Proteins

Edited by
Robert A. Meyers

Related Titles

Meyers, R. A. (ed.)

Encyclopedia of Molecular Cell Biology and Molecular Medicine

16 Volume Set

ISBN-13: 978-3-527-30542-1
ISBN-10: 3-527-30542-4

Mayer, R. J., Ciechanover, A. J., Rechsteiner, M. (eds.)

Protein Degradation

Vol. 1: Ubiquitin and the Chemistry of Life

2005
ISBN-13: 978-3-527-30837-8
ISBN-10: 3-527-30837-7

Mayer, R. J., Ciechanover, A. J., Rechsteiner, M. (eds.)

Protein Degradation

Vol. 2: Ubiquitin-Proteasome System

2006
ISBN-13: 978-3-527-31130-9
ISBN-10: 3-527-31130-0

Mayer, R. J., Ciechanover, A. J., Rechsteiner, M. (eds.)

Protein Degradation

Vol. 3: Cell Biology of the Ubiquitin-Proteasome System

2006
ISBN-13: 978-3-527-31435-5
ISBN-10: 3-527-31435-0

Janson, J.

Protein Purification: Principles, High Resolution Methods, and Applications, 3rd edition

2007
ISBN-13: 978-0-471-74661-4
ISBN-10: 0-471-74661-4

Tramontano, A.

Protein Structure Prediction

Concepts and Applications

2006
ISBN-13: 978-3-527-31167-5
ISBN-10: 3-527-31167-X

Proteins

From Analytics to Structural Genomics

Edited by
Robert A. Meyers

Volume 1

WILEY-VCH Verlag GmbH & Co. KGaA

The Editor

Dr. Robert A. Meyers
RAMTECH LIMITED
122 Escalle Lane
Larkspur, CA 94939
USA

1st Edition 2007
 1st Reprint 2008

■ All books published by Wiley-VCH are carefully produced. Nevertheless, authors, editors, and publisher do not warrant the information contained in these books, including this book, to be free of errors. Readers are advised to keep in mind that statements, data, illustrations, procedural details or other items may inadvertently be inaccurate.

Library of Congress Card No.: applied for

British Library Cataloguing-in-Publication Data: A catalogue record for this book is available from the British Library.

Bibliographic information published by the Deutsche Nationalbibliothek
The Deutsche Nationalbibliothek lists this publication in the Deutsche Nationalbibliografie; detailed bibliographic data are available in the Internet at http://dnb.d-nb.de.

© 2007 WILEY-VCH Verlag GmbH & Co. KGaA, Weinheim

All rights reserved (including those of translation into other languages). No part of this book may be reproduced in any form – by photoprinting, microfilm, or any other means – nor transmitted or translated into a machine language without written permission from the publishers. Registered names, trademarks, etc. used in this book, even when not specifically marked as such, are not to be considered unprotected by law.

Composition: Laserwords Private Ltd, Chennai, India
Printing and bookbinding: buch bücher dd ag, Birkach

Printed in the Federal Republic of Germany.
Printed on acid-free paper.

ISBN-13: 978-3-527-31608-3

ISBN-10: 3-527-31608-6

Contents

Volume 1

Preface vii

Color Plates ix

1 Structure and Function 1

1 X-Ray Diffraction of Biological Macromolecules 3
Albrecht Messerschmidt Robert Huber

2 Protein NMR Spectroscopy 81
Thomas Szyperski

3 Electron Microscopy of Biomolecules 103
Claus-Thomas Bock, Susanne Franz, Hanswalter Zentgraf, and John Sommerville

4 Circular Dichroism in Protein Analysis 129
Zhijing Dang and Jonathan D. Hirst

5 Protein Structure Analysis: High-throughput Approaches 149
Andrew P. Turnbull Udo Heinemann

6 Protein Aggregation 169
Jeannine M. Yon

7 Prions 199
Stanley B. Prusiner

8 Alzheimer's Disease 257
Jun Wang, Silva Hecimovic and Alison Goate

9 Molecular Chaperones 287
Peter Lund

Proteins. Edited by Robert A. Meyers.
Copyright © 2007 Wiley-VCH Verlag GmbH & Co. KGaA, Weinheim
ISBN: 978-3-527-31608-3

10	Motor Proteins Charles L. Asbury Steven M. Block	321
11	DNA–Protein Interactions Sylvie Rimsky and Malcolm Buckle	347

2 Modeling and Design 377

12	Protein Modeling Marian R. Zlomislic D. Peter Tieleman	379
13	Synthesis of Peptide Mimetics and their Building Blocks Bruce K. Cassels Patricio Sáez	405
14	Design and Application of Synthetic Peptides Gregory A. Grant	431
15	Synthetic Peptides: Chemistry, Biology, and Drug Design Tomi K. Sawyer	453

3 Expression, Synthesis and Degradation 485

16	Ribosome, High Resolution Structure and Function Christiane Schaffitzel Nenad Ban	487
17	Ubiquitin-Proteasome System for Controlling Cellular Protein Levels Michael H. Glickman Aaron Ciechanover	515
18	Plant-based Expression of Biopharmaceuticals Jörg Knäblein	537
19	Cell-free Translation Systems Takuya Ueda, Akio Inoue and Yoshihiro Shimizu	563

Preface

The *Proteins* two volume set was compiled from a selection of key articles from the recently published *Encyclopedia of Molecular Cell Biology and Molecular Medicine* (ISBN 978-3-527-30542-1). *Proteins* is comprised of 36 detailed articles arranged in seven sections covering structure and function; modeling and design; expression, synthesis and degradation; analytical techniques; signaling and regulation. The articles were prepared by eminent researchers from the major research institutions in the United States, Europe and around the globe including two Nobel Laureates.

Each article begins with a concise definition of the subject and its importance, followed by the body of the article and extensive references for further reading. The references are divided into secondary references (books and review articles) and primary research papers. Each subject is presented on a first-principle basis, including detailed figures, tables and drawings. Because of the self-contained nature of each article, some overlap among articles on related topics occurs. Extensive cross-referencing is provided to help the reader expand his or her range of inquiry.

The master publication, which is the basis of the *Proteins* set, is the *Encyclopedia of Molecular Cell Biology and Molecular Medicine*, which is the successor and second edition of the *VCH Encyclopedia of Molecular Biology and Molecular Medicine*, covers the molecular and cellular basis of life at a university and professional researcher level. This second edition is double the first edition in length and will comprise the most detailed treatment of both molecular and cell biology available today. The Board and I believe that there is a serious need for this publication, even in view of the vast amount of information available on the World Wide Web and in text books and monographs. We feel that there is no substitute for our tightly organized and integrated approach to selection of articles and authors and implementation of peer review standards for providing an authoritative single-source reference for undergraduate and graduate students, faculty, librarians and researchers in industry and government.

Our purpose is to provide a comprehensive foundation for the expanding number of molecular biologists, cell biologists, pharmacologists, biophysicists, biotechnologists, biochemists and physicians as well as for those entering molecular cell biology and molecular medicine from majors or careers in physics, chemistry, mathematics, computer science and engineering. For example there is an unprecedented demand for physicists, chemists and computer scientists who will work with biologists to define the genome, proteome and interactome through experimental and computational biology.

Proteins. Edited by Robert A. Meyers.
Copyright © 2007 Wiley-VCH Verlag GmbH & Co. KGaA, Weinheim
ISBN: 978-3-527-31608-3

The Board and I first divided all of molecular cell biology and molecular medicine into primary topical categories and each of these was further defined into subtopics. The following is a summary of the topics and subtopics:

- Nucleic Acids: amplification, disease genetics overview, DNA structure, evolution, general genetics, nucleic acid processes, oligonucleotides, RNA structure, RNA replication and transcription.
- Structure Determination Technologies Applicable to Biomolecules: chromatography, labeling, large structures, mapping, mass spectrometry, microscopy, magnetic resonance, sequencing, spectroscopy, x-ray diffraction.
- Proteins, Peptides and Amino Acids: analysis, enzymes, folding, mechanisms, modeling, peptides, structural genomics (proteomics), structure, types.
- Biomolecular Interactions: cell properties, charge transfer, immunology, recognition, senses.
- Molecular Cell Biology of Specific Organisms: algae, amoeba, birds, fish, insects, mammals, microbes, nematodes, parasites, plants, viruses, yeasts.
- Molecular Cell Biology of Specific Organs or Systems: excretory, lymphatic, muscular, neurobiology, reproductive, skin.
- Molecular Cell Biology of Specific Diseases: cancer, circulatory, endocrine, environmental stress, immune, infectious diseases, neurological, radiation.
- Biotechnology: applications, diagnostics, gene altered animals, bacteria and fungi, laboratory techniques, legal, materials, process engineering, nanotechnology, production of classes or specific molecules, sensors, vaccine production.
- Biochemistry: carbohydrates, chirality, energetics, enzymes, biochemical genetics, inorganics, lipids, mechanisms, metabolism, neurology, vitamins.
- Pharmacology: chemistry, disease therapy, gene therapy, general molecular medicine, synthesis, toxicology.
- Cellular Biology: developmental cell biology, diseases, dynamics, fertilization, immunology, organelles and structures, senses, structural biology, techniques.

We then selected some 340 article titles and author or author teams to cover the above topics. Each article is designed as a self-contained treatment. Each article begins with a key word section, including definitions, to assist the scientist or student who is unfamiliar with the specific subject area. The Encyclopedia includes more than 3000 key words, each defined within the context of the particular scientific field covered by the article. In addition to these definitions, the glossary of basic terms found at the back of each volume, defines the most commonly used terms in molecular and cell biology. These definitions should allow most readers to understand articles in the Encyclopedia without referring to a dictionary, textbook or other reference work.

Larkspur, July 2006

Robert A. Meyers
Editor-in-Chief

Color Plates

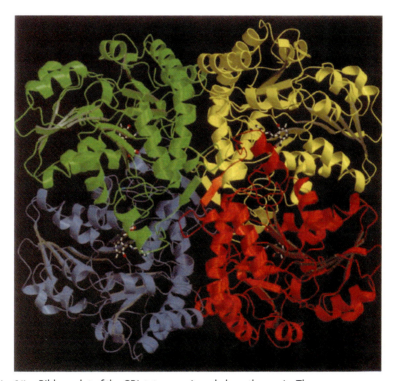

Fig. 39 (p. 64) Ribbon plot of the CBL tetramer viewed along the x-axis. The monomers are colored differently. The blue- and green-colored monomers, which are related by a crystallographic axis (horizontal, in the plane of the paper), build up one catalytic active dimer, and the yellow and red ones the other. The location of the PLP-binding site is shown in a ball-and-stick presentation; MOLSCRIPT and RASTER3D. (Reproduced by permission of Academic Press, Ltd., from Clausen, T. et al. (1996) *J. Mol. Biol.* **23**, 202–224.)

Proteins. Edited by Robert A. Meyers.
Copyright © 2007 Wiley-VCH Verlag GmbH & Co. KGaA, Weinheim
ISBN: 978-3-527-31608-3

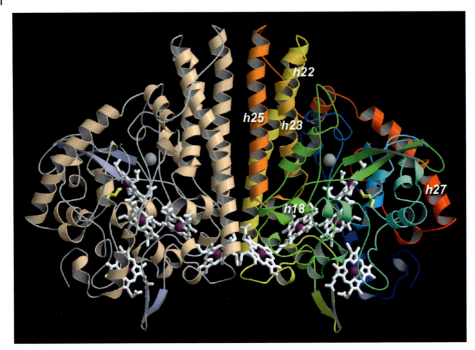

Fig. 49 (p. 73) Overall structure of the nitrite reductase dimer. A front view with the dimer axis oriented vertically, five hemes in each monomer (white), the Ca^{2+} ions (grey), and Lys133 that coordinates the active-site iron (yellow). The dimer interface is dominated by three long α-helices per monomer. All hemes in the dimer are covalently attached to the protein. (Reproduced by permission of Macmillan Magazines, Ltd., from Einsle, O. et al. (1999) *Nature (London)* **400**, 476–480.)

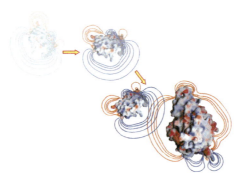

Fig. 4 (p. 396) Schematic illustration of the association of fasciculin (on the left) with acetylcholinesterase. Blue and red contour lines indicate regions of positive and negative electrostatic potential, respectively. Figure courtesy of D. Sept and A. Elcock. See Elcock, A.H., Gabdoulline, R.R., Wade, R.C., McCammon, J.A. (1999) Computer simulation of protein-protein association kinetics: acetylcholinesterase-fasciculin, *J. Mol. Biol.* **291**, 149–162 for more details.

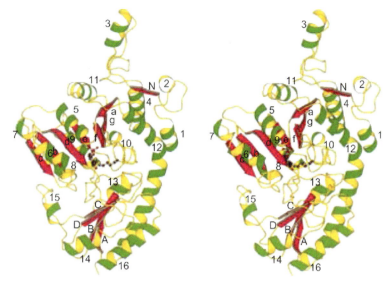

Fig. 40 (p. 65) Stereo ribbon presentation of the CBL monomer, emphasizing secondary structure elements. α-Helices are drawn as green spirals, ß-strands as magenta arrows. PLP and PLP-binding Lys210 are shown in a ball-and-stick representation; MOLSCRIPT and RASTER3D. (Reproduced by permission of Academic Press, Ltd., from Clausen, T. et al. (1996) *J. Mol. Biol.* **23**, 202–224.)

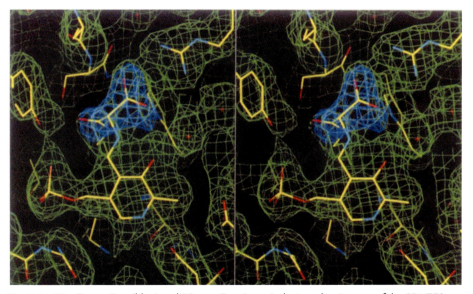

Fig. 42 (p. 66) $F_{obs} - F_{calc}$ (blue) and $2F_{obs} - F_{calc}$ (green) electron density map of the CBL/TFA complex around the active site contoured at 3.5σ and 1.0σ respectively, at 2.3-Å resolution. (Reproduced by permission of Academic Press, Ltd., from Clausen, T. et al. (1996) *J. Mol. Biol.* **23**, 202–224.)

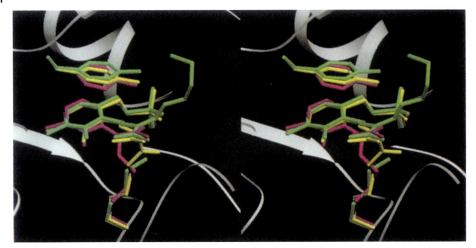

Fig. 43 (p. 67) Stereo view of a superposition of Tyr111 and the PLP derivative of the unliganded enzyme (magenta), the CBL/AVG adduct (green), and the TFA-inactivated enzyme (yellow); SETOR. (Reproduced by permission of the American Chemical Society, Clausen, T. et al. (1997) *Biochemistry* **36**, 12633–12643.)

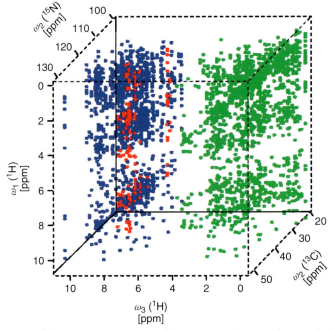

Fig. 8 (p. 94) Cube with dots representing NOEs detected in 3D heteronuclear resolved [^{1}H,^{1}H]-NOESY, which were used for calculating the NMR solution structure of YqfB shown in Fig. 9. Blue: signals detected along ω_3 on backbone amide protons; Red: detected on aromatic protons; Green: detected on aliphatic protons. Courtesy of Hanudatta Atreya.

Color Plates xiii

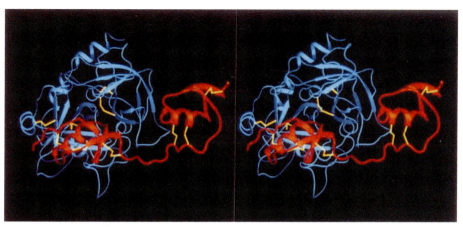

Fig. 44 (p. 68) Stereo view of the complex formed between thrombin (blue) and rhodniin (red) in the thrombin standard orientation, that is, with the active-site cleft facing the viewer and a bound inhibitor chain from left to right. Yellow connections indicate disulfide bridges. Rhodniin interacts through its N-terminal domain in a canonical manner with the active site and through its C-terminal domain with the fibrinogen recognition exosite of thrombin; SETOR. (Reproduced by permission of Oxford University Press, from van de Locht, A. et al. (1995) *EMBO J.* **14**, 5149–5157).

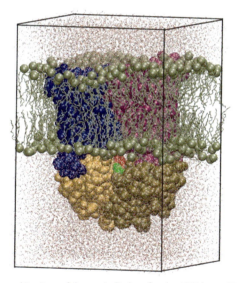

Fig. 3 (p. 393) An orthographic view of the periodic box for the ATP-bound BtuCD simulation, showing water (red and white), lipid with phosphorus atoms enlarged, and the protein. The transporter consists of two transmembrane domains (blue and purple) and two nucleotide binding domains (orange and ochre). The two docked MgATP molecules are partially visible (green and red). Figure courtesy of E. Oloo. See Oloo, E.O., Tieleman, D.P. (2004) Conformational transitions induced by the binding of MgATP to the vitamin B_{12} ABC-transporter BtuCD, *J. Biol. Chem.* **279**, 45013–45019 for more details. Rendered with VMD.

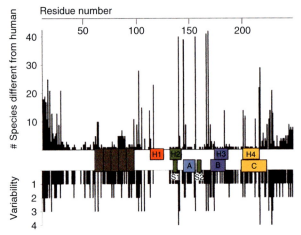

Fig. 5 (p. 214) Species variations and mutations of the prion protein gene. The x-axis represents the human PrP sequence, with the five octarepeats and H1–H4 regions of putative secondary structure shown as well as the three α-helices A, B, and C and the two β-strands S1 and S2 as determined by NMR. Vertical bars above the axis indicate the number of species that differ from the human sequence at each position. Below the axis, the length of the bars indicates the number of alternative amino acids at each position in the alignment. Data were compiled by Paul Bamborough and Fred E. Cohen

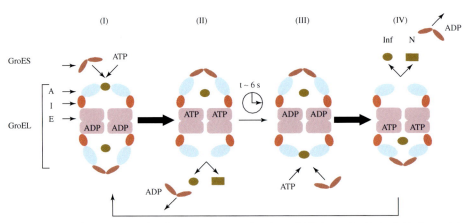

Fig. 8 (p. 183) The reaction cycle of GroEL–GroES. Inf is the unfolded protein, N the folded one, A is the apical domain, (in blue), I the intermediate domain (in red) and E the equatorial domain (in magenta). (Reproduced from Wang & Weissman (1999) *Nat. Struct. Biol.* **6**, 597, with permission.)

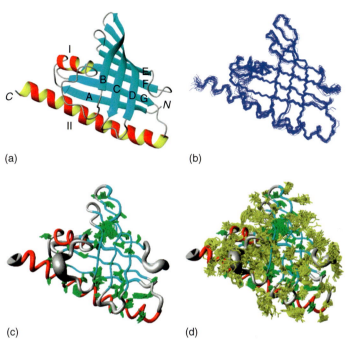

Fig. 9 (p. 95) Commonly used representations of NMR solution structures, exemplified for the 18-kDa Northeast Structural Genomics consortium target protein CC1736. (a) Ribbon drawing of the conformer exhibiting the smallest residual constraints violations. α-Helices and β-strands are depicted, respectively, in red/yellow, and cyan. (b) Superposition of the polypeptide backbone of the 20 conformers chosen to represent the NMR solution structure. The higher precision of the atomic coordinates of the regular secondary structure elements is apparent when compared with the loops. (c) The polypeptide backbone is represented by a spline function through the mean C^α coordinates, where the thickness represents the rmsd values for the C^α-coordinates after superposition of the regular secondary structure elements for minimal rmsd. The superpositions of the best defined side chains of the molecular core are shown in green to indicate the precision of their structural description. (d) Same as in (c), also showing the superposition of the more flexibly disordered side chains on the protein surface. The protein sample was provided by Drs. Acton and Montelione, Rutgers University. Courtesy of Yang Shen.

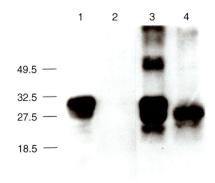

(a)

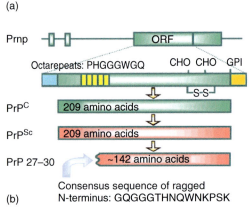

(b)

Consensus sequence of ragged
N-terminus: GQGGGTHNQWNKPSK

Fig. 1 (p. 202) Prion protein isoforms. Western immunoblot of brain homogenates from uninfected (lanes 1 and 2) and prion-infected (lanes 3 and 4) Syrian hamsters. Samples in lanes 2 and 4 were digested with 50 µg mL^{-1} of proteinase K for 30 min at 37 °C. PrP^C in lanes 2 and 4 was completely hydrolyzed under these conditions, whereas approximately 67 amino acids were digested from the N-terminus of PrP^{Sc} to generate PrP 27–30. After polyacrylamide gel electrophoresis (PAGE) and electrotransfer, the blot was developed with anti-SHaPrP R073 polyclonal rabbit antiserum. Molecular weight markers are depicted in kilodaltons (kDa). (b) Bar diagram of the SHaPrP gene that encodes a protein of 254 amino acids. After processing of the N- and C-termini, both PrP^C (green) and PrP^{Sc} (red) consist of 209 residues. After limited proteolysis, the N-terminus of PrP^{Sc} is truncated to form PrP 27–30, which is composed of approximately 142 amino acids, the N-terminal sequence of which was determined by Edman degradation.

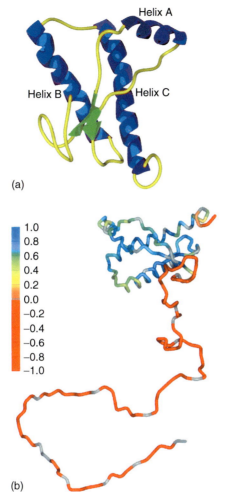

Fig. 7 (p. 218) Structures of PrP^C. (a) NMR structure of Syrian hamster (SHa) recombinant (rec) PrP(90–231). Presumably, the structure of the α-helical form of recPrP(90–231) resembles that of PrP^C. recPrP(90–231) is viewed from the interface where PrP^{Sc} is thought to bind to PrP^C. The color scheme is: α-helices A (residues 144–157), B (172–193), and C (200–227) in blue; loops in yellow; residues 129–134 encompassing strand S1 and residues 159–165 encompassing strand S2 in green. (b) Schematic diagram showing the flexibility of the polypeptide chain for PrP(29–231). The structure of the portion of the protein representing residues 90–231 was taken from the coordinates of PrP(90–231). The remainder of the sequence was hand-built for illustration purposes only. The color scale corresponds to the heteronuclear [^{1}H]-^{15}N NOE data: red for the lowest (most negative) values, where the polypeptide is most flexible, to blue for the highest (most positive) values in the most structured and rigid regions of the protein.

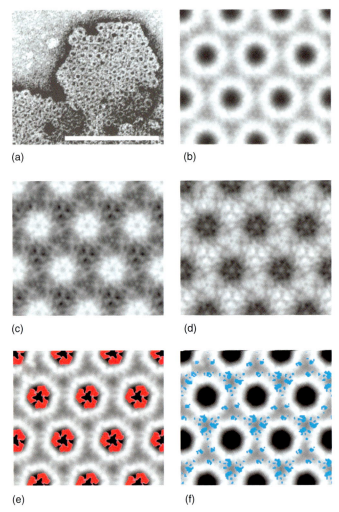

Fig. 8 (p. 220) Two-dimensional crystals of PrP^{Sc}. (a) A 2D crystal of PrP^{Sc}106 stained with uranyl acetate. Bar = 100 nm. (b) Image processing result after correlation-mapping and averaging followed by crystallographic averaging. (c and d) Subtraction maps between the averages of PrP 27–30 and PrP^{Sc}106 (panel B). (c) PrP^{Sc}106 minus PrP 27–30 and (d) PrP 27–30 minus PrP^{Sc}106, showing major differences in lighter shades. (e and f) The statistically significant differences between PrP 27–30 and PrP^{Sc}106 calculated from (c) and (d) in red and blue, respectively, overlaid onto the crystallographic average of PrP 27–30. Reprinted with permission, from Wille, H., Michelitsch, M.D., Guénebaut, V., Supattapone, S., Serban, A., Cohen, F.E., Agard, D.A. Prusiner, S.B. (2002) Structural studies of the scrapie prion protein by electron crystallography, *Proc. Natl. Acad. Sci. U. S. A.* **99**, 3563–3568 Wille et al., (2002), copyright 2002 National Academy of Sciences, USA.

Color Plates | xix

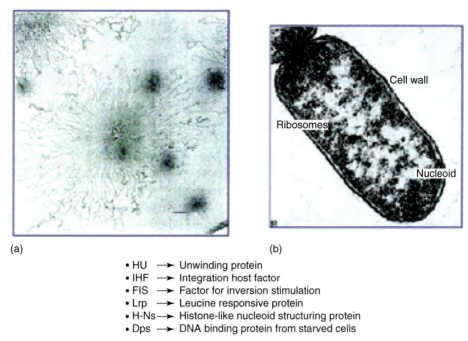

(a)   (b)

- HU ⟶ Unwinding protein
- IHF ⟶ Integration host factor
- FIS ⟶ Factor for inversion stimulation
- Lrp ⟶ Leucine responsive protein
- H-Ns ⟶ Histone-like nucleoid structuring protein
- Dps ⟶ DNA binding protein from starved cells

Fig. 1 (p. 358) Structural proteins of the bacterial nucleoid. Electron micrographs of (b) a complete *E. coli* bacterium showing the DNA condensed (the white material) within the cell) and (a) the same cell exploded after salt treatment showing the DNA now completely unwound, note that the horizontal blue bar in (a) shows the length of the original prior to disruption. The organization of the DNA is mostly carried out by the proteins listed.

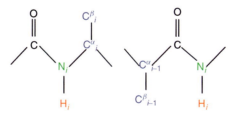

Fig. 3 (p. 86) Example of chemical shifts of nuclei (in color) correlated in triple resonance NMR experiments for obtaining sequence-specific backbone resonance assignments of proteins. A first experiment correlates the shifts of amide proton, nitrogen, α-carbon, and β-carbon within a given residue i (on the left). A second experiment correlates the shifts of amide proton and nitrogen of residue i with those of the α- and β-carbons of the preceding residue $i - 1$. Combining the two experiments allows one to obtain sequence-specific resonance assignments for the backbone (and $^{13}C^{\beta}$ shifts) of the entire polypeptide.

MKKAVINGEQIRSISDLHQTLKKELALPEYYGENLAALWDCLTGWVEYPLVLEWRQ
FEQSKQLTENGAESVLQVFREAKAEGCDITIILS

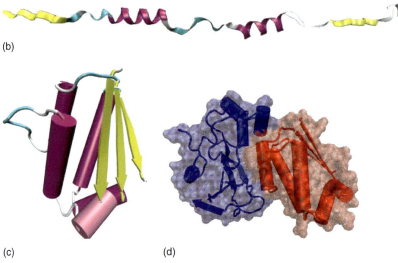

Fig. 1 (p. 383) (a) The primary sequence of the protein barstar in FASTA format. (b) The secondary structure of the first 60 residues of barstar, colored purple for helices, and yellow for β-strands. All are shown in ribbon format. (c) The complete tertiary structure of barstar in cartoon format. The helices are now bundled together, and the β-strands have formed a three-stranded parallel β-sheet.
(d) Example of quaternary structure: the barnase/barstar complex. Barstar (red) and barnase (blue) area shown in ribbon and spacefilling formats. Images created with VMD using the PDB entry 1B2U.

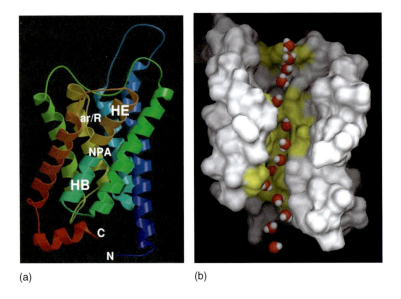

Color Plates | xxi

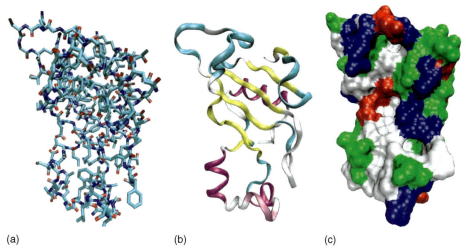

(a) (b) (c)

Fig. 2 (p. 389) Barnase rendered in three formats: (a) colored by atom-type (carbon – blue, nitrogen – navy blue, oxygen – red); (b) colored by secondary structure features in ribbon format, where β-sheets are colored yellow, helices are colored purple, and turns and random coil features are colored blue and white; (c) rendered in surface format, colored by residue type, where charged residues are colored red or blue, polar residues are green, and nonpolar residues are white. All figures are rendered with the same view of the protein, looking into the barstar binding pocket. All figures were rendered with VMD.

Fig. 5 (p. 397) (a) Ribbon diagram of the aquaporin monomer and (b) spacefilling representation illustrating a water file through the channel pore. (a) Starting at the N-terminus of the monomer, there are two transmembrane helices, followed by the coil–NPA–helix motif. A third transmembrane helix completes the first half of the protein. The second half of the monomer has two transmembrane helices, followed by the coil–NPA–helix motif, followed by another transmembrane helix. (b) The path of the water channel is through the core of the protein, following the path of the coil motifs, which meet at the NPA signature. This is more clearly illustrated in 5(b). The protein is rendered as a molecular surface in white, with the front surface of the protein cut away so that we can clearly see the pore in the middle of the protein. Surfaces colored yellow are those that interact most strongly with passing water molecules. The waterfile displayed is an overlay of a number of snapshots from the 10 ns simulation. The dipole inversion of water at the NPA motif is clearly illustrated here. Figure courtesy of B. de Groot. See De Groot, B.L., Grubmuller, H. (2001) Water permeation across biological membranes: mechanism and dynamics of aquaporin-1 and GlpF, *Science* **294**, 2353–2357. De Groot, B.L., Frigato, T., Helms, V., Grubmuller, H. (2003) The mechanism of proton exclusion in the aquaporin-1 water channel, *J. Mol. Biol.* **333**, 279–293 for more details.

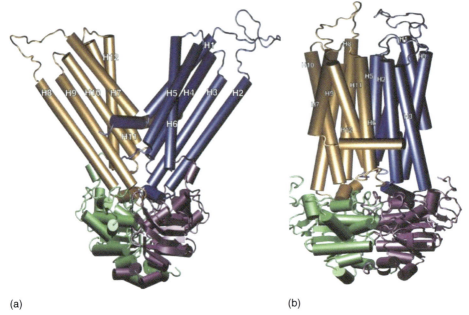

(a) (b)

Fig. 6 (p. 400) (a) Construction of a molecular model for P-glycoprotein (P-gp). Each half of P-gp was modeled by homology to the crystal structure of MsbA (PDB code 1JSQ), which had been extended to a full atom representation. The two halves of P-gp were assembled such that the NBDs adopt the ATP-dependent orientation observed in MJ0796 (PDB code 1L2T). The constituent domains are individually colored. (b) Reconciliation of cross-linking data with a P-gp model. P-gp-Model-B was generated by rotation of each NBD with respect to its cognate TMD. The final model contains a parallel TMD:TMD interface (blue and gold subunits) and a consensus NBD:NBD interface (green and purple subunits). Adapted from Stenham, D.R., Campbell, J.D., Sansom, M.S.P., Higgins, C.F., Kerr, I.D., Linton, K.J. (2003) An atomic detail model for the human ATP binding cassette transporter P-glycoprotein derived from disulphide cross-linking and homology modeling, *FASEB J.* **17**, 2287–2289.

Color Plates | xxiii

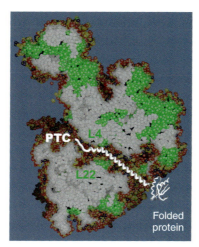

Fig. 8 (p. 504) The polypeptide exit tunnel. The 50S subunit is cut in half such that the tunnel is shown in its entire length. The ribosome atoms are shown in a space-filling representation. All RNA atoms that do not contact solvent are shown in white, and surface atoms are color-labeled with oxygen red, carbon yellow, and nitrogen blue. Proteins are depicted in green, and surface residues are in the same color code as RNA surface atoms. A model of a nascent chain polypeptide passing the tunnel is shown in white ribbon. The polypeptide is still connected to the peptidyl transferase center and can adopt an α-helical conformation in the tunnel. Outside the tunnel, the polypeptide chain can fold cotranslationally adopting its tertiary structure. The narrowest part of the tunnel is constricted by the two proteins L4 and L22.

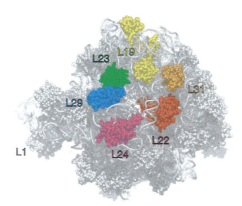

Fig. 10 (p. 504) The tunnel exit of the 50S subunit. Proteins at the rim of the exit tunnel are shown in the different colors. Some of the proteins are suggested to interact with the translocation machinery. L23 and L29 are implied to play a role in protein folding and transport across membranes based on their interaction with chaperones and SRP. The 23S rRNA and other ribosomal proteins are shown in white.

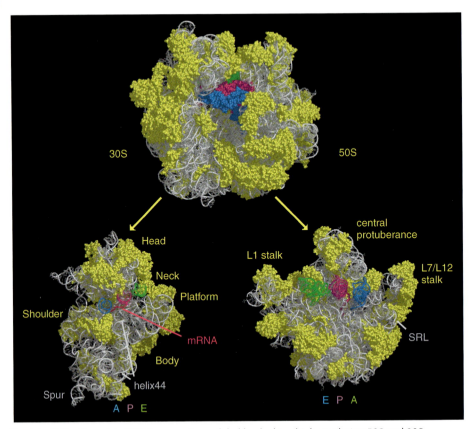

Fig. 4 (p. 494) Complete 70S ribosome is modeled by docking high-resolution 50S and 30S structures onto the low-resolution 70S structure. The 30S structure is from *T. thermophilus* and 50S structure from *H. marismortui* with tRNA positions determined from lower resolution 70S studies. Ribosomal RNA is depicted in gray and proteins in yellow. The 70S is shown from the side with the 30S subunit on the left and the 50S subunit on the right. In the subunit interface cavity, the tRNAs are visible. The A-site tRNA is depicted in blue, the P-site tRNA in magenta and the E-site tRNA in green. The anticodon arms of the tRNAs point to the 30S subunit, while the acceptor stems point into a large cleft in the large subunit. On the left-hand side, the 30S subunit is shown from the interface with the A-, P- and E-tRNA anticodon stem-loops. The anticodon loops of the A and P tRNAs contact the mRNA. Architectural characteristics and important features are labeled. On the right-hand side, the large subunit with tRNAs docked onto the *Haloarcula* 50S structure is shown from the interface as well. The acceptor stems of A- and P-site tRNAs point into the peptidyl transferase center. The sarcin–ricin loop (SRL) is a central part of the GTPase factor binding center. The three characteristic protuberances of the large subunit are labeled.

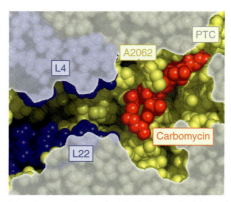

Fig. 9 (p. 504) A section of the ribosomal tunnel with Carbomycin A bound. The first and most constricted half of the tunnel is shown in a longitudinal section. All rRNA atoms are shown in yellow, and protein atoms of L4 and L22 are colored blue in a space-filling representation. The cutting plane is gray. The macrolide antibiotic Carbomycin A (red) is bound to the 50S tunnel blocking the passage of the nascent chain, and its disaccharide isobutyrate extension reaches to the peptidyl transferase center (PTC). A covalent bond is formed between N6 of A2062 of 23S rRNA and an ethylaldehyde substituent of Carbomycin A.

Phytomedics (tobacco):

- Root secretion, easy recovery
- Greenhouse contained tanks
- High density tissue
- Salts and water only
- Tobacco is well characterized
- Stable genetic system

Fig. 6 (p. 551) Secretion of the biopharmaceuticals via tobacco roots. The tobacco plants are genetically modified in such a way that the protein is secreted via the roots into the medium ("rhizosecretion"). In this example, the tobacco plant takes up nutrients and water from the medium and releases GFP (Green Fluorescent Protein). Examination of root cultivation medium by its exposure to near ultraviolet-illumination reveals the bright green-blue fluorescence characteristics of GFP in the hydroponic medium (left flask in panel lower left edge). The picture also shows a schematic drawing of the hydroponic tank, as well as tobacco plants at different growth stages, for example, callus, fully grown, and greenhouse plantation. Source: Knäblein J. (2003) *Biotech: A New Era in the New Millennium – Biopharmaceutic Drugs Manufactured in Novel Expression Systems*, DECHEMA-Jahrestagung der Biotechnologen, Munich, Germany, 21.

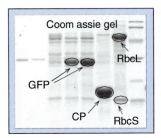

Fig. 7 (p. 552) Viral transfection of tobacco plants. This new generation platform for fast (1 to 2 weeks), high-yield (up to 5 g kg^{-1} fresh leaf weight) production of biopharmaceuticals is based on proviral gene amplification in a nonfood host. Antibodies, antigens, interferons, hormones, and enzymes could successfully be expressed with this system. The picture shows development of initial symptoms on a tobacco following the Agrobacterium-mediated infection with viral vector components that contain a GFP gene (a); this development eventually leads to a systemic spread of the virus, literally converting the plant into a sack full of protein of interest within two weeks (b). The system allows to coexpress two proteins in the same cell, a feature that allows expression of complex proteins such as full-length monoclonal antibodies. Panels (c) and (d) show the same microscope section with the same cells, expressing Green Fluorescent Protein (c) and Red Fluorescent Protein (d) at the same time. The yield and total protein concentration achievable are illustrated by a Coomassie gel with proteins in the system: GFP (protein of interest), CP (coat protein from wild-type virus), RbcS and RbcL (small and large subunit of ribulose-1,5-bisphosphate carboxylase). Source: Knäblein J. (2003) *Biotech: A New Era in the New Millennium – Biopharmaceutic Drugs Manufactured in Novel Expression Systems*, DECHEMA-Jahrestagung der Biotechnologen, Munich, Germany, 21.

Part 1
Structure and Function

Proteins. Edited by Robert A. Meyers.
Copyright © 2007 Wiley-VCH Verlag GmbH & Co. KGaA, Weinheim
ISBN: 978-3-527-31608-3

1
X-Ray Diffraction of Biological Macromolecules

Albrecht Messerschmidt and Robert Huber
Max-Planck-Institut für Biochemie, Martinsried, Germany

1	**Introduction** 7	
1.1	Crystals and Symmetry 7	
1.2	Protein Solubility 9	
1.2.1	Ionic Strength 9	
1.2.2	pH and Counterions 11	
1.2.3	Temperature 11	
1.2.4	Organic Solvents 11	
1.3	Experimental Techniques 12	
1.4	Crystallization Screenings 14	
2	**Experimental Techniques** 15	
2.1	X-ray Sources 15	
2.1.1	Conventional X-ray Generators 15	
2.1.2	Synchrotron Radiation 16	
2.1.3	Monochromators 17	
2.2	Detectors 18	
2.2.1	General Components of an X-ray Diffraction Experiment 18	
2.2.2	Image Plates 19	
2.2.3	Gas Proportional Detectors 19	
2.2.4	Charge-coupled Device-based Detectors 21	
2.3	Crystal Mounting and Cooling 21	
2.3.1	Conventional Crystal Mounting 21	
2.3.2	Cryocrystallography 22	
2.3.3	Crystal Quality Improvement by Humidity Control 24	
2.4	Data-collection Techniques 25	
2.4.1	Rotation Method 25	
2.4.2	Precession Method 27	

Proteins. Edited by Robert A. Meyers.
Copyright © 2007 Wiley-VCH Verlag GmbH & Co. KGaA, Weinheim
ISBN: 978-3-527-31608-3

3	**Principles of X-ray Diffraction by a Crystal**	**28**
3.1	Scattering of X-rays by an Atom	28
3.2	Scattering of X-rays by a Unit Cell	30
3.3	Scattering of X-rays by a Crystal	30
3.3.1	One-dimensional Crystal	30
3.3.2	Three-dimensional Crystal	31
3.4	The Reciprocal Lattice and Ewald Construction	32
3.5	The Temperature Factor	34
3.6	Symmetry in Diffraction Patterns	34
3.7	Electron Density Equation and Phase Problem	34
3.8	The Patterson Function	36
3.9	Integrated Intensity Diffracted by a Crystal	36
3.10	Intensities on an Absolute Scale	37
3.11	Resolution of the Structure Determination	37
3.12	Diffraction Data Evaluation	38
3.13	Solvent Content of Protein Crystals	39
4	**Methods for Solving the Phase Problem**	**39**
4.1	Isomorphous Replacement	39
4.1.1	Preparation of Heavy Metal Derivatives	39
4.1.2	Single Isomorphous Replacement	40
4.1.3	Multiple Isomorphous Replacement	42
4.2	Anomalous Scattering	43
4.2.1	Theoretical Background	43
4.2.2	Experimental Determination	44
4.2.3	Breakdown of Friedel's Law	45
4.2.4	Anomalous Difference Patterson Map	46
4.2.5	Phasing Including Anomalous Scattering Information	46
4.2.6	Multiwavelength Anomalous Diffraction Technique	47
4.2.7	Determination of the Absolute Configuration	50
4.3	Patterson Search Methods (Molecular Replacement)	50
4.3.1	Rotation Function	51
4.3.2	Translation Function	51
4.3.3	Computer Programs for Molecular Replacement	52
4.4	Phase Calculation	52
4.4.1	Refinement of Heavy Atom Parameters	52
4.4.2	Protein Phases	53
4.5	Phase Improvement	55
4.5.1	Solvent Flattening	55
4.5.2	Histogram Matching	56
4.5.3	Molecular Averaging	56
4.5.4	Phase Combination	57
4.6	Difference Fourier Technique	57

5	**Model Building and Refinement** 59	
5.1	Model Building 59	
5.2	Crystallographic Refinement 59	
5.3	Accuracy and Verification of Structure Determination 61	
6	**Applications** 62	
6.1	Enzyme Structure and Enzyme–Inhibitor Complex 62	
6.1.1	X-ray Structure of Cystathionine ß-Lyase 62	
6.1.2	Enzyme–Inhibitor Complex Structures of Cystathione ß-Lyase 64	
6.1.3	Crystal Structure of the Thrombin–Rhodniin Complex 67	
6.2	Metalloproteins 68	
6.2.1	Crystal Structure of the Multicopper Enzyme Ascorbate Oxidase 69	
6.2.2	Crystal Structure of Cytochrome-c Nitrite Reductase Determined by Multiwavelength Anomalous Diffraction Phasing 70	
6.3	Large Molecular Assembly 74	
6.3.1	Crystal Structure of 20S Proteasome from Yeast 74	

Bibliography 76
Books and Reviews 76
Primary Literature 76

Keywords

Anomalous scattering
If the wavelength of the incident X-ray beam is around the absorption edge of a given atom, its atomic scattering factor becomes complex with a normal, a dispersive, and an absorption component.

Atomic scattering factor
Mathematical expression for the X-ray scattering of an individual atom.

Crystallographic R-factor
Reliability factor for the X-ray structure determination, sum over the absolute values of the differences between the observed and calculated structure factors divided by the sum of the absolute values of the observed structure factors.

Electron density function
The spatial distribution of electrons in the crystal, which can be obtained by the Fourier transformation of the structure factors, represents the atomic arrangement in the crystal.

Ewald construction
Construction to visualize the diffraction of X-rays by a crystal; a sphere with a radius of the reciprocal of the incident X-ray wavelength is drawn with the origin of the reciprocal lattice of the crystal at the intersection of the incident beam with the surface of the sphere; diffraction occurs if a reciprocal lattice point intersects the surface of the so-called Ewald sphere; the direction of the diffracted beam is the connection between the center of the Ewald sphere and the intersection point of the relevant reciprocal lattice point.

MAD
Multiple anomalous diffraction, a method to exploit the anomalous scattering effect of a set of natural or artificially introduced anomalous scatterers for the solution of the phase problem.

MIR
Multiple isomorphous replacement, a method to solve the phase problem by preparing isomorphous heavy metal derivatives and measuring X-ray data sets of the native and the derivative crystals.

Phase problem
The phase of the diffracted waves is lost during the measurement as the X-ray intensities deliver the amplitude of the diffracted waves only.

Reciprocal lattice
A lattice in reciprocal space with lattice points, which are perpendicular to the corresponding set of lattice planes in the lattice in direct space with a length of the reciprocal of the lattice plane distance or its multiples.

Resolution
Minimal distance between two objects that can be resolved.

Structure factor
Mathematical expression for the X-ray wave diffracted by a crystal, which consists of amplitude and phase factor.

X-ray diffraction of biomolecules covers diffraction experiments on single crystals of small biological molecules such as oligopeptides, cofactors, steroid hormones, and so on, or biological macromolecules. But it also includes small angle scattering experiments on biological macromolecules in solution. This chapter is exclusively dedicated to the X-ray crystallography of biological macromolecules. It comprises the determination of the three-dimensional structures of proteins, nucleic acids, and other biological macromolecules at atomic resolution by diffraction of X-rays on crystals of such macromolecules. The atomic structure is obtained from the electron density distribution of the macromolecular crystal, which is the Fourier transform of the waves diffracted by the crystal. The amplitudes of the diffracted waves are determined directly from the diffraction experiment. The diffraction phases are revealed (1) from additional diffraction data of isomorphous heavy atom derivatives (multiple isomorphous replacement (MIR) technique), (2) from multiple anomalous diffraction (MAD), if suitable anomalous scatterers are in the crystal and using tunable synchrotron radiation, and (3) by Patterson search techniques (molecular replacement), if structural information on the biological macromolecule under investigation is available. X-ray crystallography of biological macromolecules is the unique method for the elucidation of the spatial structures at atomic resolution of complex biomolecules with molecular masses greater than 30 kDa. The structures represent a time and spatial average of molecules packed in a crystal. The preparation of suitable crystals may be a limiting factor.

1
Introduction

1.1
Crystals and Symmetry

In a crystal, atoms or molecules are arranged in a three-dimensionally periodic manner by translational symmetry. The crystal is formed by a three-dimensional stack of unit cells, which is called the *crystal lattice* (Fig. 1a and b). The unit cell is built up by three noncollinear vectors **a**, **b**, and **c**. In the general case, these vectors have unequal magnitudes and their mutual angles deviate from 90°. The arrangement of the molecule(s) in the unit cell may be asymmetrical, but very often it is symmetrical. This is illustrated in Fig. 2(a–e) in two-dimensional lattices for rotational symmetries.

In crystals, only 1-, 2-, 3-, 4-, and 6-fold rotations are allowed. This follows from the combination of the lattice properties with rotational operations. Other possible symmetry elements are mirror plane *m*, inversion center, and combination of rotation axis with inversion center (inversion axis). These are the point group symmetries. They can only occur among each other in a few certain combinations of angles. Other angle orientations would violate the lattice properties. The number

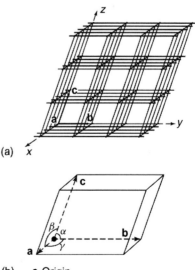

Fig. 1 (a) Crystal lattice and (b) unit cell.

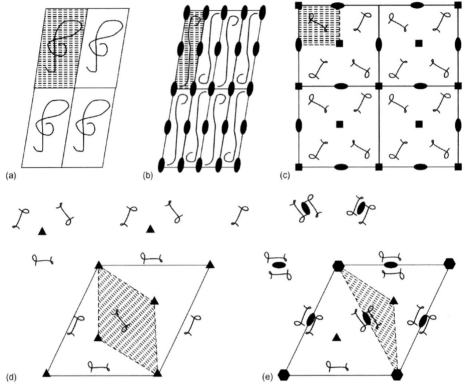

Fig. 2 Rotational symmetry elements in two-dimensional lattices: (a) 1, (b) 2, (c) 4, (d) 3, and (e) 62. The asymmetric unit is hatched.

of all possible combinations reveals the 32 point groups. The crystal morphology obeys the point group symmetries.

Adding an inversion center to the point group symmetry leads to the 11 Laue groups. These are of importance for the symmetry of X-ray diffraction patterns. Their symbols are 1, 2/m, 2/mmm, 3, 3 m, 4/m, 4/mmm, 6/m, 6/mmm, m3, and m3m. Proteins and nucleic acids are chiral molecules. Therefore, they can crystallize only in the 11 enantiomorphic point groups: 1, 2, 3, 4, 6, 23, 222, 32, 422, 622, and 432.

The combination of point group symmetries with lattices leads to seven crystal systems, triclinic, monoclinic, orthorhombic, trigonal, tetragonal, hexagonal, and cubic, with 14 different Bravais-lattice types, which can be primitive, face-centered, all-face-centered, and body-centered. Furthermore, additional symmetry elements are generated, having translational components such as screw axes or glide mirror planes. There exist 230 space groups of which 65 are enantiomorphic (for chiral molecules such as proteins). Figure 3 shows the graphical representation for the space group $P2_12_12_1$, as listed in the *International Tables for Crystallography*. The asymmetric unit is one-fourth of the unit cell and can contain one or several molecules. Multimeric molecules may have their own

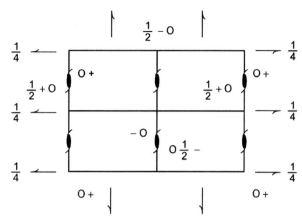

Fig. 3 Graphical representation of space group $P2_12_12_1$.

symmetries, which are called *noncrystallographic symmetries*. Here, axes that are 5-fold, 7-fold, and so on, are also allowed.

1.2 Protein Solubility

Figure 4 shows a typical phase diagram illustrating the solubility properties of a macromolecule. In the labile phase, crystal nucleation and growth compete, whereas in the metastable region, only crystal growth appears. In the unsaturated region, crystals dissolve. The solubility of proteins is influenced by several factors, as follows.

1.2.1 Ionic Strength

A protein can be considered as a polyvalent ion and, therefore, its solubility can be discussed on the basis of the Debye–Hückel theory. In aqueous solution, each ion is surrounded by an "atmosphere" of counter ions. This ionic atmosphere influences the interactions of the ion with water molecules and hence the solubility.

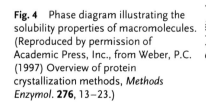

Fig. 4 Phase diagram illustrating the solubility properties of macromolecules. (Reproduced by permission of Academic Press, Inc., from Weber, P.C. (1997) Overview of protein crystallization methods, *Methods Enzymol.* **276**, 13–23.)

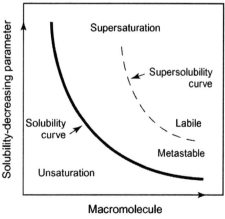

1.2.1.1 **"Salting-in"** At low ionic concentration, the "ionic atmosphere" increases the solubility as it increases the possibilities for favorable interactions with water molecules. We obtain (Eq. 1 and 2)

$$\log S - \log S_0 = \frac{AZ_+Z_-\sqrt{\mu}}{1 + aB\sqrt{\mu}} \quad (1)$$

$$\mu = \frac{1}{2}\sum c_j Z_j^2 \quad (2)$$

where μ = ionic strength, S = solubility of the salt at a given ionic strength μ, S_0 = solubility of the salt in absence of the electrolyte, Z_+, Z_- ionic charge of salt ions, A, B = constants depending on the temperature and dielectric constant, a = average diameter of ions, and c_j = concentration of the jth chemical component. Ions with higher charge are more effective for changes in solubility. Most salts and proteins are more soluble in low ionic strength than in pure water. This is termed as *salting-in* (Fig. 5).

1.2.1.2 **"Salting-out"** At higher ionic strength, the ions compete for the surrounding water. Therefore, water molecules are taken away from the dissolved agent and the solubility decreases according to Eq. (3):

$$\log S - \log S_0 = \frac{AZ_+Z_-\sqrt{\mu}}{1 + aB\sqrt{\mu}} - K_s\mu \quad (3)$$

The term $K_s\mu$ predominates at high ionic strengths, which means that "salting-out" is then proportional to the ionic strength (Fig. 5). In a medium with low ionic strength, the solubility of a protein can be decreased by increasing or decreasing the salt concentration. Salts with small, highly charged ions are more effective than those with large, lowly charged ions.

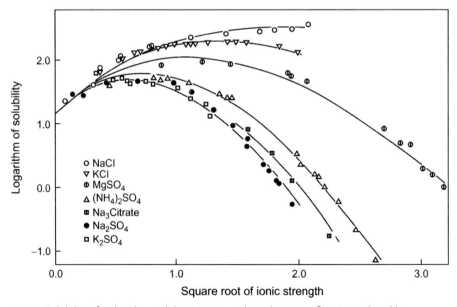

Fig. 5 Solubility of carboxyhemoglobin in various electrolytes at 25 °C. (Reproduced by permission of the American Society for Biochemistry and Molecular Biology, from Green, A.A. (1932) *J. Biol. Chem.* **95**, 47–66.)

Fig. 6 (a) Solubility of hemoglobin at different pH values in concentrated phosphate buffers; (b) extracted from (a). (Reproduced by permission of the American Society for Biochemistry and Molecular Biology, from Green, A.A. (1931) *J. Biol. Chem.* **93**, 495–516.)

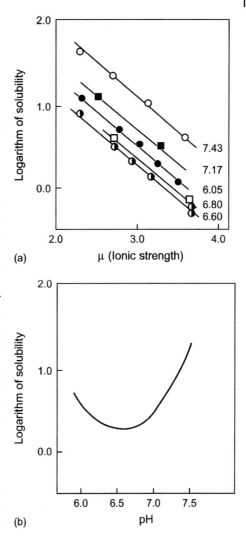

Ammonium sulfate is often used because of its high solubility.

1.2.2 pH and Counterions

A protein is more soluble when its net charge is larger. The minimum solubility is found at the isoelectric point. The net charge is zero and hence the packing in the solid state (in the crystal) is possible owing to electrostatic interactions without the accumulation of a net charge of high energy. All "salting-out" curves are parallel, K_s remains constant, and S_0 varies with pH (Fig. 6a and b). In some cases, the isoelectric point is different at low and high ionic strength owing to the interactions of the protein with counterions, which can cause a net charge at the pH of the isoelectric point.

1.2.3 Temperature

Many factors governing protein solubility are temperature dependent. The dielectric constant decreases with increasing temperature. In the solution energy, $\Delta G = \Delta H - T\Delta S$, the entropy term has an increasing influence with increasing temperature. The temperature coefficient of the solubility depends on other conditions (ionic strength, presence of organic solvents, etc.).

At high ionic strength, most proteins are less soluble at 25 °C than at 4 °C, for example the temperature coefficient is negative. The opposite is valid for low ionic strength.

1.2.4 Organic Solvents

The presence of organic solvents leads to a decrease in the dielectric constant. This causes an augmentation of the electric attraction between opposite charges on the surface of the protein molecule and hence to a reduction in solubility. In general, the solubility of a protein is reduced in the presence of an organic solvent if the temperature decreases. Often, organic solvents denature proteins. Therefore, one should work at low temperatures.

1.3
Experimental Techniques

The whole field of macromolecular crystallography has been excellently reviewed in Volumes 114 and 115 and Volumes 276 and 277 of *Methods in Enzymology*. A collection of review articles concerning the theory and practice of crystallization of biomacromolecules is given in Part A of Carter and Sweet.

A protein preparation to be used in crystallization should be "pure" or "homogeneous" at a level that established chromatographic methods are providing (protein content $\geq 95\%$). Furthermore, it should meet the requirements of "structural homogeneity." These requirements can be enumerated as follows. It is first necessary to prepare the protein in an isotypically pure state free from other cellular proteins. It may then be necessary to maintain the homogeneity of the protein preparation against covalent modification during crystallization by adding inhibitors of sulfhydryl group oxidation, by proteolysis, and by the action of reactive metals. It may be necessary to suppress the slow denaturation/aggregation of the protein and to restrict its conformational flexibility to reduce the entropic barrier to crystallization presented by extensive conformational flexibility.

For the crystallization of biomacromolecules, a broad spectrum of crystallization techniques exists. The most common techniques are described here. The oldest and simplest method is batch crystallization (Fig. 7a). In batch experiments, vials containing supersaturated protein solutions are sealed and left undisturbed. In microbatch methods, a small (2–10 µL) droplet containing both protein and precipitant is immersed in an inert oil, which prevents droplet evaporation. In the case that ideal conditions for nucleation and growth are different, it is useful to undertake the separate optimization of these processes. This can be done by seeding, a technique where crystals are transferred from nucleation conditions to those that will support only growth (Fig. 7b). For macroseeding, a single crystal is transferred to an etching solution, then to a solution of optimal growth. In microseeding experiments, a solution containing many small seed crystals, occasionally obtained by grinding a larger crystal, is transferred to a crystal growth solution.

The method of crystallization by vapor diffusion is depicted in Fig. 8(a). In this method, unsaturated precipitant-containing protein solutions are suspended over a reservoir. Vapor equilibration of the droplet and reservoir causes the protein solution to reach a supersaturation level where nucleation and initial crystal growth occur. Changes in soluble protein concentration in the droplet are likely to decrease supersaturation over the time course of the experiment. The vapor diffusion technique can be carried out as hanging drop or sitting drop method.

In crystallization by dialysis, the macromolecular concentration remains constant as in batch methods (Fig. 8b) because the molecules are forced to stay in a fixed volume. The solution composition is changed by diffusion of low-molecular-weight components through a semipermeable membrane. The advantage of dialysis is that the precipitating solution can be easily changed. Dialysis is also uniquely suited to crystallizations at low ionic strength and in the presence of volatile reagents such as alcohols.

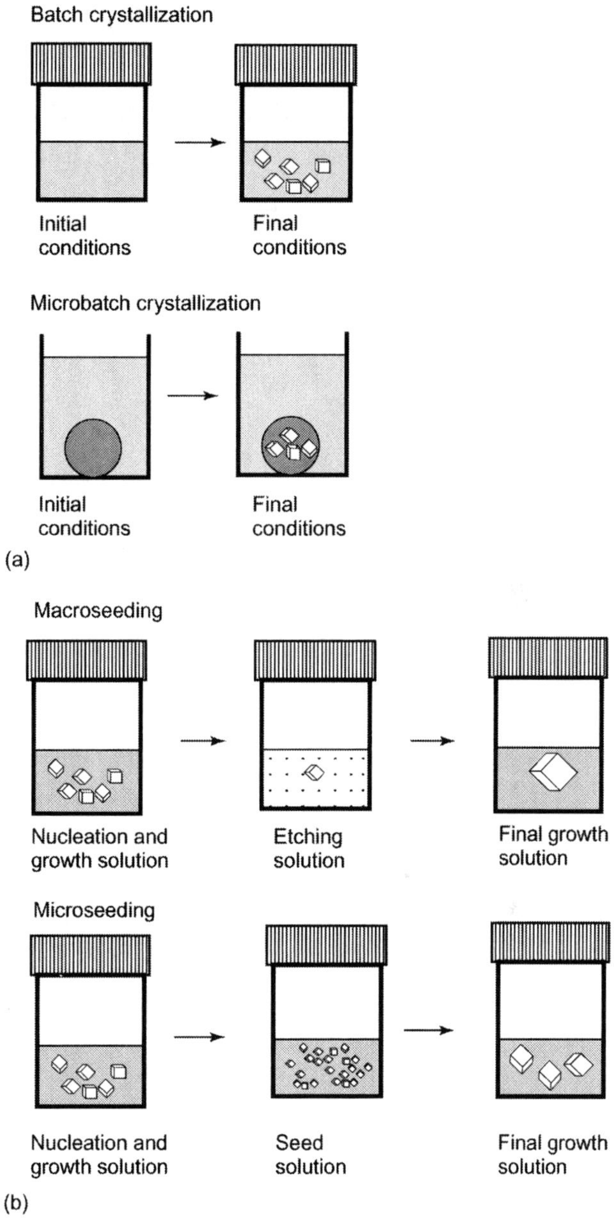

Batch crystallization

Initial conditions → Final conditions

Microbatch crystallization

Initial conditions → Final conditions

(a)

Macroseeding

Nucleation and growth solution → Etching solution → Final growth solution

Microseeding

Nucleation and growth solution → Seed solution → Final growth solution

(b)

Fig. 7 Schematic presentation of (a) batch crystallization and (b) seeding techniques. (Reproduced by permission of Academic Press, Inc., from Weber, P.C. (1997) *Methods Enzymol.* **276**, 13–23.)

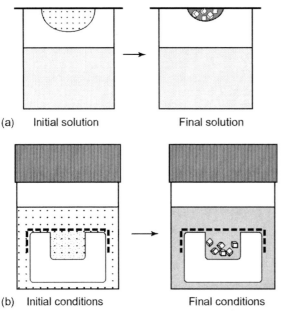

Fig. 8 Schematic representation of (a) vapor diffusion and (b) dialysis. (Reproduced by permission of Academic Press, Inc., from Weber, P.C. (1997) *Methods Enzymol.* **276**, 13–23.)

1.4
Crystallization Screenings

Screening schemes have been developed that change the most common parameters of this multiparameter problem such as protein concentration, the nature and concentration of the precipitant, pH, and temperature. Each screening can be extended by adding specific additives in low concentrations that affect the crystallization. Sparse matrix crystallization screens are widely applied. The sparse matrix formulation allows one to screen efficiently a broad range of the most popular and effective salts (e.g. ammonium sulfate, sodium and potassium phosphate, sodium citrate, sodium acetate, lithium sulfate), polymers (e.g. poly(ethyleneglycol) (PEG) of different molecular masses (from 400 to 8000)), and organic solvents (e.g. 2,4-methylpentanediol (MPD), 2-propanol, ethanol) versus a wide range of pH. Another approach is the systematic screening of the statistically most successful precipitants. A single precipitant is screened at four unique concentrations versus seven precise levels of pH between 4 and 10. Such grid screens can be done with ammonium sulfate, PEG 6000, MPD, and PEG 6000 in the presence of 1.0 M lithium chloride or sodium chloride. For the crystallization of membrane proteins, for each detergent that is necessary to make the membrane protein soluble, a whole grid screen or sparse matrix screen must be constructed. In principle, all three techniques can be applied for the different screening schemes, but mostly the vapor diffusion technique is applied because it is easy to use and the protein consumption is low. For a typical broad screening, about 2 mg of protein is sufficient. Chryschem plates (sitting drop) or Linbro plates (hanging drops) may be used for the vapor diffusion crystallization screening experiments. Once crystals have been obtained, their size and quality can be optimized by additional fine screens around the observed crystallization conditions. General rules do not exist that indicate which method for crystallization one has to use for which type of protein. Suggestions for crystallization conditions to be tested can be obtained from the Biological Macromolecule Crystallization Database.

2
Experimental Techniques

2.1
X-ray Sources

2.1.1 Conventional X-ray Generators

X-rays are produced when a beam of high-energy electrons, which have been accelerated through a voltage V in a vacuum, hit a target. An X-ray tube run at voltage V will emit a continuous X-ray spectrum with a minimum wavelength given by Eq. (4):

$$\lambda_{min} = \frac{hc}{eV} = \frac{12398}{V} \quad (4)$$

with λ in angströms ($1 \text{ Å} = 10^{-10}$ m) and V in volts. The critical voltage, V_0, which is required to excite the characteristic line of a particular element, can be calculated from the corresponding wavelength for the appropriate absorption edge. For the copper absorption edge, $\lambda_{ae} = 1.380$ Å. Hence, we have (Eq. 5)

$$V_0 = \frac{12398}{\lambda_{ae}} = 8.98 \text{ kV} \quad (5)$$

Provided that $V > V_0$, the characteristic line spectra will be produced (Fig. 9). The oldest and cheapest X-ray sources are sealed X-ray tubes. The cathode and anode are situated under vacuum in a sealed glass tube, and the heat generated at the anode is removed by a water-cooling system. For the generation of higher intensities, as needed in protein crystallography, one has to use a rotating anode (Fig. 10). Here the anode is rotated, which allows a higher power loading at the focal spot. In protein crystallography, copper targets are usually taken. The used take-off angle is near 4°, which results in apparent focal spot sizes of about 0.3×0.3 mm.

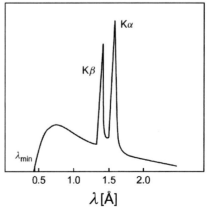

Fig. 9 X-ray spectrum emitted from a copper anode. It shows the continuous "Bremsspektrum" starting at λ_{min} and the two characteristic copper lines λ Kα = 1.5418Å (superposition of λ Kα_1 = 1.5405 Å and λ Kα_2 = 1.5443 Å) and λ Kβ = 1.3922 Å.

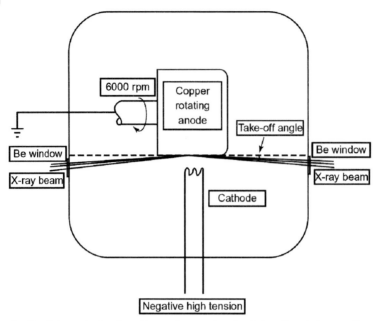

Fig. 10 Schematic drawing of a rotating anode tube. Take-off angle is near 4°. For copper, the tube is normally operated at 50-kV high tension and 100-mA cathode current.

2.1.2 Synchrotron Radiation

As electrically charged particles such as electrons or positrons of high energy are kept under the influence of magnetic fields and travel in a pseudocircular trajectory, synchrotron radiation is emitted and can be used in many different types of experiments. The particles are injected into the storage ring directly from a linear accelerator or through a booster ring. They circulate in a high vacuum for several hours at a relative constant energy. To keep the bunched particles traveling in a nearly circular path, a lattice of bending magnets is set up around the storage ring. As the particle beam traverses each magnet, the path of the beam is altered, and synchrotron radiation is emitted. The loss of energy of the particle beam is compensated by a radiofrequency input at each cycle. The synchrotron radiation can be channeled through different beamlines for use in research.

Other types of magnets – insertion devices called *wigglers* and *undulators* – can be assembled in the storage ring. Unlike the bending magnets, the primary purpose of which is to maintain the circular trajectory, wigglers and undulators are used to increase the intensity of the emitted radiation. Bending magnets and wigglers cause a continuous spectrum of radiation.

In contrast, the radiation produced by an undulator has a discontinuous spectrum, and can be tuned to various wavelengths. The importance of synchrotron radiation for macromolecular crystallography lies in the high brilliance (photons s^{-1} $mrad^{-2}$ mm^{-2} per $\Delta\lambda/\lambda$; that is, how small is the source and how well collimated are the X-rays?) of the beam, the high intensity, and the tunability of the wavelength in the relevant range from 0.5 to 3.0 Å. The time structure of the beam is of interest for time-resolved crystallography. The particles circulate in bunches with widths of 50 to 150 ps and repeat every few microseconds.

About 15 synchrotron radiation facilities equipped with beamlines for macromolecular crystallography are available throughout the world operated at energies from about 1.5 to 6 to 8 GeV for third-generation machines. An aerial view of the European Synchrotron Radiation Facility (ESRF) in Grenoble, a third-generation machine, is shown in Fig. 11. The ESRF storage ring is operated at 6 GeV and has a circumference of 844.39 m. Its critical wavelength, λ_c, is 0.6 Å.

2.1.3 Monochromators

In the majority of applied diffraction techniques, monochromatic X-rays are used. Therefore, the emitted white radiation of X-rays must be further monochromatized. With copper Kα radiation generated by a sealed or rotating anode tube, the Kβ radiation can be removed with a nickel filter. Much better results can be achieved with a monochromator. The simplest monochromator is a piece of a graphite crystal that reflects the copper Kα radiation at a Bragg angle of 13.1° and a glancing angle of 26.2°. Improved beam focusing is obtained by a double mirror system. The mirror assembly is composed of two perpendicular bent nickel-coated glass optical flats, each with translation, rotation, and slit components housed in a helium gas-flashed chamber, which is commercially available (Molecular Structure Corporation, The Woodlands, TX, USA). The prototype and basic theory in the use of this system were discussed in detail by Phillips and Rayment.

For synchrotron radiation with its much higher intensity, germanium or silicon single crystals can be applied as monochromators, which filter out a bandwidth of $\delta\lambda/\lambda$ from 10^{-4} to 10^{-5}, two orders of magnitude smaller than that with graphite. Single or double monochromators can be used, which are either flat or bent. The bent monochromators have the advantage that

Fig. 11 Aerial view of ESRF in Grenoble. Courtesy of ESRF.

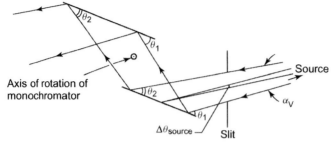

Fig. 12 Schematic drawing of a double monochromator system. (Reproduced by permission of Cambridge University Press, from Helliwell, J.R. (1992) *Macromolecular Crystallography with Synchrotron Radiation*, Cambridge University Press, Cambridge.)

they simultaneously focus the beam. The double monochromator (Fig. 12) has the advantage that the emergent monochromatic beam is parallel to and only slightly displaced from the incident synchrotron radiation beam. This makes necessary only small adjustments of the X-ray optics and detector arrangement when it is tuned to another wavelength compared to a single monochromator where the whole X-ray diffraction assembly must be moved.

2.2 Detectors

2.2.1 General Components of an X-ray Diffraction Experiment

A principal arrangement for a macromolecular X-ray diffraction experiment is depicted in Fig. 13. The primary beam leaves the X-ray source and passes the X-ray optics, which may be a simple collimator or the various types of monochromators or mirror systems described above,

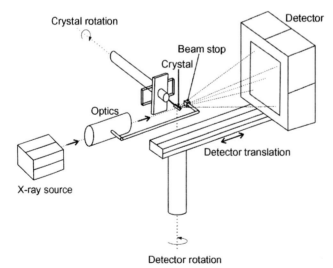

Fig. 13 Principal arrangement for a macromolecular X-ray diffraction experiment.

and terminated with a collimator. The crystal is mounted on a goniometer head either in a quartz capillary or in a cryo-loop shock-frozen at low temperature. The goniometer head is attached to a device that can perform spatial movements of the crystal around the center of the crystal. The simplest kind of such a movement is the rotation of the crystal about a spindle axis as indicated in Fig. 13. This device can be a multiple axis goniostat (2–4 axes), which allows the crystal to be brought into any spatial orientation around its center. The X-ray detector, which registers the diffracted intensities, is mounted on a device that permits the translation and rotation of the detector. If the active area of the detector is large enough to collect all generated diffracted beams at a given wavelength, detector rotation is not necessary and the detector is arranged normal to the primary beam. A small piece of lead is placed in the path of the primary beam just behind the crystal to prevent damage to the detector and superfluous gas scattering.

The classical detectors in macromolecular crystallography have been photographic films and single-photon counters. The photographic films were used on specially designed X-ray cameras and the single-photon counters on four-circle diffractometers. The main disadvantage of these detectors was their low sensitivity and with films the limited dynamic range (1 : 200). Over the last 15 years, powerful detectors have been developed, which will be discussed briefly. These new detectors have almost completely replaced photographic films and single-photon counters.

2.2.2 Image Plates

An image plate (IP) consists of a support (either a flexible plastic plate or a metal base) coated with a photostimulable phosphor (150 µm) and a protective layer (10 µm). The photostimulable phosphor is a mixture of very thin crystals of $BaF(Br,I):Eu^{2+}$ and an organic binder. This phosphor can store a fraction of the absorbed X-ray energy by electrons trapped in color centers. It emits photo-stimulated luminescence, whose intensity is proportional to the absorbed X-ray intensity, when later stimulated by visible light. The wavelength of the photostimulated luminescence ($\lambda \approx 390$ nm) is reasonably separated from that of the stimulating light ($\lambda \approx 633$ nm, in practice a red laser), allowing it to be collected by a conventional high quantum efficiency photomultiplier tube. The output of the photomultiplier is amplified and converted to a digital image, which can be processed by a computer. The residual image on the IP can be erased completely by irradiation with visible light, to allow repeated use.

IPs have several excellent performance characteristics as integrating X-ray area detectors that make them well suited in X-ray diffraction. The sensitivity is at least 10 times higher than for X-ray films and the dynamic range is much broader ($1:10^4-10^5$). Important for synchrotron radiation is their high sensitivity at shorter wavelengths (e.g. 0.65 Å). A disadvantage is the relatively long readout times for each exposure (from 45 s to several min). IP diffractometer systems are commercially available from several companies. All systems work reliably and deliver good-quality data. A photograph of the newest IP system produced by Mar Research (Norderstedt, Germany) is shown in Fig. 14.

2.2.3 Gas Proportional Detectors

As X-ray counters, gas proportional detectors provide unrivaled dynamic range and sensitivity for photons in the range important for macromolecular crystallography.

Fig. 14 Mar 345 IP diffractometer system from Mar Research, Norderstedt, Germany. The circular plate rotates for scanning and the laser is moved along a radial line. Courtesy of Mar Research.

The classical gas proportional detector is a multiwire proportional chamber (MWPC), widely used as an in-house detector with conventional X-ray sources. Two MWPC diffractometer systems are commercially available. Gas proportional detectors use as a first step the absorption of an X-ray photon in a gas mixture high in xenon or argon. This photoabsorption produces one electron–ion pair whose total energy is just the energy of the initial X-ray photon. The ion returns to its neutral state either by emission of Auger electrons or by fluorescence. Since the kinetic energy of these first electrons is far greater than the energy of the first ionization level of the xenon or argon atoms, fast collisions with atoms (or molecules) in the gas very quickly produce a cascade of new electron–ion pairs in a small region extending over a few hundred micrometers around the conversion point. The total number of primary electrons that are produced during this process is proportional to the energy of the absorbed X-ray photon and is thus a few hundred for ~10-keV photons. These primary electrons then drift to the nearest anode wire where an ionization avalanche of as many as 10 000 to 1 000 000 ion pairs results. The motion of the charged particles in this avalanche (chiefly the motion of the heavy positive ions away from the anode wire) causes a negative-going pulse on the anode wire and positive-going pulses on a few

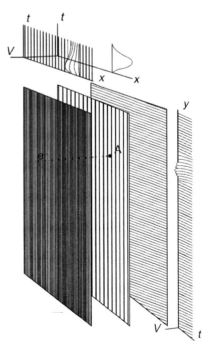

Fig. 15 Expanded view of an MWPC, showing the anode plane sandwiched between the two cathode planes. A is the position of the avalanche. The centers of the induced charge distributions are used to determine the coordinates, x and y, of the avalanche. (Reproduced by permission of Academic Press, Inc., from Kahn, R., Fourme, R. (1997) *Methods Enzymol.* **276**, 244–268.)

of the nearest wires in the back (cathode) wire plane (see Fig. 15).

Disadvantages of the MWPC detector are the limited counting rate due to the buildup of charges in the chamber and limitations in the readout electronics and the lower sensitivity at shorter wavelengths. This makes the application of MWPCs with synchrotron radiation poorly effective.

2.2.4 Charge-coupled Device-based Detectors

A remarkable development for the use with synchrotron radiation is the design and construction of charge-coupled device (CCD) detectors. CCDs were developed originally as memory devices, but the observation of localized light-induced charge accumulation in CCDs quickly led to their development as imaging sensors. These CCD detectors are integrating detectors like the conventional X-ray sensitive film, IPs, and analog electronic detectors using either silicon intensified target (SIT) or CCD sensors. Integrating detectors have virtually no upper rate limits because they measure the total energy deposited during the integration period (although individual pixels may become saturated if the signal exceeds its storage capacity).

The first commercially available analog electronic detector was the fast area television detector (FAST) detector produced by Enraf-Nonius (Delft, The Netherlands). This detector contained a SIT vidicon camera as an electronically readable sensor. The SIT vidicon exhibits higher noise than CCDs, which have therefore replaced SIT sensors during the past few years. Because of their high intrinsic noise, detectors with SIT vidicon sensors need an analog image-amplification stage and this limits the overall performance of such detectors. Several CCD detector systems have also been developed that incorporate image intensification. The most important development in detector design for macromolecular crystallography has been the incorporation of scientific-grade CCD sensors into instruments with no image intensifier. These detector designs are based on direct contact between the CCD and a fiber-optic taper. There are several commercial systems available based on this construction (Mar Research, Norderstedt, Germany; Hamlin Detector).

A schematic representation of such a detector is shown in Fig. 16. An X-ray phosphor (commonly $Gd_2O_2S:Tb$) is attached to a fiber-optic faceplate, which is tightly connected to a fiber-optic taper. The X-ray sensitive phosphor surfaces at the front convert the incident X-rays into a burst of visible-light photons. Although it is possible to permit the X-rays to strike the CCD directly, this method has several drawbacks, such as radiation damage to the CCD, signal saturation, and poor efficiency. The use of a larger phosphor as active detector area and the demagnifying fiber-optic taper is also necessary because the size of the scientific-grade CCD sensors is not as large as needed for the demands of the X-ray diffraction experiment. The fiber-optic taper is then bonded to the CCD, which is connected to the electronic readout system. The CCD must be cooled to temperatures ranging from -40 to $-90\,^\circ$C, depending on the various systems. The great advantage of CCD detectors is their short readout time, which lies in the range from 1 to a few seconds.

2.3 Crystal Mounting and Cooling

2.3.1 Conventional Crystal Mounting

The purpose of crystal mounting is to isolate a single crystal from its growth

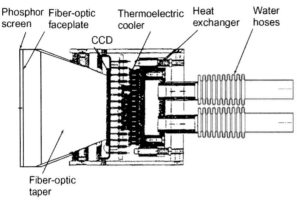

Fig. 16 Schematic representation of a CCD/taper detector. (Reproduced by permission of Academic Press, Inc., from Westbrook, E.M., Naday, I. (1997) *Methods Enzymol.* **276**, 268–288.)

medium so that it can be used in the X-ray diffraction experiment to study its diffraction properties. It is important that the manipulation of the crystal introduces as little damage as possible to its three-dimensional structure. The most important aspect of crystal mounting is to preserve the crystal in its state of hydration. This is accomplished by sealing the crystal in a thin-walled (0.001-mm thick) glass or quartz capillary tube. The important steps in conventional crystal mounting are illustrated in Fig. 17(a–c). The crystal must be dislodged from the surface on which it grew, and then it may be drawn into the capillary using suction from a small-volume (0.25 mL) syringe, micropipet, or mouth aspirator, which are connected to the funnel of the capillary by a flexible plastic hose of appropriate diameter. Next, the capillary should be inverted to allow the crystal to fall to the inner meniscus. Then, the surrounding solution may be removed using thin strips of filter paper or by a small glass pipet. The extent to which the crystal should be dried must be determined by experience. The final step is to place a small volume of mother liquor in the capillary and seal both ends. The capillary is then glued to a metal base, which can be attached to a goniometer head.

2.3.2 Cryocrystallography

Many macromolecular crystals suffer from radiation damage when exposed to X-rays with energies and intensities as used in macromolecular X-ray diffraction experiments with both conventional sources and synchrotron radiation. A possibility for reducing radiation damage of the crystal during the measurement is to cool the crystal to low temperatures, usually to 100 K. For this purpose, the crystal is flash-frozen to prevent ice formation or damage to the crystal. One method of crystal treatment is the removal of external solution by transferring the crystal in a small drop to a hydrocarbon oil and either teasing the liquid away or drawing it off with filter paper or a small pipet. The oil-coated crystal is then mounted on a glass fiber or small glass "spatula." Oil protects the crystal from drying and acts as an adhesive that hardens on cooling to hold the sample rigidly. Much more frequently used

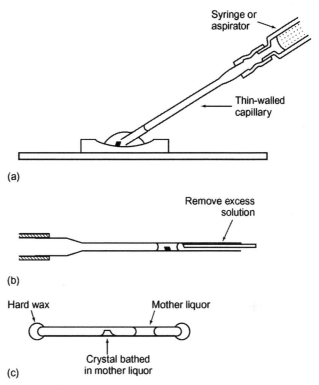

Fig. 17 Mounting of a crystal in a glass capillary. (Reproduced by permission of Academic Press, Inc., from Rayment, I. (1985) *Methods Enzymol.* **114**, 136–140.)

is a technique in which the crystal is suspended in a film of mother liquor in a small loop. This method avoids problems with damage by the oil or mechanical damage when removing the external liquid, and it has proven successful for most samples. It does, however, require the use of a cryoprotectant to prevent ice formation. The most commonly used cryoprotectants are glycerol, PEG of different molecular weights, glucose, and MPD.

The loop is produced from fine fibers that permit unobstructed data collection in nearly all sample orientations. The crystal is held within the loop, suspended in a thin film of cryoprotectant-containing harvest buffer. The loop is supported by a fine wire or pin, which itself is attached to a steel base used for placing the assembly on a goniometer head and in storing mounted crystals. Once in the loop, the crystal is cooled to a temperature at which the increasing viscosity of the liquid prevents molecular rearrangement. The rate of cooling must be rapid enough to reach this point before ice-crystal nucleation occurs. Two methods are used: cooling directly in the gas stream of a cryostat or plunging the crystal into a cryogenic liquid. The first method is explained in Fig. 18(a) and (b). The loop assembly (with crystal) is attached to the goniometer head, with the cold stream deflected. Then, the cold stream is unblocked to flash-freeze

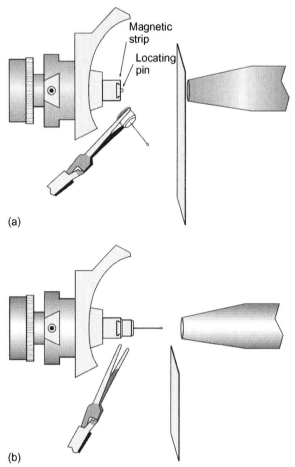

Fig. 18 Flash cooling in the direct cold gas stream of a cryostat. (Reproduced by permission of Academic Press, Inc., from Rodgers, D.W. (1997) *Methods Enzymol.* **276**, 244–268.)

the crystal. In this gas-stream position, the goniometer head must be heated to prevent ice formation on the goniometer head. Cryostats and cryocrystallographic tools are commercially available.

Cryocrystallography has had a great impact on macromolecular crystallography by dramatically increasing the lifetime of a crystal during the X-ray experiment allowing, for example, the collection of several data sets from one crystal at different wavelengths using synchrotron radiation.

2.3.3 Crystal Quality Improvement by Humidity Control

The crystal quality is of decisive significance for a successful X-ray crystal structure determination. Two principal cases must be distinguished: (1) the crystals diffract to a resolution only (>4.5 Å), which does not allow a structure determination at all and (2) the crystal quality is good enough to elucidate its 3-D structure (resolution <3.5 Å but not better than 2.5 Å) but a structure determination at higher resolution and accuracy therewith would be necessary, for example, for the characterization of a metal center in a metalloprotein. The control of the crystal packing via the solvent content of a crystal is a useful approach for the improvement of the crystal quality of biological macromolecules. For this purpose, a so-called Free Mounting System (FMS) has been developed and successfully applied. The principal construction of the FMS is shown in Fig. 19 (a). A main part of the FMS is the humidifier unit, which consists of the humidifier, the control, and power electronics. A stream of humid air with defined moisture is produced. A flexible teflon tube transports the humid air to the crystal holder, which is depicted in Fig. 19 (b). A very compact construction allows the sample to be mounted in a controlled environment with minimal restriction for the X-ray measurement. The head part is freely rotatable relative to the insert, without axial movement. A heating element and a temperature sensor are integrated into the head part. The humid air stream through the head part is adjusted to the temperature of the head part, independent of the ambient temperature. The crystal may either be mounted in a patch-clamp pipette or conventional cryo-loop (Fig 19 (c)).

In a typical experiment to increase the crystal quality, X-ray crystal diffraction is monitored at various defined crystal humidities. Usually, a positive effect is observed at lower humidity, which causes shrinkage of the unit cell volume and allows different and possibly more favorable crystal contacts. After having found the optimal condition, the crystal, mounted in a loop, can be shock-frozen for a subsequent data collection. A remarkable improvement of crystal quality by using the FMS could be observed with about 30% of different projects under investigation.

2.4
Data-collection Techniques

2.4.1 Rotation Method

Most macromolecular X-ray diffraction systems use the rotation method for data collection. For each crystal, a reciprocal lattice can be constructed, which is very useful in the interpretation of crystallographic crystal diffraction experiments. Diffraction theory (discussed later) tells us that an X-ray reflection is generated when a point of this reciprocal lattice lies on a sphere of radius $1/\lambda$ whose origin is $1/\lambda$ away from the origin of the reciprocal lattice in the direction of the primary beam (Fig. 20). The direction of such a diffracted beam is along the connection of the center of the so-called Ewald sphere (radius $1/\lambda$) and the intersection of the reciprocal lattice point on the Ewald sphere. Owing to certain factors, which will be discussed later, the apparent reciprocal lattice extends to a given radius only, which defines the resolution sphere. To bring all the reciprocal lattice points within the resolution sphere into the reflection position, the crystal must be rotated around its center. Nearly all macromolecular X-ray diffraction systems apply the rotation technique in the normal beam case where the rotation axis is normal to the incident X-ray beam. Rotating the crystal around 360° brings all reciprocal lattice points within the resolution sphere in the reflection position except for the region between the rotation axis and the Ewald sphere, which is therefore called the *blind region*. This region can be collected when the crystal has been brought into another orientation. The diffracted beams are usually registered with a flat detector at distance D from the crystal, which is also normal to the primary beam. To avoid overlapping of reflection

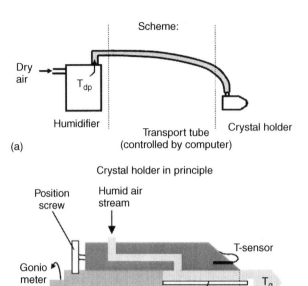

Fig. 19 The Free Mounting System: (a) principal construction; (b) schematic view of the crystal holder; (c) opened crystal holder with magnetic base and mounting loop. T_{dp}–temperature of dew point, T_g–temperature of goniometer head.

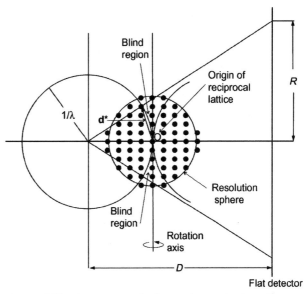

Fig. 20 Diffraction geometry in the rotation method usually applied in macromolecular X-ray diffraction systems.

spots on the detector, the crystal is rotated by rotation-angle increments, which can vary from tenths of degrees to 1 or to a few degrees, depending on the size of the crystal unit cell, crystal mosaicity, beam collimation, and other factors. Each individual exposure is processed and electronically stored in a computer. These raw data images are evaluated subsequently with relevant computer programs (discussed in some detail later) to give the intensities and geometric reference values (indices) for each collected intensity.

2.4.2 Precession Method

The rotation method delivers a distorted image of the reciprocal lattice for each geometry of the detector (flat or curved) and orientation with respect to the rotation axis. An undistorted image of the reciprocal lattice can be obtained by using the precession method. The principle of this technique is shown in Fig. 21. The detector is a flat film. During the motion of a given reciprocal lattice plane (in Fig. 21, a so-called zeroth plane going through the origin O of the reciprocal lattice), the flat detector must always be parallel to this reciprocal lattice plane to obtain an undistorted image of this plane. The normal of the reciprocal lattice plane and, consequently, also the detector are inclined with respect to the primary X-ray beam by an angle μ. When the normals of the reciprocal lattice plane and the detector carry out a concerted precession motion of angle μ around the primary X-ray beam, a circular region of the reciprocal lattice plane is registered on the detector (these regions are shown as dashed circles in Fig. 21). In a precession camera construction, the crystal and the film cassette are both held in a universal joint, which are linked so that the film and crystal move together in phase with precession angle μ. In Fig. 21, the joints are symbolized as forks and their linkage by a line. Parallel to

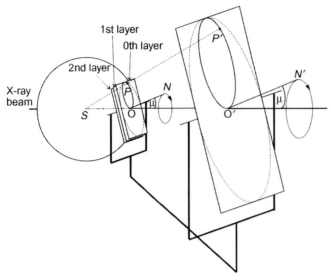

Fig. 21 Principle of the precession method (modified from Buerger, M.J. (1964) *The Precession Method in X-ray Crystallography*, John Wiley & Sons, New York).

the zeroth reciprocal lattice plane is a set of lattice planes that also carry out this precession movement. The parts of them that are swung through the Ewald sphere also give rise to an image on the flat detector (first and second reciprocal lattice layers are also indicated in Fig. 21). The images of these layers would superimpose on the detector, and the use of the technique in this way is denoted by the screenless precession method. The insertion of a screen with a suitable annular aperture between the crystal and the detector at an appropriate distance can be used to screen out the desired reciprocal layer. This screen is also inclined by the precession angle μ and is coupled to the concerted precession movement of crystal and film cassette. The strength of the precession technique is that in addition to the undistorted imaging of the reciprocal lattice planes, the indexing of the diffraction spots is straightforward and the symmetry of the diffraction pattern is readily obtained by inspection.

For this reason, the precession method has been broadly applied in macromolecular crystallography for a long time. The replacement of film by the new generation of detectors has almost completely stopped the use of the precession method. Nevertheless, the use of a precession camera has great teaching benefits in becoming familiar with the reciprocal lattice concept.

3
Principles of X-ray Diffraction by a Crystal

3.1
Scattering of X-rays by an Atom

A component of the electrical field of the incident wave has the following form (Eq. 6) in free space referred to an origin at $\mathbf{x} = 0$:

$$E(\mathbf{x}) = E_0 \exp[2\pi i(\nu t - \mathbf{x}\mathbf{s}_0)] \qquad (6)$$

where s_0 = wave vector of incident wave, $\nu = c/\lambda$ = frequency, λ = wavelength, and $s_0 = 1/\lambda$ = absolute value of incident wave vector. This wave interacts with a scattering center at position **r** (Fig. 22). The electric field component of the incident wave causes the electron at position **r** to oscillate. Together with the positively charged nucleus of the atom, which does not oscillate, this can be considered as a classical dipole oscillator. This dipole oscillator emits a spherical wave, which is denoted as a scattered wave and can be given in the following form (Eq. 7 and 8):

$$E_{SC} = CE(\mathbf{r}) \frac{\exp(-2\pi i s_0 r_1)}{r_1} \quad (7)$$

$$E_{SC} = CE_0 \exp(-2\pi i s_0 \mathbf{r})$$
$$\times \frac{\exp(2\pi i(\nu t - s_0 r_1))}{r_1} \quad (8)$$

The phase-angle-dependent factor has been omitted. The amplitude of the scattered wave is proportional to the amplitude $E(\mathbf{x})$ of the wave incident at **r**. This gives the factor $E(\mathbf{r})$. C is a proportionality factor, taking into account the peculiarities of the scattering center. The factor $1/r_1$ considers the conservation of the scattered energy flux. We add all wavelets scattered from different volume elements of an atom to get the total amplitude of the scattered wave at point **R** relative to the origin O in the atom.

Equations 9 to 12 follow from Fig. 22:

$$\mathbf{r} + \mathbf{r}_1 = \mathbf{R} \quad (9)$$

$$r_1^2 = (\mathbf{R} - \mathbf{r})^2 = R^2 + r^2$$
$$- 2rR\cos(\mathbf{r}, \mathbf{R}) \quad (10)$$

$$r_1 = R\left[1 - \frac{2r}{R}\cos(\mathbf{r}, \mathbf{R}) + \left(\frac{r}{R}\right)^2\right]^{1/2} \quad (11)$$

$$r_1 \approx R\left(1 - \frac{r}{R}\cos(\mathbf{r}, \mathbf{R}) + \ldots\right) \quad (12)$$

where $(r/R)^2$ and higher terms were neglected in the expansion of the square root. It follows that (Eq. 13)

$$r_1 \approx R - r\cos(\mathbf{r}, \mathbf{R}) \quad (13)$$

Now we can combine the spatial phase factors in the equation for E_{SC}. With the approximation for r_1, we obtain Eq. (14):

$$\exp(-2\pi i(s_0\mathbf{r} + s_0 r_1)) = \exp(-2\pi i s_0 R)$$
$$\times \exp(-2\pi i(s_0\mathbf{r} - s_0 r\cos(\mathbf{r}, \mathbf{R}))) \quad (14)$$

As $s_0 = s$ and **s** is parallel to **R**, we obtain Eq. (15).

$$s_0 r \cos(\mathbf{r}, \mathbf{R}) = s r \cos(\mathbf{r}, \mathbf{s}) = \mathbf{sr} \quad (15)$$

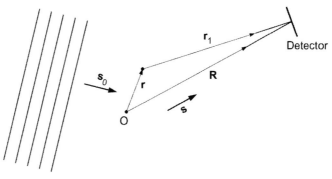

Fig. 22 Scattering of a planar X-ray wave by an electron at position **r** with respect to origin O.

It follows for the phase factor (Eq. 16):

$$\exp(-2\pi i(s_0 \mathbf{r} + s_0 \mathbf{r}_1) = \exp(-2\pi i s_0 \mathbf{R})$$
$$\times \exp(-2\pi i(s_0 \mathbf{r} - s\mathbf{r}))$$
$$= \exp(-2\pi i S_0 \mathbf{R})$$
$$\times \exp(2\pi i(\vec{\mathbf{r}}(\vec{\mathbf{S}} - \vec{\mathbf{S}}_0)) \quad (16)$$

Now we can write for the wave scattered by a center at **r** (Eq. 17):

$$E_{SC} = \left(\frac{CE_0 \exp(-2\pi i s_0 R) \exp(2\pi i \nu t)}{R} \right)$$
$$\times \exp(2\pi i \mathbf{r}(\mathbf{s} - \mathbf{s}_0)) \quad (17)$$

In a useful approximation, r_1 has been replaced by R in the denominator. In general, the scattering from an atom comes from the distribution of electrons in the atom. If the scattering of a volume element dv of the atom is proportional to the local electron density ρ (**r**), then the scattering amplitude will be proportional to the integral (Eq. 18):

$$f(\mathbf{S}) = \int_{\text{vol. of atom}} \rho(\mathbf{r}) \exp(2\pi i \mathbf{r}\mathbf{S}) dv \quad (18)$$

3.2
Scattering of X-rays by a Unit Cell

A unit cell may contain N atoms at positions of their internal origins at $\mathbf{r}_j$ ($j = 1, 2, 3, \ldots, N$) with respect to the origin of the unit cell (Fig. 23). For atom 1, we obtain Eq. (19):

$$\mathbf{f}_1 = \int_{\text{vol. of atom}} \rho(\mathbf{r}) \exp[2\pi i(\mathbf{r}_1 + \mathbf{r})\mathbf{S}] dv$$
$$= f_1 \exp(2\pi i \mathbf{r}\mathbf{S}) \quad (19)$$

with (Eq. 20)

$$f_1 = \int_{\text{vol. of atom}} \rho(\mathbf{r}) \exp 2\pi i \mathbf{r}_1 \mathbf{S} dv \quad (20)$$

where f_1 is the atomic form factor for atom 1. It reflects the characteristics of the scattering of the individual atoms and is real if the wavelength of the incident X-ray is not close to an absorption edge of the atom. The atomic form factor f is equal to Z, the ordinary number of the scattering atom, at a diffraction angle of 0° and decreases with increasing diffraction angle. For N atoms, this adds up to the total scattered wave of a unit cell $G(S)$ (Fig. 24) according to Eq. (21):

$$G(\mathbf{S}) = \sum_{j=1}^{N} f_j \exp(2\pi i \mathbf{r}_j \mathbf{S}) \quad (21)$$

3.3
Scattering of X-rays by a Crystal

3.3.1 One-dimensional Crystal

In a one-dimensional crystal, the unit cells are separated by the unit cell vector **a**. The contribution of the scattered wave

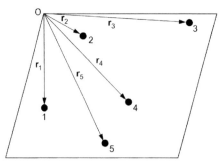

Fig. 23 Atomic positions in a unit cell.

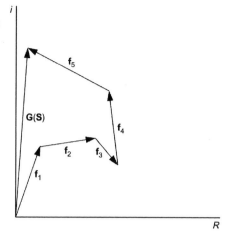

Fig. 24 Vector diagram for the total scattered wave in direction **S** added up for five atoms.

from the unit cell at the origin of the crystal is **G(S)**. All scatterers in the second unit cell are displaced by the vector **a** relative to the origin, which introduces a corresponding phase factor and reveals for the second unit cell relative to the origin, **G(S)** exp $2\pi i\mathbf{aS}$. For the nth unit cell relative to the origin, we obtain **G(S)** exp $2\pi i(n-1)\mathbf{aS}$. This sums up for the total wave to Eq. (22):

$$\mathbf{F(S)} = \sum_{n=1}^{T} \mathbf{G(S)} \exp 2\pi i(n-1)\mathbf{aS} \quad (22)$$

Generally, **F(S)** is of the same order of magnitude as **G(S)**, and no strong scattering effect is observed (Fig. 25a). However, when $2\pi \mathbf{aS} = 2\pi h$ or an integral multiple of 2π or $\mathbf{aS} = h$ (h is an integer), the waves add up constructively to a scattered wave proportional to $T|\mathbf{G(S)}|$ (Fig. 25b). $T = 10^5$ for a 1-mm long crystal with a 100-Å lattice constant. The intensity distribution of the scattered waves is concentrated around the values where **aS** is equal to an integer and depends on the number of contributing unit cells. The more unit cells are contributing, the sharper is the concentration of the intensity around these values.

3.3.2 Three-dimensional Crystal

In this case, the unit cell is spanned by the unit cell vectors **a**, **b**, and **c**, and is repeated periodically by the corresponding vector shifts in the respective spatial directions. This means that we will obtain scattered waves of measurable intensities when the three subsequent conditions are fulfilled (Eq. 23):

$$\mathbf{aS} = h;\ \mathbf{bS} = k;\ \mathbf{cS} = l \quad (23)$$

These conditions are known as *Laue equations*.

If we neglect the proportionality constant T, we obtain Eq. (24) for the total scattered wave for a three-dimensional crystal with a unit cell containing N atoms:

$$\mathbf{F(S)} = \sum_{j=1}^{N} f_j \exp 2\pi i \mathbf{r}_j \mathbf{S} \quad (24)$$

with (Eq. 25)

$$\mathbf{r}_j = \mathbf{a}x_j + \mathbf{b}y_j + \mathbf{c}z_j \quad (25)$$

Hence, we have (Eq. 26 and 27)

$$\mathbf{r}_j\mathbf{S} = x_j\mathbf{aS} + y_j\mathbf{bS} + z_j\mathbf{cS} = hx_j + hy_j + lz_j \quad (26)$$

1 X-Ray Diffraction of Biological Macromolecules

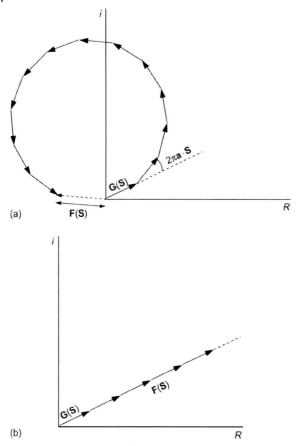

Fig. 25 Vector diagrams displaying the total wave scattered by a molecule in a crystal. (a) The phase differences between waves scattered by adjacent unit cells is $2\pi \mathbf{aS}$ and (b) the phase difference is an integral multiple of 2π (adapted from Blundell, T.L., Johnson, L.N. (1976) *Protein Crystallography*, Academic Press, New York).

(from Laue's equation) and

$$\mathbf{F}(hkl) = \sum_{j=1}^{N} f_j \exp 2\pi i(hx_j + ky_j + lz_j)$$

$$= |\mathbf{F}(hkl)| \exp i\alpha(hkl) \quad (27)$$

with $|\mathbf{F}(hkl)|$ – amplitude and α – phase angle. We obtain the intensity of the scattered wave as the structure factor $\mathbf{F}(hkl)$ multiplied by its complex conjugate value according to (Eq. 28):

$$I(hkl) = \mathbf{F}(hkl)\mathbf{F}^*(hkl)$$
$$= |\mathbf{F}(hkl)|^2 \quad (28)$$

3.4 The Reciprocal Lattice and Ewald Construction

In section 2.4, we mentioned the usefulness of the concept of the reciprocal lattice in understanding the diffraction of X-rays from a crystal. Now we have the necessary relations for the derivation of the reciprocal lattice. One can write the scattering vector **S** as Eq.(29):

$$\mathbf{S} = h_x\mathbf{a}^* + k_y\mathbf{b}^* + l_z\mathbf{c}^* \quad (29)$$

where **S** is a vector in reciprocal space with the metric $\mathbf{a}^*$, $\mathbf{b}^*$, $\mathbf{c}^*$. The relation to the direct space with metric **a**, **b**, **c** is still unknown. The vector **S** must obey the Laue equations (Eq. 30):

$$\mathbf{aS} = \mathbf{a}(h_x\mathbf{a}^* + k_y\mathbf{b}^* + l_z\mathbf{c}^*) = h$$
$$= h_x\mathbf{aa}^* + k_y\mathbf{ab}^* + l_z\mathbf{ac}^* = h \quad (30)$$

This is fulfilled only when $\mathbf{aa}^* = 1$, $h_x = h$, and $\mathbf{ab}^*$ and $\mathbf{ac}^* = 0$. Similar equations can be derived for the other two Laue conditions. Thus, vector **S** is a vector of a lattice in reciprocal space. The relation between direct and reciprocal lattice is given by the following set of nine equations (Eq. 31–39):

$$\mathbf{aa}^* = 1 \quad (31)$$
$$\mathbf{ba}^* = 0 \quad (32)$$
$$\mathbf{ca}^* = 0 \quad (33)$$
$$\mathbf{ab}^* = 0 \quad (34)$$
$$\mathbf{bb}^* = 1 \quad (35)$$
$$\mathbf{cb}^* = 0 \quad (36)$$
$$\mathbf{ac}^* = 0 \quad (37)$$
$$\mathbf{bc}^* = 0 \quad (38)$$
$$\mathbf{cc}^* = 1 \quad (39)$$

It follows from these that $\mathbf{a}^* \perp \mathbf{b}$; **c**; $\mathbf{b}^* \perp \mathbf{a}$; **c**; $\mathbf{c}^* \perp \mathbf{a}$; **b**; and vice versa. The metric relations can also be derived from these relations. It means that the inverse lattice vectors are perpendicular to the plane that is spanned by the two other noninverse lattice vectors. Bragg's law can be derived now by inspection of Fig. 26. The wave vectors for the incident wave $\mathbf{s}_0$ and the scattered wave **S** have the same absolute value of $1/\lambda$. Vector **S** must be a vector of the reciprocal lattice and its absolute value is equal to d^*. From Fig. 26, we obtain Eq. 40 to 42:

$$\sin\theta = \frac{d^*}{2}\lambda \quad (40)$$

$$\lambda = \frac{2\sin\theta}{d^*} \quad (41)$$

$$\lambda = 2d\sin\theta \quad \text{for } n = 1 \quad (42)$$

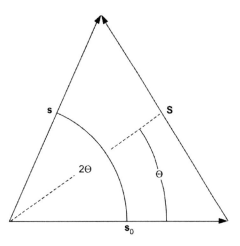

Fig. 26 Geometric representation of diffraction geometry, 2Θ, glance angle; Θ, Bragg angle.

The general equation for Bragg's law is Eq. (43):

$$2d \sin \theta = n\lambda \quad (43)$$

where n is the order of reflection and d is the interplanar distance in the direct lattice.

The Ewald construction is contained in Fig. 20. A sphere of radius $1/\lambda$ is drawn. The origin of the reciprocal lattice is located where the wave vector s_0 ends on the Ewald sphere. A diffracted beam is generated if a reciprocal lattice vector d^*_{hkl} with an absolute value of $1/d_{hkl}$ cuts the Ewald sphere. The beam is diffracted in the direction of the connection of the origin of the Ewald sphere and the intersection point of the reciprocal lattice point on the Ewald sphere. The diffraction pattern of a lattice is itself a lattice with reciprocal lattice dimensions.

3.5
The Temperature Factor

The thermal motion of the atoms causes a decrease of the scattering power by a factor of $\exp[-B(\sin^2\theta/\lambda^2)]$ with (Eq. 44)

$$B = 8\pi^2 \bar{u}^2 \quad (44)$$

where $\bar{u}$ is the mean displacement of the atoms due to the thermal motion. The atomic scattering factor f must be multiplied with this factor. In this model, the thermal motion has been assumed to be isotropic. Therefore, B is denoted as the isotropic temperature factor. In molecules, this is usually not the case and the thermal motion is described by a tensor ellipsoid. Here we obtain a set of six independent anisotropic temperature factors. In protein crystallography, isotropic B values for each atom of the molecules are used normally. The thermal motion of the atoms is one main reason for the falloff of the diffraction intensity especially at higher diffraction angles. This limits the possible recordable number of diffraction spots and, as will be seen later, the resolution of the diffraction experiment.

3.6
Symmetry in Diffraction Patterns

An X-ray diffraction data set from a crystal represents its reciprocal lattice with the corresponding diffraction intensities at the reciprocal lattice points (hkl). As the reciprocal lattice is closely related to its direct partner, it reveals symmetries, lattice properties, and other peculiarities (e.g. systematic extinctions) that are connected to the direct crystal symmetry such as unit cell dimensions and space group. A detailed discussion of this problem is given in Buerger.

In the case of real atomic scattering factors f, the diffraction intensities are centrosymmetric according to Friedel's law (Eq. 45):

$$I(hkl) = I(\overline{hkl}) \quad (45)$$

This is illustrated in Fig. 27(a,b). The square of a complex number is the product of this number by its complex conjugate. This is shown for **F** (hkl) in Fig. 27(a) and for **F**$(\overline{hkl})$ in Fig. 27(b). The resulting intensities are equal in both cases.

3.7
Electron Density Equation and Phase Problem

Inspection of the equation for the structure factor (Eq. 46)

$$\mathbf{F(S)} = \sum_{j=1}^{N} f_j \exp 2\pi i \mathbf{r}_j \mathbf{S}$$

$$= \int_{\text{vol. of unit cell}} \rho(\mathbf{r}) \exp 2\pi i \mathbf{rS} dv \quad (46)$$

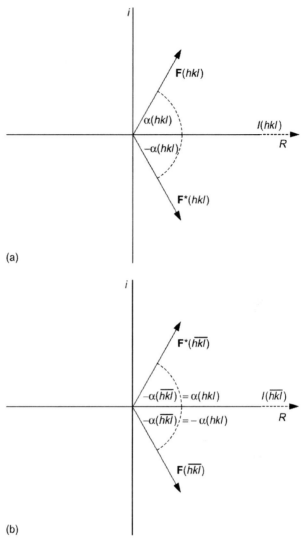

Fig. 27 Diagram illustrating the basis of Friedel's law.

shows that it is the Fourier transform of the electron density $\rho(\mathbf{r})$.

The electron density $\rho(\mathbf{r})$ is then the inverse Fourier transform of the structure factor $\mathbf{F(S)}$ according to Eq. (47):

$$\rho(\mathbf{r}) = \int_{\substack{\text{vol. of} \\ \text{diffraction space}}} \mathbf{F(S)} \exp{-2\pi i\mathbf{rS}} dv_s \quad (47)$$

The integration is replaced by summation since $\mathbf{F(S)}$ is not continuous and is nonzero only at the reciprocal lattice points. Hence, we have (Eq. 48)

$$\rho(xyz) = \frac{1}{V} \sum_{h=-\infty}^{\infty} \sum_{k=-\infty}^{\infty} \sum_{l=-\infty}^{\infty} F(hkl) \times \exp[-2\pi i(hx+ky+lz)] \quad (48)$$

Knowing the structure factors (Eq. 49)

$$F(hkl) = |F(hkl)| \exp i\alpha(hkl) \quad (49)$$

one can calculate the electron density distribution in the unit cell and thus determine the atomic positions of the scattering molecule(s). Unfortunately, the measured quantities are only the absolute values $|F(hkl)|$ of the structure factor. Information on the phase angles $\alpha(hkl)$ is lost during the diffraction experiment. The determination of these phases is the basic problem in any crystal structure determination, and we discuss the methods for solving the phase problem later.

3.8
The Patterson Function

The measured X-ray intensities are proportional to the square of the absolute value of the structure factor according to Eq. (28). Would it be possible to use the intensities directly to calculate from these a function that contains structural information? The answer is "yes." If one calculates a convolution of the electron density with itself, Patterson showed that this is just the Fourier transform of the intensities (Eq. 50):

$$P(uvw) = \int_{\substack{\text{vol. of} \\ \text{unit cell}}} \rho(xyz)$$
$$\times \rho(x+u, y+v, z+w)dv \quad (50)$$

From this, it follows that (Eq. 51)

$$P(\mathbf{u}) = \frac{1}{V} \sum_{\mathbf{h}} F_\mathbf{h}^2 \exp -2\pi i\mathbf{h}\mathbf{u} \quad (51)$$

The function $P(\mathbf{u})$ will have maxima if the positions $\mathbf{x}$ and $\mathbf{x}+\mathbf{u}$ correspond to atoms. Thus, we obtain a function that contains the interatomic vectors as maxima. We expect N^2 peaks for N atoms.

The maxima are proportional to $Z_i Z_j$. The Patterson function is a very useful tool to locate atoms when the number of atoms in the asymmetric unit of the unit cell is not too high (e.g. <20) or it contains a subset of heavy atoms among not too many (e.g. <100) light atoms such as C, N, O, or S. Here, the heavy atom–heavy atom vectors are clearly prominent. If a protein with 1000 or a multiple of that light atoms holds one or several heavy atoms per molecule, the signal resulting from the heavy atoms can no longer be resolved. However, when using the method of isomorphous replacement (discussed later), a Patterson function of the heavy atom structure can be calculated, from which it is possible to locate the heavy atoms.

3.9
Integrated Intensity Diffracted by a Crystal

Real crystals are not perfect. They can be regarded as consisting of small blocks of perfect crystals (sizes in the range of 0.1 μm), which have an average tilt angle among each other of 0.1 to 0.5° for protein crystals and which diffract independently of each other. Such a real crystal is denoted a mosaic crystal. The total energy, $E(\mathbf{h})$, in a diffracted beam for a mosaic crystal rotating with uniform angular velocity w through the reflecting position in a beam of X-rays of incident intensity I_0 is given by Darwin's equation (Eq. 52):

$$E(\mathbf{h}) = \frac{I_0}{\omega} \lambda^3 \frac{e^4}{m^2 c^4} p \frac{LAV_x}{V^2} |F(\mathbf{h})|^2 \quad (52)$$

where λ = wavelength of X-rays, e = electronic charge, m = mass of electron, c = velocity of light, p = polarization factor, L = Lorentz factor (geometrical factor taking into account the relative time each reflection spends in the reflection position), A = absorption factor, V = volume of the unit cell, and $V_x =$ irradiated crystal volume.

Owing to the mosaicity (0.1–0.5°), each reflection has a corresponding reflection width. The integrated intensity equation is valid under the assumption that apart from ordinary absorption, the incident intensity, I_0, is constant within the crystal (kinematic theory of X-ray diffraction) and the mosaic blocks are so small that no multiple scattering occurs within an individual mosaic block. The integrated intensity depends on λ to the third power. Increasing the wavelength causes appreciably stronger diffraction intensities but is accompanied by larger absorption. Copper Kα radiation with a wavelength of 1.5418 Å is an optimal choice for protein crystallography when using X-ray generator sources. Also important is the dependence of the integrated intensity on the unit cell volume V by its negative second power. Doubling of the unit cell volume with twice as many molecules, taking into account the increase in $|\mathbf{F}(\mathbf{h})|^2$ by having now $2n$ molecules per unit cell, reduces the average intensity for the reflected beams by a factor of two. In Eq. (52), $(\lambda^3/\omega V^2) \times (e^4/m^2c^4) \times V_x \times I_0$ is a constant for a given experiment. The corrected intensity on a relative scale $I(\mathbf{h})$ is obtained from Eq. (53):

$$I(\mathbf{h}) = \frac{E(\mathbf{h})}{p \times L \times A} \quad (53)$$

3.10
Intensities on an Absolute Scale

The corrected intensity on a relative scale $I(\mathbf{h})$ can be converted to an intensity given by Eq. (54)

$$I(abs, \mathbf{h}) = \mathbf{F}(\mathbf{h})\mathbf{F}(\mathbf{h})^* = |\mathbf{F}(\mathbf{h})|^2 \quad (54)$$

on an absolute scale by applying a so-called Wilson plot. The basis for this plot is an equation that connects the average intensity on an absolute scale with the average intensity on a relative scale by a scale factor C and considers the isotropic thermal motion of the scattering atoms by the temperature factor given in Eq. (44). This is written in the form of Eq. (55):

$$\ln \frac{\overline{I(\mathbf{h})}}{\sum_j (f_j)^2} = \ln C - 2B \frac{\sin^2 \theta}{\lambda^2} \quad (55)$$

This is the equation of a straight line. B, the overall temperature factor, and C, the scale factor, can be obtained by plotting $\ln \overline{I(\mathbf{h})}/\sum_j(f_j)^2$ against $(\sin^2 \theta)/\lambda^2$.

3.11
Resolution of the Structure Determination

The concept of resolution in X-ray diffraction has the same meaning as the concept in image formation in the optical microscope. After the Abbe theory, we obtain Eq. (56):

$$d_m = \frac{\lambda}{2NA} \quad (56)$$

where NA is the numerical aperture of the objective lens. In protein crystallography, the nominal resolution of an electron density map is expressed in d_m, the minimum interplanar spacing for which Fs are included in the Fourier series. The maximum attainable resolution at a given wavelength is $\lambda/2$. For Cu Kα radiation, it is 0.7709 Å and would suffice to determine protein structures at atomic resolution (the distance of a carbon–carbon single bond is about 1.5 Å). However, usually the thermal vibrations of the atoms in a protein crystal are so high that the diffraction data cannot be observed to the full theoretical resolution limit. The polypeptide chain fold can be determined at a resolution of better than 3.5 Å. A medium-resolution structure is in the resolution range of 3.0 to 2.2 Å, and makes the amino acid side chains clearly visible. A high-resolution structure has a nominal resolution better than 2.2 Å and can be as good as 1.2 Å. In such structures, the main-chain carbonyl oxygens become visible as prominent bumps and at a resolution better than 2.0 Å, aromatic side chains acquire a hole in the middle of their ring systems. For some very well diffracting crystals from small proteins, diffraction data extending to resolutions below 1.2 Å could be collected with synchrotron radiation. Such structures reveal real atomic resolution where each atom is visible as an isolated maximum in the electron density map.

3.12
Diffraction Data Evaluation

The analysis and reduction of diffraction data from a single crystal consists of seven main steps: (1) visualization and preliminary analysis of the raw, unprocessed data; (2) indexing of the diffraction patterns; (3) refinement of the crystal and detector parameters; (4) integration of the diffraction spots; (5) finding the relative scale factors between measurements; (6) precise refinement of crystal parameters using the whole data set; and (7) merging and statistical analysis of the measurements related by space-group symmetry. When using electronic area detectors with short readout times such as CCD or MWPC detectors, it is possible to collect diffraction images with small rotational increments (0.05–0.2°). In this case, the reflection profile over the crystal rotation angle can be registered, giving a three-dimensional picture of the spot. The evaluation of such diffraction data can be done with computer programs MADNES, XDS, the San Diego programs and related programs XENGEN, and X-GEN. IP systems with their longer readout times are operated in a film-like mode with rotational increments of 0.5 to 2.0°. Here, mainly the program systems MOSFLM and Denzo are applied. The most important developments in the data evaluation of macromolecular diffraction measurements are autoindexing, profile fitting, transformation of data to a reciprocal-space coordinate system, and demonstration that a single rotation image contains all of the information necessary to derive the diffraction intensities from that image.

Scaling, merging, and statistical analysis of the intensity data are either done with corresponding programs of the CCP4 program suite or with Scalepack. The principles of these operations are given in the manuals for these programs. With modern data-collection methods, the completeness should approach 100% (including the low-resolution data, which are very important for molecular replacement), the ratio $I/\sigma(I)$ should be significant even for the highest resolution shell, and undue emphasis should not be given to the reliability factor for merging the data (R-merge) unless factors such as multiplicity are taken into account. Nowadays, it is customary using synchrotron radiation techniques and fast CCD detectors to collect as much data as possible (before radiation damage becomes significant) in order to produce good statistics.

3.13
Solvent Content of Protein Crystals

The Matthews parameter V_M, which is defined according to Eq. (57):

$$V_M = \frac{V_{\text{unit cell}}}{M_{\text{Prot}}} \quad (57)$$

where $V_{\text{unit cell}}$ is the volume of the unit cell and M_{Prot} is the molecular mass of the protein in the unit cell, has values that are in the range 1.6 to 3.5 Å^3 Da^{-1} for proteins. This allows a rough estimation of the number of molecules in the unit cell. Furthermore, V_M can be used for the assessment of the solvent content of a protein crystal. Calling V_{Prot} the crystal volume occupied by the protein, V'_P its fraction with respect to the total crystal volume V, and M_{Prot} the mass of protein in the cell, we obtain Eq. (58):

$$V'_P = \frac{V_{\text{Prot}}}{V} = \frac{V_{\text{Prot}}/M_{\text{Prot}}}{M_{\text{Prot}}/V} \quad (58)$$

The first term is the specific volume of the protein, the second the reciprocal of V_M and, remembering that the molecular weight is expressed in daltons, we have Eq. (59):

$$V'_P = \frac{1.6604}{d_{\text{Prot}} V_M} \quad (59)$$

Taking 1.35 g cm^{-3} as the protein density, we obtain as a first approximation Eq. (60) and (61):

$$V'_P \approx \frac{1.23}{V_M} \quad (60)$$

$$V'_{\text{Solv}} \approx 1 - V'_P \quad (61)$$

The solvent content in a protein crystal may vary from 75 to 40%.

4
Methods for Solving the Phase Problem

4.1
Isomorphous Replacement

4.1.1 Preparation of Heavy Metal Derivatives

If one can attach one or several heavy metal atoms at defined binding site(s) to the protein molecules without disturbing the crystalline order, one can use such isomorphous heavy atom derivatives for the phase determination. The lack of isomorphism can be monitored by a change in the unit cell parameters compared with the native crystal and a deterioration of the quality of the diffraction pattern. The preparation of heavy atom derivatives is undertaken by soaking the crystals in mother liquor containing the dissolved heavy metal compound. Soaking times may be in the range from several minutes to months. Concentrations of the heavy metal compound may vary from tenths of millimolar to 50 mM. Favorite heavy atoms are Hg, Pt, U, Pb, Au, rare earth metals, and so on. Potential ligands can be classified as hard and soft ligands according to Pearson. Hard ligands are electronegative and undergo electrostatic interactions. In proteins, such ligands are glutamate, aspartate, terminal carboxylates, hydroxyls of serines and threonines, and in the buffer acetate, citrate, and phosphate. Soft ligands are polarizable and form covalent bonds such as cysteine, cystine, methionine, and histidine in proteins, and Cl$^-$, Br$^-$, I$^-$, S-ligands, CN$^-$, and imidazole in the buffer solution.

Metals are classified according to their preference for hard or soft ligands. Class (a) metals bind preferentially to hard ligands. They comprise the cations of A-metals such as alkali and alkaline earth metals, the lanthanides, some actinides, and groups IIIA, IVA, and VA of the transition metals. Class (b) metals are rather soft and polarizable and can form covalent bonds to soft ligands. They include heavy metals at the end of the transition metal groups such as Hg, Pt, and Au. Thus, in the protein, the class (b) metals Hg, Pt, and Au and complex compounds of them bind to soft ligands such as cysteine, histidine, or methionine and the class (a) metals U and Pb to hard ligands such as the carboxylate groups of glutamate or aspartate.

4.1.2 Single Isomorphous Replacement

The structure factor $\mathbf{F}_{PH}$ for the heavy atom derivative structure (Fig. 28) becomes (Eq. 62)

$$\mathbf{F}_{PH} = \mathbf{F}_P + \mathbf{F}_H \quad (62)$$

where $\mathbf{F}_P$ = structure factor of the native protein and $\mathbf{F}_H$ = contribution of the heavy atoms to the structure factor of the derivative. The isomorphous differences, $F_{PH} - F_P$, which can be calculated from experimental intensity data sets of the native and derivative protein, correspond to the distance CB in Fig. 28, and are given by Eq. (63):

$$F_{PH} - F_P = F_H \cos(\alpha_{PH} - \alpha_H)$$
$$- 2F_P \sin^2\left(\frac{\alpha_P - \alpha_{PH}}{2}\right) \quad (63)$$

If F_H is small compared with F_P and F_{PH}, the sine term will be very small and we have (Eq. 64)

$$F_{PH} - F_P \approx F_H \cos(\alpha_{PH} - \alpha_H) \quad (64)$$

When vectors $\mathbf{F}_P$ and $\mathbf{F}_H$ are collinear, then (Eq. 65)

$$|\mathbf{F}_{PH} - \mathbf{F}_P| = F_H \quad (65)$$

The square of the isomorphous differences, $F_{PH} - F_P$, can be used as coefficients in a Patterson synthesis. We get

$$(F_{PH} - F_P)^2$$
$$= 4F_P^2 \sin^4\left(\frac{\alpha_P - \alpha_{PH}}{2}\right) \quad (i)$$
$$+ F_H^2 \cos^2(\alpha_{PH} - \alpha_H) \quad (ii) \quad (66)$$
$$- 4F_P F_H \sin^2\left(\frac{\alpha_P - \alpha_{PH}}{2}\right)$$
$$\times \cos(\alpha_{PH} - \alpha_H) \quad (iii)$$

It is a theorem of Fourier theory that the Fourier transform of the sum

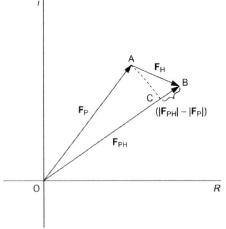

Fig. 28 Vector diagram for the vector addition of the structure factor of the native protein $\mathbf{F}_P$ and the heavy atom contribution $\mathbf{F}_H$ to the heavy atom derivative structure factor $\mathbf{F}_{PH}$.

of Fourier coefficients is equal to the sum of the Fourier transforms of the individual Fourier coefficients. Here, there are three different terms. $(\alpha_P - \alpha_{PH})$ is small if F_H is small, and term (i), which gives the protein–protein interaction will be of low weight. The transform of term (iii) is zero if sufficient terms are included. However, if $F_H \ll F_P$, $(\alpha_P - \alpha_{PH})$ is effectively random, and term (ii) will give heavy atom vectors with half the expected peak heights (Eq. 67)

$$F_H^2 \cos^2(\alpha_{PH} - \alpha_H)$$
$$= \frac{1}{2}F_H^2 + \frac{1}{2}F_H^2 \cos 2(\alpha_{PH} - \alpha_H) \quad (67)$$

with the second term on the right contributing only noise to the Patterson map because the angles α_{PH} and α_H are not correlated. Such an isomorphous heavy atom difference Patterson map allows the determination of the positions of the heavy metals on the condition of isomorphism and a not-too-large heavy atom partial structure. The interpretation of these difference Patterson maps is undertaken by vector verification routines, which are part of the CCP4 program suite. In these routines, the asymmetric unit of the unit cell is systematically scanned by calculating on each scan point the corresponding heavy atom–heavy atom vectors, determining their peak height in the Patterson map and evaluating a meaningful correlation value (e.g. the sum of the correlated maxima). Prominent heavy atom sites should show up with high correlation values.

It is important to know what intensity changes are generated by the attachment of heavy atoms to the macromolecule. According to Crick and Magdoff, the relative root-mean-square intensity change is given by Eq. (68) for centric reflections:

$$\frac{\sqrt{\overline{(\Delta I)^2}}}{\overline{I_P}} = 2 \times \sqrt{\frac{\overline{I_H}}{\overline{I_P}}} \quad (68)$$

and by Eq. (69) for acentric reflections:

$$\frac{\sqrt{\overline{(\Delta I)^2}}}{\overline{I_P}} = \sqrt{2} \times \sqrt{\frac{\overline{I_H}}{\overline{I_P}}} \quad (69)$$

where $\overline{I_H}$ is the average intensity of the reflections if the unit cell were to contain the heavy atoms only and $\overline{I_P}$ is the average intensity of the reflections of the native protein. Attaching one mercury atom ($Z = 80$) to a macromolecule with varying molecular mass and assuming 100% occupancy gives the following average relative changes in intensity: 0.51 for 14 000 Da, 0.25 for 56 000 Da, 0.18 for 112 000 Da, 0.13 for 224 000 Da, and 0.09 for 448 000 Da. From this estimation it is evident that with increasing molecular mass more heavy atoms or for large molecular masses heavy metal clusters such as $Ta_6Br_{12}^{2+}$ must be introduced to generate intensity changes that can be statistically measured (precision for intensity measurements between 5 and 10%) and which are sufficient for the phasing.

The phase calculation for single isomorphous replacement can be seen from the so-called Harker construction for this case (Fig. 29). $\mathbf{F_H}$, which can be calculated from the known heavy atom positions, is drawn in its negative direction from the origin O ending at point A. Circles are drawn with radii F_P and F_{PH} from points O and A respectively. The connections of the intersection points of both circles B and C with origin O determine two possible phases for $\mathbf{F_P}$. This means that the single isomorphous replacement leaves an ambiguity in the phase determination for the acentric reflections.

1 X-Ray Diffraction of Biological Macromolecules

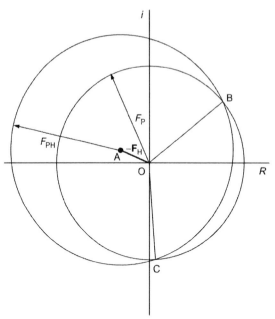

Fig. 29 Harker construction for the phase calculation by the method of single isomorphous replacement.

4.1.3 Multiple Isomorphous Replacement

The phase ambiguity can be overcome if two or more isomorphous heavy atom derivatives are used, which exhibit different heavy atom partial structures. In Fig. 30, the Harker construction for two different heavy atom derivatives is shown. In addition to Fig. 29, $-\mathbf{F}_{H2}$ is drawn from the origin O and a third circle with radius F_{PH2} is inserted around its endpoint B. The

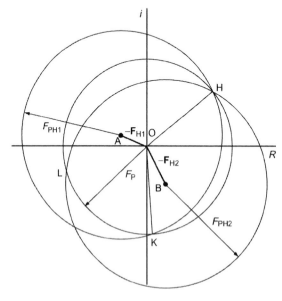

Fig. 30 Harker construction for the phase calculation by the method of MIR for two different heavy atom derivatives PH1 and PH2.

intersection point, H, of all three circles determines the protein phase, α_P. In the case of n isomorphous derivatives, there are $n+1$ circles, which have one common intersection point whose connection to origin O determines the protein phase, α_P.

4.2
Anomalous Scattering

4.2.1 Theoretical Background

So far, in the normal Thomson scattering of X-rays, the electrons in the atom have been treated as free electrons that vibrate as a dipole-oscillator in response to the incident electromagnetic radiation and generate elastic scattering of the X-rays. However, the electrons are bound to atomic orbitals in atoms, and this treatment is valid only if the frequency ω of the incident radiation is large compared to any natural absorption frequency ω_{kn} of the scattering atom. For the light atoms in biological macromolecules (H, C, N, O, S, P) with frequency ω of the used radiation (in the range of 0.4 to 3.5 Å), this condition is fulfilled and these atoms scatter normally. For heavier elements, the assumption $\omega \gg \omega_{kn}$ is no longer valid, and the frequency ω may be higher for some and lower for other absorption frequencies. If ω is equal to an absorption frequency ω_{kn}, absorption of radiation will occur, which is manifested by the ejection of a photoelectron with an energy corresponding to the ionization energy for this electron. This transition goes to a state in the continuous region because the discrete energy states are all occupied in the atom. The absorption frequencies for the K, L, or M shells are connected with the corresponding absorption edges, which are characterized by a sharp drop in the absorption curve (absorption vs λ) at the edge position. It is evident that the scattering from the electrons with their resonance frequencies close or equal to the frequency of the incident radiation will deliver a special contribution, which is called *anomalous scattering*.

The classical treatment is briefly outlined. It is assumed that the atoms scatter as if they contain electric dipole-oscillators having certain definite natural frequencies. The classical differential equation of the motion of a particle of mass m and charge e in an alternating electric field $\mathbf{E} = \mathbf{E}_0 e^{i\omega t}$ is Eq. (70):

$$\ddot{\mathbf{x}} + k\dot{\mathbf{x}} + \omega_s^2 \mathbf{x} = \frac{e\mathbf{E}_0}{m} e^{i\omega t} \quad (70)$$

where the damping factor, k, is proportional to the velocity of the displaced charge and ω_s is the natural circular frequency of the dipole if the charge is displaced. The steady state solution for this equation for the moment of the dipole that executes forced oscillations of frequency under the action of the incident wave is Eq. (71):

$$\mathbf{M} = e\mathbf{x} = \frac{e^2}{m} \frac{\mathbf{E}_0 e^{i\omega t}}{\omega_s^2 - \omega^2 + ik\omega} \quad (71)$$

The amplitude A of the scattered wave at unit distance in the equatorial plane is given by Eq. (72):

$$A = \frac{e^2}{mc^2} \frac{\omega^2 E_0}{\omega_s^2 - \omega^2 + ik\omega} \quad (72)$$

The scattering factor of the dipole, f, is now defined as the ratio of the amplitude scattered by the oscillator to that scattered by a free classical electron under the same conditions. This amplitude at unit distance and in the equatorial plane is given by Eq. (73):

$$A' = -\frac{e^2}{mc^2} E_0 \quad (73)$$

Hence, we obtain Eq. (74) for f:

$$f = \frac{\omega^2}{\omega^2 - \omega_s^2 - ik\omega} \quad (74)$$

If f is positive, the scattered wave has a phase difference of π with respect to the primary beam (introduced by the negative sign in the equation for A'). If $\omega \gg \omega_s$, f is unity. In the case of $\omega \ll \omega_s$, f is negative, and the dipole then scatters a wave in phase with the primary beam.

Equation (74) can be split into real and imaginary parts so that we obtain Eq. (75):

$$f = f' + if'' \quad (75)$$

with (Eqs. 76 and 77)

$$f' = \frac{\omega^2(\omega^2 - \omega_s^2)}{(\omega^2 - \omega_s^2)^2 + k^2\omega^2} \quad (76)$$

$$f'' = \frac{k\omega^3}{(\omega^2 - \omega_s^2)^2 + k^2\omega^2} \quad (77)$$

We now extend this for an atom consisting of s electrons, each acting as a dipole-oscillator with oscillator strength $g(s)$ and resonance frequency ω_s. We have to multiply the contribution for each electron by $g(s)$ and form the sum over all electrons. For the total real part of the atomic scattering factor, we obtain Eq. (78):

$$f' = \sum_s \frac{g(s)\omega^2}{\omega^2 - \omega_s^2} \quad (78)$$

which assumes that ω is not very nearly equal to ω_s, and a small damping. f' can be written as Eq. (79):

$$f' = f_0 + \Delta f' = \sum_s g(s) + \sum_s \frac{g(s)\omega_s^2}{\omega^2 - \omega_s^2} \quad (79)$$

For free electrons, we have $\omega_s = 0$ and $f' = f_0 = \sum_s g(s)$. The real part of the increment of the scattering factor is due to the binding of electrons. $\Delta f'$ is the dispersion component of the anomalous scattering.

If ω is comparable to ω_s but slightly greater, $ik\omega$ must not be neglected. f becomes complex (Eq. 80):

$$f = f' + if'' = f_0 + \Delta f' + i\Delta f'' \quad (80)$$

The imaginary part lags $\pi/2$ behind the primary wave, that is it is always $\pi/2$ in front of the scattered wave. $\Delta f''$ is known as the absorption component of the anomalous scattering. In the quantum mechanical treatment of the problem, the oscillator strengths are calculated from the atomic wave functions. Hönl, in theoretical work, used hydrogen-like atomic wave functions. In the frame of this approach, to each natural dipole frequency ω_s in the classical expression, there corresponds in the quantum expression a frequency ω_{kn}, which is the Bohr frequency associated with the transition of the atom from the energy state k to the state n in which it is supposed to remain during the scattering. Modern quantum mechanical calculations of anomalous scattering factors on isolated atoms, based on relativistic Dirac–Slater wave functions, have been carried out by Cromer and Liberman. It follows from the theory of the anomalous scattering of X-rays that f_0 is real, independent of the wavelength of the incident X-rays but dependent on the scattering angle. $\Delta f'$ and $\Delta f''$ depend on the wavelength, λ, of the incident radiation but are virtually independent of the scattering angle.

4.2.2 Experimental Determination

$\Delta f''$ is related to the atomic absorption coefficient μ_0 by Eq. (81):

$$\Delta f''(\omega) = \frac{mc\omega}{4\pi}\mu(\omega_0) \quad (81)$$

$\Delta f'$ can now be calculated by the Kramers–Kronig transformation (Eq. 82):

$$\Delta f'(\omega) = \frac{2}{\pi} \int_0^\infty \frac{\omega' \Delta f''(\omega')}{\omega^2 - \omega'^2} d\omega' \quad (82)$$

As fluorescence is closely related to absorption, fluorescence measurements varying the X-ray radiation frequency are used to determine the frequency dependence of the dispersive components of the different chemical elements. Instead of the radiation frequency ω, the radiation is often characterized by its wavelength, λ, or photon energy, E. The dispersion correction terms $\Delta f'$ and $\Delta f''$ are often simply denoted f' and f''. Figure 31 shows the anomalous scattering factors near the absorption K edge of selenium from a crystal of $E.\ coli$ selenomethionyl thioredoxin. The spectrum was measured with tunable synchrotron radiation. Apart from the "white line" feature at the absorption edge, f'' drops by about 4 electrons, approaching the edge from the short wavelength side; $\Delta f'$ exhibits a symmetrical drop of −8 electrons around the edge. Similar values can be observed at the K edges for Fe, Cu, Zn, and Br, whose wavelengths all lie in the range 0.9 to 1.8 Å, which is well suited for biological macromolecular X-ray diffraction experiments. For other interesting heavy atoms such as Sm, Ho, Yb, W, Os, Pt, and Hg, the LII (Sm) or LIII edges are in this range. Here, the effects are even greater. Considerably larger changes are found for several lanthanides, such as Yb, where the minimum f' is −33 electrons and the maximum f'' is 35 electrons.

4.2.3 Breakdown of Friedel's Law

Under the assumption that the crystal contains a group of anomalous scatterers, one can separate the contributions from the distinctive components of the scattering factor according to Hendrickson and Ogata to obtain Eq. (83):

$$^\lambda F(h) = {}^\circ F_N(h) + {}^\circ F_A(h) + {}^\lambda F'_A(h) + i\,{}^\lambda F''_A(h) \quad (83)$$

where ${}^\circ F_N$ is the contribution of the normal scatterers and ${}^\circ F_A$, ${}^\lambda F'_A$, and ${}^\lambda F''_A$ are the contributions for the corresponding components of the complex atomic form factor. For the centrosymmetric reflection, we obtain Eq. (84):

$$F(-h) = {}^\circ F_N(-h) + {}^\circ F_A(-h) + {}^\lambda F'_A(-h) + i\,{}^\lambda F''_A(-h) \quad (84)$$

The geometric presentation for both structure factors is given in Fig. 32. The

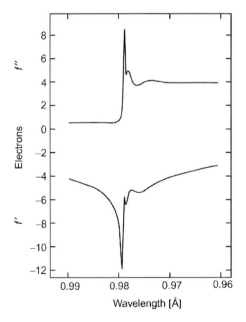

Fig. 31 Anomalous scattering factors near the absorption K edge of selenium from a crystal of E. coli selenomethionyl thioredoxin. (Reproduced by permission of Academic Press, Inc., from Hendrickson, W.A., Ogata, C.M. (1997) Methods Enzymol. **276**, 494–523.)

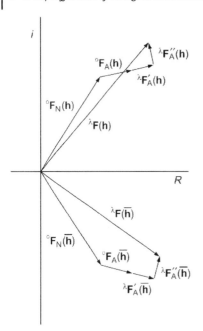

Fig. 32 Vector diagram explaining the breakdown of Friedel's law.

inversion of the sign of **h** causes a negative phase angle for all contributions where the components of the scattering factor are real. For the f''-dependent part, this is also valid, but owing to the imaginary factor i, this vector has to be constructed with a phase angle $+\pi/2$ with respect to $^\circ\mathbf{F}_A(-\mathbf{h})$ and $^\lambda F'_A(-\mathbf{h})$. The resultant absolute values for $^\lambda\mathbf{F}(\mathbf{h})$ and $^\lambda\mathbf{F}(-\mathbf{h})$ are no longer equal, which means that their intensities (square of the amplitude) are different (breakdown of Friedel's law).

4.2.4 Anomalous Difference Patterson Map

One can show that (Eq. 85)

$$^\lambda F(\mathbf{h}) - {}^\lambda F(-\mathbf{h}) \approx \frac{2}{k}[{}^\circ F_A(\mathbf{h})$$
$$+ {}^\lambda F'_A(\mathbf{h})]\sin(\alpha_\mathbf{h} - \alpha_A) \quad (85)$$

where $\alpha_\mathbf{h}$ is the phase angle of $^\lambda F(\mathbf{h})$, α_A the phase angle of the anomalous scatterers, and (Eq. 86)

$$k = \frac{{}^\circ F_A(\mathbf{h}) + {}^\lambda F'_A(\mathbf{h})}{{}^\lambda F''_A(\mathbf{h})} \quad (86)$$

As coefficients for an anomalous difference Patterson map, we obtain Eq. (87):

$$\Delta F^2_{\text{ano}} = [{}^\lambda F(\mathbf{h}) - {}^\lambda F(-\mathbf{h})]^2 \sim \frac{4}{k^2}[{}^\circ F_A(\mathbf{h})$$
$$+ {}^\lambda F'_A(\mathbf{h})]^2 \sin^2(\alpha_\mathbf{h} - \alpha_A) \quad (87)$$

The ΔF_{ano}s will be maximal if the phase angle α_A is perpendicular to the phase angle $\alpha_\mathbf{h}$ and zero if both vectors are collinear, which is opposite to the MIR case. The anomalous Patterson map contains peaks of the anomalous scatterers with heights proportional to half of $(4/k^2)[{}^\circ F_A(\mathbf{h}) + {}^\lambda F'_A(\mathbf{h})]^2$ owing to the $\sin^2$ term and is therefore suited to determine the structure of the anomalous scatterers.

4.2.5 Phasing Including Anomalous Scattering Information

The combination of anomalous scattering information with isomorphous replacement permits the unequivocal determination of the protein phases, as shown in Fig. 33. Using the anomalous scattering information alone gives two possible solutions for the protein phase characterized by the intersection points H and L in Fig. 33. Combining it with the corresponding intensities from the native protein without the anomalous scatterers leaves only one solution for the protein phase (vector O-H in Fig. 33). The case in Fig. 33 is called *single isomorphous replacement anomalous scattering (SIRAS)*. Having n isomorphous heavy atom derivatives, each with anomalous scattering contributions, the Harker

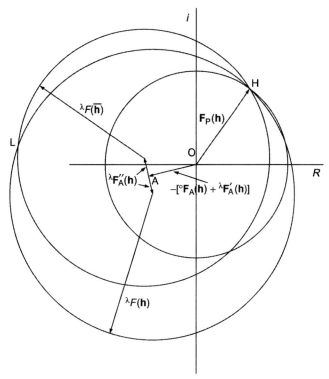

Fig. 33 Harker construction illustrating the phase determination combining information from anomalous scattering and isomorphous replacement.

construction can be extended for this situation, and the phasing method is then designated as multiple isomorphous replacement anomalous scattering (MIRAS).

4.2.6 Multiwavelength Anomalous Diffraction Technique

During the last few years, the MAD technique has matured to be a routine method and has led to a revolution in biological macromolecular crystallography. If there are one or a few anomalous scatterers in the biological macromolecule, it is possible to determine the whole spatial structure from one crystal (exact isomorphism) by the MAD technique. The anomalous scatterers may be intrinsic as in metalloproteins (e.g. Fe, Zn, Cu, Mo, Mn) or exogenous (e.g. Hg in a heavy atom derivative or Se in selenomethionyl proteins). A prerequisite for the MAD technique is well-diffracting crystals (resolution better than 2.8 Å) because the anomalous components of the atomic form factor are virtually independent of the diffraction angle and acquire increasing weight with increasing scattering angle. This advantageous property together with exact isomorphism serves for the determination of good phases down to the full resolution, and leads to the production of excellent experimental MAD-phased electron density maps. A typical MAD experiment is carried out at three different wavelengths (tunable synchrotron radiation), at minimum f' and maximum f'' at the absorption edge of the anomalous scatterer(s), and at a remote wavelength where anomalous scattering effects are small.

The basic equations for the MAD technique as formulated by Hendrickson and Ogata are as follows. Equation (83) can be written as Eq. (88):

$$^\lambda F(h) = {}^\circ F_T(h) + {}^\lambda F_A(h) + i {}^\lambda F_A''(h) \quad (88)$$

where (Eq. 89)

$$^\circ F_T = {}^\circ F_N + {}^\circ F_A \quad (89)$$

with subscript T for the totality of atoms in the structure.

Furthermore, we have Eqs. (90) to (93):

$$^\circ F_T(f^\circ) = {}^\circ F_T \exp(i {}^\circ \phi_T) \quad (90)$$

$$^\circ F_A(f^\circ) = {}^\circ F_A \exp(i {}^\circ \phi_A) \quad (91)$$

$$^\lambda F_A' = f(f') \quad (92)$$

$$^\lambda F_A'' = f(f'') \quad (93)$$

In the common case of a single kind of anomalous scatterer, we obtain Eqs. (94) and (95):

$$^\lambda F_A' = \frac{f'(\lambda)}{f^\circ} {}^\circ F_A \quad (94)$$

$$^\lambda F_A'' = \frac{f''(\lambda)}{f^\circ} {}^\circ F_A \quad (95)$$

Separating the experimentally observable squared amplitude into wavelength-dependent and wavelength-independent terms gives Eq. (96):

$$^\lambda F(\pm h)^2 = {}^\circ F_T^2 + a(\lambda) {}^\circ F_A^2$$
$$+ b(\lambda) {}^\circ F_T {}^\circ F_A \cos({}^\circ \phi_T - {}^\circ \phi_A)$$
$$\pm c(\lambda) {}^\circ F_T {}^\circ F_A \sin({}^\circ \phi_T - {}^\circ \phi_A) \quad (96)$$

with (Eqs. 97–99)

$$a(\lambda) = \frac{f'^2 + f''^2}{f^{\circ 2}} \quad (97)$$

$$b(\lambda) = 2\frac{f'}{f^\circ} \quad (98)$$

$$c(\lambda) = 2\frac{f''}{f^\circ} \quad (99)$$

The derivation of the formula for $^\lambda F(h)^2$ is illustrated in detail. $^\lambda F(h)^2$ is obtained from the triangle formed by the vectors $^\lambda F(h)$, $^\circ F_T(h)$, and $\mathbf{a}$ (Fig. 34) by use of the cosine rule. The absolute values of the vectors are represented in italics, and the relevant angle is $(180° - {}^\circ \phi_A - \delta)$. We get

$$^\lambda F(h)^2 = {}^\circ F_T^2 + \left(\frac{f'^2 + f''^2}{f^{\circ 2}}\right) {}^\circ F_A^2 - 2$$

$$\times {}^\circ F_T \times \left(\frac{f'^2 + f''^2}{f^{\circ 2}}\right)^{1/2} \times {}^\circ F_A$$

$$\times \cos(\pi - \delta - {}^\circ \phi_A + {}^\circ \phi_T) \quad (100)$$

The cosine term in Eq. (100) can be obtained using some basic trigonometry:

$$\cos(\pi + ({}^\circ \phi_T - {}^\circ \phi_A - \delta))$$
$$= -\cos({}^\circ \phi_T - {}^\circ \phi_A - \delta) \quad (101)$$

$$\cos(({}^\circ \phi_T - {}^\circ \phi_A) - \delta) = \cos({}^\circ \phi_T - {}^\circ \phi_A)$$
$$\times \cos \delta + \sin({}^\circ \phi_T - {}^\circ \phi_A) \times \sin \delta \quad (102)$$

with

$$\cos \delta = \frac{\left(\frac{f'}{f^\circ}\right) \times {}^\circ F_A}{\left(\frac{f'^2 + f''^2}{f^{\circ 2}}\right)^{1/2}}$$

$$\times {}^\circ F_A = \frac{\left(\frac{f'}{f^\circ}\right)}{\left(\frac{f'^2 + f''^2}{f^{\circ 2}}\right)^{1/2}} \quad (103)$$

1 X-Ray Diffraction of Biological Macromolecules

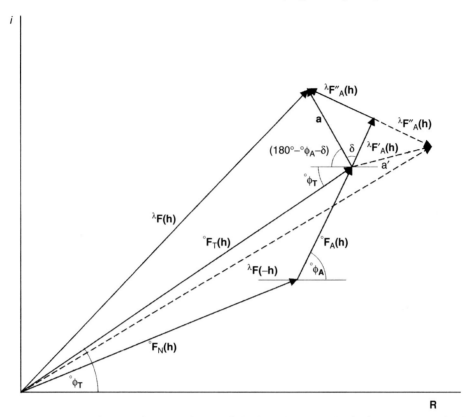

Fig. 34 Schematic drawing of structure factors of a biological macromolecule that contains one kind of anomalous scatterer.

$$\sin\delta = \frac{\left(\frac{f''}{f^\circ}\right) \times {}^\circ F_A}{\left(\frac{f'^2 + f''^2}{f^{\circ 2}}\right)^{1/2}}$$

$$\times {}^\circ F_A = \frac{\left(\frac{f''}{f^\circ}\right)}{\left(\frac{f'^2 + f''^2}{f^{\circ 2}}\right)^{1/2}} \quad (104)$$

Substituting this in Eq. 99, we obtain the expression for ${}^\lambda F(\mathbf{h})^2$. ${}^\lambda F(-\mathbf{h})^2$ is determined from the triangle formed by the vectors ${}^\lambda F(-\mathbf{h})$, ${}^\circ F_T(\mathbf{h})$, and a' (Fig. 34) using a similar approach. Maximum anomalous scattering effects can be expected in intensity differences of reflections that would be equal for exclusively normal scattering. This is the case for Friedel pairs, $\mathbf{h}$ and $-\mathbf{h}$, or their rotational symmetry partners, and the relation for such differences is given in Eq. (85). Of further interest are dispersive differences between structure amplitudes at different wavelengths, Eq. (105):

$$\Delta F_{\Delta\lambda} \equiv {}^{\lambda i}F(\mathbf{h}) - {}^{\lambda j}F(\mathbf{h}) \quad (105)$$

The anomalous or dispersive intensity differences can be used to determine the structure of the anomalous scatterers. The methods are the same as for isomorphous replacement. They include vector verification procedures of difference Patterson maps or direct methods programs such as Shake and Bake and SHELXD.

4.2.7 Determination of the Absolute Configuration

As anomalous scattering destroys the centrosymmetry of the diffraction data, this effect can be used to determine the absolute configuration of chiral biological macromolecules. The most common method is to calculate protein phases on the basis of both hands of the heavy atom or anomalous scatterer structures and check the quality of the relevant electron density map that should be better for the correct hand. Furthermore, secondary structural elements in proteins (consisting of L-amino acids) such as α-helices should be right handed.

4.3 Patterson Search Methods (Molecular Replacement)

If the structures of molecules are similar (virtually identical) or contain a major similar part, this can be used to determine the crystal structure of the related molecule if the structure of the other molecule is known. This is done by systematically exploring the Patterson function of the crystal structure to be determined with the Patterson function of the search model. Let us first consider some important features of the Patterson function. The relation between two identical molecules in the search crystal structure (Fig. 35a) can generally be formulated as Eq. (106):

$$\mathbf{X}_2 = [C]\mathbf{X}_1 + \mathbf{d} \qquad (106)$$

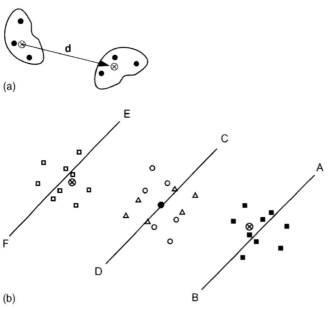

Fig. 35 Patterson function of two identical molecules separated by the spatial movement given in Eq. (106). (a) Positions of the two molecules; (b) interatomic vectors of structure in (a). △, ○, intramolecular vectors of the left and right molecule; □, ■, intermolecular vectors.

Equivalent positions X_1 in molecule 1 are at positions X_2 of molecule 2. [C] is the rotation matrix and **d** is the translation vector of the movement of the molecule. Fig. 35(b) shows the Patterson function belonging to the molecular arrangement in Fig. 35(a). It is evident that around the origin, vectors are assembled that are intramolecular, whereas the vectors around lines AB and EF are intermolecular. The intramolecular vectors depend on the molecule orientation only and, therefore, can be used for its determination. Once the orientation of the molecule(s) has been elucidated, this can be used to reveal the translation of the molecule(s) by analyzing the intermolecular vector part of the Patterson function. The distinction between intra- and intermolecular vector sets and exploiting them for orientation and translation determination was given by Hoppe. The extension to protein crystallography and the first mathematical formulation of the rotation and translation functions were given by Rossmann and Blow.

4.3.1 Rotation Function

The intramolecular vectors are arranged in a volume around the origin of the Patterson function with a radius equal to the dimension of the molecule. The rotational search is then carried out in this volume u. The search Patterson (deduced either from the search model or from the crystal Patterson itself) is rotated to any possible rotational orientation X_2, characterized by three rotational angles α, β, γ, which may be defined in different ways (polar angles, Eulerian angles, etc.). At each angular position, the actual functional values are correlated with those of the crystal Patterson all through the volume u and integrated over this volume. The correlation function may be the sum or the product of each corresponding pair of values. Rossmann and Blow proposed a product function, and the rotation function for this case is given by Eq. (107):

$$R(\alpha, \beta, \gamma) = \int_u P_2(X_2 P_1(X_1)) dX_1 \quad (107)$$

The function has maxima if the intramolecular vector sets are coincident. The calculation can be carried out in both direct and reciprocal space.

The self-rotation function is a special form of the rotation function. If an asymmetric unit contains more than one copy of a molecule, the rotation matrix between the molecules can be determined by a self-rotation function. Here, the crystal Patterson is rotated against itself, and the integration is taken over the volume u around the origin in the same manner. The identical molecules may have an arbitrary orientation to each other or they may be related by local or the so-called noncrystallographic symmetries. Searching for local rotation axes is done best in a polar angle system. The search Patterson is brought into each polar orientation and then rotated around the angle value for the local axis being sought, for example 120° for a threefold local axis.

4.3.2 Translation Function

Once the orientation of the molecule(s) has been determined, the translation of the molecule(s) can be obtained from a translation function. The model Patterson $P_2(u)$ revealed from the model in the correct orientation is calculated for different translations **t** and correlated with the crystal Patterson $P_1(u)$. The translation function proposed by Crowther and Blow has the form shown in Eq. (108):

$$T(t) = \int P_1(u) P_2(u, t) du \quad (108)$$

$T(t)$ reveals a maximum peak at the correct translation t if the center of gravity of the search model was at the origin for $t = 0$.

4.3.3 Computer Programs for Molecular Replacement

An early program for molecular replacement, working in direct space, was written by Huber. Nowadays, several program packages are available, either being exclusively dedicated to the molecular replacement technique or having integrated relevant modules. Pure molecular replacement programs are, for example, AMORE and GLRF. The rotational and translational search starting from the search model is fully automated in AMORE and includes a final rigid body refinement of each proposed solution. GLRF offers different types of rotation and translation functions, all operating in reciprocal space, and a Patterson correlation refinement. A peculiarity of the GLRF program is the locked rotation function. This function takes into account the possible noncrystallographic symmetries and is an average of n independent rotation functions with an improved peak-to-noise ratio. Frequently used program packages including molecular replacement modules are the CCP4 program suite, CNS, and PROTEIN.

4.4 Phase Calculation

4.4.1 Refinement of Heavy Atom Parameters

Before the protein phases can be calculated, it is necessary to refine the heavy atom parameters. These are the coordinates x, y, z, the temperature factor (either isotropic or anisotropic), and the occupancy. The refinement modifies the parameters in such a way that $|F_{PH}(obs)|$ becomes as close as possible to $|F_{PH}(calc)|$. Using the method of least squares, the refinement according to Rossmann minimizes Eq. (109):

$$\varepsilon = \sum_{h} w(h)[(F_{PH} - F_P)^2 - kF_{H\,calc}^2]^2 \tag{109}$$

where k is a scaling factor to correct F_{Hcalc}^2 to a theoretically more acceptable value because according to Eq. (64), $F_{PH} - F_P$ and F_H have approximately the same length when F_{PH}, F_P, and F_H point in the same direction. The probability for this case will be high if the difference between F_{PH} and F_P is large. An improvement can be obtained if the contribution from the anomalous scattering is included.

For the parameter refinement of anomalous scattering sites, the differences between the observed and calculated structure factor amplitudes for $°F_A$ are subjected to minimization. Another approach treats the anomalous or dispersive contributions as in MIR phasing.

From the refined heavy atom parameters, preliminary protein phase angles α_P can be obtained as shown in the corresponding Harker construction. A further refinement of the heavy atom parameters can be achieved by the "lack of closure" method, incorporating this knowledge. The definition of this "lack of closure" ε is illustrated in Fig. 36(a) and (b). In the case of perfect isomorphism, the vector triangle $F_P + F_H = F_{PH}$ closes exactly (Fig. 36a). In practice, this condition will not be fulfilled, and a difference ε between the observed F_{PH} and the calculated F_{PH} will remain (Fig. 36b). $F_{PH}(calc)$ can be obtained from the triangle OAB (Fig. 36b)

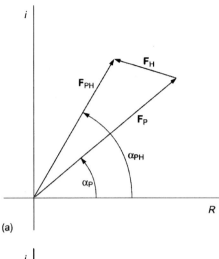

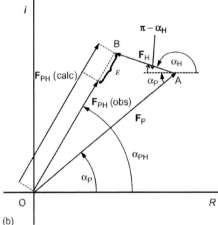

Fig. 36 Definition of "lack of closure." (a) Perfect isomorphism; (b) usually, the observed and calculated values for F_{PH} differ by the "lack of closure" ε.

is the "lack of closure" for the heavy atom derivative j, k_j is a scaling factor, and $m_\mathbf{h}$ is a weighting factor.

4.4.2 Protein Phases

As the structure factor amplitudes F_P, F_{PH}, F_H, and α_H are known, the protein phase angle α_P can be calculated. For the single isomorphous replacement situation (Fig. 28), ε is zero only for the two protein phase angles α_P where the two circles for F_P and F_{PH} intersect. In practice, all these observed quantities exhibit errors. For the treatment of these errors, it is assumed that all errors are in $\mathbf{F}_{PH}$ and that both $\mathbf{F}_H$ and $\mathbf{F}_P$ are error free. For each protein phase angle α, $\varepsilon(\alpha)$ is calculated. The smaller $\varepsilon(\alpha)$ is, the higher is the probability of a correct phase angle α. For each reflection of a derivative j, a Gaussian probability distribution is assumed for ε according to Eq. (113):

$$P(\alpha) = P(\varepsilon) = N \exp\left[-\frac{\varepsilon^2(\alpha)}{2E^2}\right] \quad (113)$$

where N is a normalization factor and E^2 is the mean square value of ε. Small values of E are related to probability curves with sharp peaks and well-determined phase angles, and the opposite is true for large E values. Such phase-angle probability curves can be calculated for each individual reflection and derivative. For single isomorphous replacement, this curve is symmetric with two high peaks corresponding to the two possible solutions for α_P. We obtain the total probability for each reflection with contributions from n heavy atom derivatives by multiplying the

with the cosine rule (Eq. 110):

$$F_{PH} = [F_P^2 + F_H^2 + 2F_P \\ \times F_H \cos(\alpha_H - \alpha_P)]^{1/2} \quad (110)$$

The function that is minimized by the least-squares method is Eq. (111):

$$E_j = \sum_\mathbf{h} m_\mathbf{h} \varepsilon_j(\mathbf{h})^2 \quad (111)$$

where (Eq. (112))

$$\varepsilon_j = k_j F_{PHj}(\text{obs}) - F_{PHj}(\text{calc}) \quad (112)$$

individual probabilities (Eq. 114):

$$P(\alpha) = \prod_{j=1}^{n} P_j(\alpha) = N' \exp\left[-\sum_j \frac{\varepsilon_j^2(\alpha)}{2E_j^2}\right] \quad (114)$$

These curves will be nonsymmetric with one or several maxima (see Fig. 37a and b).

The question arises of which phases should be taken in the electron density equation to calculate the best electron density function. An immediate guess would be to use the phases where $P(\alpha)$ has the highest value. This approach would be appropriate for unimodal distributions but not for bimodal distributions. Blow and Crick derived the phase value that must be applied under the assumption that the mean square error in electron density over the unit cell is minimal. For one reflection, this is given by Eq. (115):

$$\langle \Delta\rho^2 \rangle = \frac{1}{V^2}(\mathbf{F}_s - \mathbf{F}_t)^2 \quad (115)$$

where $\mathbf{F}_t$ is the true factor and $\mathbf{F}_s$ is the structure factor applied in the Fourier synthesis. The mean square error is then obtained as Eq. (116):

$$\langle \Delta\rho^2 \rangle = \frac{1}{V^2} \frac{\int_{\alpha=0}^{2\pi} (\mathbf{F}_s - F \exp i\alpha)^2 P(\alpha) d\alpha}{\int_{\alpha=0}^{2\pi} P(\alpha) d\alpha} \quad (116)$$

F_t has a phase probability of $P(\alpha)$ and has been given as $\mathbf{F}_t = F \exp i\alpha$. It can be shown that the numerator integral in Eq. (116) is minimal if (Eq. 117)

$$\mathbf{F}_{s(best)} = F \frac{\int_{\alpha=0}^{2\pi} \exp(i\alpha) P(\alpha) d(\alpha)}{\int_{\alpha=0}^{2\pi} P(\alpha) d\alpha}$$

$$= mF \exp(i\alpha_{best}) \quad (117)$$

Equation (117) corresponds to the center of gravity of the probability distribution with polar coordinates (mF, α_{best}), where m is defined as magnitude of $\mathbf{m}$ given by Eq. (118):

$$m = \frac{\int_{\alpha=0}^{2\pi} P(\alpha) \exp(i\alpha) d\alpha}{\int_{\alpha=0}^{2\pi} P(\alpha) d\alpha} \quad (118)$$

This magnitude of m is equivalent to a weighting function and is designated the "figure of merit." The electron density map calculated with mF and α_{best} is known

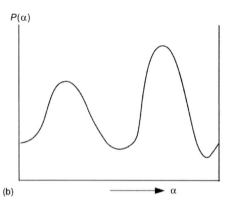

Fig. 37 Total probability curves $P(\alpha)$ for two different reflections: (a) for one derivative; (b) for more than one derivative.

as *best Fourier* and should represent a Fourier map with minimum least-squares error from the true Fourier map.

For the total error of the "best Fourier," Eq. (119) has been derived:

$$\langle \Delta \rho^2 \rangle = \frac{1}{V^2} \sum_{\mathbf{h}} F^2(\mathbf{h})(1 - m^2) \quad (119)$$

The order of magnitude of this error may be illustrated by the example of the structure determination of lysozyme. The root-mean-square error in the Fourier synthesis was 0.35 e Å^{-3}, with values of 2.0-Å resolution for the diffraction data and a mean "figure of merit" of 0.6.

The program systems CCP4 and PROTEIN contain all routines necessary to calculate protein phases according to the MIRAS technique and a number of different kinds of Fourier maps. Alternative probabilistic approaches for the phase calculation are used in programs MLPHARE and SHARP. Both programs can also carry out MAD phasing. The MADSYS program is based on the algebraic approach outlined in the MAD section of this contribution and executes all tasks of a MAD analysis from scaling to phase-angle calculation.

4.5
Phase Improvement

With the methods so far described, an experimental electron density map can be calculated, and if its quality is high enough, the atomic model can be constructed. However, there are methods for further phase improvement available, which may be applied in general or depending on the given prerequisites. Such phase improvement routines have been used routinely over the last 10 years and have had a large impact on the advancement of biological macromolecular crystallography.

4.5.1 Solvent Flattening

Protein crystals have a solvent content of 75 to 40%. In a highly refined protein crystal structure, the solvent space between the molecules is rather flat owing to the dynamic nature of this region. Usually, the initial experimental starting phases are of lower quality than the final ones and as a result, the solvent region (if the molecular boundaries can be identified) contains noise peaks. It is now obvious to set the noisy solvent space to a low constant value and calculate new improved phases by Fourier back-transforming this corrected electron density map. However, it is evident that the definition of the molecular boundaries will be tedious and depend on the quality of the electron density map. Wang has proposed an automatic procedure that smooths the electron density to define the protein region. This smoothed electron density map is traced against a threshold value that separates this map into molecule and solvent space according to their ratio of volumes in the unit cell. The space inside the molecular envelope is polished to avoid internal voids. Now, a new electron density map is calculated using the observed structure factor amplitudes and the phases revealed from the solvent corrected map. The solvent corrected map is obtained by setting all electron density values inside the molecular envelope to those of the initial map and all values outside the envelope to a low constant value. These phases from the solvent flattening procedure can be combined with the MIR or MAD phases. This procedure can be repeated in several iterative cycles because after each cycle of solvent flattening, the quality of the electron density map is improved. There are no prerequisites for the application of the method of solvent flattening. It is evident that solvent flattening is most effective for crystals with a high solvent content.

4.5.2 Histogram Matching

Histogram matching is a technique emanating from image processing. In the application to electron density maps, it is assumed that a high-quality protein crystal structure has a characteristic frequency distribution of electron density, which serves as a standard reference distribution for other electron density maps. Such maps of lower quality exhibit a frequency distribution of electron density, which deviates from the standard distribution. The electron density map of low quality is then scaled in such a way that its frequency distribution of electron density now corresponds to the standard distribution. Histogram matching is normally used together with solvent flattening and is incorporated into the density modification programs SQUASH and DM from the CCP4 program package.

4.5.3 Molecular Averaging

If there is more than one identical subunit in the asymmetric unit of the crystal, molecular averaging can be used to improve the protein phases. The spatial relations between the single identical subunits in the asymmetric unit may be determined by Patterson search methods or from the arrangement of the heavy atoms or anomalous scatterers. The spatial relation between the identical subunits can be improper (the relevant spatial movement consists of a rotation about an unsymmetrical angle value and a translation component) or proper (the spatial movements form a symmetry group that is composed of rotational symmetry elements only). Such additional symmetries are called *noncrystallographic* or *local*, and there are no limitations concerning the Zähligkeit of the symmetry axes (e.g. five-, seven-, and higherfold axes are allowed). It is evident that averaging about the different related subunits, whose electron density should be equal in each subunit, must result in an improved electron density map and therefore in improved protein phases. Molecular averaging is best done in direct space and several programs (e.g. RAVE or MAIN) are available.

The procedure of molecular averaging is composed of several steps. First, the molecular envelope must be determined from the initial electron density map or from a molecular model that, for example, has been obtained from molecular replacement. Next, the particular electron density averaging between the related subunits is performed. This is followed by the reconstitution of the complete crystal unit cell with the averaged electron density. The space outside the molecular envelope is flattened. This map is then Fourier back-transformed. The obtained phase angles can either be taken directly or combined with known phase information to calculate a new and improved electron density map. This cycle can be repeated several times until convergence of the electron density map improvement has been reached. It is very useful to refine the local symmetry operations after every macrocycle of molecular averaging. Furthermore, molecular averaging can be applied if proteins crystallize in more than one crystal form.

Molecular averaging is especially efficient if a high noncrystallographic symmetry is present as in virus structures, but the averaging over two related subunits alone (the lowest case of local symmetry) can give a considerable improvement.

In special cases where high noncrystallographic symmetry exists and the phase information extends to low resolution only, cyclic molecular averaging can be used to extend the phase angles to the full resolution of the native protein. This was first shown in the structure analysis of hemocyanin from *Panulirus interruptus*. It is extensively used in the analysis of icosahedral structures and for large molecular assemblies.

4.5.4 Phase Combination

In the course of a crystal structure analysis of a biological macromolecule, phase information from different sources may be available, such as information from isomorphous replacement, anomalous scattering, partial structures, solvent flattening, and molecular averaging. An overall phase improvement can be expected when these factors are combined, and a useful method to do this was proposed by Hendrickson and Lattman. The probability curve for each reflection is written in an exponential form as Eq. (120):

$$P_s(\alpha) = N_s \exp(K_s + A_s \cos \alpha + B_s \sin \alpha + C_s \cos 2\alpha + D_s \sin 2\alpha) \quad (120)$$

Subscript s stands for the source from which the phase information has been derived. K_s and the coefficients A_s, B_s, C_s, and D_s depend on the structure factor amplitudes and other magnitudes, for example, the estimated standard deviation of the errors in the derivative intensity, but are independent of the protein phase angles α. The overall probability function $P(\alpha)$ is obtained by a multiplication of the individual phase probabilities, and this turns out to be a simple addition of all K_s and of the related coefficients in the exponential term. We obtain Eq. (121):

$$P(\alpha) = \prod_s P_s(\alpha)$$

$$= N' \exp\left[\sum_s K_s + \left(\sum_s A_s\right)\cos\alpha \right.$$

$$+ \left(\sum_s B_s\right)\sin\alpha + \left(\sum_s C_s\right)\cos 2\alpha$$

$$\left. + \left(\sum_s D_s\right)\sin 2\alpha \right] \quad (121)$$

K_s and the coefficients A_s to D_s have special expressions for each source of phase information.

4.6 Difference Fourier Technique

Supposing that one has solved the crystal structure of a biological macromolecule and has isomorphous crystals of this macromolecule, which contain small structural changes caused by a substrate-analog or inhibitor binding, a metal removal or replacement or a local mutation of one or several amino acids. Then, these structural changes can be determined by the difference Fourier technique. The difference Fourier map is calculated with the differences between the observed structure factor amplitudes of the slightly altered molecule $F_{DERI}(obs)$ and the native molecule $F_{NATI}(obs)$ as Fourier coefficients and the phase angles of the native molecule α_{NATI} as phases according to Eq. (122):

$$\rho_{DERI} - \rho_{NATI} \cong \frac{1}{V}\sum_h m[F_{DERI}(obs)$$

$$- F_{NATII}(obs) \times \exp(i\alpha_{NATI})$$

$$\times \exp(-2\pi i\mathbf{hx})] \quad (122)$$

where m may be the figure of merit or another weighting scheme. The difference Fourier map can alternatively be calculated with coefficients $F_{DERI}(obs) - F_{DERI}(calc)$ and phases $\alpha_{DERI}(calc)$. $F_{DERI}(calc)$ and $\alpha_{DERI}(calc)$ do not include the unknown contribution of the structural change.

Figure 38(a) and (b) illustrate the relation for the structure factors involved in the difference Fourier technique. We assume that the structural change is small. If F_{NATI} is large, the structure factor amplitude of the structural change F_{SC} will be small compared to F_{NATI}, and α_{DERI} will be close to α_{NATI}. This is no longer valid if F_{NATI} is small. Now, F_{SC} is comparable to F_{NATI}, and α_{DERI} may deviate considerably from α_{NATI}. This implies the necessity to introduce a weighting scheme that scales down the contributions where the probability is high that α_{NATI} differs appreciably from the correct phase angle. Various weighting schemes have been elaborated such as those of Sim and Read. The weighting scheme of Sim has the following form (Eq. 123):

$$w = \frac{I_1(X)}{I_0(X)} \quad (123)$$

for acentric reflections and (Eq. 124)

$$w = \tanh \frac{X}{2} \quad (124)$$

for centric reflection with (Eq. 125)

$$X = \frac{2 F_{DERI} \times F_{NATI}}{\sum_{1}^{n} f_j^2} \quad (125)$$

$I_0(X)$ and $I_1(X)$ are modified Bessel functions of zeroth and first order respectively. These equations and weighting schemes can also be used for the calculation of OMIT maps (where parts of the model have been omitted from the structure factor evaluation) or when a complete structure must be developed from a known partial model. F_{DERI} must be replaced by the observed structure factor F, F_{NATI} by the structure factor of the known or included part of the model F_K, and α_{NATI} by the phase angle α_K of the known or included part of the model.

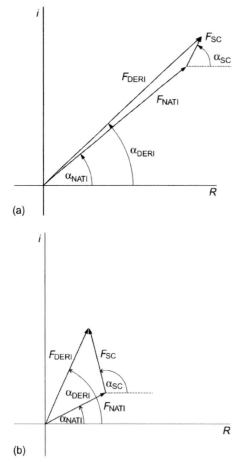

Fig. 38 Vector diagrams illustrating different situations (a) and (b) in the difference Fourier technique for the involved structure factors.

5
Model Building and Refinement

5.1
Model Building

Once the quality of the MIRAS or MAD maps is good enough, model building can be started. This is done on a computer graphics system, and the main modeling programs are O and TURBO-FRODO. An interesting alternative is the program MAIN, which additionally contains routines for molecular averaging, molecular docking, and other features. The visualization of the relevant electron density map on the computer graphics system is done as cagelike structures. For this purpose, the standard deviation from the mean value of the map is calculated, and the cagelike structure is built up for a given contour level (normally 1.0 σ). The first task in a *de novo* protein crystal structure analysis is to localize the trace of the polypeptide chain. This can be assisted by routines for automatic chain tracing such as the bones routine, an auxiliary program of O. Such automatic chain-tracing programs generate a skeleton of the electron density map. This representation was introduced by Greer. When the trace of the polypeptide chain has been identified, the atomic model can be built into the electron density. The atomic model is represented as sticks, connecting the atomic centers of bonded atoms. The individual building blocks (amino acids) of the protein molecule can be generated, interactively manipulated (e.g. linked with each other, moved, rotated, etc.), and fitted into the corresponding part of the electron density map. The geometry of the atomic model is regularized according to protein standard geometries.

The success of model building depends on the quality and resolution of the experimental electron density map. Usually, the quality of the electron density map is not so good that the complete model can be constructed in one cycle. In this case, the partial atomic model is refined crystallographically against the observed structure factor amplitudes. This phase information can be used directly to calculate a new electron density map commonly with $2F_{obs} - F_{calc}$ Fourier coefficient amplitudes. This kind of map is the sum of a normal F_{obs} Fourier and a difference Fourier synthesis. It displays the atomic model with normal weight and indicates errors in the model by its contribution of the difference Fourier map. The parallel determination and inspection of a difference Fourier map are also very helpful. As already mentioned, the model phases can be combined with phases present from other sources or incorporated in procedures of phase improvement. A further model-building cycle can be started with such new and improved electron density maps. After several cycles of model building and crystallographic refinement, the atomic model will be so well defined that the solvent structure of internally bound solvent molecules can be developed. The atomic model is complete now and the biochemical interpretation can be started.

5.2
Crystallographic Refinement

The structural model has to be subjected to a refinement procedure. In practical macromolecular crystallography, one does not always have atomic resolution. Therefore, the single atoms cannot be treated as moving independently. They must be refined using energy or stereochemistry restraints, taking care to maintain a reasonable stereochemistry of the macromolecule. There exist different approaches to structure refinement of macromolecules. The minimization of a potential energy function E together with a diffraction term D according to Eq. (126):

$$S = E + D \qquad (126)$$

where (Eqs. 127 and 128)

$$E = \sum k_b [b_{j(calc)} - b_j^0]^2$$
$$+ \sum k_\tau [\tau_{j(calc)} - \tau_j^0]^2$$
$$+ \sum k_\Theta [1 + \cos(m\Theta_k + \delta)]$$
$$+ \sum (Ar^{-12} + Br^{-6}) \qquad (127)$$

$$D = \sum_i w_i [F_{i(obs)} - kF_{i(calc)}]^2 \qquad (128)$$

is applied in the programs EREF and CNS, which are now used frequently. The four terms of the right-hand side of E describe bond, valence angle, dihedral torsion angle, and nonbonded interactions, k_b is the bond stretching constant, k_τ is the bond angle bending force constant, k_Θ is the torsional barrier, m and δ are the periodicity and the phase of the barrier, A and B are the repulsive and long-range nonbonded parameters, D is the crystallographic contribution with w_i a weighting factor, F_{obs} is the observed structure factor, F_{calc} is the calculated structure factor, and k is a scaling factor. Programs TNT and PROLSQ use stereochemically restrained least-squares refinement. For both refinement schemes, parameters are employed that were derived from small molecule crystal structures of amino acids, small peptides, nucleic acids, sugars, fatty acids, cofactors, and so on. If noncrystallographic symmetry is present, a corresponding term may be introduced in the energy or stereochemistry part of the expression to be minimized. It is possible to divide the structural model into several individual parts and refine these parts as rigid bodies. This is especially useful with solutions from molecular replacement.

A measure of the quality of the crystallographic model is calculated from the crystallographic R-factor (Eq. 129):

$$R = \frac{\sum_i |F_{obs}| - |k| F_{calc}|}{\sum_i |F_{obs}|} \qquad (129)$$

Typical R-factors are below 0.2 for a well-refined macromolecular structure.

Beside the atomic coordinates x, y, z, the atomic temperature factor B may be refined at a resolution better than 3.5 Å. This is done in most programs in a separate step where, for example, in program CNS, the target function (Eq. 130)

$$T = E_{XRAY} + E_R \qquad (130)$$

is minimized, where (Eq. 131)

$$E_R = W_B \sum_{(i,j)\text{-bonds}} \frac{(B_i - B_j)^2}{\sigma_{bonds}^2}$$
$$+ W_B \sum_{(i,j,k)\text{-angles}} \frac{(B_i - B_k)^2}{\sigma_{angles}^2} + W_B$$
$$\times \sum_{k\text{-group}} \sum_{j\text{-equivalences}} \sum_{i\text{-unique atoms}} \frac{(B_{ijk} - \overline{B_{ijk}})^2}{\sigma_{ncs}^2}$$

$$(131)$$

The last term is used only if noncrystallographic symmetry restraints should be imposed on the molecules. Normally, isotropic B-factors are applied and refined in macromolecular crystallography only. Even for a high-resolution structure (1.7 Å), the ratio of observations (observed structure factors) to parameters to be refined (x, y, z, B for each atom) is only about 3. Therefore, as many additional "observations" (energy or stereochemistry restraints) as possible are incorporated. In some cases, it is useful to refine the individual occupancy of certain atoms such as bound metal ions or solvent atoms. This must be performed in a separate step.

All of the mentioned refinement procedures are based on the least-squares method. The radius of convergence for this method is not very high because it follows a downhill path to its minimum. If the model is too far away from the correct solution, the minimization may end in a local minimum corresponding to an incorrect structure. Brünger introduced the method of molecular dynamics (MD), which is able to overcome barriers in the S-function and find the correct global minimum. MD calculations simulate the dynamic behavior of a system of particles. The basic idea of the MD refinement technique is to increase the temperature sufficiently high for the atoms to overcome energy barriers and then to cool slowly to approach the energy minimum. This MD protocol is designated as simulated annealing (SA). The crystallographic application of the MD or SA technique includes a crystallographic term D, as given in Eq. (126), treated as a pseudoenergy term. A crystallographic MD or SA refinement is capable of overcoming a high-energy barrier occurring in the flipping of a peptide plane. It can be useful in removing model bias from the system.

A new approach is the refinement of macromolecular structures by the maximum likelihood method. Programs working on the basis of this method are REFMAC and CNS. The results derived using the maximum likelihood residual are consistently better than those from least-squares refinement.

If the resolution of a biological macromolecular crystal structure is equal to or better than 1.2 Å, it is in the range of real atomic resolution, and the ratio of observations to parameters is high enough to carry out, in principle, an unrestrained crystallographic refinement. Advances in cryogenic techniques, area detectors, and the use of synchrotron radiation enabled macromolecular data to be collected to atomic resolution for an increasing number of proteins. SHELXL is a program with all tools for the crystallographic refinement of biological macromolecules at real atomic resolution.

Since the advent of structural genomics, automation of all parts of structure analysis has become of paramount importance. A first step in this direction was the compilation of the Arp/Warp program system. This system allows the building and refinement of a protein model automatically and without user intervention, starting from diffraction data extending to a resolution higher than 2.3 Å and reasonable estimates of crystallographic phases. The method is based on an iterative procedure that describes the electron density map as a set of unconnected atoms and then searches for protein-like patterns. Automatic pattern recognition (model building) combined with refinement permits a structural model to be obtained reliably within few CPU hours.

5.3
Accuracy and Verification of Structure Determination

A measure of the quality of a structure determination is the crystallographic R-factor given in Eq. (129). For a high-resolution structure, for example 1.6 Å, it should not be much larger than 0.16. As this R-factor is an overall number, it does not indicate major local errors. This can be obtained by the evaluation of a real space R-factor, which is calculated on a grid for nonzero elements according to Eq. (132):

$$R_{\text{real space}} = \frac{\sum |\rho_{\text{obs}} - \rho_{\text{calc}}|}{\sum |\rho_{\text{obs}} + \rho_{\text{calc}}|} \quad (132)$$

where ρ_{obs} is the observed and ρ_{calc} is the calculated electron density.

It has been shown that the conventional R-factor may reach rather low values in a crystallographic refinement with structural models that turned out to be wrong later. To overcome this unsatisfactory situation, Brünger proposed the additional calculation of a so-called free R-factor. For this purpose, the reflections are divided into a working set (e.g. 90%) and a test set (e.g. 10%). The reflections in the working set are used in the crystallographic refinement. The free R-factor is calculated with reflections from the test set, which were not used for the crystallographic refinement and is thus unbiased by the refinement process. There exists a high correlation between the free R-factor and the accuracy of the atomic model phases.

The accuracy of the final model expressed by the mean coordinate error can be determined alternatively by a Luzzati or σ_A plot. The mean coordinate error for a macromolecular structure determined with a resolution of 2.0 Å and a crystallographic R-factor of 0.2 is in the region of ±0.2 Å.

The stereochemistry of the final model must also be checked. The root-mean-square deviation of bond lengths and bond angles from ideal geometry should not be greater than 0.015 Å and 3.0° respectively. The conformation of the main-chain folding is verified by a Ramachandran plot. The dihedral angles Φ and Ψ are plotted against each other for each residue. The data points should lie in the allowed regions of the plot, which correspond to energetically favorable secondary structures such as α-helices, ß-sheets, and defined turn structures. Exceptions are glycine residues, which may occur at any position in the Ramachandran plot. Further stereochemical parameters to be checked are bond lengths and angles, dihedral angles (e.g. determinating side-chain conformations), noncovalent interactions, geometry of H-bonds, and interactions in the solvent structure. This can be done with the programs PROCHECK or WHAT CHECK.

Nearly all spatial structures of biological macromolecules determined either by X-ray crystallography or nuclear magnetic resonance (NMR) techniques have been and will be deposited with the RCSB Protein Data Bank at Rutgers University. The information of the structural model is in a file that contains for each individual atom of the model a record with atom number, atom name, residue type, residue name, coordinates x, y, z, B-value(s), and occupancy. The header records hold useful information such as crystal parameters, amino acid sequence, secondary structure assignments, and references.

6
Applications

6.1
Enzyme Structure and Enzyme–Inhibitor Complex

6.1.1 X-ray Structure of Cystathionine ß-Lyase

Cystathionine ß-lyase (CBL) is a member of the γ-family of pyridoxal-5'-phosphate (PLP)-dependent enzymes that cleaves Cß-S bonds of a broad variety of substrates. The crystal structure of CBL from *Escherichia coli* has been solved using

MIR phases in combination with density modification. The enzyme has been crystallized by the hanging drop vapor diffusion method using either ammonium sulfate or PEG 400 as precipitating agent. The crystals belong to the orthorhombic space group $C222_1$ with unit cell parameters $a = 60.9$ Å, $b = 154.7$ Å, and $c = 152.7$ Å. There is one dimer per asymmetric unit. A native data set has been collected using synchrotron radiation (wavelength 1.1 Å) at the wiggler beamline BW6 at the storage ring DORIS at the Deutsches Elektronensynchrotron (DESY) in Hamburg, Germany. Data sets for three heavy atom derivatives (thiomersalate, 2-mercuri-4-diazobenzoic acid and platinum(II)-2,2′-6,6″-terpyridinium chloride) were registered on an imaging plate scanner (MAR Research, Norderstedt, Germany) using graphite monochromatized Cu Kα radiation from an RU200 rotating anode generator (Rigaku, Tokyo, Japan) operating at 5.4 kW. The reflection data were processed with the MOSFLM package and scaled with programs from the CCP4 program suite. All data sets reveal satisfactory symmetry consistency factors ($R_{\text{merge}} \leq 0.08$) and completenesses (>90%). The heavy atom positions were determined from isomorphous difference Patterson maps using the vector verification routines of program PROTEIN. All derivatives have one common heavy atom–binding site at Cys72, and one of the mercury derivatives shows a second binding site at Cys229. The dimer in the asymmetric unit is related by a local twofold axis that lies parallel to the x-axis. The translation of this local axis was determined from the distribution of the heavy atom sites in the unit cell. An initial MIR map was calculated, followed by solvent flattening, twofold averaging about the local symmetry, and density modification and phase extension to 2.5 Å resolution. Phase calculations, solvent flattening, density modification, and phase extension were done with programs of the CCP4 package. Program AVE was used for the averaging. The quality of the resulting electron density map was sufficiently high to build an almost complete atomic model. Model building was performed on an ESV-30 Graphic system workstation (Evans and Sutherland, Salt Lake City, UT, USA) using program O. The atomic model was refined by energy-restrained crystallographic refinement with XPLOR.

The final model of CBL is made up of two monomers with 391 amino acids each, one cofactor and one hydrogencarbonate molecule per monomer, and 581 solvent water molecules. The final crystallographic R-factor is 0.152 for data from 8.0- to 1.83-Å resolution, and the free R-factor is 0.221. The mean positional error of the atoms as estimated from a Luzzati plot is ±0.19 Å. A homotetramer with 222 symmetry is built up by crystallographic and noncrystallographic symmetry (Fig. 39). Each monomer of CBL (Fig. 40) can be described in terms of three spatially and functionally different domains. The N-terminal domain (residues 1–60) consists of three α-helices and one ß-strand. It contributes to tetramer formation and is part of the active site of the adjacent subunit. The second domain (residues 61–256) harbors PLP and has an $\alpha/\text{ß}$ structure with a seven-stranded ß-sheet as the central part. The remaining C-terminal domain (residues 257–395), connected by a long α-helix to the PLP-binding domain, consists of four helices packed on the solvent-accessible side of an antiparallel four-stranded ß-sheet. The fold of the C-terminal and the PLP-binding domain and the location of the active site are similar to aminotransferases. Most

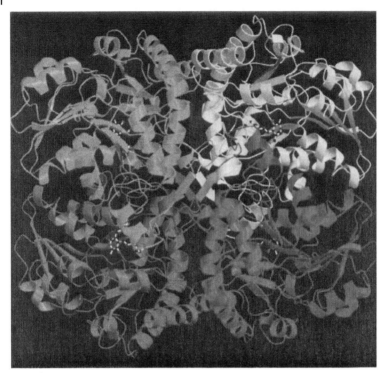

Fig. 39 Ribbon plot of the CBL tetramer viewed along the x-axis. The monomers are colored differently. The blue- and green-colored monomers, which are related by a crystallographic axis (horizontal, in the plane of the paper), build up one catalytic active dimer, and the yellow and red ones the other. The location of the PLP-binding site is shown in a ball-and-stick presentation; MOLSCRIPT and RASTER3D. (Reproduced by permission of Academic Press, Ltd., from Clausen, T. et al. (1996) *J. Mol. Biol.* **23**, 202–224.) (See color plate p. v).

of the residues in the active site are strongly conserved among the enzymes of the transulfuration pathway. Figure 41 shows the final $2F_{obs} - F_{calc}$ map superimposed with the refined atomic model around the active site of CBL. The cage-like structures for the representation of the electron density correspond to a contour level of 1.2σ. The protein part (main- and side-chain atoms), the PLP cofactor, and the hydrogencarbonate molecule are well defined in the electron density map.

The knowledge of the spatial structure of a given enzyme structure forms the basis for understanding its functional properties. It is now possible to design rational site-directed mutants or determine the structures of enzyme–substrate, enzyme–substrate analog, or enzyme–inhibitor complexes, which will deliver invaluable information in understanding the enzyme's functional properties.

6.1.2 Enzyme–Inhibitor Complex Structures of Cystathione ß-Lyase

The enzyme–inhibitor X-ray structures of ß,ß,ß-tri-fluoroalanine (TFA) and L-aminoethoxyvinylglycine (AVG) with CBL could be determined. In both cases,

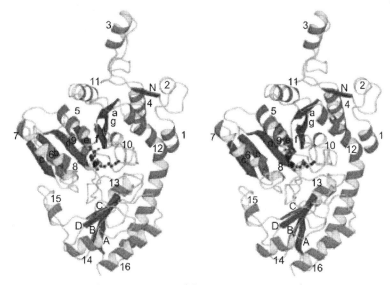

Fig. 40 Stereo ribbon presentation of the CBL monomer, emphasizing secondary structure elements. α-Helices are drawn as green spirals, ß-strands as magenta arrows. PLP and PLP-binding Lys210 are shown in a ball-and-stick representation; MOLSCRIPT and RASTER3D. (Reproduced by permission of Academic Press, Ltd., from Clausen, T. et al. (1996) *J. Mol. Biol.* **23**, 202–224.) (See color plate p. vii).

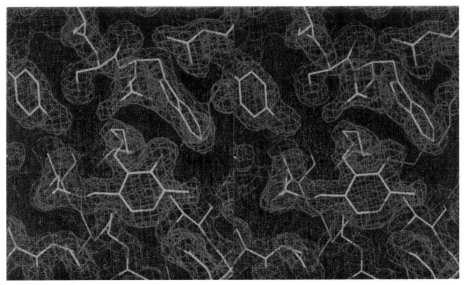

Fig. 41 Stereo plot of the electron density in the active site of CBL, superimposed with the refined model of the region around the cofactor. The $2F_{obs} - F_{calc}$ map is contoured at 1.2σ and calculated at 1.83-Å resolution. (Reproduced by permission of Academic Press, Ltd., from Clausen, T. et al. (1996) *J. Mol. Biol.* **23**, 202–224.)

crystals of the complexes were obtained by incubating the enzyme solution with inhibitor in the millimolar range and subsequent cocrystallization. The resultant crystals were isomorphous with the native enzyme, making it possible to apply the difference Fourier technique in the structure solution. The technical details for the structure analyses are given in the relevant references. The CBL/TFA complex structure was determined to substantiate that the ε-amino group of the active-site Lys210 can react with the nucleophile at the active site via Michael addition, which leads to covalent labeling and inactivation of the enzyme. The final $F_{obs} - F_{calc}$ and $2F_{obs} - F_{calc}$ electron density maps for the CBL/TFA complex around the active site are displayed in Figure 42. Clear, continuous electron density between the cofactor and Lys210 can be seen in this map, indicating a covalent lysine–inactivator–PLP product. In Fig. 42, the blue $F_{obs} - F_{calc}$ map reveals the well-defined electron density for the bound inhibitor. This binding mode of TFA to CBL corresponds to an intermediate in the reaction of TFA with CBL. The structure of the inactivation product proves that Lys210 is the active-site nucleophile reacting via Michael addition with the inactivator. It must also be the residue that transfers a proton from Cα to Sγ in the reaction with the substrates.

The CBL/AVG structure has been determined at 2.2-Å resolution and a crystallographic R-factor of 0.164. The X-ray structure shows that AVG binds to the PLP cofactor forming the external aldimine. Lys210 is no longer bound to the PLP cofactor. Figure 43 is an overlay of the atomic models of native CBL(magenta), CBL/TFA (yellow), and CBL/AVG (green). The main difference in inhibitor binding is the location of Cß and its substituents; in the

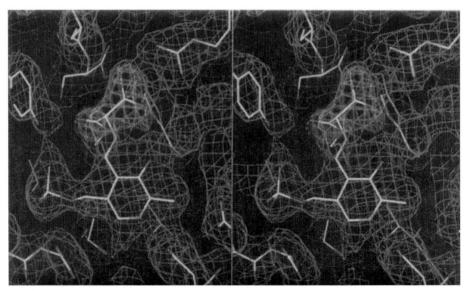

Fig. 42 $F_{obs} - F_{calc}$ (blue) and $2F_{obs} - F_{calc}$ (green) electron density map of the CBL/TFA complex around the active site contoured at 3.5σ and 1.0σ respectively, at 2.3-Å resolution. (Reproduced by permission of Academic Press, Ltd., from Clausen, T. et al. (1996) *J. Mol. Biol.* **23**, 202–224.) (See color plate p. vii).

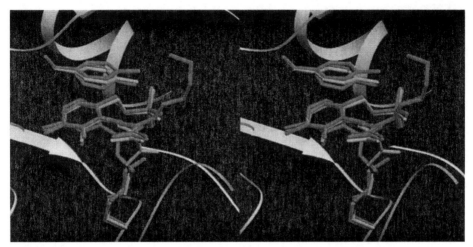

Fig. 43 Stereo view of a superposition of Tyr111 and the PLP derivative of the unliganded enzyme (magenta), the CBL/AVG adduct (green), and the TFA-inactivated enzyme (yellow); SETOR. (Reproduced by permission of the American Chemical Society, Clausen, T. et al. (1997) *Biochemistry* **36**, 12633–12643.) (See color plate p. viii).

TFA complex, the inactivator is directed toward the protein interior (the A face of the cofactor), whereas in CBL/AVG, Cß is located at the B side of the cofactor. The α-carboxylate group in CBL/AVG is located in the same position as the hydrogencarbonate in the native and the α-carboxylate group of TFA in the CBL/TFA complex. The terminal amino group of AVG is held in place mainly by interactions with the hydroxyl group of Tyr111.

The experimental determination of the external aldimine structure in the CBL/AVG complex is of high relevance because it can serve as a rational basis for modeling of substrate and inhibitor binding, leading to more effective herbicides.

6.1.3 Crystal Structure of the Thrombin–Rhodniin Complex

This complex structure is an example related to pharmaceutical research. The goal of this special application is the development of more efficient blood anticoagulants. The target enzyme is α-thrombin. This enzyme is a serine proteinase of trypsin-like specificity. α-Thrombin, the key enzyme in hemostasis and thrombosis, exhibits both enzymatic and hormonelike properties, and can be both pro- and anticoagulatory. Rhodniin is a highly specific inhibitor of thrombin isolated from the assassin bug *Rhodnius prolixus*. Such blood-sucking animals have developed various anticlotting mechanisms to prevent local clotting of the victim's blood. These natural thrombin inhibitors are polypeptides of 60 to 120 amino acid residues.

The crystal structure of the noncovalent complex between recombinant rhodniin and bovine a-thrombin has been determined at 2.6-Å resolution. Crystals were obtained by cocrystallization of thrombin with rhodniin in an approximately 1 : 1 molar ratio. The structure could be solved by molecular replacement because the spatial structure of the major constituent, bovine α-thrombin, was known. Only a diffraction data set of the complex crystal had to be collected. Rotational and translational searches for the orientation and position of the thrombin molecules in the unit cell were performed with the program AMORE. The rotational search showed two solutions with correlation values of 0.22 and 0.20 over 0.09 for the next highest peak. Translational search and rigid body fitting for these two solutions resulted in a correlation value of 0.54, with the two independent complex molecules in the asymmetric unit. The quality of the electron density map calculated from the thrombin phases was good enough in principle to build the model of the rhodniin molecule (noncrystallographic averaging was also applied). The structure was refined with XPLOR to an R-factor of 0.189 and a free R-factor of 0.262.

Figure 44 shows the structure of the complex between thrombin and rhodniin as a ribbon plot, with α-helices represented as ribbon spirals and ß-strands as arrows. The N-terminal domain binds in a substrate-like manner to the narrow active-site cleft of thrombin. The C-terminal domain, whose distorted reactive-site loop cannot adopt the canonical conformation, docks to the fibrinogen recognition exosite via extensive electrostatic interactions. The peculiarity of this complex structure is that the two KAZAL-type domains of rhodniin bind to two different sites of thrombin.

6.2
Metalloproteins

Metals bound as cofactors in proteins have a great variety of functions. They may be involved in the activation of small inorganic or organic molecules, in oxygen storage and transport, in electron transport, in regulation of biological processes, or in stabilizing a transition state during enzymatic catalysis. Their role may also be solely structural. The chemistry of metals in proteins has attracted the interest of inorganic chemists, and has led to the formation of

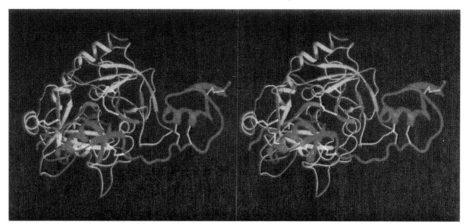

Fig. 44 Stereo view of the complex formed between thrombin (blue) and rhodniin (red) in the thrombin standard orientation, that is, with the active-site cleft facing the viewer and a bound inhibitor chain from left to right. Yellow connections indicate disulfide bridges. Rhodniin interacts through its N-terminal domain in a canonical manner with the active site and through its C-terminal domain with the fibrinogen recognition exosite of thrombin; SETOR. (Reproduced by permission of Oxford University Press, from van de Locht, A. et al. (1995) *EMBO J.* **14**, 5149–5157). (See color plate p. ix).

the field of biological inorganic chemistry. As an example of a complex metalloprotein, the multicopper oxidase ascorbate oxidase (AO) is presented.

6.2.1 Crystal Structure of the Multicopper Enzyme Ascorbate Oxidase

The blue protein AO belongs to the group of "blue" oxidases with laccase and ceruloplasmin. These are multicopper enzymes catalyzing the four-electron reduction of molecular oxygen with concomitant one-electron oxidation of the substrate. The crystal structure of AO has been solved by the MIR technique and refined to 1.9-Å resolution. The peculiarity of this structure determination is briefly described. It consists in the utilization of the information of two different crystal forms. In both crystal forms, the molecules arrange themselves as homotetramers with 222 symmetry, but in crystal form 1, one of these twofold axes is realized by a crystallographic twofold axis resulting in two subunits per asymmetric unit. In crystal form 2, one homotetramer is found per asymmetric unit. In crystal form 1, six isomorphous heavy atom derivatives could be found and interpreted. An initial MIR map was calculated, solvent flattened, and averaged about the local twofold axis. For crystal form 2, no phase information was available. From the averaged uninterpreted MIR map, a whole tetramer was selected and used for rotational and translational searches in crystal form 2. This was successful and provided the necessary phase information for crystal form 2 and, additionally, the local symmetry. Now, averaging could be performed both separately in the two crystal forms and subsequently between both crystal forms. The averaged electron density was transported into both unit cells and a new macrocycle of averaging could be started.

These macrocycles were used to extend the phases from 3.5 Å to the full attainable resolution. This structure analysis was the first example where a molecular replacement was carried out with an uninterpreted MIR electron density–based model.

AO is a homodimeric enzyme with a molecular mass of 70 kDa and 552 amino acid residues per subunit (zucchini). The three-domain structure and the location of the mononuclear centers and trinuclear copper centers in the AO monomer as derived from the crystal structure are shown in Fig. 45. The folding of all three domains is of a similar ß-barrel type. The mononuclear copper site is located in domain 3 and the trinuclear copper species is bound between domains 1 and 3. The coordination of the mononuclear copper site is depicted in Fig. 46. It has the four canonical type-1 copper ligands (His, Cys, His, Met), also found in plastocyanin and azurin. The copper is coordinated to the ND1 atoms of His445 and His512, the SG atom of the Cys507, and the SD atom of Met517 in a distorted trigonal geometry. This unusual coordination geometry confers this copper site with its blue color. The trinuclear copper site (see Fig. 47) has eight histidine ligands symmetrically supplied by domains 1 and 3 and two oxygen ligands. The trinuclear copper site may be divided into a pair of copper (CU2, CU3), with six histidine ligands in a trigonal prismatic arrangement. The pair is bridged by an OH^-, which leads to a strong antiferromagnetic coupling and makes this copper pair electron paramagnetic resonance silent. The remaining copper has two histidine ligands and an OH^- or H_2O ligand. A binding pocket for the reducing substrate, which is complementary to an ascorbate molecule, is located near the mononuclear copper site and is accessible

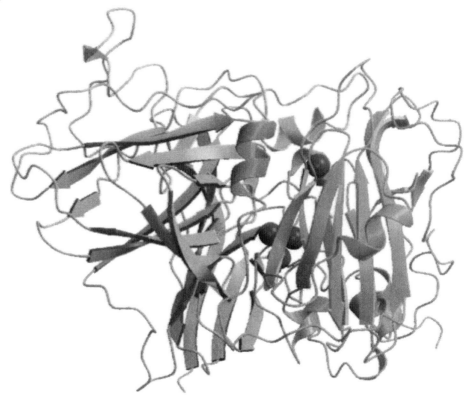

Fig. 45 Schematic representation of the monomer structure of AO. (Reproduced by permission of World Scientific Publishing Co., from Messerschmidt, A. (1997) Spatial Structures of Ascorbate Oxidase, Laccase and Related Proteins, in: Messerschmidt, A. (Ed.) *Multi Copper Oxidases*, World Scientific Publishing, Singapore.)

from solvent. A broad channel providing access from the solvent to the trinuclear copper species, which is the binding and reaction site for the dioxygen, is present in AO. During catalysis, an intramolecular electron transfer between the mononuclear copper site and the trinuclear copper cluster occurs. The distances between the mononuclear copper and the three coppers of the trinuclear center are 12.20, 12.69, and 14.87 Å, respectively. Furthermore, the crystal structures of functional derivatives of AO such as the reduced, azide, and peroxide form have been determined. They show considerable changes at the trinuclear copper site in the reduced form and the peroxide or azide binding to the trinuclear copper species in the relevant structures. The X-ray studies on AO delivered essential information in understanding the catalytic mechanism.

6.2.2 Crystal Structure of Cytochrome-c Nitrite Reductase Determined by Multiwavelength Anomalous Diffraction Phasing

The spatial structure of cytochrome-c nitrite reductase from *Sulfurospirillum deleyianum* has been solved by the MAD technique using synchrotron radiation and

1 X-Ray Diffraction of Biological Macromolecules | 71

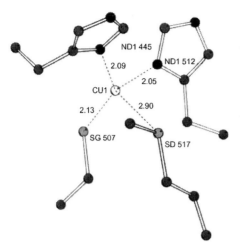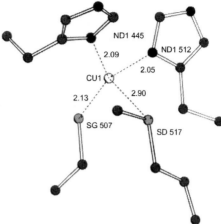

Fig. 46 Stereo drawing of the mononuclear copper site in domain 3 of AO. The displayed bond distances are for subunit A; MOLSCRIPT. (Reproduced by permission of World Scientific Publishing Co., from Messerschmidt, A. (1997) Spatial Structures of Ascorbate Oxidase, Laccase and Related Proteins, in: Messerschmidt, A. (Ed.) *Multi Copper Oxidases*, World Scientific Publishing, Singapore.)

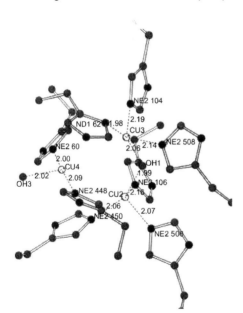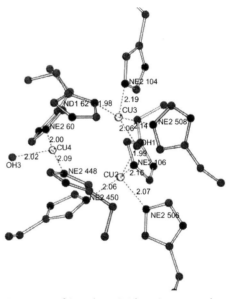

Fig. 47 Stereo drawing of the trinuclear copper site of AO. The displayed bond distances are for subunit A; MOLSCRIPT. (Reproduced by permission of World Scientific Publishing Co., from Messerschmidt, A. (1997) Spatial Structures of Ascorbate Oxidase, Laccase and Related Proteins, in: Messerschmidt, A. (Ed.) *Multi Copper Oxidases*, World Scientific Publishing, Singapore.)

refined at a resolution of 1.9 Å to a R-factor of 0.18. The data collection for the MAD phasing was performed at the wiggler beamline BW6 at the storage ring DORIS at DESY in Hamburg, Germany. At the beginning, an X-ray fluorescence spectrum around the Fe K absorption edge was registered using an NaI(Tl) scintillation counter. Evaluation of the spectrum with program DISCO gave the anomalous dispersion contributions, f' and f'', as a function of photon energy (Fig. 48). Subsequently, diffraction data were collected at three different photon energies. Two of them, 7141 and 7129 eV, correspond to maximum f'' and minimum f' respectively. A third high-resolution data set (1.90 Å resolution) was collected at a photon energy of 11 808 eV. All data sets were obtained from a single crystal that had been shock-frozen at 100 K. Anomalous difference Patterson maps were calculated from the data collected at f'' maximum of the Fe K absorption edge. Correlated peaks that appeared in all three Harker sections were chosen and used for phasing with MLPHARE. The highest peaks from the resulting Fourier map were analyzed for consistency with the anomalous difference Patterson map and, if correct, used for a new cycle of phase calculations. Thus, it was possible successively to find 15 iron positions, which divided into three groups of five. Phasing with program SHARP produced an interpretable electron density map, which was solvent flattened and threefold averaged with program AVE. The resulting electron density map at a resolution of 1.9 Å was of high quality and allowed the construction of the complete atomic model into that map.

Cytochrome-c nitrite reductase (58 kDa) catalyzes the six-electron reduction of nitrite to ammonia as one of the key steps in the biological nitrogen cycle, where it participates in the anaerobic energy metabolism of dissimilatory nitrate ammonification. The crystal structure shows that the enzyme is a homodimer

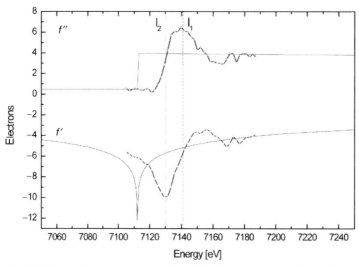

Fig. 48 Dispersion terms as a function of the photon energy as obtained from an X-ray fluorescence scan from cytochrome-c nitrite reductase sample crystal and evaluation with DISCO.

1 X-Ray Diffraction of Biological Macromolecules | 73

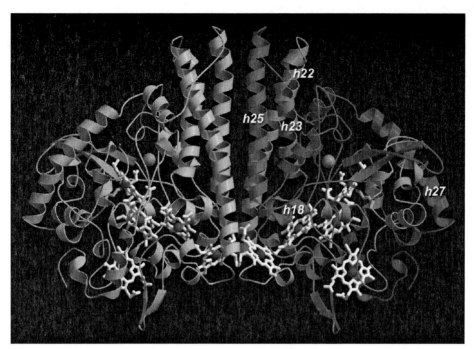

Fig. 49 Overall structure of the nitrite reductase dimer. A front view with the dimer axis oriented vertically, five hemes in each monomer (white), the Ca^{2+} ions (grey), and Lys133 that coordinates the active-site iron (yellow). The dimer interface is dominated by three long α-helices per monomer. All hemes in the dimer are covalently attached to the protein. (Reproduced by permission of Macmillan Magazines, Ltd., from Einsle, O. et al. (1999) *Nature (London)* **400**, 476–480.) (See color plate p. vi).

of ~100 Å × 80 Å × 50 Å, with 10 hemes in a remarkably close packing (Fig. 49). The protein folds into one compact domain with α-helices as the predominant secondary structural motif, ranging from short helical turns to four long helices at the C-terminal end of the peptide chain (Fig. 49). There are eight 3_{10} helices three- or four-residues long, seven of which occur within the first 200 residues, ß-sheet structures are found only in two short, antiparallel strands, where one is part of a funnel-like cavity leading to the active site. Dimer formation is mediated by helices h22 and h25, with helix h25 as the key element, as it interacts with its counterpart in the other monomer over its full length of 28 residues, corresponding to 42 Å. The five hemes in the monomer of cytochrome-c nitrite reductase are in close contact, with Fe-Fe distances of between 9 and 12.8 Å. They are arranged as a group of three, almost coplanar, hemes, with heme 1 forming the active site. Hemes 2 and 5 are farther apart and are not coplanar with hemes 1, 3, and 4. All hemes except heme 1 are bis-histidynyl-coordinated, and are linked to the peptide backbone by thioether bonds to the cysteine residues of a classical heme-binding motif for periplasmic proteins, Cys-X_1-X_2-Cys-His. The propionate side chains of heme 1 form part of the active-site cavity. The site of nitrite reduction is clearly heme

1, with NZ atom of Lys133, replacing a histidine in the classical binding motif, and an oxygen atom of a sulfate ion, present in the crystallization buffer. An additional electron density maximum detected close to the active site was assigned to Ca^{2+}.

S. deleyianum cytochrome-c nitrite reductase reduces not only nitrite to ammonia but also the potential intermediates NO and hydroxylamine. No intermediates are released during nitrite turnover. Obviously, the active site accomodates anions and uncharged molecules and releases the ammonium cation only after full six-electron reduction. The preference for anions is reflected by a positive electrostatic potential around and inside the active-site cavity. The cationic product might make use of a second channel leading to the protein surface opposite to the entry channel. It branches before reaching the protein surface and ends with both arms in areas possessing a significant electrostatic surface potential. The existence of separate pathways for substrate and product with matched electrostatic potential could also contribute to the enzyme's high specific activity compared to the siroheme-containing reductases that catalyze the same reaction.

6.3
Large Molecular Assembly

6.3.1 Crystal Structure of 20S Proteasome from Yeast

The controlled degradation of proteins in the interior of cells is of central significance for many processes that range from cell cycle control and differentiation to cellular immune response. The target protein is labeled for destruction by the covalent linkage of a small protein called *ubiquitin* and is degraded after adenosine triphosphate (ATP)-driven unfolding in the closed internal chamber of a large protein complex, which is known as *26S proteasome* (molecular mass 2 000 000 Da). The core and the proteolytic chamber of the 26S proteasome are formed by the catalytic 20S particle (molecular mass 700 000 Da). This particle is flanked at each end by a so-called 19S cap, which seems to be responsible for the recognition and unfolding of the ubiquitin-labeled target proteins. The eukaryotic catalytic 20S machinery consists of 14 different but related protein subunits that assemble to the overall structure of 28 protein chains, a monster of the molecular world. The structure of the 20S proteasome from *Saccharomyces cerevisiae* has been elucidated at atomic resolution. This is one of the most complex structures that has been determined up to now, excluding symmetrical systems such as viral capsids. It shows that complexity itself is no limit to understanding a structure on the atomic level.

The structure analysis is briefly outlined. Crystals could be obtained by the hanging drop vapor diffusion method with a final precipitant concentration of 12% MPD in the drop (pH 6.5). The crystals are monoclinic, space group $P2_1$, cell parameters $a = 135.7$ Å, $b = 301.8$ Å, $c = 144.7$ Å, and ß $= 112.6°$, and have one 20S particle per asymmetric unit. Data sets with two different inhibitors were collected at the wiggler beamline BW6 at DESY with radiation of $\lambda = 1.1$ Å and at 90 K temperature. The mean attainable resolution was 2.4 Å. A self-rotation function calculated at 5-Å resolution revealed the orientation of the local twofold axis. The atomic model of the 20S proteasome from *Thermoplasma acidophilum* could be used for molecular replacement calculations (program AMORE) at 3.5-Å resolution, and a prominent solution could be found. The resultant electron density map was cyclically averaged about the local twofold axis. The individual subunits were identified according to their characteristic amino acid sequences and were built into the map. The final model with the lactacystin inhibitor was refined at a resolution of 2.4 Å to an R-factor of 0.26, consisting of 48 888 proteins, 30 inhibitors, 18 magnesium atoms, and 1800 solvent molecules.

The structure of the eukaryotic proteasome is important because it is much more complex than its archaebacterial relative, which has two types of subunits, α and ß, only. The α-subunits do not seem to be catalytic but they may self-assemble to form sevenfold rings. In contrast, the ß-subunits show catalytic activity but are not able to assemble themselves. The two components assemble themselves to a stack of four rings, two outer α-rings, which enclose an inner pair of ß-rings. The eukaryotic proteasome retains the important property, however, that the single subunits of the sevenfold rings are different, probably reflecting the increased biological functions. The exact sevenfold symmetry of the particle is lost and a twofold symmetry remains solely to give seven different α- and seven different ß-subunits (Fig. 50a and b). In the proteasome, the unfolded protein is cut

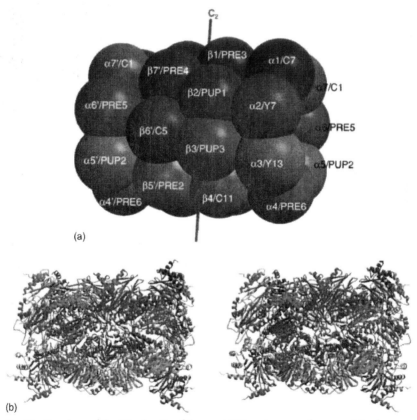

Fig. 50 Topology of the 28 subunits of the yeast 20S proteasome drawn as (a) spheres and (b) ribbon representation; MOLSCRIPT, RASTER3D. (Reproduced by permission of Macmillan Magazines Ltd., from Groll, M. et al. (1997) *Nature (London)*, **386**, 463–471.)

into peptide products in the inner chamber of the 20S particle, and these have a length distribution with a center around octa- or nonapeptides. These sizes are appropriate to bind to MHC-class-I molecules.

See also Molecular Chaperones.

Bibliography

Books and Reviews

Arndt, U.W., Wonnacott, A.J. (Eds.) (1977) *The Rotation Method in Crystallography*, North Holland, Amsterdam, The Netherlands.

Blundell, T.L., Johnson, L.N. (1976) *Protein Crystallography*, Academic Press, New York.

Brünger, A.T. (1992) *XPLOR, Version 3.1, a System for Crystallography and NMR*, Yale University Press, New Haven, CT.

Buerger, M.J. (1961) *Crystal-structure Analysis*, John Wiley & Sons, New York.

Buerger, M.J. (1964) *The Precession Method in X-ray Crystallography*, John Wiley & Sons, New York, 1964.

Carter, C.W. Jr, Sweet, R.M. (Eds.) (1997) *Macromolecular crystallography*, Part A, *Methods Enzymol.* **276**, 1–700.

Carter, C.W. Jr, Sweet, R.M. (Eds.) (1997) *Macromolecular crystallography*, Part B, *Methods Enzymol.* **277**, 1–664.

Drenth, J. (1999) *Principles of Protein X-ray Crystallography*, Springer, New York.

Glusker, J.P., Lewis, M., Rossi, M. (1994) *Crystal Structure Analysis for Chemists and Biologists*, VCH Publishers, New York.

Hahn, T. (Ed.) (1997), *International Tables for Crystallography*, Vol. A. D. Reidel, Dordrecht, The Netherlands.

Helliwell, J.R. (1992) *Macromolecular Crystallography with Synchrotron Radiation*, Cambridge University Press, Cambridge, MA.

McRee, D. (1999) *Practical Protein Crystallography*, Academic Press, San Diego, CA.

Michel, H. (1991) *Crystallization of Membrane Proteins*, CRC Press, Boca Raton, FL.

Wyckoff, H.W., Hirs, C.H.W., Timasheff, S.N. (Eds.) (1985) *Diffraction Methods for Biological Macromolecules*, Part A, *Methods Enzymol.* **114**, 1–588.

Wyckoff, H.W., Hirs, C.H.W., Timasheff, S.N. (Eds.) (1985) *Diffraction Methods for Biological Macromolecules*, Part B, *Methods Enzymol.* **115**, 1–485.

Primary Literature

Berman, H.M., Westbrook, J., Feng, Z., Gilliland, G., Bhat, T.N., Weissig, H., Shindyalov, I.N., Bourne, P.E. (2000) The protein data bank, *Nucleic Acids Res.* **28**, 235–242.

Blow, D.M., Crick, F.H.C. (1959) The treatment of errors in the isomorphous replacement method, *Acta Crystallogr.* **12**, 794–802.

Brünger, A.T. (1990) Extension of molecular replacement: new search strategy based on Patterson correlation refinement, *Acta Crystallogr.* **A46**, 46–57.

Brünger, A.T. (1992) Free R value: A novel statistical quantity for assessing the accuracy of crystal structures, *Nature (London)*, **355**, 472–475.

Brünger, A.T. (1993) Assessment of phase accuracy by cross validation: the free R value. Methods and applications, *Acta Crystallogr.* **D49**, 24–36.

Brünger, A.T., Nilges, M. (1993) Computational challenges for macromolecular structure determination by X-ray crystallography and solution NMR spectroscopy, *Q. Rev. Biophys.* **26**, 49–125.

Brünger, A.T., Kuriyan, J., Karplus, M. (1987) Crystallographic R-factor refinement by molecular dynamics, *Science* **235**, 458–460.

Brünger, A.T., Adams, P.D., Clore, G.M., Delano, W.L., Gros, P., Grosse-Kunstleve, R.W., Jiang, J.S., Kuszewski, J., Nilges, M., Pannu, N.S., Read, R.J., Rice, L.M., Simonson, T., Warren, G.L. (1998) Crystallography and NMR system – a new software suite for macromolecular structure determination, *Acta Crystallogr.* **D54**, 905–921.

CCP4, (1994) Collaborative computational project No. 4, *Acta Crystallogr.* **D50**, 760–763.

Clausen, T., Huber, R., Laber, B., Pohlenz, H.-D., Messerschmidt, A. (1996) Crystal structure of the pyridoxal-5′-phosphate dependent Cystathionine ß-Lyase from *Escherichia coli* at 1.83 Å, *J. Mol. Biol.* **262**, 202–224.

Clausen, T., Huber, R., Messerschmidt, A., Pohlenz, H.-D., Laber, B. (1997) Slow-binding inhibition of *Escherichia coli* Cystathionine ß-Lyase by L-aminoethoxyvinylglycine: a kinetic and X-ray study, *Biochemistry* **36**, 12633–12643.

Crick, F.H.C., Magdoff, B.S. (1956) The theory of the method of isomorphous replacement, *Acta Crystallogr.* **9**, 901–908.

Cromer, D.T., Liberman, D. (1970) Relativistic calculation of anomalous scattering factors for X-rays, *J. Chem. Phys.* **53**, 1891–1898.

Crowther, R.A., Blow, D.M. (1967) A method of positioning a known molecule in an unknown crystal structure, *Acta Crystallogr.* **23**, 544–548.

Dauter, Z., Lamzin, V.S., Wilson, K.S. (1997) The benefits of atomic resolution, *Curr. Opin. Struct. Biol.* **7**, 681–688.

Diamond, R. (1969) Profile analysis in single crystal diffractometry, *Acta Crystallogr.* **A25**, 43–55.

Dickerson, R.E., Weinzierl, J.E., Palmer, R.A. (1968) A least-squares refinement method for isomorphous replacement, *Acta Crystallogr.* **B24**, 997–1003.

Dodson, E., Vijayan, M. (1971) The determination and refinement of heavy-atom parameters in protein heavy-atom derivatives: some model calculations using acentric reflections, *Acta Crystallogr.* **B27**, 2402–2411.

Eichhorn, K.D. (1985) DISCO: Calculation of the Anomalous Dispersion Corrections, f' and f'', to the Atomic X-ray Form Factor from Both EXAFS and Theory, Deutsches Elektronensynchrotron DESY, Hamburg.

Einsle, O., Messerschmidt, A., Stach, P., Bourenkov, G.P., Bartunik, H.D., Huber, R., Kroneck, P.M.H. (1999) Structure of cytochrome c nitrite reductase, *Nature (London)* **400**, 476–480.

Engh, R.A., Huber, R. (1991) Accurate bond and angle parameters for X-ray protein structure refinement, *Acta Crystallogr.* **A47**, 392–400.

Evans, S.V. (1993) SETOR: Hardware lighted three-dimensional solid model representation of macromolecules, *J. Mol. Graphics* **11**, 134–138.

Ford, G. (1974) Intensity determination by profile fitting applied to precession photographs, *J. Appl. Crystallogr.* **7**, 555–564.

Gaykema, W.P.J., Hol, W.G.J., Vereijken, J.M., Soeter, N.M., Bak, H.J., Beintema, J.J. (1984) 3.2 Å structure of the copper-containing, oxygen-carrying protein *Panulirus interruptus* haemocyanin, *Nature (London)* **309**, 23–29.

Gilliland, G.L., Tung, M., Blakeslee, D.M., Ladner, J.E. (1994) Biological macromolecule crystallization database, version 3.0: new features, data and the NASA archive for protein crystal growth data, *Acta Crystallogr.* **D50**, 408–413.

Green, A.A. (1931) Studies in the physical chemistry of the proteins. VIII. The solubility of hemoglobin in concentrated salt solutions. A study of the salting out of proteins, *J. Biol. Chem.* **93**, 495–516.

Green, A.A. (1932) Studies in the physical chemistry of the proteins. X. The solubility of hemoglobin in solutions of chlorides and sulfates of varying concentration, *J. Biol. Chem.* **95**, 47–66.

Greer, J. (1974) Three-dimensional pattern recognition: an approach to automated interpretation of electron density maps of proteins, *J. Mol. Biol.* **82**, 279–301.

Groll, M., Ditzel, L., Löwe, J., Stock, D., Bochtler, M., Bartunik, H.D., Huber, R. (1997) Structure of 20S proteasome from yeast at 2.4 Å resolution, *Nature (London)* **386**, 463–471.

Hendrickson, W.A. (1985) Stereochemically restrained refinement of macromolecular structures, *Methods Enzymol.* **115**, 252–270.

Hendrickson, W.A. (1991) Determination of macromolecular structures from anomalous diffraction of synchrotron radiation, *Science* **254**, 51–58.

Hendrickson, W.A., Lattman, E.E. (1970) Representation of phase probability distributions for simplified combination of independent phase information, *Acta Crystallogr.* **B26**, 136–143.

Hendrickson, W.A., Ogata, C.M. (1997) Phase determination from Multiwavelength Anomalous Diffraction measurements, *Methods Enzymol.* **276**, 494–523.

Hendrickson, W.A., Horton, J.R., LeMaster, D.M. (1990) Selenomethionyl proteins produced for analysis by Multiwavelength Anomalous Diffraction (MAD): a vehicle for direct determination of three-dimensional structure, *EMBO J.* **9**, 1665–1672.

Hönl, H. (1933) Zur Dispersionstheorie der Röntgenstrahlen, *Z. Phys.* **84**, 1–16.

Hooft, R., Vriend, G. (1996) WHAT IF Program Manual, EMBL, Heidelberg, Chapter 23.

Hoppe, W. (1957) Die Faltmolekülmethode: Eine neue Methode zur Bestimmung der Kristallstruktur bei ganz oder teilweise bekannten Molekülstrukturen, *Z. Elektrochem.* **61**, 1076–1083.

Howard, A.J., Nielsen, C., Xuong, Ng.H. (1985) Software for a diffractometer with multiwire area detector, *Methods Enzymol.* **114**, 452–472.

Howard, A.J., Gilliland, G.L., Finzel, B.C., Poulos, T.L., Ohlendorf, D.H., Salemne, F.R. (1987) The use of an imaging proportional counter in macromolecular crystallography, *J. Appl. Crystallogr.* **20**, 383–387.

Huber, R. (1965) Die Automatisierte Faltmolekülmethode, *Acta Crystallogr.* **19**, 353–356.

Jack, A., Levitt, M. (1978) Refinement of large structures by simultaneous minimization of energy and R-factor, *Acta Crystallogr.* **A34**, 931–935.

James, R.W. (1960) *The Optical Principles of the Diffraction of X-rays*, Vol. II The Crystalline State, Bell, London, 135–167.

Jancarik, J., Kim, S.H. (1991) Sparse matrix sampling: a screening method for crystallization of proteins, *J. Appl. Crystallogr.* **24**, 409–411.

Jones, T.A. (1978) A graphics model building and refinement system for macromolecules, *J. Appl. Crystallogr.* **15**, 24–31.

Jones, T.A., Zou, J.Y., Cowan, S.W., Kjeldgaard, M. (1991) Improved methods for building protein models in electron density maps and location of errors in these models, *Acta Crystallogr.* **A47**, 110–119.

Kabsch, W. (1988) Automatic indexing of rotation diffraction patterns, *J. Appl. Crystallogr.* **21**, 67–81.

Kabsch, W. (1988) Evaluation of single-crystal X-ray diffraction data from position-sensitive detector, *J. Appl. Crystallogr.* **21**, 916–924.

Kabsch, W. (1993) Data Collection and Evaluation with Program XDS, in Data Collection and Processing, Proceedings of the CCP4 Study Weekend, 29–30 January 1993, Sawyer, L. Isaac, N. Bailey S., (Eds.) SERC Daresbury Laboratory, Warrington, UK, 63–70.

Kahn, R., Fourme, R. (1997) Gas proportional detectors, *Methods Enzymol.* **276**, 268–288.

Kiefersauer, R., Than, M.E., Dobbek, H., Gremer, L., Melero, M., Strobl, S., Dias, J.M., Soulimane, T., Huber, R. (2000) A novel free-mounting system for protein crystals: transformation and improvement of diffraction power by accurately controlled humidity changes, *J. Appl. Cryst.* **33**, 1223–1230.

Kleywegt, G.J., Jones, T.A. (1994) 'Halloween ... Masks and Bones', in *From First Map to Final Model*, Bailey, S., Hubbard, R., Waller, D., (Eds.) Proceedings of the Study Weekend, SERC Daresbury Laboratory, Warrington, UK, 59–66.

Knäblein, J., Neufeind, T., Schneider, F., Bergner, A., Messerschmidt, A., Löwe, J., Steipe, B., Huber, R. (1997) $Ta_6Br_{12}^{2+}$, a tool for phase determination of large biological assemblies by X-ray crystallography, *J. Mol. Biol.* **270**, 1–7.

Kraulis, P.J. (1991) MOLSCRIPT:A program to produce both detailed and schematic plots of protein structures, *J. Appl. Crystallogr.* **24**, 946–950.

Ladenstein, R., Schneider, M., Huber, R., Bartunik, H., Schott, K., Bacher, A. (1988) Heavy riboflavin synthase from bacillus subtilis: crystal structure analysis of the icosahedral Beta-60 capsid at 3.3 Å resolution, *J. Mol. Biol.* **203**, 1045–1070.

de La Fortelle, E., Bricogne, G. (1997) Maximum-likelihood heavy-atom parameter refinement for multiple isomorphous replacement and multiwavelength anomalous diffraction methods, *Methods Enzymol.* **276**, 472–494.

Laskowski, R.A., MacArthur, M.W., Moss, D.S., Thornton, J.M. (1993a) PROCHECK: A program to check the stereochemical quality of protein structures, *J. Appl. Crystallogr.* **26**, 283–291.

Laskowski, R.A., Moss, D.S., Thornton, J.M. (1993b) Main-chain bond lengths and bond angles in protein structures, *J. Mol. Biol.* **231**, 1049–1067.

Leslie, A. (1994) *Mosflm User Guide*, Mosflm Version 5.41', in Data Collection and Processing, Proceedings of the CCP4 Study Weekend, 29–30 January 1993, Sawyer, L., Isaac, N., Bailey, S., (Eds.) HRC Laboratory of Molecular Biology, Cambridge, UK, 44–51.

Löwe, J., Stock, D., Jap, B., Zwickl, P., Baumeister, W., Huber, R. (1995) Crystal structure of the 20S proteasome from the archaeon *Thermoplasma acidophilum* at 3.4 Å resolution, *Science* **268**, 533–539.

van de Locht, A., Lamba, D., Bauer, M., Huber, R., Friedrich, T., Kröger, B., Höffken, W., Bode, W. (1995) Two heads are better than one: crystal structure of the insect derived double domain Kazal Inhibitor rhodniin in complex with thrombin, *EMBO J.* **14**, 5149–5157.

Luzzati, V. (1952) Traitement Statistique des Erreurs dans la Determination des Structures Crystallines, *Acta Crystallogr.* **A5**, 802–810.

MacArthur, M.W., Laskowski, R.A., Thornton, J.M. (1994) Knowledge-based validation of protein-structure coordinates derived by X-ray crystallography and NMR spectroscopy, *Curr. Opin. Struct. Biol.* **4**, 731–737.

Merrit, E.A., Murphy, N.E.P. (1994) RASTER3D Version 2.0. A program for photorealistic molecular graphics, *Acta Crystallogr.* **D50**, 869–873.

Messerschmidt, A., Pflugrath, J.W. (1987) Crystal orientation and X-ray pattern prediction routines for area-detector diffractometer systems in macromolecular crystallography, *J. Appl. Crystallogr.* **20**, 306–315.

Messerschmidt, A., Luecke, H., Huber, R. (1993) X-ray structures and mechanistic implications of three functional derivatives of ascorbate oxidase from zucchini, *J. Mol. Biol.* **230**, 997–1014.

Messerschmidt, A., Ladenstein, R., Huber, R., Bolognesi, M., Avigliano, L., Petruzzelli, R., Rossi, A., Finazzi-Agro, A. Refined crystal structure of ascorbate oxidase at 1.9 Å resolution, *J. Mol. Biol.* **224**, 179–205.

Messerschmidt, A., Rossi, A., Ladenstein, R., Huber, R., Bolognesi, M., Gatti, G., Marchesini, A., Petruzzelli, R., Finazzi-Agro, A. (1989) X-ray crystal structure of the blue oxidase ascorbate oxidase from zucchini: analysis of the polypeptide fold and a model of the copper sites and ligands, *J. Mol. Biol.* **206**, 513–529.

Moffat, K. (1998) Ultrafast time-resolved crystallography, *Nature Struct. Biol., Synchrotron Suppl.* 641–643.

Murshudov, G.N., Vagin, A.A., Dodson, E.J. (1997) Refinement of macromolecular structures by the maximum likelihood method, *Acta Crystallogr.* **D53**, 240–255.

Navaza, J. (1994) AMoRe: An automated package for Molecular Replacement, *Acta Crystallogr.* **A50**, 157–163.

Otwinowski, Z. (1991) Maximum Likelihood Refinement of Heavy Atom Parameters, in *Isomorphous Replacement and Anomalous Scattering*, Wolf, W., Evans, P.R., Leslie, A.G.W., (Eds.) SERC Daresbury Laboratory, Warrington, UK, 80–86.

Otwinowski, Z. (1993) *The Denzo Program Package*, in Data Collection and Processing, Proceedings of the CCP4 Study Weekend, 29–30 January 1993, Isaac, N., Bailey, S., (Eds.) SERC Daresbury Laboratory, Warrington, UK, 56–62.

Otwinowski, Z., Minor, W. (1997) Processing of X-ray diffraction data collected in oscillation mode, *Methods Enzymol.* **276**, 307–326.

Patterson, A.L. (1934) A Fourier series method for the determination of the components of interatomic distances in crystals, *Phys. Rev.* **46**, 372–376.

Pearson, R.G. (1969) Hard and soft acids and bases, *Survey* **5**, 1–52.

Perrakis, A., Morris, R., Lamzin, V.S. (1999) Automated protein model building combined with iterative structure refinement, *Nat. Struct. Biol.* **6**, 458–463.

Phillips, W.C., Rayment, I. (1985) A systematic method for aligning double-focusing mirrors, *Methods Enzymol.* **114**, 316–329.

Ramachandran, G.N., Ramakrishnan, C., Sasisekharan, V.J. (1963) Stereochemistry of polypeptide chain configurations, *J. Mol. Biol.* **7**, 95–99.

Read, R.J. (1986) Improved Fourier coefficients for maps using phases from partial structures with errors, *Acta Crystallogr.* **A42**, 140–149.

Read, R.J. (1990) Structure-factor probabilities for selected structures, *Acta Crystallogr.* **A46**, 900–912.

Rodgers, D.W. (1997) Practical cryocrystallography, *Methods Enzymol.* **276**, 183–203.

Rossmann, M.G. (1960) The accurate determination of the position and shape of heavy-atom replacement groups in proteins, *Acta Crystallogr.* **13**, 221–226.

Rossmann, M.G. (1979) Processing oscillation diffraction data for very large unit cells with an automatic convolution technique and profile fitting, *J. Appl. Crystallogr.* **12**, 225–238.

Rossmann, M.G., Blow, D.M. (1962) The detection of subunits within the crystallographic asymmetric unit, *Acta Crystallogr.* **15**, 24–31.

Rossmann, M.G., Arnold, E., Erikson, J.W., Frankenberger, E.A., Griffith, J.P., Hecht, H.-J., Johnson, J.E., Kamer, G., Luo, M., Mosser, A.G., Rueckert, R.R., Sherry, B., Vriend, G. (1985) Structure of a human common cold virus and functional relationship to other picornaviruses, *Nature (London)* **317**, 145–153.

Roussel, A., Cambilleau, C. (1989) Turbo-Frodo in Silicon Graphics Geometry, Partners Directory, Silicon Graphics, Mountain View, CA.

Schneider, R.R., Sheldrick, G.M. (2002) Substructure solution with *SHELXD*, *Acta Crystallogr.* **D58**, 1772–1779.

Sheldrick, G.M. (1997) SHELX: High-resolution refinement, *Methods Enzymol.* **277**, 319–343.

Sim, G.A. (1959) The distribution of phase angles for structures containing heavy atoms. II. A modification of the normal heavy-atom method for noncentrosymmetrical structures, *Acta Crystallogr.* **12**, 813–815.

Steigemann, W. (1974) Die Entwicklung und Anwendung von Rechenverfahren und Rechenprogrammen zur Strukturanalyse von Proteinen am Beispiel des Trypsin-Trysininhibitorkomplexes, des freien Inhibitors und der L-Asparaginase, PhD Thesis, Technische Universität München.

Stubbs, M.T., Bode, W. (1993) A player of many parts: the spotlight falls on thrombin's structure, *Thromb. Res.* **69**, 1–58.

Teng, T.-Y. (1990) Mounting of crystals for macromolecular crystallography in a free-standing thin film, *J. Appl. Crystallogr.* **23**, 387–391.

Tong, L., Rossmann, M.G. (1990) The locked rotation function, *Acta Crystallogr.* **A46**, 783–792.

Tronrud, D.E., Ten Eyk, L.F., Matthews, B.W. (1987) An efficient general-purpose least-squares refinement program for macromolecular structures, *Acta Crystallogr.* **A43**, 489–501.

Turk, D. (1992) Weiterentwicklung eines Programmes für Molekülgrafik und Elektronendichte-Manipulation und seine Anwendung auf verschiedene Protein-Struktur-Aufklärungen, PhD Thesis, Technische Universität München.

Wang, B.-C. (1985) Resolution of phase ambiguity in macromolecular crystallography, *Methods Enzymol.* **115**, 90–112.

Westbrook, E.M., Naday, I. (1997) Charge-coupled device based area detectors, *Methods Enzymol.* **276**, 244–268.

Wilson, K.S. (1998) Illuminating crystallography, *Nat. Struct. Biol., Synchrotron Suppl.* 627–630.

Xu, H.L., Hauptmann, H.A., Weeks, C.M. (2002) Sine-enhanced *Shake-and-Bake*: the theoretical basis and applications to Se-atom substructures, *Acta Crystallogr.* **D58**, 90–96.

Yang, W., Hendrickson, W.A., Kalman, E.T., Crouch, R.J. (1990) Expression, purification, and crystallization of natural and selenomethionyl recombinant ribonuclease H from *Escherichia coli*, *J. Biol. Chem.* **265**, 13553–13559.

Zhang, K.Y.J., Main, P. (1990) The use of Sayre's equation with solvent flattening and histogram matching for phase extension and refinement of protein structures, *Acta Crystallogr.* **A46**, 377–381.

2
Protein NMR Spectroscopy

Thomas Szyperski
State University of New York, Buffalo, NY

1	Aspects of Multidimensional Protein NMR Spectroscopy	83
2	NMR Instrumentation	87
3	Experimental Protein NMR Parameters	87
4	Recently Developed Techniques	89
5	Structure Determination	92
6	Dynamics	96
7	Hydration	97
8	Folding	97
9	Structural Genomics	98

Bibliography 99
Books and Reviews 99
Primary Literature 99

Keywords

Correlation Spectroscopy
"COSY": NMR experiment devised to correlate chemical shifts via through-bond scalar nuclear spin–spin couplings.

Proteins. Edited by Robert A. Meyers.
Copyright © 2007 Wiley-VCH Verlag GmbH & Co. KGaA, Weinheim
ISBN: 978-3-527-31608-3

Fourier Transformation
Mathematical transformation that provides an analysis of the angular frequencies encoded in a given function.

Nuclear Magnetic Moment
Since atomic nuclei are positively charged, the nuclear spin is proportional to a nuclear magnetic moment.

Nuclear Magnetic Resonance
Owing to their magnetic moment, atomic nuclei interact with external magnetic fields. This enables one to pursue nuclear magnetic resonance (NMR) spectroscopy.

Nuclear Overhauser Enhancement Spectroscopy
"NOESY": NMR experiment to measure through-space dipolar interactions between the magnetic moments of protons. This allows one to estimate distances between protons in proteins.

Nuclear Spin
The nuclear spin is a nonclassical angular momentum associated with atomic nuclei possessing a spin quantum number larger than zero.

Triple Resonance NMR Experiment
Experiment devised to correlate the chemical shifts of three types of nuclei, that is, ^{1}H, ^{13}C, and ^{15}N.

Nuclear magnetic resonance (NMR) spectroscopy provides unique information about protein structure, dynamics, hydration, and folding in aqueous solution, and has become a pivotal biophysical technique to investigate biological macromolecules. Although the first application of NMR spectroscopy to study proteins date back to the 1960s, its widespread use for proteins was fostered only in the late 1970s and early 1980s by introduction of two-dimensional NMR spectroscopy conducted on spectrometers equipped with superconducting high-field magnets. Advances made in the subsequent 25 years until today have established NMR as an indispensable tool to study even large proteins with molecular masses above 100 kDa. These advances were due to new approaches for efficient production of stable isotope labeled proteins, novel spin relaxation, optimized and rapid sampling of NMR data collection strategies, and the advent of highly sensitive cryogenic probes used for signal detection at field strengths corresponding to 500–900 MHz proton resonance frequency. Recently, the outstanding value of NMR was demonstrated for high-throughput structure determination in the newly emerging field of structural genomics, which makes protein NMR spectroscopy also a key tool for systems biology.

1
Aspects of Multidimensional Protein NMR Spectroscopy

Nuclear magnetic resonance (NMR) spectroscopy is based on the existence of a nonclassical angular momentum, the *nuclear spin*. All atomic nuclei with a nonzero nuclear spin quantum number possess such a spin. Since nuclei are positively charged, the spin generates a colinearly oriented magnetic moment, which interacts with magnetic fields as compass needles interact with the earth's magnetic field. "Spin-$\frac{1}{2}$ nuclei," with a spin quantum number of $\frac{1}{2}$, exhibit neither electric dipole nor electric quadrupole moments. Hence, spin-$\frac{1}{2}$ nuclei are quite weakly coupled to the environment (often called the *lattice*) via their magnetic dipole moments only: these interact with random fluctuations of magnetic fields that arise from thermal motions. In proteins, the weak coupling enables the observation of coherent spatial reorientations of spins (briefly named *coherences*) about the axis of a strong external magnetic field, B_0, for time periods up to several hundred milliseconds. For a typical study, this allows accurate measurement of hundreds or even thousands of NMR parameters. This wealth of information is unrivaled among solution spectroscopic techniques and is a central reason for the success of modern protein NMR spectroscopy.

Protein NMR relies on the three spin-$\frac{1}{2}$ nuclei ^{1}H, ^{13}C, and ^{15}N. Since these nuclei possess the same spin quantum number of $\frac{1}{2}$, the absolute value of their spin angular momenta is equal. However, owing to varying charges (and charge distributions within the nuclei), the resulting magnetic moments differ greatly. ^{1}H possesses the largest magnetic moment, while the magnetic moments of ^{13}C and ^{15}N are fourfold and tenfold smaller, respectively. Since resonance frequencies scale accordingly, this has far-reaching consequences for the design of protein NMR experiments: highest sensitivity is achieved for experiments starting with proton polarization and ending with proton signal detection. Furthermore, only ^{1}H is 100% abundant in nature, while ^{13}C (natural abundance: ~1%; mainly naturally occurring is ^{12}C, a spin-0 nucleus which is "invisible" in NMR) and ^{15}N (~0.5%; mainly naturally occurring is ^{14}N, a spin-1 nucleus with an electric quadrupole moment that results in broad NMR lines) need to be enriched if a protein has to be studied with more demanding heteronuclear NMR spectroscopy based on ^{13}C and/or ^{15}N.

Radio frequency (rf) pulses created with an rf coil can induce the coherent precession of spin magnetic moments about the axis of an external magnetic field. This precession and its decay can be detected with an rf coil, and the resulting signal ("free induction decay") yields, after Fourier Transformation (FT), a one-dimensional (1D) NMR spectrum. For proteins, the resonance lines provide chemical shifts and, in favorable cases, estimates for scalar spin–spin couplings from resonance fine structures, and nuclear spin relaxation times from line widths. However, owing to the large number of nuclei, 1D NMR spectra of proteins and other biological macromolecules are very crowded. Hence, protein NMR primarily relies on two-, three-, and four-dimensional (2D, 3D, and 4D) spectral information. The high dimensional NMR spectra afford increased spectral resolution and correlate several chemical shifts in a single data set.

A 2D data set is obtained if the signals of a 1D spectrum are modulated by the evolution of an NMR parameter (usually a chemical shift) in an "indirect dimension" prior to acquisition (Fig. 1). Fourier transformation along the indirect dimension yields the 2D frequency domain spectrum. In the same fashion, 3D (4D) NMR data sets are obtained by recording many 2D (3D) data sets, which are modulated by an additional NMR parameter. As a result, we have the minimal measurement time, T_m, increasing steeply with dimensionality: acquiring 32 points in each indirect dimension (with one scan per 1D spectrum each second) yields $T_m(3D)$ ~0.5 h, $T_m(4D)$ ~9 h, $T_m(5D)$ ~12 days and $T_m(6D)$ ~1 year. This obstacle for acquiring the highest dimensional NMR spectral information has been named the *NMR sampling problem*.

Figure 2(a) shows a 2D $[^{15}N,^{1}H]$-correlation spectrum ("COSY"), which was recorded for a ^{15}N-labeled 21-kDa globular protein on a spectrometer operating at 900 MHz ^1H resonance

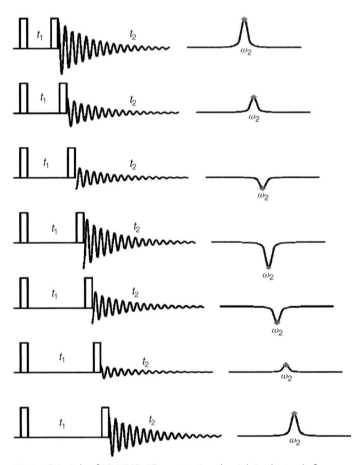

Fig. 1 Principle of 2D NMR. 1D spectra (on the right), obtained after Fourier transformation (FT) of the free induction decays along t_2, are modulated by the evolution of an NMR parameter (e.g. a chemical shift) in an indirect dimension ("t_1"). The frequency domain 2D spectrum is obtained after a second FT along t_1. Courtesy of Jeff Mills.

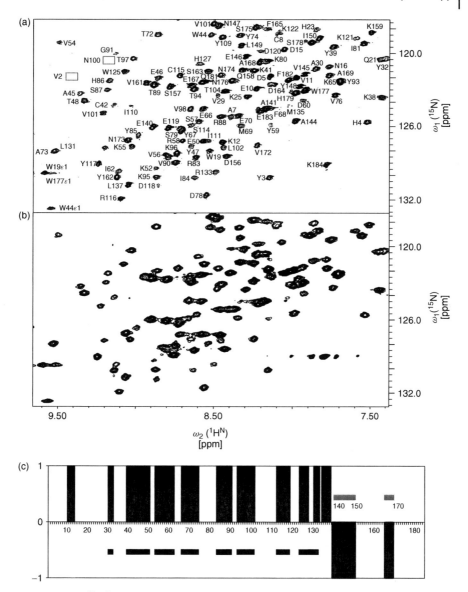

Fig. 2 (a) 2D [^{15}N,^{1}H]-COSY spectrum recorded in the transverse relaxation optimized ("TROSY") acquisition mode (see Sect. 4) on a 900-MHz Varian INOVA spectrometer (25 °C) for the uniformly ^{15}N-labeled 21-kDa protein FluA(R95K). The peaks are labeled using the one-letter code for amino acids. (b) For comparison, a 2D [^{15}N,^{1}H]-COSY spectrum recorded without "TROSY" at 600 MHz is shown. (c) Chemical shift index (CSI) consensus plot for identification of regular secondary structure elements. The eight β-strands forming a β-barrel (black bars) as well as the α-helices (gray bars) are indicated. (a and b are reproduced with permission from Liu, G., Mills, J.L., Hess, T.A., Kim, S., Skalicky, J.J., Sukumaran, D.K., Kupce, E., Skerra, A., Szyperski, T. (2003) Resonance assignments for the 21 kDa engineered fluorescein-binding lipocalin FluA, *J. Biomol. NMR* **27**, 187–188).

frequency. To generate this spectrum, 1D ^{1}H NMR spectra of amide protons are modulated with the chemical shift of the covalently attached ^{15}N nucleus, and one signal is detected for each amino acid residue (except for prolinyl residues and the N-terminal residue). Owing to high intrinsic sensitivity, one can rapidly obtain site-specific information along the entire polypeptide backbone. This feature makes 2D [^{15}N,^{1}H]-COSY (correlation spectroscopy) one of the most important experiments in modern NMR-based structural biology. First, the signal dispersion allows one to assess the degree to which the protein is folded: unfolded polypeptide segments tend to exhibit signals in a comparably narrow range attributed to "random coils." Secondly, the comparison of expected and detected number of signals enables one to investigate if parts of the protein, for example, loops connecting regular secondary structure elements, are affected by slow motional modes in the microsecond to millisecond time range. Such motions quite often lead to broadening of signals beyond detection.

Prior to using NMR spectroscopy for deriving site-specific information on structure or dynamics of proteins, the nuclear magnetic resonances must be "assigned," that is, one has to identify the chemical shifts (see Sect. 3) of spins. Since proteins contain each type of amino acid several times at various positions along the polypeptide chain, it is not sufficient to identify the type of proton (e.g. ^{1}H$^\alpha$ of glutamate) belonging to a given resonance. Instead, one has to obtain "sequence-specific" resonance assignments (e.g. ^{1}H$^\alpha$ of the glutamate in position 79). Obtaining (nearly complete) sequence-specific resonance assignments is generally considered a prerequisite for determining protein structures (see Sect. 5). This so-called "assignment problem" is comparable to the crystallographic "phase problem" in the sense that a structural interpretation of experimental data depends on solving these problems. Nowadays, sequential resonance assignment of ^{15}N/^{13}C-labeled proteins relies largely on recording ^{15}N/^{13}C/^{1}H triple resonance NMR experiments (Fig. 3). In these experiments, several proton, carbon, and nitrogen chemical shifts are correlated. These originate either from the same, or from two sequentially neighboring residues. Combining the information from the two experiments then allows one to effectively "walk" along the polypeptide chain to obtain sequence-specific resonance assignments for backbone (and often ^{13}C$^\beta$ shifts). Importantly, these chemical shifts can provide the location of regular secondary structure elements (Fig. 2c) without reference to nuclear Overhauser effects (NOEs). The

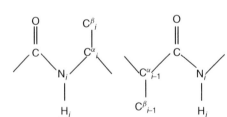

Fig. 3 Example of chemical shifts of nuclei (in color) correlated in triple resonance NMR experiments for obtaining sequence-specific backbone resonance assignments of proteins. A first experiment correlates the shifts of amide proton, nitrogen, α-carbon, and β-carbon within a given residue i (on the left). A second experiment correlates the shifts of amide proton and nitrogen of residue i with those of the α- and β-carbons of the preceding residue $i-1$. Combining the two experiments allows one to obtain sequence-specific resonance assignments for the backbone (and ^{13}C$^\beta$ shifts) of the entire polypeptide. (See color plate p. xv).

backbone NMR experiments are then combined with other experiments to assign resonances of side chains, thus yielding a *complete* protein resonance assignment.

2
NMR Instrumentation

Transitions between nuclear spin states are associated with radio frequency energy quanta that are, at ambient temperature, several orders of magnitude smaller than kT (k and T represent Boltzman constant and temperature). Hence, NMR transitions are *lowest energy* transitions. This results in both, the greatest strength and the greatest weakness of NMR spectroscopy. The strength is due to the fact that the low energy required for observing the system (using the nuclear spins as "spies") hardly affects even the subtlest conformational equilibria in their structural and dynamic manifestation. The weakness associated with the low interaction energy is due to the low sensitivity of NMR spectroscopy when compared with, for example, optical spectroscopy. This problem is further aggravated in protein NMR, since slowly tumbling macromolecules exhibit short transverse spin relaxation times and thus exhibit broad resonance lines. Hence, to alleviate this drawback, modern protein NMR depends on using large superconducting magnets (Fig. 4a). In 2005, the largest commercially available magnets have a price tag of about four million US dollars and can induce a ^{1}H resonance frequency of 900 MHz. Apart from increased sensitivity, which scales with $B_0^{3/2}$, high magnetic fields yield increased signal dispersion scaling with B_0^N, where N is the dimensionality of the NMR experiment. A second major breakthrough in hardware development was the construction of cryogenic NMR probes (Fig. 4b). In such probes, the rf coil for signal detection is cooled to about 25 K, which leads to a large reduction in thermal noise in the rf circuitry. In spite of the fact that the insulation between rf coil and NMR sample (which needs to be kept at ambient temperature) reduces the filling factor of the coil, 2 to 3 fold gains in sensitivity can be routinely achieved in protein NMR. This corresponds to about 5 to 10 fold reduced NMR data collection times, so that rapid NMR data collection techniques (see Sect. 4) are best suited to take advantage of the high sensitivity of such probes.

3
Experimental Protein NMR Parameters

Protein NMR spectroscopy aims at measurement of parameters, which provide the desired structural and/or dynamic information of a system under consideration. *Chemical shifts* are due to the site-specific shielding of the external magnetic field at the location of the nucleus. Hence, chemical shifts depend on the covalent and conformational environment in which a spin is embedded and are affected by long-range electrostatic and ring current effects. The dispersion of chemical shifts arising from the folding of a polypeptide into a tertiary structure is pivotal for using NMR: without such conformation-dependent dispersion, one could not resolve the site-specific information. Chemical shifts are routinely measured in multidimensional NMR spectra (see Sect. 1), and the resonance assignment is quite generally considered a prerequisite for the site-specific interpretation of other NMR parameters described below. Importantly, backbone and $^{13}C^\beta$ chemical shifts provide the location of the regular secondary structure elements

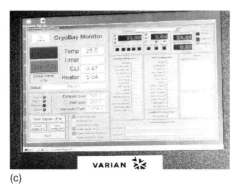

Fig. 4 Modern high-field NMR equipment used for protein NMR spectroscopy. (a) Spectrometer at the New York Structural Biology Center operating at 900 MHz ^{1}H resonance frequency (Courtesy of David Cowburn). (b) Cryogenic probe installed at the University at Buffalo high-field NMR facility (Courtesy of Dinesh Sukumaran). The probe is inserted from below in the magnet, and its rf coil is cooled to 25 K, as can be read off the monitor shown in (c).

(Fig. 2c). Moreover, chemical shifts are not isotropic, that is, the shielding of the external magnetic field depends on the orientation of the molecule relative to the field. Chemical shifts are thus accurately described by using a "chemical shift tensor."

Scalar nuclear spin–spin couplings, J, arise from the through-bond coupling of the nuclear magnetic moments. The J-coupling is the most important interaction used to devise multidimensional experiments for measurement and correlation of chemical shifts (Fig. 2a). Moreover, three-bond J-couplings are related through the "Karplus-relations" to dihedral angles and thus encode valuable structural information.

^{1}H–^{1}H NOEs arise from through-space coupling of the magnetic dipole moments of protons. Owing to the r^{-6} dependence of the dipolar coupling (with r being the distance between two protons), the measurement of NOEs in [^{1}H,^{1}H]-NOESY (nuclear Overhauser enhancement spectroscopy) allows one to estimate distances between protons: NOEs effectively constitute a "molecular ruler." Thus, having a large number of such ^{1}H–^{1}H distance constraints, one can calculate three-dimensional molecular structures (see Sect. 5).

Residual dipolar couplings (RDCs) arise if a protein is partially aligned in a dilute liquid crystalline medium. Notably, for paramagnetic metalloproteins, such alignment takes place at high magnetic fields without the presence of a liquid crystalline medium. Owing to the partial alignment, the dipolar through-space coupling between dipole moments is not completely averaged out, as in the case of molecules tumbling in an isotropic medium. RDCs encode orientational constraints, that is, they provide information regarding the orientation of the internuclear axis of the dipolarly coupled nuclei relative to the principle axes of the alignment tensor. RDC-derived orientational constraints are thus quite distinct from – and complementary to – NOE-derived distance constraints for NMR structure calculation and validation. Since RDCs depend on motional averaging, they also provide valuable information about internal motional modes.

Nuclear spin relaxation parameters, such as longitudinal/transverse spin relaxation times and heteronuclear Overhauser effects reflect dynamic features of the protein under investigation. Hence, the majority of our insights into protein dynamics, be it overall rotational reorientation or internal motions, have thus far been obtained from measurement of these parameters.

4
Recently Developed Techniques

The introduction of (1) multidimensional triple resonance NMR and heteronuclear resolved [^{1}H,^{1}H]-NOESY, and (2) efficient ^{13}C/^{15}N labeling and protein deuteration protocols until the mid 1990s, made NMR a key tool to study structure and dynamics of proteins up to about 15 to 20 kDa. During the last decade, new NMR techniques were introduced to study much larger systems and to dramatically increase NMR data collection speed.

Transverse relaxation optimized NMR spectroscopy (TROSY) takes advantage of the mutual cancellation of nuclear spin relaxation pathways in slowly tumbling macromolecules studied at high magnetic fields (Fig. 5). 2D [^{15}N,^{1}H]-TROSY (Fig. 2a), and triple resonance NMR variants thereof, are most prominent and shall be discussed. Both, the ^{15}N and the ^{1}H nucleus of a polypeptide backbone H–N moiety exhibit substantial chemical shift anisotropy (CSA). As a result, two major pathways for transverse ^{15}N and ^{1}H spin relaxation are encountered: one owing to the dipolar ^{1}H–^{15}N interaction and the other owing to the CSA. While spin relaxation due to random fluctuation of the shielding is independent of spin states, the sign of the fluctuating local B-field arising from the spatial modulation of the dipolar interaction does depend on spin states. Hence, the transverse relaxation rate of a given ^{15}N or ^{1}H spin is accelerated when both the CSA and dipolar interaction random B-field fluctuations add up, whereas the transverse rate is reduced in cases where the two random fluctuations (partially) cancel. Since short transverse relaxation times lead to broad NMR lines, this phenomenon can be directly monitored in a 2D [^{15}N,^{1}H]-COSY experiment in which the one-bond ^{1}H–^{15}N J-coupling (~95 Hz) places the various transitions into separate spectral regions (no "^{15}N–^{1}H" decoupling; Fig. 5). Out of the four transitions registered for a given ^{15}N–^{1}H moiety, one is broadened in both dimensions, two are broadened in one dimension but narrow in the other,

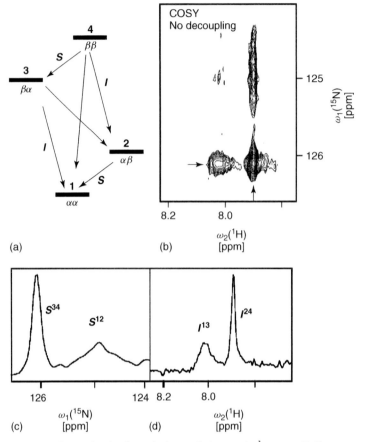

Fig. 5 (a) Energy levels of a scalarly coupled two spin-$\frac{1}{2}$ system IS (for example, $I = {}^1H$, $S = {}^{15}N$). The nuclear spin states are indicated as "α" and "β" depending on the orientation of the magnetic moments relative to the external magnetic field. (b) Cross-peak registered for a backbone N−H group in a [^{15}N,^{1}H]-COSY spectrum recorded (without decoupling of the ^{15}N−^{1}H scalar coupling interaction) for the 110-kDa protein aldolase at 20°C. In this protein preparation, aliphatic and aromatic ^{1}H were replaced by ^{2}H ("deuterated") in order to eliminate dipolar interactions between backbone amide and carbon-bound protons. (c) and (d) 1D cross sections taken along ω_1 and ω_2 at the positions indicated in (b) by arrows. Comparison of line widths of the four components reflects the different transverse relaxation rates between the transitions shown in (a). A [^{15}N,^{1}H]-TROSY experiment relies on selecting the two transitions S^{34} and I^{24}. Reproduced with permission from Pervushin, K. (2000) Impact of transverse relaxation optimized spectroscopy (TROSY) on NMR as a technique in structural biology, *Q. Rev. Biophys.* **33**, 161–197.

and one is narrow in both dimensions. TROSY relies on selecting the latter narrow component, while discarding the others.

Since CSA-based transverse relaxation scales with B_0^2, TROSY is best performed at a magnetic field strength where the two relaxation pathways cancel (nearly) entirely. Fortunately, ^{1}H and ^{15}N CSA are of comparable magnitude (in Hz), so that about the same optimal TROSY field strength is predicted for both ^{1}H and ^{15}N transitions (corresponding to 900–1100 MHz ^{1}H resonance frequency). This makes triple resonance [^{15}N,^{1}H]-TROSY feasible. In conjunction with protein deuteration, this allows one to study systems with molecular masses well above 100 kDa, that is, up to about an order of magnitude larger than what could be approached without TROSY.

Methodology for rapid acquisition of NMR data focuses on resolving the "NMR sampling problem" (see Sect. 1), that is, on making high dimensional spectral information available at short measurement times. Currently, two broader classes of rapid data sampling approaches can be distinguished.

First, *G-matrix Fourier Transform* (GFT) NMR is a projection technique, which can also serve to reconstruct the higher-dimensionality parent NMR spectra one would obtain if no projection is applied. GFT NMR is based on the joint sampling of indirect chemical shift evolution periods that leads to signals with a multiplet fine structure, encoding several chemical shifts. A "G-matrix transformation" is applied in order to sort the components of the shift multiplets into subspectra, so that the peaks in each subspectrum encode a distinct linear combination of the jointly measured chemical shifts (Fig. 6). Hence, monitoring of chemical shifts in FT NMR

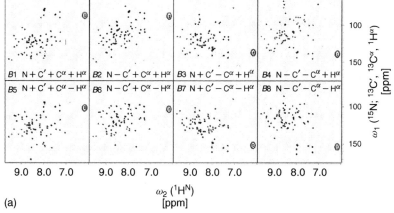

Fig. 6 15 2D planes constituting a (5,2)D HACACONHN GFT NMR experiment, which can provide the information of a 5D FT NMR experiment within less than an hour of measurement time. The type of linear combination of chemical shifts detected in a given plane is indicated. (a) The basic spectra. (b), (c) and (d) First, second, and third order central peak spectra in which linear combinations with fewer shifts are monitored to resolve assignment ambiguities. (e) Cross sections showing that signals do not broaden with an increasing number of shifts being jointly sampled. Reproduced with permission from Kim, S., Szyperski, T. (2003) GFT NMR, a new approach to rapidly obtain precise high dimensional NMR spectral information, *J. Am. Chem. Soc.* **125**, 1385–1393.

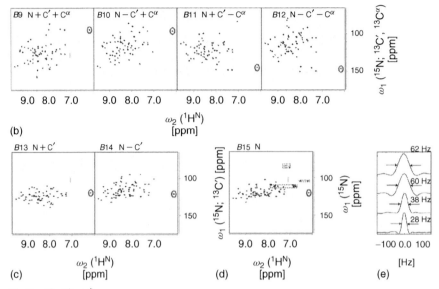

Fig. 6 (Continued)

spectroscopy is replaced in GFT NMR spectroscopy by measuring permutations of linear combinations of chemical shifts. This leads to approximately an order of magnitude reduction of measurement times *per* indirect dimension that is included in the joint sampling scheme.

Second, *single-scan acquisition* of 2D NMR spectra, named *ultrafast NMR*, was introduced. This approach constitutes an interface between high-resolution NMR and magnetic resonance imaging. The indirect chemical shift evolution is spatially encoded by use of pulse-field gradients applied along the axis of the external magnetic field B_0, and then "read out" together with the chemical shift evolution in the direct dimension in a single scan. In principle, this single-scan acquisition scheme can be extended to an arbitrary number of dimensions, that is, to the implementation of 3D and 4D single-scan acquisition. For proteins, however, reduced sensitivity currently limits this concept to the use of 2D single-scan acquisition. Importantly, ultrafast NMR is the only technique providing multidimensional NMR spectral information without sampling of indirect chemical shift evolution periods (Fig. 7). This allows one to acquire a multidimensional spectrum within the fraction of a second, and enables high-throughput acquisition of such spectra at unprecedented speed.

5
Structure Determination

NMR solution structures are usually solved in several major steps as outlined in the following. At the outset, a suitable sample (usually about 500 µL of a 1-mM protein solution) is prepared. Nowadays, both prokaryotic and eukaryotic high-yield overexpression systems serve for that purpose. If the molecular weight of a protein exceeds about 10 kDa, labeling with the NMR active spin-$\frac{1}{2}$ nuclei ^{13}C and ^{15}N is required to resolve spectral

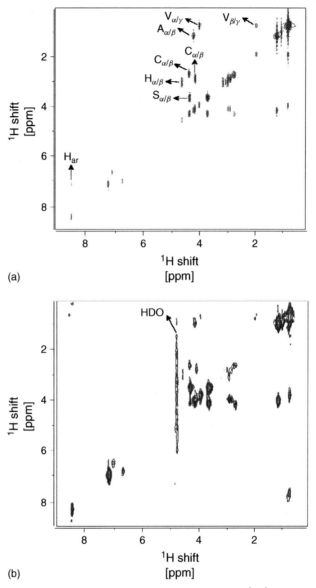

Fig. 7 Comparison of a conventionally acquired 2D [^{1}H,^{1}H] total correlation spectrum (a) acquired in 90 min for the hexapeptide CSHAVC (in the one-letter amino acid code) with the corresponding ultrafast congener recorded in 150 ms (b). In the upper panel, resonance assignments are indicated. Courtesy of Lucio Frydman.

overlap and to measure heteronuclear chemical shift correlations. For systems above about 25 kDa, one would usually consider deuterating the protein.

The sample thus obtained is used to record a set of multidimensional NMR experiments at temperatures around 30 °C. These allow (nearly) complete sequential NMR assignments to be obtained, and the conformation-dependent dispersion of the shifts enables one to derive experimental constraints for the NMR structure calculation. In most cases, structures are calculated on the basis of ^{1}H–^{1}H upper distance limit constraints derived from [^{1}H,^{1}H]-NOESY (Fig. 8).

The assignment of NOESY cross peaks and the calculation of the NMR structure is quite generally pursued in an iterative fashion. An initial set of distance constraints, which can be unambiguously obtained from the chemical shift data alone, is used to determine a low-resolution description of the protein. In turn, this initial structural model allows one to resolve the remaining NOE assignment ambiguities. This improves the precision of the structural model and allows the identification of yet additional constraints in the next round. Iterations involving structure calculations and identification of new constraints are usually pursued until (nearly) all experimentally derived constraints are in agreement with a bundle of protein conformations representing the NMR solution structure, in which conformational

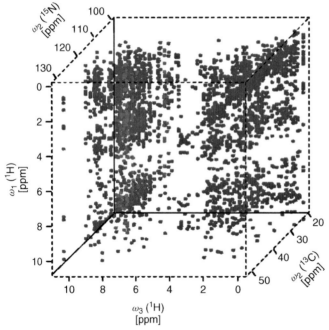

Fig. 8 Cube with dots representing NOEs detected in 3D heteronuclear resolved [^{1}H,^{1}H]-NOESY, which were used for calculating the NMR solution structure of YqfB shown in Fig. 9. Blue: signals detected along ω_3 on backbone amide protons; Red: detected on aromatic protons; Green: detected on aliphatic protons. Courtesy of Hanudatta Atreya. (See color plate p. viii).

variations reflect the precision of the NMR structure determination. Finally, the NMR structure can be refined using molecular force fields, which in essence reflect our knowledge about conformational preferences in proteins.

In order to sample the "conformational space" allowed by experimental constraints, NMR structures are usually represented by an ensemble of conformers (typically ~10–30). Various representations (Fig. 9) are used to (1) display the structural information and (2) indicate the local and global precision of the structure determination, which is reflected by rmsd values calculated for sets of atom

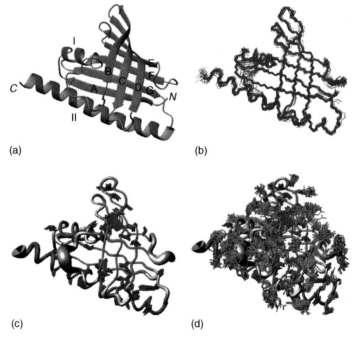

Fig. 9 Commonly used representations of NMR solution structures, exemplified for the 18-kDa Northeast Structural Genomics consortium target protein CC1736. (a) Ribbon drawing of the conformer exhibiting the smallest residual constraints violations. α-Helices and β-strands are depicted, respectively, in red/yellow, and cyan. (b) Superposition of the polypeptide backbone of the 20 conformers chosen to represent the NMR solution structure. The higher precision of the atomic coordinates of the regular secondary structure elements is apparent when compared with the loops. (c) The polypeptide backbone is represented by a spline function through the mean C^α coordinates, where the thickness represents the rmsd values for the C^α-coordinates after superposition of the regular secondary structure elements for minimal rmsd. The superpositions of the best defined side chains of the molecular core are shown in green to indicate the precision of their structural description. (d) Same as in (c), also showing the superposition of the more flexibly disordered side chains on the protein surface. The protein sample was provided by Drs. Acton and Montelione, Rutgers University. Courtesy of Yang Shen. (See color plate p. xi).

positions for the ensemble of conformers. Additional criteria, such as the number and size of residual constraint violations or the completeness of stereo-specific resonance assignments, can be used to assess the quality of an NMR structure.

6
Dynamics

Proteins are not rigid but quite dynamic biological macromolecules. In fact, the internal mobility of proteins is of key importance for their function. Nuclear spin relaxation experiments provide unique insights into protein dynamics on the sub-nanosecond to millisecond timescale. This is because the KHz to MHz random magnetic field fluctuations associated with the motions can be monitored by measuring relaxation parameters. In particular, for ^{15}N, ^{13}C, and ^{2}H-labeled proteins, the measurement of corresponding non-proton relaxation parameters can afford a rather complete coverage of the entire protein.

In a typical protein dynamics study, longitudinal (T_1) and transverse relaxation times (T_2) and heteronuclear Overhauser effects are measured. These measurements can be complemented by use of multidimensional exchange spectroscopy, which provides information regarding very slow motional modes on the ~100 ms timescale. The first key challenge that is encountered for proper interpretation of relaxation parameters is the dissection of the impact arising from the overall rotational reorientation and the internal motional modes.

While the overall tumbling of the protein is important for "finding" an interaction partner, the internal motional modes are likely the key to linking function with dynamics. Proteins are usually not of spherical shape, so that their overall tumbling can be highly anisotropic. Any failure to accurately consider this anisotropy leads to a biased picture of the internal motions. Once the overall tumbling properties of a protein are known, the proper selection of an internal motional model remains as the second key challenge. In the so-called "model-free" approach, the internal motional mode is described by a minimal set of two parameters: an order parameter reflecting spatial motional restriction and a correlation time. Importantly, the model-free approach is limited to situations in which the internal motions occur on a faster timescale than the overall tumbling. In contrast, detailed assumptions on the nature of the internal motions are made within the framework of analytical models. These are inferred from either intuition or molecular dynamics simulations.

Comparable to reaction mechanisms in chemical kinetics, motional models can never be truly "proven" by measurement of spin relaxation parameters. Instead, one can only attest consistency between model and experimental parameters. This is because the complexity of possible spatial motional modes far exceeds the complexity that can be encoded in a rather small number of scalar relaxation parameters. In the same spirit, a more extensive body of data, for example, relaxation parameters measured for various types of nuclei at different magnetic field strengths, allows one to develop more refined multiparameter motional models. Very recently, the measurement of RDCs (see Sect. 3) has been established as a valuable complement to characterize protein dynamics.

Among the currently better characterized motional modes are low-amplitude segmental motions of loops, large-amplitude flipping of aromatic rings and disulfide bonds, as well as the relative motion of entire domains in multidomain proteins. Moreover, important progress has been made in recent years to link the investigation of protein motion with functional aspects.

7
Hydration

Proteins have evolved in aqueous solution, and their surface interacts with water molecules. Hence, protein hydration plays a key role for understanding structure, dynamics, and function of proteins. In particular, macromolecular recognition involves protein surfaces: the hydration water molecules need to be "stripped off" prior to the formation of direct contacts between biological molecules. Hence, thermodynamics and kinetics of protein hydration deserve proper consideration in structural biology. Two NMR techniques provide complementary insights into protein hydration: measurement of $^1H-^1H$ NOEs between water molecules and protein, and measurement of ^{17}O and ^2H nuclear spin relaxation times at different magnetic field strengths.

For a few selected systems, such studies have shown that more than 95% of the water molecules being in contact with the protein surface are *less than* twofold motionally retarded compared to bulk water molecules. Hence, it appears that stripping off the hydration layer can hardly constitute a rate-limiting step for biomolecular interactions. This finding has evidently far-reaching consequences for understanding the kinetic properties of biological systems. In contrast to the very mobile hydration water molecules with "lifetimes" on the picosecond timescale, water molecules with longer lifetimes on the millisecond timescale have been identified and characterized. These may well play an integral structural role for proteins.

8
Folding

The rules governing the folding of a protein from an ensemble of "random coil" state into its native, functional conformation remain a major enigma of structural biology. NMR spectroscopy has been extensively employed to investigate thermodynamics and kinetics of protein folding. Measurement of chemical shifts (see Sect. 3), spin relaxation parameters (see Sect. 7), exchange spectroscopy (see Sect. 6), as well as determination of backbone amide proton–deuteron exchange rates can provide valuable insights. The majority of NMR studies focused on a few small model systems. These studies fostered the development of intriguing theoretical concepts shaping our view of how protein folding may take place, but have as yet not lead to algorithms enabling one to predict protein structure from sequence. In fact, even the existence of a concise algorithm to predict protein structure, for example, from first physical principles, could be called in question. John von Neumann noted in the framework of his automata theory, that systems may exhibit a degree of complexity, which is so high that their (empiric) description represents the easiest approach to assess them. Such complex systems have subsequently been named *von Neumann systems*, and one may ask the question whether proteins are possibly such systems. If so, it appears that the solution of the protein folding problem through structural genomics (see Sect. 9) might be the most viable way to proceed. Notably, all methods currently available for *"ab initio"* prediction of protein structures, that is, without direct reference to an experimental template, rely on our empiric knowledge of protein structures. Recently, NMR studies have also proven the common existence of intrinsically unstructured but functional proteins, and showed that folding of proteins may take place upon binding to their physiological target molecules.

9
Structural Genomics

Since the genetic code is known, genomic DNA sequences can provide the amino acid sequence of all proteins encoded in a genome. However, since the protein folding problem (see Sect. 8) is not solved, the three-dimensional structure of the proteins cannot be inferred from sequence alone. One central aim of the new discipline of "structural genomics" is to make available at least one experimentally determined protein structure for each naturally occurring family of sequence homologs. This shall allow one to (1) homologically model the remaining members of the family using an experimental X-ray or NMR structure as a template, and (2) support the functional annotation of the gene encoding the protein. One may thus argue that structural genomics corresponds to seeking a "semi-empirical" solution of the protein folding problem.

Structural genomics relies on high-throughput structure determination

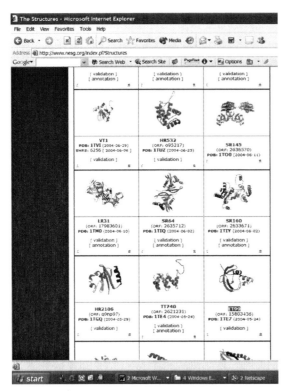

Fig. 10 "Structure gallery" accessible at the Web site of the Northeast Structural Genomics consortium in the United State (http://www.nesg.org), documenting successful implementation of a high-throughput NMR structure production pipeline.

(Fig. 10). Protein NMR spectroscopy has only very recently been shown to be a valuable technique for structural genomics (and thus also for systems biology). The ability to achieve protein structure determination (see Sect. 5) in high throughput depends primarily on: (1) sensitive high-field NMR spectrometers equipped with cryogenic probes (see Sect. 2); (2) methodology for rapid NMR data acquisition (see Sect. 4); (3) software for efficient analysis of NMR spectra and fast structure calculation; and (4) methods for automated structure quality assessment, validation, and data bank deposition. The setup of an efficient NMR-based structural genomics pipeline thus requires development of methodology, which strengthens the scientific infrastructure available for NMR-based structural biology in general.

See also Protein Modeling; Protein Structure Analysis: High-throughput Approaches.

Bibliography

Books and Reviews

Atreya, H., Szyperski, T. (2005) Rapid NMR data collection, *Methods Enzymol.* **394**, 78–108.

Bruschweiler, R. (2003) New approaches to the dynamic interpretation of NMR relaxation data from proteins, *Curr. Opin. Struct. Biol.* **13**, 175–183.

Cavanagh, J., Fairbrother, W.J., Palmer, A.G., Skelton, N.J. (1996) *Protein NMR Spectroscopy*, Academic Press, San Diego, CA.

Dyson, H.J., Wright, P.E. (2004) Unfolded proteins and protein folding studied by NMR, *Chem. Rev.* **104**, 3607–3622.

Ernst, R.R., Bodenhausen, G., Wokaun, A. (1987) *Principles of Nuclear Magnetic Resonance in One and Two Dimensions*, Clarendon Press, Oxford.

Montelione, G.T., Zheng, D., Huang, Y., Gunsalus, K.C., Szyperski, T. (2000) Protein NMR spectroscopy for structural genomics, *Nat. Struct. Biol.* **7**, 982–984.

Otting, G. (1997) NMR studies of water bound to biological molecules, *Prog. Nucl. Magn. Reson. Spectrom.* **31**, 259–285.

Pervushin, K. (2000) Impact of transverse relaxation optimized spectroscopy (TROSY) on NMR as a technique in structural biology, *Q. Rev. Biophys.* **33**, 161–197.

Sandström, J. (1982) *Dynamic NMR Spectroscopy*, Academic Press, London, UK.

Wüthrich, K. (1986) *NMR of Proteins and Nucleic Acids*, Wiley, New York.

Primary Literature

Akke, M., Palmer, A. (1996) Monitoring macromolecular motions on microsecond to millisecond scales by R(1)rho-R(1) constant relaxation time NMR spectroscopy, *J. Am. Chem. Soc.* **118**, 911–912.

Arora, A., Abildgaard, F., Bushweller, J.H., Tamm, L.K. (2001) Structure of outer membrane protein A transmembrane domain by NMR spectroscopy, *Nat. Struct. Biol.* **8**, 334–338.

Balbach, J., Forge, V., Lau, W.S., van Nuland, N.A., Brew, K., Dobson, C.M. (1996) Protein folding monitored at individual residues during a two-dimensional NMR experiment, *Science* **274**, 1161–1163.

Bax, A. (2003) Weak alignment offers new NMR opportunities to study protein structure and dynamics, *Protein Sci.* **12**, 1–16.

Bruschweiler, R., Liao, X., Wright, P.E. (1995) Long-range motional restrictions in a multidomain zinc-finger protein from anisotropic tumbling, *Science* **268**, 886–889.

Cano, K.E., Thrippleton, M.J., Keeler, J., Shaka, A.J. (2004) Cascaded z-filters for efficient single-scan suppression of zero-quantum coherence, *J. Magn. Reson.* **167**, 291–297.

Castellani, F., van Rossum, B., Diehl, A., Schubert, M., Rehbein, K., Oschkinat, H. (2002) Structure of a protein determined by solid-state magic-angle-spinning NMR spectroscopy, *Nature* **420**, 98–102.

Chou, J.J., Li, H., Salvesen, G.S., Yuan, J., Wagner, G. (1999) Solution structure of BID, an intracellular amplifier of apoptotic signaling, *Cell* **96**, 615–624.

Fiaux, J., Bertelsen, E.B., Horwich, A.L., Wuthrich, K. (2002) NMR analysis of a 900K GroEL GroES complex, *Nature* **418**, 207–211.

Frydman, L., Scherf, T., Lupulescu, A. (2002) The acquisition of multidimensional NMR spectra within a single scan, *Proc. Natl. Acad. Sci. U.S.A.* **99**, 15858–15862.

Gardner, K.H., Kay, L.E. (1998) The use of ^{2}H, ^{13}C, ^{15}N multidimensional NMR to study the structure and dynamics of proteins, *Annu. Rev. Biophys. Biomol. Struct.* **27**, 357–406.

Gutmanas, A., Billeter, M. (2004) Specific DNA recognition by the Antp homeodomain: MD simulations of specific and nonspecific complexes, *Proteins* **57**, 772–782.

Hare, B.J., Wyss, D.F., Osburne, M.S., Kern, P.S., Reinherz, E.L., Wagner, G. (1999) Structure, specificity and CDR mobility of a class II restricted single-chain T-cell receptor, *Nat. Struct. Biol.* **6**, 574–581.

Kalodimos, C.G., Biris, N., Bonvin, A.M., Levandoski, M.M., Guennuegues, M., Boelens, R., Kaptein, R. (2004) Structure and flexibility adaptation in nonspecific and specific protein-DNA complexes, *Science* **305**, 386–389.

Kim, S., Szyperski, T. (2003) GFT NMR, a new approach to rapidly obtain precise high dimensional NMR spectral information, *J. Am. Chem. Soc.* **125**, 1385–1393.

Kitahara, R., Yokoyama, S., Akasaka, K. (2005) NMR snapshots of a fluctuating protein structure: ubiquitin at 30bar-3kbar, *J. Mol. Biol.* **347**, 277–285.

Korzhev, D.M., Salvatella, X., Vendruscolo, M., DiNardo, A.A., Davidson, A.R., Dobson, C.M., Kay, L.E. (2004) Low-populated folding intermediates of Fyn SH3 characterized by relaxation dispersion NMR, *Nature* **29**, 586–590.

Kupše, E., Freeman, R. (2003) Projection-reconstruction of three-dimensional NMR spectra, *J. Am. Chem. Soc.* **125**, 13958–13959.

Lipari, G., Szabo, A. (1982) Model-free approach to the interpretation of nuclear magnetic resonance relaxation in macromolecules. 1. Theory and range of validity, *J. Am. Chem. Soc.* **104**, 4546–4559.

Liu, G., Mills, J.L., Hess, T.A., Kim, S., Skalicky, J.J., Sukumaran, D.K., Kupce, E., Skerra, A., Szyperski, T. (2003) Resonance assignments for the 21 kDa engineered fluorescein-binding lipocalin FluA, *J. Biomol. NMR* **27**, 187–188.

Luginbühl, P., Pervushin, K.V., Iwai, H., Wuthrich, K. (1997) Anisotropic molecular rotational diffusion in ^{15}N spin relaxation studies of protein mobility, *Biochemistry* **36**, 7305–7312.

Markley, J.L., Bax, A., Arata, Y., Hilbers, C.W., Kaptein, R., Sykes, B.D., Wright, P.E., Wüthrich, K. (1998) Recommendations for the presentation of NMR structures of proteins and nucleic acids, *J. Mol. Biol.* **280**, 933–952.

Modig, K., Liepinsh, E., Otting, G., Halle, B. (2003) Dynamics of protein and peptide hydration, *J. Am. Chem. Soc.* **126**, 102–114.

Monleon, D., Colson, K., Moseley, H.N.B., Anklin, C., Oswald, R., Szyperski, T., Montelione, G.T. (2002) Rapid data collection and analysis of protein resonance assignments using AutoProc, AutoPeak, and AutoAssign Software, *J. Struct. Funct. Genom.* **2**, 93–101.

Montelione, G.T. (2001) Structural genomics: an approach to the protein folding problem, *Proc. Natl. Acad. Sci. U.S.A.* **98**, 13488–13489.

Mueller, T.D., Feigon, J. (2003) Structural determinants for the binding of ubiquitin-like domains to the proteasome, *EMBO J.* **22**, 4634–4645.

Orekhov, V., Pervushin, K.V., Arseniev, A.S. (1994) Backbone dynamics of (1–71)bacterioopsin studied by two-dimensional 1H-15N NMR spectroscopy, *Eur. J. Biochem.* **219**, 887–896.

Otting, G., Wüthrich, K. (1991) Protein hydration in aqueous solution, *Science* **254**, 974–980.

Perez Canadillas, J.M., Varani, G. (2003) Recognition of GU-rich polyadenylation regulatory elements by human CstF-64 protein, *EMBO J.* **22**, 2821–2830.

Pervushin, K., Riek, R., Wider, G., Wüthrich, K. (1997) Attenuated T-2 relaxation by mutual cancellation of dipole-dipole coupling and chemical shift anisotropy indicates an avenue to NMR structures of very large biological macromolecules in solution, *Proc. Natl. Acad. Sci. U.S.A.* **94**, 12366–12371.

Peti, W., Meiler, J., Bruschweiler, R., Griesinger, C. (2002) Model-free analysis of protein backbone motion from residual dipolarcouplings, *J. Am. Chem. Soc.* **124**, 5822–5833.

Powers, R., Garrett, D.S., March, C.J., Frieden, E.A., Gronenborn, A.M., Clore, G.M. (1992)

Three-dimensional solution structure of human interleukin-4 by multidimensional heteronuclear magnetic resonance spectroscopy, *Science* **256**, 1673–1677.

Reif, B., Hennig, M., Griesinger, C. (1997) Direct measurement of angles between bond vectors in high-resolution NMR, *Science* **276**, 1230–1233.

Serber, Z., Dötsch, V. (2001) In-cell NMR spectroscopy, *Biochemistry* **40**, 14317–14323.

Skalicky, J.J., Sukumaran, D.K., Mills, J.L., Szyperski, T. (2000) Toward structural biology in supercooled water, *J. Am. Chem. Soc.* **122**, 3230–3231.

Szyperski, T., Luginbühl, P., Otting, G., Güntert, P., Wüthrich, K. (1993) Protein dynamics studied by rotating frame ^{15}N spin relaxation times, *J. Biomol. NMR* **3**, 151–164.

Tjandra, N., Bax, A. (1997) Direct measurement of distances and angles in biomolecules by NMR in a dilute liquid crystalline medium, *Science* **278**, 1111–1114.

Tolman, J.R., Flanagan, J.M., Kennedy, M.A., Prestegard, J.H. (1995) Nuclear magnetic dipole interactions in field-oriented proteins: information for structure determination in solution, *Proc. Natl. Acad. Sci. U.S.A.* **92**, 9279–9283.

Volkman, B.F., Lipson, D., Wemmer, D.E., Kern, D. (2001) Two-state allosteric behavior in a single-domain signalling protein, *Science* **291**, 2429–2433.

Von Neumann, J. (1966) *Theory of Self-Reproducing Automata*, University of Illinois Press, Urbana, IL.

Wüthrich, K. (2003) NMR studies of structure and function of biological macromolecules (Nobel lecture), *Angew. Chem., Int. Ed. Engl.* **42**, 3340–3363.

Yabuki, T., Kigawa, T., Dohmae, N., Takio, K., Terada, T., Ito, Y., Laue, E.D., Cooper, J.A., Kainosho, M., Yokoyama, S. (1998) Dual amino acid-selective and site-directed stable-isotope labeling of the human c-Ha-Ras protein by cell-free synthesis, *J. Biomol. NMR* **11**, 295–306.

Yee, A., Chang, X., Pineda-Lucena, A., Wu, B., Semesi, A., Le, B., Ramelot, T., Lee, G.M., Bhattacharyya, S., Gutierrez, P., Denisov, A., Lee, C.H., Cort, J.R., Kozlov, G., Liao, J., Finak, G., Chen, L., Wishart, D., Lee, W., McIntosh, L.P., Gehring, K., Kennedy, M.A., Edwards, A.M., Arrowsmith, C.H. (2002) An NMR approach to structural proteomics, *Proc. Natl. Acad. Sci. U.S.A.* **99**, 1825–1830.

Zheng, D., Aramini, J.M., Montelione, G.T. (2004) Validation of helical tilt angles in the solution NMR structure of the Z domain of Staphylococcal protein A by combined analysis of residual dipolar coupling and NOE data, *Protein Sci.* **13**, 549–554.

3
Electron Microscopy of Biomolecules

Claus-Thomas Bock[1], Susanne Franz[2], Hanswalter Zentgraf[2], and John Sommerville[3]
[1] Department of Molecular Pathology, Institute of Pathology, University of Tübingen, Germany
[2] German Cancer Research Center, Applied Tumor Virology, Electronmicroscopy, Heidelberg, Germany
[3] School of Biology, University of St. Andrews, St. Andrews, United Kingdom

1	**Principles** 105	
2	**Techniques** 106	
2.1	General 106	
2.2	Nucleic Acids 106	
2.3	Chromatin 109	
2.4	Proteins 113	
2.5	Macromolecular Assemblies 117	
3	**Applications** 119	
3.1	Use of Antibodies 119	
3.2	Detection of Radiolabeled Molecules 121	
3.3	Electron Spectroscopic Imaging (ESI) 123	
3.4	Image Reconstruction 124	
	Bibliography 124	
	Books and Reviews 124	
	Primary Literature 124	
	DNA Spreading 124	
	Long DNA Molecules 124	
	DNA–DNA Heteroduplex 124	
	DNA–RNA Heteroduplex 125	
	Small Chromatin Units 125	
	Large Chromatin Units (transcriptionally inactive chromatin) 125	
	Large Chromatin Units (transcriptionally active chromatin) 126	

Proteins. Edited by Robert A. Meyers.
Copyright © 2007 Wiley-VCH Verlag GmbH & Co. KGaA, Weinheim
ISBN: 978-3-527-31608-3

Negative Staining   126
Immunolabelling of Spread Preparations   126
Immunoelectron microscopy (cell biology and viruses)   127
Image Reconstruction   127
Electron Spectroscopic Imaging (ESI)   128

Keywords

Cryoelectron Microscopy
Imaging of unfixed, unstained biomolecules in a hydrated state after rapid freezing to low temperature ($-160\,°C$).

Heavy-metal Shadowing
Evaporation of a film of heavy metal (e.g. platinum–palladium or tungsten–tantalum) onto a dehydrated preparation. The deposit of metal particles around the biomolecules improves contrast and gives a shadowed, three-dimensional appearance.

High-resolution Autoradiography (EM ARG)
Detection of radiolabeled molecules by coating a stained or shadowed preparation with a film of photographic emulsion. Radioactive emissions hit silver halide crystals in the emulsion, which can be developed into silver grains.

Immunoelectron Microscopy
The application of antibodies to map specific sites on biomolecules. The antibody molecules may be visible after shadowing or negative staining, although detection can be improved by binding of a secondary antibody linked to an electron-dense tag (e.g. ferritin or colloidal gold particles).

Miller Spread
Deposition of dispersed chromatin onto a carbon-coated grid by centrifugation through a denser phase containing sucrose and formaldehyde.

Negative Staining
Instead of staining the biomolecules themselves (positive staining), a solution of heavy-metal salt (e.g. uranyl acetate or sodium phosphotungstic acid) is deposited in the hydrated spaces around and within the molecules.

Replica Casting
Adsorption of biomolecules to a mica surface followed by heavy-metal shadowing and coating of the preparation with a film of carbon. The carbon–metal replica (minus the biomolecules) is then removed from the mica and is mounted on an electron microscope grid for viewing.

Support Film
A thin film of plastic or carbon that is attached to an EM grid and provides a substrate onto which biomolecules can be adsorbed.

■ Electron microscopy (EM) is a method appropriate for examining details of the sizes and shapes of biological macromolecules and is particularly useful in studying isolated or reconstituted macromolecular assemblies. Small amounts of material (often <1 µg) can be used and prepared in a state suitable for viewing in as short a time as several minutes. The basic procedure involves adsorption of the biomolecules onto a support film, followed by staining or shadowing of the preparation and viewing of the dehydrated molecules *in vacuo* in an electron beam. A resolution of 1 to 2 nm is routine with conventional microscopes. The basic procedure can be adapted to give information about sites of specific epitopes, location of newly synthesized components, internal structures, and atomic composition.

Structures most suitable for EM analysis are nucleoprotein complexes, including ribosomes, spliceosomes, nucleosomes, and virus particles. DNA and RNA molecules are also suitable, and their lengths can be directly related to the number of base pairs or nucleotide residues determined biochemically. In general, it is difficult to visualize small proteins (<50 000 Da), but good detail can be obtained if they are isolated as multimeric complexes or if they can be induced to form filaments or crystalline arrays.

A wide range of applications are available using EM techniques, including virus identification, mapping of hybridized regions in heteroduplexes, detailing of macromolecular interaction, and analysis of the organization of molecular components in replication, transcription, splicing, and translation complexes.

1
Principles

The transmission electron microscope (TEM) consists of a metal column from which air is evacuated and through which a linear beam of electrons is accelerated and focused by electromagnetic lenses. The biomolecules are adsorbed onto a support film, stained and dehydrated (or frozen in an aqueous film on the grid), and introduced into the electron beam through an air lock. Whereas some of the electrons collide with atoms in the specimen, lose energy, and are scattered, the remaining electrons pass through the preparation and are focused to form an image on a phosphorescent screen (for direct viewing) or on a photographic plate (for later examination). Under ideal conditions, a resolution of 0.1 to 0.2 nm can be obtained; however, limitations are imposed by the naturally low masses of atoms (primarily hydrogen, carbon, and oxygen) contained in biomolecules, by distortions and artifacts created during sample preparation, and by radiation damage. Techniques are designed with the following objectives:

1. To minimize distortion by immobilization of the molecules, in their native state, onto an appropriate support.
2. To stabilize molecular complexes by suitable chemical fixation.
3. To improve contrast by staining the preparation with heavy-metal salt or by shadowing with heavy metal.

Irrespective of the method adopted, data derived from the electron microscopic examination of biomolecules should be entirely consistent with the known biochemical and biophysical properties of the particles. Whenever possible, apparent sizes should be checked by independent measurement of sedimentation rate, electrophoretic mobility, or gel filtration elution. Also, features revealed by EM of isolated molecules should be compared with observations made on them *in situ*, by EM of cell or tissue sections.

2 Techniques

2.1 General

Similar principles of sample preparation (Fig. 1) apply for nucleic acids, proteins, and nucleoprotein complexes, since most of these are less than 20 nm thick and can be adsorbed directly from solution onto the support matrix. The concentration of molecules must be high enough to permit several examples to be viewed together in one field. The efficiency of uptake onto the support is not always predictable, and a range of initial concentrations should be tried. It is important that the biomolecules be held in a solution known to maintain the proper structural features directly prior to applying to the support.

The grid consists of a fine meshwork, usually of copper, which must be coated with a thin support film of plastic (e.g. collodion, parlodion) or carbon. Carbon films are preferable because they are thin (down to 2 nm) and contribute little to the image. However, they are frequently hydrophobic and should be subjected to ionizing gases (by "glow discharging") or should be treated chemically (e.g. with Alcian blue 8GX) to render them hydrophilic before use.

2.2 Nucleic Acids

Nucleic acids are used at a concentration of 1 to 5 µg/µL, and as little as 0.1 µg is required; detailed procedures for working with these molecules may be found in the literature. The molecules can be native DNA or RNA, cloned DNA, single-stranded or double-stranded forms, partially denatured duplexes, or denatured and hybridized structures (Figs. 2–8). A

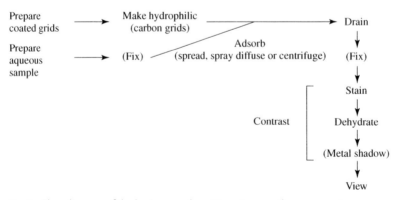

Fig. 1 Flow diagram of the basic procedure. Steps in parentheses are not always used.

key step in their preparation for EM is the spreading out of the molecules so that they can be adsorbed in an extended, nonaggregated form onto the support film. The most successful method of spreading is that devised originally by Kleinschmidt and Zahn (Table 1). The negatively charged nucleic acid is mixed with a basic protein (usually cytochrome c) in a solution appropriate for maintaining the required conformation. This spreading mix, or hyperphase, is spread, via a glass ramp, onto the surface of a second solution, the hypophase, and a molecular monolayer of nucleic acid and protein is formed at the liquid–air interface. Alternatively, the molecular monolayer can be formed by diffusion to the surface of the spreading mix contained in a small vessel or even in a droplet sitting on a hydrophobic surface (Table 2). The nucleic acid–protein complex is adsorbed through brief contact with the support film, stained with ethanolic uranyl acetate, rinsed in ethanol, and air-dried. Protein-free spreading is applied in special circumstances, for instance, to visualize very large DNA molecules that have been stabilized by polyamines (Table 3, Fig. 5). To improve contrast, spread preparations are rotary-shadowing, usually with platinum–palladium, at an angle of 6 to 9° from the plane of the metal vapor. It should be noted that although metal shadowing greatly exaggerates the thickness of the molecules, double-stranded and single-stranded regions can still be differentiated (Figs. 6 and 7).

Cytochrome c should not be used in spreading solutions when experimentally bound proteins are being studied

(a)

(b)

Fig. 2 (a) Simian virus 40 (SV40) DNA spread in the presence of cytochrome c down a glass slide and across a water hypophase. (b) *Xenopus laevis* ribosomal DNA–derived plasmid DNA spread by the cytochrome c microdiffusion droplet technique. The DNA–protein films were picked up on parlodion-coated (a) and carbon-coated (b) copper grids. The use of carbon-coated grids yields a significantly higher proportion of supercoiled molecules and results in a smoother background than is the case for parlodion-coated grids. However, the attachment of molecules to the grid is significantly higher with parlodion. For contrast enhancement, the specimens were rotary-shadowed with platinum/palladium (80:20) at an angle of 8°. Note that an additional, very brief, unidirectional shadowing in (a) leads to further contrast enhancement. The bars represent 1 μm.

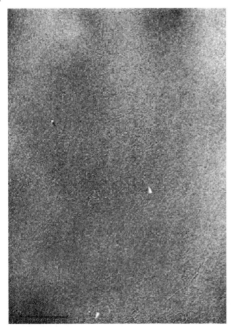

Fig. 3 Herpes simplex virus (HSV) DNA as revealed by the cytochrome c microdiffusion droplet technique, picked up on a carbon-coated grid and shadowed as described in Fig. 2a. Because shearing and stretching forces are minimal when the droplet technique is used, it is possible to spread this extremely large viral genome (150–160 kbp) to its full length. Bacteriophage PM2 circular DNA serves as size markers. The ends of the HSV genome are denoted by arrowheads. Bar represents 1 µm.

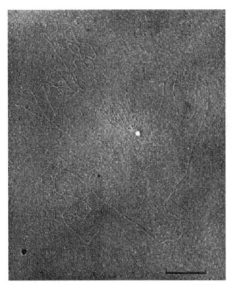

Fig. 4 The "unraveling" of the HSV genome out of the viral core as shown by cytochrome c spreading. Complete intact virus particles (see Fig. 17d) were briefly lysed in water and immediately spread onto a water hypophase. The white dotlike structure anchoring the spread DNA represents the unraveled part of the genome, surrounded by fuzzy material, most probably remnants of the viral core. Bar represents 1 µm.

because it obscures fine details of nucleic acid structure. Thus, to permit detection of proteins bound to specific regions of nucleic acid molecules, other protein-free spreading methods have been devised (Fig. 8e). These methods employ low–molecular weight inorganic substances (e.g. benzyldimethylalkylammonium chloride or ethidium bromide) in place of cytochrome c. In working with

Tab. 1 Sequence of steps for spreading nucleic acids.

1. Mix nucleic acid,[a] cytochrome c, and buffer.[b]
2. Pour hypophase solution[c] into a spreading trough.
3. Insert a glass slide in the trough with one end resting on the rim.
4. Apply spreading solution on the glass slide.
5. Wait until solution has run down the ramp and spread out.
6. Touch the film side of a grid[d] to the surface of the hypophase.
7. Stain the grid.[e]
8. Dehydrate the specimen.
9. Rotary shadow for contrast enhancement.[f]

[a]This can be dsDNA (Figs. 2a and 4), ssDNA (Fig. 6a), RNA (Fig. 6c), or heteroduplex molecules (Fig. 7).
[b]Spreading of ssDNA, RNA, or heteroduplex molecules needs denaturing agents, such as formamide and/or urea (Figs. 6 and 7).
[c]Usually, water or ammonium acetate is used; for highly denaturing conditions, formamide can be included.
[d]Normally, carbon- or parlodion-coated grids are used (Fig. 2).
[e]Ethanolic uranyl acetate is widely used.
[f]Usually, platinum/palladium (80:20) is used for rotary shadowing at an angle of 6 to 9°; specimens can also be viewed without rotary shadowing (Fig. 8c).
Note: These preparation steps are also used with slight modifications for protein-free spreading procedures (e.g. in the presence of protamines, Fig. 5, or benzyldimethylalkylammonium chloride, Fig. 8e).

Tab. 2 Sequence of steps for spreading nucleic acids: microdiffusion droplet technique.

1. Mix nucleic acid, cytochrome c, and buffer.[a]
2. Put a small drop of the solution on a hydrophobic surface.
3. Leave for 10 to 30 min for diffusion.
4. Pick up nucleic acid–protein film by allowing the grid[b] to touch the surface of the drop.
5. Stain, dehydrate, and shadow.

[a]The procedure yields best results with double-stranded nucleic acids (Figs. 2b, 3, and 8a–d). For spreading single-stranded nucleic acids, denaturing agents, such as formamide and/or urea, have to be added to the solution.
[b]Carbon- or parlodion-coated grids may be used (Fig. 2).

protein molecules bound to nucleic acids, care must be taken to stabilize the complexes before spreading by cross-linking the molecules, in solution, with glutaraldehyde (Fig. 8) or formaldehyde (for chromatin, see Section 2.3). Good detail of the association can be obtained by making a carbon–metal replica of the molecules bound to the surface of freshly cleaved mica. The use of platinum–carbon or tungsten for shadowing reveals finer detail.

2.3
Chromatin

Chromatin spreading techniques are mostly adapted from the procedure devised by Miller and coworkers (Table 4). Large chromatin units are obtained by

Tab. 3 Sequence of steps for protein-free spreading of long DNA molecules.[a]

1. Dilute DNA (0.25–1.6 Mbp) to 4 to 10 ng/μL in microinjection buffer.[b]
2. Apply a 10 μL sample[c] to a freshly glow-discharged carbon-coated grid.
3. Allow DNA to adsorb for 1 min and remove excess liquid by touching the drop with the edge of a filter paper.
4. Immediately add a 20 μL drop of 2% (w/v) aqueous uranyl acetate solution and leave to stain for 20 s.
5. Remove staining solution by touching with the edge of the filter paper and air-dry grid.
6. Rotary shadow[d] and view.

[a] Appropriate for chromosomal lengths of yeast DNA and artificial chromosomes to check molecular integrity prior to microinjection (see Fig. 6).
[b] 10 mM Tris-HCl (pH 7.5), 0.1 mM EDTA (pH 8), 100 mM NaCl, 30 mM spermine, 70 mM spermidine.
[c] Transfer of DNA can be achieved without shearing by using a plastic pipette tip with the end cut off.
[d] For instance, platinum : palladium (80 : 20) at an angle of 5°.

Tab. 4 Sequence of steps for spreading large chromatin units: Miller spreading.

1. Incubate cells, isolated nuclei, chromosomes, or chromatin under low ionic strength conditions for chromatin dispersal.
2. Keep on ice for 15 to 30 min.
3. Glow-discharge carbon-coated grids.
4. Fill the well of the centrifugation chamber with a sucrose solution containing formaldehyde.
5. Insert the carbon-coated grid, which is now hydrophilic.
6. Layer a small volume of chromatin solution on top of the sucrose cushion.
7. Centrifuge the chamber (speed and time depend on the equipment used).
8. Remove grid and dry in Photoflo detergent.
9. Stain with phosphotungstic acid and dehydrate.
10. Rotary shadow.

Note: For details see Figs. 9, 10, and 11.

Fig. 5 Full-length chromosomal DNA molecules of approximately 1 Mb isolated by pulse-field gel electrophoresis from *Saccharomyces cerevisiae* strain AB1380 and spread from a buffer containing polyamines and sodium chloride (see Table 3). The long DNA fibres may be stabilized by a coating of polyamines in the presence of sodium chloride and appear to gather into compact bundles. The two ends of a single molecule are indicated by arrows. Bar represents 0.5 μm.

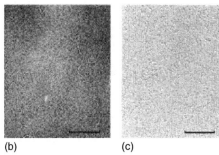

(a) (b) (c)

Fig. 6 (a) Double-stranded and single-stranded DNA of bacteriophage fd spread in 40% formamide, 1 mM EDTA, 10 mM Tris-HCl (pH 8.5) onto a water hypophase. The cytochrome c–DNA film was picked up on a parlodion-coated grid and was shadowed as described in Fig. 2b. The double-stranded DNA molecules are wider with smoother contours than the single-stranded molecules. (b) Replicative intermediate of plasmid pA2Y1, derived from a parvovirus, AAV (adeno-associated virus), spread by the microdiffusion technique in the presence of 10% formamide (parlodion-coated grids, rotary shadowing). The replication eye is denoted by an arrowhead. For contrast enhancement, (a) and (b) are printed in reverse contrast. (c) Tobacco mosaic virus (TMV) RNA was diluted in a spreading solution made from 4 M urea dissolved in "pure" formamide, 1 mM EDTA, and 10 mM Tris-HCl, (pH 8.5), heated to 80 °C (5 min), and spread on a 60 °C water hypophase, picked up on parlodion-coated grids, and shadowed. The bars represent 1 μm in (a) and 0.5 μm in (b) and (c).

Tab. 5 Sequence of steps for spreading small chromatin units.

Part A
1. Dilute chromatin for dispersal under conditions of low ionic strength.
2. Place a drop of chromatin solution on a hydrophobic surface.
3. Place a glow-discharged carbon grid on top of the drop.
4. After 15 min, remove the grid and place it on top of a drop of double-distilled water.
5. After 10 min, transfer it to a second drop of water.
6. After 10 min, remove the grid; negative-stain and air-dry the preparation.
7. Rotary shadow for contrast enhancement (see Figs. 12a, b, and d)

Part B
1. Hold a glow-discharged, carbon-coated grid with forceps.
2. Apply a drop of chromatin solution.
3. Allow the chromatin to adsorb onto the film for 1 to 2 min.
4. Stain and air-dry (alternatively shadow) (see Fig. 12c).

---

lysing cells or nuclei in a low salt concentration, high pH buffer (often referred to as "pH 9 water"). The chromatin is left to disperse and is then centrifuged at low speed, through a denser solution containing fixative (formaldehyde), onto the surface of a carbon-coated grid. The grid is dipped into a solution of wetting agent (Kodak Photoflo 200) and air-dried. The preparations can be positively stained (with ethanolic phosphotungstic acid), negatively stained (with aqueous uranyl acetate), or metal-shadowed. Chromatin spread from avian erythrocytes shows the typical nucleosomal configuration of dispersed chromatin (Fig. 9). The features of actively transcribing chromatin can be seen in nucleolar (pre-rRNA) units (Fig. 10). In contrast, nonnucleolar transcription units normally contain more widely spaced nascent transcripts (Fig. 11b).

For small chromatin units, such as chromatin fragments and viral chromatin (minichromosomes), high-speed centrifugation is required to deposit the chromatin onto the coated grid. Alternatively, small chromatin units can be adsorbed directly by floating the grid, carbon film down, on the surface of a droplet containing the sample (Table 5). The nucleosomal configuration of simian virus 40 chromatin (the SV40 minichromosome) is shown in Figs. 12a to c. Supranucleosomal

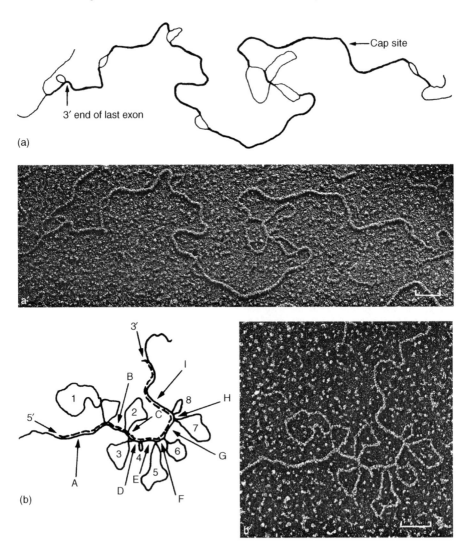

particles – for instance, those obtained by brief nuclease digestion of avian erythrocyte chromatin (Fig. 12d) – are also prepared by direct adsorption.

2.4
Proteins

The most commonly used method for contrast enhancement of proteins is negative staining (Table 6). This method is extremely rapid and involves drying of an aqueous solution of heavy-metal salt (e.g. uranyl acetate) with the protein molecules on the coated grid. The metal salt occupies hydrated regions in and around the biomolecules and, upon drying, forms a dense cast. Since the biomolecules themselves are not stained, but the background is stained, a negative image is created. This technique has been used extensively for examining protein filaments. Negatively stained intermediate filaments, in this example, glial fibrillary acidic protein (GFAP), are shown in Fig. 13a. Details of the arrangement of individual collagen molecules within the collagen fibril can be obtained from preparations like that seen in Fig. 14.

An alternative to the adsorption of proteins from droplets is to spray them, in aerosols, onto the surface of carbon-coated grids. The spraying is most easily achieved

Tab. 6 Sequence of steps for negative staining.

1. Hold a coated grid[a] with forceps.
2. Apply a drop containing biological material to the grid.
3. After 10 to 60 s, drain off excess liquid.
4. Apply a drop of negative stain[b] for 10 to 60 s.[c]
5. Remove liquid and air-dry the preparation.

[a]Carbon, parlodion, or sandwich coats (made from Formvar and carbon) may be used. Glow discharge increases the amount of biological material adsorbed onto the film; plastic films, however, are destroyed by glow discharge.
[b]Uranyl acetate and phosphotungstic acid are widely used for negative staining.
[c]Alternatively, the grid can be transferred through several drops of stain.
Note: For details on the various procedures, see Figs. 13–18.

Fig. 7 (a) A DNA:DNA heteroduplex formed between cloned mouse tyrosine aminotransferase (TAT) and rat TAT. Heteroduplexes were formed after denaturation in 0.1 M NaOH, 20 mM EDTA at 20 °C for 10 min and renaturation in the presence of 50% formamide, 20 mM Tris-HCl (pH 7.2) at 20 °C for 90 min. Heteroduplexes were then prepared from a spreading solution containing 4 M urea dissolved in "pure" formamide, 1 mM EDTA, 10 mM Tris-HCl, (pH 8.5) on water hypophase, followed by rotary shadowing. (a′) The interpretive drawing of (a): the positions of the cap and of the 3′ end of the last exon are indicated. (b) A heteroduplex formed between poly(A+) RNA from bovine muscle epidermis and a genomic clone that contains the gene coding for epidermal bovine keratin (KBIa). Heteroduplexes were formed in 70% formamide, 0.3 M NaCl, 1 mM EDTA, 20 mM Tris-HCl (pH 8.0). The sample was kept for 17 h at 62 °C, followed by incubation at 64 °C for 2 to 4 h. The hybrids were prepared for electron microscopy as described in (a). (b′) The interpretative drawing of (b): DNA is represented by a continuous line, RNA by an interrupted line; the 5′ end of mRNA is identified by a change in molecular diameter and the 3′ end by a projecting [i.e. nonhybridized, poly(A) tail]. Exons are denoted by capital letters, introns by arabic numerals. For contrast enhancement, (a) and (b) have been printed as negatives. Bars represent 100 nm.

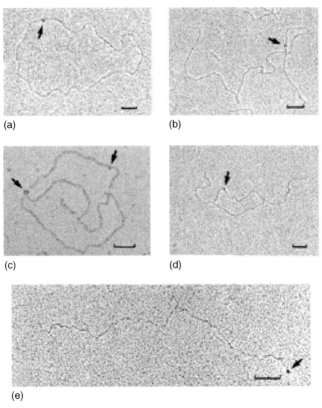

Fig. 8 (a, b) Binding of monoclonal Z-DNA antibodies to DNA from bovine papillomavirus (BPV-1). Antibodies against left-handed Z-DNA were incubated and cross-linked with 0.1% glutaraldehyde (2 h) to the pML2d-BPV-1 plasmid DNA (a). To determine the position of the binding site, the DNA was digested with the single-cutting restriction enzyme Xba I (b). Arrows indicate the position of Z-DNA antibodies on the plasmids. (c, d) Localization of large-tumor antigen (T antigen) on replicative intermediates of SV40 DNA. The molecules were immunostained with either ferritin-labeled protein A (c) or ferritin-labeled goat antimouse antibodies (d) after cross-linking with 0.1% glutaraldehyde (20 min, 37 °C). To localize T antigen on the replicated section, the complex was cleaved with the single-cutting enzyme Bgl I. Arrows indicate the ferritin-labeled T antigen on DNA. These complexes (a–d) were prepared for electron microscopy according to the cytochrome c droplet diffusion technique. The specimens were rotary-shadowed, except the one shown in (c), which is only stained by ethanolic uranyl acetate. (e) DNA-relaxing enzyme molecules linked to single-stranded SV40 DNA (arrow). After enzyme reaction, the complexes were spread from a solution containing 0.25 M NaOH, indicating that the protein dot is stable in alkali and is therefore linked via a covalent bond to the end of the single-stranded DNA. Enzyme–DNA complexes were prepared with a protein-free spreading procedure using benzyldimethylalkylammonium chloride (BAC) and were shadowed. Bars represent 100 nm.

Fig. 9 Appearance of the nucleosomal chromatin configuration of chicken erythrocytes lysed and swollen in 0.5 mM sodium borate buffer (pH 8.8) for 5 min, spread by the Miller technique, positively stained, and metal-shadowed. The dispersed chromatin shows the characteristic nucleosomal "beads-on-a-string" organization. Bar represents 1 µm.

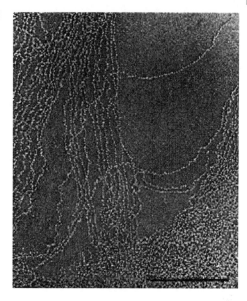

Fig. 10 Organization of nucleolar chromatin from oocytes of the newt *Pleurodeles waltlii*, after spreading using the Miller technique, positive staining, and rotary shadowing. The transcribed chromatin segments are densely packed with RNA polymerases and nascent ribonucleoprotein (RNP) fibrils. The tandemly arranged active pre-rRNA genes are interspersed by the spacer regions. The axis of the spacer is relatively thin and does not show nucleosome-sized particles. Bar represents 1 µm.

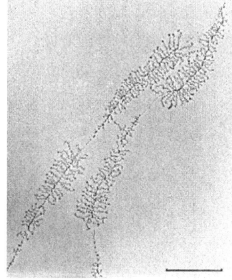

by touching the tip of a glass capillary tube, containing both sample and stain, into a stream of nitrogen. Elongate, or rod-shaped, proteins can be mixed with glycerol (and no stain) and sprayed onto the surface of a piece of freshly cleaved mica. The deposit is then shadowed finely and carbon-coated to produce a replica cast of the protein molecules. (Details are provided in the literature.) Unidirectionally

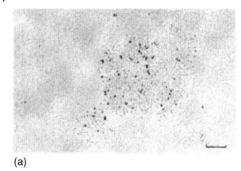

(a)

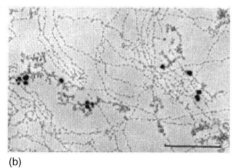

(b)

Fig. 11 High-resolution autoradiography of chromatin spread from mouse P815 cells after labeling for 5 min with [^{3}H]uridine. (a) Radioactivity (recorded as dense silver grains) associated with densely transcribing regions showing the gradient-like features of pre-rRNA transcription units (cf. Fig. 10). The background consists of many strands of nucleosomal chromatin. (b) Radioactivity associated with individual RNP fibrils characteristic of those found in nonnucleolar transcription units. In sparsely transcribing units, nucleosomes can be seen on the chromatin strand between the nascent transcripts. Both preparations were stained with phosphotungstic acid, rotary-shadowed with platinum, and exposed with Ilford L4 emulsion. Bars represent 0.5 µm. (Courtesy of S. Fakan, with permission from Springer-Verlag.)

Fig. 12 Small chromatin units such as SV40 chromatin complexes (a–c) or supranucleosomal particles from chicken erythrocyte chromatin (d) are prepared by direct adsorption to freshly glow-discharged, carbon-coated grids. The chromatin is allowed to adsorb onto the film for 1 to 2 min, stained with 2% uranyl acetate (c), and rotary-shadowed (a, b, d). Under spreading conditions involving buffers of low ionic strength (2 mM EDTA, 1 mM Tris-HCl, pH 8.4), the SV40 nucleoprotein complexes ("minichromosomes") show the typical "beads-on-a-string" morphology of the circular nucleosomal chain (a, b). The foreshortening of DNA in the SV40 "minichromosomes" due to nucleosomal organization is demonstrated by comparison with the viral DNA, which is included in the preparation (b). This packaging of the SV40 genome, at first-order level into nucleosomes, results in a foreshortening ratio of about 5.5 : 1. (d) At the second level of packaging, chromatin is organized into supranucleosomal structures. Fractions of supranucleosomal granular subunits, obtained by brief digestion of chromatin with micrococcal nuclease at physiological salt concentration, can be spread after fixation for 15 min at 4 °C with glutaraldehyde (final concentration 0.2%). (e) Nonnucleosomal forms of chromatin beside nucleosomal chains are observed in Miller spread preparations of African green monkey kidney cells (RC37) after infection with herpes simplex virus (HSV). The "bubblelike" configurations, thickly coated with a single-strand DNA-binding protein, alternate with unbranched thin intercepts (staining and shadowing as in Figs. 9 and 10). Bars represent 500 nm in (a), (d), and (e), 200 nm in (b), and 50 nm in (c).

shadowed molecules, prepared by the spray/replica technique and representing early stages in the assembly of GFAP filaments, are shown in Fig. 13b.

2.5
Macromolecular Assemblies

Multimeric enzyme complexes, ribosomes, viruses, and other particles can be treated in ways similar to those described for protein filaments or for small chromatin units. Particles in the size range of 10 to 200 nm diameter are excellent targets for negative staining (Figs. 15–18), which often reveals considerable detail of surface structure (Fig. 17). Nevertheless, the size of the particles of dried stain (or of the metal particles after shadowing) limits the resolution of structural

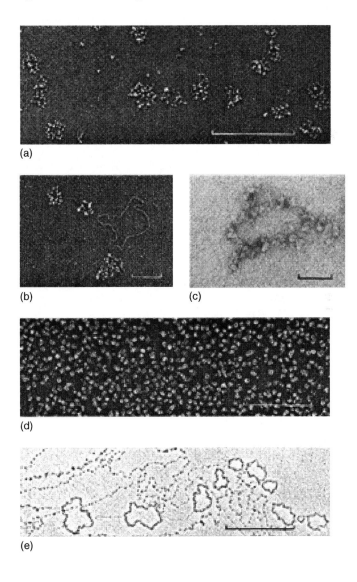

(a)

(b) (c)

(d)

(e)

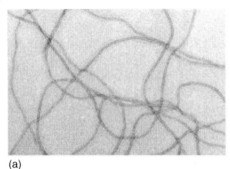

(a)

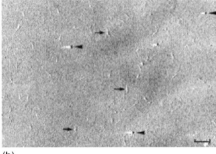

(b)

Fig. 13 Intermediate filaments as visualized by negative staining and partly assembled monomers as visualized by metal shadowing. (a) GFAP intermediate-sized filaments spread onto carbon film and negatively stained with 1% w/v uranyl acetate. GFAP was prepared from spinal cord and assembled *in vitro* by dialysis against the following buffer: 10 mM Tris-HCl (pH 7.0), 1 mM $MgCl_2$, 50 mM NaCl, and 25 mM 2-mercaptoethanol. Note that the intermediate filaments are 10 nm in width and are long with smooth edges. (b) The assembly of intermediate filaments can be arrested at various points along the assembly pathway by choosing the appropriate buffer conditions. In this preparation, GFAP molecules containing chains of four proteins were formed in 10 mM Tris-HCl (pH 8.5). A sample of this was made 50% v/v in glycerol, sprayed onto freshly cleaved mica, and unidirectionally shadowed at 10° with platinum. The replica was then floated on water and picked up onto grids. The molecules of GFAP appear as short rodlets, between 45 and 65 nm long and 2 to 3 nm wide (arrows). Latex beads demonstrate the direction of shadow and locate the position of the droplets on the replica (arrowheads). Both micrographs are shown at the same magnification; bar represents 100 nm. (Courtesy of R. Quinlan and A. M. Hutcheson.)

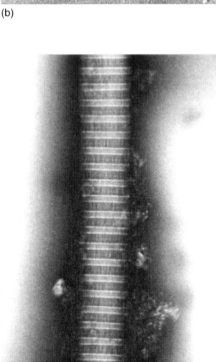

detail. A superior technique cryoelectron microscopy involves the rapid freezing of a thin (100 nm) aqueous film containing the specimen particles. Below −143 °C, water can be held in a vitrified state, which resembles the liquid state and avoids formation of ice crystals. In this condition, macromolecules and macromolecular assemblies can be viewed free of artifacts from fixation, dehydration, and staining. Under-focus phase contrast is used to produce a high-resolution image.

Fig. 14 A collagen fibril, negatively stained with 4% phosphotungstic acid, on a glow-discharged, carbon-coated grid. The staggered arrangement of the collagen molecules in the fibril results in the striated appearance after negative staining. The bars represent 100 nm.

Fig. 15 Examples of filamentous bacteriophages and rodlike viruses. The bacteriophage fd was negatively stained with either 4% phosphotungstic acid (a) or 2% uranyl acetate (b). The tobacco mosaic viruses in (c) are stained with 4% phosphotungstic acid. All preparations were made on carbon-coated, glow-discharged grids. The bars represent 0.5 µm.

(a)

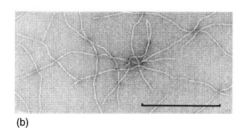
(b)

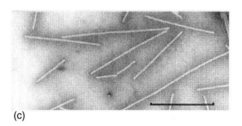
(c)

3 Applications

3.1 Use of Antibodies

Antibody molecules (IgG and IgM) are large enough to be resolved in metal-shadowed or negatively stained preparations. Monoclonal antibodies can be used to map sites on nucleic acid or protein molecules and to identify individual components within macromolecular assemblies. Antibodies can also be used to build up denser structures around small

Tab. 7 Sequence of steps for immunogold labeling: identification of individual components in macromolecular assemblies.

1. Hold a coated grid[a] with forceps.
2. Apply a drop containing biological material to the grid.
3. After 1 min, wash the grid with 20 drops of double-distilled water from a Pasteur pipette and drain off excess liquid.[b]
4. Place the grid (film side down) on a drop[c] of the primary antibody diluted in phosphate-buffered saline (PBS), 1% bovine serum albumin (BSA).[d]
5. After 30 min, remove the grid and place it on top of a drop of PBS, 1% BSA.
6. After 5 min, transfer the grid to a fresh drop of PBS, 1% BSA; repeat this washing step a third time.
7. After 5 min, transfer the grid to a drop of the gold-labeled secondary antibody[e] diluted in PBS, 1% BSA.
8. After 30 min, place the grid on top of a drop of PBS, 1% BSA and continue as described in step 6.
9. Wash the grid as described in step 3.
10. Stain the specimen with 2% aqueous uranyl acetate for 1 min and air-dry.

[a]The procedure yields best results with glow-discharged carbon-coated grids.
[b]Never drain the grids completely.
[c]Droplets are placed on a piece of Parafilm.
[d]PBS, 1% BSA; filtered through a 0.22 µm Millipore filter.
[e]Alternatively, a protein A–coated colloidal gold probe may be used.

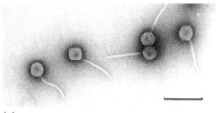

(a)

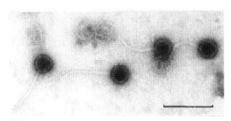

(b)

(c)

Fig. 16 Examples of tailed (a, b) and globular (c) bacteriophages. Bacteriophage λ (a) and the globular bacteriophage fr (c) were stained with 4% uranyl acetate. The bacteriophage T5 (b) was stained with 2% uranyl acetate. Note the empty phage heads in (b). All preparations were made on freshly glow-discharged, carbon-coated grids. The bars represent 200 nm.

proteins, which are themselves difficult to resolve. Secondary antibodies tagged with electron-dense markers such as ferritin (Figs. 8c and d) or colloidal gold particles are routinely used to improve detection (Table 7). For instance, gold-conjugated secondary antibodies can be used to detect a range of epitopes in recombinant viral coat particles (Fig. 18). A quite different application is the use of anti-biotin to detect biotinylated nucleic acid probes that have been hybridized to complementary sequences in chromatin preparations or *in situ* in isolated chromosomes. In the example shown in Fig. 19, two different antibody-binding sites are enhanced for EM detection by addition of a common secondary antibody tagged with colloidal gold particles. In this way, specific genes can be detected within complex chromatin masses.

Tab. 8 Sequence of steps for high-resolution autoradiography of biomolecules.

1. Radioactively label[a] living cells or subcellular fractions *in vitro*.
2. Isolate molecules or molecular complexes and prepare by direct adsorption on the grid or by spreading techniques.
3. Stain,[b] dehydrate, and rotary shadow.
4. Evaporate a thin layer of carbon (4–8 nm) on the specimen.[c]
5. Coat the specimen with a layer of photographic emulsion.
6. Follow the gold latensification–Elon–ascorbic acid (GEA) procedure.[d]
7. Transfer to Elon–ascorbic acid developer.
8. Transfer into the fixing bath.
9. Wash and air-dry.

[a] Generally, radioactive tritium is used.
[b] Negative or positive staining; shadowing with platinum/palladium.
[c] The fine coat prevents chemical interactions between specimen and emulsion.
[d] The GEA development procedure increases the sensitivity and gives rise to silver grains of rather small size.

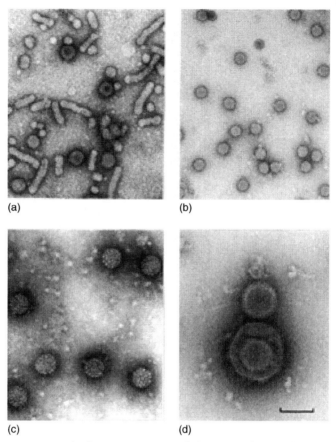

Fig. 17 Details of virus structure revealed in negatively stained preparations on carbon-coated, glow-discharged grids. (a) Hepatitis B viruses (HBV) were stained with 2% uranyl acetate. (b) Polioviruses and (c) bovine papilloma viruses (BPV) were stained with 4% phosphotungstic acid. (d) Herpes simplex virus (HSV) was stained with 2% uranyl formate. HBV was obtained from a patient's serum and reveals, in addition to the mature "Dane" particles, globular and filamentous forms of the viral S protein (a). Polioviruses show a fine-textured surface (b), while BPVs reveal a moruloid structure, which is due to the capsomere architecture (c). This organization is also visible in the HSV capsid, even when surrounded by the envelope (d). The micrographs are magnified to the same scale to give an impression of the size difference between the virus families. Bar represents 100 nm.

3.2 Detection of Radiolabeled Molecules

Although high-resolution autoradiography (EM ARG) is applied mostly to the detection of newly synthesized (radiolabeled) components *in situ* at the cellular level, application is possible with spread molecules and molecular complexes. However, EM ARG is limited to the use

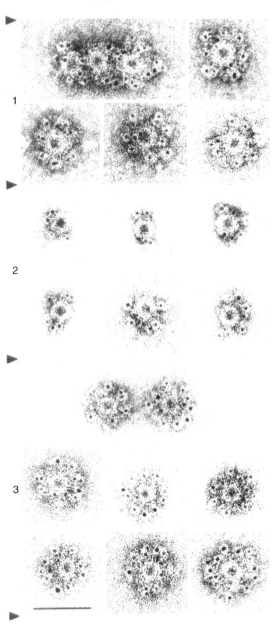

Fig. 18 Detection of different epitopes in recombinant viral core particles. The vector cassette pB–CHG–HBc (C, cytomegalovirus (CMV) promoter; H, histidine tag; G, green fluorescent protein; HBc, core protein of hepatitis B virus) was transfected into HeLa cells. The recombinant core particles were isolated by sucrose gradient centrifugation. Epitopes in the fusion protein were detected by immunoelectron microscopy after adsorption of the particles onto carbon-coated grids. All incubation and washing steps were carried out on the grids. The antibodies used were (1) rabbit anti-HBc/goat antirabbit −5 nm gold; (2) mouse anti-histidine/goat antimouse −5 nm gold; and (3) rabbit anti-GFP/goat antirabbit −5 nm gold. After immunoreactions, the grids were negatively stained with uranyl acetate. Bar represents 100 nm.

of radioisotopes emitting soft β-particles, usually ^{3}H. Molecules that have been radiolabeled either *in vivo* or *in vitro* are prepared for EM by the method most appropriate for that type of sample (Table 8). After staining, the preparation is covered, first with a thin protective coat of carbon and then with a layer of photographic emulsion. Ideally, the emulsion should consist of a homogeneous monolayer of

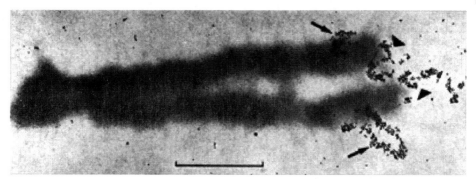

Fig. 19 Whole-chromosome mount from *Xenopus* culture cells after *in situ* hybridization with a mixture of biotinylated DNA encoding tRNA and oocyte-specific 5S RNA. The chromosomes are partially unfolded, and the sites of hybridization are detected using anti-biotin and secondary antibodies tagged with colloidal gold. The gold particles are seen to decorate DNA loops containing tRNA genes (arrows) and 5S RNA genes (arrowheads). The bar represents 2 µm. (Courtesy of S. Narayanswami and B. A. Hamkalo, with permission from Oxford University Press.)

the silver halide crystals. After being kept in the dark, to allow a sufficient amount of radioactive decay, the emulsion is developed, leaving silver grains above the sites of radiolabeling in the molecules. EM ARG is subject to two severe limitations:

1. The accuracy with which the developed silver grain can be located to the source of radiation in the specimen molecule. This depends on the thickness of the preparation and the diameter of the silver halide crystal in the emulsion and results in scattering of grains with a half distance of about 100 nm.
2. The efficiency in detecting radioactive disintegrations. Even molecules labeled to high specific activities require exposure times of generally more than a few weeks.

In spite of these limitations, EM ARG has been used successfully in studying sites of replication in isolated DNA molecules and the location of active transcription complexes in spread chromatin (Fig. 11).

3.3
Electron Spectroscopic Imaging (ESI)

By adapting the electron microscope to include an imaging electron energy filter, positional information can be recovered from electrons that have lost a specific and characteristic amount of energy in colliding with atoms in the specimen molecules. This approach has been used successfully in the location of phosphorus atoms, particularly those in the phosphodiester bonds of nucleic acids. The resulting electron spectroscopic imaging (ESI) has been used to delineate the path of the sugar–phosphate backbone in nucleosomes and the configuration of SRP–RNA (7SL RNA) within the signal recognition (ribonucleoprotein) particle. Along with related techniques in element mapping within biomolecules and their complexes, ESI is becoming increasingly sophisticated and promises many diverse applications.

3.4
Image Reconstruction

Individual molecules, under optimum conditions of spreading and staining, give weak and ill-defined images. To improve on the amount of structural detail, information from many molecules can be combined to smooth out random variations between images. To do this, it is necessary to use molecules (e.g. protein filaments), which contain regular, repeating arrays of subunits. Alternatively, the biomolecules can be induced to form crystalline arrays of regular, tightly packed, oriented units. Electron micrographs of arrays of these types can then be used for image processing to produce enhanced structural detail.

Bibliography

Books and Reviews

Bazett-Jones, D.P. (1992) Electron spectroscopic imaging of chromatin and other nucleoprotein complexes, *Electron Microsc. Rev.* **5**, 37–58.

Dubochet, J., Adrian, M., Chang, J., Homo, J.-C., Lepault, J., McDowall, A.W., Schultz, P. (1988) Cryoelectron microscopy of vitrified specimens, *Q. Rev. Biophys.* **21**, 129–228.

Harris, J.R. (Ed.) (1991) *Electron Microscopy in Biology: A Practical Approach*, Oxford University Press, London, New York.

McLean, D.M., Wong, K.K. (1984) *Same-Day Diagnosis of Human Virus Infections*, CRC Press, Boca Raton.

Ottensmeyer, F.P. (1986) Elemental mapping by energy filtration: advantages, limitations and compromises, *Ann. N. Y. Acad. Sci.* **483**, 339–353.

Skepper, J.N. (2000) Immunocytochemical strategies for electron microscopy: choice or compromise, *J. Microsc.* **199**, 1–36.

Sommerville, J. (1981) Immunolocalization and Structural Organization of Nascent RNP, in: Busch, H. (Ed.) *The Cell Nucleus*, Vol. 8, Academic Press, New York, pp. 1–57.

Sommerville, J., Scheer, U. (Eds.) (1987) *Electron Microscopy in Molecular Biology: A Practical Approach*, Oxford University Press, London, New York.

van Tuinen, E.J. (1996) Immunoelectron microscopy, *Methods Mol. Biol.* **53**, 407–422.

Primary Literature

DNA Spreading

Bock, C.-T., Zentgraf, H. (1993) Detection of minimal amounts of DNA by electron microscopy using simplified spreading procedures, *Chromosoma* **102**, 249–252.

Egel-Mitani, M., Egel, R. (1972) A rapid visualization of Kleinschmidt-type DNA preparations by phosphotungstic acid, *Z. Naturforsch.* **27**, 480–481.

Kleinschmidt, A.K. (1968) Monolayer techniques in electron microscopy of nucleic acid molecules, *Methods Enzymol.* **12**, 361–377.

Lang, D., Mitani, M. (1970) Simplified quantitative electron microscopy of biopolymers, *Biopolymers* **9**, 373–379.

Spiess, E., Lurz, R. (1988) Electron microscopic analysis of nucleic acids and nucleic acid-protein complexes, *Methods Microbiol.* **20**, 293–321.

Vollenweider, H.H., Sogo, J.M., Koller, T. (1975) A routine method for protein-free spreading of double- and single-stranded nuclei acid molecules, *Proc. Natl. Acad. Sci. USA* **72**, 83–87.

Long DNA Molecules

Montoliu, L., Bock, C.-T., Schütz, G., Zentgraf, H. (1995) Visualization of large DNA molecules by electron microscopy with polyamines: application to the analysis of yeast endogenous and artificial chromosomes, *J. Mol. Biol.* **246**, 486–492.

DNA–DNA Heteroduplex

Davis, R.W., Davidson, N. (1968) Electron microscopic visualization of deletion mutations, *Proc. Natl. Acad. Sci. USA* **60**, 243–250.

Davis, R.W., Simon, M., Davidson, N. (1971) Electron microscope heteroduplex methods for mapping regions of base sequence homology in nucleic acids, *Methods Enzymol.* **21**, 413–428.

Scherer, G., Bock, C.T., Zentgraf, H. (1990) Heteroduplex between an EMBL3 genomic clone and a gt11 cDNA clone provides internal size standards, *Nucleic Acids Res.* **18**, 4944.

Walfield, A.M., Storb, U., Selsing, E., Zentgraf, H. (1980) Comparison of different rearranged immunoglobulin kappa genes of a myeloma by electron microscopy and restriction mapping of cloned DNA's: implications for allelic exclusion, *Nucleic Acids Res.* **8**, 4689–4707.

Westmoreland, B.C., Szybalski, W., Ris, H. (1969) Mapping of deletions and substitutions in heteroduplex DNA molecules of bacteriophage lambda by electron microscopy, *Science* **163**, 1343–1351.

DNA–RNA Heteroduplex

Chow, L.T., Roberts, J.M., Lewis, J.B., Broker, T.R. (1977) A map of cytoplasmic RNA transcripts from lytic adenovirus type 2, determined by electron microscopy of RNA: DNA hybrids, *Cell* **11**, 819–825.

Hsu, M.-T., Ford, J. (1977) A novel sequence arrangement of SV40 late RNA, *Cold Spring Harbor Symp. Quant. Biol.* **42**, 571–576.

May, E., Maizel, J.V., Salzman, N. (1977) Mapping of transcription sites of simian virus 40-specific late 16S and 19S mRNA by electron microscopy, *Proc. Natl. Acad. Sci. USA* **71**, 496–500.

Müller, G., Scherer, G., Zentgraf, H., Ruppert, S., Herrmann, B., Lehrach, H., Schütz, G. (1985) Isolation, characterization and chromosomal mapping of the mouse tyrosin aminotransferase gene, *J. Mol. Biol.* **184**, 367–373.

Schmid, W., Scherer, G., Danesch, U., Zentgraf, H., Matthias, P., Strange, C.M., Röwekamp, W., Schütz, G. (1982) Isolation and characterization of the rat tryptophan oxygenase gene, *EMBO J.* **10**, 1287–1293.

Thomas, M., White, R.L., Davies, R.W. (1976) Hybridization of RNA to double-stranded DNA: formation of R-loops, *Proc. Natl. Acad. Sci. USA* **73**, 2294–2301.

White, R.L., Hogness, D.S. (1977) R-loop mapping of the 18S and 28S sequences in the long and short repeating units of *Drosophila melanogaster* DNA, *Cell* **10**, 177–185.

Small Chromatin Units

Bellard, M., Oudet, P., Germond, J.E., Chambon, P. (1976) Subunit structure of simian virus 40 minichromosomes, *Eur. J. Biochem.* **70**, 543–549.

Bock, C.T., Schwinn, S., Locarnini, S., Fyfe, J., Manns, M.P., Trautwein., C., Zentgraf, H. (2001) Structural organization of the hepatitis B virus minichromosome, *J. Mol. Biol.* **307**, 183–196.

Cremisi, C., Pignatti, P.F., Croissant, O., Yaniv, M. (1976) Chromatin-like structures in polyoma virus and simian virus 40 lytic cycle, *J. Virol.* **17**, 204–211.

Griffith, J.D. (1975) Chromatin structure: deduced from a minichromosome, *Science* **187**, 1202–1203.

Keller, W., Müller, U., Eicken, I., Wendel, I., Zentgraf, H. (1977) Biochemical and ultrastructural analysis of SV40 chromatin, *Cold Spring Harbor Symp. Quant. Biol.* **42**, 227–244.

Müller, U., Zentgraf, H., Eicken, I., Keller, W. (1978) Higher-order structure of simian virus 40 chromatin, *Science* **201**, 406–415.

Scheer, U., Zentgraf, H. (1978) Nucleosomal and supranucleosomal organization of transcriptional inactive rDNA circles in *Dytiscus* oocytes, *Chromosoma* **69**, 243–254.

Strätling, W.H., Müller, U., Zentgraf, H. (1978). The higher-order repeat-structure of chromatin is built up of globular particles containing eight nucleosomes. *Exp. Cell Res.* **117**, 301–311.

Zentgraf, H., Trendelenburg, M.F., Spring, H., Scheer, U., Franke, W.W., Müller, U., Drury, K.C., Runger, D. (1979) Mitochondrial DNA arranged into chromatin-like structures after injection into amphibian oocyte nuclei, *Exp. Cell Res.* **122**, 363–375.

Large Chromatin Units (transcriptionally inactive chromatin)

Finch, J.T., Klug, A. (1976) Solenoidal model for superstructure in chromatin, *Proc. Natl. Acad. Sci. USA* **73**, 1897–1900.

Müller, U., Schröder, C.H., Zentgraf, H., Franke, W.W. (1980) Coexistence of nucleosomal and various non-nucleosomal chromatin configurations in cells infected with herpes simplex virus, *Eur. J. Cell Biol.* **23**, 197–203.

Olins, A.L., Olins, D.E. (1974) Spheroid chromatin units (ν bodies), *Science* **183**, 330–332.

Oudet, P., Gross-Bellard, M., Chambon, P. (1975) Electron microscopic and biochemical evidence that chromatin structure is a repeating unit, *Cell* **4**, 281–300.

Woodcock, C.L.F. (1973) Ultrastructure of inactive chromatin, *J. Cell Biol.* **59**, 368a.

Zentgraf, H., Müller, U., Franke, W.W. (1980) Supranucleosomal organization of sea urchin sperm chromatin in regularly arranged 40–50 mm large granular subunits, *Eur. J. Cell Biol.* **20**, 254–264.

Zentgraf, H., Müller, U., Franke, W.W. (1980) Reversible in vitro packing of nucleosomal filaments into globular supranucleosomal units in chromatin of whole chick erythrocyte nuclei, *Eur. J. Cell Biol.* **23**, 171–188.

Large Chromatin Units (transcriptionally active chromatin)

Foe, V.E., Wilkinson, L.E., Laird, C.D. (1976) Comparative organization of active transcription units in Oncopeltus fasciatus, *Cell* **9**, 131–146.

Franke, W.W., Scheer, U., Trendelenburg, M., Spring, H., Zentgraf, H. (1976) Absence of nucleosomes in transcriptionally active chromatin, *Cytobiologie* **13**, 401–434.

Lamb, M.M., Daneholt, B. (1979) Characterization of active transcription units in Balbiani rings of Chironomus tentans, *Cell* **17**, 835–848.

McKnight, S.L., Sullivan, N.L., Miller, O.L. (1976) Visualization of the silk fibroin transcription unit and nascent silk fibroin molecules on polyribosomes of Bombyx mori, *Prog. Nucleic Acid Res. Mol. Biol.* **19**, 313–318.

Miller, O.L., Bakken, A.H. (1972) Morphological studies of transcription, *Acta endocrinol. Suppl.* **168**, 155–177.

Miller, O.L., Beatty, B.R. (1969) Visualization of nucleolar genes, *Science* **164**, 955–957.

Puvion-Dutilleul, F., Bernadac, A., Puvion, E., Bernard, W. (1977) Visualization of nulear transcriptional complexes in rat liver cells, *J. Ultrastr. Res.* **58**, 107–117.

Samuel, C., Mackie, J., Sommerville, J. (1981) Macronuclear chromatin organization in *Paramecium primaurelia*, *Chromosoma* **83**, 481–492.

Scheer, U. (1978) Changes of nucleosome frequency in nucleolar and non-nucleolar chromatin as a function of transcription: an electron microscopic study, *Cell* **13**, 535–549.

Scheer, U., Franke, W.W., Trendelenburg, M.F., Spring, H. (1976) Classification of loops of lampbrush chromosomes according to the arrangement of transcriptional complexes, *J. Cell Sci.* **22**, 503–520.

Scheer, U., Trendelenburg, M.F., Franke, W.W. (1976) Regulation of transcription of genes of ribosomal RNA during amphibian oogenesis, *J. Cell Biol.* **69**, 465–489.

Scheer, U., Zentgraf, H., Sauer, H.W. (1981) Different chromatin structures in *Physarum polycephalum*: a special form of transcriptionally active chromatin devoid of nucleosomal particles, *Chromosoma* **84**, 279–290.

Trendelenburg, M.F., Spring, H., Scheer, U., Franke, W.W. (1974) Morphology of nucleolar cistrons in a plant cell, *Acetabularia mediterrane*, *Proc. Natl. Acad. Sci. USA* **71**, 3626–3630.

Negative Staining

Almeida, J.D., Howatson, A.F. (1963) A negative staining method for cell associated virus, *J. Cell Biol.* **16**, 616–623.

Baumeister, W. (1978) Biological horizons in molecular microscopy, *Cytobiologie* **17**, 246–297.

Brenner, S., Horne, R.W. (1959) A negative staining method for high resolution electron microscopy of viruses, *Biochim. Biophys. Acta* **34**, 103–110.

Harris, R., Horne, R.W. (1991) Negative Staining, in: Harris, R. (Ed.) *Electron Microscopy in Biology: A Practical Approach*, Oxford University Press, London, New York, pp. 203–228.

Horne, R.W., Wildy, P. (1979) An historical account of the development and applications of the negative staining technique to the electron microscopy of viruses, *J. Microsc.* **117**, 103–122.

Smith, K.O. (1967) Identification of Viruses by Electron Microscopy, in: Busch, H. (Ed.) *Methods in Cancer Research*, Academic Press, New York, pp. 545–568.

Spiess, E., Zimmermann, H.-P., Lünsdorf, H. (1987) Negative Staining of Protein Molecules and Filaments, in: Sommerville, J., Scheer, U. (Eds.) *Electron Microscopy in Molecular Biology: A Practical Approach*, Oxford University Press, London, New York, pp. 147–166.

Immunolabelling of Spread Preparations

Bock, C.-T., Schwinn, S., Zentgraf, H. (1995) Diheteroduplex formation using gold labelled single-stranded PCR fragments and its application in electron microscopy, *Chromosoma* **103**, 653–657.

Broker, T.R., Angerer, L.M., Yen, P.H., Hershey, N.D., Davidson, N. (1978) Electron visualization of tRNA genes with ferritin-avidin: biotin labels, *Nucleic Acids Res.* **5**, 363–383.

Delius, H., van Heerikhuizen, H., Clarke, J., Koller, B. (1985) Separation of complementary strands of plasmid DNA using the biotin-avidin system and its application to heteroduplex formation and RNA/DNA hybridizations in electron microscopy, *Nucleic Acids Res.* **13**, 5457–5469.

Rösl, F., Waldeck, W., Zentgraf, H., Sauer, G. (1986) Properties of intracellular bovine papillomavirus chromatin, *J. Virol.* **58**, 500–507.

Stahl, H., Dröge, P., Zentgraf, H., Knippers, R. (1985) A large-tumor-antigen-specific monoclonal antibody inhibits DNA replication of simian virus 40 minichromosomes in an *in vitro* elongation system, *J. Virol.* **54**, 473–482.

Immunoelectron microscopy (cell biology and viruses)

An, K., Paulsen, A.Q., Tilley, M.B., Consigli, R.A. (2000) Use of electron microscopic and immunogold labeling techniques to determine polyomavirus recombinant VP1 capsid-like particles entry into mouse 3T6 cell nucleus, *J. Virol. Methods* **90**, 91–97.

Arlucea, J., Andrade, R., Alonso, R., Arechaga, J. (1998) The nuclear basket of the nuclear pore complex is part of a higher-order filamentous network that is related to chromatin, *J. Struct. Biol.* **124**, 51–58.

Chang, P., Giddings, T.H. Jr., Winey, M., Stearns, T. (2003) Epsilon-tubulin is required for centriole duplication and microtubule organization, *Nat. Cell Biol.* **5**, 71–76.

Kirkham, M., Muller-Reichert, T., Oegema, K., Grill, S., Hyman, A.A. (2003) SAS-4 is a C. elegans centriolar protein that controls centrosome size, *Cell* **112**, 575–587.

Nermut, M.V., Zhang, W.H., Francis, G., Ciampor, F., Morikawa, Y., Jones, I.M. (2003) Time course of Gag protein assembly in HIV-1-infected cells: a study by immunoelectron microscopy, *Virology* **305**, 219–227.

Portoles, M., Faura, M., Renau-Piqueras, J., Iborra, F.J., Saez, R., Guerri, C., Serratosa, J., Rius, E., Bachs, O. (1994) Nuclear calmodulin/62 kDa calmodulin-binding protein complexes in interphasic and mitotic cells, *J. Cell Sci.* **107**, 3601–3614.

Ueki, S., Citovsky, V. (2002) The systemic movement of a tobamovirus is inhibited by a cadmium-ion-induced glycine-rich protein, *Nat. Cell Biol.* **4**, 478–486.

Urban, S., Schwarz, C., Marx, U.C., Zentgraf, H., Schaller, H., Multhaup, G. (2000) Receptor recognition by a hepatitis B virus reveals a novel mode of high affinity virus-receptor interaction, *EMBO J.* **19**, 1217–1227.

Weiland, E., Bolz, S., Weiland, F., Herbst, W., Raamsman, M.J., Rottier, P.J., De Vries, A.A. (2000) Monoclonal antibodies directed against conserved epitopes on the nucleocapsid protein and the major envelope glycoprotein of equine arteritis virus, *J. Clin. Microbiol.* **38**, 2065–2075.

Image Reconstruction

Andel, F. III, Ladurner, A.G., Inouye, C., Tjian, R., Nogales, E. (1999) Three-dimensional structure of the human TFIID-IIA-IIB complex, *Science* **286**, 2153–2156.

Bottcher, B., Wynne, S.A., Crowther, R.A. (1997) Determination of the fold of the core protein of hepatitis B virus by electron cryomicroscopy, *Nature* **386**, 88–91.

Mayo, K., Huseby, D., McDermott, J., Arvidson, B., Finlay, L., Barklis, E. (2003) Retrovirus capsid protein assembly arrangements, *J. Mol. Biol.* **325**, 225–237.

Nandhagopal, N., Simpson, A.A., Gurnon, J.R., Yan, X., Baker, T.S., Graves, M.V., Van Etten, J.L., Rossmann, M.G. (2002) The structure and evolution of the major capsid protein of a large, lipid-containing DNA virus, *Proc. Natl. Acad. Sci. USA* **99**, 14758–14763.

Smith, C.L., Horowitz-Scherer, R., Flanagan, J.F., Woodcock, C.L., Peterson, C.L. (2003) Structural analysis of the yeast SWI/SNF chromatin remodeling complex, *Nat. Struct. Biol.* **10**, 141–145.

Sperling, R., Koster, A.J., Melamed-Bessudo, C., Rubinstein, A., Angenitzki, M., Berkovitch-Yellin, Z., Sperling, J. (1997) Three-dimensional image reconstruction of large nuclear RNP (lnRNP) particles by automated electron tomography, *J. Mol. Biol.* **267**, 570–583.

Trus, B.L., Roden, R.B., Greenstone, H.L., Vrhel, M., Schiller, J.T., Booy, F.P. (1997) Novel structural features of bovine papillomavirus capsid revealed by a three-dimensional reconstruction to 9 Å resolution, *Nat. Struct. Biol.* **4**, 413–420.

Zhou, Z.H., He, J., Jakana, J., Tatman, J.D., Rixon, F.J., Chiu, W. (1995) Assembly of VP26 in herpes simplex virus-1 inferred from structures of wild-type and recombinant capsids, *Nat. Struct. Biol.* **2**, 1026–1030.

Electron Spectroscopic Imaging (ESI)

Bazett-Jones, D.P., Brown, M.L. (1989) Electron microscopy reveals that transcription factor TFIIIA bends 5S DNA, *Mol. Cell Biol.* **9**, 336–341.

Bazett-Jones, D.P., Kimura, K., Hirano, T. (2002) Efficient supercoiling of DNA by a single condensin complex as revealed by electron spectroscopic imaging, *Mol. Cell* **9**, 1183–1190.

Boisvert, F.M., Hendzel, M.J., Bazett-Jones, D.P. (2000) Promyelocytic leukemia (PML) nuclear bodies are protein structures that do not accumulate RNA, *J. Cell Biol.* **148**, 283–292.

Czarnota, G.J., Ottensmeyer, F.P. (1996) Structural states of the nucleosome, *J. Biol. Chem.* **271**, 3677–3683.

Grohovaz, F., Bossi, M., Pezzati, R., Meldolesi, J., Tarelli, F.T. (1996) High resolution ultrastructural mapping of total calcium: electron spectroscopic imaging/electron energy loss spectroscopy analysis of a physically/chemically processed nerve-muscle preparation, *Proc. Natl. Acad. Sci. USA* **93**, 4799–4803.

Hendzel, M.J., Boisvert, F., Bazett-Jones, D.P. (1999) Direct visualization of a protein nuclear architecture, *Mol. Biol. Cell* **10**, 2051–2062.

Locklear, L. Jr., Ridsdale, J.A., Bazett-Jones, D.P., Davie, J.R. (1990) Ultrastructure of transcriptionally competent chromatin, *Nucleic Acids Res.* **18**, 7015–7024.

Malecki, M., Hsu, A., Truong, L., Sanchez, S. (2002) Molecular immunolabeling with recombinant single-chain variable fragment (scFv) antibodies designed with metal-binding domains, *Proc. Natl. Acad. Sci. USA* **99**, 213–218.

4
Circular Dichroism in Protein Analysis

Zhijing Dang and Jonathan D. Hirst
School of Chemistry, University of Nottingham, University Park, Nottingham, UK

1	**Chirality and Polarized Light** 131	
2	**Electronic Circular Dichroism** 132	
2.1	Far Ultraviolet 133	
2.2	Near Ultraviolet 135	
3	**Vibrational Circular Dichroism** 135	
4	**Estimation of Secondary Structure Content** 136	
5	**Experimental Aspects** 137	
5.1	Instrumentation 137	
5.2	Measurements 138	
5.3	Solvents and Temperature Effects 139	
5.4	Stopped-flow 139	
5.5	Synchrotron Radiation 140	
5.6	Fast Time-resolved Techniques 140	
6	**Spectral Characteristics of Elements of Protein Structures** 141	
6.1	α-helix 141	
6.2	3_{10}-helix 142	
6.3	β-sheet 142	
6.4	β-turn 143	
6.5	Random Coil 143	
6.6	Polyproline Helices 144	
6.7	Extrinsic Chromophores 144	
6.8	Membrane Proteins 144	
7	**Conclusion and Outlook** 145	

Proteins. Edited by Robert A. Meyers.
Copyright © 2007 Wiley-VCH Verlag GmbH & Co. KGaA, Weinheim
ISBN: 978-3-527-31608-3

Bibliography 145
Books and Reviews 145
Primary Literature 146

Keywords

Chiral
Possessing a nonsuperposable mirror image (like left and right hands).

Chromophore
Chemical moiety that absorbs light through excitation of one of its electrons.

Circular Dichroism
The differential absorption of left- and right-circularly polarized light.

Circularly Polarized Light
The resultant from combining two beams of linearly polarized light (*vide infra*) of equal intensity, but orthogonal polarizations, with a phase difference of a quarter of a wavelength.

Linearly Polarized Light
A beam of light in which the electric fields of all the photons oscillate in the same plane (also called *plane-polarized light*).

Optical Rotation
The rotation of plane-polarized light by a chiral compound (equivalent to circular birefringence).

■ Proteins are chiral objects. Elements of local, regular structure, called *secondary structure*, within proteins have a distinct handedness. For example, α-helices are mostly right-handed. Proteins absorb ultraviolet light, exciting electrons into high-energy states. Plane-polarized light (such as that created by Polaroid sunglasses) is light that propagates in only a single plane. It is possible to generate circularly polarized light. Left- and right-handed circularly polarized light may be considered as two components of plane-polarized light. When plane-polarized light passes through a chiral medium, its plane of polarization is rotated. If the light excites an electronic transition, a chiral molecule will absorb the left- and right-handed circularly polarized components differently. This differential absorption is termed *circular dichroism*. Circular dichroism spectroscopy may be used to estimate quantitatively the secondary structure content of proteins, giving the fraction of residues in α-helices, β-sheets, turns, and in coil conformations. It is widely used to characterize proteins under equilibrium conditions and to measure the kinetics of protein folding and unfolding.

1
Chirality and Polarized Light

At a molecular level, much of matter is asymmetrical, like our left and right hands, which are mirror images. In the world of organic chemistry, many compounds have chiral centers, typically carbon atoms bonded to four different atoms or groups. Chiral compounds have two forms (left- and right-handed), which cannot be superimposed on each other by rotation, although they have the same chemical composition. The asymmetric character of chiral compounds induces different interactions with electromagnetic radiation, which gives rise to optical activity.

A beam of natural light is a set of electromagnetic waves propagating in a single direction, but with all possible orientations perpendicular to the direction of propagation. When it passes through a linear polarizer, only a single orientation is selected and we get linearly polarized light. It may be divided into two chiral components of equal intensity, left- and right-circularly polarized light (Fig. 1). The electric vector of each circularly polarized component rotates along the direction of propagation. The overall effect of these two vectors is a planar sinusoidal wave.

Circular dichroism (CD) spectroscopy is the differential absorption of left- and right-circularly polarized light by a chiral medium. When linearly polarized light passes through achiral molecules, the absorption of left- and right-circularly polarized light is equal and no CD spectrum results. However, when the medium comprises chiral molecules, such as peptides, proteins, or DNA, there is differential absorption. The resultant of these two components is an ellipse with a specific optical rotational angle (Fig. 1). CD spectroscopy is based on the observation of optical rotation or circular birefringence, which can reflect detailed information about the three-dimensional conformation of biomolecules. The differential absorption arises because the left- and right-handed circularly polarized components experience different refractive indices on passing through chiral molecules.

The absorbance of light by a sample is given by the Beer–Lambert law:

$$A = \log\left(\frac{I_0}{I}\right) = \varepsilon c l \quad (1)$$

where A is the absorption, c is the concentration in mol dm^{-3}, I_0 is the intensity of the incident light, l is the pathlength in cm, and ε is the molar absorption coefficient in units of dm^3 mol^{-1} cm^{-1}. The differential absorption is then

$$\Delta A = A_L - A_R = (\varepsilon_L - \varepsilon_R)cl = \Delta\varepsilon cl \quad (2)$$

where $\Delta\varepsilon$ is the difference in the molar absorption coefficients for left- and right-circularly polarized components. This may also be described, removing the linear dependence on pathlength and solute concentration, by

$$\Delta\varepsilon = \varepsilon_L - \varepsilon_R = \frac{\theta}{cl} \quad (3)$$

where θ is the ellipticity in degrees. The ellipticity is equal to the ratio of the minor radius to the major radius of the now elliptically polarized light, that is A/B, as shown in Fig. 1, and it is related to the differential absorption by

$$\theta = 32.98\, \Delta A \quad (4)$$

For a peptide or protein, the mean residue ellipticity in units of degree cm^2 dmol^{-1} is defined as

$$[\theta] = 100\,\frac{\theta}{cl} = 3298\,\Delta\varepsilon$$

$$= \theta \times 0.1 \times \frac{MRW}{cl} \quad (5)$$

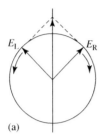

Fig. 1 (a) shows linearly polarized light with equally intense left- and right-polarized components (E_L and E_R). (b) Shows circularly polarized light with unequal left- and right-circularly polarized components. θ is the ellipticity and is the ratio of the minor axis to the major axis of the ellipse (A/B); α is called the *optical rotation*.

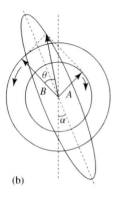

where c is the concentration in mg ml^{-1}; MRW is the mean residue weight, which is equal to molecular mass/$(N-1)$; N is the number of amino acid units and is typically one more than the number of peptide bonds, giving rise to the factor of $N-1$. The mean residue ellipticity is independent of the pathlength and concentration.

Although optical rotation is a dispersive measurement requiring the measurement over the complete spectrum, experimentally the measurements are confined to a relatively narrow wavelength range. In principle, CD can be measured with any frequency of electromagnetic radiation. However, in practice, most CD spectra are recorded in the ultraviolet–visible (UV–vis) region arising from electronic transitions. Since the early 1980s, new developments in optical and electronic technology have enabled optical activity measurements to be extended into the vibrational spectrum using both infrared and Raman techniques. We will focus on electronic CD, which is more commonly used, and will give only a brief description of vibrational optical activity.

2
Electronic Circular Dichroism

CD spectra arise from transitions that are electrically and magnetically allowed. When an electron of a molecule is excited from the ground state to a higher energy state, there is a change in the electric field of the radiation or the magnetic field or both. The change of the electric field causes a linear rearrangement of the electrons. The net linear displacement of charge during the transition is called the *electric transition dipole moment*, μ. The magnetic field change induces a circular rearrangement of electron density. The net circulation of charge is termed the *magnetic transition dipole moment*, m. In electronic absorption spectroscopy, bands arise from transitions that are either electrically allowed and magnetically forbidden or magnetically allowed and electrically forbidden. The oscillator strength, f, is proportional to the square of the electric transition dipole moment, μ^2, and corresponds to the area under a band in an electronic absorption spectrum. The rotational strength of a band in a CD spectrum is analogous to the oscillator strength in electronic absorption spectroscopy. It is related to the product of the electric and magnetic transition moments and corresponds to the area under the band in the CD spectrum.

Electronic CD spectra may be divided into the far UV region (180–250 nm), the near UV region (250–300 nm), and the UV–vis region (300–700 nm). Absorption in the far UV region is dominated by the excitation of electrons on amide chromophores. It furnishes information on secondary structure (Fig. 2). In the near UV region, the CD arises from side-chain chromophores, such as aromatic groups and disulfide bonds. In this region, CD spectra can provide information on protein tertiary structure. The UV–vis region reflects the conformation and environment of the protein. Transitions in this region arise from chromophoric prosthetic groups, such as the heme group.

2.1
Far Ultraviolet

Peptide bonds are the key chromophores in proteins. To appreciate the origin of the CD of proteins, one needs to understand the electronic structure of the peptide bond. The valence shell of the peptide bond comprises four π electrons in three π orbitals (π_+, π_0, and the unoccupied π^*) and two electron pairs in two lone-pair orbitals (n and n'). The $n\pi^*$ and $\pi_0\pi^*$ electronic transitions of the peptide chromophore occur in the far UV region. The lowest energy transition is the $n\pi^*$ transition, which is polarized along the carbonyl bond. It is an excitation of an electron from a lone pair on the oxygen atom to an antibonding π orbital and occurs at 210 to 230 nm. It is magnetically allowed and electrically forbidden (Fig. 3). The transition energy is sensitive to the extent of hydrogen-bonding to the oxygen atom. In the lower symmetry of

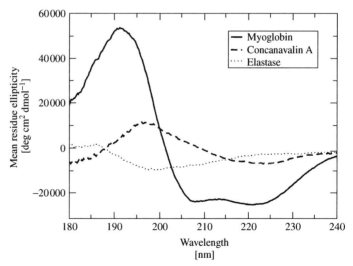

Fig. 2 The standard CD spectra for α-helix, β-sheet, and random coil, as typified by the proteins myoglobin, concanavalin A, and elastase respectively; the spectra are plotted from the freely available experimental data.

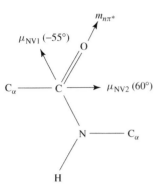

Fig. 3 The $n\pi^*$ and $\pi\pi^*$ transitions of the peptide chromophore in the far UV region. μ is the electrical transition dipole moment. m is the magnetic transition dipole moment.

a protein, the $n\pi^*$ transition is not strictly forbidden, but it is very weak. In solvents that lack hydrogen-bond donors, the $n\pi^*$ transition appears near 230 nm, while in aqueous solution, it is around 210 nm. The $n'\pi^*$ transition occurs at ~140 nm and is polarized perpendicular to the carbonyl bond.

The $\pi\pi^*$ transitions at 190 and 140 nm are electrically allowed. The first $\pi\pi^*$ transition, the NV_1 band, involves excitation of an electron from the nonbonding π orbital, π_0, to the π^* orbital. In secondary amides, the NV_1 band is observed around 185 to 190 nm, whereas in tertiary amides, it is closer to 200 nm. The second $\pi\pi^*$ transition, the NV_2 band, at 140 nm involves electronic excitation from the bonding π orbital, π_+, to the π^* orbital. The transition dipole moments of the two $\pi\pi^*$ transitions in secondary amides are approximately parallel (NV_1) and perpendicular (NV_2) to the C–N bond. The electric transition dipole moment of the $\pi_0\pi^*$ transition (NV_1) is about 3.1 D and that of the $\pi_+\pi^*$ transition (NV_2) is about 1.8 D. In an α-helix, the electric dipole coupling of the $\pi_0\pi^*$ (NV_1) transitions on neighboring residues results in two bands in the CD spectrum, one at ~190 nm and the other, a long-wavelength component, at ~208 nm. Interactions among the amide $n\pi^*$ and $\pi\pi^*$ transitions in a peptide are relatively delocalized and are significantly affected by environmental or local perturbations.

The arrangement of peptide bonds in a protein largely determines its CD spectrum in the far UV. The main-chain dihedral angles define the relative orientation of the backbone chromophores. If the dihedral angles for each amino acid residue were the same, the protein would form a perfectly regular helix (a β-strand may be viewed as a very narrow helix). Alternatively, an arbitrary set of (sterically allowed) dihedral angles would generate a random coil conformation. In reality, most secondary structures lie between these two extremes. The origins of the electronic CD of proteins can be understood by dividing the biopolymer into subunits, which do not have significant electronic exchange among them. This basic concept underlies theoretical calculations of protein CD spectra. Protein CD spectra arise from three types of interactions between $\pi\pi^*$ and $n\pi^*$ transitions of amides. Exciton interactions occur between degenerate or nearly degenerate $\pi\pi^*$ transitions on different peptide groups. An $n\pi^*$ transition and a $\pi\pi^*$ transition on different peptide groups may mix; this is also referred to as $\mu-m$ coupling. Finally, the $n\pi^*$ and $\pi\pi^*$

transitions on a single peptide group may mix – an interaction known as the *one-electron effect*. In an infinite helix, exciton interactions of $\pi\pi^*$ transitions on different peptide groups lead to three allowed transitions, one polarized parallel to the helix axis and two degenerate (equal energy) bands polarized perpendicular to the helix axis.

2.2
Near Ultraviolet

The near UV region, 250 to 300 nm, is also called the *aromatic region*. Because the peptide absorption in this area is at least an order of magnitude weaker than in the far UV region, a small number of aromatic chromophores have a distinct advantage. Bands in this region are due to $\pi\pi^*$ transitions from the aromatic side chains of histidine, phenylalanine, tyrosine, and tryptophan and $n\sigma^*$ transitions from disulfide groups. The indole of tryptophan has two or more transitions in the range of 240 to 290 nm with a combined maximum extinction coefficient $\varepsilon_{max}(279\,\text{nm}) \sim 5600\,\text{dm}^3\,\text{cm}^{-1}\,\text{mol}^{-1}$; tyrosine has one transition with $\varepsilon_{max}(274\,\text{nm}) \sim 1400\,\text{dm}^3\,\text{cm}^{-1}\,\text{mol}^{-1}$ and phenylalanine has one transition with $\varepsilon_{max}(258\,\text{nm}) \sim 190\,\text{dm}^3\,\text{cm}^{-1}\,\text{mol}^{-1}$. The cystine disulfide bond absorbs from 250 to 270 nm with $\varepsilon_{max} \sim 300\,\text{dm}^3\,\text{cm}^{-1}\,\text{mol}^{-1}$. The near UV CD also reflects the environment of aromatic groups. For example, the most red-shifted tyrosine feature should be due to tyrosine(s) in a highly polarizable environment, while the most blue-shifted are due to exposure to solvent. Although tryptophan has the most intense transition in this region, usually proteins have few tryptophans compared to the other aromatic groups, so the region tends not to be dominated by tryptophan transitions.

The disulfide group of cystine in many extracellular proteins has two well-characterized $n\sigma^*$ electronic transitions in the near UV region. The wavelengths depend on the dihedral angle of the disulfide band. This dihedral angle is normally approximately 90° in proteins, and the two $n\sigma^*$ transitions are degenerate, giving a single broad absorption band near 260 nm. As the dihedral angle deviates from 90°, the degeneracy is broken and one transition shifts to higher energy, with the other moving to lower energy. Disulfide contributions to a near UV CD spectrum are distinguishable from those of aromatic side chains, because the former are much broader.

3
Vibrational Circular Dichroism

Vibrational optical activity consists of two complementary spectroscopic areas, vibrational circular dichroism, and vibrational Raman optical activity. Both spectra show the differential response to left- versus right-circularly polarized radiation due to a vibrational transition in a chiral molecule. Vibrational circular dichroism (VCD) arises from simultaneous changes in the electric and magnetic dipole moments of the molecule due to the nuclear motion. The vibrational transition is between the g_0 and g_1 vibrational sublevels of the ground electronic state. The transition dipole coupling is highly dependent on peptide secondary structure. Because the interpretation of VCD depends on band shape, it is less susceptible to error from frequency shifts arising from nonstereochemical sources.

The amide I band (amide I′ in D_2O), due to primary C–O stretch, is the most characteristic and the easiest to identify with VCD. Its frequency is relatively free of overlapping transitions from other parts of the molecule. An α-helix has a major positive peak at $\sim$1640 cm^{-1}. An antiparallel β-sheet has a split amide I' band, with a major, sharp positive peak at $\sim$1615 to 1620 cm^{-1} and a weaker positive one at $\sim$1690 cm^{-1}. Random coil conformations give a broad positive peak at $\sim$1650 cm^{-1} and a positive trough at $\sim$1620 cm^{-1}, which is quite universal in band shape and intensity in most oligo- and polypeptides.

The amide II band is a mixture of N−H deformation and C−N stretch. The differences among α-helix, β-sheet, and random coil are less clear than those with the amide I band. The distinct feature of the amide II VCD for the α-helix is a dominant negative band; for β-sheet one observes a negative couplet and for random coil there is a weaker intensity positive band. The amide III band of an α-helix is mostly positive in the region of $\sim$1350 to 1250 cm^{-1}, and for β-sheet it is mostly negative in the same region. The amide III band is more sequence dependent, which affects its ultimate utility for peptide backbone conformational analyses.

VCD provides complementary data to electronic CD. The spectra are richer in information, but the instrumentation is not as developed or as widely used as that for electronic CD. The rotational strengths of vibrational transitions are much less intense than those of electronic transitions, which makes it a challenge to achieve the necessary signal/noise ratio. The technique has clear promise, but for the remainder of our discussion, we shall focus on electronic CD.

4
Estimation of Secondary Structure Content

The three-dimensional conformation of a protein can be specified by the dihedral angles of the backbone. Hydrogen-bonding patterns and particular values of dihedral angles lead to regular, local elements of structure, known as *secondary structure*. Secondary structures can be grouped into α-helices, β-sheet, β-turn, and random coil. As most peptides and proteins are mixtures of these four basic secondary structures, the overall mean residue ellipticity at a specific wavelength, $[\theta(\lambda)]$, is the sum of contributions from each secondary structure. It can be calculated by the equation:

$$[\theta(\lambda)] = f_\alpha[\theta(\lambda)]_\alpha + f_\beta[\theta(\lambda)]_\beta + f_\gamma[\theta(\lambda)]_\gamma$$
$$+ f_R[\theta(\lambda)]_R + \text{noise} \quad (6)$$

which is based on the fraction of each secondary structure type: α-helix (f_α), β-sheet (f_β), β-turn (f_γ), and random coil (f_R). As a first approximation, the CD spectrum of a protein is the sum of the appropriate percentages of each component spectrum, but the relative arrangements of structural units and motifs may also contribute to the observed CD spectrum.

There are many computer programs that can analyze CD spectra to estimate protein secondary structure content. The major methods are either algorithm-based or rely on databases. The former includes constrained least squares, normalized least squares, singular value deconvolution, parameterized fit and neural networks. Frequently used software packages are SELCON3, CONTIN, and Varslc. Database approaches are available in SELCON3, CONTIN, and SELCON. Most of these facilities can be found on the web site: http://www.cryst.bbk.ac.uk/cdweb.

The variable selection-self-consistent program SELCON provides analysis based on a database of 17 proteins. In SELCON, the proteins in the database are arranged in an order of increasing root-mean-square difference from the spectrum; the spectra least like the spectrum of interest are deleted to increase the speed. Then the program utilizes the observation that prediction improves when the protein is included in the basis set. The initial guess of the structure of the protein is made, and the secondary structure of the protein is then determined. The solution replaces the initial guess and the process is repeated until self-consistency is attained. This program gives a fairly accurate estimation of α-helix, β-sheet, and β-turn of globular proteins based on CD spectra between 240 and 200 nm. However, it overestimates α-helix and underestimates β-sheet when polypeptides have high β-sheet content.

The neural network program, K2D, uses a learnt mapping between spectra and secondary structure content to analyze new experimental spectra. It was trained on input data, which were CD spectra represented as 41 inputs corresponding to the intensities at the wavelengths between 240 and 200 nm. The output units formed a self-organized map, which could be analyzed to give fractional weights of protein secondary structures. In the training phase, the weighting of each wavelength is altered, until the error between the calculated and actual secondary structure content for the training data is minimized. For a new spectrum, the data are fed through the network using the adjusted weights to give an estimate of the secondary structure content. The K2D program predicts the secondary structure content fairly well, but does not give a matched spectrum. A different type of neural network was used, in which the output units gave directly the fractional secondary structure content: α-helical, parallel and antiparallel β-sheet, β-turn, and others. The network was trained on 13 spectra, each represented as 83 inputs corresponding to the intensities from 178 to 260 nm.

Each of these methods and programs has its advantages and disadvantages. Even when a method provides a good match between the calculated spectrum and the experimental data, it may not necessarily provide the best estimation of protein secondary structure content. Among these methods, SELCON may perform better than the others. Most of these programs usually estimate α-helical content well, but underestimate β-sheet and β-turn contents.

5
Experimental Aspects

5.1
Instrumentation

Nowadays, measuring CD is routine work in many laboratories. There are four manufacturers of CD instruments. Aviv and Associates (Lakewood, New Jersey) has Cary 61 and 62 CD instruments with a photoelastic modulator, modernized electronics, and a computer control system. JASCO (Easton, Maryland and Tokyo) makes J710 and J720 spectrometers. Jobin-Yvon (Longjumeau, France) manufactures the JY Mark VI and OLIS (On-Line Instrument Systems, Bogart, Georgia) offers the Cary 61 and 62 CD instruments. CD accessories for stopped-flow instruments are available from Applied Photophysics Ltd. (Leatherhead, UK) and Biologic (Grenoble, France).

Most instruments use a single beam. In Eq. (2), the intensity of the incident light, I_0, does not appear for calculation of absorbance, so there is no need for a reference beam in measurements. However, a new double-beam instrument made by OLIS using sample and reference beams permits direct calculation of the CD signal. This technique reduces drift over time and factors such as lamp fluctuations are negated, since they affect both beams equally.

The light source is usually a 150 or 450 W xenon arc, which provides a continuous spectrum from the edge of the vacuum UV into the infrared region. Linearly polarized light from a monochromator passes through a modulating device, a photoelastic modulator, which converts the plane-polarized light to circularly polarized light, alternating between left- and right-circularly polarized light at the modulation frequency. In the detection end of the spectrometer, the light intensity is also modulated at the same frequency as the incident light.

A different modulation method, digital subtraction, is used by OLIS. The unique aspect of this method is that both of the two diverging orthogonally polarized beams are rotated by 180°. Then, the digitized signals from the two photomultiplier tubes are combined in such a way as to yield the CD signal directly and to cancel the contributions of noise. So, in this system, the absolute CD is measured and no calibration is required. However, one shortcoming is that the two beams pass through different parts of the sample and the cuvette; the significance of this is still uncertain.

The cooling system of the instruments is usually water or a narrow bore central heating system pump with car radiator cooling fluid. If it is possible, an air-cooled lamp may give better performance. Nitrogen flow is needed to avoid the creation of ozone, which reacts with surfaces of the mirrors. Nitrogen may also keep oxygen from entering the sample compartment, as it absorbs the incident light.

High quality quartz cuvettes, which transmit the full wavelength range of UV–vis light, are required for CD experiments. Cuvettes have rectangular and cylindrical shapes. A cylindrical cuvette has lower birefringence than a rectangular shaped one. However, rectangular cells are more often used especially in titration experiments. Different pathlength cuvettes are used for measurements at different wavelengths. Ten and 1 mm pathlengths cannot be used for wavelengths under 200 and 190 nm respectively. For shorter wavelengths, and for more strongly absorbing solvents at longer wavelengths, cells of 0.1 or 0.05 mm are necessary. For these cells, as the total volume is small, a higher concentration is demanded to keep absorbance of more than 1.0.

5.2
Measurements

For measurements in the far UV, solvent absorbance is not negligible, even with aqueous solutions. It is necessary to check the CD signal as a function of concentration to have a maximum absorbance of less than 1.5. If a linear relationship of absorbance as a function of concentration is not there, then there are solute–solute interactions in the sample. A baseline spectrum should be collected of solvent and buffer under the same conditions as the sample spectrum.

The spectrometer scans from longer to shorter wavelengths. Initially, the examining wavelength should be set 20 nm longer than that demanded by the experiment and it should be established that in this region the absorbance is zero. Otherwise, it may be necessary to clean the cuvette and test again. The signal-to-noise ratio depends on many factors: it increases as $n^{1/2}$ (where n is the number of measurements), $I^{1/2}$ (where I is the intensity of light beam), and $\tau^{1/2}$ (where τ is the measuring time for each point). The CD instrument must be calibrated regularly using a standard sample. The most commonly used standard for visible and UV measurements is (+)-10-camphorsulfonic acid (CSA).

5.3
Solvents and Temperature Effects

The amount of sample required for CD experiments is relatively small compared to other absorption spectroscopic techniques. Water is the most frequently used solvent for CD measurements because of its excellent UV transparency. Fluorinated alcohols, such as 2,2,2-trifluoroethanol (TFE) and 1,1,1,3,3,3-hexafluoro-2-propanol (HFIP), are good solvents as well. However, secondary structure, especially of peptides, can be influenced by solvents, for example, trifluoroethanol promotes α-helical conformations and nonpolar solvents, such as lipids and destabilize α-helices. MeCN is a good organic solvent. Phosphate buffers are also useful due to their low absorbance. However, some common salts and buffer components do significantly interfere with the far UV spectra and are best avoided.

Temperature may have a significant effect on CD spectra. When the free energy, difference between two conformations of a molecule, ΔG, is not too much higher than RT in the range of accessible temperatures (where R is the universal gas constant), a mixture of both conformers can be expected in solution. In this case, modifications of the experimental conditions can lead to a change in the measured rotational strength and can be calculated by

$$R_{obs} = \frac{R_1 + R_2 \exp(-\Delta G/RT)}{1 + \exp(-\Delta G/RT)} \quad (7)$$

where R_1 and R_2 are the rotational strengths of these two conformers. Even when there is no a temperature-sensitive conformational change, there are temperature effects on CD spectra. For short helices, the temperature coefficient is small, but it increases significantly with helical length.

5.4
Stopped-flow

Stopped-flow CD is used for kinetic measurements of conformational transitions in proteins and in protein folding studies. It has been applied in the near UV and the far UV. The principle of stopped-flow is straightforward. The apparatus uses a drive motor to mix together rapidly two solutions, contained in separate drive syringes, into a mixing cell. The solutions then flow into the observation cell displacing the previous contents with freshly mixed reactants. A stop syringe is used to limit the volume of solution expended with each experiment and abruptly stops the flow. The flow of solution into the stop syringe causes the plunger to move back and trigger data collection. Then CD spectra may be measured as a function of time.

The primary objective of the stopped-flow experiments is to investigate transient changes in chiroptical properties associated with chemical, biochemical, and biophysical reactions, such as structural changes undergone during protein folding on a millisecond (ms) time scale and refolding. In many cases, secondary structure is formed rapidly within 2 to 5 ms, and tertiary structure is formed slowly. The two-state folding of a small protein may occur within 3 ms. Stopped-flow measurements require relatively few combined shots to provide folding kinetics data. It can be made over a wide wavelength range, allowing the collection of time-resolved CD spectra. These can be analyzed to provide time-courses of the development of individual secondary structure types. However, the limitation of stopped-flow is the "dead-time" of the mixing process, which is typically more than 2 ms. In order to study faster folding kinetics, temperature-jump methods can be employed to shorten the observing time to $10 \sim 20$ ns with laser-induced rapid heating (discussed in Sect. 5.6). In general, stopped-flow only provides average information of the secondary structure of protein; it cannot specify precisely where the secondary structure is formed.

5.5
Synchrotron Radiation

The X-ray source for measuring an X-ray absorption spectrum is synchrotron radiation. It covers all the wavelengths of the electromagnetic spectrum with an intensity over 100 times higher than the conventional X-ray tubes. When charged particles (electrons or positrons) from a linear accelerator with a speed close to that of light are injected into a storage ring under high vacuum, the magnetic fields from the strong bending magnets around the ring force the accelerated electrons to follow the ring in the curved sections. The high-energy particles hit the curved parts and lose part of their energy as synchrotron radiation, which is emitted tangential to these curved sections.

Synchrotron radiation CD has been developed for more than 20 years, but it has been used only recently in protein secondary and tertiary structure analysis. Synchrotron radiation CD provides higher photon fluxes of linearly polarized light particularly at short wavelengths than conventional sources. There is no need of a polarizer and the measuring limit extends to the cut-off of the photoelastic modulators (usually 120–130 nm) with much higher signal-to-noise ratio. So, CD spectra can be collected to a lower wavelength limit, allowing more accurate secondary structure content on folding motifs and a high-intensity light source may work well in the presence of salts, buffers, and detergents as well. Collection times for synchrotron radiation CD spectra can be considerably shorter than with conventional instruments.

5.6
Fast Time-resolved Techniques

In order to study protein folding on short time scales, fast time-resolved CD has been developed. It is now possible to resolve motions of chromophores well into the subpicosecond time scale. Time-resolved spectroscopy determines the spectral properties of a material at various instants of time after a stimulus has been applied. There are two main approaches: signal gating by sampling the output of a fast-responding photodetector and probing with a short-duration pulse of light. The limitations of the former are that few detectors operate at long wavelengths and the signal-to-noise ratio would be unacceptable. The pump-probe technique overcomes these limitations. It provides high temporal resolution of repeatable, photostimulated events.

Time-resolved CD can provide information on structural rearrangement and folding for both smaller chiral species and large macromolecules in different environments. Conventional time-resolved CD spectroscopy is limited by the time scale of the polarization modulator employed. Now, with shorter-pulse He–Ne lasers and a photoelastic modulator to alternate the pump laser polarizations without limiting the time response of the measurement, it is possible to detect the inherently weak difference between two large absorption signals, which are ellipsometric and are subject to significant optical artifacts. The limitation is the pulse length of the grating forming (pump) laser. Measurements can be made on microsecond, nanosecond, and picoseconds timescales. Although picosecond resolution is achievable, there are some technical problems, including intensity instability, polarization scrambling, and tunability limitations.

6
Spectral Characteristics of Elements of Protein Structures

CD is commonly used to probe the secondary conformations of proteins to study protein folding and to investigate interactions of proteins with small molecules such as achiral molecules whose induced CD is due to their interaction with the protein. Changes in CD can be used to provide evidence for conformational changes and to determine equilibrium constants. A few examples of conformational transitions are the α-helix to β-sheet conformational switch in the prion protein as the concentration of sodium dodecyl sulfate is altered; the unfolding of the coiled-coil structure of fibrinogen binding protein as temperature increases; the increasing helical content of troponin-C on binding Ca^{2+} and the binding of molybdate by the protein ModE, which gives large changes in the Trp signal at 292 nm. Far UV and near UV CD can be used to monitor protein unfolding upon the alteration of pH or the addition of a denaturant, such as urea or guanidinium chloride (GdmCl). CD can also be used to examine the interactions between multiple domains of proteins, for example, the flavin domain and the heme domain of flavocytochrome P450-BM3 give characteristic changes in the environment of the cofactors when they are linked in the intact enzyme.

6.1
α-helix

The right-handed α-helix (Fig. 4) has mainchain hydrogen bonds between amino acid residues and positions i and $i+4$ along the backbone chain. There is a 0.15-nm translation and a 100° rotation between two consecutive peptide units. Its helical pitch (the number of residues multiplied by the distance between α-carbon atoms on neighboring residues) is 0.54 nm. The α-helix has three distinctive bands in the far

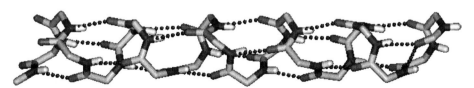

Fig. 4 The main chain of a polypeptide in an α-helical conformation, with hydrogen bonds denoted by dashed lines.

UV CD spectrum: two negative bands with separate minima of similar magnitude at 222 nm (the $n\pi^*$ transition) and 208 nm (the parallel $\pi\pi^*$ exciton band) and a positive band, the NV_1 band, at 190 nm due to the antiparallel $\pi\pi^*$ exciton band. The ratio $[\theta]_{208}/[\theta]_{222}$ is sensitive to the backbone dihedral angles.

The average fractional helicity, f_H, of an N-residue peptide may be related to the observed mean residue molar ellipticity at 222 nm by

$$f_H = \frac{[\theta]_{222}}{[\theta_{H\infty}]_{222}(1-k/N)} \quad (8)$$

where $[\theta_{H\infty}]_{222}$ is the ellipticity of an infinite, completely helical peptide, k is an end-effect correction, which is ~3, and N is the number of residues. Estimated values for $[\theta]_{222}$ of an infinite, completely helical peptide range from $-37\,000$ to $-44\,000$ deg cm^2 dmol^{-1}. However, anomalously intense bands have been observed in some peptide models under certain experimental conditions.

6.2
3_{10}-helix

The 3_{10}-helix has $(i, i+3)$ main-chain hydrogen bonds. There are three representative dihedral angles (φ, ψ) for 3_{10}-helices, type I $(-60°, -30°)$; type II $(-54°, -28°)$, and type III $(-44°, -33°)$. The three types of 3_{10}-helices are anticipated to have different CD spectra based on first-principles calculations. The CD spectrum for a type I 3_{10}-helix is similar to that for an α-helix, but a little weaker band at 222 nm relative to the 208 nm band, and the $[\theta]_{222}/[\theta]_{208}$ ratio ranges from about 0.3 to 0.85; this ratio is greater than 1 for an α-helix. Type II has the strongest experimental foundation; type I and III have weaker and stronger $n\pi^*$ transitions than type II, respectively. It may be possible to distinguish reliably between 3_{10}-helix and α-helix by CD, but it is still a matter of debate.

6.3
β-sheet

β-sheets are composed of β-strands, which can be thought of as helices with two residues per turn. Typical backbone dihedral angles are $\phi = -120°$ and $\psi = +120°$ producing a translation of 3.2 to 3.4 Å per residue for residues in antiparallel and parallel strands respectively. An antiparallel arrangement is shown in Fig. 5. The CD spectra for β-sheet proteins are less intense and have fewer features than those of α-helical proteins, with a negative band near 217 nm, a positive band near 195 nm, and another negative band near 180 nm. However, the CD spectra for β-sheet proteins have much greater variability than those for α-helices: for example, the intensities

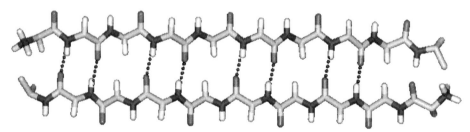

Fig. 5 The main chain of a polypeptide in an antiparallel two-stranded β-sheet, with hydrogen bonds depicted by dashed lines.

of the two long-wavelength bands, their ratio, and the wavelength of the positive band. These can be affected by side chains, solvents, and other environmental factors. Structural factors may also be important; β-sheets can be antiparallel, parallel, or mixed, intra- or intermolecular, and are also twisted to varying extents. Theoretical studies suggest that the extent of twisting of β-sheets is probably more important than the distinction between antiparallel and parallel sheets.

The wavelength at which the intensity of the CD spectrum changes sign is the most convenient criterion for differentiating parallel and antiparallel β-sheets. Crossover at ∼178 nm indicates antiparallel β-sheet, at ∼192 nm, parallel β-sheet, and at ∼187 nm, a mixture of parallel and antiparallel β-sheets. The effects of twisting on β-sheet CD are predicted to include large differences in the magnitudes of the two long-wavelength bands, and a red-shift of both the positive maximum and the short-wavelength crossover from positive to negative CD.

Unlike α-helix content, it is not possible to estimate β-sheet content in proteins reliably from CD spectra because of the twisted and planar β-sheet showing quantitative differences between parallel and antiparallel β-sheet. One method is to calculate the difference between the CD at 217 nm and 195 nm. For 100% β-sheet, it is $50–55 \times 10^3$ deg cm^2 dmol^{-1}. This type of estimate may be reasonable, as long as the β-sheet is not strongly twisted.

6.4
β-turn

There are eight types of β-turn, three of which are common in proteins. The CD spectra of β-turn are rather varied. The type II β-turn has a strong negative band between 180 and 190 nm, a strong positive band between 200 and 205 nm, and a weak, red-shifted band at 225 nm due to the nπ* transition. Generally, type II CD spectra are similar to β-sheet CD, except that the maximum is red-shifted by 5 to 10 nm. The CD spectra for type I and III turns are similar to that of the α-helix, with a negative nπ* band and a negative ππ* couplet.

6.5
Random Coil

CD spectroscopy is sensitive to precise protein conformation, but most random coils are very flexible. So, there are less common features in their CD spectra, except a strong negative band near 200 nm and a weak band at longer wavelengths, but the latter band may be either positive or negative. Poly(Glu) and poly(Lys) are frequently used models of unordered polypeptides. Their neutral aqueous solution spectra show two characteristic features: a strong negative band near 200 nm ($[\theta]_{max} \sim -40\,000$ deg cm^2 dmol^{-1}) and a significant band shift between 200 nm and 191 nm. When the pH is set to neutralize the side chains, or the ionic strength is increased by added salt, or the temperature is increased, the very weak negative band in the 235- to 240-nm region is observable. Urea and guanidinium chloride (GuCl) can have a complicated effect on CD spectra. At low concentrations, GuCl causes a decrease in the long-wavelength CD bands of the charged polypeptides by shielding the charges on the side chains and decreasing the electrostatic stabilization of the helical peptide.

There are two different explanations for unordered polypeptides CD spectra. In the conventional view, random coil polypeptides exist as an ensemble of an enormous number of different conformers, with the relative geometry of nearest neighbors determined statistically. Another opinion is that fully ionized poly(Glu) and poly(Lys) at low temperature and low ionic strengths are in an extended helix conformation, which is similar to the left-handed 3_1 helix of poly(Pro)II helix. Theoretical calculations suggest that the parallel-polarized $\pi\pi^*$ transition occurs as a positive peak at 209 nm and that the perpendicular-polarized component is at 201 nm with a negative rotational strength. In general, random coil structures with different sequences and under different conditions can give a variety of CD curves. However, it seems likely that unordered polypeptides that have a positive band near 218 nm may have a substantial amount of poly(Pro)II helical conformation.

6.6
Polyproline Helices

The homopolymer poly(Pro) can adopt two different conformations. Poly(Pro) I is a right-handed helix with *cis*-peptide bonds throughout and it is stable only in relatively nonpolar solvents such as *n*-propanol. Poly(Pro)II is a left-handed 3_1-fold helix with trans residues. These two types of polyproline helices have different CD spectra. Poly(Pro)I has a strong positive band at ∼215 nm and a slightly weaker negative at ∼195 nm. Poly(Pro)II has a weak positive band at ∼230 nm and a strong negative band at ∼205 nm of ellipticity ∼$-60\,000$ deg cm^2 dmol^{-1}.

6.7
Extrinsic Chromophores

The CD spectra above 300 nm can be used to study transitions in chromophoric prosthetic groups, metal ions, inhibitors, or substrate analogues. Proteins lacking prosthetic groups do not exhibit absorption or CD bands above 300 nm, except possibly for the tail of a disulfide transition or a tryptophan band just above 300 nm. The CD spectra of disulfides have a characteristic tail stretching beyond 300 nm. There are three mechanisms contributing to the CD of chromophores bound to proteins. An extrinsic chromophore may be inherently chiral. There may be coupling between electronic transitions on the extrinsic chromophore and chromophores in the protein. It is also possible that transitions of differing symmetry on the extrinsic chromophore may mix because of the electrostatic field of the protein. The last two mechanisms depend on both the extrinsic chromophore geometry and also the structure of the protein.

6.8
Membrane Proteins

There are two kinds of membrane proteins: extrinsic (or peripheral) and intrinsic (or integral). Extrinsic membrane proteins may be removed from the membrane, or solubilized, by mild treatment, such as shaking with a dilute salt solution. Intrinsic membrane proteins cannot be removed from the membrane without treatment that destroys the membrane structure, such as dissolving it with detergent. Membrane proteins are difficult to study in NMR or X-ray crystallography, since the use of detergents may affect the conformation. So, CD spectroscopy is a useful method for investigating the structure of membrane proteins.

Membrane proteins tolerate mildly disruptive detergents without loss of activity. To analyze the CD spectra of membrane proteins requires the known structures of membrane proteins (determined by X-ray diffraction), of which there are relatively few. Nevertheless, it is possible to estimate secondary structures content from CD spectra of membrane proteins using approaches such as convex constraint analysis. Five component spectra have been identified. They are two different types of α-helices (the α-helix in the soluble domain and the α_T-helix for transmembrane α-helix), β-sheet, β-turn, and unordered conformation. The characteristics of CD spectra of α-helix, β-sheet, β-turn, and random coil have been discussed above. The α_T-helix has a positive band in the range of 195 to 200 nm. The intensity of the 208-nm band is slightly more negative than that of the 222-nm band. α_T-helix has a larger rotational strength than an α-helix in aqueous conditions. Membrane proteins are immersed in a much lower dielectric medium than soluble proteins, which may explain some differences. Another important factor may be that the average chain length of an α_T-helix is 25 residues, which is about twice that of an α-helix in a soluble domain.

CD spectra can give much information about membrane proteins, such as secondary structure, fold motifs, conformational changes, environmental effects, folding, and insertion into membranes. However, there are potential artifacts of differential scattering, absorption flattening, and wavelength shifts, which may affect the CD spectra. CD spectra of membrane proteins often exhibit various degrees of distortions in shapes, intensities, and/or positions of the CD bands, and shifts in crossover points. There have been several experimental and theoretical approaches for differentiating the CD bands from the artifacts of the membrane protein CD spectra.

7
Conclusion and Outlook

CD spectroscopy is a useful technique for detecting protein secondary structure quickly and quantitatively. It is also useful in following protein folding/unfolding processes with stopped-flow or faster time-resolved techniques. The development of transient VCD may present a significant advance in quantifying the timescales of motions of biological systems. Currently, there is ongoing research to improve the accuracy of estimates of protein secondary structure content. Making such estimates more quantitative remains a challenge, as there are many factors that may affect the intensity and wavelength of CD spectra. Temperature and solvents are important external factors; dihedral angles and hydrogen-bonding patterns are key internal factors. All of these can change the relative orientation of $n\pi^*$ and $\pi\pi^*$ transitions, which in turn can lead to different CD spectra. In general, CD spectroscopy is a rapid technique, which is sensitive to protein secondary structure. More reliable analysis of CD spectra may emerge from ongoing experimental and theoretical studies.

Bibliography

Books and Reviews

Berova, N., Nakanishi, K., Woody, R.W. (1994) *Circular Dichroism: Principles and Applications*, VCH Publishers Inc., New York.

Fasman, G.D. (1996) *Circular Dichroism and the Conformational Analysis of Biomolecules*, Plenum Press, New York.

Greenfield, N.J. (1996) Methods to estimate the conformation of proteins and polypeptides from circular dichroism data, *Anal. Biochem.* **235**, 1–10.

Rodger, A., Norden, B. (1997) *Circular Dichroism and Linear Dichroism*, Oxford University Press, Oxford.

Woody, R.W. (1995) Circular dichroism, *Methods Enzymol.* **246**, 34–71.

Primary Literature

Andrade, M.A., Chacon, P., Merelo, J.J., Moran, F. (1993) Evaluation of secondary structure of proteins from UV circular dichroism spectra using an unsupervised learning neural network, *Protein Eng.* **6**, 383–390.

Besley, N.A., Hirst, J.D. (1999) Theoretical studies toward quantitative protein circular dichroism calculations, *J. Am. Chem. Soc.* **121**, 9636–9644.

Böhm, G., Muhr, R., Jaenicke, R. (1992) Quantitative analysis of protein for UV circular dichroism spectra by neural networks, *Protein Eng.* **5**, 191–195.

Brahms, S., Brahms, J. (1980) Determination of protein secondary structure in solution by vacuum ultraviolet circular dichroism, *J. Mol. Biol.* **138**, 149–178.

Chang, C.T., Wu, C.S.C., Yang, J.T. (1978) Circular dichroic analysis of protein conformation: inclusion of the beta-turns, *Anal. Biochem.* **91**, 13–31.

Chin, D.H., Woody, R.W., Rohl, C.A., Baldwin, R.L. (2002) Circular dichroism spectra of short, fixed-nucleus alanine helices, *Proc. Natl. Acad. Sci. U.S.A.* **99**, 15416–15421.

Compton, L.A., Johnson, W.C. Jr. (1986) Analysis of protein circular dichroism spectra for secondary structure using a simple matrix multiplication, *Anal. Biochem.* **155**, 155–167.

Dang, Z., Hirst, J.D. (2001) Short hydrogen bonds, circular dichroism, and over-estimates of peptide helicity, *Angew. Chem., Int. Ed.* **40**, 3619–3621.

Fasman, G.D. (1995) The measurement of transmembrane helices by the deconvolution of CD spectra of membrane proteins: a review, *Biopolymers* **37**, 339–362.

Fitts, D.D., Kirkwood, J.G. (1957) The optical rotatory dispersion of the α-helix, *Proc. Natl. Acad. Sci. U.S.A.* **43**, 1046–1052.

Holzwarth, G., Doty, P. (1965) The ultraviolet circular dichroism of polypeptides, *J. Am. Chem. Soc.* **87**, 218–228.

Johnson, W.C. Jr. (1990) Protein secondary structure and circular dichroism: a practical guide, *Proteins: Struct., Funct., Genet.* **7**, 205–214.

Lewis, J.W., Goldbeck, R.A., Kliger, D.S., Xie, X., Dunn, R.C., Simon, J.D. (1992) Time-resolved circular dichroism spectroscopy: experiment, theory, and applications to biological systems, *J. Phys. Chem.* **96**, 5243–5254.

Lobley, A., Whitmore, L., Wallace, B.A. (2002) Dichroweb: an interactive website for the analysis of protein secondary structure from circular dichroism spectra, *Bioinformatics* **18**, 211–212.

Luchins, J., Beychok, S. (1978) Far-ultraviolet stopped-flow circular dichroism, *Science* **199**, 425–426.

Luo, P., Baldwin, R.L. (1997) Mechanism of helix induction by trifluoroethanol: a framework for extrapolating the helix-forming properties of peptides from trifluoroethanol/water mixtures back to water, *Biochemistry* **36**, 8413–8421.

Manning, M.C., Illangasekare, M., Woody, R.W. (1988) Circular dichroism studies of distorted α-helices, twisted β-sheets, and β-turns, *Biophys. Chem.* **31**, 77–86.

Manning, M.C., Woody, R.W. (1991) Theoretical CD studies of polypeptide helices: examination of important electronic and geometric factors, *Biopolymers* **31**, 569–586.

Moffitt, W. (1956) Optical rotatory dispersion of helical polymers, *J. Chem. Phys.* **25**, 467–478.

Nitta, K., Segawa, T., Kuwajima, K., Sugai, S. (1977) Application of stopped-flow circular dichroism to the study of the unfolding of proteins, *Biopolymers* **16**, 703–706.

Pancoska, P., Bitto, E., Janota, V., Urbanova, M., Gupta, V.P., Keiderling, T.A. (1995) Comparison of and limits of accuracy for statistical analyses of vibrational and electronic circular dichroism spectra in terms of correlations to and predictions of protein secondary structure, *Protein Sci.* **4**, 1384–1401.

Pancoska, P., Keiderling, T.A. (1991) Systematic comparison of statistical analyses of electronic and vibrational circular dichroism for secondary structure prediction of selected proteins, *Biochemistry* **30**, 6885–6895.

Provencher, S.W., Glockner, J. (1981) Estimation of protein secondary structure from circular dichroism, *Biochemistry* **20**, 33–37.

Schellman, J.A. (1968) Symmetry rules for optical rotation, *Acc. Chem. Res.* **1**, 144–151.

Sreerema, N., Venyaminov, S.Y., Woody, R.W. (1999) Estimation of the number of helical and strand segments in proteins using CD spectroscopy, *Protein Sci.* **8**, 370–380.

Sreerema, N., Woody, R.W. (1993) A self-consistent method for the analysis of protein secondary structure from circular dichroism, *Anal. Biochem.* **209**, 32–44.

Sreerama, N., Woody, R.W. (2000) Estimation of protein secondary structure from CD spectra: comparison of CONTIN, SELCON and CDSSTR methods with an expanded reference set, *Anal. Biochem.* **282**, 252–260.

Toniolo, C., Polese, A., Formaggio, F., Crisma, M., Kamphuis, J. (1996) Circular dichroism spectrum of a peptide 3_{10}-helix, *J. Am. Chem. Soc.* **118**, 2744–2745.

Wallace, B.A., Janes, R.W. (2001) Synchrotron radiation circular dichroism spectroscopy of proteins: secondary structure, fold recognition and structural genomics, *Curr. Opin. Chem. Biol.* **5**, 567–571.

Wallimann, P., Kennedy, R.J., Kemp, D.S. (1999) Large circular dichroism ellipticities for N-templated helical polypeptides are inconsistent with currently accepted helicity algorithms, *Angew. Chem., Int. Ed.* **38**, 1290–1292.

Woody, R.W. (1992) Circular dichroism and conformation of unordered polypeptides, *Adv. Biophys. Chem.* **2**, 37–79.

Yang, J.T., Wu, C.S., Martinez, H.M. (1986) Calculation of protein conformation from circular dichroism, *Methods Enzymol.* **130**, 208–269.

5
Protein Structure Analysis: High-throughput Approaches

Andrew P. Turnbull and Udo Heinemann
Max Delbrück Center for Molecular Medicine, Berlin, Germany

1	**Bioinformatics** 151	
1.1	Structure-to-function Approaches 151	
1.2	Identification of Disordered Regions in a Protein 154	
1.3	Protein-ligand Complexes 154	
2	**Protein Production** 155	
2.1	Yeast 156	
2.2	Baculovirus–insect Cell 156	
2.3	*Leishmania tarentolae* (Trypanosomatidae) 156	
2.4	Cell-free Expression Systems 157	
3	**Purification** 157	
4	**Structure Determination by X-ray Crystallography** 158	
4.1	Crystallization 158	
4.2	Data Collection 160	
4.3	Phasing 160	
4.4	Automated Structure Determination 161	
4.5	Molecular Replacement 162	
5	**Structure-based Drug Design** 162	
6	**Structure Determination by NMR** 162	
	Bibliography 164	
	Books and Reviews 164	
	Primary Literature 164	

Proteins. Edited by Robert A. Meyers.
Copyright © 2007 Wiley-VCH Verlag GmbH & Co. KGaA, Weinheim
ISBN: 978-3-527-31608-3

Keywords

Active Site
Part of an enzyme molecule made up of amino acid residues, and involved in substrate binding and catalysis.

Anomalous Diffraction
Diffraction of X-rays near the absorption edge of a scattering atom, where a phase shift occurs and the atomic scattering factor becomes a complex quantity; This phenomenon is exploited in phasing X-ray diffraction data.

Isotope Labeling
Introduction of ^{13}C, ^{15}N and other nonradioactive isotopes into a protein for structure determination by nuclear magnetic resonance (NMR) spectroscopy.

Phasing
Reconstitution of the phase relations of diffracted X-rays as part of a crystal structure analysis, using anomalous diffraction and other techniques.

Posttranslational Modification
Covalent modification of a protein molecule after translation by phosphorylation, glycosylation, acetylation and so on. This is rare in prokaryotic proteins, and common in proteins from eukaryotes.

Protein Disorder
Lack of defined three-dimensional structure in segments of proteins or complete proteins in the absence of stabilizing binding partners.

Protein Domain
Compact folding unit of a protein, recognizable on the sequence and/or three-dimensional structure level.

Protein Fold
Recurrent pattern of three-dimensional structure in a protein or protein domain. There are far fewer protein folds than sequences or sequence families.

Selenomethionine
Nonnatural amino acid that can be incorporated into proteins by gene technology in place of methionine. Its selenium atom is used as an anomalous scatterer for phasing the X-ray diffraction pattern.

Structural Genomics
Large-scale project to determine the shapes of all proteins and other important biomolecules encoded by the genomes of key organisms.

Structural Proteomics
Large-scale project to determine the shapes and functions of all proteins encoded by the genomes of key organisms.

Synchrotron Radiation
Electromagnetic radiation emitted by subatomic particles (electrons or positrons) traveling at high velocity in storage rings. X-rays produced at synchrotrons are used in crystal structure determination.

> Developments in the high-throughput analysis of protein structure have been primarily driven by worldwide structural genomics initiatives that are aimed at determining the three-dimensional structures of all proteins and other important biomolecules encoded by the genomes of key organisms. Structural genomics requires a large number of procedural steps in order to convert sequence information into a three-dimensional structure; this has led to new high-throughput methods for protein production, characterization, and structure determination. Over the past decade, the most notable technological advances have been in the fields of X-ray crystallography and NMR – the principal tools of structural genomics – which have facilitated high-throughput, rapid, and cost-effective structure determinations. In recent years, these developments have resulted in an exponential increase in the number of structures being deposited in the Protein Data Bank (PDB; http://www.rcsb.org/pdb/), in which the total number currently exceeds 30 000. Major developments in the fields of bioinformatics, protein production, and structure determination, and the impact of these on high-throughput protein structure analysis will be discussed in what follows.

1
Bioinformatics

Primary sequence analysis, such as similarity searches against protein sequence databases, protein domain architecture determination, identification of specialized local structural motifs, and prediction of protein structure are possible with a variety of homology-based modeling methods (Fig. 1). Sequence database searches are particularly useful in selecting targets for structural genomics initiatives and, where possible, generating homologous probes for determining structures by the molecular replacement method. For example, the Web-based program 3D-PSSM (http://www.sbg.bio.ic.ac.uk/~3dpssm/) is a fast method for predicting the protein fold from the primary amino acid sequence, and SWISS-MODEL (http://swissmodel.expasy.org/) is a fully automated protein structure homology-modeling server.

1.1
Structure-to-function Approaches

There are a number of bioinformatics resources available that are aimed

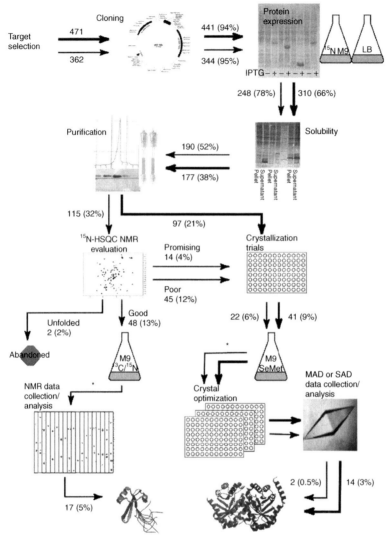

Fig. 1 Schematic flow diagram of the strategies employed in structural genomics initiatives, using *Methanobacterium thermoautotrophicum* as an example. The number of protein targets after each step and the percentage relative to the number of starting targets are indicated in brackets. Thin arrows and italicized numbers are for smaller molecular-weight proteins, and wide arrows and bold numbers are for larger molecular-weight proteins. Diagram taken from Yee, A., Pardee, K., Christendat, D., Savchenko, A., Edwards, A.M., Arrowsmith, C.H. (2003) Structural proteomics: toward high-throughput structural biology as a tool in functional genomics, *Acc. Chem. Res.* **36**, 183–189. Picture with permission from Prof. Cheryl Arrowsmith.

at identifying a protein's biochemical function from its three-dimensional structure (Fig. 2). For example, Dali (http://www2.ebi.ac.uk/dali) and VAST (Vector Alignment Search Tool; http://www.ncbi.nlm.nih.gov:80/Structure/VAST/vastsearch.html) offer Web-based servers for automatically comparing the fold of a newly determined structure against known folds, as represented by the protein structures in the PDB. Such comparisons can often reveal striking similarities between proteins that are not evident from sequence analysis alone, and that can provide important insights into biological function even in the absence of any other biochemical or functional data. However, computer-based approaches fail to assign functions to proteins that adopt novel folds. Enzymes are a notable exception to this rule, because the groupings of residues constituting their active sites tend to be highly conserved in their spatial disposition even in cases where there is no overall similarity in sequence or fold. For example, the relative positioning of the Ser-His-Asp catalytic triad of the serine proteases is highly conserved even when found in protein structures adopting different folds. Hence, screening a new protein structure against a database of enzyme active-site templates such as PROCAT (http://www.biochem.ucl.ac.uk/bsm/PROCAT/PROCAT.html) can be used for detecting key functional residues. The spatial patterns of residues can be automatically generated using various techniques including graph theory (ASSAM) and "fuzzy pattern matching" (RIGOR). Alternative approaches to identifying enzymes

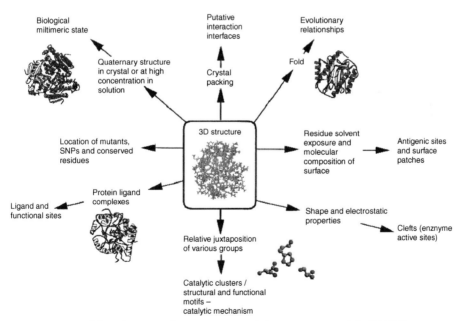

Fig. 2 Summary of the information deriving from the three-dimensional structure of a protein, relating to its biological function. Taken from Thornton, J.M., Todd, A.E., Milburn, D., Borkakoti, N., Orengo, C.A. (2000) From structure to function: approaches and limitations, *Nat. Struct. Biol.* **7**(Suppl.), 991–994.

on the basis of their three-dimensional structure and predicting their functions, have recently been reported. Here, a vector machine-learning algorithm is used, based on the secondary structure of proteins the propensities of amino acids, and surface properties, in order to discriminate enzymes from nonenzymes. Another approach analyses protein surface charges to identify conserved residues that can serve as catalytic sites. Once a general class of biochemical function of a protein has been proposed, experimental screening of enzymatic activity can be used to derive the precise biochemical function. For example, after the structure determination of BioH (an enzyme involved in biotin biosynthesis in *E. coli*) the protein structure was screened against a library of enzyme active sites, a Ser/His/Asp catalytic triad was identified, and subsequent hydrolase assays showed BioH to be a carboxylesterase.

1.2
Identification of Disordered Regions in a Protein

The occurrence of regions in proteins that lack any fixed tertiary structure is increasingly being observed in structural studies. These disordered regions or "random coils" are inherently flexible and are involved in a variety of functions, including the modulation of the specificity/affinity of protein-binding interactions, activation by cleavage, and DNA recognition. During the target selection process, it is important to consider any intrinsic protein disorder, because it can often lead to problems with the expression of protein-coding genes, protein stability, purification, and crystallization. PONDR, DisEMBL, and GlobPlot are useful tools for predicting potential disordered regions within a protein sequence that can be used to help design constructs corresponding to globular proteins or domains. PONDR (Predictor of Naturally Disordered Regions; http://www.pondr.com) and DisEMBL use methods based on artificial neural networks, whereas GlobPlot (http://globplot.embl.de) relies on a novel, propensity-based disorder-prediction algorithm. These methods can also be used to predict inherently flexible regions in protein sequences. For example, PONDR predicted that the linker between the DNA-operator-binding central domain of the transcriptional regulator KorB (KorB-O) and the KorB dimerization domain (KorB-C) is flexible, which was indeed observed in crystal structures and is thought to facilitate complex formation on circular plasmids (Fig. 3).

1.3
Protein-ligand Complexes

Protein-ligand complexes are the most useful in terms of providing functional information, because they reveal the nature of the ligand, the site at which it is bound to the protein, the location of the active site, and of the catalytic machinery (if the protein is an enzyme). There are several examples of structural analyses in which an unexpected protein-bound ligand or cofactor derived from the cloning organism was discovered. For example, the structure of the trimeric human protein p14.5 was found to have picked up benzoate molecules from the crystallization buffer at its inter-subunit tunnels, which most likely mark a hydrolytic active site (Fig. 4). When such data are available at high resolution, proposing a biological function for the protein can be relatively straightforward, because these data identify the nature of the ligand, the

5 Protein Structure Analysis: High-throughput Approaches

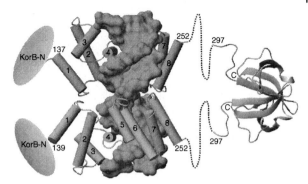

Fig. 3 Natively disordered regions in the bacterial transcriptional regulator and partitioning protein KorB. The KorB DNA-binding domains (KorB-O, center) are connected by flexible linkers to N-terminal domains of unknown structure and function (KorB-N, left), and the KorB dimerization domains (KorB-C, right). Picture taken from Khare et al., 2004.

Fig. 4 Crystal structure of the trimeric human protein, hp14.5. Benzoate molecules picked up from the crystallization buffer bind in the inter-subunit tunnels and mark putative hydrolytic active sites (Manjasetty et al., 2004). Picture with permission from Dr. B.A. Manjasetty.

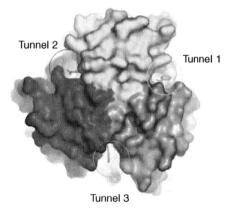

ligand-binding site, and the arrangement of catalytic residues from which a catalytic mechanism can be postulated.

2
Protein Production

Protein expression and purification play a central role in high-throughput protein structure analysis. Cloning using restriction enzymes is impractical for high-throughput approaches, because of the complications of selecting compatible and appropriate restriction enzymes for each cloning procedure, and the multiple steps of experimental refinement and treatment that must be performed. Therefore, high-throughput cloning requires procedures based on the polymerase chain reaction (PCR). High-throughput cloning and expression methods are now being developed in many laboratories, enabling the generation and testing of up to hundreds of DNA constructs for high-level expression in parallel, using rapid and generic protocols. To generate expression vector clones, generic cloning systems can be used; the Gibco/Life Technologies GATEWAY™ system is one such example, streamlining the expression and cloning process by alleviating recloning steps and avoiding the use of restriction enzymes in the cloning and subcloning processes.

High-throughput approaches may rely on both prokaryotic and eukaryotic hosts. *Escherichia coli* expression systems are advantageous for many reasons, most notably because the overproduced protein is usually obtained without any posttranslational modification heterogeneity, and protein expression is cheaper and faster than with eukaryotic systems. However, it is not always possible to obtain soluble protein for many eukaryotic proteins, in particular, human proteins, by using the heterologous expression of eukaryotic genes, in which the codon usage for a cDNA could be suboptimal in *E. coli*. Furthermore, eukaryotic systems are necessary for the expression of proteins that require posttranslational modifications for their correct folding and activity. In situations in which a protein cannot be synthesized in *E. coli*, the eukaryotic yeasts (*Saccharomyces cerevisiae* and *Pichia pastoris*), baculovirus-infected insect cells, *Leishmania tarentolae*, and the cell-free wheat germ systems that have been developed for high-throughput approaches can be used. Furthermore, in these and other eukaryotic expression systems, the codon usage is closer to that in humans. The various merits of each system are discussed below.

2.1
Yeast

The yeasts *S. cerevisiae* and *P. pastoris* can be used for the routine production of recombinant proteins. More recently, *P. pastoris* has emerged as the preferred yeast, because of its strong, highly inducible promoter system resulting in higher yields of recombinant protein, stable genomic integration and posttranslational modifications, such as phosphorylation, which may be important for both the structure and function of some human proteins. In comparison with *S. cerevisiae*, the distribution and chain length of N-linked oligosaccharides are significantly shorter, and therefore this system represents a suitable alternative for the extracellular expression of human proteins. An additional advantage of the methylotrophic *P. pastoris* expression system is that it makes it possible to introduce ^{13}C label into recombinant proteins (by feeding ^{13}C-labeled methanol) for nuclear magnetic resonance (NMR) structural analyses.

2.2
Baculovirus–insect Cell

The recombinant baculovirus–insect cell expression system accomplishes most posttranslational modifications, including phosphorylation, N- and O-linked glycosylation, acylation, disulfide cross-linking, oligomeric assembly, and subcellular targeting, which may all be critical for the accurate production and function of human proteins. In contrast to bacterial expression systems, the recombinant baculovirus expression system usually produces soluble proteins without the need for induction or specific temperature conditions.

2.3
Leishmania tarentolae (Trypanosomatidae)

The use of the parasitic Trypanosomatidae species, *L. tarentolae*, as the host for *in vitro* protein production has recently been reported, and can achieve high levels of protein expression. The Trypanosomatidae species of parasites naturally produce large amounts of glycoproteins, which is an advantage in the production of sialylated heterologous glycosylated proteins. Furthermore, given the natural auxotrophy of *L. tarentolae* for methionine, it has been suggested that such a system could prove to be useful for the production of selenomethionine-labeled proteins, for high-throughput X-ray crystallographic structure determination using single- or multiple-wavelength anomalous diffraction (SAD and MAD) phasing techniques.

2.4
Cell-free Expression Systems

Cell-free systems facilitate the parallel expression of many protein-coding genes, and are therefore suitable for high-throughput protein production. Obtaining the protein directly in a form suitable for X-ray crystallography or NMR spectroscopy is another prerequisite for high-throughput structure analysis. For the purpose of X-ray analysis, anomalous diffraction techniques (SAD or MAD) necessitate the substitution of methionine residues with selenomethionine. Additionally, NMR structure analysis often requires the labeling of proteins with ^{13}C and/or ^{15}N, which can be introduced through cell growth on media containing these isotopes in the form of ^{13}C-glucose and $^{15}NH_4Cl$. In this respect, the *E. coli* cell-free protein synthesis system of Kigawa et al. (1999) permits the straightforward incorporation of isotopes for NMR analysis. However, this system is not always suitable for the expression of some eukaryotic proteins and can result in aggregation, formation of insoluble inclusion bodies, and degradation of the expression product. Furthermore, in the case of multidomain proteins, which are found more often in eukaryotes, correct folding occurs more frequently in eukaryotic than in prokaryotic translation systems. Another limitation with the use of *E. coli* systems for high-throughput cell-free expression is that PCR-generated fragments are not transcribed and translated efficiently in these systems: contamination by the mRNA- and DNA-degradation enzymes originating from the cell decreases the stability of the templates and reduces yields. By contrast, in eukaryotic cell-free systems, the added mRNAs are stable for long periods of time, and therefore these systems overcome many of the limitations associated with *E. coli* cell-free systems. Furthermore, cell-free systems can produce high yields of correctly folded proteins and, unlike *in vivo* systems, facilitate the expression of proteins that would otherwise interfere with the host cell physiology. Recently, the synthesis and screening of gene products based on the cell-free system prepared from eukaryotic wheat embryos has been reported; this method bypasses many of the time-consuming cloning steps involved in conventional expression systems and lends itself to the high-throughput expression of proteins, using automated robotic systems.

3
Purification

Protein purification has also seen significant improvements because of the use of affinity tags fused to the protein of interest, so that it can be separated from the host cell proteins rapidly by using standardized purification schemes. N-terminal tags range from large tags, such as glutathione-S-transferase (GST), maltose-binding protein (MBP), thioreductase, and the chitin-binding protein, to fairly small tags such as His_6, and various epitope tags. In some systems, the use of large tags can lead to the production of fusion products that are too large for high-level expression, but the use of such tags can improve the correct folding and stability of the overexpressed protein. His_6 is the most commonly used purification tag, because it is easier to incorporate into expression constructs and it allows a generic one-step purification using automated methods based on nickel-nitrilotriacetic acid (Ni-NTA) or other immobilized metal-affinity chromatography resins. The use of an additional C-terminal StrepII tag enables dual-affinity column purification, which ensures that only full-length gene products are separated. For crystallization to occur, it is generally agreed that the fusion tags must be removed, because they can introduce flexible regions into the protein, which can interfere with crystallization or lead to various forms of microheterogeneity. In a recent study, the compatibility of small peptide affinity tags with protein crystallization was assessed, in which the N-terminus of the chicken spectrin SH3 domain was labeled with a His_6 tag and a StrepII tag, fused to the N- and C-termini, respectively. The resulting protein, His_6-SH3-StrepII, comprised 83 amino acid residues, 23 of which originated from the tags (accounting for 23% of the total fusion protein mass). In contrast to the general consensus that the presence of affinity tags is detrimental in structural studies, this study demonstrated that the fused affinity tags did not interfere with crystallization or structure analysis, and did not change the protein structure. This suggests that, in some cases, protein constructs utilizing both N- and C-terminal peptide tags may lend themselves to structural investigations in high-throughput regimes.

4
Structure Determination by X-ray Crystallography

4.1
Crystallization

Crystallization is regarded as a major bottleneck in structure determination by X-ray crystallography and can be divided into two stages: coarse screening for initial crystallization conditions, followed by optimization of the conditions in order to produce diffraction-quality single crystals. The field of crystallization has recently been revolutionized by significant developments in automation, miniaturization, and process integration. These developments have led to the availability of robotic liquid-handling systems that are capable of rapidly and efficiently screening thousands of crystallization conditions, in which different parameters such as ionic strength, precipitant concentration, additives, pH, and temperature are altered. High-throughput screening begins with the automated preparation and/or reformatting of precipitant solutions into crystallization microplates. The latest nanoliter robotic liquid-dispensing systems are capable of dispensing very small drops (typically containing between 25 and 100 nL of protein), which reduces the amount of protein required for screening conditions and the overhead on protein production. Furthermore, smaller drops equilibrate faster than larger drops, leading to a more rapid appearance of crystals. The crystallization experiments are then regularly monitored using an imaging robot, and the collected images can either be analyzed visually or by using automatic crystal recognition systems. HomeBase™ (The Automation Partnership, UK) represents an integrated system combining high-density plate storage and imaging.

A recent development in protein crystallization has been the use of a microfluidic system for crystallizing proteins, using the free-interface diffusion method at the nanoliter scale. This system is capable of screening hundreds of crystallization conditions in which droplets, each containing solutions of protein, precipitants, and additives in various ratios, are formed in the flow of immiscible fluids inside a polydimethylsiloxane (PDMS)/glass capillary composite microfluidic device (Fig. 5). The system is capable of performing multidimensional screening (mixing 5–10 solutions) and therefore explores more of the crystallization space than the conventional method of vapor diffusion, increasing the likelihood of obtaining crystals.

Furthermore, the capillary containing protein crystals can be directly exposed to the synchrotron X-ray beam, eliminating the need to manually manipulate the crystal. These features also mean that it has the potential to serve as the basis for future high-throughput automated crystallization systems.

Membrane proteins represent the most persistent bottleneck for all preparatory procedures and analytical methods, because they are water soluble only in the presence of detergents, and are difficult to overproduce in the quantities required for structural studies. Membrane proteins, which constitute up to 30% of the protein repertoire of an organism, represent the targets for more than 50% of the drugs that are being currently used and tested. An adapted robotic system has recently been reported that enables high-throughput crystallization of membrane proteins using lipidic mesophases.

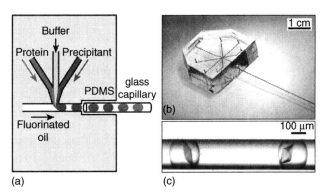

Fig. 5 (a) Schematic illustration of the droplet-based microfluidic system for protein crystallization. (b) Photograph of a polydimethylsiloxane (PDMS)/glass capillary composite microfluidic device. (c) A micrograph of thaumatin crystals grown in droplets, produced by the method outlined in panel (a), within a capillary. Taken from Zheng, B., Tice, J.D., Roach, L.S., Ismagilov, R.F. (2004) A droplet-based, composite PDMS/glass capillary microfluidic system for evaluating protein crystallization conditions by microbatch and vapor-diffusion methods with on-chip X-ray diffraction, *Angew. Chem., Int. Ed. Engl.* **43**, 2508–2511. Picture with permission from Dr. Rustem Ismagilov.

4.2
Data Collection

X-ray data collection has been revolutionized over the past decade, with the development of improved X-ray sources and detectors and the universal adoption of flash-freezing techniques that greatly reduce crystal radiation damage. New third-generation synchrotrons are now available across the world, which provide more intense and stable X-ray beams and, combined with new, faster, larger, and more sensitive X-ray detectors, allow higher-quality data to be collected much more rapidly, leading to a dramatic increase in the success rate of structure determination. Furthermore, specialized, highly collimated microfocus beamlines, such as ID13 at the ESRF (Grenoble, France; http://www.esrf.fr/exp_facilities/ID13/index.html) and the protein crystallography beamline at the Swiss Light Source (SLS), are specifically tailored to study biological crystals with small physical dimensions (microcrystals from 5 to 50 μm in size) or very large unit cell dimensions. Future developments in detector design, such as the MarResearch GmbH (Norderstedt, Germany) flat-panel detector, promise larger continuous active areas, higher spatial resolution, very low noise, and fast readout times (∼1 s). High-throughput X-ray data collection has also led to the development of automated robotic sample changers that store and mount crystals sequentially while maintaining the samples at liquid-nitrogen temperatures (100 K). A novel system has recently been described, using a proprietary crystallization plate that facilitates the preliminary investigation of the diffraction properties of crystals *in situ* in the drop, by direct exposure to the X-ray beam. The BIOXHIT project (Biocrystallography (X) on a Highly Integrated Technology Platform for European Structural Genomics; http://www.bioxhit.org), comprising more than 20 research groups from all over Europe, aims to develop an integrated technology platform for synchrotron beamlines by promoting new approaches to crystallization, and fully automating diffraction data collection and structure determination.

Recent developments in technology have not been limited exclusively to synchrotron sources. The latest generation of high-intensity X-ray generators has revolutionized X-ray sources to such an extent that the latest "in-house" systems from Rigaku/MSC (e.g. the FR-E SuperBright; http://www.rigakumsc.com/) and Bruker AXS (http://www.bruker-axs.com) are comparable in intensity to first-generation synchrotron sources and, coupled with automated sample changers, make high-throughput crystallography possible in the laboratory. Furthermore, the development of the Compact Light Source (Lyncean Technologies, Inc., Palo Alto, CA, USA) promises to have a huge impact on protein structure determination, offering the possibility of a "synchrotron beamline" for home laboratory applications. This tunable, tabletop X-ray source combines an electron beam with a laser beam to generate an intense X-ray beam and, as a next-generation X-ray source, directly addresses the increasing demand for high-throughput protein crystallography.

4.3
Phasing

The central problem in X-ray crystallography is the determination of the protein phases. X-ray data collected from a crystal consist of structure factor amplitudes, but there is no way of directly measuring the phase associated with each amplitude. Recent advances in macromolecular phasing have simplified and further automated this crucial stage of X-ray structure determination to such an extent that the eventual success of a project is usually assured, if well-diffracting crystals are available. The techniques of isomorphous replacement, anomalous scattering, molecular replacement, and single (SAD) and multiple (MAD) anomalous dispersion are commonly used to solve the phase problem. However, phase determination has been dramatically facilitated by the widespread adoption of SAD and MAD phasing techniques by the crystallographic community, primarily as a consequence of the availability of stable and tunable synchrotron sources. These allow the optimal exploitation of the anomalous effect as a source of phase information by delivering X-ray energies corresponding to absorption maxima of the anomalous scatterers, very often selenium introduced as selenomethionine through substitution of methionine. Additionally, heavy-atom labels (mercury, platinum, and others) may be bound to crystalline proteins by soaking to yield phase information by anomalous diffraction techniques.

Modern phasing techniques, such as fast halide soaks and sulfur-SAD, hold promise of simpler and faster protein structure determination than traditional methods. Fast halide soaks using bromide and iodide that diffuse rapidly into the crystal and display significant anomalous scattering signals can be used to quickly derivatize protein crystals. Heavy-atom reagents can also be incorporated into the crystal in a relatively short time if a concentration greater than 10 mM is used. These derivatives then display better isomorphism and diffraction qualities than those obtained after a standard, prolonged soak. The availability of stable synchrotron beamlines and improvements in data processing programs make it possible to collect extremely accurate diffraction data, and to determine structures using the very weak anomalous signals from atoms such as sulfur and phosphorus that are inherently present in macromolecules or nucleic acids; several novel structures have been determined by using sulfur-SAD phasing.

4.4
Automated Structure Determination

High-throughput crystallographic structure determination requires software that is automated and designed for minimum user intervention. There has been considerable development in the direct-method programs, SHELXD and SnB, which can automatically determine heavy-atom substructures from a very small signal. HKL2MAP connects several programs from the SHELX suite to guide the user from analyzing scaled diffraction data (SHELXC), through substructure solution (SHELXD) and phasing (SHELXE), to displaying an electron density map (Xfit). There are a number of other automated software systems, such as ACrS (automated crystallographic system) and PHENIX (Python-based hierarchical environment for integrated Xtallography) that are currently being developed to meet the requirements of high-throughput structure determination by combining multiple structure-determination software packages into one intuitive interface.

Finally, new algorithms for interpreting electron density maps and for automated model building, such as, SOLVE/RESOLVE, AUTOSHARP/

SHARP, and ARP/wARP, enable rapid construction of protein models without the need for significant manual intervention. At present, the success rates for these programs are dependent on the resolution of the diffraction data (typically, 2.5-Å resolution or higher is necessary for automatic chain fitting).

4.5
Molecular Replacement

When an approximate structural model of a protein under investigation is available, either from NMR, a homologous X-ray structure, or from homology modeling, initial phases can be obtained using molecular replacement where the homologous probe structure is fitted to the experimental data using three rotational and three translational parameters. The programs AMoRe and MolRep that are integrated into the CCP4i GUI simplify the problem of positioning a molecule in the asymmetric unit by running sequential rotational and translational searches. Advances in molecular replacement include the implementation of the maximum likelihood–based algorithms in BEAST, and the six-dimensional evolutionary search algorithm in EPMR. As the number of protein structures increases, it is anticipated that molecular replacement will become the standard method for structure determination, using a generalized search of all unique protein domains present in the PDB.

5
Structure-based Drug Design

Knowledge of the three-dimensional structure of proteins can play a key role in the development of small-molecule drugs, because being able to verify how lead compounds bind to their targets accelerates drug development and is more cost-effective. Notable drugs that have been successfully designed using protein three-dimensional structure information include the HIV protease inhibitors, Viracept™ (Agouron, USA and Eli Lilly, USA), and Agenerase™ (Vertex, USA; Kissei, Japan; Glaxo Wellcome, UK). However, it is usually necessary to screen a large number of protein–ligand complex structures in the iterative process of rational structure-based drug design. Hence, high-throughput protein X-ray crystallography offers an unprecedented opportunity for facilitating drug discovery. Recent advances in rapid binding-site analysis of *de novo* targets using virtual ligand (*in silico*) screening and small-molecule cocrystallization methodologies, in combination with the miniaturization and automation of structural biology, enable more rapid lead compound identification and faster optimization, providing a framework for direct integration into the drug discovery process. In the high-throughput structure determination of protein–ligand complexes, it is desirable to use tools that can locate, build, and refine the structure of the bound ligand with minimal human intervention. One such tool is X-LIGAND, part of the QUANTA software package (Accelrys, San Diego, CA, USA) that automatically searches for unoccupied regions of electron density in the structure of the protein–ligand complex in which it tries to fit the ligand.

6
Structure Determination by NMR

Solution-state NMR spectroscopy can serve as a technique complementary to X-ray crystallography in protein structure analysis, particularly in the context of structural genomics initiatives, where many protein targets either do not crystallize or do not form crystals suitable for crystallographic studies (owing to small crystal size or poor diffraction quality). NMR measurements are performed in aqueous solution, obviating the need for growing crystals. This technique is applicable primarily to small proteins (<30 kDa) that are highly soluble (millimolar concentrations), and is particularly useful in the study of proteins that are partially unfolded in the absence of their appropriate binding partners. The additional technique of solid-state NMR is useful in providing structural information for some integral membrane proteins that may not be accessible using crystallographic methods. Furthermore, chemical shift perturbation studies can be used to validate proposed biochemical functions, to map ligand-binding epitopes, and to screen for small-molecule ligands in drug development.

High-throughput, NMR-based structure determination requires rapid and automated data acquisition and analysis methods. The major challenges in realizing this have been those of increasing instrumentation sensitivity (signal-to-noise ratio) and reducing the time required for data collection. These technical issues have been addressed by constructing new high-field magnets and by the recent introduction of cryogenic probes that operate at low temperatures (~25 K), permitting the investigation of proteins that have either low solubility or low yields from purification. Additionally, the application of TROSY (transverse relaxed optimized spectroscopy), a novel spectroscopic concept based on the selection of slowly relaxing NMR transitions, has provided significant sensitivity enhancements for large proteins.

The most time-consuming aspects of structure determination by NMR are the long data collection times necessary for independently sampling three or more indirect dimensions along with the time taken to interpret the correspondingly large number of spectra from ^{13}C/^{15}N-isotope-labeled samples. Rapid resonance assignment is a prerequisite for high-throughput NMR structure determination; techniques such as reduced-dimensionality ^{13}C,^{15}N,^{1}H-triple resonance NMR also avoid the sampling limited regime through the simultaneous frequency-labeling of two spin types in a single indirect dimension. Heteronuclear multidimensional data reduce complications arising from interspectral variations by maximizing the dimensionality of the spectra and decrease signal overlap of data sets sufficiently for the data to be analyzed automatically. Recent approaches to automated structure elucidation from NMR spectra include NOESY-Jigsaw, in which sparse and unassigned NMR data can be used to reasonably and accurately assess secondary structure and align it. The information thus retrieved is useful for quick structural assays for assessing folds before full structural determination and can therefore assist in fold prediction. Additionally, the program ATNOS (automated NOESY peak picking) enables automated peak picking and NOE signal identification in homonuclear 2D and heteronuclear 3D [^{1}H, ^{1}H]-NOESY spectra during *de novo* protein structure determination.

See also Protein NMR Spectroscopy.

Bibliography

Books and Reviews

Blundell, T.L., Mizuguchi, K. (2000) Structural genomics: an overview, *Prog. Biophys. Mol. Biol.* **73**, 289–295.

Dauter, Z. (2002) New approaches to high-throughput phasing, *Curr. Opin. Struct. Biol.* **12**, 674–678.

Guntert, P. (2004) Automated NMR structure calculation with CYANA, *Methods Mol. Biol.* **278**, 353–378.

Laskowski, R.A., Watson, J.D., Thornton, J.M. (2003) From protein structure to biochemical function? *J. Struct. Funct. Genomics* **4**, 167–177.

Norin, M., Sundstrom, M. (2002) Structural proteomics: developments in structure-to-function predictions, *Trends Biotechnol.* **20**, 79–84.

Stevens, R.C. (2000) Design of high-throughput methods of protein production for structural biology, *Struct. Fold. Des.* **8**, R177–R185.

Stewart, L., Clark, R., Behnke, C. (2002) High-throughput crystallization and structure determination in drug discovery, *Drug Discov. Today* **7**, 187–196.

Tickle, I., Sharff, A., Vinkovic, M., Yon, J., Jhoti, H. (2004) High-throughput protein crystallography and drug discovery, *Chem. Soc. Rev.* **33**, 558–565.

Primary Literature

Abola, E., Kuhn, P., Earnest, T., Stevens, R.C. (2000) Automation of X-ray crystallography, *Nat. Struct. Biol.* **7**, 973–977.

Adams, P., Gopal, k., Grossekunstleve, R.W., Hung, L.-W., Ioerger, T.R., McCoy, A.J., Moriarty, N.W., Pai, R.K., Read, R.J., Romo, T.D., Sacchettini, J.C., Sauter, N.K., Storoni, L.C., Terwilliger, T.C. (2004) Recent developments in the PHENIX software for automated crystallographic structure determination, *J. Synchrotron Radiat.* **11**, 53–55.

Albala, J.S., Franke, K., McConnell, I.R., Pak, K.L., Folta, P.A., Rubinfeld, B., Davies, A.H., Lennon, G.G., Clark, R. (2000) From genes to proteins: high-throughput expression and purification of the human proteome, *J. Cell. Biochem.* **80**, 187–191.

Bailey-Kellogg, C., Widge, A., Kelley, J.J., Berardi, M.J., Bushweller, J.H., Donald, B.R. (2000) The NOESY jigsaw: automated protein secondary structure and main-chain assignment from sparse, unassigned NMR data, *J. Comput. Biol.* **7**, 537–558.

Bate, P., Warwicker, J. (2004) Enzyme/non-enzyme discrimination and prediction of enzyme active site location using charge-based methods, *J. Mol. Biol.* **340**, 263–276.

Breitling, R., Klingner, S., Callewaert, N., Pietrucha, R., Geyer, A., Ehrlich, G., Hartung, R., Muller, A., Contreras, R., Beverley, S.M., Alexandrov, K. (2002) Non-pathogenic trypanosomatid protozoa as a platform for protein research and production, *Protein Expr. Purif.* **25**, 209–218.

Brenner, S.E. (2000) Target selection for structural genomics, *Nat. Struct. Biol.* **7**(Suppl.), 967–969.

Brown, J., Walter, T.S., Carter, L., Abrescia, G.A., Aricescu, A.R., Batuwangala, T.D., Bird, L.E., Brown, N., Chamberlain, P.P., Davis, S.J., Dubinina, E., Endicott, J., Fennelly, J.A., Gilbert, R.J.C., Harkiolaki, M., Hon, W.-C., Kimberley, F., Love, C.A., Mancini, E.J., Manso-Sancho, R., Nichols, C.E., Robinson, R.A., Sutton, G.C., Schueller, N., Sleeman, M.C., Stewart-Jones, G.B., Vuong, M., Welburn, J., Zhang, Z., Stammers, D.K., Owens, R.J., Jones, E.Y., Harlos, K., Stuart, D.I. (2003) A procedure for setting up high-throughput nanoliter crystallization experiments. II. Crystallization results, *J. Appl. Crystallogr.* **36**, 315–318.

Bruel, C., Cha, K., Reeves, P.J., Getmanova, E., Khorana, H.G. (2000) Rhodopsin kinase: expression in mammalian cells and a two-step purification, *Proc. Natl. Acad. Sci. U.S.A.* **97**, 3004–3009.

Brunzelle, J.S., Shafaee, P., Yang, X., Weigand, S., Ren, Z., Anderson, W.F. (2003) Automated crystallographic system for high-throughput protein structure determination, *Acta Crystallogr.* **D59**, 1138–1144.

Cereghino, J.L., Cregg, J.M. (2000) Heterologous protein expression in the methylotrophic yeast *Pichia pastoris*, *FEMS Microbiol. Rev.* **24**, 45–66.

Cherezov, V., Peddi, A., Muthusubramaniam, L., Zheng, Y.F., Caffrey, M. (2004) A robotic system for crystallizing membrane and soluble proteins in lipidic mesophases, *Acta Crystallogr.* **D60**, 1795–1807.

Collaborative Computational Project, N. (1994) The CCP4 suite: programs for protein crystallography, *Acta Crystallogr.* **D50**, 760–763.

Dauter, Z., Dauter, M., Rajashankar, K.R. (2000) Novel approach to phasing proteins: derivatization by short cryo-soaking with halides, *Acta Crystallogr.* **D56**, 232–237.

del Val, C., Mehrle, A., Falkenhahn, M., Seiler, M., Glatting, K.H., Poustka, A., Suhai, S., Wiemann, S. (2004) High-throughput protein analysis integrating bioinformatics and experimental assays, *Nucleic Acids Res.* **32**, 742–748.

Delbruck, H., Ziegelin, G., Lanka, E., Heinemann, U. (2002) An Src homology 3-like domain is responsible for dimerization of the repressor protein KorB encoded by the promiscuous IncP plasmid RP4, *J. Biol. Chem.* **277**, 4191–4198.

Dobson, P.D., Doig, A.J. (2003) Distinguishing enzyme structures from non-enzymes without alignments, *J. Mol. Biol.* **330**, 771–783.

Endo, Y., Sawasaki, T. (2004) High-throughput, genome-scale protein production method based on the wheat germ cell-free expression system, *J. Struct. Funct. Genomics* **5**, 45–57.

Fortelle, Edl., Bricogne, G. (1997) Maximum-likelihood heavy-atom parameter refinement for multiple isomorphous replacement and multiwavelength anomalous diffraction methods, *Methods enzymol.* **276**, 472–494.

Garman, E. (1999) Cool data: quantity AND quality, *Acta Crystallogr.* **D55**, 1641–1653.

Garner, E., Cannon, P., Romero, P., Obradovic, Z., Dunker, A.K. (1998) Predicting disordered regions from amino acid sequence: common themes despite differing structural characterization, *Genome Inf. Ser. Worksh. Genome Inf.* **9**, 201–213.

Garner, E., Romero, P., Dunker, A.K., Brown, C., Obradovic, Z. (1999) Predicting binding regions within disordered proteins, *Genome Inf. Ser. Worksh. Genome Inf.* **10**, 41–50.

Goodwill, K.E., Tennant, M.G., Stevens, R.C. (2001) High-throughput x-ray crystallography for structure-based drug design, *Drug Discov. Today* **15**(Suppl.), 113–118.

Grinna, L.S., Tschopp, J.F. (1989) Size distribution and general structural features of N-linked oligosaccharides from the methylotrophic yeast, *Pichia pastoris*, *Yeast* **5**, 107–115.

Heinemann, U., Frevert, J., Hofmann, K., Illing, G., Maurer, C., Oschkinat, H., Saenger, W. (2000) An integrated approach to structural genomics, *Prog. Biophys. Mol. Biol.* **73**, 347–362.

Herrmann, T., Guntert, P., Wuthrich, K. (2002) Protein NMR structure determination with automated NOE-identification in the NOESY spectra using the new software ATNOS, *J. Biomol. NMR* **24**, 171–189.

Holm, L., Sander, C. (1993) Protein structure comparison by alignment of distance matrices, *J. Mol. Biol.* **233**, 123–138.

Jhoti, H. (2001) High-throughput structural proteomics using X-rays, *Trends Biotechnol.* **19**, S67–S71.

Kaldor, S.W., Kalish, V.J., Davies, J.F. II, Shetty, B.V., Fritz, J.E., Appelt, K., Burgess, J.A., Campanale, K.M., Chirgadze, N.Y., Clawson, D.K., Dressman, B.A., Hatch, S.D., Khalil, D.A., Kosa, M.B., Lubbehusen, P.P., Muesing, M.A., Patick, A.K., Reich, S.H., Su, K.S., Tatlock, J.H. (1997) Viracept (nelfinavir mesylate, AG1343): a potent, orally bioavailable inhibitor of HIV-1 protease, *J. Med. Chem.* **40**, 3979–3985.

Kelley, L.A., MacCallum, R.M., Sternberg, M.J. (2000) Enhanced genome annotation using structural profiles in the program 3D-PSSM, *J. Mol. Biol.* **299**, 499–520.

Khare, D., Ziegelin, G., Lanka, E., Heinemann, U. (2004) Sequence-specific DNA binding determined by contacts outside the helix-turn-helix motif of the ParB homolog KorB, *Nat. Struct. Mol. Biol.* **11**, 656–663.

Kim, E.E., Baker, C.T., Dwyer, M.D., Murcko, M.A., Rao, B.G., Tung, R.D., Navia, M.A. (1995) Crystal structure of HIV-1 protease in complex with VX-478, a potent and orally bioavailable inhibitor of the enzyme, *J. Am. Chem. Soc.* **117**, 1181–1182.

Kissinger, C.R., Gehlhaar, D.K., Fogel, D.B. (1999) Rapid automated molecular replacement by evolutionary search, *Acta Crystallogr.* **D55**, 484–491.

Kleywegt, G.J. (1999) Recognition of spatial motifs in protein structures, *J. Mol. Biol.* **285**, 1887–1897.

Kolb, V.A., Makeyev, E.V., Spirin, A.S. (2000) Co-translational folding of an eukaryotic multidomain protein in a prokaryotic translation system, *J. Biol. Chem.* **275**, 16597–16601.

Kuhn, P., Wilson, K., Patch, M.G., Stevens, R.C. (2002) The genesis of high-throughput structure-based drug discovery using protein crystallography, *Curr. Opin. Chem. Biol.* **6**, 704–710.

Lamzin, V.S., Perrakis, A. (2000) Current state of automated crystallographic data analysis, *Nat. Struct. Biol.* **7**(Suppl.), 978–981.

Li, X., Obradovic, Z., Brown, C.J., Garner, E.C., Dunker, A.K. (2000) Comparing predictors of disordered protein, *Genome Inf. Ser. Worksh. Genome Inf.* **11**, 172–184.

Li, X., Romero, P., Rani, M., Dunker, A.K., Obradovic, Z. (1999) Predicting protein disorder for N-, C-, and internal regions, *Genome Inf. Ser. Worksh. Genome Inf.* **10**, 30–40.

Linding, R., Russell, R.B., Neduva, V., Gibson, T.J. (2003) GlobPlot: exploring protein sequences for globularity and disorder, *Nucleic Acids Res.* **31**, 3701–3708.

Linding, R., Jensen, L.J., Diella, F., Bork, P., Gibson, T.J., Russell, R.B. (2003) Protein disorder prediction: implications for structural proteomics, *Structure* **11**, 1453–1459.

Linial, M., Yona, G. (2000) Methodologies for target selection in structural genomics, *Prog. Biophys. Mol. Biol.* **73**, 297–320.

Madej, T., Gibrat, J.F., Bryant, S.H. (1995) Threading a database of protein cores, *Proteins* **23**, 356–369.

Manjasetty, B.A., Delbruck, H., Pham, D.T., Mueller, U., Fieber-Erdmann, M., Scheich, C., Sievert, V., Bussow, K., Niesen, F.H., Weihofen, W., Loll, B., Saenger, W., Heinemann, U., Neisen, F.H. (2004) Crystal structure of Homo sapiens protein hp14.5, *Proteins* **54**, 797–800.

Montelione, G.T., Zheng, D., Huang, Y.J., Gunsalus, K.C., Szyperski, T. (2000) Protein NMR spectroscopy in structural genomics, *Nat. Struct. Biol.* **7**(Suppl.), 982–985.

Mueller, U., Bussow, K., Diehl, A., Bartl, F.J., Niesen, F.H., Nyarsik, L., Heinemann, U. (2003) Rapid purification and crystal structure analysis of a small protein carrying two terminal affinity tags, *J. Struct. Funct. Genomics* **4**, 217–225.

Navaza, J. (2001) Implementation of molecular replacement in AMoRe, *Acta Crystallogr.* **D57**, 1367–1372.

Netzer, W.J., Hartl, F.U. (1997) Recombination of protein domains facilitated by co-translational folding in eukaryotes, *Nature* **388**, 343–349.

Novotny, M., Madsen, D., Kleywegt, G.J. (2004) Evaluation of protein fold comparison servers, *Proteins* **54**, 260–270.

Oldfield, T.J. (2001) X-LIGAND: an application for the automated addition of flexible ligands into electron density, *Acta Crystallogr.* **D57**, 696–705.

Pandey, N., Ganapathi, M., Kumar, K., Dasgupta, D., Das Sutar, S.K., Dash, D. (2004) Comparative analysis of protein unfoldedness in human housekeeping and non-housekeeping proteins, *Bioinformatics* **20**, 2904–2910.

Pape, T., Schneider, T.R. (2004) HKL2MAP: a graphical user interface for phasing with SHELX programs, *J. Appl. Crystallogr.* **37**, 843–844.

Perrakis, A., Morris, R., Lamzin, V.S. (1999) Automated protein model building combined with iterative structure refinement, *Nat. Struct. Biol.* **6**, 458–463.

Potterton, E., Briggs, P., Turkenburg, M., Dodson, E. (2003) A graphical user interface to the CCP4 program suite, *Acta Crystallogr.* **D59**, 1131–1137.

Prinz, B., Schultchen, J., Rydzewski, R., Holz, C., Boettner, M., Stahl, U., Lang, C. (2004) Establishing a versatile fermentation and purification procedure for human proteins expressed in the yeasts Saccharomyces cerevisiae and Pichia pastoris for structural genomics, *J. Struct. Funct. Genomics* **5**, 29–44.

Read, R.J. (2001) Pushing the boundaries of molecular replacement with maximum likelihood, *Acta Crystallogr.* **D57**, 1373–1382.

Sali, A., Glaeser, R., Earnest, T., Baumeister, W. (2003) From words to literature in structural proteomics, *Nature* **422**, 216–225.

Sanchez, R., Pieper, U., Melo, F., Eswar, N., Marti-Renom, M.A., Madhusudhan, M.S., Mirkovic, N., Sali, A. (2000) Protein structure modeling for structural genomics, *Nat. Struct. Biol.* **7**(Suppl.), 986–990.

Sanishvili, R., Yakunin, A.F., Laskowski, R.A., Skarina, T., Evdokimova, E., Doherty-Kirby, A., Lajoie, G.A., Thornton, J.M., Arrowsmith, C.H., Savchenko, A., Joachimiak, A., Edwards, A.M. (2003) Integrating structure, bioinformatics, and enzymology to discover function: BioH, a new carboxylesterase from Escherichia coli, *J. Biol. Chem.* **278**, 26039–26045.

Sawasaki, T., Ogasawara, T., Morishita, R., Endo, Y. (2002) A cell-free protein synthesis system for high-throughput proteomics, *Proc. Natl. Acad. Sci. U.S.A.* **99**, 14652–14657.

Schneider, T.R., Sheldrick, G.M. (2002) Substructure solution with SHELXD, *Acta Crystallogr.* **D58**, 1772–1779.

Schwede, T., Kopp, J., Guex, N., Peitsch, M.C. (2003) SWISS-MODEL: an automated protein homology-modeling server, *Nucleic Acids Res.* **31**, 3381–3385.

Spriggs, R.V., Artymiuk, P.J., Willett, P. (2003) Searching for patterns of amino acids in 3D protein structures, *J. Chem. Inf. Comput. Sci.* **43**, 412–421.

Stevens, R.C., Yokoyama, S., Wilson, I.A. (2001) Global efforts in structural genomics, *Science* **294**, 89–92.

Sun, P.D., Radaev, S. (2002) Generating isomorphous heavy-atom derivatives by a quick-soak method. Part II: phasing of new structures, *Acta Crystallogr.* **D58**, 1099–1103.

Sun, P.D., Radaev, S., Kattah, M. (2002) Generating isomorphous heavy-atom derivatives by a quick-soak method. Part I: test cases, *Acta Crystallogr.* **D58**, 1092–1098.

Szyperski, T., Yeh, D.C., Sukumaran, D.K., Moseley, H.N., Montelione, G.T. (2002) Reduced-dimensionality NMR spectroscopy for high-throughput protein resonance assignment, *Proc. Natl. Acad. Sci. U.S.A.* **99**, 8009–8014.

Terwilliger, T.C. (2000) Maximum-likelihood density modification, *Acta Crystallogr.* **D56**, 965–972.

Terwilliger, T.C., Berendzen, J. (1999) Automated MAD and MIR structure solution, *Acta Crystallogr.* **D55**, 849–861.

Thornton, J.M., Todd, A.E., Milburn, D., Borkakoti, N., Orengo, C.A. (2000) From structure to function: approaches and limitations, *Nat. Struct. Biol.* **7**(Suppl.), 991–994.

Vagin, A., Teplyakov, A. (1997) MOLREP: an automated program for molecular replacement, *J. Appl. Crystallogr.* **30**, 1022–1025.

Voss, S., Skerra, A. (1997) Mutagenesis of a flexible loop in streptavidin leads to higher affinity for the Strep-tag II peptide and improved performance in recombinant protein purification, *Protein Eng.* **10**, 975–982.

Walter, T.S., Diprose, J., Brown, J., Pickford, M., Owens, R.J., Stuart, D.I., Harlos, K. (2003) A procedure for setting up high-throughput nanoliter crystallization experiments. I. Protocol design and validation, *J. Appl. Crystallogr.* **36**, 308–314.

Watanabe, N., Murai, H., Tanaka, I. (2002) Semi-automatic protein crystallization system that allows in situ observation of X-ray diffraction from crystals in the drop, *Acta Crystallogr.* **D58**, 1527–1530.

Watson, J.D., Todd, A.E., Bray, J., Laskowski, R.A., Edwards, A., Joachimiak, A., Orengo, C.A., Thornton, J.M. (2003) Target selection and determination of function in structural genomics, *IUBMB Life* **55**, 249–255.

Xu, H., Hauptman, H., Weeks, C.M. (2002) Sine-enhanced shake-and-bake: the theoretical basis for applications to Se-atom substructures, *Acta Crystallogr.* **D58**, 90–96.

Yee, A., Pardee, K., Christendat, D., Savchenko, A., Edwards, A.M., Arrowsmith, C.H. (2003) Structural proteomics: toward high-throughput structural biology as a tool in functional genomics, *Acc. Chem. Res.* **36**, 183–189.

Zheng, B., Roach, L.S., Ismagilov, R.F. (2003) Screening of protein crystallization conditions on a microfluidic chip using nanoliter-size droplets, *J. Am. Chem. Soc.* **125**, 11170–11171.

Zheng, B., Tice, J.D., Roach, L.S., Ismagilov, R.F. (2004) A droplet-based, composite PDMS/glass capillary microfluidic system for evaluating protein crystallization conditions by microbatch and vapor-diffusion methods with on-chip X-ray diffraction, *Angew. Chem., Int. Ed. Engl.* **43**, 2508–2511.

6
Protein Aggregation

Jeannine M. Yon
Université de Paris-sud, Orsay, France

1	Introduction 171	
2	**Protein Folding, Misfolding, and Aggregation** 171	
2.1	The New View of Protein Folding 171	
2.2	Detection of Aggregates during the Refolding Process 173	
2.2.1	Transient Aggregation 173	
2.2.2	Irreversible Aggregation 175	
2.3	Mechanisms of Protein Aggregation 176	
3	**Protein Folding in the Cellular Environment** 180	
3.1	Molecular Crowding in the Cells 180	
3.2	The Role of Molecular Chaperones 180	
4	**Protein Aggregation in the Cellular Environment** 186	
4.1	The Formation of Inclusion Bodies 186	
4.1.1	Occurrence of Inclusion Bodies 186	
4.1.2	Characteristics of Inclusion Bodies 187	
4.1.3	Strategies for Refolding Inclusion Body Proteins 189	
4.2	The Formation of Amyloid Fibrils and its Pathological Consequences 191	
	Bibliography 195	
	Books and Reviews 195	
	References of Primary Literature 195	

Proteins. Edited by Robert A. Meyers.
Copyright © 2007 Wiley-VCH Verlag GmbH & Co. KGaA, Weinheim
ISBN: 978-3-527-31608-3

Keywords

Aggregates
The association of nonnative protein molecules through intermolecular hydrophobic interactions.

Amyloid Fibrils
Ordered aggregates.

Inclusion Bodies
Insoluble, amorphous, disordered aggregates.

Molecular Chaperones
Proteins that assist protein folding within cells.

Protein Folding
The process by which polypeptide chains acquire their three-dimensional and functional structure.

Three dimensional–Domain Swapping
An aggregation mechanism in which one domain in a multidomain protein is swapped with the same domain of another molecule.

Protein misfolding and aggregation are frequent phenomena that occur under different conditions *in vivo* as well as *in vitro*. Aggregation is a serious problem affecting both the production of proteins in the biotechnology and pharmaceutical industries and human health. The aggregates are formed from nonnative proteins through intermolecular interactions that compete with intramolecular interactions. There is thus a kinetic competition between proper folding and misfolding, which can generate aggregates.

Recent evidence for transient association of intermediates during *in vitro* refolding has been obtained for several monomeric proteins. Irreversible and insoluble aggregates are formed in an off-pathway folding process; their formation is concentration dependent and could be prevented by using very small protein concentrations. These aggregates can dissociate and dissolve only in the presence of high concentrations of denaturant. The mechanisms involved in these aggregation processes will be discussed in light of the so-called *new view* of protein folding.

The environmental conditions within cells are markedly different from those used in *in vitro* refolding studies. In the production of recombinant proteins in foreign hosts, the formation of disordered aggregates, that is, inclusion bodies, is often observed. However, aggregation can also result in the formation of amyloid fibrils, which are ordered aggregates. These amyloid formations are at the origin of serious diseases.

1
Introduction

Protein misfolding and aggregation have been recognized for many years as common processes. Aggregation can occur under various conditions. The aggregation of which we are speaking is very different from the precipitation of a native protein at the isoelectric point or upon salting out, which can be reversed under appropriate conditions. In the precipitate, the protein remains in a native conformation. The aggregates, however, are formed from partially folded intermediates and result from intermolecular interactions, which compete with intramolecular interactions. Thermal denaturation of proteins is frequently accompanied by the formation of aggregates leading to the irreversibility of the process. As early as 1931, Wu, in a review on protein denaturation, distinguished between aggregation and precipitation. The aggregated species are not in equilibrium with the soluble species, complicating experimental approaches.

Aggregation has been reported to occur during the *in vitro* refolding of monomeric as well as oligomeric proteins, lowering the refolding yield. As mentioned above, the use of very low protein concentrations could prevent protein aggregation. However, during the folding of nascent polypeptide chains biosynthesized within prokaryotic and eukaryotic cells, aggregates can accumulate. The overexpression of genes in foreign hosts often result in aggregated nonnative proteins called *inclusion bodies*, which are disordered aggregates, leading to serious limitation in the production of recombinant proteins. It is a real problem needing a lot of effort to fully exploit the sequence information contained in the genome projects. Ordered aggregates resulting in amyloid fibrils lead to a number of serious human diseases such as Alzheimer's disease and the transmissible spongiform encephalopathies. The formation of amyloid aggregates has also been reported in *in vitro* experiments.

Experimental and theoretical studies together have provided significant insights into the mechanisms of protein folding, also allowing a better understanding of the aggregation processes.

The following different aspects of protein aggregation must be considered:

1. Theoretical and methodological aspects of protein folding, misfolding, and aggregation including the detection of aggregates and the mechanisms of aggregation processes.
2. Protein aggregation in the cellular environment including the folding into the cell, the role of molecular chaperones, and the formation of different aggregate morphologies, as well as the pathological consequences.

2
Protein Folding, Misfolding, and Aggregation

2.1
The New View of Protein Folding

The question of the mechanisms of protein folding has intrigued scientists for many decades. As early as the 1930s, attempts to refold denatured proteins were published, but significant progress began to be made when Anfinsen successfully refolded, denatured, and reduced ribonuclease into the fully active enzyme. In 1973, he stated the fundamental principle of protein folding referred to as the Anfinsen postulate: "all the information necessary to achieve the native conformation of a protein in a given environment is contained in its amino acid sequence." The thermodynamic control of protein folding was considered to be a corollary of the Anfinsen postulate, meaning that the native structure is at a minimum of the Gibbs free energy. This statement was discussed by Levinthal in a consideration of the short time required for the folding process *in vitro* as well as *in vivo*. It was concluded that a random search of the native conformation among all possible ones would require an astronomic time and is therefore unrealistic. Thus, it is clear that evolution has found an effective solution to this combinatorial problem. This is referred to as the Levinthal paradox and has dominated discussions for the last three decades.

In order to understand how the polypeptide chain could overcome the Levinthal paradox, different folding models were proposed and submitted to experimental tests. Kinetic studies were carried out to follow the folding pathway. A considerable number of experiments were performed to detect and characterize the folding intermediates. A stepwise sequential and hierarchical folding process in which several stretches of structure are formed and assembled at different levels following a unique route was supported by a majority of scientists for many years. According to this view, misfolded species could be formed from folding intermediates leading to the formation of aggregates in a kinetic competition with the correct folding.

Progressively, with the development of computers, theoretical studies have approached the folding problem, using simplified models to take into account the computational limitations in simulations of the folding from the random coil to the native structure. Different methods were developed using either lattice models or molecular dynamics simulations. In the lattice model, the polypeptide chain is represented as a string of beads on a two-dimensional square lattice or on a three-dimensional cubic lattice. The interactions between residues (the beads) provide the energy function for Monte Carlo simulations. In such simplified models, the essential features of proteins, that is, the heterogeneous character (hydrophobic or polar) of the interactions and the existence of long-range interactions, were included to explore the general characteristics of the possible folds. Lattice models were first applied to protein folding by Go and coworkers while simple exact models were initiated by Dill and his group, and have been used by several theoreticians. From the lattice simulations, insights into possible folding scenarios have been obtained, providing a basis for exploring the general characteristics of folding for real proteins. The exploration of such models supplies useful information that can be submitted to experimental tests.

The so-called "new view" has evolved during the past 10 years from both experiment and theory with the use of simplified models. It is illustrated by the metaphor of the folding funnel introduced in 1995 by Wolynes and coworkers. The model is represented in terms of an energy landscape and describes the thermodynamic and kinetic behavior of the transformation of an ensemble of unfolded molecules to a predominantly native state as illustrated in Fig. 1. According to this model, there are several micropathways, each individual polypeptide chain following its own route. Toward the bottom of the funnel, the number of protein conformations decreases as does the protein entropy. The steeper the slope, the faster the folding. As written by Wolynes et al., "To fold, a protein navigates with remarkable ease

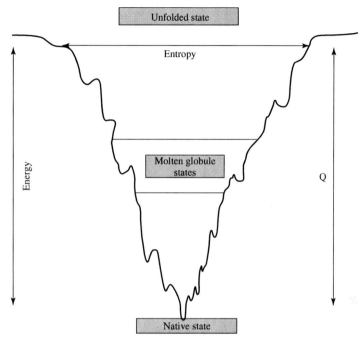

Fig. 1 Schematic representation of the folding funnel. Q is the number of native interactions.

through a complicated energy landscape." Thus, a wide variety of folding behaviors emerge from the energy landscape, depending on the energetic parameters and conditions. The folding rate could be slowed by ripples in the energy landscape corresponding to local minima populated by transiently stable intermediates. In a rugged energy landscape with kinetic traps formed by energy barriers, the folding will be even slower. When local energy barriers are high enough, protein molecules could be trapped and possibly aggregate.

The new view has progressively replaced the classical one of a unique sequential pathway and is now quite generally accepted. It is similar to the jigsaw puzzle model proposed in 1986 by Harrison and Durbin, suggesting the possibility of multiple folding routes to reach a unique solution. Many experimental results are consistent with this view. There is an increasing amount of evidence showing that the extended polypeptide chain folds through a heterogeneous population of partially folded intermediates in fluctuating equilibrium. Several alternative folding pathways have been observed for different proteins. From the convergence of theoretical and experimental studies, a unified view of the folding process has progressively emerged, also providing an explanation for the aggregation processes.

2.2
Detection of Aggregates during the Refolding Process

2.2.1 Transient Aggregation

Several observations indicate that transient aggregation could occur during *in vitro* protein refolding. Direct evidence for the transient association of intermediates has been obtained from small angle X-ray scattering, in the case of apomyoglobin by Doniach and his group, and in the case of carbonic anhydrase by Semisotnov and Kuwajima, and by Silow et al. During the refolding of phosphoglycerate kinase, rapidly transient multimeric species (dimers, trimers, and tetramers) yielding to the native monomeric protein have been detected by Pecorari et al. These species are not in equilibrium, but are formed rapidly and disappear in the slow folding step. Unlike classical aggregates, their distribution does not depend on protein concentration, and they are produced at concentrations as low as 0.05 µM. The distribution of the oligomers is completely established at the end of the fast refolding step. To take into account all these observations, a model, which is formally similar to a reaction of copolymerization between two types of monomers, has been proposed. In this model, the refolding of the protein produces two types of intermediate conformers that can associate with the same or the other type. In the latter case, the association cannot be extended further (Fig. 2). Transient multimeric species have also been observed during the refolding of the isolated N-terminal domain under conditions in which neither the whole native protein nor the folded isolated N-domain associate. However, they cannot transform to the native form in the absence of the interactions with the complementary

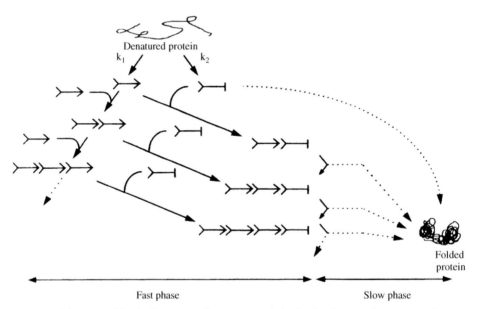

Fig. 2 Model proposed for the formation of transient multimeric species during the refolding of yeast phosphoglycerate kinase. Two types of conformers are produced in the early step of folding. One of these can be directionally extended by association with either the same conformer or another type of conformer. In this last case, the association cannot be further extended. The distribution of species results from a kinetic competition between two kinetic processes. (Reproduced from Pecorari, F., Minard, P., Desmadril, M., Yon, J.M. (1996) Occurrence of transient multimeric species during the refolding of a monomeric protein, *J. Biol. Chem.* **271**, 5270–5276.)

domain indicating the importance of long-range interactions in directing the correct folding. Such species have not been observed with the C-terminal fragment. Thus, the occurrence of transient multimeric species arising from partially folded intermediates through hydrophobic interactions does not prevent the correct folding of a monomeric protein.

2.2.2 Irreversible Aggregation

Thermal unfolding of proteins is frequently accompanied by the formation of aggregates and therefore behaves as an irreversible process. It occurs at temperatures that vary widely according to the protein, since the temperature of optimum stability depends on the balance between hydrogen bonds and hydrophobic interactions. Generally, the products of thermal denaturation are not completely unfolded and retain some structured regions. At the end of the thermal transition, the addition of a denaturant such as urea or GdnHCl frequently induces further unfolding.

An apparent irreversibility at a critical concentration of denaturant has been observed during the refolding of monomeric as well as oligomeric proteins. It was reported for the first time by M.Goldberg and coworkers for the refolding of β-galactosidase, and for tryptophanase. It was also observed for a two-domain protein, horse muscle phosphoglycerate kinase by Yon and coworkers. In the latter study, when the enzyme activity was used as a conformational probe of the native structure, an irreversibility was observed for a critical concentration of denaturant equal to $0.7\ M \pm 0.1\ M$ GdnHCl, a concentration very close to the end of the transition curve. Such irreversibility was found to be concentration dependent. For protein concentrations higher than 30 μM, restoration of enzyme activity was practically null.

The formation of irreversible nonnative species was found to be temperature dependent; it was practically abolished at 4 °C, suggesting that aggregation occurs through hydrophobic interactions. The aggregation also depends on the time of exposure of the protein to the denaturant. When the unfolding–refolding process was observed using structural signals such as fluorescence or circular dichroism, it appeared completely reversible whatever the final denaturant concentration.

Another example is provided by rhodanese, a two-domain monomeric protein. During refolding at low denaturant concentration, an intermediate accumulates with partially structured domains and apolar surfaces exposed to the solvent, leading to the formation of aggregates. The aggregation can be prevented by refolding the protein in the presence of lauryl maltoside.

Most of the examples discussed above are related to multidomain proteins. Another degree of complexity appears in the folding of oligomeric proteins. It is generally accepted that the early steps of the process are practically identical to the folding of monomeric proteins. In the last step, subunit association and subsequent conformational readjustments yield the native and functional oligomeric protein. The correct recognition of subunit interfaces is required to achieve the process. The overall process of the folding of oligomeric proteins was extensively studied by Jaenicke and his coworkers for several enzymes and described in reviews. As with monomeric proteins, the formation of aggregates is concentration dependent. The kinetics of aggregation are complex and multiphasic, indicating that several rate-limiting reactions are involved in the process. In an attempt to characterize these aggregates, it was shown that noncovalent interactions occur between monomeric species with partially restored secondary structures. The aggregates formed by either heat or pH denaturation can be disrupted in 6 M GdnHCl into monomeric unfolded species and then renatured under optimal conditions to yield an active enzyme. Only strong denaturants such as high concentrations of guanidine hydrochloride are efficient in this disruption process.

The presence of covalent cross-links such as disulfide bridges in a protein molecule can complicate the refolding of the denatured and reduced protein resulting in the formation of incorrect and intramolecular disulfide bridges leading to further aggregation. The first well-documented studies were performed by Anfinsen and his group on the refolding of reduced ribonuclease. The authors showed that the reoxidation of the enzyme produces a great number of species with incorrectly paired disulfide bonds. This scrambled ribonuclease is capable of regaining its native structure in a slow step, a process that is accelerated by the addition of a small quantity of reducing reagent such as β-mercaptoethanol yielding about 100% of active enzyme. The reshuffling of a protein's disulfide bonds takes place through a series of redox equilibria according to either an intramolecular or an intermolecular exchange. To prevent a wrong pairing of half-cystine and further aggregation, the addition of small amounts of reducing reagents or redox mixture is frequently used as investigated by Wetlaufer.

The detection and characterization of aggregates represent an important aspect of folding studies. The aggregation phenomenon can occur without precipitation. Indeed, the degree of association of protein intermediates during folding might be small, depending on the intermolecular interactions, and does not necessarily lead to a visible insolubility. The association state may be determined in several ways. The most common methods, available in any biochemistry laboratory, are gel permeation and sodium dodecyl sulfate polyacrylamide gel electrophoresis (SDS-PAGE), used both with and without cross-linking. The detection of aggregates can also be monitored by other hydrodynamic methods such as analytical ultracentrifugation or classical light scattering. The latter method also gives information on the size of the aggregates. Quasi-elastic light scattering is a dynamic technique that can be used to determine macromolecule diffusion coefficients as a function of time, that is, to follow the kinetics of aggregation. Neutron scattering can also be used to detect protein aggregates, and mass spectrometry has become a useful tool as well.

2.3
Mechanisms of Protein Aggregation

A substantial body of information supports the idea that protein aggregation arises from partially folded intermediates through hydrophobic interactions. The formation of aggregates has often been considered as a trivial phenomenon, a nonspecific association of partially folded polypeptide chains to form a disordered precipitate. However, several analyses indicate that aggregation occurs by specific intramolecular associations involving the recognition of a sequence partner in another molecule rather than in the same molecule during the folding process. Analyses of the aggregation mechanisms of various proteins, such as bovine growth hormone and phosphoglycerate kinase, has permitted the identification of specific sites that are critical in the association.

An elegant demonstration of the specificity of aggregation was provided by King and coworkers. The authors showed that during the *in vitro* refolding of a mixture of two proteins, tailspike endorhamnidase and coat protein from phage P22, no heterogeneous aggregates were formed. Tailspike endorhamnidase is a thermostable trimer whose folding intermediates are thermolabile and either undergo productive folding or form multimeric aggregates (Fig. 3). The P22 coat protein, which comprises the capsid shell of phage P22, yields either a correct fold or "off-pathway" aggregates upon refolding. Both proteins were intensively studied by King and coworkers who first denatured the two proteins in urea and then chose refolding conditions such that aggregation competes with correct folding. Folding and soluble aggregates of the two proteins were characterized either separately or mixed together. No heterogeneous aggregates were found, clearly indicating that only self-association of transient refolding molecules occurs in the formation of soluble multimers.

One mechanism that accounts for the formation of aggregates during refolding of multidomain proteins is domain swapping. This was first suggested by Monod and later proposed by Goldberg and colleagues to account for the formation of aggregates during the refolding of tryptophanase. The concept was foreshadowed by the results of Crestfield and coworkers in 1962. From their experiments based on chemical modification of bovine pancreatic ribonuclease, the authors proposed that the dimer is formed by exchanging the N-terminal fragments. The term *3D domain swapping* was introduced in 1994 by Bennett and coworkers to describe the structure of a diphtheria toxin dimer. The mechanism involves the replacement of one domain of a monomeric protein by the same domain of an identical neighboring

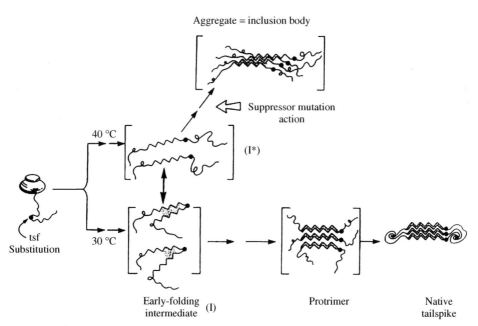

Fig. 3 The folding pathway of the P22 tailspike protein. (From Mitraki, A., King, J. (1992) *FEBS Lett.* **307**, 20–25; reproduced with permission.)

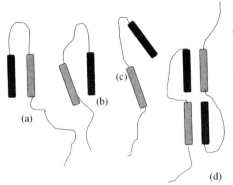

Fig. 4 Schematic representation of domain swapping. (a) monomeric protein, (b) and (c) partially unfolded monomers, (d) domain-swapped dimer.

molecule, thus resulting in an intertwined dimer or oligomer, as defined by Eisenberg and colleagues (Fig. 4). When the exchange is reciprocated, domain-swapped dimers are formed. However, if the exchange is not reciprocated but propagated along multiple polypeptide chains, higher order assemblies or aggregates may form. Domain-swapped oligomers are divided into two types, open and closed. The open oligomers are linear and have one closed interface (closed in the monomer) exposed to the solvent, whereas closed oligomers are cyclic and do not expose a closed interface. Eisenberg and coworkers have defined the structure of the monomer as the "closed monomer" and the conformation of the polypeptide chain in the domain-swapped oligomer as the "open monomer."

The ability of monomeric proteins to swap structural elements requires the presence of a hinge or linker region that permits the protein to attain the native fold with parts of two polypeptide chains. In fact, domain-swapped structures reveal regions of protein structure that are flexible. Bergdoll and coworkers have suggested that a proline in the linker region, by rigidifying the hinge region in intermediate states, might facilitate domain swapping. Baker and colleagues proposed that strain in a hairpin loop might predispose a protein to domain swapping. The possible role of 3D domain swapping in the evolution of oligomeric proteins has been discussed in several reviews. In the past years, the number of known domain-swapped proteins has increased and today about 40 such structures are solved. One common feature of these proteins is that all the swapped domains are from either the N-terminus or the C-terminus of the polypeptide chain. In this regard, an interesting example arises from the work of Eisenberg and his group on the dimerization of ribonuclease A. This protein forms two types of dimers upon concentration in mild acid. The minor dimer is formed by swapping of its N-terminal α-helix with that of an identical molecule. The major dimer results from the swapping of its C-terminal β-strand. RNase A was also reported to form trimers. On the basis of the structure of the N- and C-terminal swapped dimers, a model was proposed (Fig. 5). This indicates that two types of swapping can occur simultaneously in the same oligomer. Further biochemical studies have supported this model. A less abundant trimer in which only the C-terminal β-strand is swapped and exhibits a cyclic structure was also found. RNase represents the

Fig. 5 Domain swapping in ribonuclease. Ribbon diagram of the structures of (a) the ribonuclease A monomer (2.0 Å), (b) the N-terminal swapped dimer (2.1 Å), (c) the C-terminal swapped dimer (1.75 Å), (d) the N- and C-terminal trimer model, and (e) the cyclic C-terminal swapped trimer (2.2 Å) (reproduced from Liu et al. Prot. Sci. **11**, 371, 2002 with permission).

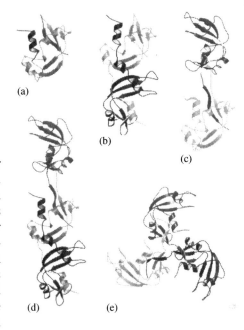

first protein found to form both linear and cyclic domain-swapped oligomers. This protein also was described to form tetramers. Models based on the structures of dimers and trimers were proposed for these tetramers. Two linear models exhibit both types of swapping that occur in one molecule, and a cyclic tetramer shows the swapping of the C-terminal β-strand only. A trimeric domain-swapped barnase was obtained at low pH and high protein concentration. Crystallographic studies revealed a structure suggesting a probable folding intermediate. Domain swapping was described for the cell cycle regulatory protein p13suc1, a small protein of 113 amino acids.

Folding studies as well as molecular dynamics simulations have shown that domain swapping occurs in the unfolded state. Eisenberg and his colleagues have proposed a free energy diagram for the pathway of domain swapping. The free energy difference between the closed monomer and domain-swapped oligomer is small since they share the same structures except at the hinge loop, but the energy barrier can be reduced under certain conditions making domain swapping more favorable. Several molecular or environmental events may favor the formation of extended domain-swapped polymers. Genetic mutations introducing a deletion in the hinge loop can destabilize the monomeric form of a protein. The replacement of only one amino acid can also favor the polymerization of the mutated protein. Three-dimensional domain-swapped oligomers are expected to be increasingly favored as the protein concentration increases. Thus, a metabolic change that increases the concentration of a protein will favor aggregation. Charge effects, caused either by mutations or by pH change or salt concentration, can induce domain swapping; for example, in RNase A, a decrease in pH, by protonating the residues involved in hydrogen bonds and in salt bridges, lowers the energy barrier of the formation of the open monomer, hence inducing domain swapping.

There is great diversity of swapped domains, with different sizes and sequences. They can consist of entire tertiary domains or smaller structural elements made of several residues. No specific sequence motif seems to be involved among the swapped domains. Three-dimensional domain swapping has also been proposed as a mechanism for amyloid formation. This aspect will be discussed in Sect. 4.2.

As can be seen here, several mechanisms exist, which lead to the formation of aggregates. It is recognized that aggregation results from the association of incompletely or incorrectly folded intermediates through hydrophobic interactions. In the energy landscape of protein folding, the presence of local minima separated by an energy barrier allows the accumulation of intermediates. If the barrier is high enough, these intermediates cannot easily reach the native state, and kinetic competition thus favors the formation of aggregates.

3
Protein Folding in the Cellular Environment

3.1
Molecular Crowding in the Cells

The main rules of protein folding have been deduced from a considerable body of *in vitro* and *in silico* studies. It has been accepted that the same mechanisms are involved in *in vitro* refolding and in the folding of a nascent polypeptide chain in the cell. However, the intracellular environment differs markedly from that of the test tube where low protein concentrations are used. The interior of a cell is highly crowded with macromolecules. The concentration is so high that a significant proportion of the volume is occupied. As mentioned by Ellis, in general, 20 to 30% volume of the interior of the cells are occupied by macromolecules; for example, the concentration of total protein inside cells ranges from 200 to 300 g L^{-1}. The total concentration of proteins and RNA inside *Escherichia coli* ranges from 300 to 400 g L^{-1} depending on the growth phase. Polysaccharides also contribute to the crowding. It can be predicted practically that diffusion coefficients will be reduced by factors up to 10-fold due to crowding. Since the average time for a molecule to move a certain distance varies by D^{-2}, D being the diffusion coefficient, it will take 100 times longer to move this distance in the cell as would be necessary under low concentration conditions. Another prediction indicates that equilibrium constants for macromolecular associations may be increased by two to three orders of magnitude.

Molecular crowding inside cells also has consequences for protein folding, favoring the association of partly folded polypeptide chains into aggregates. This could explain why cells contain molecular chaperones, even though most denatured proteins refold spontaneously in the test tube.

3.2
The Role of Molecular Chaperones

The discovery of a ubiquitous class of proteins mediating the correct folding in cellular environment has led to a reconsideration of the mechanism of protein folding *in vivo*. Historically, the term *molecular chaperone* was introduced by Laskyard and coworkers in 1987 to describe the function of nucleoplasmin, which mediates the *in vitro* assembly of nucleosomes from separated histones and DNA. The concept was further extended by Ellis to define a class of proteins whose function is to ensure the correct folding and assembly of proteins through a transient association with the nascent polypeptide chain. Studies on heat-shock proteins have widely contributed to the development of this concept.

Today, more than 20 protein families have been identified as molecular chaperones. Molecular chaperones comprise several highly conserved families of related proteins. They can be divided into two classes according to their size. Small chaperones are less than 200 kDa, whereas large chaperones are more than 800 kDa. During the past few years, a large amount of biochemical, biophysical, and low- and high-resolution structural data have provided mechanistic insights into the machinery of protein folding as assisted by molecular chaperones.

Molecular chaperones are involved in diverse cellular functions. The constitutive members of the heat-shock protein family (Hsp70) can stabilize nascent polypeptide chains during their elongation in ribosomes. The large cylindrical chaperonins GroEL in bacteria, mitochondria, and chloroplasts and the corresponding TriC in eukaryotes and archaebacteria provide a sequestered environment for productive folding. Several chaperones are stress-dependent; their expression is induced under conditions such as high temperatures, which provoke protein unfolding and aggregation. The members of the Hsp90 and Hsp100 families, as well as small Hsp, play a role in preventing protein aggregation under stress. Chaperone interactions are also important for the translocation of polypeptide chains into membranes.

Within cells, the nascent polypeptide chain is synthesized sequentially on the ribosome by a vectorial process. For many proteins, the rate of this process is slower than the rate of folding. Synthesis times range from 20 s for a 400 residue–polypeptide chain in E. coli at 37 °C to 10 times as long for such a chain in an eukaryotic cell. Many unfolded proteins refold completely in 20 s under the same conditions. Thus, there is the possibility for the elongating polypeptide either to misfold before completion or to be degraded by proteolytic enzymes. Chaperones prevent such unfavorable events by protecting the nascent chain. Hsp70 and its prokaryotic homolog DnaK recognize extended hydrophobic regions of the elongating polypeptides. These interactions are not specific. Hsp70 and DnaK interact with most unfolded polypeptide chains that expose hydrophobic residues. They do not recognize folded proteins. Binding and release of unfolded proteins from Hsp70 are ATP-dependent and require the presence of various cochaperones such as DnaJ and GrpE. The basic mechanism of Hsp70 (DnaK in E. coli) is represented in Fig. 6. In E. coli, DnaJ binds the nascent unfolded polypeptide, U; then the complex binds to the ATP-bound state of DnaK. ATP is hydrolyzed in the ternary complex

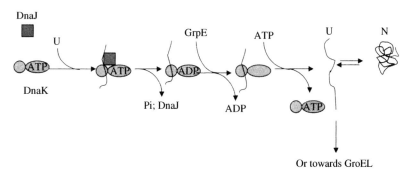

Fig. 6 Schematic representation of the basic mechanism of DnaK (see text).

allowing the release of DnaJ and P_i. In the following step, GrpE acts as an exchange factor to regenerate the ATP-bound state of DnaK. The unfolded polypeptide chain is released into the bulk solution. Thus, Hsp70 systems bind and release the polypeptide in an unfolded conformation. The unfolded protein has the possibility either to fold or to be transferred to the GroEL system, as illustrated in Fig. 6. Significant insights into this mechanism were obtained from structural data. The three-dimensional structures of Hsp70 and DnaJ as well as those of a complex DnaK–polypeptide and a complex of GrpE with the ATP binding domain of DnaK are known. DnaK and its homologs are composed of two domains, a C-terminal domain that binds ATP and an N-terminal domain that binds peptides. GrpE is a tight homodimer associated along two long helices. It binds DnaK–ATPase domain through its proximal monomer. DnaJ activates the ATP hydrolysis by DnaK. It was shown that a conformational change may occur upon ATP binding, opening the polypeptide binding cleft in the polypeptide binding domain of DnaK. The closed state may correspond to the ADP-bound conformation. The ADP-bound state of DnaK binds the peptide tightly. Peptide release requires the dissociation of ADP, which is mediated by GrpE. DnaK then rebinds ATP.

The GroEL–GroES system acts by a different mechanism in which the unfolded protein is sequestered. The chaperonins are large cylindrical protein complexes. The crystal structure of *E. coli* chaperonin GroEL was determined in 1994 and that of the asymmetric GroEL–GroES–(ADP)7 complex in 1997 by Sigler and his group. GroEL consists of two heptameric rings of 58-kDa subunits stacked back to back with a dyad symmetry and forming a porous cylinder (Fig. 7). Each subunit is organized in three structural domains. A large equatorial domain forms the foundation of the assembly and holds the rings together. It contains the nucleotide binding site. A large apical domain forms the end of the cylinder. The apical domain contains a number of hydrophobic

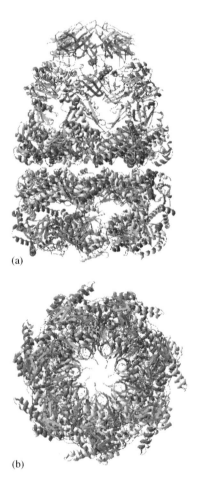

(a)

(b)

Fig. 7 Crystal structure of GroEL–GroES–(ADP)7 complex determined by Sigler et al. (a) view along the axis and (b) view from the top of the complex. (Reproduced from the PDB web site.)

residues exposed to the solvent. A small intermediate domain connects the two large domains. The intermediate segments have some flexibility allowing a hinge-like opening of the apical domains, which occurs upon nucleotide binding. These movements are large and have been visualized by three-dimensional reconstruction from cryoelectron microscopy by Sebil and her group.

GroES is a heptamer of 10 kDa subunits forming a flexible dome-shaped structure with an internal cavity large enough to accommodate proteins up to 70 kDa. Each subunit is folded into a single domain containing β-sheets and flexible loop regions. The loop regions are critical for the interactions between GroEL and GroES. It was deduced from electron microscopy studies that GroES binding to GroEL induces large movements in the apical GroEL domains. This provokes a significant increase in the volume of the central cavity in which protein folding proceeds. NMR coupled with the study of hydrogen-exchange techniques has indicated that small proteins are essentially unfolded in their GroEL-bound states. Mass spectroscopy has revealed the presence of fluctuating elements of secondary structure for several proteins. In a way, the GroEL–GroES system recognizes nonnative proteins.

The reaction cycle of the GroEL–GroES system is represented in Fig. 8. The nonnative protein binds to the apical domain of the upper ring of GroEL through hydrophobic interactions. Then, the equatorial domain of the same ring binds ATP, and GroES caps the upper ring, sequestering the protein inside the internal chamber in which the protein folding proceeds. The binding of GroES induces a conformational change in GroEL and ATP hydrolysis, which is a cooperative process that produces a conformational change in the lower ring, allowing it to bind a nonnative protein molecule. This promotes subsequent binding of ATP and GroES in the lower ring, and the dissociation of the upper complex, releasing the protein and ejecting GroES. If the protein has not reached the native state, it is subjected to a new cycle.

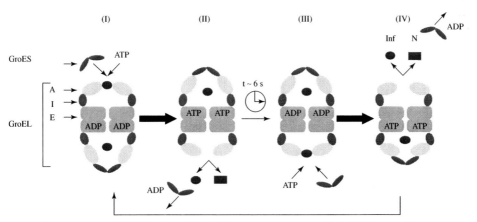

Fig. 8 The reaction cycle of GroEL–GroES. Inf is the unfolded protein, N the folded one, A is the apical domain, (in blue), I the intermediate domain (in red) and E the equatorial domain (in magenta). (Reproduced from Wang & Weissman (1999) Nat. Struct. Biol. **6**, 597, with permission.) (See color plate p. x).

The hydrolysis of ATP by GroEL is used only to induce conformational changes of the chaperone, which permits the release of the folded protein. The molecular chaperones, by their transient association through hydrophobic interactions with nascent, stress-destabilized, or translocated proteins, have a role in preventing improper folding and subsequent aggregation. They do not interact with folded proteins. They do not carry information capable of directing a protein to assume a structure different from that dictated by its amino acid sequence. Therefore, molecular chaperones assist the folding in the cells without violation of the Anfinsen postulate. They increase the yield but not the rate of folding reactions; in this respect they do not act as catalysts. Furthermore, the majority of newly synthesized polypeptide chains in both bacterial and eukaryotic cells fold spontaneously without the assistance of molecular chaperones.

Many proteins from prokaryotic and eukaryotic organisms are produced with an amino-terminal propeptide, which is removed by limited proteolysis during the activation process. Several of these propeptides consist of a long polypeptide chain; for example, there are 174 amino acids in the propeptide of pro-α-lytic protease, 91 in that of procarboxypeptidase Y, and 77 in that of prosubtilisin. Several studies have shown that the propeptide is required for proper folding of these proteins. The mature enzymes are not able to refold correctly. They seem to have kinetic stability only, whereas the proenzymes have thermodynamic stability. Since propeptides perform the function of mediating protein folding, they have been classified as intramolecular chaperones. However, this terminology is not appropriate since the nascent protein is the proenzyme, not the enzyme that has undergone proteolytic cleavage. Thus, it is not surprising that the proenzyme refolds spontaneously, whereas the mature protein does not. Indeed, the information is contained in the totality of the proenzyme sequence.

Two other classes of proteins play the role of helpers during protein folding *in vivo*: protein disulfide isomerases (PDIs) and peptidyl–prolyl cis–trans isomerases. Protein disulfide isomerase is an abundant component of the lumen of the endoplasmic reticulum in secretory cells. The enzyme was discovered independently in 1963 by two research groups: in rat and ox by Anfinsen and coworkers, and in chicken and pigeon pancreas by Straub and coworkers. Proteins destined to be secreted enter the endoplasmic reticulum in an unfolded state. In this environment, the folding process is associated with the formation of disulfide bonds, which is catalyzed by PDI through thiol–disulfide interchange. The first PDI cDNA was sequenced in 1985 by Edman et al. It displays sequence homologies implying a multidomain architecture. PDI consists of four structural domains arranged in the order a, b, b', a', with the b' and a' domains being connected by a linker region. Furthermore, it possesses an acidic C-terminal extension. The a and a' domains contain the active site motif – W-C-G-H-C-. They display significant sequence identity to thioredoxin, a small cytoplasmic protein involved in several redox functions, and they have a similar active site sequence.

Recombinants of the a and b domains have been obtained and studied by high-resolution NMR. The a domain has the same overall fold as thioredoxin, an α/β fold with a central core made up of a five-stranded β-sheet surrounded by four helices. As in thioredoxin, the active site is located at the N-terminus of helix

α_2. Preliminary NMR data of the a' domain confirm its structural similarity to the a domain. The b and b' domains have significant sequence similarity to each other, but no similarity with the a domain. Nevertheless, NMR studies of the b domain have indicated a similar overall fold. From its sequence, it could be inferred that b' also has the same fold. Neither b nor b' contain the active site.

The folding pathway of disulfide-bound proteins involves isomerizations between a number of species containing disulfide bonds. *In vitro* experimental studies were performed using the isolated a and a' domains, and the results were compared with those obtained with the holoenzyme. It was concluded that the activity of long length PDI is not simply the sum of the activities of the isolated a and a' domains. Using a series of constructs including nearly every linear combination of domains, the contribution of each domain was investigated. It was determined that the thiol-disulfide chemistry requires only the a and a' domains, and that simple isomerization requires one of these in a linear combination including b', whereas complex isomerization involving large conformational changes requires all the PDI domains except the C-terminal extension. Thus, it appears that the b' domain is the principal peptide binding site, but all domains contribute to the binding of larger polypeptide chains holding them in a partially unfolded conformation while the catalytic sites acts synergistically to perform the thiol-disulfide exchange. Since PDI has binding properties, it has been proposed that it acts as a molecular chaperone. However, as underlined by Freedman and coworkers, this property does not represent a chaperone activity and instead reflects its role as a catalyst to accelerate the formation of native disulfide bridges during protein folding.

Several gene products with similarity to PDI have been identified in higher eukaryotes. All are probably localized in the endoplasmic reticulum and have thiol-disulfide exchange activity.

In prokaryotes, the disulfide formation occurs in the periplasm and is catalyzed by a protein called DsbA, which exchanges its Cys30–Cys33 to a pair of thiols in the target protein, leaving DsbA in its reduced state. The crystal structure of oxidized DsbA displays a domain with a thioredoxin-like fold and another domain, which caps the thioredoxin-like active site C30-P31-H32-C33, located at the domain interface. Reoxidation of DsbA is catalyzed by a cytoplasmic membrane protein called DsbB, which contains four cysteine residues essential for catalysis. DsbB transfers the electrons from the reduced DsbA to membrane embedded quinones. The reduced quinones are then oxidized enzymatically either aerobically or anaerobically. Thus, DsbA is found in normal cells in its oxidized state. *E. coli* also has a complex reductive system including another periplasmic protein DsbC, which is a homodimer. The molecule consists of two thioredoxin-like domains with a CxxC motif, joined via hinged linker helices to an N-terminal dimerization domain. The hinge regions allow movement of the active site, and a broad hydrophobic cleft between the two domains may bind the polypeptide chain. Its function consists of reducing proteins with incorrect disulfide bonds. DsbC is maintained in its reduced form by a membrane protein called DsbD, which contains six essential cysteine residues. Then, the electrons are transferred to thioredoxin and ultimately to NADPH by thioredoxin reductase.

All these enzymes, which catalyze the pairing of cysteine residues in disulfide-bridged proteins, have functional domains pertaining to the thioredoxin superstructure.

Another type of enzyme, peptidyl–prolyl cis–trans isomerases, facilitates the folding of some proteins by catalyzing the cis–trans isomerization of X-Pro peptide bonds. Two classes of unrelated proteins demonstrate this activity, those that bind cyclosporin, which are known as *cyclophilins*, and those that bind FK506. The cellular function of these enzymes is important, since cyclosporin and FK506 are potent immunosuppressors that regulate T-cells activation. Both classes of peptidyl–prolyl isomerases are ubiquitous, and abundant in prokaryotes and eukaryotes. The sequences of several members of each family are known, and the three-dimensional structures of at least one member of each family have been elucidated by X-ray crystallography and multidimensional NMR. Their role is to accelerate the cis-trans isomerization of X-pro peptide bonds when this process is the rate-limiting step in protein folding. Although they do not present structural similarity, both exhibit a hydrophobic binding cleft favoring the rotamase activity by excluding water molecules.

4
Protein Aggregation in the Cellular Environment

4.1
The Formation of Inclusion Bodies

The overexpression of genes introduced in foreign hosts frequently results in aggregated nonnative proteins called *inclusion bodies*. In cells, inclusion bodies appear as unordered amorphous aggregates clearly separated from the rest of the cytoplasm; they form a highly refractive area when observed microscopically. A great variety of experimental studies indicates that the formation of inclusion bodies results from partially folded intermediates in the intracellular folding pathway and not from either totally unfolded or native proteins.

4.1.1 Occurrence of Inclusion Bodies

Inclusion bodies were first identified in the blood cells of patients with abnormal hemoglobins, the resulting pathology being anemia. Pathological point mutants of hemoglobin aggregate into inclusion bodies; this is the case for hemoglobin Köln (Val98Met on the β chain) and hemoglobin Sabine (Leu91Pro on the β chain). Similar deposits have been described in studies on the metabolism of abnormal proteins subjected to covalent modification in *E. coli*. The formation of aggregates also occurs when cells are subjected to heat shock.

The *in vivo* folding pathway of tailspike endorhamnosidase of Salmonella phage 22 is a well-documented system studied by J.King's group. Furthermore, it is one of the few systems in which the *in vivo* folding pathway has been compared with the *in vitro* refolding pathway. The protein is a trimer of 666 amino acids. The secondary structure is predominantly β-sheet. Newly synthesized polypeptide chains released from the ribosome generate an early partially folded intermediate. This intermediate further evolves into a species sufficiently structured for chain–chain recognition. In the following step, an incompletely folded trimer is formed upon close association with the latter species. The protrimer is then transformed into the native tailspike. A clear difference between the physicochemical properties of the intermediates and the native state has allowed their identification. Figure 3 illustrates the folding pathway of the protein. The native protein is highly thermostable with a T_m of 88 °C; it is also resistant to detergents and proteases. During the *in vivo* folding process, the intermediates are sensitive to these factors, allowing their identification. At low temperature, almost 100% of the newly synthesized chains reach the native trimer conformation. When the temperature increases in the cells, the number of polypeptide chains achieving the native state decreases. At 39 °C, the maturation proceeds with 30% efficiency, while the remainder aggregates into inclusion bodies. It has been shown that the aggregation does not result from an intracellular denaturation of the native protein, but is generated from an early thermolabile intermediate. The aggregated chains cannot recover their proper folding by lowering the temperature. But when polypeptide chains that have been synthesized at high temperatures are shifted to low temperature early enough, they can refold correctly.

A set of mutations that alter protein folding without modifying the properties and stability of native P22 tailspike has been identified; they are referred to as temperature-sensitive folding (tsf) mutants. These mutations have been supposed to destabilize the already thermolabile intermediate and are located at more than 30 sites in the central region of the polypeptide chain. Starting from mutants kinetically blocked in their folding, a second set of mutants capable of correcting the folding defects was selected, and the sequences surrounding the suppressor mutations were identified. Only two substitution positions on the 666 amino acids of the polypeptide chain were sufficient to prevent inclusion body formation. Thus, single temperature mutations that affect the folding pathway but not the native conformation of a protein are efficient in preventing off-pathway and subsequent aggregation. A similar result has been found for heterodimeric luciferase. For recombinant proteins such as interferon-γ and interleukin 1β, as well as for P22 tailspike, amino acid substitutions that can decrease or increase the formation of inclusion bodies without alteration of the functional structure were found by Wetzel and coworkers.

The formation of inclusion bodies is frequently observed in the production of recombinant proteins. High levels of expression of these proteins result in the formation of inactive amorphous aggregates, and has been reported for proteins expressed in *E. coli* and also in several host cells, gram-negative as well as gram-positive bacteria, and eukaryotic cells such as *Saccharomyces cerevisiae*, insect cells, and even animal cells. The production of recombinant proteins, among them human insulin, interferon-γ, interleukin 1β, β-lactamase, prochymosin, tissue plasminogen activator, basic fibroblast growth hormone, and somatotropin, gives rise to inclusion bodies.

4.1.2 Characteristics of Inclusion Bodies

Inclusion bodies can form in the cytoplasm and in the periplasmic space of *E. coli*. Wild-type β-lactamase expressed in *E. coli* results in the formation of inclusion bodies in the periplasm, whereas the protein expressed without its signal sequence aggregates in the cytoplasm.

The characteristics of the aggregates depend on how the protein is expressed. Different sizes and morphologies have been observed. Generally, inclusion bodies appear as dense isomorphous aggregates of nonnative proteins separated from the rest of the cytoplasm, but not surrounded

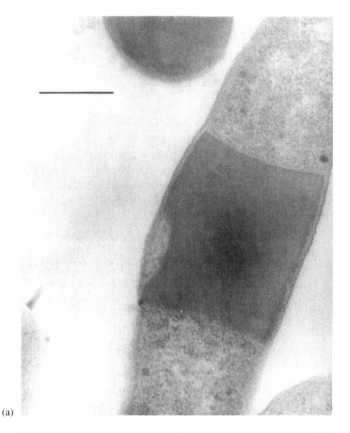

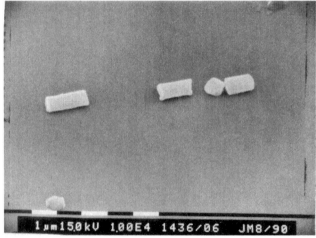

Fig. 9 Electron micrographs of (a) cytoplasmic β-lactamase inclusion bodies in E. coli RB791(pGB1) and (b) purified inclusion bodies from the same origin (courtesy of G.A. Bowden, A.M. Paredes & G. Georgiou).

by a membrane (Fig. 9). They look like refractile inclusions, which can be easily recognized by phase contrast microscopy when large enough. For prochymosin expressed in *E. coli*, the lack of birefringence indicates that inclusion bodies are not crystalline. The size distribution of inclusion bodies has been studied for prochymosin and interferon-γ, and Marston reported the mean size of particles to be 0.81 and 1.28 µm respectively, with a relatively high void fraction. The void volume was about 70% of the total volume for interferon-γ and 85% for prochymosin. Structural characterization studies using ATR-FTIR (attenuated total reflectance Fourier transformed infrared spectroscopy) have shown that the insoluble nature of inclusion bodies may be due to their increased levels of nonnative intramolecular β-sheet content.

Inclusion bodies consist mostly of the overexpressed recombinant protein, and can contain little contaminating molecules. Thus, they can be used as a source of relatively pure misfolded protein when refolding yields the active protein. However, some amorphous bodies incorporate other molecules, for example, inclusion bodies from *E. coli* cells overexpressing β-lactamase contain only between 35 and 95% intact β-lactamase. The rest consists of a variety of intracellular proteins, some lipids, and a small amount of nucleic acids. Homogeneous inclusion bodies were obtained by expressing β-lactamase without its leader peptide. Under these conditions, aggregation occurs within the cytoplasm. The extent of incorporation of other macromolecules in inclusion bodies depends upon the overexpressed protein.

The formation of inclusion bodies generally appears to be a disadvantage, since it requires the dissolving of the aggregates in denaturant and subsequent refolding of the protein. However, when the recovery of the active product can be obtained with a sufficient yield, certain advantages may accrue. Indeed, aggregation generally prevents proteolytic attack, except when the protein coaggregates with a protease. The formation of inclusion bodies is also an advantage for the production of proteins that are toxic for the host cells. Furthermore, these aggregates contain a great quantity of the overexpressed protein.

4.1.3 Strategies for Refolding Inclusion Body Proteins

The recovery of the active protein from inclusion bodies is crucial for industrial purposes. In structural proteomics today, efficient production of genetically engineered proteins is a prerequisite for exploiting the information contained in the genome sequences. The strategy to recover active proteins involves several steps of purification. The first step, the separation of the inclusion bodies from the cell, consists of cell lysis monitored either by high-pressure homogeneization, or by a combination of mechanical, chemical, and enzymatic techniques such as the use of EDTA and lysozyme. The lysates are then treated by low-speed centrifugation or filtration to remove the soluble fraction from the pellet containing inclusion bodies and cell debris. The most difficult task is to remove the contaminants; this is achieved by the washing steps, which commonly utilize EDTA and low concentrations of denaturants or detergents such as Triton X-100, deoxycholate, or octylglucoside. Using centrifugation in a sucrose gradient, it is generally possible to remove cell debris and membrane proteins. When the accumulation levels of aggregates are very high, inclusion bodies may be directly solubilized by treatment in a high concentration of denaturant, eliminating the need for gradient centrifugation. In this case, the costs of production are considerably reduced.

A variety of techniques are available to solubilize purified inclusion bodies. The most commonly used solubilizing reagents are strong denaturants such as guanidine hydrochloride and urea. Generally, high denaturant concentrations are employed, 4 to 6 M for guanidine hydrochloride, and 5 to 10 M for urea to allow the disruption of noncovalent intermolecular interactions. Conditions may differ somewhat according to the denaturant and the protein. Lower denaturant concentrations have been used to solubilize cytokines from E. coli inclusion bodies. The purity of the solubilized protein was much higher at 1.5 to 2 M guanidinium chloride than at 4 to 6 M guanidinium chloride. At higher denaturant concentrations, contaminating proteins were also released from the particulate fractions.

Extremes of pH have also been used to solubilize inclusion bodies and for growth hormone, proinsulin, and some antifungal recombinant peptides. However, exposure to very low or very high pH may not be applicable to many proteins and may cause irreversible chemical modifications.

Detergents such as sodium dodecylsulfate (SDS) and n-cetyl trimethylammonium bromide (CTAB), have also been used to solubilize inclusion bodies. Extensive washing may then be needed to remove the solubilizing detergents. They also may be extracted from the refolding mixture by using cyclodextrins, linear dextrins, or cycloamylose. Recent developments include the use of high hydrostatic pressure (1–2 kbar) for solubilization and renaturation. For proteins with disulfide bonds, the addition of a reducing reagent such as dithiothreitol or β-mercaptoethanol is necessary to disrupt the incorrectly paired disulfide bonds. The concentrations generally used are 0.1 M for dithiothreitol and 0.1 to 0.3 M for β-mercaptoethanol.

When expression levels are very high, an *in situ* solubilization method can be used. It consists of adding the solubilizing reagent directly to the cells at the end of the fermentation process. The main disadvantage of this technique concerns the release of contaminants.

The last step is the recovery of the active protein. When inclusion bodies have been solubilized, the refolding is achieved by removal of the denaturant. This can be done by different techniques including dilution, dialysis, diafiltration, gel filtration, chromatography, or immobilization on a solid support. Dilution has been extensively used. It considerably reduces concentrations of both denaturant and protein. This procedure, however, cannot be applied to the commercial scale refolding of recombinant proteins, because large downstream processing volumes increase the cost of products. Although dialysis through semipermeable membranes has been used successfully to refold several proteins, it is not employed in large-scale processes. This is because it requires very long processing times, and there is the risk that during dialysis, the protein will remain too long at a critical concentration of denaturant and aggregate. The removal of the denaturant may be accomplished through gel filtration. However, here again, a possible aggregation could lead to flow restriction within the column. Dialfiltration through a semipermeable membrane allows the removal of denaturant and other small molecules and retains the protein. This procedure has been used for large-scale processing and was particularly efficient in the refolding of prorennin and interferon-β.

During the refolding process, the formation of incorrectly folded species and aggregates usually decreases the refolding yield. For disulfide-bridged proteins, the renaturation buffer must contain redox-shuffling mixture to allow the formation of correctly paired disulfide bridges. Stabilizing reagents may be added to improve the refolding yield. An efficient strategy is the addition of small molecules to suppress intermolecular interactions leading to aggregation. Sugar, alcohols, polyols (including sucrose, glycerol, polyethylene glycol, isopropanol), cyclodextrin, laurylmaltoside, sulfobetains, L-arginine, and low concentrations of denaturants and detergents, have been used to increase the refolding yield. L-arginine at a concentration ranging from 0.4 M to 0.8 M is the most widely used additive today.

Another important factor in the refolding process is the rate of removal of the denaturant. Since there is kinetic competition between the correct folding and the formation of aggregates from a folding intermediate, conditions that favor folding over the accumulation of aggregates must be found. To optimize this selection, Vilick and de Bernadez–Clark developed a strategy for achieving high protein refolding yields. They start from a model of refolding, develop the equations of refolding kinetics, characterize the rate-limiting step of the process, determine the influence of various environmental parameters, and finally optimize the system of equations in a scheme involving diafiltration to remove the denaturant. The approach was evaluated in the refolding of carbonic anhydrase from 8 M urea. The yield obtained after three diafiltration experiments was 69% whereas the model predicted a yield of 73%.

The properties of molecular chaperones have also been utilized to increase the refolding yield. Altamiro and coworkers have developed a system for refolding chromatography that utilizes GroEL, DsbA, and peptidyl–prolyl isomerase immobilized on an agarose gel. Kohler and coworkers have built a chaperone-assisted bioreactor; however, it could only be used for three cycles of refolding and needs to be improved. Another strategy consists of the co-overproduction of the DnaK–DnaJ or GroEL–GroES chaperones with the desired protein; this can greatly increase the soluble yield of aggregation-prone proteins. Fusion proteins have also been used to minimize aggregation.

The recovery of active proteins from inclusion bodies is a rather complex process. Although some general strategies have been developed, optimal conditions have to be determined for each protein. Recently, genetic strategies to improve recovery processes for recombinant proteins have been introduced. They consist of the introduction of combinatorial protein engineering to generate molecules highly specific to a particular ligand. Such methods, which allow efficient recovery of a recombinant protein, will be increasingly used in industrial scale bioprocesses as well.

4.2
The Formation of Amyloid Fibrils and its Pathological Consequences

The formation of amyloid fibrils plays a key role in the origin of several neurodegenerative pathologies, such as spongiform encephalopathies and Alzheimer's disease. Historically, the term amyloid was introduced to describe fibrillar protein deposits associated with diseases known as *amyloidoses* that involve the extracellular deposition of amyloid fibrils and plaques with the aspect of starch. For many of these diseases, the major fibrillar protein component has been identified. In the 1970s, it was demonstrated that lysosomal proteins under acidic conditions could form amyloid fibrils. It was generally accepted at this time that proteolysis was the amyloidogenic determinant. Twenty years later, it was shown that purified transthyretin is converted into amyloid fibrils via an acid-induced conformational change *in vitro*, demonstrating that conformational changes alone were responsible for producing an intermediate generating amyloid structure. These aberrant protein self-assemblies are at the origin of more than hundred human amyloid diseases, some of them being lethal.

Twenty unrelated protein precursors are known to form amyloid fibrils, among them transthyretin, lysozyme, immunoglobulin light chain, β_2 microglobulin, Alzheimer $A\beta 1-40$ and $A\beta 1-42$ peptides, the mammalian prion protein, and the yeast prion-like proteins (Table 1). Since they are subjects of another chapter, prion proteins will not be discussed here. Although they have no homology in sequence and structure, all form amyloid fibrils with a similar overall structure, suggesting a common self-assembly

Tab. 1 Amyloidogenic proteins and the corresponding diseases.

Clinical syndrome	Precursor protein	Fibril component
Alzeimer's disease	APP	β-peptide 1–40 to 1–43
Primary systemic amyloidosis	Immunoglobulin light chain	Intact light chain or fragments
Secondary systemic amyloidosis	Serum amyloid A	Amyloid A (76-residue fragment)
Senile systemic amyloidosis	Transthyretin	Transthyretin or fragments
Familial amyloid polyneuropathy I	Transthyretin	Over 45 transthyretin variants
Hereditary cerebral amyloid angiopathy	Cystatin C	Cystatin C minus 10 residues
Hemodialysis-related amyloidosis	β_2-microglobulin	β_2-microglobulin
	Apolipoprotein A1	Fragments of Apolipoprotein A1
Familial amyloid polyneuropathy III	Gelsosin	71-amino acid fragment of gelsosin
Finnish hereditary systemic amyloidosis	Islet amyloid polypeptide (IAPP)	Fragment of IAPP
Type II diabetes	Calcitonin	Fragments of calcitonin
Medullary carcinoma of the thyroid		
Spongiform encephalopathies	Prion	Prion or fragments thereof
Atrial amyloidosis	Atrial natriuretic factor (ANF)	ANF
Hereditary nonneuropathic systemic amyloidosis	Lysozyme	Lysozyme or fragments thereof
Injection-localized amyloidosis	Insulin	Insulin
Hereditary renal amyloidosis	Fibrinogen	Fibrinogen fragments
Parkinson disease	α-synuclein*	

Source: (According to Kelly, J.W. (1996) Alternative conformations of amyloidogenic proteins govern their behavior, *Curr. Opin. Struct. Biol.* **6**, 11–17); *From J.C. Rochet & P.T. Lansbury (2000) *Curr. Opin. Struct. Biol.* **10**, 60–68.

pathway. In all proteins known to form amyloid fibrils, there is a conversion of α- to β-structure. Amyloid fibrils are abnormal, insoluble, and generally protease-resistant structures. They were first recognized by their staining properties. The most commonly used method to detect amyloid is staining by Congo Red, which exhibits a green birefringence. Amyloid fibrils are generally 60 to 100 Å in diameter and of variable length. X-ray diffraction data on fibrils, solid-state NMR studies, cryoelectron microscopy, and infrared Fourier transform experiments have shown that amyloid fibrils are made of two or more β-sheet filaments wound around one another. They have a characteristic cross-β repeat structure, the individual β-strands being oriented perpendicular to the long axis of the fibril.

Recently, progress has been made in the knowledge of the mechanisms involved in the formation of amyloid fibrils. Oligomeric prefibrillar intermediates have been extensively characterized with respect to their structure and temporal evolution. A well-documented example is provided by the studies on transthyretin. The biological role of this protein is the transport of thyroxin by direct binding and the transport of retinol via the retinol binding protein. The wild-type protein is very stable at neutral pH. In certain individuals, however, it is converted into amyloid fibrils, and this is associated with the disease, senile systemic amyloidosis. Several variants are associated with familial polyneuropathies. *In vitro* biophysical studies have identified conditions leading to amyloid formation. The three-dimensional structure of the protein is known. The wild-type protein is a tetramer at pH ranging between 5 and 7; the tetramer dissociates into a monomer when the pH decreases. The dissociation is the rate-limiting step of the process. The monomer exhibits an altered tertiary structure, which aggregates in amyloid

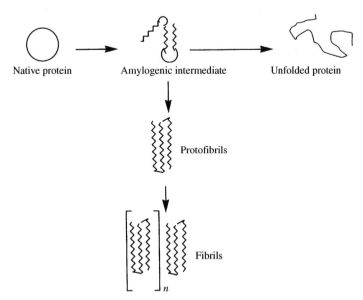

Fig. 10 Schematic representation of the formation of amyloid fibrils from a partially folded intermediate.

protofilaments, and then forms amyloid fibrils. This formation is at its maximum at pH 4.4. Using deuterium–proton exchange monitored by two-dimensional NMR spectroscopy on transthyretin at pH 5.75 and 4.5, Liu et al. have shown a selective destabilization of one half of the β-sandwich structure of the protein, increasing the mobility of this region. These studies have identified the residues that undergo increased conformational fluctuations under amyloidogenic conditions. The mutations in the pathological variants responsible for familial amyloid polyneuropathies are localized in this region. A strategy to delay the formation of amyloid fibrils proposed by Saccheti & Kelly was to develop molecules capable of stabilizing the tetramer.

Two variants of human lysozyme, Ile56Thr and Asp67His have been reported to be amyloidogenic; they are responsible for fatal amyloidoses. Pepys and colleagues have determined the precise structures and properties of these mutants. The native fold of the two amyloidogenic variants, as resolved by X-ray crystallography, is similar to that of the wild-type protein. Both variants are enzymatically active, but have been shown to be unstable. The replacement of an aspartate by a histidine suppresses a hydrogen bond formed in the wild-type protein with a tyrosine in a neighboring β-strand. This rupture opens a large gap between two β-strands. In the other variant, the replacement of an isoleucine by a threonine suppresses a van der Waals contact with a neighboring helix. Consequently, changes in the interface between the α- and β-domains occur in both variants, destabilizing the molecule.

The mutations leading to amyloid fibril formation are observed to result in a decreased stability of the native state. In all cases, the formation of fibrils occurs from a partially structured molecule via nucleation-dependent oligomerization. It was observed for several proteins that fibrillation takes place only after a lag phase, which is abolished upon seeding. Nucleation is followed by the formation of protofibrils whose characteristics have been determined (Fig. 10). Atomic force microscopy and fluorescence correlation spectroscopy have been used to monitor transitions among the different types of assemblies.

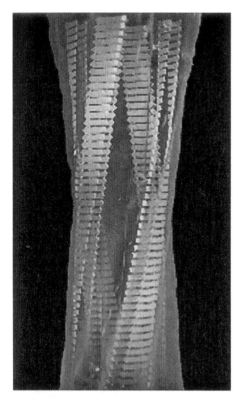

Fig. 11 Molecular model of an amyloid fibril derived from cryoelectron microscopy analysis of fibrils grown from an SH3 domain by incubation of a solution containing the protein at low pH (reproduced from Dobson, C. (1999) TIBS 24, 331, with permission).

Recent observations from Dobson and his group have shown that several proteins unrelated to amyloid diseases are able to aggregate *in vitro* into amyloid fibrils when exposed to mild denaturing conditions. These fibrils are indistinguishable from those found in pathological conditions. It was demonstrated for different proteins such as normal lysozyme, an SH_3 domain of a phosphatidyl inositol protein kinase (Fig. 11), an acyl phosphatase, and an α-helical protein, myoglobin suggesting a common mechanism for the formation of amyloid. These findings clearly indicate that amyloid formation is a general property of polypeptide chains rather than one restricted to definite sequences as occurs with chameleon sequences capable of adopting either a β- or an α-helicoidal structure depending on their environment. Furthermore, these aggregates exhibited an inherent toxicity when incubated with mouse fibroblasts. Several groups suggest that oligomeric intermediates rather than fibrils themselves are responsible for pathogenicity.

Significant progress has been made in understanding the mechanisms involved in the formation of amyloid fibrils. This is an important step in guiding research into the discovery of molecules with therapeutic efficiency.

See also Circular Dichroism in Protein Analysis.

Bibliography

Books and Reviews

Dobson, C.M., Šali, A., Karplus, M. (1998) Protein folding. A perspective from theory and experiment, *Angew. Chem. Int. Ed.* **37**, 868–893.

Ellis, R.J. (1991) Molecular chaperones, *Ann. Rev. Biochem.* **60**, 321–347.

Ellis, R.J., Hartl, F.U. (1999) Principles of protein folding in the cellular environment, *Curr. Opin. Struct. Biol.* **9**, 102–110.

Georgiou, G., de Bernadez-Clark, E. (Eds.) (1991) *Protein Refolding*, ACS Symposium Series 470, American Chemical Society, Washington, DC.

Ghélis, C., Yon, J.M. (1982) *Protein folding*, Academic Press, New York.

Jaenicke, R. (1987) Folding and association of proteins, *Prog. Biophys. Mol. Biol.* **49**, 117–237.

Kim, P.S., Baldwin, R.L. (1990) Intermediates in protein folding of small proteins, *Ann. Rev. Biochem.* **59**, 631–660.

Schlunegger, M.P., Bennett, M.J., Eisenberg, D. (1997) Oligomer formation by 3D domain swapping: a model for protein assembly and disassembly, *Adv. Protein Chem.* **50**, 61–122.

Wetlaufer, D.B., Ristow, S. (1973) Acquisition of the three-dimensional structure of proteins, *Ann. Rev. Biochem.* **42**, 135–158.

References of Primary Literature

Altamiro, M.M., Garcia, C., Possani, L.D., Fersht, A.R. (1999) Oxidative refolding chromatography: folding of the scorpion toxin Cn5, *Nat. Biotechnol.* **17**, 187–191.

Anfinsen, C.B. (1973) Principles that govern the folding of polypeptide chains, *Science* **181**, 223–230.

Anfinsen, C.B., Haber, E., Sela, M., White, F.H. (1961) The kinetics of formation of native ribonuclease during oxidation of the reduced polypeptide chain, *Proc. Natl. Acad. Sci. USA* **47**, 1309–1314.

Baneyx, F. (1999) Recombinant protein expression in Escherichia coli, *Curr. Opin. Biotechnol.* **10**, 411–421.

Bennett, M.J., Choe, S., Eisenberg, D. (1994) Domain swapping: Entangling alliance between proteins, *Proc. Natl. Acad. Sci. USA* **91**, 3127–3131.

Bergdoll, M., Remy, M.H., Capron, C., Masson, J.M., Dumas, P. (1997) Proline-dependent oligomerization with arm exchange, *Structure* **5**, 391–401.

de Bernadez-Clark, E. (2001) Protein refolding for industrial processes, *Curr. Opin. Biotechnol.* **12**, 202–207.

de Bernadez-Clark, E., Schwarz, E., Rudolph, R. (1999) Inhibition of aggregation side-reactions during in vitro protein folding, *Methods Enzymol.* **309**, 217–236.

Booth, D.R., Suide, M., Belotti, V., Robinson, C.V., Hutchinson, W.L., Fraser, P.E., Hawkins, P.N., Dobson, C.M. Radford, S.E., Blake, C.C.F., Pepys, M.B. (1997) Instability, unfolding, and aggregation of human lysozyme variants underlying amyloid fibrillogenesis, *Nature* **385**, 787–793.

Bucciantini, M., Giannoni, E., Chiti, F., Baroni, F., Formigli, L., Zurdo, J., Taddei, N., Ramponi, G., Dobson, C.M., Stefani, M. (2002) Inherent toxicity of aggregates implies a common mechanism for protein misfolding diseases, *Nature* **416**, 507–511.

Chiti, F., Wabster, P., Taddei, N., Clark, A., Stefani, M., Ramponi, G., Dobson, C.M. (1999) Designing conditions for in vitro formation of amyloid protofilaments and fibrils, *Proc. Natl. Acad. Sci. USA* **96**, 3342–3344.

Crestfield, A.M., Stein, W.H., Moore, S. (1962) On the aggregation of bovine pancreatic ribonuclease, *Arch. Biochem. Biophys.* **1**(Suppl.), 217–222.

Dijkstra, K., Karvonen, P., Pirneskoski, A., Koivunen, P., Kivirikko, K.I., Darby, N.J., van Staaden, M., Sheek, R.M., Kemmink, J. (1999) Assignment of ^{1}H, ^{13}C and ^{15}N resonances of the a domain of protein disulfide isomerase, *J. Biomol. NMR* **14**, 194–196.

Dill, K.A. (1985) Theory for folding and stability of globular proteins, *Biochemistry* **24**, 1501–1509.

Dill, K.A., Bromberg, S., Yue, K., Fiebig, K.H., Thomas, P., Chan, H.S. (1995) Principles of protein folding: a perspective from simple exact models, *Protein Sci.* **4**, 561–602.

van Duyne, G.D., Standaert, R.F., Karplus, P.A., Stuart, L., Shreiber, S.L., Clardy, J. (1993) Atomic structure of the human immunophilin FKBP12 complexes with FK506 and rapamycin, *J. Mol. Biol.* **229**, 105–124.

Edman, J.C., Ellis, L., Blacher, R.W., Roth, R.A., Rutter, W.J. (1985) Sequence of protein disulfide isomerase: implications of its relationship to thioredoxin, *Nature* **317**, 267–270.

Edwards, K.J., Ollis, D.L., Dixon, N.E. (1997) Crystal structure of cytoplasmic Escherichia coli peptidyl-prolyl isomerase: evidence for decrease mobility of loops upon complexation, *J. Mol. Biol.* **271**, 258–265.

Eliezer, D., Chiba, K., Tsurata, H., Doniach, S., Hodgson, K.O., Kihara, H. (1993) Evidence of an associative intermediate on the myoglobin refolding pathway, *Biophys. J.* **65**, 912–917.

Eliezer, D., Jennings, P.A., Wright, P.E.H. Doniach, S., Hodgson, K.O., Tsurata, H. (1995) The radius of gyration of an apomyoglobin folding intermediate, *Science* **270**, 487–488.

Ellis, R.J. (2001) Molecular crowding: an important but neglected aspect of the cellular environment, *Curr. Opin. Struct. Biol.* **11**, 114–119.

Ellis, R.J., Hartl, F.U. (1999) Principles of protein folding in the cellular environment, *Curr. Opin. Struct. Biol.* **9**, 102–110.

Fandrich, M., Flechter, M.A., Dobson, C.M. (2001) Amyloid fibrils from muscle myoglobin, *Nature* **410**, 163–166.

Fink, A.L. (1998) Protein aggregation, folding aggregates, inclusion bodies and amyloid, *Fold. Des.* **3**, R9–R23.

Freedman, R.B., Klappa, P., Ruddock, L.W. (2002) Protein disulfide isomerases exploit synergy between catalytic and specific binding domains, *EMBO Rep.* **31**, 136–140.

Georgiou, G., Valax, P. (1999) Isolating inclusion bodies from bacteria, *Methods Enzymol.* **309**, 48–58.

Goldberg, M.E. (1973) L'état natif est-il l'état fondamental? in: Sadron, C. (Ed.) *Dynamic Aspect of Conformational Changes in Biological Macromolecules*, Reidel, Dordrecht.

Guijarro, J.L., Sunde, M., Jones, J.A., Campbell, I.D., Dobson, C.M. (1998) Amyloid fibril by an SH3 domain, *Proc. Natl. Acad. Sci. USA* **95**, 4224–4228.

Harrison, S.C., Durbin, R. (1985) Is there a single pathway for the folding of a polypeptide chain? *Proc. Natl. Acad. Sci. USA* **82**, 4028–4030.

Horowitz, P. Criscimagna, N.L. (1986) Low concentrations of guanidinium chloride expose apolar surfaces and cause differential perturbation in catalytic intermediates of rhodanese, *J. Biol. Chem.* **261**, 15652–15658.

Jaenicke, R. (1991) Protein folding: local structures, domains, subunits and assemblies, *Biochemistry* **30**, 3147–3161.

Kadokura, H., Beckwith, J. (2002) Four cysteines of the membrane protein DsbB act in concert to oxidize its substrate DsbA, *EMBO J.* **21**, 2354–2363.

Karplus, M. Šali, A. (1996) Theoretical studies of protein folding and unfolding, *Curr. Opin. Struct. Biol.* **5**, 58–73.

Kelly, J.W. (1996) Alternative conformations of amyloidogenic proteins govern their behavior, *Curr. Opin. Struct. Biol.* **6**, 11–17.

Kelly, J.W. (1998) The alternative conformations of amyloidogenic proteins and their multi-step assembly pathways, *Curr. Opin. Struct. Biol.* **8**, 101–106.

Kelly, J.W. (2002) Towards an understanding of amyloidogenesis, *Nat. Struct. Biol.* **9**, 323–325.

Kohler, R.J., Preuss, M., Miller, A.D. (2000) Design of a molecular chaperone-assisted protein folding bioreactor, *Biotechnol. Prog.* **16**, 671–675.

Kuhlman, B., O'Neill, J.M., Kim, D.E., Zhang K.Y., Baker, D. (2001) Conversion of a monomeric protein L to an obligate dimer by computational protein design, *Proc. Natl. Acad. Sci. USA* **98**, 10687–10691.

Langer, T., Lu, C., Flanagan, J., Hayer, A.K., Hartl, F.U. (1994) Successive action of DnaK, DnaJ and GroEL along the pathway of chaperone-mediated protein folding, *Nature* **356**, 683–689.

Levinthal, C. (1968) Are there pathways for protein folding? *J. Chim. Phys.* **65**, 44–45.

Liu, K. Cho, H.S., Lashuel, H.A., Kelly, J.W., Wemmer, D.E. (2000) A glimpse of a possible amyloidogenic intermediate of transthyretin, *Nat. Struct. Biol.* **7**, 754–757.

Liu, Y., Gotte, G. Libonati, M., Eisenberg, D. (2001) A domain-swapped ribonuclease A dimer with implication for amyloid formation, *Nat. Struct. Biol.* **8**, 282–284.

Liu, Y., Gotte, G. Libonati, M., Eisenberg, D. (2002) Structures of the two 3D domain-swapped ribonuclease A trimers, *Protein Sci.* **11**, 371–380.

London, J., Skrzynia C., Goldberg, M.E. (1974) Renaturation of Escherichia coli tryptophanase after exposure to 8 M urea. Evidence for the existence of nucleation centers, *Eur. J. Biochem.* **47**, 409–415.

Marston, F.A.O. (1986) The purification of eukaryotic polypeptides synthesized in Escherichia coli, *Biochem. J.* **240**, 1–12.

Martin, J.L., Bardwell, J.C., Kuriyan, J. (1993) Crystal structure of DsbA protein required for disulfide bond formation, *Nature* **365**, 464–468.

McCarthy, A.A., Haebel, P.W., Torronen, A., Rybin, V., Baker, E.N., Metcalf, P. (2000) Crystal structure of the protein disulfide bond isomerase, DsbC, from Escherichia coli, *Nat. Struct. Biol.* **7**, 196–199.

Mitraki, A., Betton, J.M., Desmadril, M., Yon, J.M. (1987) Quasi-irreversibility in the unfolding-refolding transition of phosphoglycerate kinase induced by guanidine hydrochloride, *Eur. J. Biochem.* **163**, 29–34.

Morozova-Roche, L.A., Zurdo, J., Spencer, A., Noppe, W., Receveur, V., Archer, D.B., Joniau, M., Dobson, C.M. (2000) Amyloid fibril formation and seeding by wild-type lysozyme and its disease-related mutational variants, *J. Struct. Biol.* **130**, 339–351.

Pecorari, F., Minard P., Desmadril, M., Yon, J.M. (1996) Occurrence of transient multimeric species during the refolding of a monomeric protein, *J. Biol. Chem.* **271**, 5270–5276.

Rochet, J.C., Lansbury, P.T. (2000) Amyloid fibrillogenesis: themes and variations, *Curr. Opin. Struct. Biol.* **10**, 60–68.

Roseman, A.M., Chen, S., White, H., Sebil, H.R. (1996) The chaperonin ATPase cycle: mechanism of allosteric switching and movements of substrate binding domains in GroEL, *Cell* **87**, 241–251.

Saccheti, J.C., Kelly, J.W. (2002) Therapeutic strategies for human amyloid diseases, *Nat. Rev. Drug Disc.* **1**, 267–275.

Semisotnov, G.V., Kuwajima, K. (1996) Protein globularization during folding- a study by synchrotron small-angle X-ray scattering, *J. Mol. Biol.* **262**, 559–574.

Silow, M., Tan, Y.J., Fersht, A.N., Oliveberg, M. (1999) Formation of short-lived protein aggregates directly from the coil in two-state folding, *Biochemistry* **38**, 13006–13012.

Speed, M.A., Wang, D.I.C., King, J. (1996) Specific aggregation of partially folded polypeptide chains. The molecular basis of inclusion body composition, *Nat. Biotech.* **14**, 1283–1287.

Taketomi, H., Kano, F., Gô, N. (1988) The effect of amino acid substitutions on protein folding and unfolding studied by computer simulation, *Biopolymers* **27**, 527–560.

Wetzel, R., Perry, L.J., Veilleux, C. (1991) Mutations in human interferon γ affecting inclusion body formation by a general immunochemical screen, *Biotechnology* **9**, 731–737.

Wolynes, P.G., Onuchic, J.N., Thirumalai, D. (1995) Navigating the folding routes, *Science* **267**, 1618–1620.

Xu, Z., Horwich, A.L., Sigler, P.B. (1997) Crystal structure of the asymmetric GroEL-GroES-(ADP)$_7$ chaperonin complex, *Nature* **388**, 741–750.

Yon, J.M. (1996) The specificity of protein aggregation, *Nat. Biotech.* **14**, 1231.

Zhang, X., Beuron, F., Freemont, P.S. (2002) Machinery of protein folding and unfolding, *Curr. Opin. Struct. Biol.* **12**, 231–238.

7
Prions

Stanley B. Prusiner
University of California, San Francisco, CA

1	**Prions** 202	
2	**Prions are Distinct from Viruses** 203	
3	**Disease Paradigms** 204	
4	**Nomenclature** 205	
5	**Discovery of the Prion Protein** 208	
6	**Prion Protein Isoforms** 209	
6.1	Cell Biology of PrP^{Sc} Formation 209	
7	**Rodent Models of Prion Disease** 210	
8	**PrP Gene Structure and Organization** 210	
8.1	Expression of the PrP Gene 211	
8.2	Overexpression of Wild-type PrP Transgenes 211	
8.3	PrP Gene Dosage Controls the Incubation Time 212	
8.4	PrP-deficient Mice 212	
8.5	Species Variations in the PrP Sequence 213	
8.6	N-terminal Sequence Repeats 213	
8.7	Conserved Ala-Gly Region 216	
9	**Structures of PrP Isoforms** 216	
9.1	NMR Structures of recPrP 217	
9.2	Electron Crystallography of PrP^{Sc} 219	

Proteins. Edited by Robert A. Meyers.
Copyright © 2007 Wiley-VCH Verlag GmbH & Co. KGaA, Weinheim
ISBN: 978-3-527-31608-3

10	**Prion Replication** 221	
10.1	Mechanism of Prion Propagation 222	
10.2	Template-assisted Prion Formation 222	
10.3	Evidence for Protein X 223	
10.4	Dominant-negative Inhibition 224	
11	**Sporadic Human Prion Diseases** 225	
12	**Heritable Human Prion Diseases** 225	
12.1	GSS and Genetic Linkage 226	
12.2	fCJD Caused by Octarepeat Inserts 226	
12.3	fCJD in Libyan Jews 226	
12.4	Penetrance of fCJD 226	
12.5	Fatal Familial Insomnia 227	
12.6	Human PrP Gene Polymorphisms 227	
13	**Infectious Human Prion Diseases** 228	
13.1	Human Growth Hormone 228	
13.2	Variant Creutzfeldt–Jakob Disease (vCJD) 228	
13.3	Transmission of vCJD Prions by Blood Transfusion 229	
14	**Strains of Prions** 229	
14.1	Isolation of New Strains 230	
14.2	Interplay between the Species and Strains of Prions 231	
15	**Prion Diseases of Animals** 231	
15.1	PrP Polymorphisms in Sheep, Cattle, and Elk 231	
15.2	Bovine Spongiform Encephalopathy 232	
15.3	Monitoring Cattle for BSE Prions 233	
15.4	Compelling Evidence of Transmission of Bovine Prions to Humans 233	
15.5	Chronic Wasting Disease 234	
16	**Fungal Prions** 235	
16.1	Yeast Prion Domains 235	
16.2	Dependence of Yeast Prions on Molecular Chaperones 235	
16.3	Differences between Yeast and Mammalian Prions 236	
17	**Prion Diseases are Disorders of Protein Conformation** 236	
18	**Prevention and Therapeutics for Prion Diseases** 237	
18.1	Prion Therapeutics 237	
18.1.1	Quinacrine and other Acridine Derivatives 237	
18.1.2	Anti-PrP Antibodies 238	
18.2	Dominant-negative Sheep 238	

19	**Some Principles of Prion Biology** 239
19.1	Implications for Common Neurodegenerative Diseases 239

Bibliography 239
Books and Reviews 239
Primary Literature 241

Keywords

Prion
Proteinaceous infectious particle.

PrP^C
Normal isoform of the prion protein.

PrP^{Sc}
Disease-causing isoform of the prion protein.

■ Prions are infectious proteins that have been described in both mammals and fungi. That prions are composed solely of proteins makes them unprecedented infectious pathogens. Prions are proteins that can adopt at least two different conformations; they multiply by forcing the precursor protein to acquire a second conformation. Different conformations of proteins in the prion state encipher distinct strains and are prone to aggregation. In mammals, prions accumulate to high levels in the nervous system where they cause dysfunction and fatal degeneration.

Both mammalian and fungal prions have been produced in cell-free systems. Synthetic prion protein (PrP) peptides and recombinant PrP fragments have been used to form mammalian prions while N-terminal regions, called *prion domains* that are rich in glutamine and asparagine, have been used to form fungal prions.

Prions cause a group of invariably fatal, neurodegenerative diseases. Prion diseases may present as genetic, infectious, or sporadic disorders, all of which involve modification of PrP. The tertiary structure of PrP is profoundly altered as prions are formed, and as such, prion diseases represent disorders of protein conformation. Creutzfeldt-Jakob disease (CJD) generally presents as a progressive dementia in humans while scrapie of sheep and bovine spongiform encephalopathy (BSE) usually manifest as ataxic illnesses.

1
Prions

In mammals, prions reproduce by recruiting the normal, cellular isoform of the prion protein (PrP^C) and stimulating its conversion into the disease-causing isoform (PrP^{Sc}). PrP^C has a high α-helical content and little β-sheet structure, whereas PrP^{Sc} has less α-helical structure and a high β-sheet content. Comparisons of secondary structures of PrP^C and PrP^{Sc} were performed on proteins purified from Syrian hamster (SHa) brains. Limited proteolysis of PrP^{Sc} produces PrP 27–30, which retains prion infectivity; under these conditions, PrP^C is completely hydrolyzed (Fig. 1).

Solution structures of recombinant SHa and mouse (Mo) PrPs produced in bacteria showed three α-helices denoted A, B, and C as well as two short β-strands using nuclear magnetic resonance (NMR) imaging. The atomic structure of PrP^{Sc} has not been determined because of the insolubility of the protein. Two-dimensional crystals of PrP 27–30 have been used to constrain computational models of PrP^{Sc} suggesting it contains α-helix C and a portion of α-helix B while α-helix A, the two β-strands, and the surrounding segments form a β-helix. From these studies as well as from ionizing radiation inactivation analyses, it is likely that the infectious monomer is a trimer of PrP^{Sc} molecules. Recombinant

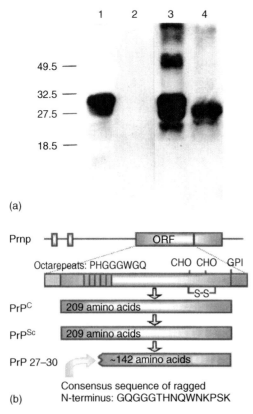

Fig. 1 Prion protein isoforms. Western immunoblot of brain homogenates from uninfected (lanes 1 and 2) and prion-infected (lanes 3 and 4) Syrian hamsters. Samples in lanes 2 and 4 were digested with 50 µg mL^{-1} of proteinase K for 30 min at 37 °C. PrP^C in lanes 2 and 4 was completely hydrolyzed under these conditions, whereas approximately 67 amino acids were digested from the N-terminus of PrP^{Sc} to generate PrP 27–30. After polyacrylamide gel electrophoresis (PAGE) and electrotransfer, the blot was developed with anti-SHaPrP R073 polyclonal rabbit antiserum. Molecular weight markers are depicted in kilodaltons (kDa). (b) Bar diagram of the SHaPrP gene that encodes a protein of 254 amino acids. After processing of the N- and C-termini, both PrP^C (green) and PrP^{Sc} (red) consist of 209 residues. After limited proteolysis, the N-terminus of PrP^{Sc} is truncated to form PrP 27–30, which is composed of approximately 142 amino acids, the N-terminal sequence of which was determined by Edman degradation. (See color plate p. xii).

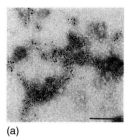 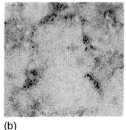 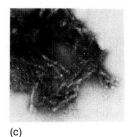

(a) (b) (c)

Fig. 2 Electron micrographs of negatively stained and immunogold-labeled prion proteins. (a) PrP^C and (b) PrP^{Sc}. Neither PrP^C nor PrP^{Sc} forms recognizable, ordered polymers. (c) Prion rods composed of PrP 27–30 were negatively stained. The prion rods are indistinguishable from many purified amyloids. Bar = 100 nm. Reprinted with permission, from (Pan, K-M., Baldwin, M., Nguyen, J., Gasset, M., Serban, A., Groth, D., Mehlhorn, I., Huang, Z., Fletterick, R.J., Cohen, F.E., Prusiner, S.B. (1993) Conversion of α-helices into β-sheets features in the formation of the scrapie prion proteins, Proc. Natl. Acad. Sci. U. S. A. **90**, 10962–10966), copyright 1993 National Academy of Sciences, USA.

antibody fragments have been used to map the surfaces of PrP^C and PrP^{Sc} and those results are consistent with the foregoing structural data. The structural transition from an α-helical protein into a β sheet–rich molecule seems to be the fundamental event underlying the formation of nascent prions.

Limited proteolysis truncates the N-terminus of PrP^{Sc} to produce PrP 27–30 consisting of the C-terminal ~142 amino acids. PrP 27–30 polymerizes into amyloid fibrils that are indistinguishable from fibrils found in amyloid plaques of the brains of mammals with prion disease. When full-length PrP^{Sc} was purified from SHa brain, only amorphous aggregates were seen by electron microscopy. While limited digestion of purified PrP^{Sc} with proteinase K produced PrP 27–30, the ultrastructural appearance of the samples remained unchanged (Fig. 2). Addition of a detergent, such as Sarkosyl, either before or after limited proteolysis provoked the assembly of PrP 27–30 into rod-shaped structures with the ultrastructural and tinctorial properties of amyloid.

2
Prions are Distinct from Viruses

Major features that distinguish prions from viruses are: (1) the ability to create prion infectivity by modifying the conformation of a polypeptide devoid of nucleic acid and (2) PrP is encoded by a chromosomal gene, designated *PRNP* in humans and *Prnp* in mice. The PrP gene is located on the short arm of chromosome 20 in humans and the syntenic region of chromosome 2 in mice.

Prions differ from viruses and viroids in that they lack a nucleic acid genome that directs the synthesis of their progeny. Many investigators argued for a nucleic acid genome within the infectious prion particle while others contended for a small, noncoding polynucleotide of either foreign or cellular origin. No nucleic acid has been found despite intensive searches using a wide variety of techniques and approaches. On the basis of a wealth of evidence, it is reasonable to assert that such a nucleic acid has not been found because it does not exist. Prions are composed of an alternative isoform of a cellular protein, whereas most viral proteins are encoded by a viral genome and viroids are devoid of protein.

In contrast to viruses, prions are non-immunogenic. Prions do not elicit an immune response because the host is tolerant to PrP^C, which prevents the host from mounting an immune response to PrP^{Sc}. In contrast, foreign proteins of viruses that are encoded by the viral genome often elicit a profound immune response. Thus, it seems unlikely that vaccination, which has been so effective in preventing many viral illnesses, will be a useful strategy for preventing or treating prion diseases.

When prions and viruses are passaged from one host species to another, the consequences are very different. The passage of prions from one host to another is accompanied by the acquisition of a new PrP amino acid sequence encoded by the genome of the new host. The crossing of prions from one species to another is restricted by what has been called the "species barrier". The closer the evolutionary relationship is between the host in which the prions last replicated and the newly infected animal, the more likely replication will occur in the new host. However, particular strains of prions can change this rule. For example, variant CJD prions from humans replicate much more readily in transgenic (Tg) mice expressing bovine (Bo) PrP than in Tg mice expressing human or chimeric mouse-human PrP.

The issue of prion strains posed a profound conundrum for many years. How could an infectious pathogen composed only of protein encipher biological information? This riddle was solved when prion strains with different physical properties were isolated. Subsequently, new strains of prions have been isolated either by passage through mice expressing artificial PrP transgenes or by forming synthetic prions from PrP produced in *Escherichia coli*.

3
Disease Paradigms

Despite some similarities between prion and viral illnesses, these disorders are very different. Prion diseases are uniformly fatal. No human or animal has ever recovered from a prion disease once neurologic dysfunction is manifest. No host defenses are mounted in response to prion infection: no humoral immunity, no cellular immunity, and no interferons are elicited to the replicating prion.

While prions can be spread from one host to another, the most common form of prion disease is spontaneous or sporadic. In these illnesses, prions arise endogenously and replicate. Recent experiments with synthetic prions demonstrate that only PrP^C is required to generate prion infectivity. These findings contrast with viruses in which exogenous infection is required except in the case of latent retroviral genomes. For example, after infection with exogenous HIV, the virus may disappear but its RNA genome may be reverse-transcribed into DNA and the DNA copies may remain dormant for years.

The dramatically different principles that govern prion biology have often been poorly understood. This lack of understanding has led to some regrettable decisions of great economic, political, and possibly public health importance. For example, scrapie and BSE have different names, yet they are the same disease in two different species. Scrapie and BSE differ in only two respects: first, the PrP sequence in sheep differs from that of cattle at seven or eight positions of 270 amino acids, which results in different PrP^{Sc} molecules. Second, some aspects of each disease are determined by the particular prion strain that infects the respective host.

Understanding prion strains and the "species barrier" is of paramount importance with respect to the BSE epidemic in Britain, in which more than 180 000 cattle have died over the past decade. Brain extracts from eight cattle with BSE resulted in similar incubation times and patterns of vacuolation in the neuropil when inoculated into a variety of inbred mice. Incubation times and profiles of neuronal vacuolation have been used for three decades to study prion strains. Brain extracts prepared from three domestic cats, one nyala, and one kudu, all of which died of a neurologic illness, produced incubation times and lesion profiles indistinguishable from those found in the BSE cattle. Cats and exotic ungulates (such as the kudu) presumably developed prion disease from eating food containing bovine prions.

4
Nomenclature

Although the prions that cause transmissible mink encephalopathy (TME) and BSE are referred to as *TME prions* and *BSE prions*, this may be unjustified because both are thought to originate from the oral consumption of scrapie prions in sheep-derived foodstuffs and because many lines of evidence argue that the only difference among various prions is the sequence of PrP, which is dictated by the host and not the prion itself. Human (Hu) prions present a similar semantic conundrum. Transmission of Hu prions to laboratory animals produces prions carrying PrP molecules with sequences dictated by the PrP gene of the last host, not that of the inoculum.

To simplify the terminology, the generic term PrP^{Sc} was suggested in place of such terms as PrP^{CJD}, PrP^{BSE}, and PrP^{res}. To distinguish PrP^{Sc} found in humans and cattle from that found in other animals, "$HuPrP^{Sc}$" and "$BoPrP^{Sc}$" are suggested instead of PrP^{CJD} and PrP^{BSE}, respectively. Once human prions, or $HuPrP^{Sc}$ molecules, have been passaged into animals, the prions and PrP^{Sc} are no longer of the human species unless they were formed in an animal expressing a HuPrP transgene.

The "Sc" superscript of PrP^{Sc} was initially derived from the term "scrapie" because scrapie was the prototypic prion disease. Because all of the known prion diseases (Table 1) of mammals involve aberrant metabolism of PrP similar to that observed in scrapie, the "Sc" superscript was suggested for all abnormal, pathogenic PrP isoforms. In this context, the "Sc" superscript is used to designate the disease-causing isoform of PrP (Table 2). The development of the conformation-dependent immunoassay (CDI) for PrP^{Sc} led to the discovery of a protease-sensitive form of PrP^{Sc}, designated $sPrP^{Sc}$. The CDI is based on measuring antibody binding to an epitope that is exposed in PrP^{C} but buried in PrP^{Sc}. Under conditions of limited proteolysis in which the protease-resistant form of PrP^{Sc}, designated $rPrP^{Sc}$, is converted into PrP 27–30, $sPrP^{Sc}$ is completely hydrolyzed. Whether $sPrP^{Sc}$ is an intermediate in the formation of infectious prions remains to be established.

In the case of mutant PrP, mutations and polymorphisms can be denoted in parentheses following the PrP isoform.

Tab. 1 The prion diseases.

Disease	Host	Mechanism of pathogenesis
Kuru	Fore people	Infection through ritualistic cannibalism
Iatrogenic CJD	Humans	Infection from prion-contaminated HGH, dura mater grafts, etc.
Variant CJD	Humans	Infection from bovine prions?
Familial CJD	Humans	Germline mutations in PrP gene
GSS	Humans	Germline mutations in PrP gene
FFI	Humans	Germline mutations in PrP gene (D178N,M129)
Sporadic CJD	Humans	Somatic mutation or spontaneous conversion of PrP^C into PrP^{Sc}?
sFI		Somatic mutation or spontaneous conversion of PrP^C into PrP^{Sc}?
Scrapie	Sheep	Infection in genetically susceptible sheep
Bovine spongiform encephalopathy	Cattle	Infection with prion-contaminated MBM
Transmissible mink encephalopathy	Mink	Infection with prions from sheep or cattle
Chronic wasting disease	Mule deer, elk	Unknown
Feline spongiform encephalopathy	Cats	Infection with prion-contaminated bovine tissues or MBM
Exotic ungulate encephalopathy	Greater kudu, nyala, oryx	Infection with prion-contaminated MBM

Notes: CJD: Creutzfeldt–Jakob disease; FFI: fatal familial insomnia; sFI: sporadic fatal insomnia; GSS: Gerstmann–Sträussler–Scheinker disease; HGH: human growth hormone; MBM: meat and bone meal.

For fatal familial insomnia (FFI), in which it might be important to identify the mutation, the prions would be designated $HuPrP^{Sc}(D178N,M129)$ (Table 3).

Parentheses following PrP^{Sc} can also be used to notate a particular prion strain. For example, prions from Syrian hamsters inoculated with Sc237 or 139H prion strains can be designated $SHaPrP^{Sc}(Sc237)$ or $SHaPrP^{Sc}(139H)$, respectively. Similarly, sheep inoculated with scrapie or BSE prions produce $OvPrP^{Sc}(Sc)$ or $OvPrP^{Sc}(BSE)$, respectively.

The terms "PrP^{res}" and "PrP-res" were derived to describe the protease resistance of PrP^{Sc} and have sometimes been used interchangeably with PrP^{Sc}. The use of "PrP^{res}" and "PrP-res" became particularly problematic with the discovery of $sPrP^{Sc}$. Protease resistance, insolubility, and high β-sheet content should be considered only as surrogate markers of PrP^{Sc} infectivity because not all characteristics may be present. For example, $MoPrP^{Sc}(P101L)$ from Tg mice that express high levels of MoPrP(P101L) is transmissible to Tg(MoPrP,P101L)196/$Prnp^{0/0}$ mice, but these prions are sensitive to proteolytic digestion at 37 °C. When digestions were performed at 4 °C or the CDI was used to detect PrP^{Sc}, then $MoPrP^{Sc}(P101L)$ was detected. In contrast, when Tg196 mice were inoculated with mouse RML prions, the resulting $MoPrP^{Sc}(P101L)$ was resistant to proteolytic digestion at 37 °C. These

Tab. 2 Glossary of prion terminology.

Term	Description
Prion	A proteinaceous infectious particle that lacks nucleic acid. Prions are composed largely, if not entirely, of PrP^{Sc} molecules.
PrP^{Sc}	Abnormal, pathogenic isoform of the prion protein that causes illness. This protein is the only identifiable macromolecule in purified preparations of prions.
PrP^{C}	Cellular isoform of the prion protein.
PrP 27–30	N-terminally truncated PrP^{Sc}, generated by digestion with proteinase K.
PRNP	Human PrP gene located on chromosome 20.
Prnp	Mouse PrP gene located on syntenic chromosome 2. *Prnp* controls the length of the prion incubation time and is congruent with the incubation-time genes *Sinc* and *Prn-i*. PrP-deficient (*Prnp*$^{0/0}$) mice are resistant to prion infection.
PrP amyloid	Fibril of PrP fragments derived from PrP^{Sc} by proteolysis. Plaques containing PrP amyloid are found in the brains of some mammals with prion disease.
Prion rod	An amyloid polymer composed of PrP 27–30 molecules. Created by detergent extraction and limited proteolysis of PrP^{Sc}.
Protein X	A hypothetical macromolecule that is thought to act as a molecular chaperone in facilitating the conversion of PrP^{C} into PrP^{Sc}.

Tab. 3 Examples of human PrP gene mutations found in the inherited prion diseases.

Inherited prion disease	PrP gene mutation
Gerstmann–Sträussler–Scheinker disease	P102L[a]
Gerstmann–Sträussler–Scheinker disease	A117V
Familial Creutzfeldt–Jakob disease	D178N, V129
Fatal familial insomnia	D178N, M129[a]
Gerstmann-Sträussler-Scheinker disease	F198S[a]
Familial Creutzfeldt–Jakob disease	E200K[a]
Gerstmann–Sträussler–Scheinker disease	Q217R
Familial Creutzfeldt–Jakob disease	octarepeat insert[a]

[a] Signifies genetic linkage between the mutation and the inherited prion disease.

findings emphasize the ambiguities that may arise from simply assessing resistance to proteolytic digestion. Whether PrP^{res} is useful in denoting PrP molecules that have been subjected to procedures modifying resistance to proteolysis but that neither convey infectivity nor cause disease remains questionable. Perhaps PrP^{res} is best reserved for PrP molecules that exhibit resistance to limited proteolysis after binding to PrP^{Sc}.

The term "PrP*" has been used in two different ways. First, it has been used to identify a fraction of PrP^{Sc} molecules that are infectious. Such a designation is thought to be useful because there are $\sim 10^5$ PrP^{Sc} molecules per infectious (ID$_{50}$) unit. Second, PrP* has been used to designate a metastable intermediate of PrP^C that is bound to a putative conversion cofactor, provisionally designated protein X. It is noteworthy that neither a subset of biologically active PrP^{Sc} molecules nor a metastable intermediate of PrP^C has been identified, to date.

In mice, *Prnp* is now known to be identical to two genes, *Sinc* and *Prn-i*, that are known to control the length of the incubation time in mice inoculated with prions. A gene, designated *Pid-1*, on mouse chromosome 17 also appears to influence experimental CJD and scrapie incubation times but information on this locus is limited.

Distinguishing among CJD, FFI, and Gerstmann–Sträussler–Scheinker syndrome (GSS) has grown increasingly difficult with the recognition that familial (f) CJD, GSS, and FFI are autosomal-dominant diseases caused by mutations in *PRNP* (Table 3). Initially, it was thought that each *PRNP* mutation was associated with a particular cliniconeuropathologic phenotype, but more exceptions are being recognized. Multiple examples of variations in the cliniconeuropathologic phenotype have been recorded within a single family in which all affected members carry the same *PRNP* mutation. Most patients with a *PRNP* mutation at codon 102 present with ataxia and have PrP amyloid plaques; such patients are generally given the diagnosis of GSS, but some individuals present with dementia, a clinical characteristic that is usually associated with CJD. For most inherited prion disease, the disease is specified by the respective mutation, such as fCJD(E200K) and GSS(P101L). In the case of FFI, describing the D178N mutation and M129 polymorphism seems unnecessary because this is the only known mutation–polymorphism combination that results in the FFI phenotype. The sporadic form of fatal insomnia is denoted sFI.

5
Discovery of the Prion Protein

The discovery of PrP 27–30 in SHa brain fractions progressively enriched for scrapie infectivity transformed research on scrapie and related diseases. PrP 27–30 was so named because it has an apparent molecular weight (M_r) of 27 kDa to 30 kDa (Fig. 1). PrP 27–30 is derived from the larger PrP^{Sc} by N-terminal truncation; both PrP 27–30 and PrP^{Sc} are infectious. PrP 27–30 not only provided a molecular marker that is specific for prion disorders but it was later shown to be the sole component of the prion particle.

The molecular biology and genetics of prions began with the purification of PrP 27–30 that allowed determination of its N-terminal, amino acid sequence. Multiple signals in each cycle of the Edman degradation suggested that either multiple proteins were present in these "purified fractions" or a single protein with a ragged N-terminus was present. When the signals in each cycle were grouped according to their strong, intermediate, and weak intensities, it became clear that a single protein with a ragged N-terminus was being sequenced (Fig. 3). Determination of a single, unique sequence for the N-terminus of PrP 27–30 permitted the synthesis of isocoding mixtures of oligonucleotides that were subsequently used to identify incomplete PrP cDNA clones from hamster and mouse. cDNA clones encoding the entire open reading

Relative amount	Amino acid sequence*
1	G-Q-G-G-G-T-H-N-Q-W-N-K-P-S-K
0.4	X-X-X-T-H-N-X-W-X-K-P
0.2	X-X-P-W-X-Q-X-X-X-T-H-X-Q-W

*Single-letter amino acid code. X = amino acid not determined at that cycle.

Fig. 3 Interpreted sequence, shown by single-letter amino acid codes, of the N-terminus of PrP 27–30. The "ragged ends" of PrP 27–30 are shown. X = amino acid was not detected at that cycle of the Edman degradation.

frames (ORFs) of SHaPrP and MoPrP were eventually recovered.

6
Prion Protein Isoforms

In Syrian hamsters, PrP^C and PrP^{Sc} are 209-residue proteins that are anchored to the cell surface by a glycosylphosphatidyl inositol (GPI) moiety and have the same covalent structure (Fig. 1). The N-terminal sequencing, the deduced amino acid sequences from PrP cDNA, and immunoblotting studies argue that PrP 27–30 is a truncated protein of approximately 142 residues, which is derived from PrP^{Sc} by limited proteolysis of the N-terminus.

In general, $\sim 10^5$ PrP^{Sc} molecules correspond to one ID_{50} unit (U) using the most sensitive bioassay. PrP^{Sc} is probably best defined as the alternative or abnormal isoform of PrP, which stimulates conversion of PrP^C into nascent PrP^{Sc}, accumulates, and causes disease. Although resistance to limited proteolysis has proved to be a convenient tool for detecting PrP^{Sc}, not all PrP^{Sc} molecules possess protease resistance, as discussed above. Some investigators equate protease resistance with PrP^{Sc} and this erroneous view has been compounded by the use of the term "PrP-res".

Although insolubility and protease resistance were used in initial studies to differentiate PrP^{Sc} from PrP^C, subsequent investigations showed that these properties are only surrogate markers of infectivity, as are high β-sheet content and polymerization into amyloid. When these surrogate markers are present, they are useful, but their absence does not establish a lack of prion infectivity. PrP^{Sc} is usually not detected by Western immunoblotting if fewer than 10^5 ID_{50} U mL^{-1} are present in a sample. Furthermore, PrP^{Sc} from different species or prion strains may exhibit different degrees of protease resistance.

6.1
Cell Biology of PrP^{Sc} Formation

In scrapie-infected cells, PrP^C molecules destined to become PrP^{Sc} exit to the cell surface prior to conversion into PrP^{Sc}. Like other GPI-anchored proteins, PrP^C appears to reenter the cell through a subcellular compartment bounded by cholesterol-rich, detergent-insoluble membranes, which might be caveolae or early endosomes. Within this cholesterol-rich, nonacidic compartment, GPI-anchored PrP^C can be either converted into PrP^{Sc} or partially degraded. Subsequently, PrP^{Sc} is trimmed at the N-terminus in an acidic compartment in scrapie-infected cultured cells to form PrP 27–30. In contrast, N-terminal truncation of PrP^{Sc} is minimal in the brain, where little PrP 27–30 is found.

7
Rodent Models of Prion Disease

Mice and hamsters are commonly used in experimental studies of prion disease. The shortest incubation times are achieved with intracerebral inoculation of prions with a sequence identical to that of the host animal; under these conditions, all animals develop prion disease within a narrow interval for a particular dose. When the PrP sequence of the donor prion differs from that of the recipient host, the incubation time is prolonged, and can be quite variable; often, many of the inoculated animals do not develop disease. This phenomenon is generally called the "species barrier".

8
PrP Gene Structure and Organization

Prnp is a member of the *Prn* gene family. The second member of this family to be identified is the *Prnd* gene that lies approximately 19 kb downstream from the PrP locus and encodes the doppel (Dpl) protein. The respective genes that encode PrP and Dpl appear to represent ancient gene duplication that occurred an prior to the speciation of mammals. The sequences are approximately 25% identical but the structures of the two proteins are highly conserved (Fig. 4). In contrast to PrP, which is expressed in many different tissues, Dpl expression is confined to the testis. Both Dpl and PrP are found on the surface of sperm but their functions are unknown. In contrast to PrP-deficient ($Prnp^{0/0}$) mice, Dpl-deficient ($Prnd^{0/0}$) mice are sterile. The knockout of both PrP and Dpl genes resulted in a sterile phenotype.

The entire ORF of all known mammalian and avian PrP genes resides within a single exon, which eliminates the possibility that PrP^{Sc} arises from alternative RNA splicing. The two exons of the SHaPrP gene are separated by a 10-kb intron; exon 1 encodes a portion of the 5' untranslated leader sequence. The PrP

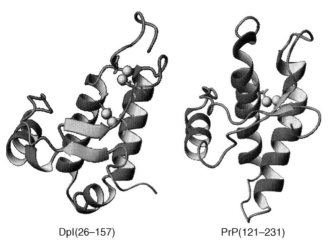

Fig. 4 Comparison of the NMR structures of mouse Dpl and PrP. Backbone topology of mouse Dpl(26–157) and mouse PrP(121–231). Figure prepared by Jane Dyson and Peter Wright.

genes of Syrian hamster, mouse, sheep, and rat contain three exons, with exon 3 encoding the ORF and 3′ untranslated region. The promoters of both the SHaPrP and MoPrP genes contain multiple copies of G-C–rich repeats and are devoid of TATA boxes. These G-C nonamers represent a motif that may function as a canonical binding site for the transcription factor Sp1.

Like the PrP gene, the entire ORF of the Dpl gene is encoded within a single exon. In some lines of $Prnp^{0/0}$ mice, high levels of Dpl expression were found in the brain. The expression of Dpl in the CNS was due to intergenic splicing of the nontranslated exons of the PrP gene with the translated exon of Dpl. That Dpl expression is neurotoxic was subsequently demonstrated by construction of Tg mice expressing Dpl in the brain.

8.1
Expression of the PrP Gene

Although PrP mRNA is constitutively expressed in the brains of adult animals, it is highly regulated during development. In the septum, levels of PrP mRNA and choline acetyltransferase were found to increase in parallel during development. In other brain regions, PrP gene expression occurs at an earlier age. *In situ* hybridization studies show that the highest levels of PrP mRNA are found in neurons.

Because no antibodies are currently available that clearly distinguish between PrP^C and PrP^{Sc}, PrP^C is generally measured in tissues from uninfected control animals, in which no PrP^{Sc} is found. PrP^{Sc} staining was minimal in the regions that intensely stained for PrP^C. A similar relationship between PrP^C and PrP^{Sc} was found in the amygdala. In contrast, PrP^{Sc} accumulated in the medial habenular nucleus, the medial septal nuclei, and the diagonal band of Broca; these areas were virtually devoid of PrP^C. In the white matter, bundles of myelinated axons contained PrP^{Sc} but were devoid of PrP^C. These findings suggest that prions are transported along axons, which is consistent with earlier findings in which scrapie infectivity migrated in a pattern consistent with retrograde transport. While the rate of PrP^{Sc} synthesis appears to be a function of the level of PrP^C expression in Tg mice, the level of PrP^{Sc} accumulation appears to be independent of PrP^C concentration.

8.2
Overexpression of Wild-type PrP Transgenes

Mice expressing different levels of the wild-type (wt) SHaPrP transgene were constructed. Inoculation of these Tg(SHaPrP) mice with SHa prions demonstrated abrogation of the species barrier, resulting in abbreviated incubation times due to a nonstochastic process. The length of the incubation time after inoculation with SHa prions was inversely proportional to the level of $SHaPrP^C$ in the brains of Tg(SHaPrP) mice. Bioassays of brain extracts from clinically ill Tg(SHaPrP) mice inoculated with Mo prions revealed that Mo prions, but no SHa prions, were produced. Conversely, inoculation of Tg(SHaPrP) mice with SHa prions led to the synthesis of only SHa prions.

During transgenetic studies, uninoculated older mice harboring high copy numbers of wt PrP transgenes derived from Syrian hamsters, sheep, and $Prnp^b$ mice spontaneously developed truncal ataxia, hind-limb paralysis, and tremors. These Tg mice exhibited profound necrotizing myopathy involving skeletal muscle, demyelinating polyneuropathy, and focal vacuolation of the CNS. Development of disease was dependent on transgene dosage. For example, Tg(SHaPrP$^{+/+}$)7 mice homozygous for the SHaPrP transgene array regularly developed disease between 450 and 600 days of age, while hemizygous Tg(SHaPrP$^{+/0}$)7 mice developed disease after >650 days. Attempts to demonstrate spontaneous generation of SHa prions in Tg(SHaPrP$^{+/+}$)7 mice have been unsuccessful.

8.3
PrP Gene Dosage Controls the Incubation Time

Incubation times have been used to isolate prion strains inoculated into sheep, goats, mice, and hamsters. The *Sinc* gene is a major determinant of incubation periods in mice. Once molecular clones of *Prnp* became available, a study was performed to determine if control of the length of the incubation time is genetically linked to the PrP gene. Because the availability of VM mice with prolonged incubation times that were used to define *Sinc* was restricted, I/LnJ mice were used in the crosses. Indeed, the incubation-time locus, designated *Prn-i*, was found to be either congruent with or closely linked to *Prnp*.

Although the amino acid substitutions in PrP that distinguish *Prnp^a* from *Prnp^b* mice argued for the congruency of *Prnp* and *Prn-i*, experiments with *Prnp^a* mice expressing *Prnp^b* transgenes demonstrated a "paradoxical" shortening of incubation times. These Tg mice were predicted to exhibit a prolongation of the incubation time after inoculation with RML prions based on (*Prnp^a* × *Prnp^b*) F1 mice, which exhibit long incubation times. Those findings were described as a "paradoxical" shortening because we and others had believed for many years that long incubation times are dominant traits. From studies of congenic and transgenic mice expressing different numbers of the *a* and *b* alleles of *Prnp*, these findings were discovered not to be paradoxical; indeed, they resulted from increased PrP gene dosage. When the RML isolate was inoculated into congenic and transgenic mice, increasing the number of copies of the *a* allele was found to be the major determinant in reducing the incubation time; however, increasing the number of copies of the *b* allele also reduced the incubation time, but not to the same extent as that seen with the *a* allele. Gene-targeting studies established that the *Prnp* gene controls the incubation time and as such, is congruent with both *Prn-i* and *Sinc*.

8.4
PrP-deficient Mice

Ablation of the PrP gene in mice did not affect development of these animals. $Prnp^{0/0}$ mice remain healthy for more than two years. Acute suppression of PrP expression by addition of doxycycline to the drinking water of bigenic mice with an inducible PrP transgene under the control of the tetracycline promoter did not result in any untoward effects in adult mice. Similarly, bigenic mice, in which neuronal PrP expression was terminated in adulthood, showed no discernable deficits.

In two $Prnp^{0/0}$ lines, Purkinje cell loss was accompanied by ataxia beginning at approximately 70 weeks of age. Crossing one of these $Prnp^{0/0}$ lines with Tg mice overexpressing MoPrP rescued the ataxic phenotype. With the discovery of Dpl, it became clear, as described above, that Dpl expression in the brains of these $Prnp^{0/0}$ mice provoked cerebellar degeneration.

$Prnp^{0/0}$ mice inoculated with prions are resistant to infection. $Prnp^{0/0}$ mice were sacrificed 5 days, 60 days, 120 days, and 315 days after inoculation with RML prions and brain extracts were bioassayed in Swiss CD-1 mice. Except for residual infectivity from the inoculum detected at 5 days after inoculation, no infectivity was found in the brain extracts, as the Swiss CD-1 mice did not develop disease.

$Prnp^{0/0}$ mice crossed with Tg(SHaPrP) mice were rendered susceptible to SHa prions but remained resistant to Mo prions. Because the absence of PrP^C expression does not provoke disease, it is likely that scrapie and other prion diseases are a consequence of PrP^{Sc} accumulation rather than inhibition of PrP^C function. Such an interpretation is consistent with the dominant inheritance of familial prion diseases.

Mice heterozygous ($Prnp^{0/+}$) for ablation of the PrP gene had prolonged incubation times when inoculated with Mo prions and developed signs of neurologic dysfunction at 400 to 460 days after inoculation. These findings are in accord with studies on Tg(SHaPrP) mice in which increased SHaPrP expression was accompanied with shortened incubation times.

Because $Prnp^{0/0}$ mice do not express PrP^C, we reasoned that they might more readily produce α-PrP antibodies. $Prnp^{0/0}$ mice immunized with Mo or SHa prion rods produced α-PrP antisera that bound MoPrP, SHaPrP, and HuPrP. These findings contrast with earlier studies in which α-MoPrP antibodies could not be produced in mice presumably because the mice had been rendered tolerant by the presence of $MoPrP^C$. That $Prnp^{0/0}$ mice readily produce α-PrP antibodies is consistent with the hypothesis that the lack of an immune response in prion diseases is due to the fact that PrP^C and PrP^{Sc} share many epitopes.

8.5
Species Variations in the PrP Sequence

PrP is posttranslationally processed to remove a 22-amino acid, N-terminal signal peptide. The C-terminal 120 amino acids contain two conserved disulfide-bonded cysteines and a sequence that beckons addition of a GPI anchor. Twenty-three residues are removed during the addition of this GPI moiety, which anchors the protein to the cell membrane. Contributing to the mass of the protein are two Asn side chains linked to large oligosaccharides with multiple structures that have been shown to be complex and diverse. Although many species variants of PrP have now been sequenced, only the chicken sequence has been found to differ greatly from the human sequence. The alignment of the translated sequences from more than 40 PrP genes shows a striking degree of conservation between the mammalian sequences and is suggestive of the retention of some important function through evolution (Fig. 5). Cross-species conservation of PrP sequences makes it difficult to draw conclusions about the functional importance of many of the individual residues in the protein.

8.6
N-terminal Sequence Repeats

The N-terminal domain of mammalian PrP contains five copies of a P(H/Q)GGG (G)WGQ octarepeat sequence, occasionally more, as in the case of one sequenced bovine allele, which has six copies. These repeats are remarkably conserved between species, which implies a functionally

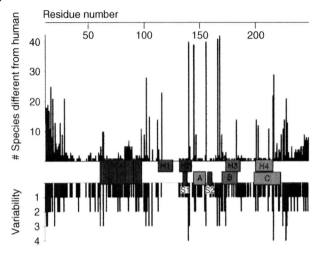

Fig. 5 Species variations and mutations of the prion protein gene. The x-axis represents the human PrP sequence, with the five octarepeats and H1–H4 regions of putative secondary structure shown as well as the three α-helices A, B, and C and the two β-strands S1 and S2 as determined by NMR. Vertical bars above the axis indicate the number of species that differ from the human sequence at each position. Below the axis, the length of the bars indicates the number of alternative amino acids at each position in the alignment. Data were compiled by Paul Bamborough and Fred E. Cohen (see color plate p. x).

important role. The chicken sequence contains a different repeat, PGYP(H/Q)N. Although insertions of extra repeats have been found in patients with familial prion disease, naturally occurring deletions of single octarepeats do not appear to cause disease, and deletion of all these repeats does not prevent PrP^C from undergoing a conformational transition into PrP^{Sc}.

It was suggested that the histidine residues in the octarepeats might bind metal ions and that immobilized metal ion affinity chromatography (IMAC) might facilitate the purification of PrP; indeed, IMAC did prove to be useful in the purification of PrP^C. The availability of purified PrP^C allowed studies comparing the secondary structures of PrP^C and PrP^{Sc}. The metal ion-peptide complexes were found to be much more soluble than the metal ions bound to full-length PrP. Using full-length recombinant (rec) PrP, Cu^{2+} was found to bind with a much higher avidity than any other metal ion, but the concentration for half-maximal binding for Cu^{2+} was 14 μM at pH 6.0, indicating a rather low affinity of PrP for Cu^{2+}. At neutral pH, Cu^{2+}–PrP complexes tended to form large aggregates and precipitate, so, synthetic peptides containing the octarepeats were used to study more extensively the interaction of Cu^{2+} with PrP. The concentration for half-maximal binding for Cu^{2+} to a peptide (51–75) containing two octarepeats was 6 μM at pH 7.5. When a peptide (58–91) containing four octarepeats was studied, the binding of Cu^{2+} was found to be cooperative and highly pH-dependent (Fig. 6). The midpoint of the pH-dependence transition was pH

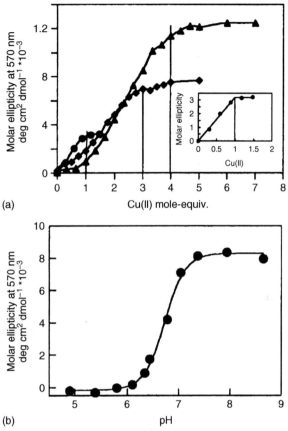

Fig. 6 Binding of copper ions to synthetic peptides containing the octarepeats of PrP. (a) Cu(II) binding curves: molar ellipticity at 570 nm with increasing amounts of Cu(II), pH 7.5. Filled circles, 2-His peptide, PrP(51–75) (0.34 mM). Filled diamonds, 3-His peptide, PrP(66–91) (0.021 mM). Filled triangles, 4-His peptide, PrP(58–91) (0.033 mM). (b) pH dependence of the ellipticity at 570 nm for PrP(58–91). The pH-dependence curve has been fitted to the following equation: $\Delta\varepsilon_{obs} = [\Delta\varepsilon_{acid}[H^+]^n + \Delta\varepsilon_{base}[H^+]K_a]/[H^+]^n + K_a]$, in which n = Hill coefficient and K_a = acid dissolution constant for the transition. The midpoint of the transition is pH 6.7. Reprinted with permission, from (Viles, J.H., Cohen, F.E., Prusiner, S.B., Goodin, D.B., Wright, P.E., Dyson, H.J. (1999) Copper binding to the prion protein: structural implications of four identical cooperative binding sites, *Proc. Natl. Acad. Sci. U. S. A.* **96**, 2042–2047), copyright 1999 National Academy of Sciences, USA.

6.7, suggesting that the binding of Cu^{2+} occurred through the imidazole nitrogens of histidine residues.

In studies of full-length recPrP, Cu^{2+} was found to catalyze the oxidation of the histidine residues within the octarepeats. These findings argue that PrP may function as Cu^{2+}-binding protein. Consistent with studies on recPrP, membranes isolated from the brains of $Prnp^{0/0}$ mice were reported to have substantially lower levels of Cu^{2+} than wt mice. However, attempts to confirm this relationship between brain Cu^{2+} and PrP levels were unsuccessful. Interestingly, the copper-chelating reagent cuprizone administered to rodents causes spongiform degeneration resembling that induced by prions.

8.7
Conserved Ala-Gly Region

In addition to the octarepeat, the other region of notable conservation is in the sequence at the C-terminal end of the last octarepeat. Here, an unusual glycine- and alanine-rich region from A113 to Y128 is found (Fig. 5). Although no differences have been found in this part of the sequence between species, a single point mutation A117V is linked to GSS. The conservation of structure suggests an important role in the function of PrP^C; in addition, this region is likely to be important in the conversion of PrP^C into PrP^{Sc}.

9
Structures of PrP Isoforms

Mass spectrometry and gas phase sequencing were used to search for posttranslational chemical modifications that might explain the differences in the properties of PrP^C and PrP^{Sc}. No modifications differentiating PrP^C from PrP^{Sc} were found. These observations forced the consideration of the possibility that a conformational change distinguishes the two PrP isoforms.

When the secondary structures of the PrP isoforms were compared by optical spectroscopy, they were found to be markedly different. Fourier transform infrared (FTIR) and circular dichroism (CD) spectroscopy studies showed that PrP^C contains approximately 40% α-helix and little β-sheet while PrP^{Sc} is composed of approximately 30% α-helix and 45% β-sheet. That the two PrP isoforms have the same amino acid sequence runs counter to the widely accepted view that the amino acid sequence specifies only one biologically active conformation of a protein. Like PrP^{Sc}, PrP 27–30 has a high β-sheet content, which is consistent with the earlier finding that PrP 27–30 polymerizes into amyloid fibrils. Denaturation of PrP 27–30 under conditions that reduced infectivity resulted in a concomitant diminution of β-sheet content.

Prior to comparative studies on the structures of PrP^C and PrP^{Sc}, metabolic labeling studies showed that the acquisition of protease resistance in PrP^{Sc} is a posttranslational process. In a search for chemical differences that would distinguish PrP^{Sc} from PrP^C, ethanolamine was identified in hydrolysates of PrP 27–30, which signaled the possibility that PrP might contain a GPI anchor. Both PrP isoforms were found to carry GPI anchors. PrP^C was found on the surface of cells, where it could be released by cleavage of the anchor. Subsequent studies showed that PrP^{Sc} formation occurs after PrP^C reaches the cell surface and localizes to caveolae-like domains (CLDs).

9.1
NMR Structures of recPrP

The NMR structures of more than 10 PrPs from different species have been determined to be quite similar (Table 4). The recPrP molecules were produced in *E. coli* and isotopically labeled. When refolded into proteins with a high α-helical content, these recPrPs appear to be representative of PrP^C. It is noteworthy that recPrPs produced in *E. coli* have neither N-linked sugar chains nor a GPI anchor, both of which are attached to PrP^C. However, synthetic prions produced from recMoPrP(89–230) demonstrated that neither N-linked sugar chains nor a GPI anchor is required for PrP^{Sc} formation. The structure of MoDpl

Tab. 4 Structures of PrP, Dpl, and Ure2p[a].

Species	Protein	Determined by	PDB accession codes
Mouse	recPrP(121–231)	NMR	1AG2
Mouse	recPrP(23–231)	NMR	data not released?
Syrian hamster	recPrP(90–231)	NMR	1B10, updated 1999
Syrian hamster	recPrP(29–231)	NMR	data not released?
Syrian hamster	synPrP(104–113) in complex with 3F4	NMR	1CU4 in complex with 3F4
Syrian hamster	recPrP(90–231)	NMR	used to update 1B10
Human	recPrP(23–230)	NMR	1QLX and 1QLZ
Human	recPrP(90–230)	NMR	1QM0 and 1QM1
Human	recPrP(121–230)	NMR	1QM2 and 1QM3
Cattle	recPrP(121–230)	NMR	1DWY and 1DWZ
Cattle	recPrP(23–230)	NMR	1DX0 and 1DX1
Human	recPrP(121–230), M166V	NMR	1E1G and 1E1J
Human	recPrP(121–230), S170N	NMR	1E1P and 1E1S
Human	recPrP(121–230), R220K	NMR	1E1U and 1E1W
Human	recPrP(90–231), E200K	NMR	1QM0, 1FKC, and 1FO7
Human	recPrP(90–231)	X ray	1I4M
Human	recPrP(121–230), M166C, E221C	NMR	1H0L
Sheep	synPrP(145–169)	NMR	1G04
Sheep	synPrP(145–169)	NMR	1M25
Mouse	recDpl(26–157)	NMR	1I17
Human	recDpl(24–152)	NMR	1LG4
S. cerevisiae	recUre2p(95–354)	X ray	1G6W and 1G6Y
S. cerevisiae	recUre2p(95–354) Glutathion complex	X ray	1JZR, 1K0B, and 1K0D
S. cerevisiae	recUre2p(95–354) S-P-Nitrobenzylglutathione complex	X ray	1K0C
S. cerevisiae	recUre2p(95–354) S-Hexylglutathione complex	X ray	1K0A
S. cerevisiae	recUre2p(97–354)	X ray	1HQO

[a]Data compiled by Holger Wille.

is similar to that of MoPrP even though the two proteins share only 25% sequence similarity (Fig. 4).

The NMR structure of recSHaPrP (90–231) was determined after the protein was purified and refolded (Fig. 7a). Residues 90–112 are not shown because marked conformational heterogeneity was found in this region while residues 113–126 constitute the conserved hydrophobic region that also displays some structural plasticity. Although some features of the structure of recSHaPrP(90–231) are similar to those reported earlier for the smaller recMoPrP(121–231) fragment, substantial differences were found. For example, the loop at the N-terminus of helix B is defined in recSHaPrP(90–231) but is disordered in recMoPrP(121–231); in addition, helix C is composed of residues 200–227 in recSHaPrP(90–231) but extends only from residues 200–217 in recMoPrP(121–231). The loop and the C-terminal portion of

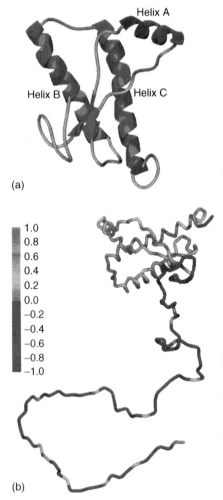

Fig. 7 Structures of PrP^C. (a) NMR structure of Syrian hamster (SHa) recombinant (rec) PrP(90–231). Presumably, the structure of the α-helical form of recPrP(90–231) resembles that of PrP^C. recPrP(90–231) is viewed from the interface where PrP^{Sc} is thought to bind to PrP^C. The color scheme is: α-helices A (residues 144–157), B (172–193), and C (200–227) in blue; loops in yellow; residues 129–134 encompassing strand S1 and residues 159–165 encompassing strand S2 in green. (b) Schematic diagram showing the flexibility of the polypeptide chain for PrP(29–231). The structure of the portion of the protein representing residues 90–231 was taken from the coordinates of PrP(90–231). The remainder of the sequence was hand-built for illustration purposes only. The color scale corresponds to the heteronuclear [^{1}H]-^{15}N NOE data: red for the lowest (most negative) values, where the polypeptide is most flexible, to blue for the highest (most positive) values in the most structured and rigid regions of the protein. (See color plate. p. xiii)

helix C are particularly important, as described below. Whether the differences between the two recPrP fragments are due to (1) their different lengths, (2) species-specific differences in sequences, or (3) the conditions used for solving the structures remains to be determined.

NMR studies of recMoPrP(23–231) and recSHaPrP(29–231) showed that like the shorter fragments described above, full-length PrP is also a three helix–bundle protein with two short antiparallel β-strands. While the three helices form a globular C-terminal domain, the N-terminal domain is highly flexible and lacks identifiable secondary structure under the experimental conditions employed (Fig. 7b). Studies of recSHaPrP(29–231) indicate transient interactions between the C-terminal end of helix B and the highly flexible, N-terminal random-coil containing the octarepeats (residues 29–125).

9.2
Electron Crystallography of PrP^{Sc}

Because the insolubility of PrP^{Sc} has frustrated structural studies by X-ray crystallography or NMR spectroscopy, electron crystallography was used to characterize the structure of two infectious variants of PrP. Isomorphous, two-dimensional crystals of PrP 27–30 and a miniprion (PrP^{Sc}106) were identified by negative-stain electron microscopy. Image processing allowed the extraction of limited structural information to 7-Å resolution. Comparing projection maps of PrP 27–30 and PrP^{Sc}106, the 36-residue internal deletion of the miniprion was visualized and the N-linked sugars were localized (Fig. 8). The dimensions of the monomer and the locations of the deleted segment and sugars were used as constraints in the construction of models for PrP^{Sc}. Only models featuring parallel β-helices as the key element could satisfy the constraints.

A single antiparallel β-sheet as previously suggested was not consistent with the observed densities in the projection maps obtained from the 2D crystals. Specifically, the sheets were far too wide to fit into the observed hexameric arrangement. Efforts to adjust the sheet morphology to fit the density required the use of shorter strands. The amount of β-structure in these altered β-sheets was no longer compatible with the amounts of β-sheet observed by FTIR spectroscopy. Furthermore, antiparallel β-sheets typically have a twist of ~20° per strand.

From a study of 119 all-β folds observed in globular proteins, it was determined that if PrP^{Sc} follows a known protein fold, it adopts either a β-sandwich or parallel β-helical architecture. With increasing evidence arguing for a parallel β-sheet organization in amyloids, it was argued that the sequence of PrP is compatible with a parallel left-handed β-helical fold. Left-handed β-helices readily form trimers, providing a natural template for a trimeric model of PrP^{Sc} (Fig. 9a). This trimeric model accommodates the PrP sequence from residues 89–175 in a β-helical conformation with the C-terminus (residues 176–227) retaining the disulfide-linked α-helical conformation observed in PrP^{C}. In addition, the proposed model matches the structural constraints of the crystals, positioning residues 141–176 and the N-linked sugars appropriately. The parallel left-handed β-helical model provides a coherent framework that is consistent with the stacking of trimeric units of PrP 27–30 to form PrP amyloid fibrils (Fig. 9b).

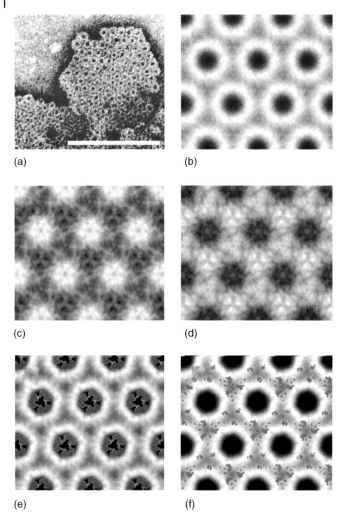

Fig. 8 Two-dimensional crystals of PrP^{Sc}. (a) A 2D crystal of PrP^{Sc}106 stained with uranyl acetate. Bar = 100 nm. (b) Image processing result after correlation-mapping and averaging followed by crystallographic averaging. (c and d) Subtraction maps between the averages of PrP 27–30 and PrP^{Sc}106 (panel B). (c) PrP^{Sc}106 minus PrP 27–30 and (d) PrP 27–30 minus PrP^{Sc}106, showing major differences in lighter shades. (e and f) The statistically significant differences between PrP 27–30 and PrP^{Sc}106 calculated from (c) and (d) in red and blue, respectively, overlaid onto the crystallographic average of PrP 27–30. Reprinted with permission, from Wille, H., Michelitsch, M.D., Guénebaut, V., Supattapone, S., Serban, A., Cohen, F.E., Agard, D.A. Prusiner, S.B. (2002) Structural studies of the scrapie prion protein by electron crystallography, *Proc. Natl. Acad. Sci. U. S. A.* **99**, 3563–3568 Wille et al., (2002), copyright 2002 National Academy of Sciences, USA. (See color plate. p. xiv)

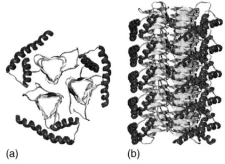

Fig. 9 Left-handed β-helical models of PrP 27–30. (a) Trimeric model of PrP 27–30 built by superimposing three monomeric models onto the coordinates of the C'_αs of the structure of trimeric carbonic anhydrase from *Methanosarcina thermophila* (PDB ID 1THJ). (b) A model of the PrP 27–30 fiber, constructed by stacking five trimeric discs. Figure prepared by Cédric Govaerts.

10
Prion Replication

The mechanism of prion replication is not well understood. The formation of nascent prions requires that PrP^C be converted into PrP^{Sc}. Some investigators have argued that this process involves a nucleation–polymerization (NP) reaction while others have contended that a more likely mechanism employs a template-assisted process.

In an uninfected cell, PrP^C with the wt sequence is likely to exist in equilibrium in its monomeric α-helical, protease-sensitive state or bound to some other protein, such as protein X (Fig. 10). The conformation of PrP^C that is bound to protein X is denoted PrP*; this conformation is likely to be different from that determined under aqueous conditions for monomeric recPrP. The PrP*–protein X complex will bind PrP^{Sc}, thereby creating a replication-competent assembly. Order-of-addition experiments demonstrate that protein X binding to PrP^C precedes productive PrP^{Sc} interactions. A conformational change takes place wherein PrP*, in a shape competent for binding to protein X and PrP^{Sc}, represents the initial phase in the formation of infectious PrP^{Sc}.

Although many lines of evidence reviewed below argue for a template-assisted process in the formation of nascent PrP^{Sc}

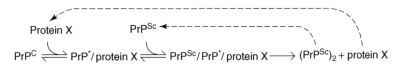

Fig. 10 Hypothetical scheme for template-assisted PrP^{Sc} formation. In the initial step, PrP^C binds to protein X to form the PrP*–protein X complex. Next, PrP^{Sc} binds to PrP* that has already formed a complex with protein X. When PrP* is transformed into a nascent molecule of PrP^{Sc}, protein X is released and a dimer of PrP^{Sc} remains. The inactivation target size of an infectious prion suggests that it is composed of a dimer of PrP^{Sc}. In the model depicted here, a fraction of infectious PrP^{Sc} dimers dissociates into uninfectious monomers as the replication cycle proceeds, while a majority of the dimers accumulates in accord with the increase in prion titer that occurs during the incubation period. Another fraction of PrP^{Sc} is cleared presumably by cellular proteases. The precise stoichiometry of the replication process remains uncertain. Reprinted with permission, from (Prusiner, S.B., Scott, M.R., DeArmond, S.J., Cohen, F.E. (1998) Prion protein biology, *Cell* **93**, 337–348), copyright 1998 Cell Press.

in mammalian cells, the formation of synthetic prions from PrP(89–143,P101L) as well as from recPrP(89–230) establishes that infectious prions can be generated *de novo* without the aid of auxiliary proteins.

10.1
Mechanism of Prion Propagation

In attempting to probe the mechanism of prion formation, it is reasonable to determine the rate-limiting step in prion formation. First, the impact of the concentration of PrP^{Sc} in the inoculum, which is inversely proportional to the length of the incubation time, must be considered. Second, the sequence of PrP^{Sc} that forms an interface with PrP^C must be taken into account. When the sequences of the two isoforms are identical, the shortest incubation times are observed. Third, the strain-specific conformation of PrP^{Sc} must be considered. Some prion strains exhibit longer incubation times than others; the mechanism underlying this phenomenon is not understood. From these considerations, a set of conditions exist under which initial PrP^{Sc} concentrations can be rate limiting. These effects presumably relate to the stability of the PrP^{Sc}, its targeting to the correct cells and subcellular compartments, and its ability to be cleared. Once infection in a cell is initiated and endogenous PrP^{Sc} production is operative, then the following discussion of PrP^{Sc} formation seems most applicable. If the assembly of PrP^{Sc} into a specific dimeric or multimeric arrangement is difficult, then an NP formalism would be relevant. In NP processes, nucleation is the rate-limiting step and elongation, or polymerization, is facile. These conditions are frequently observed in peptide models of aggregation phenomena; however, studies in ScN2a cells and with Tg mice expressing foreign PrP genes suggest that a different process occurs.

10.2
Template-assisted Prion Formation

From investigations with mice expressing both the SHaPrP transgene and the endogenous MoPrP gene, designated Tg(SHaPrP) mice, it is clear that PrP^{Sc} provides a template for directing prion replication; a template is defined as a catalyst that leaves its imprint on the product of the reaction. Inoculation of these mice with SHaPrP^{Sc} leads to the production of nascent SHaPrP^{Sc} and not MoPrP^{Sc}. Conversely, inoculation of the Tg(SHaPrP) mice with MoPrP^{Sc} results in MoPrP^{Sc} and not SHaPrP^{Sc} formation.

Even stronger evidence for templating has emerged from studies of prion strains passaged in Tg(MHu2M)*Prnp*$^{0/0}$ mice, which express a chimeric mouse–human gene, as described in more detail below. Even though the conformational templates were initially generated with PrP^{Sc} molecules having different sequences from patients with inherited prion diseases, these templates are sufficient to direct replication of distinct PrP^{Sc} molecules when the amino acid sequences of the substrate PrPs are identical.

Another line of evidence for template-assisted prion replication comes from studies of FI. In FFI, the protease-resistant fragment of PrP^{Sc} after deglycosylation has an M_r of 19 kDa when measured by immunoblotting of brain extracts from humans with FFI or from Tg(MHu2M)*Prnp*$^{0/0}$ mice that were inoculated with FFI prions. In both humans and inoculated Tg(MHu2M)*Prnp*$^{0/0}$ mice on both first and second passage, PrP^{Sc} is confined largely to the thalamus. These findings argue that the conformation of

PrP^{Sc} that yields a 19-kDa polypeptide after deglycosylation is propagated both in humans and in mice expressing an artificial chimeric transgene; in addition, prion accumulation is confined to the thalamus. Additional evidence supporting these assertions comes from a patient who died after developing a clinical disease similar to FFI. Because both PrP alleles encoded the wt PrP sequence and methionine at position 129, we labeled this sFI. At autopsy, the spongiform degeneration, reactive astrocytic gliosis, and PrP^{Sc} deposition were confined to the thalamus. Moreover, the PrP^{Sc} both in the patient's brain and in the brains of inoculated Tg(MHu2M)$Prnp^{0/0}$ mice yielded a 19-kDa fragment after deglycosylation. These findings argue that the clinicopathologic phenotype is determined by the conformation and not the amino acid sequence of PrP^{Sc}. Additionally, FI prions can be replicated with mutant human PrP(D178N), wt human PrP, or chimeric human–mouse PrP. With these different PrPs as substrates, the conformation of PrP^{Sc} must be faithfully copied; template-assisted prion replication provides for such a mechanism.

10.3
Evidence for Protein X

Protein X was postulated to explain the results of the transmission of Hu prions to Tg mice (Table 5). Mice expressing both MoPrP^C and HuPrP^C were resistant while those expressing only HuPrP^C were susceptible to Hu prions. These results argue that MoPrP^C inhibited transmission of Hu prions, that is, the formation of nascent HuPrP^{Sc}. In contrast, mice expressing both MoPrP^C and chimeric MHu2MPrP^C were susceptible to Hu prions and mice expressing MHu2MPrP^C alone were only slightly more susceptible. These findings contend that MoPrP^C has only a minimal effect on the formation of chimeric MHu2MPrP^{Sc}.

When the data on Hu prion transmission to Tg mice were considered together, they suggested that MoPrP^C prevented the conversion of HuPrP^C into PrP^{Sc} by binding to another Mo protein (protein X) with a higher affinity than HuPrP^C does. It was postulated that MoPrP^C has little effect on the formation of PrP^{Sc} from MHu2MPrP^C (Table 5) because MoPrP^C

Tab. 5 Mouse PrP^C inhibits transmission of human sCJD prions to transgenic mice expressing human PrP^C [a].

Transgene	MoPrP genotype	Incubation time (Mean days ± SEM) (n/n$_0$)
Non-Tg FVB	+/+	701 (1/10)
HuPrP(V129)	+/+	721 (1/10)
HuPrP(V129)	0/0	263 ± 2 (6/6)
MHu2M	+/+	238 ± 3 (8/8)
MHu2M	0/0	191 ± 3 (10/10)
MHu2M(M165V,E167Q)	0/0	106 ± 2 (13/13)
MHu2M(M165V,E167Q)b	0/0	196 ± 1 (40/40)

[a] Inocula from sCJD(MM1) case designated RG. All MHu2M transgenes encode M129. n, number of sick mice; n$_0$, number of inoculated mice.
[b] Stainless steel suture wire coated with sCJD(MM1) brain homogenate were implanted into the brains of Tg mice.

and MHu2MPrP^C share the same amino acid sequence at the C-terminus. This also suggested that MoPrP^C only weakly inhibited transmission of SHa prions to Tg(SHaPrP) mice because SHaPrP is more closely related to MoPrP than to HuPrP.

The search for protein X has been frustrating because many proteins are known to bind to PrP^C but none have been shown to participate in PrP^{Sc} formation. Whether protein X is one protein or a complex of proteins remains to be established. It is reasonable to expect that mice deficient in protein X would not replicate prions or do so extremely slowly, resulting in prolonged incubation times.

10.4
Dominant-negative Inhibition

To extend the foregoing findings in Tg mice, an expression vector with an insert encoding a chimeric mouse–hamster PrP, designated MHM2PrP, was transfected into ScN2a cells. MHM2PrP carries two Syrian hamster residues, which create an epitope that is recognized by the anti-SHaPrP 3F4 mAb that does not react with mouse PrP. Substitution of a Hu residue at position 214 or 218 in MHM2PrP prevented formation of MHM2PrP^{Sc} in ScN2a cells. The side chains of these residues protrude from the same surface of the C-terminal α-helix, forming a discontinuous epitope with residues 167 and 171 in an adjacent loop. Like MHM2PrP(Q218K), substitution of a basic residue at position 167 or 171 prevented PrP^{Sc} formation. When MHM2PrP and MHM2PrP(Q218K) were coexpressed, the conversion of MHM2PrP^C into PrP^{Sc} was inhibited, arguing that MHM2PrP(Q218K) was acting as a dominant negative. Similar results were obtained when studies were performed with cells expressing MHM2PrP(Q167R).

One interpretation of dominant-negative inhibition of prion propagation is that mutant PrP binds more avidly to protein X than does wt PrP. In this scenario, mutant PrP binds to protein X and prevents binding of wt PrP. This explanation is also consistent with the protective effects of basic polymorphic residues in PrP of humans and sheep. The E219K substitution seems to render humans resistant to sCJD and Q171R renders sheep resistant to scrapie. In MHM2, the Q218K mutation corresponds to E219K in humans and the Q167R mutation corresponds to Q171R in sheep (Fig. 11).

To determine whether dominant-negative inhibition of prion formation occurs *in vivo*, Tg mice expressing PrP with either the Q167R or Q218K substitution alone or in combination with wt PrP were produced. Tg(MoPrP,Q167R) $Prnp^{0/0}$ mice expressing mutant PrP at levels equal to non-Tg FVB mice were inoculated with prions and remained healthy for more than 550 days, indicating that inoculation did not initiate a process sufficient to cause disease. Immunoblots of brain homogenates and histologic analysis revealed no abnormalities. Tg(MoPrP,Q167R) $Prnp^{+/+}$ mice expressing both mutant and wt PrP exhibited neurologic dysfunction at $\sim$450 days after inoculation; the brains of three of these mice that were sacrificed at 300 days revealed low levels of PrP^{Sc} as well as numerous vacuoles and severe astrocytic gliosis. Both Tg(MoPrP,Q218K) $Prnp^{0/0}$ and Tg(MoPrP,Q218K) $Prnp^{+/+}$ mice expressing the transgene product at 16× remained healthy for more than $\sim$450 days after inoculation. Neither PrP^{Sc} deposition nor neuropathology was found. Tg mice expressing MoPrP(Q218K) at 32× developed spontaneous neurodegeneration

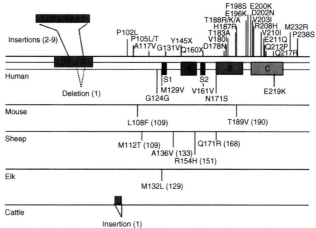

Fig. 11 Mutations and polymorphisms of the prion protein gene. PrP mutations causing inherited human prion disease and PrP polymorphisms found in humans, mice, sheep, elk, and cattle. Above the line of the human sequence are mutations that cause inherited prion disease. Below the lines are polymorphisms, some but not all of which are known to influence the onset as well as the phenotype of disease. Residue numbers in parentheses correspond to the human codons. Much of the data was compiled by J.-L. Laplanche.

at ~450 days of age. These studies demonstrate that although dominant-negative inhibition of wt PrP^{Sc} formation occurs, expression of dominant-negative PrP at the same level as wt PrP does not completely prevent prion formation. However, expression of dominant-negative PrP alone had no deleterious effects on the mice and did not support prion propagation.

11
Sporadic Human Prion Diseases

In most patients with CJD, there is neither an infectious nor heritable etiology. How prions arise in patients with sCJD is unknown; hypotheses include: (1) horizontal transmission from humans or animals; (2) somatic mutation of the *PRNP* ORF; (3) spontaneous conversion of PrP^C into PrP^{Sc}; and (4) the accumulation of PrP^{Sc}, which is normally present at very low levels. Numerous attempts to establish an infectious link between sCJD and a preexisting prion disease in animals or humans have been unrewarding. Studies demonstrating the formation of synthetic prions from PrP(89–143,P101L) as well as from recPrP(89–230) show that only PrP^C is needed for an animal for producing infectious prions. Thus, the spontaneous conversion of PrP^C to PrP^{Sc} seems reasonable and any mammal expressing PrP^C is capable of forming prions.

12
Heritable Human Prion Diseases

The recognition that 10 to 15% of CJD cases are familial led to the suspicion that genetics plays a role in this disease. As with scrapie, the relative contributions of genetic and infectious etiologies in the human prion diseases remain puzzling.

More than 30 different mutations of the PrP gene have been shown to segregate with the heritable human prion diseases. Five of these mutations have been genetically linked to the inherited human prion diseases (Table 3). Virtually all cases of GSS and FFI appear to be caused by germline mutations in the PrP gene. The brains of humans dying of inherited prion disease contain infectious prions that have been transmitted to experimental animals.

12.1
GSS and Genetic Linkage

The discovery that GSS, which is a familial disease, could be transmitted to apes and monkeys was first reported when many still thought that scrapie, CJD, and related disorders were caused by viruses. With the discovery that the P102L mutation of the human PrP gene was genetically linked to GSS, prion diseases were concluded to have both genetic and infectious etiologies. In that study, the P102L mutation was linked to development of GSS with a logarithm of the odds (LOD) score exceeding 3, demonstrating a tight association between the altered genotype and disease phenotype (Fig. 3B). This mutation may be caused by the deamination of a methylated CpG in a germline PrP gene, which results in the substitution of a leucine for proline. This mutation has been found in many families in numerous countries, including the first family identified to have GSS.

12.2
fCJD Caused by Octarepeat Inserts

An insert of 144 bp containing six octarepeats at codon 53, in addition to the five that are normally present, was described in patients with CJD from four families residing in southern England. Genealogic investigations showed that all four families are related, arguing for a single founder born more than two centuries ago. The LOD score for this extended pedigree exceeds 11. Studies from several laboratories have demonstrated that inserts of two, four, five, six, seven, eight, or nine octarepeats in addition to the normal five are found in individuals with inherited CJD (Fig. 11).

12.3
fCJD in Libyan Jews

The unusually high incidence of CJD among Israeli Jews of Libyan origin was thought to be due to the consumption of lightly cooked sheep brain or eyeballs. Molecular genetic investigations revealed that Libyan and Tunisian Jews with fCJD have a PrP gene point mutation at codon 200, resulting in a Glu-to-Lys substitution (Fig. 11). The E200K mutation has been genetically linked to fCJD, with an LOD score exceeding 3; the same mutation has also been found in patients from Orava in North Central Slovakia, in a cluster of familial cases in Chile, and in a large German family living in the United States.

Most patients are heterozygous for the mutation and thus, express both mutant and wt PrP^C. In the brains of patients who die of fCJD(E200K), mutant PrP^{Sc} is both insoluble and protease-resistant, whereas much of wt PrP differs from both PrP^C and PrP^{Sc} in that it is insoluble but readily digested by proteases. Whether this form of PrP is an intermediate in the conversion of PrP^C into PrP^{Sc} remains to be established.

12.4
Penetrance of fCJD

Life-table analyses of carriers harboring the codon 200 mutation exhibit complete penetrance. Therefore, if the carriers live long enough, they will all eventually develop prion disease. Some investigators have argued that the inherited prion diseases are not fully penetrant and thus an environmental factor, such as the ubiquitous "scrapie virus," is required for illness to be manifest, but no viral pathogen has been found to date.

12.5
Fatal Familial Insomnia

The D178N mutation has been linked to the development of FFI, with an LOD score exceeding 5. More than 30 families worldwide with FFI have been recorded. Studies of inherited human prion diseases demonstrate that the amino acid at polymorphic residue 129 with the D178N pathogenic mutation alters the clinical and neuropathologic phenotype. The D178N mutation combined with M129 results in FFI. In this disease, adults generally over age 50 years present with a progressive sleep disorder and usually die within one year. In their brains, deposition of PrP^{Sc} is confined largely within the anteroventral and the dorsal medial nuclei of the thalamus. In contrast, the same D178N mutation with V129 produces fCJD, in which the patients present with dementia and widespread deposition of PrP^{Sc} is found postmortem. The first family to be recognized with CJD was found to carry the D178N mutation.

12.6
Human PrP Gene Polymorphisms

Several polymorphisms in the PrP gene may affect the development and expression of prion disease. A Met/Val polymorphism at PrP codon 129 has been identified (Fig. 11). This polymorphism appears to influence expression of the inherited as well as the iatrogenic and sporadic forms of prion disease. A second polymorphism resulting in an amino acid substitution at codon 219 (Glu/Lys) has been reported to occur with a frequency of approximately 12% in the Japanese population but not in Caucasians. In people heterozygous for the lysine substitution, this expression seems to act as a dominant negative and protects against CJD by the avid binding of mutant PrP^C to protein X. By sequestering protein X, $PrP^C(K219)$ prevents $PrP^C(E219)$ from being converted into PrP^{Sc}. In people homozygous for the lysine substitution, it seems likely that $PrP^C(K219)$ is not converted into PrP^{Sc}. A third polymorphism results in an amino acid substitution at codon 171 (Asn/Ser), which lies adjacent to the protein X binding site. This polymorphism has been found in Caucasians but it has not been studied extensively and it is not known to influence the binding of PrP^C to protein X. A fourth polymorphism is the deletion of a single octarepeat (24 bp), which has been found in 2.5% of Caucasians. In another study of over 700 individuals, this single octarepeat deletion was found in 1.0% of the population.

Studies of Caucasian patients with sCJD have shown that most are homozygous for Met or Val at codon 129. This contrasts with the general Caucasian population, in which frequencies for the codon 129 polymorphism are 12% Val/Val, 37% Met/Met, and 51% Met/Val. In contrast, the frequency of the valine allele in the Japanese population is much lower. Heterozygosity at codon 129 is more frequent in Japanese CJD patients (18%) than in the general Japanese population, in which the polymorphism frequencies are 0% Val/Val, 92% Met/Met, and 8% Met/Val.

While no specific mutations have been identified in the PrP gene of patients with sCJD, homozygosity at codon 129 is consistent with the results of Tg mouse studies. The finding that homozygosity at codon 129 predisposes to sCJD supports a model of prion production that favors PrP interactions between homologous proteins, as appears to occur in Tg(SHaPrP) mice inoculated with either hamster prions or mouse prions, as well as in Tg(MHM2) mice inoculated with "artificial" prions.

13
Infectious Human Prion Diseases

Prions from different sources have infected humans. Human prions have been transmitted to others both by ritualistic cannibalism and iatrogenic means. Kuru in the highlands of New Guinea was transmitted by ritualistic cannibalism, as people in the region attempted to immortalize their dead relatives by eating their brains. Iatrogenic transmissions include prion-tainted human growth hormone (HGH) and gonadotropin, dura mater grafts, and corneal transplants from people who died of CJD. In addition, CJD cases have been recorded after neurosurgical procedures in which ineffectively sterilized depth electrodes or instruments were used.

13.1
Human Growth Hormone

More than 165 young adults have been diagnosed with iatrogenic (i) CJD between 4 and 30 years after receiving HGH or gonadotropin from cadaveric pituitaries. The longest incubation periods (20 to 30 years) are similar to those associated with more recent cases of kuru. Since 1985, recombinant HGH produced in *E. coli* has been used in place of cadaveric HGH. With recombinant HGH, no cases of iCJD have been identified.

13.2
Variant Creutzfeldt–Jakob Disease (vCJD)

In 1994, the first cases of CJD in teenagers and young adults that were eventually labeled variant (v) CJD occurred in Great Britain. More than 150 teenagers and young adults have died of vCJD in Britain, Ireland, Italy, and France. While the average age of vCJD patients is 26 years of age, the youngest patient was 12 years old and the oldest was 74 years of age. The median duration of illness was 13 months, with the range from 6 to 69 months.

In addition to the young age of these patients, vCJD is characterized by numerous PrP amyloid plaques surrounded by a halo of intense spongiform degeneration in the brain. These unusual neuropathologic changes have not been seen in CJD cases in the United States, Australia, or Japan. Both macaque monkeys and marmosets developed neurologic disease several years after inoculation with bovine prions, but only the macaques exhibited numerous PrP plaques similar to those found in vCJD.

The majority of vCJD patients presented with psychiatric symptoms, including dysphoria, withdrawal, anxiety, insomnia, and loss of interest. Generally, neurologic deficits did not appear until at least four months later; these neurologic changes consisted of memory loss, paresthesias, sensory deficits, gait disturbances, and dysarthria. To date, all vCJD cases have been reported from Britain, except for six cases in France, one in Italy, one in Ireland, and one in the United States. The US case is a 23-year-old woman, who is thought to have been exposed to bovine prions while living in Britain during the first 12 years of her life. From both epidemiologic and experimental studies, evidence is now quite compelling that vCJD is the result of prions being transmitted from cattle with BSE to humans through consumption of contaminated beef products.

Studies of the prion diseases have taken on new significance with the identification of vCJD. The number of cases of vCJD caused by bovine prions that will occur in the years ahead is unknown. Until more time passes, it will be impossible to assess the magnitude of this problem. These tragic cases have generated a continuing discourse concerning mad cows, prions, and the safety of the human and animal food supplies throughout the world. Untangling politics and economics from the science of prions seems to have been difficult in disputes between Great Britain and other European countries over the safety of beef and lamb products.

13.3
Transmission of vCJD Prions by Blood Transfusion

vCJD has been identified in two patients who received blood transfusions from donors that later died of vCJD. In one case, the recipient was a 69-year-old male who was transfused 6.5 years before the onset of neurologic dysfunction. His PrP gene encoded Met/Met at position 129. Many details of the second case are not published but the patient is known to have died of a non-neurologic disease. While vCJD prions were found in the spleen and cervical lymph nodes of this patient, none were found in the brain. The patient's PrP gene encoded Met/Val at position 129 and as such, is the first heterozygous codon 129 human identified with vCJD prions.

A glimpse of future vCJD cases caused by prion-tainted transfused blood may come from a survey of tissues collected during appendectomies and tonsillectomies. This UK survey reports that of the 12 674 appendectomy specimens examined, three were positive for PrP^{Sc} by immunohistochemistry. This finding argues that as many as 3800 people in the United Kingdom may be replicating vCJD prions in their lymphoid tissues. Considering that immunohistochemistry is less sensitive than Western immunoblotting, the number of people with vCJD prions in their lymphoid tissues may be substantially greater than the above estimate.

14
Strains of Prions

That goats with scrapie can manifest two different syndromes, one in which the goats became hyperactive and the other in which they became drowsy, raised the possibility that strains of prions might exist. Subsequent studies with mice documented the existence of multiple strains through careful measurements of incubation times and the distribution of vacuoles in the CNS.

The search for an explanation of how biological information encrypting the disease phenotype could be enciphered within the prion posed a conundrum. Many investigators kept arguing for a small nucleic acid but none could be found. Others argued that this information must be enciphered within the structure of PrP^{Sc}. The first evidence supporting the hypothesis that strain-specific information is enciphered in PrP^{Sc} came from studies on prions causing TME, which were passaged into Syrian hamsters. On serial passaging, two strains emerged: one strain (HY) produced hyperactivity in Syrian hamsters and the other (DY) was manifest as a drowsy syndrome, as was seen with scrapie strains in goats. The HY strain was similar to the Sc237 prion strain with respect to protease resistance and sedimentation properties, while the DY strain showed minimal protease resistance and was less readily sedimented.

Because the HY and DY strains emerged after passaging TME prions in Syrian hamsters, concern over the origin of these strains made the conclusions somewhat questionable. Because of such concerns, a more convincing investigation came from the transmission of FFI and fCJD to Tg mice. In this study, two different strains of prions were generated *de novo* in patients with PrP gene mutations and propagated in mice expressing the chimeric MHu2MPrP transgene. Brain homogenates of FFI and fCJD(E200K) patients transmitted disease to Tg(MHu2M) mice approximately 200 days after inoculation. The FFI inoculum induced formation of a 19-kDa PrP^{Sc} fragment, as measured by SDS-PAGE after limited proteolysis and removal of Asn-linked carbohydrates; in contrast, fCJD(E200K) produced a 21-kDa PrP^{Sc} fragment. On second passage, Tg(MHu2M) mice inoculated with FFI prions showed an incubation time of ~130 days and a 19-kDa PrP^{Sc} fragment while those inoculated with fCJD(E200K) prions exhibited an incubation time of ~170 days and a 21-kDa PrP^{Sc} fragment. The experimental data demonstrated that MHu2MPrP^{Sc} can exist in two different conformations, based on the sizes of the protease-resistant fragments, within an invariant amino acid sequence. The results of these studies argue that PrP^{Sc} acts as a template for the conversion of PrP^{C} into nascent PrP^{Sc}. Imparting the size of the protease-resistant fragment of PrP^{Sc} through conformational templating provides a mechanism for both the generation and propagation of prion strains.

In another set of studies, prions from cattle with BSE and humans with vCJD were used to demonstrate distinct prion strains in humans expressing wt PrP, in contrast to the studies described above with prions from patients harboring the E200K mutation. Both BSE and vCJD prions transmitted to Tg mice expressing BoPrP but not to mice expressing either HuPrP or MHu2MPrP. Moreover, sCJD, iCJD, and fCJD(E200K) prions failed to transmit disease to Tg(BoPrP) mice after more than 500 days. These findings demonstrate the prions that cause BSE and vCJD are similar, but distinct from those responsible for sCJD, iCJD, and fCJD(E200K).

14.1
Isolation of New Strains

Additional evidence implicating PrP in the phenomenon of prion strains comes from studies on the interspecies transmission of strains. Such studies were especially revealing when mice expressing chimeric hamster–mouse transgenes were used. Evidence was obtained showing that new strains pathogenic to Syrian hamsters could be obtained from prion strains that had been previously replicated by limiting dilution in mice. Both the generation and propagation of prion strains result from interactions between PrP^{Sc} and PrP^{C}. Additionally, strains that were isolated from different breeds of sheep with scrapie and thought to be distinct were shown to have identical properties. Such findings argue for the convergence of some strains and raise the issue of the limits of prion diversity.

Studies with different strains of prions propagated in Syrian hamsters and subsequently in Tg mice expressing MH2MPrP are most illustrative with respect to the isolation of new prion strains. These Tg mice were inoculated with the two prion strains, Sc237 and DY that had been serially passaged in Syrian hamsters. On first passage in Tg(MHM2)$Prnp^{0/0}$ mice, the SHa(Sc237) prions exhibited prolonged incubation times, typical of a species barrier. On subsequent passage in Tg(MHM2)$Prnp^{0/0}$ mice, the MH2M(Sc237) strain showed a profound shortening of incubation time. Moreover, PrP^{Sc} of the MH2M(Sc237) strain possesses different structural properties from PrP^{Sc} of the SHa(Sc237) strain, as demonstrated by relative conformational stability measurements. Conversely, transmission of SHa(DY) prions to Tg(MHM2)$Prnp^{0/0}$ mice did not encounter a species barrier and the MH2M(DY) strain retained the conformational and phenotypic properties of SHa(DY). These results extend the findings described above for FFI and fCJD(E200K) prions and contend that a change in PrP^{Sc} conformation is intimately associated with the emergence of a new prion strain.

14.2
Interplay between the Species and Strains of Prions

The recent advances in our understanding of the role of the primary and tertiary structures of PrP in the transmission of disease have given new insights into the pathogenesis of the prion diseases. The amino acid sequence of PrP encodes the species of the prion (Table 5), and the prion derives its PrP^{Sc} sequence from the last mammal in which it was passaged. While the primary structure of PrP is likely to be the most important or even sole determinant of the tertiary structure of PrP^C, existing PrP^{Sc} seems to function as a template in determining the tertiary structure of nascent PrP^{Sc} molecules as they are formed from PrP^C. In turn, prion diversity appears to be enciphered in the conformation of PrP^{Sc} and prion strains represent different conformers of PrP^{Sc}. The total number of prion strains, that is, different conformations of PrP^{Sc}, in existence remains to be established.

15
Prion Diseases of Animals

The prion diseases of animals include scrapie of sheep and goats, BSE, TME, chronic wasting disease (CWD) of mule deer and elk, feline spongiform encephalopathy, and exotic ungulate encephalopathy (Table 1).

15.1
PrP Polymorphisms in Sheep, Cattle, and Elk

In 1962, Parry argued that host genes were responsible for the development of scrapie in sheep. He was convinced that natural scrapie is a genetic disease that could be eradicated by proper breeding protocols. He considered its transmission by inoculation of importance primarily for laboratory studies and communicable infection of little consequence in nature. Other investigators viewed natural scrapie as an infectious disease and argued that host genetics only modulates susceptibility to an endemic infectious agent.

In sheep, polymorphisms at codons 136, 154, and 171 of the PrP gene that produce amino acid substitutions have been studied with respect to the incidence of scrapie (Fig. 11). Studies of natural scrapie in the United States have shown that 85% of afflicted sheep are of the Suffolk breed. Only those Suffolk sheep homozygous for Gln (Q) at codon 171 developed scrapie although healthy controls with Gln/Gln, Gln/Arg, and Arg/Arg genotypes were also found. These results argue that susceptibility in Suffolk sheep is governed by the PrP codon 171 polymorphism. As with the Suffolk breed, the PrP codon 171 polymorphism in Cheviot sheep has a profound influence on susceptibility to scrapie, and codon 136 also modulates susceptibility but less so than codon 171.

In contrast to sheep, different breeds of cattle have no specific PrP polymorphisms. The only polymorphism recorded in cattle is a variation in the number of octarepeats: most cattle, like humans, have five octarepeats but some have six. However, the presence of six octarepeats does not seem to be overrepresented in BSE.

In studies of CWD, the susceptibility of elk, but not deer, seems to be modulated by codon 132, which corresponds to codon 129 in humans (Fig. 11). Elk with CWD consistently express Met/Met at position 132; no elk with CWD expressing leucine at this residue have been found.

15.2
Bovine Spongiform Encephalopathy

Prion strains and the species barrier are of paramount importance in understanding the BSE epidemic in Britain, in which it is estimated that between one and two million cattle were infected with prions. The mean incubation time for BSE is approximately 5 years. Therefore, most cattle did not manifest disease because they were slaughtered between 2 and 3 years of age.

Nevertheless, more than 180 000 cattle, primarily dairy cows, have died of BSE over the past decade. BSE is a massive common-source epidemic caused by meat and bone meal (MBM) fed primarily to dairy cows. MBM was prepared from the offal of sheep, cattle, pigs, and chickens as a high-protein nutritional supplement. In the late 1970s, the hydrocarbon-solvent extraction method used in the rendering of offal began to be abandoned, resulting in MBM with a much higher fat content. It is now thought that this change allowed scrapie prions from sheep to survive the rendering process and to be passed into cattle. Alternatively, bovine prions may have been present at low levels prior to modification of the rendering process and with the processing change, survived in sufficient numbers to initiate the BSE epidemic when reintroduced into cattle through ingestion of MBM. Perhaps, a particular conformation of BoPrP^{Sc} selected for heat-resistance during the rendering process and then reselected multiple times as cattle infected by ingesting prion-contaminated MBM were slaughtered and their offal rendered into more MBM. Against the latter hypothesis is the widespread geographical distribution of the initial 17 cases of BSE throughout England, which occurred almost simultaneously.

In July 1988, the practice of feeding MBM to sheep and cattle was banned. Although statistical analyses demonstrate that the epidemic is disappearing as a result of this ruminant feed ban, reminiscent of the disappearance of kuru in the Fore people of New Guinea, it is unclear how many current cases of BSE are due to infection and how many arise spontaneously. In 2003, there were 1390 cases of BSE; how many were due to tainted feed is unknown. Although evidence of a preexisting prion disease of cattle, either in Great Britain or elsewhere, is scant, an outbreak of TME in Wisconsin has been cited as evidence for sporadic BSE. Moreover, the formation of synthetic prions from wt PrP argues that any mammal expressing PrP^C can produce prions spontaneously.

The origin of BSE prions cannot be determined by examining the amino acid sequence of PrP^{Sc} from cattle with BSE because it has the bovine sequence regardless of whether the initial prions originated from sheep or cattle. The bovine PrP sequence differs from that of sheep at seven or eight positions.

Brain extracts from cattle with BSE cause disease in cattle, sheep, mice, pigs, and mink after intracerebral inoculation, but prions in brain extracts from sheep with scrapie fed to cattle produced an illness substantially different from BSE. However, no exhaustive effort has been made to test different strains of sheep prions or to examine the disease following bovine-to-bovine passage.

15.3
Monitoring Cattle for BSE Prions

Although no reliable, specific test for prion disease in live animals is available, immunoassays for PrP^{Sc} in the brainstems of cattle currently provide the best available approach for preventing BSE prions from entering the human food supply. Determining how early in the incubation period PrP^{Sc} can be detected by immunological methods is now possible because a reliable bioassay is now available with the creation of Tg(BoPrP)$Prnp^{0/0}$ mice. Prior to the construction of Tg(BoPrP)$Prnp^{0/0}$ mice, cattle were used for bioassays of bovine prions. In a very limited study using cattle bioassays, bovine prions were undetectable in the obex of the brainstem until 26 months after oral inoculation. In these studies, prion infectivity was detected much earlier in the lymphoid tissue of the distal ileum.

Prior to development of Tg(BoPrP) $Prnp^{0/0}$ mice, non-Tg mice inoculated intracerebrally with BSE brain extracts required more than 300 days to develop disease. Generally, many of the inoculated mice fail to develop prion disease. It is noteworthy that a highly stable, mouse prion strain denoted 301V was derived by passage of BSE inocula into non-Tg $Prnp^{b/b}$ mice.

Both Western blots and ELISA tests are currently used in the postmortem evaluation of bovine brainstems. As configured, these tests are quite insensitive. The CDI, whose sensitivity is superior to conventional Western blots and ELISAs, offers a new approach to measuring PrP^{Sc} in tissues. The sensitivity of the CDI for BSE prions was found to be similar to that of bioassays using Tg(BoPrP)$Prnp^{0/0}$ mice. These studies suggest that the CDI may be able to detect BSE prions in asymptomatic cattle even when titers are quite low.

15.4
Compelling Evidence of Transmission of Bovine Prions to Humans

The restricted geographical occurrence and chronology of vCJD raise the possibility that BSE prions have been transmitted to humans. That only ~150 vCJD cases have been recorded and the incidence has remained relatively constant made establishing the origin of vCJD difficult. No set of dietary habits distinguishes vCJD patients from apparently healthy people. Moreover, there is no explanation for the predilection of vCJD for teenagers and young adults. Why have older individuals not developed vCJD-based neuropathologic criteria? It is noteworthy that epidemiologic studies over the past three decades have failed to find evidence of transmission of sheep prions to humans. Attempts to predict the future number of cases of vCJD, assuming exposure to bovine prions prior to the offal ban, have been uninformative because so few cases of vCJD have occurred. Are we at the beginning of a prion disease epidemic in Britain, as seen for BSE and kuru, or will the number of vCJD cases remain small, as seen with iCJD caused by cadaveric HGH?

Recent studies of PrP^{Sc} from brains of patients who died of vCJD show a PrP glycoform pattern different from those found for sCJD and iCJD. The utility of measuring PrP glycoforms is questionable in trying to relate BSE to vCJD because PrP^{Sc} is formed after the protein is glycosylated and enzymatic deglycosylation of PrP^{Sc} requires denaturation. Moreover, synthetic prions have been formed from unglycosylated recMoPrP(89–230), demonstrating that glycosylation is unnecessary for prion formation.

Prior to the investigations recorded below laboratory studies were unconvincing in establishing a relationship between the conformations of PrP^{Sc} from cattle with BSE and those from humans with vCJD. Perhaps slightly more persuasive is the relationship demonstrated between vCJD and BSE based on similar incubation times in non-Tg RIII mice, of ~310 days after inoculation with human vCJD or bovine BSE prions. But such studies suffer from transmission of both BSE and vCJD prions to a heterologous host, that is, non-Tg mice expressing MoPrP^C. Using Tg mice, as was done for strains generated in the brains of patients with FFI or fCJD, compelling evidence for the transmission of bovine prions to humans was found. BSE prions transmitted to all inoculated Tg(BoPrP) mice after ~240 days but not to Tg mice expressing either HuPrP or MHu2MPrP. On second passage to Tg(BoPrP)$Prnp^{0/0}$ mice, the incubation time was unaltered, demonstrating the complete absence of a species barrier. Similar to BSE prions, vCJD prions transmitted readily to Tg(BoPrP)$Prnp^{0/0}$ mice, with a slightly longer incubation time of ~270 days, but poorly to Tg(HuPrP) and Tg(MHu2M) mice. Moreover, sCJD, iCJD, and fCJD(E200K) prions failed to transmit disease to Tg(BoPrP) mice after more than 500 days. On second passage of vCJD prions to Tg(BoPrP)$Prnp^{0/0}$ mice, the incubation time was reduced to ~225 days, demonstrating a small but expected species barrier. These findings argue that the strain-specific PrP^{Sc} conformations from BSE and vCJD prions are quite similar despite substantial differences in the amino acid sequences of BoPrP and HuPrP. Clearly, the conformation of HuPrP^{Sc}(vCJD) makes these prions much more readily transmittable to Tg(BoPrP)$Prnp^{0/0}$ mice than to either Tg(HuPrP)$Prnp^{0/0}$ or Tg(MHu2M)$Prnp^{0/0}$ mice.

15.5
Chronic Wasting Disease

Mule deer, white-tailed deer, and elk have been reported to develop CWD, which is unique among the prion diseases because it seems to be far more communicable than scrapie, BSE, or TME; moreover, it is the only prion disease known in free-ranging animals. CWD was first described in 1967 and reported to be a spongiform encephalopathy in 1978 based on histopathology in the brain. CWD has been found in the United States, Canada, and South Korea. In the United States,

CWD has been reported in Colorado, Wyoming, South Dakota, Nebraska, Oklahoma, Montana, New Mexico, Minnesota, and Wisconsin. In captive cervid herds, up to 90% of mule deer have been reported to be positive for prions and up to 60% of elk in Colorado and Wyoming develop CWD. Moreover, the incidence of CWD in cervids living in the wild has been estimated to be as high as 15%. The mode of transmission of CWD prion among mule deer, white-tailed deer, and elk is unresolved but contamination of grass with prions excreted in fecal matter seems to be a likely source. The high content of PrP^{Sc} in the intestinal lymphoid tissue of cervids with CWD supports such a scenario.

Brain homogenates from mule deer with CWD have transmitted disease to 4 of 13 cattle after intracerebral inoculation. These findings are particularly important because there is great concern that CWD prions might be transmitted to cattle grazing in contaminated pastures. In addition, CWD has been transmitted to ferrets, mink, squirrel monkeys, goats, and mice after intracerebral inoculation; however, only mule deer demonstrate efficient transmission of CWD prions by intracerebral inoculation. To date, endpoint titrations of CWD prions have not been performed in mule deer and Tg mice susceptible to CWD prions have not been developed, but the CDI has been adapted to measure CWD prions rapidly.

16
Fungal Prions

Although prions were originally defined in the context of an infectious pathogen, it is now becoming widely accepted that prions are elements that impart and propagate variability through multiple conformers of a normal, cellular protein. It is likely that such a mechanism will not be restricted to a single class of transmissible pathogens. Indeed, it is probable that this original definition will need to be extended to encompass other situations in which a similar mechanism of information transfer occurs. Two notable prion-like determinants, [URE3] and [PSI], have been described in yeast and another prion-like determinant has been reported in other fungi.

16.1
Yeast Prion Domains

In both [PSI] and [URE3], the "functional" determinants have been mapped to the C-terminal region of the protein, distinct from the "prion" domain, which comprises the N-terminal residues 65 and 114 of Ure2p and Sup35p, respectively. Although neither of the prion domains displays sequence identity either to each other or to PrP, the N-terminal regions of both Sup35p and mammalian PrP contain short repeated sequence elements: PQGGYQQYN in Sup35p and PHGGG-WGQ in PrP. Interestingly, when the prion domains of both Sup35 and Ure2p are expressed in *E. coli* and purified, they polymerize spontaneously into amyloid-like fibrils. Polymerization of the prion domains of both Sup35 and Ure2p are able to create [PSI] and [URE3] states in yeast, respectively.

16.2
Dependence of Yeast Prions on Molecular Chaperones

The intrinsic power of the yeast genetic system has provided striking evidence of the involvement of chaperones in the propagation of yeast [PSI] "prions." A genetic screen for factors that suppress the [PSI] phenotype resulted in the isolation of a single suppressor plasmid, which was found to contain the chaperone Hsp104. Furthermore, propagation of [PSI] was eliminated by either overproduction or absence of Hsp104, and treatment of cells with guanidine or UV light led to induction of Hsp104. The significance of Hsp104 is unclear because no published data exist to indicate that [URE3] utilizes Hsp104; furthermore, overexpression of Sup35 at high levels can induce [PSI] in the absence of Hsp104.

16.3
Differences between Yeast and Mammalian Prions

A wealth of studies has established the concept of fungal prions. While fungal prions are produced in the cytoplasm, mammalian prions are produced within cholesterol-rich microdomains on the surface of cells called *rafts* or *CLDs*. Whether the process of fungal and mammalian prion replication are fundamentally different or quite similar remains to be established. It seems likely that many more similarities than differences will emerge because distinct strains of both fungal and mammalian prions seem to represent different conformations of a particular protein. This finding necessitates a mechanism by which a particular conformation can be templated and reproduced with a high degree of fidelity.

Interestingly, the prion state in yeast is proposed to be functionally inert in the case of both [PSI] and [URE3] and produces the same phenotype as inactivation of the maintenance gene. In contrast, prion diseases in mammals cannot be explained simply by the loss of function of PrP because ablation of the PrP gene does not as yet have a detectable deleterious effect.

17
Prion Diseases are Disorders of Protein Conformation

The study of prions has followed several unexpected directions over the past three decades. The discovery that prion diseases in humans are uniquely genetic and infectious has greatly strengthened and extended the prion concept. To date, more than 30 different mutations in the human PrP gene, all resulting in nonconservative substitutions have been found either to be linked genetically to or to segregate with the inherited prion diseases (Fig. 11).

Understanding how PrP^C unfolds and refolds into PrP^{Sc} will be of paramount importance in transferring advances in the prion diseases to studies of other degenerative illnesses. The mechanism by which PrP^{Sc} is formed must involve a templating process in which existing PrP^{Sc} directs the refolding of PrP^C into nascent PrP^{Sc} with the same conformation. Undoubtedly, molecular chaperones of some type participate in a process that appears to be associated with CLDs on the cell surface.

Studies of prions in fungi have been extremely helpful in establishing the prion concept. In the case of yeast, Sup35 and Ure2p fold into alternative conformations that create the new metabolic states [PSI] and [URE3], respectively. Whether PrP^{Sc} in mammals represents a misfolded protein or an alternatively folded protein, as with yeast prions, remains to be established.

18
Prevention and Therapeutics for Prion Diseases

Because people at risk for inherited prion diseases can now be identified decades before neurologic dysfunction is evident, the development of an effective therapy for these fully penetrant disorders is imperative. Although it is difficult to predict the number of individuals who may develop neurologic dysfunction from bovine prions in the future, seeking an effective therapy now seems most prudent. Interfering with the conversion of PrP^C into PrP^{Sc} would seem to be the most attractive therapeutic target.

Defining the pathogenesis of prion disease is an important issue with respect to developing an effective therapy. The issue of whether large aggregates of misprocessed proteins or misfolded monomers (or oligomers) cause CNS degeneration has been addressed in several studies of prion diseases in humans as well as in Tg mice. In humans, the frequency of PrP amyloid plaques varies from 100% in GSS and vCJD to ~70% in kuru and ~10% in sCJD, arguing that these plaques are a nonobligatory feature of the disease. In Tg mice expressing both MoPrP and SHaPrP, animals inoculated with hamster prions produced hamster prions and developed amyloid plaques composed of SHaPrP^{Sc}. In contrast, Tg mice inoculated with mouse prions did not develop plaques even though they produced mouse prions and died of prion disease.

18.1
Prion Therapeutics

Various compounds have been proposed as potential therapeutics for treatment of prion diseases; these include polysulfated anions, dextrans, Congo red dye, oligonucleotides, and cyclic tetrapyrroles, all of which have been shown to increase survival time when given prior to prion infection in rodents, but not when administered a month or more after infection has been established.

Besides studies in rodents, ScN2a cells chronically infected with scrapie prions have been used to identify several candidate antiprion drugs but none have been shown to be effective in halting prion diseases in either animals or humans. Structure-based drug design based on dominant-negative inhibition of prion formation has produced several lead compounds. Prion replication depends on protein–protein interactions and a subset of these interactions gives rise to dominant-negative phenotypes produced by single residue substitutions. A particularly interesting set of drugs is the branched polyamines, or dendrimers, which enhance the clearance of PrP^{Sc} from cells. While these compounds cure cultured cells of prion infection, they have not been successfully deployed in mice because of difficulties in delivering such highly charged compounds to the CNS.

18.1.1 Quinacrine and other Acridine Derivatives

Tricyclic derivatives of acridine exhibit half-maximal inhibition of PrP^{Sc} formation at effective concentrations (IC$_{50}$) between 0.3 µM and 3 µM in cultured ScN2a cells. The IC$_{50}$ for chlorpromazine was 3 µM while quinacrine was 10 times more potent. A variety of 9-substituted, acridine-based analogs of quinacrine were synthesized, which demonstrated variable potencies similar to chlorpromazine and emphasized the importance of the side chain in mediating the inhibition of PrP^{Sc}

formation. These studies showed that tricyclic compounds with an aliphatic side chain at the middle ring moiety constitute a new class of antiprion agents. Because quinacrine and chlorpromazine have been used for many years in humans as antimalarial and antipsychotic drugs, respectively, and are known to penetrate the blood-brain barrier (BBB), these compounds became immediate candidates for the treatment of CJD and other human prion diseases.

An asymmetric carbon in the side chain of quinacrine creates two stereoisomers. (S)-Quinacrine is two to three times more potent than the (R)-quinacrine enantiomer in cultured cells. Whether the use of (S)-quinacrine in place of the racemic mixture will allow significantly more drug to be given remains to be established. Studies performed almost six decades ago in humans demonstrated that (S)-quinacrine is selectively metabolized and that (R)-quinacrine remains intact. Whether (R)-quinacrine or the metabolites of (S)-quinacrine are responsible for the side effects recorded in patients receiving the racemic mixture is unknown. Side effects of quinacrine include liver toxicity, cardiomyopathy, and toxic psychoses.

18.1.2 Anti-PrP Antibodies

A panel of recombinant antibody fragments (recFabs), recognizing different epitopes on PrP, was studied with respect to inhibition of prion propagation in cultured ScN2a cells. Recombinant Fabs binding to PrP^C on the cell surface inhibited PrP^{Sc} formation in a dose-dependent manner. In ScN2a cells treated with the most potent recFab D18, prion replication was completely abolished and preexisting PrP^{Sc} rapidly cleared, suggesting that this antibody may cure established infection. The activity of recFab D18 is associated with its ability to recognize more completely the total population of PrP^C molecules on the cell surface than other recFabs and with the location of its epitope on PrP^C. In other studies, a monoclonal antibody, pH4, which is thought to bind to the same region of PrP as recFab D18, was found to inhibit prion accumulation in ScN2a cells.

Whether antibodies or Fabs can be effectively administered to humans for the prevention and treatment of prion diseases is unclear. Neither antibodies nor Fabs cross the BBB in high concentration so that delivery of these proteins to the CNS remains a critical issue.

It is notable that anti-PrP antibodies delivered by transgenetics or by injection were successful in blocking prion replication in mice inoculated intraperitoneally. The same anti-PrP antibodies were ineffective in treating prion infection initiated by intracerebral inoculation.

18.2 Dominant-negative Sheep

The production of domestic animals that do not replicate prions may also prove to be a practical way to prevent prion disease. Sheep encoding Arg/Arg at polymorphic position 171 seem resistant to scrapie; presumably, this was the genetic basis of Parry's scrapie eradication program in Great Britain 30 years ago. A more effective approach applying dominant-negative inhibition for producing prion-resistant domestic animals, including sheep and cattle, is probably the expression of PrP transgenes encoding K219 and/or R171 (Fig. 11). Such an approach has been evaluated in Tg mice and shown to be effective. In fact, the replacement of all sheep with Arg/Arg at codon 171 is underway in the Netherlands. Such an approach can be instituted by artificial insemination using sperm from males homozygous for Arg/Arg at residue 171.

In comparison, a less practical approach is the production of PrP-deficient cattle and sheep. Although such animals would not be susceptible to prion disease, they might suffer some deleterious effects from the ablation of the PrP gene.

19
Some Principles of Prion Biology

Many principles of prion replication are clearly unprecedented in biology. As such, it is not surprising that some of these principles have been slow to be embraced. Although prion replication resembles viral replication superficially, the underlying principles are quite different. For example, in prion replication, the substrate is a host-encoded protein, PrP^C, which undergoes modification to form PrP^{Sc}, the only known component of the infectious prion particle. In contrast, viruses carry a DNA or RNA genome that is copied and directs the synthesis of most, if not all, of the viral proteins. The mature virion consists of a nucleic acid genome surrounded by a protein coat, whereas a prion appears to be composed of a dimer of PrP^{Sc}. When viruses pass from one species to another, they often replicate without any structural modification, whereas prions undergo a profound change. The prion adopts a new PrP sequence, which is encoded by the PrP gene of the current host. That change in amino acid sequence can result in a restriction of transmission for some species while making the new prion permissive to others. In viruses, different properties exhibited by distinct strains are encoded in the viral genome, whereas in prions, strain-specific properties are enciphered in the conformation of PrP^{Sc}.

19.1
Implications for Common Neurodegenerative Diseases

Understanding how PrP^C unfolds and refolds into PrP^{Sc} may also open new approaches to deciphering the causes of and to developing effective therapies for some common neurodegenerative diseases, including Alzheimer's disease, Parkinson's disease, and amyotrophic lateral sclerosis (ALS). Whether or not therapies designed to prevent the conversion of PrP^C into PrP^{Sc} will be effective in these more common neurodegenerative diseases is unknown. Alternatively, developing a therapy for the prion diseases might provide a blueprint for designing somewhat different drugs for these common disorders. Like the inherited prion diseases, subsets of Alzheimer's disease and ALS are caused by mutations that result in nonconservative amino acid substitutions in proteins expressed in the CNS.

As knowledge about prions continues to expand, our understanding of how prions replicate and cause disease will undoubtedly evolve. It is important to add that many of the basic principles of prion biology are becoming increasingly well understood.

See also Alzheimer's Disease.

Bibliography

Books and Reviews

Aguzzi, A., Heppner, F.L., Heikenwalder, M., Prinz, M., Mertz, K., Seeger, H., Glatzel, M. (2003) Immune system and peripheral nerves in propagation of prions to CNS, *Br. Med. Bull.* **66**, 141–159.

Alpers, M.P. (1968) Kuru: Implications of its Transmissibility for the Interpretation of its Changing Epidemiological Pattern, in: Bailey, O.T., Smith, D.E. (Eds.) *The Central Nervous System: Some Experimental Models of Neurological Diseases*, Williams and Wilkins Company, Baltimore, MD, pp. 234–251.

Behrens, A. (2003) Physiological and pathological functions of the prion protein homologue Dpl., *Br. Med. Bull.* **66**, 35–42.

Brandner, S. (2003) CNS pathogenesis of prion diseases, *Br. Med. Bull.* **66**, 131–139.

Bruce, M., Chree, A., McConnell, I., Foster, J., Pearson, G., Fraser, H. (1994) Transmission of bovine spongiform encephalopathy and scrapie to mice: strain variation and the species barrier, *Philos. Trans. R. Soc. Lond. B. Biol. Sci.* **343**, 405–411.

Bruce, M.E. (1996) Strain Typing Studies of Scrapie and BSE, in: Baker, H.F., Ridley, R.M. (Eds.) *Methods in Molecular Medicine: Prion Diseases*, Humana Press, Totowa, NJ, pp. 223–236.

Bruce, M.E. (2003) TSE strain variation: An investigation into prion disease diversity, *Br. Med. Bull.* **66**, 99–108.

Budka, H. (2003) Neuropathology of prion diseases, *Br. Med. Bull.* **66**, 121–130.

Caughey, B. (2003) Prion protein conversions: insight into mechanisms, TSE transmission barriers and strains, *Br. Med. Bull.* **66**, 109–120.

Cervenáková, L., Brown, P., Piccardo, P., Cummings, J.L., Nagle, J., Vinters, H.V., Kaur, P., Ghetti, B., Chapman, J., Gajdusek, D.C., Goldfarb, L.G. (1996) 24-Nucleotide Deletion in the *PRNP* Gene: Analysis of Associated Phenotypes, in: Court, L., Dodet, B. (Eds.) *Transmissible Subacute Spongiform Encephalopathies: Prion Diseases*, Elsevier, Paris, France, pp. 433–444.

Chesebro, B. (2003) Introduction to the transmissible spongiform encephalopathies or prion diseases, *Br. Med. Bull.* **66**, 1–20.

Chien, P., Weissman, J.S., DePace, A.H. (2004) Emerging principles of conformation-based prion inheritance, *Annu. Rev. Biochem.* **73**, 617–656.

Collinge, J., Palmer, M.S. (1997) Human Prion Diseases, in: Collinge, J., Palmer, M.S. (Eds.) *Prion Diseases*, Oxford University Press, Oxford, UK, pp. 18–56.

DeArmond, S.J., Prusiner, S.B. (1997) Prion Diseases, in: Lantos, P., Graham, D. (Eds.) *Greenfield's Neuropathology*, 6th edition, Edward Arnold, London, UK, pp. 235–280.

Dickinson, A.G., Outram, G.W. (1988) Genetic Aspects of Unconventional Virus Infections: the Basis of the Virino Hypothesis, in: Bock, G., Marsh, J. (Eds.) *Novel Infectious Agents and the Central Nervous System. Ciba Foundation Symposium 135*, John Wiley and Sons, Chichester, UK, pp. 63–83.

Diringer, H. (1991) Transmissible spongiform encephalopathies (TSE) virus-induced amyloidoses of the central nervous system (CNS), *Eur. J. Epidemiol.* **7**, 562–566.

Dormont, D. (2003) Approaches to prophylaxis and therapy: An investigation into prion disease diversity, *Br. Med. Bull.* **66**, 281–292.

Erdtmann, R., Sivitz, L.B. (Eds.) (2004) *Advancing Prion Science: Guidance for the National Prion Research Program*, National Academies Press, Washington, DC

Gambetti, P., Kong, Q., Zou, W., Parchi, P., Chen, S.G. (2003) Sporadic and familial CJD: classification and characterisation, *Br. Med. Bull.* **66**, 213–239.

Gambetti, P., Parchi, P., Petersen, R.B., Chen, S.G., Lugaresi, E. (1995) Fatal familial insomnia and familial Creutzfeldt-Jakob disease: clinical, pathological and molecular features, *Brain Pathol.* **5**, 43–51.

Goodman, L.S., Gilman, A. (Eds.) (1970) *The Pharmacological Basis of Therapeutics; A Textbook of Pharmacology, Toxicology, and Therapeutics for Physicians and Medical Students*, Macmillan, New York.

Harris, D.A. (2003) Trafficking, turnover and membrane topology of PrP: protein function in prion disease, *Br. Med. Bull.* **66**, 71–85.

Harris, D.A. (Ed.) (2004) *Mad Cow Disease and Related Spongiform Encephalopathies*, Current Topics in Microbiology and Immunology, Springer-Verlag, Berlin, Germany.

Hill, A.F., Collinge, J. (2003) Subclinical prion infection in humans and animals, *Br. Med. Bull.* **66**, 161–170.

Hunter, N. (2003) Scrapie and experimental BSE in sheep, *Br. Med. Bull.* **66**, 171–183.

Kimberlin, R.H. (1996) Speculations on the Origin of BSE and the Epidemiology of CJD, in: Gibbs Jr., C.J. (Ed.) *Bovine Spongiform Encephalopathy: The BSE Dilemma*, Springer Verlag, New York, pp. 155–175.

Kübler, E., Oesch, B., Raeber, A.J. (2003) Diagnosis of prion diseases, *Br. Med. Bull.* **66**, 267–279.

Lasmézas, C.I. (2003) Putative functions of PrP^C, *Br. Med. Bull.* **66**, 61–70.

Malmgren, R., Kurland, L., Mokri, B., Kurtzke, J. (1979) The Epidemiology of Creutzfeldt-Jakob Disease, in: Prusiner, S.B. Hadlow, W.J. (Eds.) *Slow Transmissible Diseases of the Nervous System*, Vol. 1, Academic Press, New York, pp. 93–112.

Parry, H.B. (1983) *Scrapie Disease in Sheep*, Academic Press, New York.

Pattison, I.H. (1965) Experiments with Scrapie with Special Reference to the Nature of the Agent and the Pathology of the Disease, in: Gajdusek, D.C. Gibbs Jr., C.J., Alpers, M.P. (Eds.) *Slow, Latent and Temperate Virus Infections, NINDB Monograph 2*, U.S. Government Printing, Washington, DC, pp. 249–257.

Prusiner, S.B. (1989) Scrapie prions, *Annu. Rev. Microbiol.* **43**, 345–374.

Prusiner, S.B. (1991) Molecular biology of prion diseases, *Science* **252**, 1515–1522.

Prusiner, S.B. (1997) Prion diseases and the BSE crisis, *Science* **278**, 245–251.

Prusiner, S.B. (Ed.) (2004) *Prion Biology and Diseases*, Cold Spring Harbor Laboratory Press, Cold Spring Harbor, New York.

Prusiner, S.B., Scott, M.R., DeArmond, S.J., Cohen, F.E. (1998) Prion protein biology, *Cell* **93**, 337–348.

Prusiner, S.B., Baron, H., Carlson, G., Cohen, F.E., DeArmond, S.J., Gabizon, R., Gambetti, P., Hope, J., Kitamoto, T., Kretzschmar, H.A., Laplanche, J.-L., Tateishi, J., Telling, G., Will, R. (2000) Prions, in: van Regenmortel, M.H.V., Fauquet, C.M., Bishop, D.H.L., Carstens, E.B., Estes, M.K., Lemon, S.M., Maniloff, J., Mayo, M.A., McGeoch, D.J., Pringle, C.R., Wickner, R.B. et al., *Virus Taxonomy – Classification and Nomenclature of Viruses*, Academic Press, San Diego, CA, 1032–1039.

Riesner, D. (2003) Biochemistry and structure of PrP^C and PrP^{Sc}, *Br. Med. Bull.* **66**, 21–33.

Sigurdson, C.J., Miller, M.W. (2003) Other animal prion diseases, *Br. Med. Bull.* **66**, 199–212.

Smith, P.G., Bradley, R. (2003) Bovine spongiform encephalopathy (BSE) and its epidemiology, *Br. Med. Bull.* **66**, 185–198.

Solassol, J., Crozet, C., Lehmann, S. (2003) Prion propagation in cultured cells, *Br. Med. Bull.* **66**, 87–97.

Tabrizi, S.J., Elliott, C.L., Weissmann, C. (2003) Ethical issues in human prion diseases, *Br. Med. Bull.* **66**, 305–316.

Tateishi, J., Kitamoto, T. (1993) Developments in diagnosis for prion diseases, *Br. Med. Bull.* **49**, 971–979.

Taylor, D.M. (2003) Preventing accidental transmission of human transmissible spongiform encephalopathies, *Br. Med. Bull.* **66**, 293–303.

Tuite, M.F., Lindquist, S.L. (1996) Maintenance and inheritance of yeast prions, *Trends Genet.* **12**, 467–471.

Wadsworth, J.D., Hill, A.F., Beck, J.A., Collinge, J. (2003) Molecular and clinical classification of human prion disease, *Br. Med. Bull.* **66**, 241–254.

Weissmann, C. (1991) A "unified theory" of prion propagation, *Nature* **352**, 679–683.

Weissmann, C., Flechsig, E. (2003) PrP knockout and PrP transgenic mice in prion research, *Br. Med. Bull.* **66**, 43–60.

Wells, G.A.H., Wilesmith, J.W. (1995) The neuropathology and epidemiology of bovine spongiform encephalopathy, *Brain Pathol.* **5**, 91–103.

Wickner, R.B., Liebman, S.W., Saupe, S.J. (2004) Prions of Yeast and Filamentous Fungi: [URE3], [PSI$^+$], [PIN$^+$], and [Het-s], in: Prusiner, S.B. (Ed.) *Prion Biology and Diseases*, Cold Spring Harbor Laboratory Press, Cold Spring Harbor, pp. 305–372.

Will, R.G. (2003) Acquired prion disease: iatrogenic CJD, variant CJD, kuru, *Br. Med. Bull.* **66**, 255–265.

Will, R.G., Alpers, M.P., Dormont, D., Schonberger, L.B. (2004) Infectious and Sporadic Prion Diseases, in: Prusiner, S.B. (Ed.) *Prion Biology and Diseases*, Cold Spring Harbor Laboratory Press, Cold Spring Harbor, pp. 629–671.

Williams, E.S., Miller, M.W. (2002) Chronic wasting disease in deer and elk in North America, *Rev. Sci. Tech.* **21**, 305–316.

Primary Literature

Anderson, R.M., Donnelly, C.A., Ferguson, N.M., Woolhouse, M.E.J., Watt, C.J., Udy, H.J., MaWhinney, S., Dunstan, S.P., Southwood, T.R.E., Wilesmith, J.W., Ryan, J.B.M.,

Hoinville, L.J., Hillerton, J.E., Austin, A.R., Wells, G.A.H. (1996) Transmission dynamics and epidemiology of BSE in British cattle, *Nature* **382**, 779–788.

Anfinsen, C.B. (1973) Principles that govern the folding of protein chains, *Science* **181**, 223–230.

Asante, E.A., Linehan, J.M., Desbruslais, M., Joiner, S., Gowland, I., Wood, A.L., Welch, J., Hill, A.F., Lloyd, S.E., Wadsworth, J.D., Collinge, J. (2002) BSE prions propagate as either variant CJD-like or sporadic CJD-like prion strains in transgenic mice expressing human prion protein, *EMBO J.* **21**, 6358–6366.

Baker, H.F., Ridley, R.M., Wells, G.A.H. (1993) Experimental transmission of BSE and scrapie to the common marmoset, *Vet. Rec.* **132**, 403–406.

Barry, R.A., Prusiner, S.B. (1986) Monoclonal antibodies to the cellular and scrapie prion proteins, *J. Infect. Dis.* **154**, 518–521.

Basler, K., Oesch, B., Scott, M., Westaway, D., Wälchli, M., Groth, D.F., McKinley, M.P., Prusiner, S.B., Weissmann, C. (1986) Scrapie and cellular PrP isoforms are encoded by the same chromosomal gene, *Cell* **46**, 417–428.

Bastian, F.O., Foster, J.W. (2001) Spiroplasma sp. 16S rDNA in Creutzfeldt-Jakob disease and scrapie as shown by PCR and DNA sequence analysis, *J. Neuropathol. Exp. Neurol.* **60**, 613–620.

Bateman, D., Hilton, D., Love, S., Zeidler, M., Beck, J., Collinge, J. (1995) Sporadic Creutzfeldt-Jakob disease in a 18-year-old in the UK (Lett.), *Lancet* **346**, 1155–1156.

Behrens, A., Genoud, N., Naumann, H., Rülicke, T., Janett, F., Heppner, F.L., Ledermann, B., Aguzzi, A. (2002) Absence of the prion protein homologue Doppel causes male sterility, *EMBO J.* **21**, 3652–3658.

Bellinger-Kawahara, C.G., Kempner, E., Groth, D.F., Gabizon, R., Prusiner, S.B. (1988) Scrapie prion liposomes and rods exhibit target sizes of 55 000 Da, *Virology* **164**, 537–541.

Belt, P.B., Muileman, I.H., Schreuder, B.E.C., Ruijter, J.B., Gielkens, A.L.J., Smits, M.A. (1995) Identification of five allelic variants of the sheep PrP gene and their association with natural scrapie, *J. Gen. Virol.* **76**, 509–517.

Bertoni, J.M., Brown, P., Goldfarb, L., Gajdusek, D., Omaha, N.E. (1992) Familial Creutzfeldt-Jakob disease with the PRNP codon 200^{lys} mutation and supranuclear palsy but without myoclonus or periodic EEG complexes, *Neurology* **42**(4), Suppl. 3, 350 [Abstr].

Bessen, R.A., Marsh, R.F. (1994) Distinct PrP properties suggest the molecular basis of strain variation in transmissible mink encephalopathy, *J. Virol.* **68**, 7859–7868.

Billette de Villemeur, T., Deslys, J.-P., Pradel, A., Soubrié, C., Alpérovitch, A., Tardieu, M., Chaussain, J.-L., Hauw, J.-J., Dormont, D., Ruberg, M., Agid, Y. (1996) Creutzfeldt-Jakob disease from contaminated growth hormone extracts in France, *Neurology* **47**, 690–695.

Borchelt, D.R., Scott, M., Taraboulos, A., Stahl, N., Prusiner, S.B. (1990) Scrapie and cellular prion proteins differ in their kinetics of synthesis and topology in cultured cells, *J. Cell Biol.* **110**, 743–752.

Britton, T.C., Al-Sarraj, S., Shaw, C., Campbell, T., Collinge, J. (1995) Sporadic Creutzfeldt-Jakob disease in a 16-year-old in the UK (Lett.), *Lancet* **346**, 1155.

Brown, D.R., Qin, K., Herms, J.W., Madlung, A., Manson, J., Strome, R., Fraser, P.E., Kruck, T., von Bohlen, A., Schulz-Schaeffer, W., Giese, A., Westaway, D., Kretzschmar, H. (1997) The cellular prion protein binds copper *in vivo*, *Nature* **390**, 684–687.

Brown, P., Cathala, F., Raubertas, R.F., Gajdusek, D.C., Castaigne, P. (1987) The epidemiology of Creutzfeldt-Jakob disease: conclusion of a 15-year investigation in France and review of the world literature, *Neurology* **37**, 895–904.

Bruce, M., Chree, A., McConnell, I., Foster, J., Fraser, H. (1993) Transmissions of BSE, scrapie and related diseases to mice (Abstr.), *Proceedings of the IXth International Congress of Virology*, Glasgow, Scotland, Aug. 9–12, 93.

Bruce, M.E., Dickinson, A.G. (1987) Biological evidence that the scrapie agent has an independent genome, *J. Gen. Virol.* **68**, 79–89.

Bruce, M.E., McConnell, I., Fraser, H., Dickinson, A.G. (1991) The disease characteristics of different strains of scrapie in *Sinc* congenic mouse lines: implications for the nature of the agent and host control of pathogenesis, *J. Gen. Virol.* **72**, 595–603.

Bruce, M.E., Will, R.G., Ironside, J.W., McConnell, I., Drummond, D., Suttie, A., McCardle, L., Chree, A., Hope, J., Birkett, C., Cousens, S., Fraser, H., Bostock, C.J. (1997) Transmissions to mice indicate that 'new variant' CJD is caused by the BSE agent, *Nature* **389**, 498–501.

Büeler, H., Raeber, A., Sailer, A., Fischer, M., Aguzzi, A., Weissmann, C. (1994) High prion and PrP^{Sc} levels but delayed onset of disease in scrapie-inoculated mice heterozygous for a disrupted PrP gene, *Mol. Med.* **1**, 19–30.

Büeler, H., Aguzzi, A., Sailer, A., Greiner, R.-A., Autenried, P., Aguet, M., Weissmann, C. (1993) Mice devoid of PrP are resistant to scrapie, *Cell* **73**, 1339–1347.

Büeler, H., Fisher, M., Lang, Y., Bluethmann, H., Lipp, H.-P., DeArmond, S.J., Prusiner, S.B., Aguet, M., Weissmann, C. (1992) Normal development and behaviour of mice lacking the neuronal cell-surface PrP protein, *Nature* **356**, 577–582.

Burns, C.S., Aronoff-Spencer, E., Legname, G., Prusiner, S.B., Antholine, W.E., Gerfen, G.J., Peisach, J., Millhauser, G.L. (2003) Copper coordination in the full-length, recombinant prion protein, *Biochemistry* **42**, 6794–6803.

Buschmann, A., Pfaff, E., Reifenberg, K., Müller, H.M., Groschup, M.H. (2000) Detection of cattle-derived BSE prions using transgenic mice overexpressing bovine PrP^C, *Arch. Virol. [Suppl].* **16**, 75–86.

Carlson, G.A., Westaway, D., DeArmond, S.J., Peterson-Torchia, M., Prusiner, S.B. (1989) Primary structure of prion protein may modify scrapie isolate properties, *Proc. Natl. Acad. Sci. U. S. A.* **86**, 7475–7479.

Carlson, G.A., Kingsbury, D.T., Goodman, P.A., Coleman, S., Marshall, S.T., DeArmond, S., Westaway, D., Prusiner, S.B. (1986) Linkage of prion protein and scrapie incubation-time genes, *Cell* **46**, 503–511.

Carlson, G.A., Ebeling, C., Yang, S.-L., Telling, G., Torchia, M., Groth, D., Westaway, D., DeArmond, S.J., Prusiner, S.B. (1994) Prion isolate specified allotypic interactions between the cellular and scrapie prion proteins in congenic and transgenic mice, *Proc. Natl. Acad. Sci. U. S. A.* **91**, 5690–5694.

Caughey, B., Race, R.E. (1992) Potent inhibition of scrapie-associated PrP accumulation by Congo red, *J. Neurochem.* **59**, 768–771.

Caughey, B., Raymond, G.J. (1991) The scrapie-associated form of PrP is made from a cell-surface precursor that is both protease- and phospholipase-sensitive, *J. Biol. Chem.* **266**, 18217–18223.

Caughey, B., Kocisko, D.A., Raymond, G.J., Lansbury Jr., P.T. (1995) Aggregates of scrapie-associated prion protein induce the cell-free conversion of protease-sensitive prion protein to the protease-resistant state, *Chem. Biol.* **2**, 807–817.

Caughey, B., Raymond, G.J., Ernst, D., Race, R.E. (1991) N-terminal truncation of the scrapie-associated form of PrP by lysosomal protease(s): implications regarding the site of conversion of PrP to the protease-resistant state, *J. Virol.* **65**, 6597–6603.

Caughey, B., Neary, K., Butler, R., Ernst, D., Perry, L., Chesebro, B., Race, R.E. (1990) Normal and scrapie-associated forms of prion protein differ in their sensitivities to phospholipase and proteases in intact neuroblastoma cells, *J. Virol.* **64**, 1093–1101.

Caughey, B.W., Dong, A., Bhat, K.S., Ernst, D., Hayes, S.F., Caughey, W.S. (1991) Secondary structure analysis of the scrapie-associated protein PrP 27–30 in water by infrared spectroscopy, *Biochemistry* **30**, 7672–7680.

Centers for Disease Control (1996) Surveillance for Creutzfeldt-Jakob disease – United States, *MMWR Morb. Mortal. Wkly. Rep.* **45**, 665–668.

Chapman, J., Ben-Israel, J., Goldhammer, Y., Korczyn, A.D. (1994) The risk of developing Creutzfeldt-Jakob disease in subjects with the *PRNP* gene codon 200 point mutation, *Neurology* **44**, 1683–1686.

Chernoff, Y.O., Lindquist, S.L., Ono, B., Inge-Vechtomov, S.G., Liebman, S.W. (1995) Role of the chaperone protein Hsp104 in propagation of the yeast prion-like factor [*psi*$^+$], *Science* **268**, 880–884.

Chesebro, B. (1998) Prion diseases: BSE and prions: uncertainties about the agent, *Science* **279**, 42–43.

Chesebro, B., Race, R., Wehrly, K., Nishio, J., Bloom, M., Lechner, D., Bergstrom, S., Robbins, K., Mayer, L., Keith, J.M., Garon, C., Haase, A. (1985) Identification of scrapie prion protein-specific mRNA in scrapie-infected and uninfected brain, *Nature* **315**, 331–333.

Clousard, C., Beaudry, P., Elsen, J.M., Milan, D., Dussaucy, M., Bounneau, C., Schelcher, F., Chatelain, J., Launay, J.-M., Laplanche, J.-L. (1995) Different allelic effects of the codons 136 and 171 of the prion protein gene in sheep with natural scrapie, *J. Gen. Virol.* **76**, 2097–2101.

Cohen, F.E., Pan, K.-M., Huang, Z., Baldwin, M., Fletterick, R.J., Prusiner, S.B. (1994) Structural clues to prion replication, *Science* **264**, 530–531.

Collinge, J., Palmer, M.S., Dryden, A.J. (1991) Genetic predisposition to iatrogenic Creutzfeldt-Jakob disease, *Lancet* **337**, 1441–1442.

Collinge, J., Sidle, K.C.L., Meads, J., Ironside, J., Hill, A.F. (1996) Molecular analysis of prion strain variation and the aetiology of "new variant" CJD, *Nature* **383**, 685–690.

Collinge, J., Whittington, M.A., Sidle, K.C., Smith, C.J., Palmer, M.S., Clarke, A.R., Jefferys, J.G.R. (1994) Prion protein is necessary for normal synaptic function, *Nature* **370**, 295–297.

Collinge, J., Palmer, M.S., Sidle, K.C., Hill, A.F., Gowland, I., Meads, J., Asante, E., Bradley, R., Doey, L.J., Lantos, P.L. (1995) Unaltered susceptibility to BSE in transgenic mice expressing human prion protein, *Nature* **378**, 779–783.

Cousens, S.N., Vynnycky, E., Zeidler, M., Will, R.G., Smith, P.G. (1997) Predicting the CJD epidemic in humans, *Nature* **385**, 197–198.

Cousens, S.N., Harries-Jones, R., Knight, R., Will, R.G., Smith, P.G., Matthews, W.B. (1990) Geographical distribution of cases of Creutzfeldt-Jakob disease in England and Wales 1970–84, *J. Neurol. Neurosurg. Psychiatry* **53**, 459–465.

Dawson, M., Wells, G.A.H., Parker, B.N.J. (1990) Preliminary evidence of the experimental transmissibility of bovine spongiform encephalopathy to cattle, *Vet. Rec.* **126**, 112–113.

Dawson, M., Wells, G.A.H., Parker, B.N.J., Scott, A.C. (1990) Primary parenteral transmission of bovine spongiform encephalopathy to the pig, *Vet. Rec.* **127**, 338.

DeArmond, S.J., McKinley, M.P., Barry, R.A., Braunfeld, M.B., McColloch, J.R., Prusiner, S.B. (1985) Identification of prion amyloid filaments in scrapie-infected brain, *Cell* **41**, 221–235.

Derkatch, I.L., Chernoff, Y.O., Kushnirov, V.V., Inge-Vechtomov, S.G., Liebman, S.W. (1996) Genesis and variability of [PSI] prion factors in *Saccharomyces cerevisiae*, *Genetics* **144**, 1375–1386.

Dickinson, A.G., Fraser, H., Outram, G.W. (1975) Scrapie incubation time can exceed natural lifespan, *Nature* **256**, 732–733.

Dickinson, A.G., Meikle, V.M.H., Fraser, H. (1968) Identification of a gene which controls the incubation period of some strains of scrapie agent in mice, *J. Comp. Pathol.* **78**, 293–299.

Dickinson, A.G., Meikle, V.M., Fraser, H. (1969) Genetical control of the concentration of ME7 scrapie agent in the brain of mice, *J. Comp. Pathol.* **79**, 15–22.

Dickinson, A.G., Young, G.B., Stamp, J.T., Renwick, C.C. (1965) An analysis of natural scrapie in Suffolk sheep, *Heredity* **20**, 485–503.

Doel, S.M., McCready, S.J., Nierras, C.R., Cox, B.S. (1994) The dominant $PNM2^-$ mutation which eliminates the ψ factor of *Saccharomyces cerevisiae* is the result of a missense mutation in the *SUP35* gene, *Genetics* **137**, 659–670.

Doh-ura, K., Iwaki, T., Caughey, B. (2000) Lysosomotropic agents and cysteine protease inhibitors inhibit scrapie-associated prion protein accumulation, *J. Virol.* **74**, 4894–4897.

Doh-ura, K., Kitamoto, T., Sakaki, Y., Tateishi, J. (1991) CJD discrepancy, *Nature* **353**, 801–802.

Doh-ura, K., Tateishi, J., Sasaki, H., Kitamoto, T., Sakaki, Y. (1989) Pro→Leu change at position 102 of prion protein is the most common but not the sole mutation related to Gerstmann-Sträussler syndrome, *Biochem. Biophys. Res. Commun.* **163**, 974–979.

Donne, D.G., Viles, J.H., Groth, D., Mehlhorn, I., James, T.L., Cohen, F.E., Prusiner, S.B., Wright, P.E., Dyson, H.J. (1997) Structure of the recombinant full-length hamster prion protein PrP(29–231): the N terminus is highly flexible, *Proc. Natl. Acad. Sci. U. S. A.* **94**, 13452–13457.

Edenhofer, F., Rieger, R., Famulok, M., Wendler, W., Weiss, S., Winnacker, E.-L. (1996) Prion protein PrP^C interacts with molecular chaperones of the Hsp60 family, *J. Virol.* **70**, 4724–4728.

Ehlers, B., Diringer, H. (1984) Dextran sulphate 500 delays and prevents mouse scrapie by impairment of agent replication in spleen, *J. Gen. Virol.* **65**, 1325–1330.

Enari, M., Flechsig, E., Weissmann, C. (2001) Scrapie prion protein accumulation by scrapie-infected neuroblastoma cells abrogated by exposure to a prion protein antibody, *Proc. Natl. Acad. Sci. U. S. A.* **98**, 9295–9299.

Endo, T., Groth, D., Prusiner, S.B., Kobata, A. (1989) Diversity of oligosaccharide structures linked to asparagines of the scrapie prion protein, *Biochemistry* **28**, 8380–8388.

Fink, J.K., Peacock, M.L., Warren, J.T., Roses, A.D., Prusiner, S.B. (1994) Detecting prion protein gene mutations by denaturing gradient gel electrophoresis, *Hum. Mutat.* **4**, 42–50.

Fischer, M., Rülicke, T., Raeber, A., Sailer, A., Moser, M., Oesch, B., Brandner, S., Aguzzi, A., Weissmann, C. (1996) Prion protein (PrP) with amino-proximal deletions restoring susceptibility of PrP knockout mice to scrapie, *EMBO J.* **15**, 1255–1264.

Foster, J., Goldmann, W., Parnham, D., Chong, A., Hunter, N. (2001) Partial dissociation of PrP^{Sc} deposition and vacuolation in the brains of scrapie and BSE experimentally affected goats, *J. Gen. Virol.* **82**, 267–273.

Fraser, H., Dickinson, A.G. (1973) Scrapie in mice. Agent-strain differences in the distribution and intensity of grey matter vacuolation, *J. Comp. Pathol.* **83**, 29–40.

Fraser, H., McConnell, I., Wells, G.A.H., Dawson, M. (1988) Transmission of bovine spongiform encephalopathy to mice, *Vet. Rec.* **123**, 472.

Fraser, H., Bruce, M.E., Chree, A., McConnell, I., Wells, G.A.H. (1992) Transmission of bovine spongiform encephalopathy and scrapie to mice, *J. Gen. Virol.* **73**, 1891–1897.

Furukawa, H., Kitamoto, T., Tanaka, Y., Tateishi, J. (1995) New variant prion protein in a Japanese family with Gerstmann-Sträussler syndrome, *Mol. Brain Res.* **30**, 385–388.

Gabizon, R., Telling, G., Meiner, Z., Halimi, M., Kahana, I., Prusiner, S.B. (1996) Insoluble wild-type and protease-resistant mutant prion protein in brains of patients with inherited prion disease, *Nat. Med.* **2**, 59–64.

Gabizon, R., Rosenmann, H., Meiner, Z., Kahana, I., Kahana, E., Shugart, Y., Ott, J., Prusiner, S.B. (1993) Mutation and polymorphism of the prion protein gene in Libyan Jews with Creutzfeldt-Jakob disease (CJD), *Am. J. Hum. Genet.* **53**, 828–835.

Gabriel, J.-M., Oesch, B., Kretzschmar, H., Scott, M., Prusiner, S.B. (1992) Molecular cloning of a candidate chicken prion protein, *Proc. Natl. Acad. Sci. U. S. A.* **89**, 9097–9101.

Gajdusek, D.C. (1977) Unconventional viruses and the origin and disappearance of kuru, *Science* **197**, 943–960.

Gajdusek, D.C., Gibbs Jr., C.J., Asher, D.M., Brown, P., Diwan, A., Hoffman, P., Nemo, G., Rohwer, R., White, L. (1977) Precautions in medical care of, and in handling materials from, patients with transmissible virus dementia (Creutzfeldt-Jakob disease), *N. Engl. J. Med.* **297**, 1253–1258.

Gambetti, P., Parchi, P. (1999) Insomnia in prion diseases: sporadic and familial, *N. Engl. J. Med.* **340**(21), 1675–1677.

Gasset, M., Baldwin, M.A., Fletterick, R.J., Prusiner, S.B. (1993) Perturbation of the secondary structure of the scrapie prion protein under conditions that alter infectivity, *Proc. Natl. Acad. Sci. U. S. A.* **90**, 1–5.

Gauczynski, S., Peyrin, J.M., Haik, S., Leucht, C., Hundt, C., Rieger, R., Krasemann, S., Deslys, J.P., Dormont, D., Lasmézas, C.I., Weiss, S. (2001) The 37-kDa/67-kDa laminin receptor acts as the cell-surface receptor for the cellular prion protein, *EMBO J.* **20**, 5863–5875.

Ghani, A.C., Ferguson, N.M., Donnelly, C.A., Anderson, R.M. (2000) Predicted vCJD mortality in Great Britain, *Nature* **406**, 583–584.

Glasse, R.M. (1967) Cannibalism in the kuru region of New Guinea, *Trans. NY Acad. Sci. [Ser. 2]* **29**, 748–754.

Glover, J.R., Kowal, A.S., Schirmer, E.C., Patino, M.M., Liu, J.-J., Lindquist, S. (1997) Self-seeded fibers formed by Sup35, the protein determinant of [PSI$^+$], a heritable prion-like factor of S. cerevisiae, *Cell* **89**, 811–819.

Goldfarb, L., Brown, P., Goldgaber, D., Garruto, R., Yanaghiara, R., Asher, D., Gajdusek, D.C. (1990) Identical mutation in unrelated patients with Creutzfeldt-Jakob disease, *Lancet* **336**, 174–175.

Goldfarb, L.G., Mitrova, E., Brown, P., Toh, B.H., Gajdusek, D.C. (1990) Mutation in codon 200 of scrapie amyloid protein gene in two clusters of Creutzfeldt-Jakob disease in Slovakia, *Lancet* **336**, 514–515.

Goldfarb, L.G., Haltia, M., Brown, P., Nieto, A., Kovanen, J., McCombie, W.R., Trapp, S., Gajdusek, D.C. (1991) New mutation in scrapie amyloid precursor gene (at codon 178) in Finnish Creutzfeldt-Jakob kindred, *Lancet* **337**, 425.

Goldfarb, L.G., Brown, P., Goldgaber, D., Asher, D.M., Rubenstein, R., Brown, W.T., Piccardo, P., Kascsak, R.J., Boellaard, J.W., Gajdusek, D.C. (1990) Creutzfeldt-Jakob disease and kuru patients lack a mutation consistently found in the Gerstmann-Sträussler-Scheinker syndrome, *Exp. Neurol.* **108**, 247–250.

Goldfarb, L.G., Brown, P., McCombie, W.R., Goldgaber, D., Swergold, G.D., Wills, P.R., Cervenakova, L., Baron, H., Gibbs, C.J.J., Gajdusek, D.C. (1991) Transmissible familial Creutzfeldt-Jakob disease associated with five, seven, and eight extra octapeptide coding repeats in the *PRNP* gene, *Proc. Natl. Acad. Sci. U. S. A.* **88**, 10926–10930.

Goldfarb, L.G., Petersen, R.B., Tabaton, M., Brown, P., LeBlanc, A.C., Montagna, P., Cortelli, P., Julien, J., Vital, C., Pendelbury, W.W. (1992) Fatal familial insomnia and familial Creutzfeldt-Jakob disease: disease phenotype determined by a DNA polymorphism, *Science* **258**(5083), 806–808.

Goldfarb, L.G., Brown, P., Mitrova, E., Cervenakova, L., Goldin, L., Korczyn, A.D., Chapman, J., Galvez, S., Cartier, L., Rubenstein, R., Gajdusek, D.C. (1991) Creutzfeldt-Jacob disease associated with the *PRNP* codon 200^{Lys} mutation: an analysis of 45 families, *Eur. J. Epidemiol.* **7**, 477–486.

Goldgaber, D., Goldfarb, L.G., Brown, P., Asher, D.M., Brown, W.T., Lin, S., Teener, J.W., Feinstone, S.M., Rubenstein, R., Kascsak, R.J., Boellaard, J.W., Gajdusek, D.C. (1989) Mutations in familial Creutzfeldt-Jakob disease and Gerstmann-Sträussler-Scheinker's syndrome, *Exp. Neurol.* **106**, 204–206.

Goldmann, W., Hunter, N., Manson, J. Hope, J. (1990) The PrP gene of the sheep, a natural host of scrapie, *Proceedings of the VIIIth International Congress of Virology*, Berlin, Germany, Aug. 26–31, 284.

Goldmann, W., Hunter, N., Benson, G., Foster, J.D., Hope, J. (1991) Different scrapie-associated fibril proteins (PrP) are encoded by lines of sheep selected for different alleles of the *Sip* gene, *J. Gen. Virol.* **72**, 2411–2417.

Goldmann, W., Hunter, N., Martin, T., Dawson, M., Hope, J. (1991) Different forms of the bovine PrP gene have five or six copies of a short, G-C-rich element within the protein-coding exon, *J. Gen. Virol.* **72**, 201–204.

Goldmann, W., Hunter, N., Smith, G., Foster, J., Hope, J. (1994) PrP genotype and agent effects in scrapie: change in allelic interaction with different isolates of agent in sheep, a natural host of scrapie, *J. Gen. Virol.* **75**, 989–995.

Goldmann, W., Hunter, N., Foster, J.D., Salbaum, J.M., Beyreuther, K., Hope, J. (1990) Two alleles of a neural protein gene linked to scrapie in sheep, *Proc. Natl. Acad. Sci. U. S. A.* **87**, 2476–2480.

Gorodinsky, A., Harris, D.A. (1995) Glycolipid-anchored proteins in neuroblastoma cells form detergent-resistant complexes without caveolin, *J. Cell Biol.* **129**, 619–627.

Govaerts, C., Wille, H., Prusiner, S.B., Cohen, F.E. (2004) Evidence for assembly of prions with left-handed β-helices into trimers, *Proc. Natl. Acad. Sci. U. S. A.* **101**, 8342–8347.

Graner, E., Mercadante, A.F., Zanata, S.M., Forlenza, O.V., Cabral, A.L.B., Veiga, S.S., Juliano, M.A., Roesler, R., Walz, R., Mineti, A., Izquierdo, I., Martins, V.R., Brentani, R.R. (2000) Cellular prion protein binds laminin and mediates neuritogenesis, *Mol. Brain Res.* **76**, 85–92.

Grassi, J., Comoy, E., Simon, S., Creminon, C., Frobert, Y., Trapmann, S., Schimmel, H., Hawkins, S.A., Moynagh, J., Deslys, J.P., Wells, G.A. (2001) Rapid test for the preclinical postmortem diagnosis of BSE in central nervous system tissue, *Vet. Rec.* **149**, 577–582.

Hamir, A.N., Cutlip, R.C., Miller, J.M., Williams, E.S., Stack, M.J., Miller, M.W., O'Rourke, K.I., Chaplin, M.J. (2001) Preliminary findings on the experimental transmission of chronic wasting disease agent of mule deer to cattle, *J. Vet. Diagn. Invest.* **13**, 91–96.

Hammick, D.L., Chambers, W.E. (1945) Optical activity of excreted mepacrine, *Nature* **155**, 141.

Haraguchi, T., Fisher, S., Olofsson, S., Endo, T., Groth, D., Tarantino, A., Borchelt, D.R., Teplow, D., Hood, L., Burlingame, A., Lycke, E., Kobata, A., Prusiner, S.B. (1989) Asparagine-linked glycosylation of the scrapie and cellular prion proteins, *Arch. Biochem. Biophys.* **274**, 1–13.

Harries-Jones, R., Knight, R., Will, R.G., Cousens, S., Smith, P.G., Matthews, W.B. (1988) Creutzfeldt-Jakob disease in England and Wales, 1980–1984: a case-control study of potential risk factors, *J. Neurol. Neurosurg. Psychiatry* **51**, 1113–1119.

Harris, D.A., Falls, D.L., Walsh, W., Fischbach, G.D. (1989) Molecular cloning of an acetylcholine receptor-inducing protein, *Soc. Neurosci.* **15**, 70. 7.

Heppner, F.L., Musahl, C., Arrighi, I., Klein, M.A., Rülicke, T., Oesch, B., Zinkernagel, R.M., Kalinke, U., Aguzzi, A. (2001) Prevention of scrapie pathogenesis by transgenic expression of antiprion protein antibodies, *Science* **294**, 178–182.

Hill, A.F., Desbruslais, M., Joiner, S., Sidle, K.C.L., Gowland, I., Collinge, J., Doey, L.J., Lantos, P. (1997) The same prion strain causes vCJD and BSE, *Nature* **389**, 448–450.

Hilton, D.A., Ghani, A.C., Conyers, L., Edwards, P., McCardle, L., Ritchie, D., Penney, M., Hegazy, D., Ironside, J.W. (2004) Prevalence of lymphoreticular prion protein accumulation in UK tissue samples, *J. Pathol.* **203**, 733–739.

Hornshaw, M.P., McDermott, J.R., Candy, J.M. (1995) Copper binding to the N-terminal tandem repeat regions of mammalian and avian prion protein, *Biochem. Biophys. Res. Commun.* **207**, 621–629.

Hornshaw, M.P., McDermott, J.R., Candy, J.M., Lakey, J.H. (1995) Copper binding to the N-terminal tandem repeat region of mammalian and avian prion protein: structural studies using synthetic peptides, *Biochem. Biophys. Res. Commun.* **214**, 993–999.

Hsiao, K., Baker, H.F., Crow, T.J., Poulter, M., Owen, F., Terwilliger, J.D., Westaway, D., Ott, J., Prusiner, S.B. (1989) Linkage of a prion protein missense variant to Gerstmann-Sträussler syndrome, *Nature* **338**, 342–345.

Hsiao, K., Meiner, Z., Kahana, E., Cass, C., Kahana, I., Avrahami, D., Scarlato, G., Abramsky, O., Prusiner, S.B., Gabizon, R. (1991) Mutation of the prion protein in Libyan Jews with Creutzfeldt-Jakob disease, *N. Engl. J. Med.* **324**, 1091–1097.

Hsiao, K.K., Cass, C., Schellenberg, G.D., Bird, T., Devine-Gage, E., Wisniewski, H., Prusiner, S.B. (1991) A prion protein variant in a family with the telencephalic form of Gerstmann-Sträussler-Scheinker syndrome, *Neurology* **41**, 681–684.

Hsiao, K.K., Groth, D., Scott, M., Yang, S.-L., Serban, H., Rapp, D., Foster, D., Torchia, M., DeArmond, S.J., Prusiner, S.B. (1994) Serial transmission in rodents of neurodegeneration from transgenic mice expressing mutant prion protein, *Proc. Natl. Acad. Sci. U. S. A.* **91**, 9126–9130.

Huang, Z., Prusiner, S.B., Cohen, F.E. (1995) Scrapie prions: a three-dimensional model of an infectious fragment, *Fold. Des.* **1**, 13–19.

Hundt, C., Peyrin, J.M., Haik, S., Gauczynski, S., Leucht, C., Rieger, R., Riley, M.L., Deslys, J.P., Dormont, D., Lasmézas, C.I., Weiss, S. (2001) Identification of interaction domains of the prion protein with its 37-kDa/67-kDa laminin receptor, *EMBO J.* **20**, 5876–5886.

Hunter, N., Foster, J.D., Benson, G., Hope, J. (1991) Restriction fragment length polymorphisms of the scrapie-associated fibril protein (PrP) gene and their association with susceptiblity to natural scrapie in British sheep, *J. Gen. Virol.* **72**, 1287–1292.

Hunter, N., Goldmann, W., Smith, G., Hope, J. (1994) Frequencies of PrP gene variants in healthy cattle and cattle with BSE in Scotland, *Vet. Rec.* **135**, 400–403.

Hunter, N., Goldmann, W., Benson, G., Foster, J.D., Hope, J. (1993) Swaledale sheep affected by natural scrapie differ significantly in PrP genotype frequencies from healthy sheep and those selected for reduced incidence of scrapie, *J. Gen. Virol.* **74**, 1025–1031.

Hunter, N., Moore, L., Hosie, B.D., Dingwall, W.S., Greig, A. (1997) Association between natural scrapie and PrP genotype in a flock of Suffolk sheep in Scotland, *Vet. Rec.* **140**, 59–63.

Hunter, N., Cairns, D., Foster, J.D., Smith, G., Goldmann, W., Donnelly, K. (1997) Is scrapie solely a genetic disease? *Nature* **386**, 137.

Ikeda, T., Horiuchi, M., Ishiguro, N., Muramatsu, Y., Kai-Uwe, G.D., Shinagawa, M. (1995) Amino acid polymorphisms of PrP with reference to onset of scrapie in Suffolk and Corriedale sheep in Japan, *J. Gen. Virol.* **76**, 2577–2581.

Ingrosso, L., Ladogana, A., Pocchiari, M. (1995) Congo red prolongs the incubation period in scrapie-infected hamsters, *J. Virol.* **69**, 506–508.

Ironside, J.W. (1997) The new variant form of Creutzfeldt-Jakob disease: a novel prion protein amyloid disorder (Editorial), *Amyloid: Int. J. Exp. Clin. Invest.* **4**, 66–69.

James, T.L., Liu, H., Ulyanov, N.B., Farr-Jones, S., Zhang, H., Donne, D.G., Kaneko, K., Groth, D., Mehlhorn, I., Prusiner, S.B., Cohen, F.E. (1997) Solution structure of a 142-residue recombinant prion protein corresponding to the infectious fragment of the scrapie isoform, *Proc. Natl. Acad. Sci. U. S. A.* **94**(19), 10086–10091.

Jeffrey, M., Wells, G.A.H. (1988) Spongiform encephalopathy in a nyala *(Tragelaphus angasi)*, *Vet. Pathol.* **25**, 398–399.

Jendroska, K., Heinzel, F.P., Torchia, M., Stowring, L., Kretzschmar, H.A., Kon, A., Stern, A., Prusiner, S.B., DeArmond, S.J. (1991) Proteinase-resistant prion protein accumulation in Syrian hamster brain correlates with regional pathology and scrapie infectivity, *Neurology* **41**, 1482–1490.

Kahana, E., Milton, A., Braham, J., Sofer, D. (1974) Creutzfeldt-Jakob disease: focus among Libyan Jews in Israel, *Science* **183**, 90–91.

Kaneko, K., Zulianello, L., Scott, M., Cooper, C.M., Wallace, A.C., James, T.L., Cohen, F.E., Prusiner, S.B. (1997) Evidence for protein X binding to a discontinuous epitope on the cellular prion protein during scrapie prion propagation, *Proc. Natl. Acad. Sci. U. S. A.* **94**, 10069–10074.

Kaneko, K., Ball, H.L., Wille, H., Zhang, H., Groth, D., Torchia, M., Tremblay, P., Safar, J., Prusiner, S.B., DeArmond, S.J., Baldwin, M.A., Cohen, F.E. (2000) A synthetic peptide initiates Gerstmann-Sträussler-Scheinker (GSS) disease in transgenic mice, *J. Mol. Biol.* **295**(4), 997–1007.

Kascsak, R.J., Rubenstein, R., Merz, P.A., Tonna-DeMasi, M., Fersko, R., Carp, R.I., Wisniewski, H.M., Diringer, H. (1987) Mouse polyclonal and monoclonal antibody to scrapie-associated fibril proteins, *J. Virol.* **61**, 3688–3693.

Kellings, K., Prusiner, S.B., Riesner, D. (1994) Nucleic acids in prion preparations: unspecific background or essential component? *Philos. Trans. R. Soc. Lond. B. Biol. Sci.* **343**, 425–430.

Kellings, K., Meyer, N., Mirenda, C., Prusiner, S.B., Riesner, D. (1992) Further analysis of nucleic acids in purified scrapie prion preparations by improved return refocussing gel electrophoresis (RRGE), *J. Gen. Virol.* **73**, 1025–1029.

Keshet, G.I., Bar-Peled, O., Yaffe, D., Nudel, U., Gabizon, R. (2000) The cellular prion protein colocalizes with the dystroglycan complex in the brain, *J. Neurochem.* **75**, 1889–1897.

Kimberlin, R.H. (1982) Reflections on the nature of the scrapie agent, *Trends Biochem. Sci.* **7**, 392–394.

Kimberlin, R.H. (1990) Scrapie and possible relationships with viroids, *Semin. Virol.* **1**, 153–162.

Kimberlin, R.H., Walker, C.A. (1983) The antiviral compound HPA-23 can prevent scrapie when administered at the time of infection, *Arch. Virol.* **78**, 9–18.

Kimberlin, R.H., Collis, S.C., Walker, C.A. (1976) Profiles of brain glycosidase activity in cuprizone-fed Syrian hamsters and in scrapie-affected mice, rats, Chinese hamsters and Syrian hamsters, *J. Comp. Pathol.* **86**, 135–142.

Kimberlin, R.H., Field, H.J., Walker, C.A. (1983) Pathogenesis of mouse scrapie: evidence for spread of infection from central to peripheral nervous system, *J. Gen. Virol.* **64**, 713–716.

Kimberlin, R.H., Walker, C.A., Fraser, H. (1989) The genomic identity of different strains of mouse scrapie is expressed in hamsters and preserved on reisolation in mice, *J. Gen. Virol.* **70**, 2017–2025.

King, C.-Y., Tittman, P., Gross, H., Gebert, R., Aebi, M., Wüthrich, K. (1997) Prion-inducing domain 2–114 of yeast Sup35 protein transforms *in vitro* into amyloid-like filaments, *Proc. Natl. Acad. Sci. U. S. A.* **94**, 6618–6622.

Kingsbury, D.T., Kasper, K.C., Stites, D.P., Watson, J.D., Hogan, R.N., Prusiner, S.B. (1983) Genetic control of scrapie and Creutzfeldt-Jakob disease in mice, *J. Immunol.* **131**, 491–496.

Kirkwood, J.K., Cunningham, A.A., Wells, G.A.H., Wilesmith, J.W., Barnett, J.E.F. (1993) Spongiform encephalopathy in a herd of greater kudu (*Tragelaphus strepsiceros*): epidemiological observations, *Vet. Rec.* **133**, 360–364.

Kitamoto, T., Tateishi, J. (1994) Human prion diseases with variant prion protein, *Philos. Trans. R. Soc. Lond. B. Biol. Sci.* **343**, 391–398.

Klatzo, I., Gajdusek, D.C., Zigas, V. (1959) Pathology of kuru, *Lab. Invest.* **8**, 799–847.

Klitzman, R.L., Alpers, M.P., Gajdusek, D.C. (1984) The natural incubation period of kuru and the episodes of transmission in three clusters of patients, *Neuroepidemiology* **3**, 3–20.

Koch, T.K., Berg, B.O., DeArmond, S.J., Gravina, R.F. (1985) Creutzfeldt-Jakob disease in a young adult with idiopathic hypopituitarism. Possible relation to the administration of cadaveric human growth hormone, *N. Engl. J. Med.* **313**, 731–733.

Kocisko, D.A., Come, J.H., Priola, S.A., Chesebro, B., Raymond, G.J., Lansbury Jr., P.T., Caughey, B. (1994) Cell-free formation of protease-resistant prion protein, *Nature* **370**, 471–474.

Korth, C., May, B.C.H., Cohen, F.E., Prusiner, S.B. (2001) Acridine and phenothiazine derivatives as pharmacotherapeutics for prion disease, *Proc. Natl. Acad. Sci. U. S. A.* **98**, 9836–9841.

Korth, C., Kaneko, K., Groth, D., Heye, N., Telling, G., Mastrianni, J., Parchi, P., Gambetti, P., Will, R., Ironside, J., Heinrich, C., Tremblay, P., DeArmond, S.J., Prusiner, S.B. (2003) Abbreviated incubation times for human prions in mice expressing a chimeric mouse – human prion protein transgene, *Proc. Natl. Acad. Sci. U. S. A.* **100**, 4784–4789.

Korth, C., Stierli, B., Streit, P., Moser, M., Schaller, O., Fischer, R., Schulz-Schaeffer, W., Kretzschmar, H., Raeber, A., Braun, U., Ehrensperger, F., Hornemann, S., Glockshuber, R., Riek, R., Billeter, M., Wuthrick, K., Oesch, B. (1997) Prion (PrP^{Sc})-specific epitope defined by a monoclonal antibody, *Nature* **389**, 74–77.

Kretzschmar, H.A., Neumann, M., Stavrou, D. (1995) Codon 178 mutation of the human prion protein gene in a German family (Backer family): sequencing data from 72 year-old celloidin-embedded brain tissue, *Acta Neuropathol.* **89**, 96–98.

Kretzschmar, H.A., Prusiner, S.B., Stowring, L.E., DeArmond, S.J. (1986) Scrapie prion proteins are synthesized in neurons, *Am. J. Pathol.* **122**, 1–5.

Kretzschmar, H.A., Honold, G., Seitelberger, F., Feucht, M., Wessely, P., Mehraein, P., Budka, H. (1991) Prion protein mutation in family first reported by Gerstmann, Sträussler, and Scheinker, *Lancet* **337**, 1160.

Kurschner, C., Morgan, J.I. (1995) The cellular prion protein (PrP) selectively binds to Bcl-2 in the yeast two-hybrid system, *Mol. Brain Res.* **30**, 165–168.

Laplanche, J.-L., Chatelain, J., Launay, J.-M., Gazengel, C., Vidaud, M. (1990) Deletion in prion protein gene in a Moroccan family, *Nucleic Acids. Res.* **18**, 6745.

Laplanche, J.-L., Chatelain, J., Beaudry, P., Dussaucy, M., Bounneau, C., Launay, J.-M. (1993) French autochthonous scrapied sheep without the 136Val PrP polymorphism, *Mamm. Genome* **4**, 463–464.

Lasmézas, C.I., Deslys, J.-P., Demaimay, R., Adjou, K.T., Hauw, J.-J., Dormont, D. (1996) Strain specific and common pathogenic events in murine models of scrapie and bovine spongiform encephalopathy, *J. Gen. Virol.* **77**, 1601–1609.

Lasmézas, C.I., Deslys, J.-P., Demaimay, R., Adjou, K.T., Lamoury, F., Dormont, D., Robain, O., Ironside, J., Hauw, J.-J. (1996) BSE transmission to macaques, *Nature* **381**, 743–744.

Lasmézas, C.I., Deslys, J.-P., Robain, O., Jaegly, A., Beringue, V., Peyrin, J.-M., Fournier, J.-G., Hauw, J.-J., Rossier, J., Dormont, D. (1997) Transmission of the BSE agent to mice in the absence of detectable abnormal prion protein, *Science* **275**(17), 402–405.

Legname, G., Baskakov, I.V., Nguyen, H.-O.B., Riesner, D., Cohen, F.E., DeArmond, S.J., Prusiner, S.B. (2004) Synthetic mammalian prions, *Science* **305**, 673–676.

Li, G., Bolton, D.C. (1997) A novel hamster prion protein mRNA contains an extra exon: increased expression in scrapie, *Brain Res.* **751**, 265–274.

Liu, H., Farr-Jones, S., Ulyanov, N.B., Llinas, M., Marqusee, S., Groth, D., Cohen, F.E., Prusiner, S.B., James, T.L. (1999) Solution structure of Syrian hamster prion protein rPrP(90–231), *Biochemistry* **38**(17), 5362–5377.

Lledo, P.-M., Tremblay, P., DeArmond, S.J., Prusiner, S.B., Nicoll, R.A. (1996) Mice deficient for prion protein exhibit normal neuronal excitability and synaptic transmission in the hippocampus, *Proc. Natl. Acad. Sci. U. S. A.* **93**, 2403–2407.

Llewelyn, C.A., Hewitt, P.E., Knight, R.S., Amar, K., Cousens, S., Mackenzie, J., Will, R.G. (2004) Possible transmission of variant Creutzfeldt-Jakob disease by blood transfusion, *Lancet* **363**, 417–421.

Locht, C., Chesebro, B., Race, R., Keith, J.M. (1986) Molecular cloning and complete sequence of prion protein cDNA from mouse brain infected with the scrapie agent, *Proc. Natl. Acad. Sci. U. S. A.* **83**, 6372–6376.

Lugaresi, E., Medori, R., Montagna, P., Baruzzi, A., Cortelli, P., Lugaresi, A., Tinuper, P., Zucconi, M., Gambetti, P. (1986) Fatal familial insomnia and dysautonomia with selective degeneration of thalamic nuclei, *N. Engl. J. Med.* **315**, 997–1003.

Maddelein, M.L., Dos Reis, S., Duvezin-Caubet, S., Coulary-Salin, B., Saupe, S.J. (2002) Amyloid aggregates of the HET-s prion protein are infectious, *Proc. Natl. Acad. Sci. U. S. A.* **99**, 7402–7407.

Mallucci, G., Dickinson, A., Linehan, J., Klohn, P.C., Brandner, S., Collinge, J. (2003) Depleting neuronal PrP in prion infection prevents disease and reverses spongiosis, *Science* **302**, 871–874.

Manson, J.C., Clarke, A.R., Hooper, M.L., Aitchison, L., McConnell, I., Hope, J. (1994) 129/Ola mice carrying a null mutation in PrP that abolishes mRNA production are developmentally normal, *Mol. Neurobiol.* **8**, 121–127.

Manson, J.C., Jameison, E., Baybutt, H., Tuzi, N.L., Barron, R., McConnell, I., Somerville, R., Ironside, J., Will, R., Sy, M.-S., Melton, D.W.,

Hope, J., Bostock, C. (1999) A single amino acid alteration (101L) introduced into murine PrP dramatically alters incubation time of transmissible spongiform encephalopathy, *EMBO J.* **18**, 6855–6864.

Marsh, R.F., Bessen, R.A., Lehmann, S., Hartsough, G.R. (1991) Epidemiological and experimental studies on a new incident of transmissible mink encephalopathy, *J. Gen. Virol.* **72**, 589–594.

Martins, V.R., Graner, E., Garcia-Abreu, J., de Souza, S.J., Mercadante, A.F., Veiga, S.S., Zanata, S.M., Neto, V.M., Brentani, R.R. (1997) Complementary hydropathy identifies a cellular prion protein receptor, *Nat. Med.* **3**, 1376–1382.

Masison, D.C., Wickner, R.B. (1995) Prion-inducing domain of yeast Ure2p and protease resistance of Ure2p in prion-containing cells, *Science* **270**, 93–95.

Masters, C.L., Gajdusek, D.C., Gibbs Jr., C.J. (1981) Creutzfeldt-Jakob disease virus isolations from the Gerstmann-Sträussler syndrome, *Brain* **104**, 559–588.

Masters, C.L., Harris, J.O., Gajdusek, D.C., Gibbs Jr., C.J., Bernouilli, C., Asher, D.M. (1978) Creutzfeldt-Jakob disease: patterns of worldwide occurrence and the significance of familial and sporadic clustering, *Ann. Neurol.* **5**, 177–188.

Mastrianni, J., Nixon, F., Layzer, R., DeArmond, S.J., Prusiner, S.B. (1997) Fatal sporadic insomnia: fatal familial insomnia phenotype without a mutation of the prion protein gene, *Neurology* **48** (Suppl.), A296.

Mastrianni, J.A., Nixon, R., Layzer, R., Telling, G.C., Han, D., DeArmond, S.J., Prusiner, S.B. (1999) Prion protein conformation in a patient with sporadic fatal insomnia, *N. Engl. J. Med.* **340**(21), 1630–1638.

McKinley, M.P., Meyer, R.K., Kenaga, L., Rahbar, F., Cotter, R., Serban, A., Prusiner, S.B. (1991) Scrapie prion rod formation in vitro requires both detergent extraction and limited proteolysis, *J. Virol.* **65**, 1340–1351.

McKnight, S., Tjian, R. (1986) Transcriptional selectivity of viral genes in mammalian cells, *Cell* **46**, 795–805.

Medori, R., Montagna, P., Tritschler, H.J., LeBlanc, A., Cortelli, P., Tinuper, P., Lugaresi, E., Gambetti, P. (1992) Fatal familial insomnia: a second kindred with mutation of prion protein gene at codon 178, *Neurology* **42**, 669–670.

Meggendorfer, F. (1930) Klinische und genealogische Beobachtungen bei einem Fall von spastischer Pseudosklerose Jakobs, *Z. Gesamte Neurol. Psychiatr.* **128**, 337–341.

Meyer, R.K., McKinley, M.P., Bowman, K.A., Braunfeld, M.B., Barry, R.A., Prusiner, S.B. (1986) Separation and properties of cellular and scrapie prion proteins, *Proc. Natl. Acad. Sci. U. S. A.* **83**, 2310–2314.

Miller, M.W., Wild, M.A., Williams, E.S. (1998) Epidemiology of chronic wasting disease in captive Rocky Mountain elk, *J. Wildl. Dis.* **34**, 532–538.

Miller, M.W., Williams, E.S., McCarty, C.W., Spraker, T.R., Kreeger, T.J., Larsen, C.T., Thorne, E.T. (2000) Epizootiology of chronic wasting disease in free-ranging cervids in Colorado and Wyoming, *J. Wildl. Dis.* **36**, 676–690.

Mills, J.L., Schonberger, L.B., Wysowski, D.K., Brown, P., Durako, S.J., Cox, C., Kong, F., Fradkin, J.E. (2004) Long-term mortality in the United States cohort of pituitary-derived growth hormone recipients, *J. Pediatr.* **144**, 430–436.

Minor, P., Newham, J., Jones, N., Bergeron, C., Gregori, L., Asher, D., Van Engelenburg, F., Stroebel, T., Vey, M., Barnard, G., Head, M. (2004) Standards for the assay of Creutzfeldt-Jakob disease specimens, *J. Gen. Virol.* **85**, 1777–1784.

Miyazono, M., Kitamoto, T., Doh-ura, K., Iwaki, T., Tateishi, J. (1992) Creutzfeldt-Jakob disease with codon 129 polymorphism (Valine): a comparative study of patients with codon 102 point mutation or without mutations, *Acta Neuropathol.* **84**, 349–354.

Mo, H., Moore, R.C., Cohen, F.E., Westaway, D., Prusiner, S.B., Wright, P.E., Dyson, H.J. (2001) Two different neurodegenerative diseases caused by proteins with similar structures, *Proc. Natl. Acad. Sci. U. S. A.* **98**, 2352–2357.

Mobley, W.C., Neve, R.L., Prusiner, S.B., McKinley, M.P. (1988) Nerve growth factor increases mRNA levels for the prion protein and the β-amyloid protein precursor in developing hamster brain, *Proc. Natl. Acad. Sci. U. S. A.* **85**, 9811–9815.

Moore, R.C., Hope, J., McBride, P.A., McConnell, I., Selfridge, J., Melton, D.W., Manson, J.C. (1998) Mice with gene targeted prion protein alterations show that *Prnp*, *Sinc* and *Prni* are congruent, *Nat. Genet.* **18**, 118–125.

Moore, R.C., Mastrangelo, P., Bouzamondo, E., Heinrich, C., Legname, G., Prusiner, S.B., Hood, L., Westaway, D., DeArmond, S.J., Tremblay, P. (2001) Doppel-induced cerebellar degeneration in transgenic mice, *Proc. Natl. Acad. Sci. U. S. A.* **98**, 15288–15293.

Moore, R.C., Lee, I.Y., Silverman, G.L., Harrison, P.M., Strome, R., Heinrich, C., Karunaratne, A., Pasternak, S.H., Chishti, M.A., Liang, Y., Mastrangelo, P., Wang, K., Smit, A.F.A., Katamine, S., Carlson, G.A., Cohen, F.E., Prusiner, S.B., Melton, D.W., Tremblay, P., Hood, L.E., Westaway, D. (1999) Ataxia in prion protein (PrP)-deficient mice is associated with upregulation of the novel PrP-like protein doppel, *J. Mol. Biol.* **292(4)**, 797–817.

Mouillet-Richard, S., Ermonval, M., Chebassier, C., Laplanche, J.-L., Lehmann, S., Launay, J.M., Kellermann, O. (2000) Signal transduction through prion protein, *Science* **289**, 1925–1928.

Nathanson, N., Wilesmith, J., Griot, C. (1997) Bovine spongiform encephalopathy (BSE): cause and consequences of a common-source epidemic, *Am. J. Epidemiol.* **145**, 959–969.

Nishida, N., Tremblay, P., Sugimoto, T., Shigematsu, K., Shirabe, S., Petromilli, C., Erpel, S.P., Nakaoke, R., Atarashi, R., Houtani, T., Torchia, M., Sakaguchi, S., DeArmond, S.J., Prusiner, S.B., Katamine, S. (1999) A mouse prion protein transgene rescues mice deficient for the prion protein gene from Purkinje cell degeneration and demyelination, *Lab. Invest.* **79(6)**, 689–697.

Oesch, B., Teplow, D.B., Stahl, N., Serban, D., Hood, L.E., Prusiner, S.B. (1990) Identification of cellular proteins binding to the scrapie prion protein, *Biochemistry* **29**, 5848–5855.

Oesch, B., Westaway, D., Wälchli, M., McKinley, M.P., Kent, S.B.H., Aebersold, R., Barry, R.A., Tempst, P., Teplow, D.B., Hood, L.E., Prusiner, S.B., Weissmann, C. (1985) A cellular gene encodes scrapie PrP 27–30 protein, *Cell* **40**, 735–746.

O'Rourke, K.I., Holyoak, G.R., Clark, W.W., Mickelson, J.R., Wang, S., Melco, R.P., Besser, T.E., Foote, W.C. (1997) PrP genotypes and experimental scrapie in orally inoculated Suffolk sheep in the United States, *J. Gen. Virol.* **78**, 975–978.

O'Rourke, K.I., Besser, T.E., Miller, M.W., Cline, T.F., Spraker, T.R., Jenny, A.L., Wild, M.A., Zebarth, G.L., Williams, E.S. (1999) PrP genotypes of captive and free-ranging Rocky Mountain elk (Cervus elaphus nelsoni) with chronic wasting disease, *J. Gen. Virol.* **80**, 2765–2769.

Owen, F., Poulter, M., Collinge, J., Crow, T.J. (1990) Codon 129 changes in the prion protein gene in Caucasians, *Am. J. Hum. Genet.* **46**, 1215–1216.

Owen, F., Poulter, M., Lofthouse, R., Collinge, J., Crow, T.J., Risby, D., Baker, H.F., Ridley, R.M., Hsiao, K., Prusiner, S.B. (1989) Insertion in prion protein gene in familial Creutzfeldt-Jakob disease, *Lancet* **1**, 51–52.

Paisley, D., Banks, S., Selfridge, J., McLennan, N.F., Ritchie, A.M., McEwan, C., Irvine, D.S., Saunders, P.T., Manson, J.C., Melton, D.W. (2004) Male infertility and DNA damage in Doppel knockout and prion protein/Doppel double-knockout mice, *Am. J. Pathol.* **164**, 2279–2288.

Palmer, M.S., Dryden, A.J., Hughes, J.T., Collinge, J. (1991) Homozygous prion protein genotype predisposes to sporadic Creutzfeldt-Jakob disease, *Nature* **352**, 340–342.

Palmer, M.S., Mahal, S.P., Campbell, T.A., Hill, A.F., Sidle, K.C.L., Laplanche, J.-L., Collinge, J. (1993) Deletions in the prion protein gene are not associated with CJD, *Hum. Mol. Genet.* **2**, 541–544.

Pan, K.-M., Stahl, N., Prusiner, S.B. (1992) Purification and properties of the cellular prion protein from Syrian hamster brain, *Protein Sci.* **1**, 1343–1352.

Pan, K.-M., Baldwin, M., Nguyen, J., Gasset, M., Serban, A., Groth, D., Mehlhorn, I., Huang, Z., Fletterick, R.J., Cohen, F.E., Prusiner, S.B. (1993) Conversion of α-helices into β-sheets features in the formation of the scrapie prion proteins, *Proc. Natl. Acad. Sci. U. S. A.* **90**, 10962–10966.

Paramithiotis, E., Pinard, M., Lawton, T., LaBoissiere, S., Leathers, V.L., Zou, W.Q., Estey, L.A., Lamontagne, J., Lehto, M.T., Kondejewski, L.H., Francoeur, G.P., Papadopoulos, M., Haghighat, A., Spatz, S.J., Head, M., Will, R., Ironside, J., O'Rourke, K., Tonelli, Q., Ledebur, H.C., Chakrabartty, A., Cashman, N.R. (2003) A prion protein epitope selective for the pathologically misfolded conformation, *Nat. Med.* **9**, 893–899.

Parchi, P., Capellari, S., Chin, S., Schwarz, H.B., Schecter, N.P., Butts, J.D., Hudkins, P., Burns,

D.K., Powers, J.M., Gambetti, P. (1999) A subtype of sporadic prion disease mimicking fatal familial insomnia, *Neurology* **52**, 1757–1763.

Parchi, P., Capellari, S., Chen, S.G., Petersen, R.B., Gambetti, P., Kopp, P., Brown, P., Kitamoto, T., Tateishi, J., Giese, A., Kretzschmar, H. (1997) Typing prion isoforms (Lett.), *Nature* **386**, 232–233.

Parry, H.B. (1962) Scrapie: a transmissible and hereditary disease of sheep, *Heredity* **17**, 75–105.

Patino, M.M., Liu, J.-J., Glover, J.R., Lindquist, S. (1996) Support for the prion hypothesis for inheritance of a phenotypic trait in yeast, *Science* **273**, 622–626.

Pattison, I.H., Millson, G.C. (1961) Scrapie produced experimentally in goats with special reference to the clinical syndrome, *J. Comp. Pathol.* **71**, 101–108.

Pattison, I.H., Jebbett, J.N. (1971) Clinical and histological observations on cuprizone toxicity and scrapie in mice, *Res. Vet. Sci.* **12**, 378–380.

Peden, A.H., Head, M.W., Ritchie, D.L., Bell, J.E., Ironside, J.W. (2004) Preclinical vCJD after blood transfusion in a *PRNP* codon 129 heterozygous patient, *Lancet* **364**, 527–529.

Peretz, D., Scott, M., Groth, D., Williamson, A., Burton, D., Cohen, F.E., Prusiner, S.B. (2001) Strain-specified relative conformational stability of the scrapie prion protein, *Protein Sci.* **10**, 854–863.

Peretz, D., Williamson, R.A., Legname, G., Matsunaga, Y., Vergara, J., Burton, D., DeArmond, S.J., Prusiner, S.B., Scott, M.R. (2002) A change in the conformation of prions accompanies the emergence of a new prion strain, *Neuron* **34**, 921–932.

Peretz, D., Williamson, R.A., Matsunaga, Y., Serban, H., Pinilla, C., Bastidas, R.B., Rozenshteyn, R., James, T.L., Houghten, R.A., Cohen, F.E., Prusiner, S.B., Burton, D.R. (1997) A conformational transition at the N-terminus of the prion protein features in formation of the scrapie isoform, *J. Mol. Biol.* **273**, 614–622.

Peretz, D., Williamson, R.A., Kaneko, K., Vergara, J., Leclerc, E., Schmitt-Ulms, G., Mehlhorn, I.R., Legname, G., Wormald, M.R., Rudd, P.M., Dwek, R.A., Burton, D.R., Prusiner, S.B. (2001) Antibodies inhibit prion propagation and clear cell cultures of prion infectivity, *Nature* **412**, 739–743.

Perrier, V., Wallace, A.C., Kaneko, K., Safar, J., Prusiner, S.B., Cohen, F.E. (2000) Mimicking dominant-negative inhibition of prion replication through structure-based drug design, *Proc. Natl. Acad. Sci. U. S. A.* **97**, 6073–6078.

Perrier, V., Kaneko, K., Safar, J., Vergara, J., Tremblay, P., DeArmond, S.J., Cohen, F.E., Prusiner, S.B., Wallace, A.C. (2002) Dominant-negative inhibition of prion replication in transgenic mice, *Proc. Natl. Acad. Sci. U. S. A.* **99**, 13079–13084.

Peters, J., Miller, J.M., Jenny, A.L., Peterson, T.L., Carmichael, K.P. (2000) Immunohistochemical diagnosis of chronic wasting disease in preclinically affected elk from a captive herd, *J. Vet. Diagn. Invest.* **12**, 579–582.

Petersen, R.B., Tabaton, M., Berg, L., Schrank, B., Torack, R.M., Leal, S., Julien, J., Vital, C., Deleplanque, B., Pendlebury, W.W., Drachman, D., Smith, T.W., Martin, J.J., Oda, M., Montagna, P., Ott, J., Autilio-Gambetti, L., Lugaresi, E., Gambetti, P. (1992) Analysis of the prion protein gene in thalamic dementia, *Neurology* **42**, 1859–1863.

Phillips, N.A., Bridgeman, J. Ferguson-Smith, M. (2000) Findings & Conclusions. *The BSE Inquiry*, Vol. 1, Stationery Office, London, UK.

PHS (1997) Report on Human Growth Hormone and Creutzfeldt-Jakob Disease, Public Health Service Interagency Coordinating Committee. **14**: 1–11.

Poulter, M., Baker, H.F., Frith, C.D., Leach, M., Lofthouse, R., Ridley, R.M., Shah, T., Owen, F., Collinge, J., Brown, G., Hardy, J., Mullan, M.J., Harding, A.E., Bennett, C., Doshi, R., Crow, T.J. (1992) Inherited prion disease with 144 base pair gene insertion. 1. Genealogical and molecular studies, *Brain* **115**, 675–685.

Priola, S.A., Raines, A., Caughey, W.S. (2000) Porphyrin and phthalocyanine antiscrapie compounds, *Science* **287**, 1503–1506.

Proske, D., Gilch, S., Wopfner, F., Schätzl, H.M., Winnacker, E.L., Famulok, M. (2002) Prion-protein-specific aptamer reduces PrP^{Sc} formation, *Chembiochem* **3**, 717–725.

Prusiner, S.B. (1982) Novel proteinaceous infectious particles cause scrapie, *Science* **216**, 136–144.

Prusiner, S.B. (2004) Detecting mad cow disease, *Sci. Am.* **291**, 86–93.

Prusiner, S.B., Groth, D.F., Bolton, D.C., Kent, S.B., Hood, L.E. (1984) Purification and structural studies of a major scrapie prion protein, *Cell* **38**, 127–134.

Prusiner, S.B., Bolton, D.C., Groth, D.F., Bowman, K.A., Cochran, S.P., McKinley, M.P. (1982) Further purification and characterization of scrapie prions, *Biochemistry* **21**, 6942–6950.

Prusiner, S.B., McKinley, M.P., Bowman, K.A., Bolton, D.C., Bendheim, P.E., Groth, D.F., Glenner, G.G. (1983) Scrapie prions aggregate to form amyloid-like birefringent rods, *Cell* **35**, 349–358.

Prusiner, S.B., Groth, D., Serban, A., Koehler, R., Foster, D., Torchia, M., Burton, D., Yang, S.-L., DeArmond, S.J. (1993) Ablation of the prion protein (PrP) gene in mice prevents scrapie and facilitates production of anti-PrP antibodies, *Proc. Natl. Acad. Sci. U. S. A.* **90**, 10608–10612.

Prusiner, S.B., Fuzi, M., Scott, M., Serban, D., Serban, H., Taraboulos, A., Gabriel, J.-M., Wells, G., Wilesmith, J., Bradley, R., DeArmond, S.J., Kristensson, K. (1993) Immunologic and molecular biological studies of prion proteins in bovine spongiform encephalopathy, *J. Infect. Dis.* **167**, 602–613.

Prusiner, S.B., Scott, M., Foster, D., Pan, K.-M., Groth, D., Mirenda, C., Torchia, M., Yang, S.-L., Serban, D., Carlson, G.A., Hoppe, P.C., Westaway, D., DeArmond, S.J. (1990) Transgenetic studies implicate interactions between homologous PrP isoforms in scrapie prion replication, *Cell* **63**, 673–686.

Raymond, G.J., Hope, J., Kocisko, D.A., Priola, S.A., Raymond, L.D., Bossers, A., Ironside, J., Will, R.G., Chen, S.G., Petersen, R.B., Gambetti, P., Rubenstein, R., Smits, M.A., Lansbury Jr., P.T., Caughey, B. (1997) Molecular assessment of the potential transmissibilities of BSE and scrapie to humans, *Nature* **388**, 285–288.

Requena, J.R., Groth, D., Legname, G., Stadtman, E.R., Prusiner, S.B., Levine, R.L. (2001) Copper-catalyzed oxidation of the recombinant SHa(29–231) prion protein, *Proc. Natl. Acad. Sci. U. S. A.* **98**, 7170–7175.

Rieger, R., Edenhofer, F., Lasmézas, C.I., Weiss, S. (1997) The human 37-kDa laminin receptor precursor interacts with the prion protein in eukaryotic cells, *Nat. Med.* **3**, 1383–1388.

Riek, R., Hornemann, S., Wider, G., Billeter, M., Glockshuber, R., Wüthrich, K. (1996) NMR structure of the mouse prion protein domain PrP(121–231), *Nature* **382**, 180–182.

Robinson, M.M., Hadlow, W.J., Knowles, D.P., Huff, T.P., Lacy, P.A., Marsh, R.F., Gorham, J.R. (1995) Experimental infection of cattle with the agents of transmissible mink encephalopathy and scrapie, *J. Comp. Path.* **113**, 241–251.

Rogers, M., Yehiely, F., Scott, M., Prusiner, S.B. (1993) Conversion of truncated and elongated prion proteins into the scrapie isoform in cultured cells, *Proc. Natl. Acad. Sci. U. S. A.* **90**, 3182–3186.

Rogers, M., Serban, D., Gyuris, T., Scott, M., Torchia, T., Prusiner, S.B. (1991) Epitope mapping of the Syrian hamster prion protein utilizing chimeric and mutant genes in a vaccinia virus expression system, *J. Immunol.* **147**, 3568–3574.

Rossi, D., Cozzio, A., Flechsig, E., Klein, M.A., Rülicke, T., Aguzzi, A., Weissmann, C. (2001) Onset of ataxia and Purkinje cell loss in PrP null mice inversely correlated with Dpl level in brain, *EMBO J.* **20**, 694–702.

Ryou, C., Legname, G., Peretz, D., Craig, J.C., Baldwin, M.A., Prusiner, S.B. (2003) Differential inhibition of prion propagation by enantiomers of quinacrine, *Lab. Invest.* **83**, 837–843.

Saeki, K., Matsumoto, Y., Hirota, Y., Matsumoto, Y., Onodera, T. (1996) Three-exon structure of the gene encoding the rat prion protein and its expression in tissues, *Virus Genes* **12**, 15–20.

Safar, J., Roller, P.P., Gajdusek, D.C., Gibbs, C.J.J. (1993) Thermal-stability and conformational transitions of scrapie amyloid (prion) protein correlate with infectivity, *Protein Sci.* **2**, 2206–2216.

Safar, J., Wille, H., Itri, V., Groth, D., Serban, H., Torchia, M., Cohen, F.E., Prusiner, S.B. (1998) Eight prion strains have PrP^{Sc} molecules with different conformations, *Nat. Med.* **4**(10), 1157–1165.

Safar, J.G., Scott, M., Monaghan, J., Deering, C., Didorenko, S., Vergara, J., Ball, H., Legname, G., Leclerc, E., Solforosi, L., Serban, H., Groth, D., Burton, D.R., Prusiner, S.B., Williamson, R.A. (2002) Measuring prions causing bovine spongiform encephalopathy or chronic wasting disease by immunoassays and transgenic mice, *Nat. Biotechnol.* **20**, 1147–1150.

Sakaguchi, S., Katamine, S., Nishida, N., Moriuchi, R., Shigematsu, K., Sugimoto, T., Nakatani, A., Kataoka, Y., Houtani, T., Shirabe, S., Okada, H., Hasegawa, S., Miyamoto, T., Noda, T. (1996) Loss of cerebellar Purkinje cells in aged mice homozygous for a disrupted PrP gene, *Nature* **380**, 528–531.

Schätzl, H.M., Da Costa, M., Taylor, L., Cohen, F.E., Prusiner, S.B. (1995) Prion protein gene variation among primates, *J. Mol. Biol.* **245**, 362–374.

Schmitt-Ulms, G., Legname, G., Baldwin, M.A., Ball, H.L., Bradon, N., Bosque, P.J., Crossin, K.L., Edelman, G.M., DeArmond, S.J., Cohen, F.E., Prusiner, S.B. (2001) Binding of neural cell adhesion molecules (N-CAMs) to the cellular prion protein, *J. Mol. Biol.* **314**, 1209–1225.

Scott, M., Foster, D., Mirenda, C., Serban, D., Coufal, F., Wälchli, M., Torchia, M., Groth, D., Carlson, G., DeArmond, S.J., Westaway, D., Prusiner, S.B. (1989) Transgenic mice expressing hamster prion protein produce species-specific scrapie infectivity and amyloid plaques, *Cell* **59**, 847–857.

Scott, M., Groth, D., Foster, D., Torchia, M., Yang, S.-L., DeArmond, S.J., Prusiner, S.B. (1993) Propagation of prions with artificial properties in transgenic mice expressing chimeric PrP genes, *Cell* **73**, 979–988.

Scott, M.R., Köhler, R., Foster, D., Prusiner, S.B. (1992) Chimeric prion protein expression in cultured cells and transgenic mice, *Protein Sci.* **1**, 986–997.

Scott, M.R., Groth, D., Tatzelt, J., Torchia, M., Tremblay, P., DeArmond, S.J., Prusiner, S.B. (1997) Propagation of prion strains through specific conformers of the prion protein, *J. Virol.* **71**, 9032–9044.

Scott, M.R., Will, R., Ironside, J., Nguyen, H.-O.B., Tremblay, P., DeArmond, S.J., Prusiner, S.B. (1999) Compelling transgenetic evidence for transmission of bovine spongiform encephalopathy prions to humans, *Proc. Natl. Acad. Sci. U. S. A.* **96**, 15137–15142.

Scott, M.R., Safar, J., Telling, G., Nguyen, O., Groth, D., Torchia, M., Koehler, R., Tremblay, P., Walther, D., Cohen, F.E., DeArmond, S.J., Prusiner, S.B. (1997) Identification of a prion protein epitope modulating transmission of bovine spongiform encephalopathy prions to transgenic mice, *Proc. Natl. Acad. Sci. U. S. A.* **94**, 14279–14284.

Serban, D., Taraboulos, A., DeArmond, S.J., Prusiner, S.B. (1990) Rapid detection of Creutzfeldt-Jakob disease and scrapie prion proteins, *Neurology* **40**, 110–117.

Sethi, S., Lipford, G., Wagner, H., Kretzschmar, H. (2002) Postexposure prophylaxis against prion disease with a stimulator of innate immunity, *Lancet* **360**, 229–230.

Shibuya, S., Higuchi, J., Shin, R.-W., Tateishi, J., Kitamoto, T. (1998) Protective prion protein polymorphisms against sporadic Creutzfeldt-Jakob disease, *Lancet* **351**, 419.

Sigurdson, C.J., Williams, E.S., Miller, M.W., Spraker, T.R., O'Rourke, K.I., Hoover, E.A. (1999) Oral transmission and early lymphoid tropism of chronic wasting disease PrP^{res} in mule deer fawns (*Odocoileus hemionus*), *J. Gen. Virol.* **80**, 2757–2764.

Somerville, R.A., Chong, A., Mulqueen, O.U., Birkett, C.R., Wood, S.C., Hope, J. (1997) Biochemical typing of scrapie strains, *Nature* **386**, 564.

Sparkes, R.S., Simon, M., Cohn, V.H., Fournier, R.E.K., Lem, J., Klisak, I., Heinzmann, C., Blatt, C., Lucero, M., Mohandas, T., DeArmond, S.J., Westaway, D., Prusiner, S.B., Weiner, L.P. (1986) Assignment of the human and mouse prion protein genes to homologous chromosomes, *Proc. Natl. Acad. Sci. U. S. A.* **83**, 7358–7362.

Spencer, M.D., Knight, R.S., Will, R.G. (2002) First hundred cases of variant Creutzfeldt-Jakob disease: retrospective case note review of early psychiatric and neurological features, *Br. Med. J.* **324**, 1479–1482.

Spudich, S., Mastrianni, J.A., Wrensch, M., Gabizon, R., Meiner, Z., Kahana, I., Rosenmann, H., Kahana, E., Prusiner, S.B. (1995) Complete penetrance of Creutzfeldt-Jakob disease in Libyan Jews carrying the E200K mutation in the prion protein gene, *Mol. Med.* **1**, 607–613.

Stahl, N., Baldwin, M.A., Burlingame, A.L., Prusiner, S.B. (1990) Identification of glycoinositol phospholipid linked and truncated forms of the scrapie prion protein, *Biochemistry* **29**, 8879–8884.

Stahl, N., Borchelt, D.R., Hsiao, K., Prusiner, S.B. (1987) Scrapie prion protein contains a phosphatidylinositol glycolipid, *Cell* **51**, 229–240.

Stahl, N., Baldwin, M.A., Teplow, D.B., Hood, L., Gibson, B.W., Burlingame, A.L., Prusiner, S.B. (1993) Structural analysis of the scrapie prion protein using mass spectrometry and amino acid sequencing, *Biochemistry* **32**, 1991–2002.

Stekel, D.J., Nowak, M.A., Southwood, T.R.E. (1996) Prediction of future BSE spread, *Nature* **381**, 119.

Stöckel, J., Safar, J., Wallace, A.C., Cohen, F.E., Prusiner, S.B. (1998) Prion protein selectively binds copper (II) ions, *Biochemistry* **37**(20), 7185–7193.

Sulkowski, E. (1985) Purification of proteins by IMAC, *Trends Biotechnol.* **3**, 1–7.

Supattapone, S., Nguyen, H.-O.B., Cohen, F.E., Prusiner, S.B., Scott, M.R. (1999) Elimination of prions by branched polyamines and implications for therapeutics, *Proc. Natl. Acad. Sci. U. S. A.* **96**(25), 14529–14534.

Supattapone, S., Wille, H., Uyechi, L., Safar, J., Tremblay, P., Szoka, F.C., Cohen, F.E., Prusiner, S.B., Scott, M.R. (2001) Branched polyamines cure prion-infected neuroblastoma cells, *J. Virol.* **75**, 3453–3461.

Supattapone, S., Bosque, P., Muramoto, T., Wille, H., Aagaard, C., Peretz, D., Nguyen, H.-O.B., Heinrich, C., Torchia, M., Safar, J., Cohen, F.E., DeArmond, S.J., Prusiner, S.B., Scott, M. (1999) Prion protein of 106 residues creates an artificial transmission barrier for prion replication in transgenic mice, *Cell* **96**, 869–878.

Taraboulos, A., Scott, M., Semenov, A., Avrahami, D., Laszlo, L., Prusiner, S.B. (1995) Cholesterol depletion and modification of COOH-terminal targeting sequence of the prion protein inhibits formation of the scrapie isoform, *J. Cell Biol.* **129**, 121–132.

Tatzelt, J., Maeda, N., Pekny, M., Yang, S.-L., Betsholtz, C., Eliasson, C., Cayetano, J., Camerino, A.P., DeArmond, S.J., Prusiner, S.B. (1996) Scrapie in mice deficient in apolipoprotein E or glial fibrillary acidic protein, *Neurology* **47**, 449–453.

Taylor, D.M., Fernie, K., Steele, P.J., McConnell, I., Somerville, R.A. (2002) Thermostability of mouse-passaged BSE and scrapie is independent of host PrP genotype: implications for the nature of the causal agents, *J. Gen. Virol.* **83**, 3199–3204.

Telling, G.C., Haga, T., Torchia, M., Tremblay, P., DeArmond, S.J., Prusiner, S.B. (1996) Interactions between wild-type and mutant prion proteins modulate neurodegeneration in transgenic mice, *Genes Dev.* **10**, 1736–1750.

Telling, G.C., Scott, M., Mastrianni, J., Gabizon, R., Torchia, M., Cohen, F.E., DeArmond, S.J., Prusiner, S.B. (1995) Prion propagation in mice expressing human and chimeric PrP transgenes implicates the interaction of cellular PrP with another protein, *Cell* **83**, 79–90.

Telling, G.C., Parchi, P., DeArmond, S.J., Cortelli, P., Montagna, P., Gabizon, R., Mastrianni, J., Lugaresi, E., Gambetti, P., Prusiner, S.B. (1996) Evidence for the conformation of the pathologic isoform of the prion protein enciphering and propagating prion diversity, *Science* **274**, 2079–2082.

Telling, G.C., Scott, M., Hsiao, K.K., Foster, D., Yang, S.-L., Torchia, M., Sidle, K.C.L., Collinge, J., DeArmond, S.J., Prusiner, S.B. (1994) Transmission of Creutzfeldt-Jakob disease from humans to transgenic mice expressing chimeric human-mouse prion protein, *Proc. Natl. Acad. Sci. U. S. A.* **91**, 9936–9940.

Ter-Avanesyan, M.D., Dagkesamanskaya, A.R., Kushnirov, V.V., Smirnov, V.N. (1994) The *SUP35* omnipotent suppressor gene is involved in the maintenance of the non-Mendelian determinant [psi^+] in the yeast *Saccharomyces cerevisiae*, *Genetics* **137**, 671–676.

Tobler, I., Gaus, S.E., Deboer, T., Achermann, P., Fischer, M., Rülicke, T., Moser, M., Oesch, B., McBride, P.A., Manson, J.C. (1996) Altered circadian activity rhythms and sleep in mice devoid of prion protein, *Nature* **380**, 639–642.

Tremblay, P., Ball, H.L., Kaneko, K., Groth, D., Hegde, R.S., Cohen, F.E., DeArmond, S.J., Prusiner, S.B., Safar, J.G. (2004) Mutant PrP^{Sc} conformers induced by a synthetic peptide and several prion strains, *J. Virol.* **78**, 2088–2099.

Tremblay, P., Meiner, Z., Galou, M., Heinrich, C., Petromilli, C., Lisse, T., Cayetano, J., Torchia, M., Mobley, W., Bujard, H., DeArmond, S.J., Prusiner, S.B. (1998) Doxycycline control of prion protein transgene expression modulates prion disease in mice, *Proc. Natl. Acad. Sci. U. S. A.* **95**, 12580–12585.

Vey, M., Pilkuhn, S., Wille, H., Nixon, R., DeArmond, S.J., Smart, E.J., Anderson, R.G., Taraboulos, A., Prusiner, S.B. (1996) Subcellular colocalization of the cellular and scrapie prion proteins in caveolae-like membranous domains, *Proc. Natl. Acad. Sci. U. S. A.* **93**, 14945–14949.

Viles, J.H., Cohen, F.E., Prusiner, S.B., Goodin, D.B., Wright, P.E., Dyson, H.J. (1999) Copper binding to the prion protein: structural implications of four identical cooperative binding sites, *Proc. Natl. Acad. Sci. U. S. A.* **96**, 2042–2047.

Vnencak-Jones, C.L., Phillips, J.A. (1992) Identification of heterogeneous PrP gene deletions in controls by detection of allele-specific heteroduplexes (DASH), *Am. J. Hum. Genet.* **50**, 871–872.

Waggoner, D.J., Drisaldi, B., Bartnikas, T.B., Casareno, R.L., Prohaska, J.R., Gitlin, J.D. Harris, D.A. (2000) Brain copper content and cuproenzyme activity do not vary with prion protein expression level, *J. Biol. Chem.* **275**(11), 7455–7458.

Weissmann, C. (1991) Spongiform encephalopathies – the prion's progress, *Nature* **349**, 569–571.

Wells, G.A., Hawkins, S.A., Austin, A.R., Ryder, S.J., Done, S.H., Green, R.B., Dexter, I., Dawson, M., Kimberlin, R.H. (2003) Studies of the transmissibility of the agent of bovine spongiform encephalopathy to pigs, *J. Gen. Virol.* **84**, 1021–1031.

Wells, G.A.H. (2002) Report on TSE infectivity distribution in ruminant tissues (State of Knowledge, December 2001), European Commission: 10–37.

Westaway, D., Cooper, C., Turner, S., Da Costa, M., Carlson, G.A., Prusiner, S.B. (1994) Structure and polymorphism of the mouse prion protein gene, *Proc. Natl. Acad. Sci. U. S. A.* **91**, 6418–6422.

Westaway, D., Goodman, P.A., Mirenda, C.A., McKinley, M.P., Carlson, G.A. Prusiner, S.B. (1987) Distinct prion proteins in short and long scrapie incubation period mice, *Cell* **51**, 651–662.

Westaway, D., Zuliani, V., Cooper, C.M., Da Costa, M., Neuman, S., Jenny, A.L., Detwiler, L., Prusiner, S.B. (1994) Homozygosity for prion protein alleles encoding glutamine-171 renders sheep susceptible to natural scrapie, *Genes Dev.* **8**, 959–969.

Westaway, D., Mirenda, C.A., Foster, D., Zebarjadian, Y., Scott, M., Torchia, M., Yang, S.-L., Serban, H., DeArmond, S.J., Ebeling, C., Prusiner, S.B., Carlson, G.A. (1991) Paradoxical shortening of scrapie incubation times by expression of prion protein transgenes derived from long incubation period mice, *Neuron* **7**, 59–68.

White, A.R., Enever, P., Tayebi, M., Mushens, R., Linehan, J., Brandner, S., Anstee, D., Collinge, J., Hawke, S. (2003) Monoclonal antibodies inhibit prion replication and delay the development of prion disease, *Nature* **422**, 80–83.

Wickner, R.B. (1994) [URE3] as an altered URE2 protein: evidence for a prion analog in *Saccharomyces cerevisiae, Science* **264**, 566–569.

Wilesmith, J.W. (1991) The epidemiology of bovine spongiform encephalopathy, *Semin. Virol.* **2**, 239–245.

Wilesmith, J.W., Ryan, J.B.M. Atkinson, M.J. (1991) Bovine spongiform encephalopathy – Epidemiologic studies on the origin, *Vet. Rec.* **128**, 199–203.

Will, R.G., Ironside, J.W., Zeidler, M., Cousens, S.N., Estibeiro, K., Alperovitch, A., Poser, S., Pocchiari, M., Hofman, A., Smith, P.G. (1996) A new variant of Creutzfeldt-Jakob disease in the UK, *Lancet* **347**, 921–925.

Wille, H., Zhang, G.-F., Baldwin, M.A., Cohen, F.E. Prusiner, S.B. (1996) Separation of scrapie prion infectivity from PrP amyloid polymers, *J. Mol. Biol.* **259**, 608–621.

Wille, H., Michelitsch, M.D., Guénebaut, V., Supattapone, S., Serban, A., Cohen, F.E., Agard, D.A. Prusiner, S.B. (2002) Structural studies of the scrapie prion protein by electron crystallography, *Proc. Natl. Acad. Sci. U. S. A.* **99**, 3563–3568.

Williams, E.S., Young, S. (1980) Chronic wasting disease of captive mule deer: a spongiform encephalopathy, *J. Wildl. Dis.* **16**, 89–98.

Williamson, R.A., Peretz, D., Smorodinsky, N., Bastidas, R., Serban, H., Mehlhorn, I., DeArmond, S.J., Prusiner, S.B., Burton, D.R. (1996) Circumventing tolerance to generate autologous monoclonal antibodies to the prion protein, *Proc. Natl. Acad. Sci. U. S. A.* **93**: 7279–7282.

Wyatt, J.M., Pearson, G.R., Smerdon, T.N., Gruffydd-Jones, T.J., Wells, G.A.H., Wilesmith, J.W. (1991) Naturally occurring scrapie-like spongiform encephalopathy in five domestic cats, *Vet. Rec.* **129**, 233–236.

Yehiely, F., Bamborough, P., Costa, M.D., Perry, B.J., Thinakaran, G., Cohen, F.E., Carlson, G.A., Prusiner, S.B. (1997) Identification of candidate proteins binding to prion protein, *Neurobiol. Dis.* **3**, 339–355.

Zulianello, L., Kaneko, K., Scott, M., Erpel, S., Han, D., Cohen, F.E., Prusiner, S.B. (2000) Dominant-negative inhibition of prion formation diminished by deletion mutagenesis of the prion protein, *J. Virol.* **74**(9), 4351–4360.

8
Alzheimer's Disease

Jun Wang, Silva Hecimovic and Alison Goate
Department of Psychiatry, Washington University School of Medicine,
St. Louis, MO, USA

1	**Pathology and Epidemiology of Alzheimer's Disease**	**260**
2	**The Genetics of Alzheimer's Disease**	**262**
2.1	The Genetics of Familial Alzheimer's Disease	262
2.1.1	β-amyloid Precursor Protein Mutations	262
2.1.2	Presenilin Mutations	263
2.2	The Genetics of Sporadic Alzheimer's Disease	265
2.2.1	Apolipoprotein E	265
2.2.2	The Search for New Alzheimer's Disease Susceptibility Genes	266
3	**Molecular and Cellular Mechanisms Underlying Alzheimer's Disease**	**267**
3.1	β-amyloid Precursor Protein	267
3.2	Production of Aβ – Proteolytic Processing of APP	267
3.3	β-secretase – BACE1	268
3.4	γ-secretase – Presenilins and Cofactors	269
3.4.1	Presenilins	269
3.4.2	Cofactors	270
3.5	Neurofibrillary Tangles and Tau Protein	271
3.6	Apolipoprotein E	272
4	**Transgenic Mouse Models of Alzheimer's Disease**	**273**
4.1	APP Transgenic Mice	273
4.2	Presenilin 1/2 Transgenic Mice	274
4.3	Tau Transgenic Mice	275
4.4	Apolipoprotein E Transgenic Mice	276
4.5	Modulating and Enhancing the Phenotype: Transgenic Crosses	276

Proteins. Edited by Robert A. Meyers.
Copyright © 2007 Wiley-VCH Verlag GmbH & Co. KGaA, Weinheim
ISBN: 978-3-527-31608-3

5	Potential Treatments	277
5.1	Secretase Inhibitors	278
5.2	Vaccine Approaches	279
5.3	Other Therapeutic Approaches	279
6	Summary	280

Bibliography 280
Books and Reviews 280
Primary Literature 280

Keywords

Association Study
Tests for differences in allele or genotype frequencies between unrelated, affected, and unaffected individuals.

Autosomal Dominant Trait
A phenotype that is expressed in a heterozygote, which maps to a chromosome other than the sex chromosomes.

β-secretase
Also known as BACE1; a transmembrane aspartyl protease involved in processing of APP-generating N-terminal end of $A\beta$ peptides.

Candidate Gene Study
An association study performed on a particular gene(s) selected according to the biology of the disease and/or their location within chromosome regions found to be implicated in disease (by linkage or association study).

Full Genome Scan
A linkage or association analysis performed between the disease and polymorphic markers that are spaced throughout the genome.

γ-secretase
A multiprotein complex involved in intramembranous cleavage of APP-generating C-terminal end of $A\beta$ peptides. It contains four proteins: the catalytic component (presenilins) and three cofactors (nicastrin, APH-1, and PEN-2).

Neurofibrillary Tangles (NFTs)
Abnormal intraneuronal deposits abundant in the brains of patients with AD and some other neurodegenerative diseases. The main constituent of NFTs is hyperphosphorylated tau protein.

Senile Plaques
Complex extracellular deposits abundant in the brains of AD patients. The main proteinaceous component of plaques is amyloid-β peptide.

Susceptibility Locus/Gene
A region of a chromosome or a specific gene that may influence the risk to a certain condition/disease.

Familial Alzheimer's Disease
Transmitted as an autosomal dominant trait; usually has an age of onset <60 years.

Sporadic Alzheimer's Disease
Shows modest familial clustering and probably results from the synergistic action of genetic and environmental factors. Usually has an age of onset >60 years.

Abbreviations

Aβ	β-amyloid peptide
AD	Alzheimer's disease
AICD	APP Intracellular domain
APLP	amyloid precursor-like protein
APOE	apolipoprotein E
APP	β-amyloid precursor protein
BBB	blood brain barrier
CAA	cerebral amyloid angiopathy
CNS	central nervous system
CTF	C-terminal fragment
FAD	familial Alzheimer's disease
FTDP-17	frontotemporal dementia with parkinsonism linked to chromosome 17
KPI	Kunitz-type protease inhibitor
MT	microtubule
NFT	neurofibrillary tangles
NSAID	nonsteroidal anti-inflammatory drugs
NTF	N-terminal fragment
PDGF	platelet-derived growth factor
PHF	paired helical filaments
PS	presenilin
TM	transmembrane

■ Alzheimer's disease (AD) is the most common cause of dementia in the elderly. It is characterized by a progressive loss of memory, reasoning, judgment, and orientation, by the presence of large numbers of extracellular senile plaques and intracellular neurofibrillary tangles, as well as substantial cell loss in the brain. Genetic studies in early onset families have identified mutations in three genes that cause AD, while genetic studies in sporadic AD have identified the apolipoprotein E4 allele as a risk factor for disease. *In vitro* cell biology and transgenic mouse studies implicate all of these genes in the biology of $A\beta$ metabolism. A consistent effect of these genetic factors is an increase in $A\beta$ deposition. Furthermore, cell biology studies have identified targets for possible treatment of AD. Transgenic models are now being used to test the validity of these targets and therapeutic approaches to the treatment of AD.

1
Pathology and Epidemiology of Alzheimer's Disease

The German psychiatrist Alois Alzheimer first described the clinical and pathological features of Alzheimer's disease (AD) in a 55-year-old female patient; he published the discovery as a case report in 1907. It was thought that the "presenile dementia" described by Alzheimer was distinct from "senile dementia" seen in older patients, but it is now widely accepted that the two conditions are the same disease, differing primarily in the age of onset.

Alzheimer's disease is the most common cause of dementia in the western world, and the fourth leading cause of deaths in the United States. The disease is confined mainly to aged individuals and is progressive. It affects 1% of people in the developed nations and is likely to become a major problem in developing countries as the proportion of elderly persons increases in these populations. The incidence of AD rises with increasing age: 1 to 5% of people aged 65 years are affected, while 10 to 20% of those over the age of 80 have AD.

The first clinical manifestation is usually a loss of short-term memory; this is followed, over the next 6 to 20 years, by a progressive loss of memory, reasoning, judgment, and orientation (dementia). Death occurs, on average, about 10 years after the onset of symptoms. Many epidemiological surveys have been carried out to identify risk factors. There is a clear correlation between AD and a positive family history in about one-third of cases; the risk of AD is also elevated in individuals with Down syndrome. Severe head trauma with loss of consciousness increases risk for AD, particularly in those individuals who also have an apolipoprotein E4 allele. In contrast, the use of nonsteroidal anti-inflammatory agents or cholesterol-lowering agents, such as the statins, may reduce risk for AD. Another factor that may potentially decrease risk for AD is the years of education; the more the years of education, the lower the risk of disease.

Alzheimer's disease is characterized neuropathologically by the presence of large numbers of neuritic plaques and neurofibrillary tangles (NFT) within the brain cortex (Fig. 1). In addition, there is massive

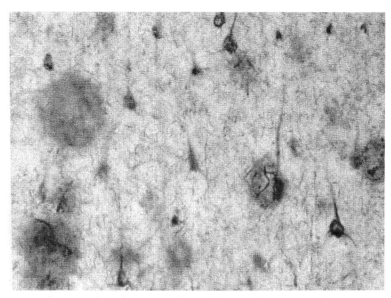

Fig. 1 Plaques and tangles in an AD brain. Shown is a brain section from an individual with mild AD. The section was double-stained with antibody PHF-1 specific to tangles and 10D5 specific to β-amyloid. Courtesy of Dr. Joel Price.

neuronal cell loss, and amyloid deposits that occur in the walls of the blood vessels of the meninges and cerebral cortex. Neither plaques nor tangles are specific to AD – they occur in intellectually normal, elderly people and in certain other diseases. However, they are more numerous and more widely distributed throughout the brains of Alzheimer's patients. Senile plaques are extracellular deposits, while tangles initially appear within the degenerating neurons. The major proteinaceous constituent of plaques is a 39 to 43 amino acid peptide called β-amyloid, or Aβ. The microtubule-associated protein, tau, is a major constituent of tangles. In the AD brain, there is extensive activation of astrocytes and microglia, the two primary mediators of inflammation in the central nervous system (CNS). A large number of inflammatory mediators and complement components are elevated in the AD brain, and colocalized with both plaques and tangles. The inflammatory response might be induced by increased Aβ levels or by plaque formation.

The deposition of Aβ has also been observed in association with several other pathological conditions. Individuals with Down syndrome, who live into their late thirties and beyond, develop a neuropathology indistinguishable from that of AD. Temporal studies in Down syndrome suggest that Aβ deposition is the earliest of the known pathological markers to appear, and that neurofibrillary tangles and neuronal cell loss occur decades after the first Aβ deposition. However, dementia correlates more closely with neurofibrillary tangles and neuron loss than with plaques. According to one hypothesis based on these observations, amyloid deposition precedes and induces tangle formation; tangles (and perhaps plaques) cause neuronal death, and the resultant neuronal damage destroys important neurotransmission pathways in the brain, producing dementia.

2
The Genetics of Alzheimer's Disease

There are two forms of AD: familial Alzheimer's disease (FAD), in which the disease is transmitted as an autosomal dominant trait; and sporadic AD, which shows modest familial clustering and probably results from the synergistic action of genetic and environmental factors. Sporadic AD can have an early (<60 yrs) or late (>60 yrs) age of onset. FAD does not appear to be clinically or neuropathologically different from the more common sporadic form of AD, except that it generally has an earlier age of onset.

2.1
The Genetics of Familial Alzheimer's Disease

The existence of families in which AD segregates as a fully penetrant, autosomal, dominant trait presents the most striking evidence for the involvement of genetic factors in the etiology of AD. Mutations in three distinct genes, β-amyloid precursor protein (APP), *PS1* (presenilin 1) and *PS2* (presenilin 2) have been identified to cause FAD.

2.1.1 β-amyloid Precursor Protein Mutations

The main component of the AD-associated senile plaques, Aβ, is derived by the proteolytic processing of APP. The gene coding for APP is located on the long arm of chromosome 21. It contains 18 exons, three of which can be alternatively spliced to produce a variety of mRNA species.

The first mutation shown to cause FAD was a valine to isoleucine substitution at residue 717 of APP (Fig. 2). To date, 22 APP mutations have been reported, of which 17 are linked to early-onset AD (www.alzforum.org). Most mutations lie within the transmembrane (TM) domain of APP in the region that is involved in the generation of the C-terminal end of Aβ. All TM domain FAD-APP mutations result in an increased production of the longer, more amyloidogenic form of Aβ (Aβ42/43). The most severe mutation (T714I) causes a 11-fold increase in the Aβ42/Aβ40 ratio; it causes both a decrease in Aβ40 and an increase in Aβ42 production. Patients carrying the T714I FAD-APP mutation die in their 40s. FAD mutations in APP can differentially affect Aβ-peptide formation by either increasing Aβ42 or decreasing Aβ40, or both. The Swedish mutation (K670N, M671L) lies in the region outside of the TM domain in which cleavage by β-secretase generates the N-terminal end of the Aβ, resulting in an increase in both Aβ40 and Aβ42 levels. Five mutations are located within the Aβ sequence but outside the TM domain: Flemish (A692G), Dutch (E693Q), Arctic (E693G), Italian (E693K), and Iowa (D694N). Patients carrying these mutations deposit Aβ in senile plaques and/or in cerebral vascular walls (cerebral amyloid angiopathy – CAA) and present clinically with either AD or hemorrhagic strokes, or both. For example, Flemish A692G patients present with both AD and strokes, while most Dutch E693Q patients suffer from strokes followed by a progressive multi-infarct dementia (HCHWA-D, hereditary cerebral hemorrhage with amyloidosis Dutch type). Pathologically, Flemish A692G patients demonstrate a classical AD pathology although with a strong CAA component, while Dutch APP patients

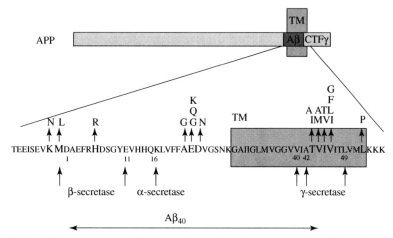

Fig. 2 FAD mutations in APP. Shown is a schematic representation of the APP protein showing the regions coding for Aβ peptide and C-terminal intracellular domain-CTFγ/AICD. The location of each FAD mutation is shown. Letters in bold are FAD-mutation sites, letters in gray are FAD substitutions. Major sites of APP processing, α-, β-, and γ-secretase sites, are also labeled. The amino acid numbering below the letters is according to the Aβ sequence.

have mainly vascular Aβ in the absence of neurofibrillary pathology. Recent analysis of pathogenic effect on Aβ levels of Arctic, Italian, Dutch, and Flemish APP mutations revealed that all, except the Flemish mutation, cause a decrease in both Aβ40 and Aβ42. In contrast, the Flemish mutation increases both Aβ40 and Aβ42 levels.

2.1.2 Presenilin Mutations

The discovery that mutations within the presenilin 1 *(PS1)* gene located on chromosome 14 (14q24.3) were responsible for most FAD cases opened up a whole new area of research into the molecular mechanisms of the pathogenesis of AD. Shortly after the discovery of *PS1*, DNA sequence database searches led to the identification of a homologous gene named presenilin 2 *(PS2)*, located on chromosome 1 (1q41.2). Mutations in *PS2* also result in FAD.

Pathogenic mutations in both of these genes have been identified; overall, mutations in *PS1*, *PS2*, and *APP* genes account for the majority of FAD cases. Although the proteins encoded by the two presenilin genes are 63% homologous at the amino acid level, intensive mutation and epidemiological analysis have revealed some interesting differences between the two proteins. The number of mutations identified in *PS1* is far greater than those in *PS2*. To date, more than 107 individual point mutations in *PS1* have been identified (Fig. 3, http://molgen-www.uia.ac.be/ADMutations) and found to be responsible for the majority of the early onset FAD. Most of these alterations are missense mutations that result in single amino acid changes in residues that are conserved between the PS1 and PS2 proteins. An intronic mutation (referred as ΔE9) results in the disruption of a splice acceptor site leading to the in-frame deletion of exon 9/10. This is the only *PS1*-FAD mutation that results in pathologically active protein that does not undergo proteolysis. Endoproteolytic

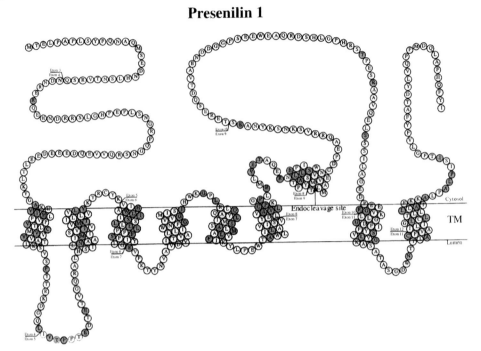

Fig. 3 FAD mutations in presenilin-1. Shown is a schematic representation of human presenilin 1 protein. The protein is predicted to have eight transmembrane domains. Arrow indicates the endocleavage site. Shaded residues are those associated with FAD mutations.

processing of the presenilins is an important aspect of the biochemistry of both PS1 and PS2 proteins, since it has been implicated in the formation of the functionally active complex (see Sect. 3). The most consistent effect mediated by the presenilin FAD mutations is increased production of Aβ42. This has been confirmed both in transfected cells and in transgenic mice. Although the mechanism by which FAD mutations in *PS1* and *PS2* cause AD is not fully understood, it is suggested that FAD mutations may alter γ-secretase activity toward cleavage of the more amyloidogenic peptide of Aβ, Aβ42/43. Although point mutations are distributed throughout the *PS1* gene, the majority of them are clustered within the TM domains and large hydrophilic loop between TM6 and TM7 (Fig. 3). The mean age of onset in all *PS1* families is about 45 years with a range of 28 to 62 years, suggesting that mutations in *PS1* result in a more aggressive form of FAD than APP mutations.

In contrast to *PS1*, only eight point mutations have been identified in *PS2* (http://molgen-www.uia.ac.be/ADMutations). The first *PS2* mutation to be identified was the N141I mutation that is responsible for disease in a large Volga German kindred. Whether the preponderance of *PS1* versus *PS2* mutations is due to the differing genomic environment of the two genes or the factors related to differing biological functions of the two molecules, remains unclear. Mutation analysis of the two proteins has also led to the identification of a phenotypic difference. Overall, the mutations in *PS1* result in relatively early and constant age of onset while those in *PS2* lead to a somewhat later and more variable age of onset. The identification of a *PS1* N135I mutation, a position and substitution that corresponds to the *PS2* Volga German mutation, and the finding that the affected members of this pedigree have a relatively constant age of onset suggests that the observed variations in age of onset are dependent on the molecule itself as opposed to the specific position or nature of the amino acid substitution.

2.2
The Genetics of Sporadic Alzheimer's Disease

While the genetics of FAD is fairly well understood, our understanding of the more common sporadic AD remains much less complete. The majority of cases have onset >60 years. Some families have stronger clustering of AD cases than others, suggesting that the genetic component of the disease can be quite variable. The estimated cumulative risk to first-degree relatives of AD-affected probands approaches 50% by age 90 compared to a disease risk of 10 to 15% in the general population. To date, only the Apolipoprotein E4 (*APOE4*) allele has been linked to increased risk for sporadic AD. While it is clear that *APOE* is a major risk factor for AD, epidemiological studies estimate that 42 to 68% of AD cases do not have an *APOE4* allele, indicating that additional genetic and environmental factors are involved in this form of the disease. The balance of these risk factors most likely determines both the age of onset of the disease and the number of affected first-degree relatives.

2.2.1 Apolipoprotein E

Linkage analysis in late-onset AD pedigrees led to the observation that these families inherit a genetic risk factor located on chromosome 19. The defect involved in these particular pedigrees is a susceptibility defect as opposed to a causative defect such as those present in the gene coding for APP and the presenilins. The inheritance of the susceptibility gene defect alone is not sufficient to cause the disease, and one or more secondary events, either environmental or genetic, must accompany the inheritance of the primary defect to cause the disease. Association studies have shown that the chromosome 19 susceptibility gene (localized to 19q13.2) is within the *APOE* gene locus.

APOE protein is a lipid-transport molecule and the primary apolipoprotein observed in the CNS. It is encoded by a polymorphic gene that exists as three alleles designated *APOE2*, *APOE3*, and *APOE4*. These genetic variations result in amino acid substitutions (arginine or cysteine) at positions 112 and 158 of the protein. Thus, whereas the *APOE2* polypeptide has a cysteine at both of these positions, *APOE3* has a cysteine at 112 and an arginine at 158, and *APOE4* has arginine at both positions. In most Caucasian populations, the *APOE3* is the most common allele (frequency 0.78); *APOE4* (frequency 0.14) and *APOE2* (frequency 0.08) are considered variants.

When the association of the three common *APOE* alleles with AD was examined in late-onset AD pedigrees, it was found that the *APOE4* allele frequency was 52%, compared to 16% for age-matched controls. In addition, in sporadic AD cases, the *APOE4* allele frequency was 40%. The *APOE4* association with AD was rapidly and widely confirmed in patients with familial and sporadic AD. These multiple confirmations established the *APOE4* allele as the most important genetic marker of risk for the disease identified so far, accounting for approximately 50% of the genetic component of AD. The frequency of *APOE4* allele among AD and control cases can vary in different populations, however, the association with AD is consistent.

Individuals inheriting two copies of the *APOE4* allele have an increased risk and earlier age of onset of AD than individuals with one *APOE4* allele. Likewise, individuals with one *APOE4* allele have a greater risk and earlier onset of disease than those with no *APOE4* alleles. These data demonstrate that the effect of *APOE4* on risk and age of onset is dose-related. In contrast to the effect of *APOE4* allele, the *APOE2* allele has a protective effect on AD, as evidenced by a decreased frequency of this allele in AD cases. The effect of the most commonly inherited allele, *APOE3*, on risk and age of onset falls between *APOE4* and *APOE2*. There also appears to be an interaction of *APOE* genotypes with the age of onset in FAD. When survival analysis was used in a large Colombian kindred carrying a presenilin mutation, the presence of an *APOE4* allele was observed to decrease age of onset, while presence of an *APOE2* allele increased age of onset. Although it is still not clear how an *APOE4* allele modifies risk for AD, several studies suggest that APOE may act via an $A\beta$-dependent mechanism.

2.2.2 The Search for New Alzheimer's Disease Susceptibility Genes

There are two genetic approaches to finding susceptibility loci: full genome scans and candidate gene studies. In candidate gene studies, genes are selected on the basis of the known biology of the disease or their location within chromosomal regions implicated in risk for late-onset AD, and are assessed individually in order to determine whether variants in each candidate are associated with disease. An alternative approach, a genome scan, does not select genes *a priori*, but instead tests for linkage or association between the disease and polymorphic markers that are spaced evenly through the genome. Association methods use unrelated case-control samples or siblings who are discordant for disease, and test for differences in allele frequencies between affected and unaffected individuals. In contrast, linkage studies use extended families or affected sibling pairs and look for regions of the genome in which there is an increased allele sharing between affected relatives.

To date, only five late-onset AD genome-wide screens have been reported. The most consistent findings among these studies are evidences for AD susceptibility loci on chromosomes 19 (*APOE*), 9, 10, and 12. Since FAD is closely associated with elevated $A\beta42$ levels, some investigators have hypothesized that loci, which modify plasma $A\beta42$, might also modify risk for AD. Using this approach, evidence for linkage was observed in the same region of chromosome 10 as was observed in the AD-linkage studies. Over the past few years, a large number of biologically interesting candidate genes as well as candidate genes that map within chromosome regions implicated in risk for late-onset AD have been evaluated in case-control populations. The genes that have received most attention are those encoding *APOE* on chromosome 19, $\alpha2$-macroglobulin and the LDL-receptor related protein on chromosome 12 and insulin degrading enzyme and urokinase on chromosome 10. All of these genes have been closely linked to $A\beta$ metabolism. However, with the exception of the *APOE* gene, whose association with

AD has been universally replicated, the relative significance of these associations is still in question.

3 Molecular and Cellular Mechanisms Underlying Alzheimer's Disease

3.1 β-amyloid Precursor Protein

β-Amyloid precursor protein (APP) is a large, type I transmembrane protein that is expressed in almost all cell types except erythrocytes. It consists of a large extracellular N-terminal region, a transmembrane domain, and a short intracellular C-terminal tail. The Aβ-peptide sequence begins close to the membrane on the extracellular part of APP and ends within the membrane-spanning domain. At least five different mRNAs arise as a result of alternative splicing of the primary transcript of the APP gene: APP770, 751, 714, 695, and 563. Transcript APP695 lacks the Kunitz-type protease inhibitor (KPI) domain and is the major form expressed in neurons, whereas APP751, which contains the KPI domain, is the major transcript in non-neuronal cell types. The APP transcripts appear to be developmentally regulated, and all cell types so far tested contain at least one of the transcripts.

In addition to differential RNA splicing, APP undergoes posttranslational modification, including phosphorylation, N- and O-linked glycosylation, sulfation and proteolytic processing. It is not clear how these different events modify APP metabolism or function. A detailed knowledge of posttranslational events is undoubtedly crucial to understanding amyloid plaque formation and to elucidating the biological functions of APP and Aβ.

The normal function of APP is still unclear, but it has been implicated in neuroprotective action. APP is a member of a multigene family that contains at least two other homologs known as amyloid precursor-like protein 1 and 2 (APLP1 and APLP2). Both $APP^{-/-}$ and $APLP2^{-/-}$ mice are viable while APP/APLP2 double knockout mice are early postnatal lethal suggesting that APP and APLPs may share important physiological functions and the APLPs may compensate for the function of APP.

3.2 Production of Aβ – Proteolytic Processing of APP

It is clear that Aβ production does occur in normal cells, suggesting that excessive Aβ production in AD is due to enhanced or altered processing of APP by normal pathways that generate Aβ, rather than by an abnormal route specific to Alzheimer's cells. APP undergoes a series of proteolytic processing events (Fig. 4) by several different proteases called secretases. It is first cleaved by either α- or β-secretase. Two disintegrin metalloproteases, ADAM 10 and TACE (TNF-α converting enzyme), are involved in α-cleavage of APP. This cleavage produces a large soluble N-terminal fragment (NTF), APPsα and an 83-amino acid membrane-bound C-terminal fragment, C83 (CTFα). Alternatively, cleavage by the β-secretase results in an N-terminal fragment APPsβ and a 99-amino acid membrane-bound fragment, C99 (CTFβ). The β-secretase has recently been identified as a transmembrane aspartyl protease called BACE1, for β-site APP-cleaving enzyme. Both C83 and C99 are substrates for γ-secretase, a mysterious enzyme that carries out an unusual proteolysis in the middle of the transmembrane domain of

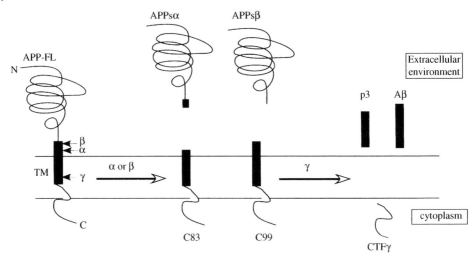

Fig. 4 Proteolytic processing of APP. APP is initially cleaved by either α- or β-secretase to generate membrane-bound C-terminal fragments, C83 or C99, which are subsequently cleaved by the γ-secretase (in the transmembrane domain) to produce p3 or Aβ.

APP, resulting in the release and secretion of the 4-kD Aβ from C99 and the 3-kD p3, a nonpathogenic peptide, from C83. γ-secretase cleavage is promiscuous and produces a spectrum of Aβ peptides varying from 39 to 43 residues in length. Over 90% of the Aβ in the brain is Aβ40 and the remaining 5 to 10% is the longer species Aβ42/43.

The C-terminal product of the γ-secretase cleavage is called CTFγ or AICD (from APP Intracellular Domain). Interestingly, CTFγ, purified from brain tissue, is not C59/57, which would be the predicted length of the C-terminal product of Aβ cleavage. Instead, sequence analysis reveals that it is C50/49, several amino acids shorter. The mechanism that accounts for such differential cleavages is currently under investigation. Several reports have suggested a role of CTFγ in APP signaling to the nucleus. Indeed, it was shown that CTFγ, when together with Fe65 protein, translocates to the nucleus and that this complex can activate transcription of a reporter gene *in vitro*.

3.3 β-secretase – BACE1

In late 1999, four groups almost simultaneously reported the discovery of BACE1, a novel aspartyl protease that exhibited all the known functional characteristics of the β-secretase. BACE1 is a type I transmembrane protein with an open reading frame of 501 amino acids. It is most closely related to the pepsin family of aspartyl proteases. BACE1 is expressed in most tissues and cell types at modest levels, but shows much higher expression levels in neurons. Structurally, it contains an N-terminal 21 amino acid signal peptide, a propeptide domain, a large lumenal domain followed by a transmembrane domain, and a short cytosolic C-terminal tail. BACE1 contains two aspartyl protease active site motifs in the lumenal domain: DTGS (residues 93–96) and DSGT (residues 289–292). Both aspartic acid residues have been shown to be essential for enzymatic activity. BACE1 contains intrinsic proteolytic activity; as a purified recombinant, BACE1 can directly cleave APP *in vitro*. Overexpression of BACE1 in cell lines dramatically increases β-secretase activity and β-cleavage products: APPsβ and C99. In addition, mice deficient in BACE1 show a normal phenotype, but Aβ generation is abolished, further supporting that BACE1 is indeed a β-secretase.

BACE1 is initially synthesized in the ER as an immature precursor protein, proBACE1, which undergoes rapid maturation involving N-glycosylation as well as removal of the propeptide domain by a furin-like convertase in the Golgi apparatus. Unlike most zymogens, proBACE1 possesses β-secretase activity. The propeptide domain does not appear to significantly inhibit protease activity, but rather acts as a chaperone to assist proper folding of the protease domain. Phosphorylation of BACE1 at Ser498 appears to be important for the intracellular trafficking of the protein between early endosomes and late endosomes/TGN compartments.

A homologous aspartyl protease, BACE2 was shortly identified by searching the EST database with the BACE1 sequence. BACE1 and BACE2 share $\sim$64% amino acid similarity. Together they define a novel family of transmembrane aspartyl proteases. Brain expression of BACE2 is very low and most evidence suggests that it does not play a major role in β-cleavage of APP.

3.4
γ-secretase – Presenilins and Cofactors

The enzyme that is clearly central to AD pathogenesis is the γ-secretase, which cleaves C99 and produces Aβ. Evidence suggests that it is an aspartyl protease with the unusual property of cleaving substrates within the transmembrane domain. Independent from its role in processing APP, γ-secretase also mediates the intramembranous cleavage of several other type I transmembrane proteins including Notch and ErbB-4. γ-secretase cleavage of Notch releases the Notch intracellular domain, which translocates to the nucleus and acts as a cofactor to modify transcription of target genes.

The identity of the γ-secretase is rather enigmatic. It appears to be a multiprotein, high molecular weight complex that consists of presenilins (PS) and three other cofactors: nicastrin, APH-1 and PEN-2.

3.4.1 Presenilins

Presenilins are a family of polytopic transmembrane proteins that are implicated in AD pathogenesis. PS homologs have been found in species as diverse as *Drosophila melanogaster*, *Caenorhabditis elegans* and *Arabidopsis*, but not in yeast. The function of PS appears to be conserved. Deficiency in sel-12, one of the nematode PS homolog, causes egg-laying defects. This phenotype can be rescued by expressing human PS1 in the mutant worms. In mammals, there are two PS genes, *PS1* and *PS2*, that share 65% identity. *PS1* is believed to account for the majority of PS functions, while *PS2* seems to play a complementary role.

Presenilins are found primarily within intracellular membranes including the ER and Golgi as well as the plasma membrane. Structurally, they are predicted to have eight transmembrane domains with both the amino and carboxyl termini oriented toward the cytoplasm (Fig. 3). A large cytoplasmic loop is postulated between TM6 and TM7. PS are synthesized as single polypeptides that rapidly undergo endoproteolysis within the cytoplasmic loop, generating a 30-kD N-terminal fragment (NTF) and a 20-kD C-terminal fragment (CTF). The NTF and CTF form a stable complex, which appears to be the functional form of presenilins. PS endoproteolysis is tightly regulated such that overexpression of PS leads to the accumulation of a full-length protein, but does not result in an increase in the fragments. The NTF/CTF complex, but not the holoprotein, are components of a high molecular weight complex, estimated at ~250 kD or even larger. This high molecular weight complex is believed to be the γ-secretase complex.

Many lines of evidence indicate that *PS1* is the catalytic component of the γ-secretase. The two membrane-embedded aspartyl residues in *PS1* (D257 in TM6 and D385 in TM7) are essential for γ-secretase activity. The mutation of either residue abolishes γ-secretase processing of APP and Notch. The knockout of the PS genes blocks the Aβ production and Notch function. While in *PS1* single K/O cells, there is residual Aβ production, in PS1/2 double knockout cells, there is no detectable Aβ, demonstrating that there is no PS-independent γ-secretase activity. Inhibitors designed to be transition-state analogs of the γ-secretase bind to presenilin fragments. Recently, a family of signal peptide peptidases was discovered, which are apparently *bona fide* intramembranous aspartyl proteases, with similar active site motifs and similar topology to the presenilins. This finding further supports the hypothesis that presenilin is an aspartyl protease.

3.4.2 Cofactors

Nicastrin is a large type I transmembrane glycoprotein that was immunopurified from PS1 immunoprecipitates. It has no significant sequence homology to other functionally characterized proteins. Nicastrin is a 709 amino acid protein containing a putative signal peptide, a long N-terminal hydrophilic domain with numerous glycosylation sites, a TM domain and a short hydrophilic C-terminal tail of 20 residues. The mature protein migrates at ~150 kD owing to heavy glycosylation. There is evidence suggesting that the association of nicastrin with γ-secretase is tightly regulated via glycosylation. Nicastrin binds to presenilins and the active γ-secretase complex. Nicastrin deficiency in *Drosophila* abolishes Notch signaling and γ-secretase cleavage of APP. In addition, presenilin stability was compromised in the absence of nicastrin. All these data support that nicastrin is a component of the γ-secretase complex and an important regulator of the γ-secretase activity.

Genetic screens in *C. elegans* identified two proteins, APH-1 and PEN-2, which interact with worm homolog of *PS1* and nicastrin. Loss-of-function mutations of either protein produces phenotypes that resemble that of null mutants in nicastrin and presenilins. APH-1 is an integral protein containing seven predicted TM domains. There are two mammalian *APH-1* genes (*mAPH-1a* and *mAPH-1b*). *mAPH-1a* has at least two splice variants: *mAPH-1a^L* has seven exons and encodes a longer protein of 265 residues; mAPH-1a^S has six exons and encodes a shorter protein of 247 residues. All three variants are widely expressed, although their abundance seems to be coordinately regulated. To date, no functional difference among these variants has been documented. PEN-2 is a small protein of 101 amino acids containing two TM domains. Topology study suggests that both the N- and C-terminals of the protein face the lumen of the ER.

Human APH-1 and PEN-2 can partially rescue the *C. elegans* mutants, suggesting functional conservation among eukaryotes. The human genes must be present together to rescue the phenotypes and adding PS1 improves the rescue, indicating that these proteins act in concert. Consistent with the genetic studies, biochemical studies showed that reduction of APH-1 and PEN-2 expression in mammalian cells by RNA inhibition reduced γ-secretase activity.

While the exact functions of these cofactors remain to be elucidated, the data obtained so far suggest that all four proteins, presenilins, nicastrin, APH-1, and PEN-2, are necessary components of the γ-secretase. They appear to regulate one another for protein stability, proper maturation, and functional γ-secretase complex formation. It is also very likely that the least abundant component becomes the cellular-limiting factor that controls overall γ-secretase activity.

3.5
Neurofibrillary Tangles and Tau Protein

Neurofibrillary tangles are intracellular deposits that consist of aggregates of a fibrous substance, the paired helical filaments (PHF). These filaments are highly insoluble, even resistant to strong detergents such as SDS and guanidine hydrochloride. A major component of PHF is tau, a microtubule (MT)-associated phosphoprotein. Tau is abundantly expressed in the brain. At least six isoforms of tau are present in the brain, resulting from alternative mRNA splicing of a single gene on chromosome 17q.

Unlike APP, whose functions are not clear, tau has a biological role that is reasonably well understood. One function of tau is to promote microtubule assembly, a process controlled by phosphorylation of tau. In addition, it has been proposed that tau has a specific role in the generation of axon morphology. Increased phosphorylation of tau decreases its ability to promote tubulin polymerization to microtubules.

Tau in PHF is hyperphosphorylated, and is insoluble under conditions in which normal tau is soluble. It is believed that tangles are formed when tau becomes hyperphosphorylated and dissociates from microtubules and spontaneously forms insoluble PHF. Over 25 phosphorylation sites have been identified in PHF tau purified from AD brains. Most of the phosphorylation sites are located in the flanking region of the MT-binding domain. A variety of protein kinases, including glycogen synthase kinase-3β (GSK3β), cyclin-dependent kinase 5 (cdk5), mitogen-activated protein kinase (MAPK) and protein kinase A, have been found to be involved in hyperphosphorylation of tau. This suggests that multiple phosphorylation cascades may be activated in NFT-bearing neurons. However, it remains unclear whether these cascades are activated independently or whether one or more kinases are primarily responsible for initiating the cascade *in vivo*.

In addition to their presence in AD, filamentous tau inclusions accompanied by extensive gliosis and loss of neurons are the neuropathological hallmarks of other neurodegenerative diseases that are sometimes designated as tauopathies. There has been considerable debate in the AD literature concerning the primacy of Aβ versus NFT in the pathogenesis of AD. This was largely resolved by the identification of mutations in the tau gene in families with frontotemporal dementia with parkinsonism (FTDP-17). FTDP-17 is characterized clinically by the presence of prominent behavioral changes early in the disease. The neuropathology associated with FTDP-17 is characterized by large numbers of NFT but no senile plaques. Mutations in the tau gene in FTDP-17 can alter splicing and produce a shift from the short tau isoform with three repeats to the longer isoform, which contains four repeats. Several other point mutations have been discovered in the coding region of tau that impact the functional integrity of the tau protein itself, P301L being the most common. So far, genetic analysis has not linked tau mutation directly to AD, but tau dysfunction may still have a significant role to play in the disease, if not in its initiation, then in its progression.

3.6
Apolipoprotein E

Apolipoprotein E is a 299 amino acid glycoprotein with a molecular weight of ~34 kD. Its N-terminal domain consists of four amphipathic α-helices and the C-terminal domain contains most of the lipid-binding activity. APOE is a major apolipoprotein that regulates cholesterol uptake and release. The CNS contains high levels of APOE with abundance only second to the liver. In the CNS, APOE is synthesized predominantly by astrocytes and is secreted into the CSF, where it is a major lipid carrier.

APOE appears to have multiple functions in the brain – any or all of which could potentially influence AD pathogenesis. The molecular mechanisms by which APOE protein is involved in AD pathogenesis are unclear because reports on the allele-specific effects of APOE are controversial. Nevertheless, numerous studies have demonstrated that APOE can affect AD pathogenesis either by influencing $A\beta$ deposition or by a direct effect on neuronal survival. APOE is a major component of senile plaques and specifically binds to soluble $A\beta$s. Allele-specific differences in $A\beta$ binding, aggregation, and fibrillogenesis have been reported. Most histopathologic studies demonstrated a correlation between senile plaque density and *APOE4* allele dose in AD. Animal studies also show that APOE has isoform-specific effects on $A\beta$ deposition and plaque formation (see Sect. 4.4).

There is also evidence that different alleles of *APOE* have different effects on neurite extension, neuronal survival, and the cell's response to oxidative stress. *In vitro* studies with primary neurons and neuronal cell cultures show that lipid-bound E3 promotes neurite extension, while E4 is neutral or inhibitory. These *in vitro* findings are consistent with reported defects in neuronal remodeling in AD brains carrying an E4 allele. APOE can have allele-specific differences on both neurotoxic or neuroprotective activities. High levels of APOE or some APOE synthetic peptides were shown to cause neurite degeneration or neuronal death in cultured cells, with the E4 having the most neurotoxic effects. On the other hand, APOE appears to play a role in protection against oxidative stress, a prominent phenomenon in AD pathogenesis. Apoe -/- mice exhibit increased markers of lipid peroxidation within the brain and appear to have increased susceptibility to neuronal damage. Different *APOE* alleles are shown to have isoform-specific neuroprotective effects (E2 > E3 > E4) on oxidative cytotoxicity caused by hydrogen peroxide, or $A\beta$ peptides. A possible mechanism for the different antioxidant activities of the *APOE* alleles is their ability to detoxify 4-hydroxynonenal, a neurotoxic lipid-peroxidation product that is believed to play a key role in neuronal death in

AD. 4-hydroxynonenal covalently binds to APOE, with more binding to E2 and E3 than E4, probably because of the cysteine residues that are present in E2 and E3. E2 and E3 may therefore protect neurons by binding to, and thereby detoxifying, 4-hydroxynonenal.

4
Transgenic Mouse Models of Alzheimer's Disease

A major hindrance to basic research into the molecular pathogenesis of AD has been the absence of a mouse or rat model. As a result of the identification of mutations that cause AD in humans, this has begun to change. Although initial efforts to make transgenic models of AD failed, more recent efforts have recapitulated at least some of the key features of the disease. Many genetically altered mice have been designed to reproduce the neuropathology of AD, however, the majority of them share only some of the neuropathological and/or cognitive impairment characteristics of AD (http://www.alzforum.org). These include mice expressing human forms of the APP, the PS, APOE and, more recently, tau. Transgenic mice expressing human APP develop amyloid plaques, but neurodegeneration and neurofibrillary tangles are not observed. The Presenilin transgenic mice produced only increased Aβ42/43 levels and did not develop signs of AD pathology, even though the same mutations caused some of the earliest forms of inherited AD in humans. Strategies emphasizing tau resulted in increased phosphorylation of tau and tangle formation, although amyloid plaques were absent. Nevertheless, crossing transgenic animals expressing mutated tau and APP has produced a mouse that closely recapitulates the neuropathology of AD. This section provides a review of the various murine models and their role in understanding the pathogenesis of AD.

4.1
APP Transgenic Mice

The most extensively studied APP transgenic mouse lines are known as PDAPP and Tg2576. The PDAPP transgenic mouse expresses a human APP770 mini gene containing the V717F FAD-APP mutation, while Tg2576 expresses human APP695 that contains the Swedish FAD-APP mutation. In PDAPP, expression of the *APP* gene is under the control of the human platelet–derived growth factor (PDGF)-β chain neuronal promoter that targets expression preferentially to neurons in the cortex, hippocampus, hypothalamus, and cerebellum of the transgenic animals. This mouse was made on a mixed-strain background (C57BL/6, DBA/2, and Swiss-Webster), while Tg2576 was made on a single C57B6/SJL background. Expression of the *APP* gene in Tg2576 mice is under the control of the prion protein promoter resulting in a less-limited expression of APP both in the CNS and the periphery. The success of both models was due to the high level of APP expression achieved (>10-fold overexpression of human APP in PDAPP mouse and >6-fold higher expression of APP in Tg2576 compared to endogenous murine APP levels). Cortical and limbic amyloid deposits begin slightly earlier in the PDAPP mouse: by 3 months of age in homozygotes and at 6 to 9 months in heterozygotes, while in Tg2576 mouse they occur between 9 to 12 months of age, elevated Aβ production is observed as early as 3 months of age. Both models develop β-amyloid deposits (in hippocampus and neocortex but limited in striatum and cerebellum), and the deposits are associated with dystrophic neurites, punctate immunoreactivity to hyperphosphorylated tau, astrocytosis, microgliosis, and vascular amyloidosis. The elevation of Aβ in the Tg2576 mouse correlates with the appearance of memory and learning deficits in the oldest group of transgenic mice as well. However, both APP transgenic models fail to show all the pathological features of human AD; no significant neuronal loss and no formation of NFTs were observed in either line.

Several other APP transgenic models were generated in order to obtain a complete or more comprehensive picture of the hallmarks of AD. The TgAPP22 transgenic mice overexpress a double mutant APP with Swedish (K670N, M671L) and London (V717I) FAD mutations in cis, while TgAPP23 only carries the Swedish mutation. In both models, the human APP751 isoform is used and its expression is under the control of the murine Thy-1 promoter. The TgAPP22 mice show twofold overexpression of the transgene over endogenous APP, and Aβ deposits were detected in the neocortex and hippocampus at 18 months of age, while the TgAPP23 has higher, sevenfold overexpression of human APP and typical Aβ plaques appear at the age of six months. Interestingly, these two murine AD models showed differences in plaque type; the majority of amyloid deposits in TgAPP22 mice brains are of the "diffuse" type, while in the TgAPP23 mice almost all extracellular amyloid deposits were fibrillar. Both substantial neurodegeneration and a reduction of neuron numbers were apparent in TgAPP23. In 14–18 months old TgAPP23 neuronal loss was 14% and reached 25% in mice with high Aβ plaque load. Additional pathological features included dystrophic neurites surrounding the plaques, hyperphosphorylated tau, but no NFTs developed.

Each of the models described above is remarkable in that the anatomical pattern of plaque formation parallels that seen in human AD. Furthermore, the morphology of amyloid plaques in aged APP-transgenic mice recapitulates amyloid pathology in human AD: the plaques span a continuum from diffuse Aβ deposits to compact core plaques with inflammation and neuritic dystrophy. However, none of these models reflects a complete picture of the neuropathology of AD.

4.2
Presenilin 1/2 Transgenic Mice

Several PS1 transgenic mice have been created to study presenilin biology *in vivo*. These mice differ in the promoter and the strains of mice used, but are similar in achieving high levels of protein production (1–3 fold over endogenous) in neuronal regions of the brain. Knockin transgenes were created expressing either the wild-type human PS1 or PS1 containing the FAD mutation A246E under the transcriptional control of the human Thy-1 promoter in PS1 null background. Both mice rescued the PS1 knockout mouse from embryonic lethality, confirming that PS1 mutation does not cause a loss of function of PS1 protein but has similar physiological properties as wild-type PS1. A second PS1 transgenic model expressing human mutant (M146L or M146V) and wild-type PS1 was generated under the control of the PDGF promoter. All three PS1 transgenics showed similar results: overexpressing mutant PS1 in the brains of transgenic mice lead to the elevation of

Aβ42/43, but not Aβ40. This effect was a direct result of the mutation and not overexpression of the human protein as the overexpression of the wild-type PS1 did not have any significant effect on Aβ levels. The physiological significance of the specific elevation of Aβ42/43 appears to be the observation that AD patients with PS1 mutations show plaques composed primarily of Aβ42/43. However, all three PS1 transgenic mice, in contrast to APP transgenics, revealed no Aβ deposition or other AD-associated pathology.

Recently, a PS1 mouse model carrying a PS1-P264L FAD mutation was developed using gene-targeting approach. The gene-targeted models are distinct from transgenic models because the mutant gene is expressed at normal levels, in the absence of the wild-type protein. PS1$^{P264L/P264L}$ mice had normal expression of PS1 mRNA, but levels of the N- and C-terminal protein fragments of PS1 were reduced while levels of the holoprotein were increased. APP$^{sw/sw}$/PS1$^{P264L/P264L}$ double gene-targeted mice had elevated levels of Aβ42, sufficient to cause Aβ deposition beginning at six months of age. This was the first animal model that exhibited Aβ deposition without overexpression of APP.

Three transgenic mice that express PS2-FAD mutation N141I under the control of PDGFβ chain promoter, chicken β-actin promoter or the neuron-specific enolase promoter were designed. The first two PS2 transgenic mice showed increased levels of Aβ42 in the brain of 12-month old mice. The third model utilizing neuron-specific enolase promoter did not show any difference in Aβ42 levels compared to the PS2 wild-type transgenic mice and both had no obvious differences in AD phenotypes (both had amyloid deposits).

4.3
Tau Transgenic Mice

One of the deficits of most AD transgenic models is the lack of tau pathology. APP and APP/PS transgenic mice, which exhibit extensive amyloid deposition also develop hyperphosphorylated tau around the amyloid deposits but it does not progress to resemble human NFTs. This suggests that some disruption of tau does occur in response to amyloid accumulation, but it does not progress to resemble human tauopathy in the mouse model. The idea that a species barrier prevents this progression has been proposed suggesting that primates and rodents differ in their response to injected amyloid not only in terms of neurodegeneration, but also in the accumulation of abnormal tau around the injected amyloid.

The most exciting development in the tau field is the creation of a transgenic mouse model with robust and reproducible tangle pathology. This transgenic mouse line, designated JNPL3, express human four repeat tau containing the most common FTDP-17 mutation, P301L, under the control of the mouse prion promoter. The mice develop motor and behavioral deficits that are associated with age- and gene-dosage-dependent development of congophilic NFT. Tau-immunoreactive pretangles are found in cortex, hippocampus, and basal ganglia, but at lower levels than commonly found in human disease. The tangles are associated with neurodegeneration, especially in the spinal cord where motor neurons were reduced by approximately 48%. This mouse has recently been used in breeding double and triple transgenic mouse models to produce a model system that more accurately recapitulates the hallmark pathology of AD (see further in the text). Recently, another

P301L tau transgenic model has been described that develops fivefold increase in the numbers of NFT soon after 18 days after the intracerebral injection of Aβ42 into the CA1 region of the hippocampus. These experiments indicate that an interaction of β-amyloid with the P301L tau mutation accelerates NFT formation.

4.4
Apolipoprotein E Transgenic Mice

A consistent consequence of carrying the *APOE4* allele is an increased number of amyloid plaques in the brain and more abundant amyloid deposition in the cerebral vasculature. The mechanism by which *APOE4* contributes to the development of neurodegeneration remains unknown, although the evidence suggests that this may be linked to the ability of APOE to interact with the amyloid β peptide and influence its concentration and structure. Recent studies with transgenic mice that overexpress the human APOE isoforms APOE2, APOE3, and APOE4 in a PDAPP hemizygous mouse (Apoe-/- background) show that isoforms that are known to increase the risk of AD enhance amyloid load and increase the neuritic pathology associated with fibrillar plaque development in aged animals (by 15 months of age APOE4 expressing mice having a 10-fold greater amyloid burden than APOE3 mice). The time course of Aβ deposition, Aβ levels, structure, and anatomic and subcellular distribution showed profound age-, species-, and isoform-dependent effect of APOE. It was also found that APOE not only impacts on the nature of aggregated Aβ, but also on the clearance of soluble Aβ from the brain across the blood–brain barrier (BBB). Aβ40 was rapidly cleared from the brain, while Aβ42 was cleared much less effectively, supporting the idea that Aβ42 production may favor amyloid deposition due to a reduced clearance across the BBB. However, Aβ clearance was not APOE isoform specific.

These recent findings are consistent with earlier work on Apoe null mice. When Tg2576 and PDAPP mice were crossed with Apoe null mice it was revealed that the absence of Apoe altered the quantity, character, and distribution of Aβ deposits in the transgenic animals, confirming that APOE affects the neuropathological phenotype in AD brains. The Aβ deposition is significantly reduced and is not thioflavine-S-positive.

4.5
Modulating and Enhancing the Phenotype: Transgenic Crosses

One of the great advantages of transgenic animals is that different mice can be mated together so that the effect of multiple transgenes on the disease phenotype can be observed. In particular, since most transgenic AD models do not exhibit all the pathological features of AD, different crosses have been made in order to design a model that will more closely describe the pathology, neurodegeneration, and memory loss seen in AD patients.

Double transgenic mice that carry *PS1-A246E* and *APPsw* FAD mutations show an elevated Aβ42/Aβ40 ratio in brain homogenates compared to the ratio observed in transgenic mice expressing *APPsw* alone or transgenic mice coexpressing wild-type human *PS1* and *APPsw*. In addition, double transgenic mice *APPswxPS1-A246E* developed numerous amyloid deposits much earlier (9 months of age) than age-matched mice expressing *APPsw* (18 months of age) and wild-type *huPS1* and *APPsw* alone. Interestingly, the majority of Aβ deposits in the double transgenic mice were not immunoreactive to $A\beta42$ but instead were stained with antisera to $A\beta40$. Similarly, crossing the Tg2576 transgenic mice, which express mutant APPsw, with mice transgenic for *PS1-M146L* (PDPS1) revealed 41% increase in $A\beta42/43$ levels and formation of $A\beta$ deposits in the cortex and hippocampus as early as 12 weeks of age, compared to development of $A\beta$ deposits in Tg2576 mice of 9–12 months of age and approximately 1.5-fold elevation of $A\beta42/43$ from birth in PDPS1 line. The early $A\beta$ deposits were found to be primarily composed of fibrillar $A\beta$ and resembled compact amyloid plaques. As the mice aged, these fibrillar deposits did not increase substantially beyond the 12 months of age. Interestingly, the diffuse deposits appeared only until later, this is opposite to the general perception that in AD the compact deposits are formed by condensation of the diffuse material.

Studies on APPxPS1 double transgenics, therefore, revealed that marginal increases in $A\beta42/43$ levels, evoked by PS1 mutant transgene, can accelerate the deposition process by several months. As these mice have a severe $A\beta$ pathology for an extended time relative to singly transgenic APP mice, other features of the disease such as tau abnormalities or major cell loss may become apparent as the animals age.

Double mutants produced from crossing JNPL3 transgenic mice expressing mutant P301L tau with Tg2576 mice expressing the APPsw mutation offered proof that $A\beta$ influences the development of NFTs. The double mutant exhibited NFT pathology that was substantially enhanced in the limbic system and the olfactory cortex. These results suggest that APP or $A\beta$ augments the formation of NFTs in the regions of the brain vulnerable to the formation of these lesions. Recently, a triple transgenic mouse was generated expressing APPsw, PS1-M146L, and P301L tau mutant protein that exhibited both plaques and NFT. However, other characteristics of AD such as neurodegeneration and memory loss have yet to be examined.

Although most of the transgenic models for AD do not completely recreate the disease phenotype, they have shown considerable utility, both for studying the disease mechanisms and for the preliminary testing of therapeutic agents, particularly those that are designed to modulate $A\beta$ deposition.

5 Potential Treatments

The ultimate goal of AD research is to prevent disease and/or develop treatments for this devastating disease. Unfortunately, there is still no effective treatment available. Currently, the only FDA-approved therapy for AD is cholinesterase inhibitors, which can enhance cholinergic activity and temporarily improve cognitive function in some individuals in the early stages of disease. However, these drugs only treat the symptoms but have no impact on progression of the disease.

Recent advances in dissecting the molecular and cellular mechanisms that are involved in AD pathogenesis have provided substantial knowledge for determining novel targets for drug development. Several new therapeutic approaches targeted at distinct aspects of the disease are currently being pursued. Potential therapies include: anti-inflammatory agents, cholesterol-lowering drugs, antioxidants, hormonal therapy, and approaches that inhibit $A\beta$ production or increase $A\beta$ degradation and clearance. Strategies targeted directly at $A\beta$ may be most effective given the fact that Aβ deposition is the central and fundamental event that causes AD. Other approaches affecting downstream events in the neurodegenerative process are likely to be used as supplementary treatments.

5.1
Secretase Inhibitors

Two prime targets for drug development are β- and γ-secretases, the two enzymes that process APP to Aβ. Inhibitors to these two proteases are expected to have the greatest Aβ-lowering effect since inhibiting either enzyme can completely inhibit Aβ generation.

γ-secretase is considered to be central to AD pathogenesis because altering its activity invariantly favors Aβ42 production and causes AD. The major concern in developing a drug that targets the γ-secretase is its potential for side effects. It is clear that γ-secretase is important for several physiological functions through its role as the protease that mediates the intramembranous cleavage of a number of TM proteins, including Notch. One possible solution is to develop an inhibitor that selectively lowers Aβ (Aβ42 in particular) production without significantly affecting the cleavage of other γ-secretase substrates. The development of such inhibitors has been reported. Another possible approach is partial inhibition of γ-secretase activity, at inhibitor concentrations that reduce Aβ production without affecting Notch signaling. Indeed, it has been shown that compounds with no reported selectivity allow significant Aβ reduction without changing the Notch-related function. Some inhibitors, however, are reported to preferentially increase Aβ42 production when used at low concentrations.

Most evidence suggests that β-secretase (BACE1) may be a better therapeutic target than γ-secretase. BACE1 knockout mice do not generate Aβ, but are otherwise healthy and fertile without obvious deficits in the basal, neurological, and physiological functions. In addition, the X-ray structure of the BACE1 protease domain has been solved and should provide valuable knowledge for inhibitor design.

Although the knockout data are reassuring that the detrimental effects of inhibiting β-secretase may be minimal, concerns regarding the possible side effects remain. First, there are likely other β-secretase substrates since it is counterintuitive to assume that BACE1 has evolved just to generate Aβ. Indeed, one recent study suggests that β-secretase may be responsible for the cleavage and secretion of a Golgi resident sialyltransferase. It is very important to identify these substrates because they are valuable for predicting the possible side effects of BACE1 inhibitors. It will also be important to know the phenotype of BACE1/2 double knockout mice. Although BACE2 is not primarily involved in Aβ generation, it may have an important physiological role and inhibitors of BACE1 may also inhibit BACE2 because of the high degree of homology between these two proteins.

While inhibitors to both secretases may potentially be effective in AD treatment, the challenges facing drug development are significant. The inhibitors need to be highly selective for the target enzyme and target substrate to reduce possible side effects to a minimum. The inhibitors need to be able to efficiently cross the BBB. Potent peptide-based inhibitors to BACE1 have been developed. However, such large compounds are not viable drug candidates because they will not penetrate the BBB to a sufficient extent. Instead, the enzyme inhibitors with therapeutic potential are preferably small organic molecules with high specificity. The X-ray structure of BACE1 suggests that its active site is more open and less hydrophobic than that of other aspartyl proteases. It may pose challenges for the development of small-molecule inhibitors.

5.2
Vaccine Approaches

Another novel approach for Alzheimer therapy is $A\beta$ immunization. Many studies have shown that direct immunization with $A\beta$ peptide or anti-$A\beta$ antibodies can greatly reduce plaque formation, neuritic dystrophy, and other AD-like pathology in APP transgenic mice. One study also shows that such an immunization strategy is also effective in reversing behavioral deficits in the transgenic mouse model. It appears that antibodies to $A\beta$ can enhance $A\beta$ clearance, prevent $A\beta$ fibril formation, disrupt $A\beta$ fibrils, as well as block $A\beta$ toxicity.

Despite the great effects in mice, there are serious concerns about the safety of this approach when used in humans. One concern is that autoimmunity may occur in humans, although it has not been reported in mice. Another major concern is the toxicity of $A\beta42$, which can cross the BBB and seed fibril formation in the brain, and therefore may actually promote plaque formation. In addition, $A\beta42$ may cause inflammation and neurotoxicity. A phase II clinical trial using $A\beta42$ vaccination was eventually terminated because of cerebral inflammation observed in several patients. The first autopsy result from this trial was just published. It shows exactly the two sides of this approach. Vaccination has such a powerful effect that the patient's cerebral cortex was almost cleared of $A\beta$ deposition, but the side effect of cerebral inflammation was so severe that the patient developed meningoencephalitis, which led to death.

Safer therapeutic approaches including using nonamyloidogenic/nontoxic $A\beta$ derivatives as an immunogen or passive immunization with $A\beta$ antibodies are being developed. One study shows that passive immunization with (Fab')$_2$ fragments of $A\beta$-specific antibody can clear $A\beta$ in a mouse model. This is promising because (Fab')$_2$ fragments do not interact with Fc receptors and so will not activate the cellular-immune response which has been the major cause for the severe side effects.

5.3
Other Therapeutic Approaches

Several FAD-approved nonsteroidal anti-inflammatory drugs (NSAIDs), including ibuprofen, indomethacin, and sulindac have been shown to selectively reduce $A\beta42$ levels in cultured cells. In addition, treatment of APP transgenic mice with ibuprofen reduces $A\beta42$ levels in the brain and suppresses plaque pathology. It appears that NSAIDs can subtly alter γ-secretase activity so that cleavage is shifted from $A\beta42$ to $A\beta38$.

Many studies also indicate that cholesterol influences $A\beta$ metabolism and cholesterol-reducing drugs may have a beneficial effect on AD. Individuals taking cholesterol-lowering drugs such as statins show greatly reduced risk for developing AD, whereas individuals with elevated cholesterol are at a higher risk. Statins and other cholesterol-lowering drugs have been shown to reduce $A\beta$ levels and $A\beta$ deposition in both cell culture systems and animal models. The role of cholesterol in $A\beta$ metabolism appears to be quite complex and the mechanisms are not clear.

Nevertheless, treatment with statins and other cholesterol-reducing drugs may have a significant clinical benefit in the prevention of AD.

6
Summary

We have seen a dramatic increase in our knowledge of the underlying molecular mechanisms of AD pathology during the last ten years. Understanding the effects of causative mutations in *APP* and the presenilins and the genetic risk factor, *APOE4* has led to the conclusion that changes in Aβ metabolism and deposition are central to the disease process. These studies suggest that similar mechanisms underlie both FAD and sporadic AD. Transgenic models of AD now provide a new hope for the rapid development in rational treatments for AD.

Bibliography

Books and Reviews

Duff, K. (2001) Transgenic mouse models of Alzheimer's disease: phenotype and mechanisms of pathogenesis, *Biochem. Soc. Symp.* **67**, 195–202.

Golde, T.E. (2003) Alzheimer disease therapy: Can the amyloid cascade be halted? *J. Clin. Invest.* **11**, 11–18.

Hardy, J., Selkoe, D.J. (2002) The amyloid hypothesis of Alzheimer's disease: progress and problems on the road to therapeutics, *Science* **297**, 353–356.

Myers, A., Goate, A.M. (2001) The genetics of late-onset Alzheimer's disease, *Curr. Opin. Neurol.* **14**, 433–440.

Richardson, J.A., Burns, D.K. (2002) Mouse models of Alzheimer's disease: a quest for plaques and tangles, *ILAR J.* **43**, 89–99.

Selkoe, D.J., Podlisny, M.B. (2002) Deciphering the genetic basis of Alzheimer's disease, *Annu. Rev. Genomics Hum. Genet.* **3**, 67–99.

Tandon, A., Fraser, P. (2002) The presenilins, *Genome Biol.* **3**, 3014.1–3014.9; reviews.

Vassar, R. (2002) β-Secretase (BACE) as a drug target for alzheimer's disease, *Adv. Drug Delivery Rev.* **54**, 1589–1602.

Primary Literature

Alzheimer, A. (1907) Über eine eigenartige Erkrankung der Hirnrinde, *Allg. Zeitsch. Psychiatrie Psychisch-gerichtliche Med.* **64**, 146–148.

Biernat, J., Gustke, N., Drewes, G., Mandelkow, E.M., Mandelkow, E. Phosphorylation of Ser262 strongly reduces binding of tau to microtubules: distinction between PHF-like immunoreactivity and microtubule binding, *Neuron* **11**, 153–163.

Blacker, D., Bertram, L., Saunders, A.J., Moscarillo, T.J., Albert, M.S., Wiener, H., Perry, R.T., Collins, J.S., Harrell, L.E., Go, R.C., Mahoney, A., Beaty, T., Fallin, M.D., Avramopoulos, D., Chase, G.A., Folstein, M.F., McInnis, M.G., Bassett, S.S., Doheny, K.J., Pugh, E.W., Tanzi, R.E. (2003) Results of a high-resolution genome screen of 437 Alzheimer's disease families, *Hum. Mol. Genet.* **12**, 23–32.

Borchelt, D.R., Ratovitski, T., van Lare, J., Lee, M.K., Gonzales, V., Jenkins, N.A., Copeland, N.G., Price, D.L., Sisodia, S.S. (1997) Accelerated amyloid deposition in the brains of transgenic mice coexpressing mutant presenilin 1 and amyloid precursor proteins, *Neuron* **19**, 939–945.

Borchelt, D.R., Thinakaran, G., Eckman, C.B., Lee, M.K., Davenport, F., Ratovitsky, T., Prada, C.M., Kim, G., Seekins, S., Yager, D., Slunt, H.H., Wang, R., Seeger, M., Levey, A.I., Gandy, S.E., Copeland, N.G., Jenkins, N.A., Price, D.L., Younkin, S.G., Sisodia, S.S. (1996) Familial Alzheimer's disease-linked presenilin 1 variants elevate Aβ1-42/1-40 ratio *in vitro* and *in vivo*, *Neuron* **17**, 1005–1013.

Braak, H., Braak, E. (1991) Neuropathological stageing of Alzheimer-related changes, *Acta Neuropathol. (Berl)* **82**, 239–259.

Brendza, R.P., Bales, K.R., Paul, S.M., Holtzman, D.M. (2002) Role of apoE/Aβ interactions in Alzheimer's disease: insights from transgenic mouse models, *Mol. Psychiatry* **7**, 132–135.

Cai, X.D., Golde, T.E., Younkin, S.G. (1993) Release of excess amyloid β protein from a mutant amyloid β protein precursor, *Science* **259**, 514–516.

Chung, H.M., Struhl, G. (2001) Nicastrin is required for Presenilin-mediated transmembrane cleavage in *Drosophila*, *Nat. Cell. Biol.* **3**, 1129–1132.

Citron, M., Oltersdorf, T., Haass, C., McConlogue, L., Hung, A.Y., Seubert, P., Vigo-Pelfrey, C., Lieberburg, I., Selkoe, D.J. (1992) Mutation of the β-amyloid precursor protein in familial Alzheimer's disease increases β-protein production, *Nature* **360**, 672–674.

Citron, M., Westaway, D., Xia, W., Carlson, G., Diehl, T., Levesque, G., Johnson-Wood, K., Lee, M., Seubert, P., Davis, A., Kholodenko, D., Motter, R., Sherrington, R., Perry, B., Yao, H., Strome, R., Lieberburg, I., Rommens, J., Kim, S., Schenk, D., Fraser, P., St. George Hyslop, P., Selkoe, D.J. (1997) Mutant presenilins of Alzheimer's disease increase production of 42-residue amyloid β-protein in both transfected cells and transgenic mice, *Nat. Med.* **3**, 67–72.

Corder, E.H., Saunders, A.M., Risch, N.J., Strittmatter, W.J., Schmechel, D.E., Gaskell, P.C. Jr., Rimmler, J.B., Locke, P.A., Conneally, P.M., Schmader, K.E., et al. (1994) Protective effect of apolipoprotein E type 2 allele for late onset Alzheimer disease, *Nat. Genet.* **7**, 180–184.

Corder, E.H., Saunders, A.M., Strittmatter, W.J., Schmechel, D.E., Gaskell, P.C., Small, G.W., Roses, A.D., Haines, J.L., Pericak-Vance, M.A. (1993) Gene dose of apolipoprotein E type 4 allele and the risk of Alzheimer's disease in late onset families, *Science* **261**, 921–923.

DeMattos, R.B., Bales, K.R., Cummins, D.J., Dodart, J.C., Paul, S.M., Holtzman, D.M. (2001) Peripheral anti-Aβ antibody alters CNS and plasma Aβ clearance and decreases brain Aβ burden in a mouse model of Alzheimer's disease, *Proc. Natl. Acad. Sci. U.S.A.* **98**, 8850–8855.

DeMattos, R.B., Bales, K.R., Cummins, D.J., Paul, S.M., Holtzman, D.M. (2002) Brain to plasma amyloid-β efflux: a measure of brain amyloid burden in a mouse model of Alzheimer's disease, *Science* **295**, 2264–2267.

De Strooper, B., Annaert, W., Cupers, P., Saftig, P., Craessaerts, K., Mumm, J.S., Schroeter, E.H., Schrijvers, V., Wolfe, M.S., Ray, W.J., Goate, A., Kopan, R.A. (1999) presenilin-1-dependent γ-secretase-like protease mediates release of Notch intracellular domain, *Nature* **398**, 518–522.

De Strooper, B., Saftig, P., Craessaerts, K., Vanderstichele, H., Guhde, G., Annaert, W., von Figura, K., Van Leuven, F. (1998) Deficiency of presenilin-1 inhibits the normal cleavage of amyloid precursor protein, *Nature* **391**, 387–390.

Duff, K., Eckman, C., Zehr, C., Yu, X., Prada, C.M., Perez-Tur, J., Hutton, M., Buee, L., Harigaya, Y., Yager, D., Morgan, D., Gordon, M.N., Holcomb, L., Refolo, L., Zenk, B., Hardy, J., Younkin, S. (1996) Increased amyloid-β42(43) in brains of mice expressing mutant presenilin 1, *Nature* **383**, 710–713.

Edbauer, D., Winkler, E., Haass, C., Steiner, H. (2002) Presenilin and nicastrin regulate each other and determine amyloid β-peptide production via complex formation, *Proc. Natl. Acad. Sci. U.S.A.* **99**, 8666–8671.

Ertekin-Taner, N., Graff-Radford, N., Younkin, L.H., Eckman, C., Baker, M., Adamson, J., Ronald, J., Blangero, J., Hutton, M., Younkin, S.G. (2000) Linkage of plasma Aβ42 to a quantitative locus on chromosome 10 in late-onset Alzheimer's disease pedigrees, *Science* **290**, 2303–2304.

Esch, F.S., Keim, P.S., Beattie, E.C., Blacher, R.W., Culwell, A.R., Oltersdorf, T., McClure, D., Ward, P.J. (1990) Cleavage of amyloid β peptide during constitutive processing of its precursor, *Science* **248**, 1122–1124.

Estus, S., Golde, T.E., Younkin, S.G. (1992) Normal processing of the Alzheimer's disease amyloid β protein precursor generates potentially amyloidogenic carboxyl-terminal derivatives, *Ann. N. Y. Acad. Sci.* **674**, 138–148.

Fagan, A.M., Watson, M., Parsadanian, M., Bales, K.R., Paul, S.M., Holtzman, D.M. (2002) Human and murine ApoE markedly alters Aβ metabolism before and after plaque formation in a mouse model of Alzheimer's disease, *Neurobiol. Dis.* **9**, 305–318.

Farrer, L.A., Bowirrat, A., Friedland, R.P., Waraska, K., Korczyn, A.D., Baldwin, C.T. (2003) Identification of multiple loci for Alzheimer disease in a consanguineous Israeli-Arab community, *Hum. Mol. Genet.* **12**, 415–422.

Flood, D.G., Reaume, A.G., Dorfman, K.S., Lin, Y.G., Lang, D.M., Trusko, S.P., Savage, M.J., Annaert, W.G., De Strooper, B., Siman, R.,

Scott, R.W. (2002) FAD mutant PS-1 gene-targeted mice: increased Aβ42 and Aβ deposition without APP overproduction, *Neurobiol. Aging* **23**, 335–348.

Francis, R., McGrath, G., Zhang, J., Ruddy, D.A., Sym, M., Apfeld, J., Nicoll, M., Maxwell, M., Hai, B., Ellis, M.C., Parks, A.L., Xu, W., Li, J., Gurney, M., Myers, R.L., Himes, C.S., Hiebsch, R., Ruble, C., Nye, J.S., Curtis, D. (2002) aph-1 and pen-2 are required for Notch pathway signaling, γ-secretase cleavage of βAPP, and presenilin protein accumulation, *Dev. Cell.* **3**, 85–97.

Games, D., Adams, D., Alessandrini, R., Barbour, R., Berthelette, P., Blackwell, C., Carr, T., Clemens, J., Donaldson, T., Gillespie, F. (1995) Alzheimer-type neuropathology in transgenic mice overexpressing V717F β-amyloid precursor protein, *Nature* **373**, 523–527.

Glenner, G.G., Wong, C.W. (1984) Alzheimer's disease: initial report of the purification and characterization of a novel cerebrovascular amyloid protein, *Biochem. Biophys. Res. Commun.* **120**, 885–890.

Goate, A., Chartier-Harlin, M.C., Mullan, M., Brown, J., Crawford, F., Fidani, L., Giuffra, L., Haynes, A., Irving, N., James, L. (1991) Segregation of a missense mutation in the amyloid precursor protein gene with familial Alzheimer's disease, *Nature* **349**, 704–706.

Goedert, M., Wischik, C.M., Crowther, R.A., Walker, J.E., Klug, A. (1988) Cloning and sequencing of the cDNA encoding a core protein of the paired helical filament of Alzheimer disease: identification as the microtubule-associated protein tau, *Proc. Natl. Acad. Sci. U.S.A.* **85**, 4051–4055.

Goldgaber, D., Lerman, M.I., McBride, O.W., Saffiotti, U., Gajdusek, D.C. (1987) Characterization and chromosomal localization of a cDNA encoding brain amyloid of Alzheimer's disease, *Science* **235**, 877–880.

Gómez-Isla, T., Price, J.L., McKeel, D.W., Morris, J.C., Growdon, J.H., Hyman, B.T. (1996) Profound loss of layer II entorhinal cortex neurons occurs in very mild Alzheimer's disease, *J. Neurosci.* **16**, 4491–4500.

Gotz, J., Chen, F., van Dorpe, J., Nitsch, R.M. (2001) Formation of neurofibrillary tangles in P301l tau transgenic mice induced by Aβ42 fibrils, *Science* **293**, 1491–1495.

Goutte, C., Tsunozaki, M., Hale, V.A., Priess, J.R. (2002) APH-1 is a multipass membrane protein essential for the Notch signaling pathway in *Caenorhabditis elegans* embryos, *Proc. Natl. Acad. Sci. U.S.A.* **99**, 775–779.

Grundke-Iqbal, I., Iqbal, K., Quinlan, M., Tung, Y.C., Zaidi, M.S., Wisniewski, H.M. (1986) Microtubule-associated protein tau. A component of Alzheimer paired helical filaments, *J. Biol. Chem.* **261**, 6084–6089.

Haass, C., Koo, E.H., Mellon, A., Hung, A.Y., Selkoe, D.J. (1992) Targeting of cell-surface β-amyloid precursor protein to lysosomes: alternative processing into amyloid-bearing fragments, *Nature* **357**, 500–503.

Haass, C., Schlossmacher, M.G., Hung, A.Y., Vigo-Pelfrey, C., Mellon, A., Ostaszewski, B.L., Lieberburg, I., Koo, E.H., Schenk, D., Teplow, D.B. (1992) Amyloid β-peptide is produced by cultured cells during normal metabolism, *Nature* **359**, 322–325.

Holcomb, L., Gordon, M.N., McGowan, E., Yu, X., Benkovic, S., Jantzen, P., Wright, K., Saad, I., Mueller, R., Morgan, D., Sanders, S., Zehr, C., O'Campo, K., Hardy, J., Prada, C.M., Eckman, C., Younkin, S., Hsiao, K., Duff, K. (1998) Accelerated Alzheimer-type phenotype in transgenic mice carrying both mutant amyloid precursor protein and presenilin 1 transgenes, *Nat. Med.* **4**, 97–100.

Holtzman, D.M. (2002) Aβ conformational change is central to Alzheimer's disease, *Neurobiol. Aging* **23**, 1085–1088.

Hong, C.S., Caromile, L., Nomata, Y., Mori, H., Bredesen, D.E., Koo, E.H. (1999) Contrasting role of presenilin-1 and presenilin-2 in neuronal differentiation *in vitro*, *J. Neurosci.* **19**, 637–643.

Hsiao, K., Chapman, P., Nilsen, S., Eckman, C., Harigaya, Y., Younkin, S., Yang, F., Cole, G. (1996) Correlative memory deficits, Aβ elevation, and amyloid plaques in transgenic mice, *Science* **274**, 99–102.

Hu, Y., Ye, Y., Fortini, M.E. (2002) Nicastrin is required for γ-secretase cleavage of the *Drosophila* Notch receptor, *Dev. Cell.* **2**, 69–78.

Hutton, M., Lendon, C.L., Rizzu, P., Baker, M., Froelich, S., Houlden, H., Pickering-Brown, S., Chakraverty, S., Isaacs, A., Grover, A., Hackett, J., Adamson, J., Lincoln, S., Dickson, D., Davies, P., Petersen, R.C., Stevens, M., de Graaff, E., Wauters, E., van Baren, J., Hillebrand, M., Joosse, M., Kwon, J.M., Nowotny, P., Heutink, P. (1998) Association of missense and 5′-splice-site mutations in tau with the inherited dementia FTDP-17, *Nature* **393**, 702–705.

Iqbal, K., Zaidi, T., Thompson, C.H., Merz, P.A., Wisniewski, H.M. (1984) Alzheimer paired helical filaments: bulk isolation, solubility, and protein composition, *Acta Neuropathol. (Berl)* **62**, 167–177.

Kang, J., Lemaire, H.G., Unterbeck, A., Salbaum, J.M., Masters, C.L., Grzeschik, K.H., Multhaup, G., Beyreuther, K., Müller-Hill, B. (1987) The precursor of Alzheimer's disease amyloid A4 protein resembles a cell-surface receptor, *Nature* **325**, 733–736.

Leissring, M.A., Murphy, M.P., Mead, T.R., Akbari, Y., Sugarman, M.C., Jannatipour, M., Anliker, B., Muller, U., Saftig, P., De Strooper, B., Wolfe, M.S., Golde, T.E., LaFerla, F.M. (2002) A physiologic signaling role for the γ-secretase-derived intracellular fragment of APP, *Proc. Natl. Acad. Sci. U.S.A.* **99**, 4697–4702.

Levy, E., Carman, M.D., Fernandez-Madrid, I.J., Power, M.D., Lieberburg, I., van Duinen, S.G., Bots, G.T., Luyendijk, W., Frangione, B. (1990) Mutation of the Alzheimer's disease amyloid gene in hereditary cerebral hemorrhage, Dutch type, *Science* **248**, 1124–1126.

Levy-Lahad, E., Wasco, W., Poorkaj, P., Romano, D.M., Oshima, J., Pettingell, W.H., Yu, C.E., Jondro, P.D., Schmidt, S.D., Wang, K. (1995) Candidate gene for the chromosome 1 familial Alzheimer's disease locus, *Science* **269**, 973–977.

Lewis, J., Dickson, D.W., Lin, W.L., Chisholm, L., Corral, A., Jones, G., Yen, S.H., Sahara, N., Skipper, L., Yager, D., Eckman, C., Hardy, J., Hutton, M., McGowan, E. (2001) Enhanced neurofibrillary degeneration in transgenic mice expressing mutant tau and APP, *Science* **293**, 1487–1491.

Lewis, J., McGowan, E., Rockwood, J., Melrose, H., Nacharaju, P., Van Slegtenhorst, M., Gwinn-Hardy, K., Paul Murphy, M., Baker, M., Yu, X., Duff, K., Hardy, J., Corral, A., Lin, W.L., Yen, S.H., Dickson, D.W., Davies, P., Hutton, M. (2000) Neurofibrillary tangles, amyotrophy and progressive motor disturbance in mice expressing mutant (P301L) tau protein, *Nat. Genet.* **25**, 402–405.

Lopez-Schier, H., St. Johnston, D. (2002) *Drosophila* nicastrin is essential for the intramembranous cleavage of notch, *Dev. Cell.* **2**, 79–89.

Masters, C.L., Simms, G., Weinman, N.A., Multhaup, G., McDonald, B.L., Beyreuther, K. (1985) Amyloid plaque core protein in Alzheimer disease and Down syndrome, *Proc. Natl. Acad. Sci. U.S.A.* **82**, 4245–4249.

Mayeux, R., Lee, J.H., Romas, S.N., Mayo, D., Santana, V., Williamson, J., Ciappa, A., Rondon, H.Z., Estevez, P., Lantigua, R., Medrano, M., Torres, M., Stern, Y., Tycko, B., Knowles, J.A. (2002) Chromosome-12 mapping of late-onset Alzheimer disease among Caribbean Hispanics, *Am. J. Hum. Genet.* **70**, 237–243.

Murrell, J., Farlow, M., Ghetti, B., Benson, M.D. (1991) A mutation in the amyloid precursor protein associated with hereditary Alzheimer's disease, *Science* **254**, 97–99.

Nicoll, J.A., Wilkinson, D., Holmes, C., Steart, P., Markham, H., Weller, R.O. (2003) Neuropathology of human Alzheimer disease after immunization with amyloid-β peptide: a case report, *Nat. Med.* **9**, 448–452.

Nilsberth, C., Westlind-Danielsson, A., Eckman, C.B., Condron, M.M., Axelman, K., Forsell, C., Stenh, C., Luthman, J., Teplow, D.B., Younkin, S.G., Naslund, J., Lannfelt, L. (2001) The 'Arctic' APP mutation (E693G) causes Alzheimer's disease by enhanced Aβ protofibril formation, *Nat. Neurosci.* **4**, 887–893.

Price, J.L., Davis, P.B., Morris, J.C., White, D.L. The distribution of tangles, plaques and related immunohistochemical markers in healthy aging and Alzheimer's disease, *Neurobiol. Aging* **12**, 295–312.

Rogaev, E.I., Sherrington, R., Rogaeva, E.A., Levesque, G., Ikeda, M., Liang, Y., Chi, H., Lin, C., Holman, K., Tsuda, T. (1995) Familial Alzheimer's disease in kindreds with missense mutations in a gene on chromosome 1 related to the Alzheimer's disease type 3 gene, *Nature* **376**, 775–778.

St. George-Hyslop, P.H., Tanzi, R.E., Polinsky, R.J., Haines, J.L., Nee, L., Watkins, P.C., Myers, R.H., Feldman, R.G., Pollen, D., Drachman, D. (1987) The genetic defect causing familial Alzheimer's disease maps on chromosome 21, *Science* **235**, 885–890.

Schenk, D., Barbour, R., Dunn, W., Gordon, G., Grajeda, H., Guido, T., Hu, K., Huang, J., Johnson-Wood, K., Khan, K., Kholodenko, D., Lee, M., Liao, Z., Lieberburg, I., Motter, R., Mutter, L., Soriano, F., Shopp, G., Vasquez, N., Vandevert, C., Walker, S., Wogulis, M., Yednock, T., Games, D., Seubert, P. (1999) Immunization with amyloid-β attenuates

Alzheimer-disease-like pathology in the PDAPP mouse, *Nature* **400**, 173–177.

Scheuner, D., Eckman, C., Jensen, M., Song, X., Citron, M., Suzuki, N., Bird, T.D., Hardy, J., Hutton, M., Kukull, W., Larson, E., Levy-Lahad, E., Viitanen, M., Peskind, E., Poorkaj, P., Schellenberg, G., Tanzi, R., Wasco, W., Lannfelt, L., Selkoe, D., Younkin, S. (1996) Secreted amyloid β-protein similar to that in the senile plaques of Alzheimer's disease is increased *in vitro* by the presenilin 1 and 2 and APP mutations linked to familial Alzheimer's disease, *Nat. Med.* **2**, 864–870.

Seubert, P., Oltersdorf, T., Lee, M.G., Barbour, R., Blomquist, C., Davis, D.L., Bryant, K., Fritz, L.C., Galasko, D., Thal, L.J. (1993) Secretion of β-amyloid precursor protein cleaved at the amino terminus of the β-amyloid peptide, *Nature* **361**, 260–263.

Seubert, P., Vigo-Pelfrey, C., Esch, F., Lee, M., Dovey, H., Davis, D., Sinha, S., Schlossmacher, M., Whaley, J., Swindlehurst, C. (1992) Isolation and quantification of soluble Alzheimer's β-peptide from biological fluids, *Nature* **359**, 325–327.

Sherrington, R., Rogaev, E.I., Liang, Y., Rogaeva, E.A., Levesque, G., Ikeda, M., Chi, H., Lin, C., Li, G., Holman, K. (1995) Cloning of a gene bearing missense mutations in early-onset familial Alzheimer's disease, *Nature* **375**, 754–760.

Shoji, M., Golde, T.E., Ghiso, J., Cheung, T.T., Estus, S., Shaffer, L.M., Cai, X.D., McKay, D.M., Tintner, R., Frangione, B. (1992) Production of the Alzheimer amyloid β protein by normal proteolytic processing, *Science* **258**, 126–129.

Sinha, S., Anderson, J.P., Barbour, R., Basi, G.S., Caccavello, R., Davis, D., Doan, M., Dovey, H.F., Frigon, N., Hong, J., Jacobson-Croak, K., Jewett, N., Keim, P., Knops, J., Lieberburg, I., Power, M., Tan, H., Tatsuno, G., Tung, J., Schenk, D., Seubert, P., Suomensaari, S.M., Wang, S., Walker, D., John, V. (1999) Purification and cloning of amyloid precursor protein β-secretase from human brain, *Nature* **402**, 537–540.

Skovronsky, D.M., Zhang, B., Kung, M.-P., Kung, H.F., Trojanowski, J.Q., Lee, V.M. (2000) *In vitro* detection of amyloid plaques in a mouse model of Alzheimer's disease, *Proc. Natl. Acad. Sci. U.S.A.* **97**, 7609–7614.

Stenh, C., Nilsberth, C., Hammarback, J., Engvall, B., Naslund, J., Lannfelt, L. (2002) The Arctic mutation interferes with processing of the amyloid precursor protein, *Neuro Report* **13**, 1857–1860.

Strittmatter, W.J., Saunders, A.M., Schmechel, D., Pericak-Vance, M., Enghild, J., Salvesen, G.S., Roses, A.D. (1993) Apolipoprotein E: high-avidity binding to β-amyloid and increased frequency of type 4 allele in late-onset familial Alzheimer disease, *Proc. Natl. Acad. Sci. U.S.A.* **90**, 1977–1981.

Sturchler-Pierrat, C., Abramowski, D., Duke, M., Wiederhold, K.H., Mistl, C., Rothacher, S., Ledermann, B., Burki, K., Frey, P., Paganetti, P.A., Waridel, C., Calhoun, M.E., Jucker, M., Probst, A., Staufenbiel, M., Sommer, B. (1997) Two amyloid precursor protein transgenic mouse models with Alzheimer disease-like pathology, *Proc. Natl. Acad. Sci. U.S.A.* **94**, 13287–13292.

Sturchler-Pierrat, C., Staufenbiel, M. (2000) Pathogenic mechanisms of Alzheimer's disease analyzed in the APP23 transgenic mouse model, *Ann. N. Y. Acad. Sci.* **920**, 134–139.

Suzuki, N., Cheung, T.T., Cai, X.D., Odaka, A., Otvos, L., Eckman, C., Golde, T.E., Younkin, S.G. (1994) An increased percentage of long amyloid β protein secreted by familial amyloid β protein precursor (β APP717) mutants, *Science* **264**, 1336–1340.

Tanzi, R.E., Gusella, J.F., Watkins, P.C., Bruns, G.A., St. George-Hyslop, P., Van Keuren, M.L., Patterson, D., Pagan, S., Kurnit, D.M., Neve, R.L. (1987) Amyloid β protein gene: cDNA, mRNA distribution, and genetic linkage near the Alzheimer locus, *Science* **235**, 880–884.

Terry, R.D., Masliah, E., Salmon, D.P., Butters, N., DeTeresa, R., Hill, R., Hansen, L.A., Katzman, R. (1991) Physical basis of cognitive alterations in Alzheimer's disease: synapse loss is the major correlate of cognitive impairment, *Ann. Neurol.* **30**, 572–580.

Thinakaran, G., Borchelt, D.R., Lee, M.K., Slunt, H.H., Spitzer, L., Kim, G., Ratovitsky, T., Davenport, F., Nordstedt, C., Seeger, M., Hardy, J., Levey, A.I., Gandy, S.E., Jenkins, N.A., Copeland, N.G., Price, D.L., Sisodia, S.S. (1996) Endoproteolysis of presenilin 1 and accumulation of processed derivatives *in vitro*, *Neuron* **17**, 181–190.

Thinakaran, G., Harris, C.L., Ratovitski, T., Davenport, F., Slunt, H.H., Price, D.L., Borchelt, D.R., Sisodia, S.S. (1997) Evidence that levels of presenilins (PS1 and PS2) are coordinately regulated by competition for limiting cellular factors, *J. Biol. Chem.* **272**, 28415–28422.

Van Nostrand, W.E., Schmaier, A.H., Farrow, J.S., Cunningham, D.D. (1990) Protease nexin-II (amyloid β-protein precursor): a platelet α-granule protein, *Science* **248**, 745–748.

Vassar, R., Bennett, B.D., Babu-Khan, S., Kahn, S., Mendiaz, E.A., Denis, P., Teplow, D.B., Ross, S., Amarante, P., Loeloff, R., Luo, Y., Fisher, S., Fuller, J., Edenson, S., Lile, J., Jarosinski, M.A., Biere, A.L., Curran, E., Burgess, T., Louis, J.C., Collins, F., Treanor, J., Rogers, G., Citron, M. (1999) B-secretase cleavage of Alzheimer's amyloid precursor protein by the transmembrane aspartic protease BAC, *Science* **286**, 735–741.

Weggen, S., Eriksen, J.L., Das, P., Sagi, S.A., Wang, R., Pietrzik, C.U., Findlay, K.A., Smith, T.E., Murphy, M.P., Bulter, T., Kang, D.E., Marquez-Sterling, N., Golde, T.E., Koo, E.H. (2001) A subset of NSAIDs lower amyloidogenic Aβ42 independently of cyclooxygenase activity, *Nature* **414**, 212–216.

Weihofen, A., Binns, K., Lemberg, M.K., Ashman, K., Martoglio, B. (2002) Identification of signal peptide peptidase, a presenilin-type aspartic protease, *Science* **296**, 2215–2218.

Wolfe, M.S., Xia, W., Ostaszewski, B.L., Diehl, T.S., Kimberly, W.T., Selkoe, D.J. (1999) Two transmembrane aspartates in presenilin-1 required for presenilin endoproteolysis and γ-secretase activity, *Nature* **398**, 513–517.

Yu, G., Nishimura, M., Arawaka, S., Levitan, D., Zhang, L., Tandon, A., Song, Y.Q., Rogaeva, E., Chen, F., Kawarai, T., Supala, A., Levesque, L., Yu, H., Yang, D.S., Holmes, E., Milman, P., Liang, Y., Zhang, D.M., Xu, D.H., Sato, C., Rogaev, E., Smith, M., Janus, C., Zhang, Y., Aebersold, R., Farrer, L.S., Sorbi, S., Bruni, A., Fraser, P., St. George-Hyslop, P. (2000) Nicastrin modulates presenilin-mediated notch/glp-1 signal transduction and βAPP processing, *Nature* **407**, 48–54.

9
Molecular Chaperones

Peter Lund
School of Biosciences, University of Birmingham, Birmingham B15 2TT, UK

1	The Definition of a Molecular Chaperone	289
2	The Need for Molecular Chaperones: Protein Folding *In Vitro* and *In Vivo*	290
3	Classification of Molecular Chaperones	293
4	Mechanisms and Roles of the Major Molecular Chaperones	294
4.1	The Hsp60 Family	294
4.1.1	Introduction to the Hsp60 Family	294
4.1.2	Cellular Roles of the Hsp60 Family	294
4.1.3	Mechanisms of Action of the Hsp60 Family	297
4.1.4	Structure and Function of the Hsp60 Family	299
4.2	The Hsp70 Family	301
4.2.1	Introduction to the Hsp70 Family	301
4.2.2	Cellular Roles of the Hsp70 Family	301
4.2.3	Mechanisms of Action of the Hsp70 Family	302
4.2.4	Structure and Function of the Hsp70/Hsp40 Family	304
4.3	The Hsp90 Family	304
4.3.1	Introduction to the Hsp90 Family	304
4.3.2	Cellular Roles of the Hsp90 Family	304
4.3.3	Mechanisms of Action of the Hsp90 Family	305
4.3.4	Structure and Function of the Hsp90 Family	306
4.4	The Small Heat Shock Proteins	307
4.4.1	Introduction to the Small Heat Shock Proteins	307
4.4.2	Cellular Roles of sHSPs	307
4.4.3	Mechanism of Action of sHSPs	307
4.4.4	Structure and Function of sHSPs	308

Proteins. Edited by Robert A. Meyers.
Copyright © 2007 Wiley-VCH Verlag GmbH & Co. KGaA, Weinheim
ISBN: 978-3-527-31608-3

5	**Chaperones and Cellular Processes** 308
5.1	Chaperones and Redox Potential 308
5.1.1	Disulfide Bond Formation and Isomerization in Prokaryotes and Eukaryotes 308
5.1.2	Hsp33: A Chaperone with a Redox Switch 310
5.2	Chaperone Networks and Protein Quality Control in the Cell 311
5.2.1	Chaperone Networks and Protein Quality Control in the *E. coli* Cytosol 311
5.2.2	Protein Quality Control in the Endoplasmic Reticulum 313
5.3	Examples of Substrate-specific Chaperones 314
5.3.1	Hsp47 and Collagen Assembly 314
5.3.2	PapD and the Assembly of Bacterial Pili 315
6	**Conclusion** 316
	Acknowledgments 317
	Bibliography 317
	Books and Reviews 317
	Primary Literature 317

Keywords

Chaperonin
One of a member of the family of molecular chaperones that are homologous to the *E. coli* GroEL chaperone, and the mitochondrial and chloroplast Hsp60 chaperones.

Cochaperone
A protein that, while not possessing intrinsic chaperone properties itself, is required for a particular chaperone to function.

Heat Shock
The exposure of cells or organisms to temperatures a few degrees above their normal growth temperature, widely used as an experimental model for inducing cellular stress.

HSP
Abbreviation for "heat shock protein." HSPs are proteins that are induced *de novo* or are upregulated as a result of heat shock.

Molecular Chaperone
Proteins that assist other proteins to reach their final active conformation, but that are not themselves part of this final conformation.

Unfolded Protein
Any protein that is not in its fully folded state. This term encompasses a large range of possible states, from a protein that is nearly but not completely native to one that is a random coil.

▎ Some proteins in the cell are needed to help other proteins reach their final active conformations: these are referred to as molecular chaperones. Molecular chaperones can be grouped into families by sequence similarity, and they act through a variety of mechanisms. They are found in all cell types and in most locations within the cell. Many are induced by stresses such as heat shock. This chapter will review the reasons why molecular chaperones are required in the cell, and then discuss each of the major families of molecular chaperones in turn, in terms of structure, mechanism of action, and cellular role. Most molecular chaperones act in networks with other chaperones, and some of these will be reviewed. Many molecular chaperones interact with a broad range of different substrates, but some are specific to particular proteins, and examples of these will be discussed.

1
The Definition of a Molecular Chaperone

The number and range of proteins now classified as molecular chaperones are very large, and a definition that encompasses all of them is likely to be so vague as to be uninformative. Molecular chaperones must, however, have two properties in order to be defined as such, which are as follows:

1. They must assist other molecules to reach their final active state, by blocking or reversing unfavorable events that would otherwise prevent this state from being reached.
2. They must not be part of the final active structure whose existence they have helped to create.

The name *molecular chaperone* was originally coined to refer to the protein nucleoplasmin, which is required for the correct assembly of nucleosomes, and the choice of the term "chaperone" is based on the conventional use of the word, to refer to a person who prevents inappropriate interactions from occurring that could hinder an otherwise desirable outcome. (The etymology of the word "chaperone" is of interest: it is derived from the French verb "chaperonner," which means "to protect." This, in turn, was originally derived from the Old French noun "chaperon," which meant "a protective hood"). The justification for this choice of name is that for biological molecules such as proteins, getting from the state where they are first synthesized to the state where they are finally active (folded, and in the correct cellular compartment) is a complex process during which many unfavorable interactions can take place. These, if they were allowed to occur, would drastically lower the efficiency with which active biological material could be formed. Molecular chaperones prevent or reverse such interactions, which explains why they are so important for normal cellular function. Although the original chaperone (nucleoplasmin) is the key to chromatin assembly, the vast majority of chaperones are involved in protein folding and assembly, and it is only these that will be considered in this chapter.

In order for a given protein to be classified as a molecular chaperone, the chaperone activity of the protein should be experimentally demonstrable. Ideally, this demonstration would be made both *in vitro* and *in vivo*, through a combination of biochemical and genetic methods, but in practice this is not always feasible. The kinds of experiments that can demonstrate molecular chaperone activity are discussed below, in the sections on individual classes of molecular chaperone.

2
The Need for Molecular Chaperones: Protein Folding *In Vitro* and *In Vivo*

All life depends upon the ability of proteins to fold into specific conformations. An understanding of the process by which proteins succeed in folding from a linear chain of amino-acyl residues to a functioning tertiary or quaternary structure is thus central to an appreciation of all aspects of cell and organismal biology. The fact that this process occasionally goes wrong, and in doing so can lead to many pathological states, makes the need for such an understanding even more important. Most of the initial studies in this field were done using highly purified proteins, and these gave us many important insights into the essential features of protein folding. Studies of this nature are still continuing. More recently, however, it has been realized that protein folding inside the cell is a more complex process than that studied in the test tube, and many key discoveries have been made by looking at cellular protein folding in more detail. These include the discovery of molecular chaperones, which act in many ways to help proteins overcome some of the intrinsic problems caused by having to fold up in the complex internal environment of the cell. This initial section will thus briefly review our current understanding of protein folding both *in vitro* and *in vivo*, and will discuss in general terms some of the areas where molecular chaperones have been shown to be important or essential.

The earliest work in the area of protein folding was done by Anfinsen, who demonstrated that proteins will spontaneously adopt their native, active conformation in aqueous solution, thus proving that folded proteins are more thermodynamically stable than unfolded proteins. This conclusion was reached by taking purified proteins, denaturing them in suitable solvents that caused them to adopt an unfolded conformation, and then removing the denaturants and observing recovery of activity of the protein in question. No energy source is required in these types of experiments for proteins to refold, thus proving that the folded and active protein must be more stable than the unfolded form.

Although this was an elegant demonstration of the thermodynamically spontaneous nature of protein folding, it was subsequently pointed out by Levinthal that there was an apparent kinetic problem in protein folding. Because the number of possible conformations that can be adopted by a protein is very high indeed, and it takes a finite time for proteins to sample these different conformations, it would in theory take, on an average, billions of years of searching through all the different conformational options for a given protein molecule to find the most stable conformation. It is known, however, that proteins can fold from the denatured state over a timescale of microseconds to seconds. To resolve this apparent paradox, Levinthal proposed that folding pathways exist, with a limited number of possible conformations existing along the pathways.

This view is still thought to be essentially correct, but the view of a folding pathway has been refined to that of an energy landscape. According to this view, the folding of a protein is analogous to the trajectory taken by a ball that is rolling around on a rubber sheet, which has been deformed by a heavy weight. The ball will eventually finish up at the bottom of the dip caused by the heavy weight; this is equivalent to the protein in its lowest energy conformation. The precise route that the ball takes to get to the bottom of the well depends on a number of factors, such as its initial starting point and the shape of the distorted sheet. Similarly, within a given population of proteins, some molecules may fold very rapidly and others more slowly. The fact that they all finish up in the same final conformation does not imply that they all have to follow precisely the same pathway to get there.

This view of protein folding can be further expanded to consider what events might lead to some proteins failing to reach a fully folded and active conformation. At least two such events can be envisaged, leading respectively to protein misfolding and to protein aggregation. In the case of protein misfolding, the polypeptide chain may reach a metastable state where it is indeed folded but has not reached its correct and active conformation, and cannot proceed to it on a biologically relevant timescale. In the case of protein aggregation, the protein may have formed interactions with other partially folded protein chains such that all these proteins are now held together but are inactive because they have not reached their fully folded states. Interestingly, aggregation appears to be specific in that, when it occurs in mixtures of proteins, similar or identical proteins will tend to aggregate with each other. Aggregation is particularly likely to be a problem where protein concentrations are high, since the likelihood of two proteins interacting with each other before they fold into the native state is greater. Aggregation is potentially a particularly serious problem for proteins, because the forces that drive protein folding (hydrophobic forces) are the same as the forces that lead to protein aggregation. It is also a problem inside the cell, where protein concentrations may be extremely high.

Aggregation and misfolding are not the only problems that may be faced by proteins as they fold within the cell (Fig. 1). Another difficulty for proteins is that many targeting events require proteins to cross membranes into different cellular compartments, and yet it is hard to envisage a more effective barrier to a folded protein than a lipid bilayer. In order to cross membranes, proteins generally need to be prevented from folding but protected from other folding proteins with which they might interact both before and after they have crossed the membrane. Proteins also often need to form interactions with a large array of cofactors and effectors, and may require the presence of these in order to adopt their active conformations. In this event, proteins would need to be synthesized but kept in a receptive state until the cofactor becomes available. Proteins on the secretory pathway in

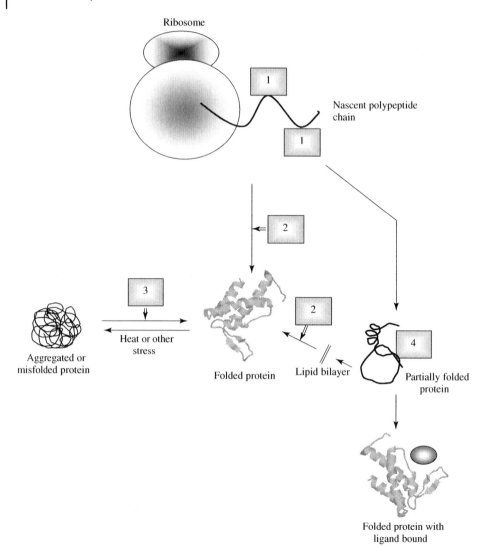

Fig. 1 The general roles of molecular chaperones. Molecular chaperones are involved in many different cellular processes, including protection of nascent proteins on the ribosome (1); folding of newly synthesized or newly membrane-translocated proteins (2); protection and refolding of misfolded, partially unfolded, or aggregated proteins (such as after heat shock (3); and maintenance of proteins in a partially unfolded state (4) for binding effectors or crossing membranes. Other roles are discussed further in the text.

eukaryotic cells have to be modified before leaving the endoplasmic reticulum (ER), and molecular chaperones play an important role here in quality control, ensuring that only those proteins with the correct modifications proceed further down the pathway. Finally, proteins are not particularly stable entities, and hence have a propensity to unfold under certain nonoptimal conditions such as a small increase in temperature. Unfolding of proteins reintroduces all the potential problems of misfolding and aggregation that exist when the proteins are first synthesized.

Molecular chaperones are proteins that have evolved to deal with all these inherent barriers to the cell's need to have the right proteins in the right place in their active conformations. As later sections will describe, they have been shown to participate in all the examples given above, and in their absence many of these essential cellular functions cannot take place at all. Some chaperones appear to act predominantly by binding to and preventing the aggregation of proteins that have become unfolded; others are able to actively promote folding of proteins that otherwise cannot fold; and yet others can act on aggregated proteins to unfold them in order to allow them another opportunity to refold.

3
Classification of Molecular Chaperones

Molecular chaperones can be described in terms of their cellular locations, their mode of expression, their mode of action, and the proteins or other molecules upon which they act, but they are best classified in terms of sequence similarity. Chaperones that belong to a particular group defined thus turn out, unsurprisingly, to have similar structures and act on similar substrates through related mechanisms. The mechanisms of action and structures of the members of different groups are often quite distinct. The degree of sequence similarity within a group varies between different groups: some molecular chaperones are among the most highly conserved of all proteins, whereas others show very high heterogeneity between members of the same group. It must be remembered that sequence similarity alone is only an indicator of potential molecular chaperone activity, not proof of it, and that direct experimental evidence should always be sought to confirm whether or not a given protein is indeed acting as a chaperone.

The classification of molecular chaperones is complicated by a number of factors, some of which are as follows:

- First, certain proteins have been discovered that do not themselves act directly on protein substrates, but are required for the full chaperone reaction to take place. These are generally referred to as "cochaperones," and as they are central to the chaperoning process they are usually discussed in the same context as molecular chaperones.
- Second, the fact that often proteins from the same chaperone family have been discovered and studied in different organisms or several different cellular locations at roughly the same time has led to a proliferation of names for proteins within a given family, which can be confusing.
- Third, many chaperones are heat-inducible, and as they were often first discovered by virtue of this property, they are referred to as heat shock proteins (HSPs). Unfortunately, this means many chaperone families are labeled as HSPs, even though not all the proteins within that family may be heat shock–inducible. An alternative approach is to use the names of the best studied member to label the group, but this is equally unsatisfactory as the name given to this protein may refer to

the phenotype through which the protein was first discovered, and which may be quite specific to a particular organism.
- Fourth, the name "molecular chaperone" is used in a very catholic sense to refer to any protein with the properties listed in Sect. 1 above, but these may range from extremely broad-spectrum chaperones, which act on a large range of substrates to those that have evolved to act with only one specific substrate.
- Fifth, many proteins are unfortunately described as molecular chaperones with only minimal supporting evidence from *in vitro* and *in vivo* studies.

The Table 1 is not intended to be comprehensive, but illustrates the major known families of molecular chaperones, together with the names of some of their better-studied members, their cellular locations, and a very brief description of their properties. The individual molecular chaperone families shown in the Table are discussed more fully in the following sections.

4
Mechanisms and Roles of the Major Molecular Chaperones

4.1
The Hsp60 Family

4.1.1 Introduction to the Hsp60 Family

The Hsp60 family is one of the best characterized of the molecular chaperone families. It is a large family consisting of highly similar proteins, all with a subunit molecular mass of around 60 kDa. Sequence comparisons have enabled the division of the Hsp60 family into two groups: Group I and Group II. The Group I proteins are found in nearly all bacteria, and also in mitochondria and chloroplasts. The Group II proteins are found in archaea, and in the eukaryotic cytosol. All these proteins have a remarkable multimeric structure: Group I proteins assemble into a "double doughnut" structure with seven subunits in each ring, and the Group II proteins assemble into a similar structure but with eight or nine subunits in each ring, depending on the particular protein and organism. Each ring encloses a cavity or cage. Proteins in the Hsp60 family are often referred to generically as "chaperonins."

4.1.2 Cellular Roles of the Hsp60 Family

The Group I family of the Hsp60 proteins has been the most intensively studied, particularly the protein from *Escherichia coli*, which is referred to as GroEL. Genetic evidence shows that these proteins are essential, both in prokaryotes and eukaryotes, and they are found in all organisms so far studied, with the exception of some mycoplasmas. They are present at fairly high levels in cells (*E. coli* is estimated to contain around 2000 GroEL complexes per cell under normal growth conditions) and are further induced by heat shock. All of them appear to require a cochaperone, called Hsp10, which is a smaller protein of typically around 10 kDa, which itself assembles into a ring with sevenfold symmetry. In *E. coli*, this protein is called GroES, and in *E. coli* and most other bacteria it is encoded in the same operon as the GroEL protein and is also essential.

If the levels of GroEL become too low to sustain cellular growth, or if a temperature-sensitive mutant is shifted to the nonpermissive temperature, a large subset of cellular proteins fails to fold

Tab. 1 Selected examples of the major chaperone families in the cell.

Name of family	Prokaryotic examples	Eukaryotic examples	Cellular location	Known cochaperones	Cellular functions
Hsp60	GroEL		Cytosol	GroES	Folding of a subset of newly synthesized proteins, folding of a subset of proteins damaged by heat shock and other stresses
	CCT (archaea)		Cytosol	GimC	Unknown; induced by heat shock
		Hsp60	Mitochondria	Hsp10	Import of proteins synthesized in cytosol, folding of proteins, protection against heat shock
		RuBP (rubisco subunit binding protein)	Chloroplasts	Cpn21	Import and folding of proteins synthesized in cytosol, and folding of proteins synthesized within chloroplasts, particularly the large subunit of rubisco
		CCT	Cytosol	GimC	Folding of actin and tubulin, folding of other cytosolic proteins
Hsp70 with Hsp40	DnaK/DnaJ		Cytosol	GrpE	Binding to nascent polypeptides, protection of protein from aggregation at high temperatures, dissolution of aggregates (with ClpB), regulation of heat shock response in E. coli
		Hsp70/Hsp40	Cytosol	Fes1, Bag1	Multiple proteins and multiple roles including binding of nascent proteins, protein import into organelles, uncoating of clathrin cages
		BiP/Sec63	ER	SLS1, BAP	Binding of incompletely folded secretory proteins, facilitating retrotranslocation to cytosol for proteasome degradation; note: several different proteins with J-domains interact with BiP
		mtHsp70/Mdj1	Mitochondria	Mt-GrpE	Import and folding of proteins, heat shock protection

(continued overleaf)

Tab. 1 (continued)

Name of family	Prokaryotic examples	Eukaryotic examples	Cellular location	Known cochaperones	Cellular functions
Hsp90	HtpG		Cytosol		Unknown, heat shock induced, but dispensable
		Hsp90	Cytosol	Hop, FKBP51, p23	Binding to proteins that require activation with ligand such as steroid hormone, protection against heat shock? Also buffers against genetic variation.
		Grp94	ER		Unknown, probably a role in quality control of protein transport
		TRAP1	Mitochondria		Unknown
Hsp100	ClpB		Cytosol		Works with DnaK to dissipate protein aggregates
		Hsp104	Cytosol		Works with Hsp70 to dissipate protein aggregates
sHSPs	IbpA, IbpB		Cytosol		Probably act to hold denatured proteins for other chaperones to refold
		Hsp25	Cytosol		Probably act to hold denatured proteins for other chaperones to refold
		α-B-crystallin	Eye lens		Keep eye lens proteins in solution
Redox chaperones	DsbA/DsbC		Periplasm		Respectively introduce and isomerize disulfide bonds; require DsbB and DsbD respectively to regenerate active forms
	Hsp33		Cytosol		Activated by oxidation, may protect proteins damaged by oxidation in cytosol
		Protein disulfide isomerase	ER		Introduce and isomerize disulfide bonds; requires Ero1 to regenerate active form
ER chaperones		Calnexin, calreticulin	ER		Act to prevent incorrectly glycosylated proteins from entering the secretory pathway

correctly. Coimmunoprecipitation experiments with antibodies against GroEL have also shown that a large number of different proteins (about 10–15% of all proteins in E. coli) bind to GroEL shortly after their synthesis is complete. It is thus thought that GroEL, together with GroES, acts to promote the correct folding of a subset of the proteins present in the bacterial cytosol that would otherwise not fold correctly. After heat shock, an increased association of proteins with GroEL is seen, and this implies that GroEL also acts to help refold proteins, which become wholly or partly denatured by an increase in temperature. The Hsp60 proteins present in mitochondria and chloroplasts are also involved in promoting the folding of some of the proteins that have been newly imported into the organelle after synthesis in the cytosol as well as some of those that are synthesized within the organelle. It was the recognition (in the late 1980s) of the unexpectedly high sequence homology between the chloroplast Hsp60 protein (the major function of which is the folding of the large subunit of rubisco, encoded by the chloroplast genome) and the E. coli GroEL protein (identified originally as a protein required for the folding of a component of bacteriophage lambda), which helped to galvanize research into the role and mechanism of action of these proteins.

The role of the Group II proteins is less clear. In archaea, the proteins are heat shock inducible, but not in eukaryotes. Archaea encode one, two, or three proteins in the Group II family, and these assemble into large complexes of two rings with eight- or ninefold symmetry. Eukaryotes encode eight proteins in the Group II family, which assemble into a doubling-ring complex with eight-fold symmetry. All eight genes are essential in yeast. The protein complex has been referred to variously as CCT, TRiC, and TCP-1. The principal role of this complex in eukaryotes appears to be the folding of the cytoskeletal proteins actin and tubulin, and genetic evidence shows that when the function of the CCT complex is compromised with conditional mutants, cells show disorganized actin, defects in morphogenesis and cell division, and acute sensitivity to microtubule-depolymerizing compounds. Although many other proteins are also substrates for the complex, the complete range of cellular substrates has yet to be defined. Moreover, actin and tubulin are not found in archaea, and the substrates for the archaeal Hsp60 proteins are completely unknown. The Group II Hsp60 proteins do not require a cofactor homologous to the Hsp10 protein used by the Group I proteins, but substrates are delivered to them by a totally unrelated chaperone called GimC or prefoldin, which again is found in both archaea and eukaryotes but not in bacteria.

4.1.3 Mechanisms of Action of the Hsp60 Family

The mechanism of action of the Group I proteins, as typified by the E. coli GroEL protein, has been very intensively studied, and yet there remain considerable uncertainties about exactly how the protein works. In vitro experiments show that if certain proteins are denatured (by chemical or heat treatment) and are then allowed to refold, the yield of refolded protein is very low, but can be increased to nearly 100% by incubation with GroEL, GroES, and ATP (Fig. 2). This process requires a complex reaction cycle, which is discussed more fully in the next section. During the reaction cycle, the unfolded protein is transiently bound in the cavity formed on one ring by the capping of the ring by the GroES protein. There are

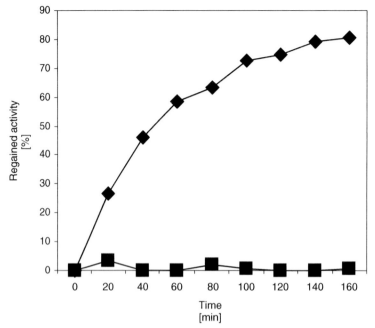

Fig. 2 Refolding of a denatured protein *in vitro* by GroEL, GroES, and ATP. In the example shown here, a substrate protein has been denatured in buffer using heat, and then returned to the temperature at which it is normally active. At suitable time points, the protein is assayed for regain of enzyme activity. The square symbols show the effect in buffer alone, and demonstrate that this particular substrate cannot effectively renature under the conditions shown. The diamond symbols show the effect of having GroEL, GroES, and ATP present in the buffer, with the GroEL and GroES at the same molar concentration as the substrate protein. As can be seen, this addition leads to restoration of roughly 80% of the initial activity of the protein.

at least three different models to explain how this process helps the protein to fold, and these are not mutually exclusive. The first (referred to sometimes as the "Anfinsen cage" model) proposes that the enclosed cavity simply provides a passive and protected environment where the protein can fold without interacting with other folding proteins with which it might otherwise aggregate. A refinement of this model (the "assisted folding" model) suggests that the environment of the cage is important, as the interior surface of the cage is rich in hydrophilic residues that tend to favor the burial of hydrophobic residues in the encapsulated substrate protein, thus accelerating folding. A third model (the "iterative annealing" model) suggests a very different role for GroEL. In this model, the principal role of GroEL is to *unfold* proteins that have become trapped in nonproductive misfolded states, and to redeliver them to the start of the protein folding pathway. It is quite possible that elements of all three models are correct, and that the extent to which any one applies varies with different substrate proteins. Further refinements to these models may be required following recent demonstrations that GroEL can help fold some proteins that are too large to fit into the cavity of a single ring.

The mechanism of action of the Group II proteins is much less well understood. Actin and tubulin have both been shown to interact with specific subunits of the CCT complex (which are always present in the same order in the complex), and there is evidence that the substrates remain bound throughout the folding cycle, unlike the case in GroEL in which they are discharged into the cavity. It may be that the large domain movements of the complex that take place during the folding cycle act to force the bound unfolded protein into a more compact folded form.

4.1.4 Structure and Function of the Hsp60 Family

It is the remarkable structure of the Hsp60 proteins that has been in part responsible for the degree of interest that they have raised. Members of both the Group I and Group II families have been studied by different structural techniques, notably X-ray crystallography and cryoelectron microscopy, and more recently in the case of GroEL by NMR. Hsp60 proteins assemble into double-ring structures with several subunits in each ring; in the case of CCT from eukaryotes, the relative position of each of these subunits is unique, so that each always has the same neighbors. The individual subunits of proteins from both families consist of three distinct domains: an equatorial domain, which is responsible for the contacts between the two rings and for ATP-binding; a flexible intermediate domain; and an apical domain, which contains the region in which protein binding initially occurs. Moreover, both structures undergo large conformational changes as they pass through the different stages of the protein folding cycle. Again, it is the GroEL protein that is best understood.

The sequence of events that is thought to take place during the folding cycle of the GroEL protein is depicted in Fig. 3. Partially folded protein binds to one ring via a series of hydrophobic contacts with residues at the top of the apical domain. GroES binds to some of the same residues, and when it binds (which requires the presence of bound nucleotide) it displaces the bound peptide into the center of the cavity and simultaneously causes a very large conformational change in the ring to which it binds, the effect of which is to nearly double the size of the cavity. This structure slowly turns over the ATP, and until this reaction is complete, no protein or nucleotide can bind to the opposite ring. Once the ATP has been turned over to ADP, unfolded protein can bind to the opposite ring, followed by ATP and GroES, the net effect of which is to displace the bound substrate, GroES, and ADP, from the opposite ring to which binding first occurred. Thus, both rings are active in binding unfolded proteins, but do not do so at the same time as each other.

The models for action of GroEL discussed above can now be seen in the light of this mechanistic cycle. The Anfinsen cage and assisted folding models predict that the key event in the cycle is the encapsidation of the unfolded protein, which may lead to either the passive or active involvement of the cavity in promoting protein folding. The iterated annealing model sees the main event as being the large conformational change that takes place in the apical domains of GroEL as they move to bind GroES. It is proposed that these act to pull apart misfolded protein, before releasing it again at the start of the protein folding pathway.

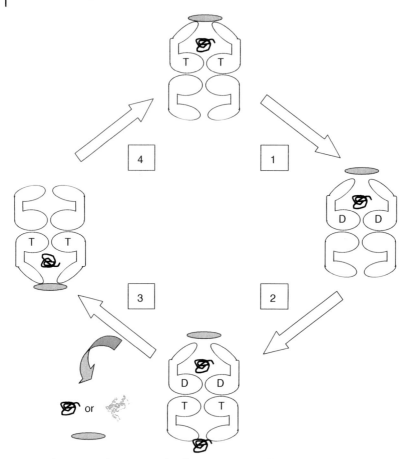

Fig. 3 The proposed reaction cycle of GroEL. GroEL is shown in cross-section, and GroES is shown as a filled ellipse. The complex at the apex of the cycle consists of substrate protein encapsidated in the top (cis) ring of the GroEL complex, which has the ATP (shown as T) bound, and is capped by GroES. The first reaction is the slow hydrolysis of ATP to ADP (reaction 1), which has the effect of weakening the GroES-GroEL interaction. During the time that this reaction occurs, folding of the protein may take place within the cavity. In reaction 2, ATP and another molecule of substrate protein bind to the bottom (trans) ring. This binding results in the release of GroES and ADP from the cis ring (reaction 3), and also the release of the bound substrate, which may now be either folded or still in a partially folded state. The new complex is now essentially the same as the starting one, as can be seen by flipping the complex (step 4) except that the former cis ring is the new trans ring, and vice versa.

The Group II proteins lack a functional GroES cochaperone, but in its place they possess a large helical protrusion at the top of the apical domain that forms a lid to the complex in a way similar to that formed by GroES. It is clear that large domain movements occur in the ATP-binding and hydrolysis cycle with the Group II proteins, but their relationship to the ability of this protein complex to promote the folding of proteins is not yet understood.

4.2 The Hsp70 Family

4.2.1 Introduction to the Hsp70 Family

The Hsp70 proteins comprise a large family, often with many representatives in a given organism (the yeast *Saccharomyces cerevisiae* has 14 Hsp70 homologs). They are found in all eubacteria, all eukaryotes, and most archaea. The Hsp70 proteins are all approximately 70 kDa in size, and show a high degree of sequence similarity. They act in concert with another family of chaperones, the Hsp40 proteins. These are more heterogeneous in size, and include both soluble and integral membrane proteins. All of the Hsp40 family members possess a particular domain (referred to as a J domain, named after DnaJ which is one of the *E. coli* homologs of this protein), which is key to their activity. In eukaryotic cells, both proteins are found in many different cellular environments, including the cytosol, nucleus, mitochondria, chloroplasts, and the lumen of the endoplasmic reticulum. They have a multiplicity of roles, all of which are related to their ability to bind to short stretches of hydrophobic residues in unfolded proteins. As with the Hsp60 proteins, their reaction cycle is driven by the binding and hydrolysis of ATP, but their mode of action appears to be very different. They utilize a wide variety of cochaperones and other factors, and are integrated into a network of interactions with other chaperones in the cell, as described below. Although originally described as heat shock proteins, some members of the family are synthesized constitutively and show no heat shock induction.

4.2.2 Cellular Roles of the Hsp70 Family

The Hsp70 proteins, together with the Hsp40 proteins, fulfill many functions inside the cell, as expected from the large number of homologs that can exist for these proteins in even a single organism. It is clear that not all of these functions have yet been defined. Some of the cytosolic members of the family are involved in binding nascent polypeptides as they emerge from the ribosome, probably thereby preventing them from folding prematurely before the whole chain has been synthesized. Other roles in the cytosol include the uncoating of clathrin from clathrin-coated vesicles, and the delivery of proteins to the mitochondrial and chloroplast import complexes. In the endoplasmic reticulum, the Hsp70 homolog Bip has a key role in ensuring that only correctly folded proteins are delivered to the secretory pathway, and that incorrectly folded proteins are returned to the cytosol for destruction by the proteasome. In the mitochondria and chloroplasts, the organellar homologs of the Hsp70 and Hsp40 proteins have a key role in the uptake of proteins that are synthesized in the cytosol and targeted to organelles, as well as in assisting the folding of proteins encoded by the organellar genomes.

That these roles vary in their importance can be seen from genetic studies. The main Hsp70 homolog in *E. coli*, DnaK, can be deleted, and although the resultant cells show reduced viability and no ability to grow at high or low temperatures, they are capable of growth at normal temperatures. The loss of viability is in part due to the fact that DnaK has a role in regulation of the heat shock response, and if the same deletion is made in *Bacillus subtilis*, an organism where DnaK does not have this role, the effect on viability is reduced. Two interpretations of these data are possible: one is that DnaK fulfills roles of relatively minor importance; the other is that the roles of DnaK are so important that other proteins have evolved to back them up should DnaK become nonfunctional. At least one of the roles for DnaK, namely, that of binding proteins as they emerge from the ribosome, has been shown to fall into the latter category. Another chaperone, called trigger factor, also has this role, and although loss of either trigger factor or DnaK alone is not lethal, a double mutant is not viable under normal growth conditions (this is referred to as *synthetic lethality*). In yeast, some of the Hsp70 homologs are essential and others are not, and again synthetic lethality is seen for some combinations of mutants in different *hsp70* genes, proving that some of the cellular functions carried out by Hsp70 are essential. Trigger factor does not appear to have a homolog in eukaryotic cells, but another protein called NAC (nascent-polypeptide association complex) is thought to fulfill a similar role.

Another important role that has been demonstrated in *E. coli* and *S. cerevisiae* is not only the prevention of protein aggregation at heat shock temperatures but also the breaking down of any aggregates that may occur. For this activity, another chaperone is required: this is the protein referred to as ClpB in *E. coli* and Hsp104 in *S. cerevisiae*. The structure and precise mechanism of action of this chaperone is not known, but it clearly interacts specifically with Hsp70 in mediating protein disaggregation and subsequent refolding. This interaction is specific: yeast Hsp104 cannot substitute for *E. coli* ClpB. The role of ClpB will be considered further below (Sect. 5.2.1).

Hsp40 proteins share many properties with Hsp70 proteins. In all cases in which an Hsp70 protein is known to act, an Hsp40 partner protein is also involved.

4.2.3 Mechanisms of Action of the Hsp70 Family

Both Hsp70 and Hsp40 act by binding to short, exposed regions of predominantly hydrophobic stretches in unfolded proteins. Proteins bound to Hsp40 or Hsp70 do not appear to undergo any further folding, unlike the situation with the Hsp60 proteins, where the protein released from the complex may be in a very different conformation to the one that was first bound by it. Rather, it appears that the transient binding of these regions is enough to reduce the risk of their aggregation to other hydrophobic patches on proteins in the neighborhood. The importance of this is particularly obvious in those cases in which proteins have to be prevented from folding until the entire protein sequence is available to fold, namely, proteins being synthesized on ribosomes, and proteins being transported across membranes. Similarly, in the case of protein folding in the ER, proteins that still have regions exposed and are hence not ready for transport to the Golgi will bind to the Hsp70 homolog BiP. This protein has an ER retention signal at its C-terminus, and consequently is constantly cycled back to the ER from the early steps of the secretory pathway, together with any unfolded proteins that are bound to it.

Although the nature and role of binding of unfolded protein to the Hsp70 and Hsp40 proteins are different from those pertaining to the Hsp60 proteins, the reaction cycle has some similarities. In particular, as with the Hsp60 proteins, the reaction cycle relies upon the chaperones existing in two states, one with a high affinity for unfolded protein and the other with a low affinity, with the binding of nucleotide mediating the transition between those states. The reaction cycle, shown in Fig. 4 for the case of the *E.*

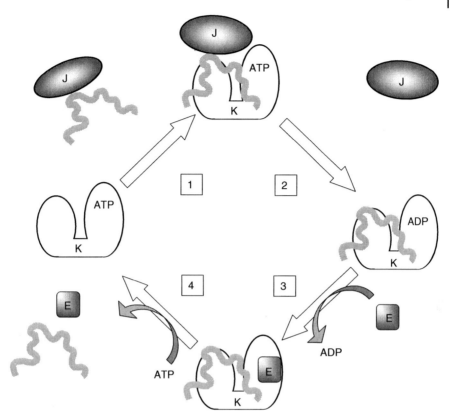

Fig. 4 The proposed DnaK/DnaJ/GrpE reaction cycle. Substrate protein is initially bound to DnaJ (shown as J), which delivers it to DnaK-ATP (step 1). The substrate is transferred to DnaK, and the presence of DnaJ stimulates the ATPase of DnaK to form a stable DnaK-ADP-substrate complex; DnaJ is probably lost at this point (step 2). This complex is stable until the cochaperone GrpE catalyzes the exchange of ADP for ATP, shown here as a two-step reaction where GrpE initially displaces bound ADP (step 3) and is in turn displaced by ATP (step 4). This causes DnaK to release the bound peptide, and the cycle can begin again. It is important to note that several molecules of DnaJ or DnaK may bind one protein. The reaction cycle is best characterized for the *E. coli* proteins shown; in eukaryotes, many different cofactors are involved with cycles involving Hsp70 homologs.

coli proteins DnaK and DnaJ, is thought to begin with the binding of unfolded protein to the Hsp40 partner, from where it is transferred to Hsp70, which is initially in a low-affinity binding state (with ATP bound). The formation of the Hsp70-Hsp40-unfolded protein complex stimulates the Hsp70 ATPase, and the ATP is hydrolyzed to ADP, which in turn has the effect of increasing the affinity of Hsp70 for the unfolded protein, but weakening it for Hsp40, which then dissociates. The resultant Hsp70-ADP-unfolded protein complex is now quite stable. In order for the unfolded protein to be discharged from its bound state, the ADP must be exchanged for ATP to lower the Hsp70 affinity for the unfolded protein. This exchange is a slow reaction but it is speeded up considerably by a cochaperone, termed GrpE in E. coli, which catalyzes the exchange reaction and allows the cycle to be completed with the release of bound protein and the regeneration of the Hsp70-ATP form. GrpE homologs are found in archaea, mitochondria, and chloroplasts, while other proteins that catalyze the same kind of nucleotide exchange, such as Bag-1, are found in eukaryotes.

4.2.4 Structure and Function of the Hsp70/Hsp40 Family

Hsp70 is a two-domain protein, with one domain containing the ATP binding site and the other domain containing the peptide binding site. To date, no structure for the whole protein is available, but structures of the individual domains are known. The peptide binding domain has been crystallized with a bound peptide in place, and this structure shows that the peptide binds to a hydrophobic groove in the chaperone. The bound peptide is held in place by an α-helix, the position of which is thought to change depending on the nucleotide-binding state of the other domain, thus explaining how Hsp70 can change between a low-affinity and a high-affinity state depending on the presence of ATP. Hsp40 proteins can also bind peptides with high affinity, and structural studies on the conserved J domain shared by all these proteins show that again this binding is to a hydrophobic groove. The features that determine affinity of peptides for the grooves on these two proteins are essentially the same – consistent with the model that Hsp40 binds peptides (or, in the cell, stretches of unfolded protein with the appropriate amino acids present) first and then transfers them to Hsp70. The complete structural details of the complex formed by these two proteins are not known, however, so the fine details of this mechanism are not yet understood.

4.3 The Hsp90 Family

4.3.1 Introduction to the Hsp90 Family

The two chaperone families we have looked at so far have benefited from extensive study using simple prokaryotic systems, and many important features of the chaperones are conserved in both prokaryotic and eukaryotic organisms. The situation is somewhat different with the Hsp90 family, a family of well-conserved proteins of approximately 90 kDa, in that studies on prokaryotes show that loss of the *hsp90* gene does not in general produce a strong phenotype, whereas in eukaryotes the gene is essential. In eukaryotes, there are several genes encoding members of this family: the minimum number in complex organisms seems to be two very closely related forms in the cytosol, one in mitochondria, and one in the ER, although in *S. cerevisiae* the ER homolog is not present. Hsp90 proteins are very abundant, and are further induced by heat shock, and they have a rather broad range of cellular functions. Many of these are related to the ability of Hsp90 to hold proteins in a state that is not completely folded or active but which is competent to become so on the reception of a suitable signal.

4.3.2 Cellular Roles of the Hsp90 Family

Many substrates for eukaryotic Hsp90 proteins have been identified, and of these a large proportion are involved in some form of signaling. The best characterized of these are the steroid hormone receptors (SHRs), which are proteins that, when bound to their substrate (steroid hormones), enter the nucleus and stimulate the transcription of specific genes. Hsp90 holds these proteins in an inactive form until the hormone is available, whereupon it binds to the SHR, which then dissociates from Hsp90, dimerizes, and enters the nucleus. Other proteins, such as the eIF-2α kinase, are bound by Hsp90 and released to become fully folded only after phosphorylation. The fact that Hsp90 is also a heat shock protein suggests that it may also have an important role in protecting cells, and it has been shown *in vitro* that it can bind to and prevent the aggregation of proteins under thermal stress.

The high levels and essential nature of cytosolic Hsp90 point to some other specialized function, and a clue as to what this might be has come from recent remarkable experiments that have shown that Hsp90 may be able to buffer potential phenotypic variation that is present, but not expressed, in all cells. When levels of Hsp90 activity are intentionally lowered, either genetically or by various external treatments, it is found in both plant and animal models that the degree of phenotypic variability seen is increased. Different alterations in phenotype are seen even between organisms that are genetically identical. Thus, it appears that Hsp90 may act to reduce "noise" that is present when organisms develop due to chance fluctuations in environmental conditions or small genetic differences, and this noise can be revealed when Hsp90 activity is compromised. It has even been speculated that this noise suppression may be a mechanism that has evolved to increase the amount of phenotypic variation present and expressed in a population at times of stress, where rapid evolution may be important for survival.

4.3.3 Mechanisms of Action of the Hsp90 Family

The way in which Hsp90 proteins bind to their substrates, and subsequently release them under the control of a suitable signal, is not well understood. What is clear is that the reaction mechanism in the cell is very distinct from those considered above, in that many more cellular cochaperones are involved. The folding cycle for SHRs has been particularly closely studied, and appears to begin with the partially folded SHR binding to Hsp70. Hsp70 can itself associate with a number of different cofactors *in vivo*, and one of these, referred to as Hop or Sti1 depending on the organism, mediates the binding of Hsp70 to Hsp90. The SHR is then transferred to Hsp90 and the Hsp70 and Hop components leave the complex. The SHR is at this stage still not competent to bind steroid hormone. Before it can become so, at least two other components associate with the Hsp90: a protein called p23, and one of a number of proteins that are peptidyl–prolyl cis–trans isomerases (PPIases). Both of these proteins have chaperone activities of their own, and the combined effect of all these proteins coming together is to alter the conformation of the bound SHR such that it can now bind steroid hormone. These reactions are shown in outline in Fig. 5.

A good deal of attention has focused on these latter stages in the binding cycle, as there was for some time a controversy about the role of ATP in the action of Hsp90. Hsp90 tested *in vitro* does not appear to require either binding or hydrolysis of ATP in order to act as a chaperone, but nevertheless contains an ATP binding site, deletion of which is lethal *in vivo*. It now appears that Hsp90 cannot bind to the PPIases or p23 until ATP is bound. As Hsp90 has a potential role in the cell cycle, promoting the folding of some of the kinases needed for cell cycle control, its inhibition in rapidly proliferating cells has been seen as a potential anticancer strategy. Some drugs

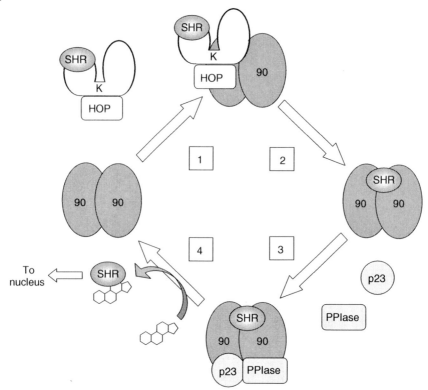

Fig. 5 Outline of the reaction cycle for Hsp90 binding to steroid hormone receptor (SHR). The SHR is delivered to the Hsp90 dimer by Hsp70 in a complex with Hop protein (step 1). SHR is transferred to Hsp90 (step 2), and p23 and PPIase associate (step 3). The precise order of reactions and the stoichiometry of the different proteins are uncertain. SHR can now bind to steroid hormone, which causes it to dissociate from Hsp90 and enter the nucleus to activate gene transcription (step 4).

that have been shown to inhibit Hsp90, in particular, the compound geldanamycin, appear to act as competitive inhibitors of ATP-binding, blocking the access of ATP and hence blocking the later stages in the Hsp90 reaction cycle.

4.3.4 Structure and Function of the Hsp90 Family

The structure of the complete Hsp90 protein is not known, but the structure of the N-terminal domain, which is the main site for binding of ATP, has been solved. As with other chaperones there is evidence of large conformational changes occurring on ATP-binding, but the structural details of these are unknown. Hsp90 is active as a dimer, with the dimerization domain in the C-terminal part of the protein, while the N-terminal regions may have a clamp activity, perhaps with the binding of a nucleotide causing the opening and closing of the clamp. Remarkably, it appears that there are likely to be peptide binding sites and ATP binding sites in both the N- and the C-terminal regions of the protein, but the functional significance of this is not understood.

4.4 The Small Heat Shock Proteins

4.4.1 Introduction to the Small Heat Shock Proteins

The small heat shock proteins (sHSPs) are a much more diverse group of chaperones than those considered above, and are correspondingly less characterized. They range in size from 15 kDa to greater than 40 kDa, and the degree of sequence similarity between them is lower than that for other chaperone families. All possess a conserved domain of approximately 90 amino-acyl residues near the C-terminal end, and all assemble into large oligomers, although again the sizes of these can vary considerably, even for a single sHSP. Organisms often contain numerous genes for sHSPs (plants are particularly abundant in them) and in eukaryotes they have been found in the cytosol, the ER lumen, and in mitochondria and chloroplasts.

4.4.2 Cellular Roles of sHSPs

The precise cellular role of the sHSPs has proven hard to pin down. Strains of *E. coli* or *S. cerevisiae* with all their sHSP genes deleted show very little significant phenotype unless the deletions are combined with other mutations, although in some organisms there is a reduction in thermotolerance when sHSPs are deleted. This lack of phenotype is a surprise, given that the sHSPs are often among the most strongly induced genes by heat shock. Overexpression of sHSPs has been shown to enhance thermotolerance in some cases, which does support the hypothesis that sHSPs have a role to play in survival of temperature stress. They are often found associated with protein aggregates, which have arisen either from the overexpression of proteins that failed to fold correctly, through heat stress or through disease states. There is also evidence that they may play a role in aspects of cytoskeletal and intermediate filament assembly.

A particularly interesting example of the sHSPs is the α-crystallin group of proteins, which are very abundant in the eye lens. Given that the eye lens is effectively a concentrated protein solution, which has to be kept clear for the lifetime of the organism (as cells laid down in fetal development remain in the center of the lens), the avoidance of protein aggregation is of particular importance. Loss of acuity of vision with age, and cataract formation, in particular, are associated with a breakdown in the ability of the α-crystallins to act effectively as chaperones. But defects in α-crystallin function have other consequences as well, showing that its importance is not limited to the eye lens. For example, the inherited disease desmin-related myopathy, which leads to muscle weakness, has been found in some cases to be associated with mutations in one of the α-crystallins, and probably results from defects in intermediate filament assembly.

4.4.3 Mechanism of Action of sHSPs

Despite the lack of a clear *in vivo* role, sHSPs have been demonstrated to have chaperone activity *in vitro*, in that they are able to bind and stabilize unfolded proteins and prevent their aggregation. The range of proteins that sHSPs can bind is very large, with no apparent limits on the size of the protein. There is no clear evidence that sHSPs are able to assist in folding *per se*, and the current view is that they act predominantly as reservoirs of unfolded protein, holding it in a nonaggregated state until conditions more favorable for folding are restored. Refolding of the bound proteins may occur either after the proteins are released into free solution, or transferred to another chaperone system, such as the Hsp70/Hsp40 system. This is discussed further below, in Sect. 5.2.1. ATP is not required for any of this to occur, further supporting the idea that sHSPs are "holders" rather than "folders." The way in which protein is discharged is not clear, but it may be related to changes in oligomerization and hydrophobicity that occur at different temperatures.

4.4.4 Structure and Function of sHSPs

The sHSPs form oligomers of variable size with fewer than 10 to more than 50 subunits. Moreover, when complexed with bound substrate proteins, their structure and degree of oligomerization may be significantly different from when they have no substrate bound. Several structures have been obtained of free sHSP, and all show oligomers formed roughly into a hollow sphere. The binding sites for unfolded proteins on the sHSPs have not been determined, and there is no detailed structural information for an sHSP-substrate complex.

5
Chaperones and Cellular Processes

A complete understanding of the role of chaperones in the cell is still some way off, but recent work has improved our understanding not only of the individual families of chaperones but also of how they interact with one another and in what way their activities are required for the survival and growth of healthy cells, both under normal and under stressed conditions. The following sections will examine some examples of chaperones and chaperone networks, which are employed for various different processes in the cell. This is not by any means a comprehensive list, but should give at least a flavor of the ways in which chaperones are involved in many different aspects of cell growth.

5.1
Chaperones and Redox Potential

Chaperones have a key role in enabling the cell to maintain active proteins under a variety of redox states, in at least two ways. First, they are directly involved in ensuring that proteins that require disulfide bonds for their activity form those bonds quickly and efficiently. Proteins that catalyze the reactions that make or break disulfide bonds in proteins are very important in maximizing the yield of active folded protein in cells, and many also can act as chaperones on substrates that lack cysteine residues. Second, chaperones exist that are activated by anomalous redox states to help chaperone the folding of those proteins that may themselves have become nonfunctional due to exposure to these states.

5.1.1 Disulfide Bond Formation and Isomerization in Prokaryotes and Eukaryotes

The formation of disulfide bonds in prokaryotes takes place in normal circumstances in the periplasm, the space between the inner and outer membranes in Gram-negative bacteria. (Gram-positive bacteria that have only a single membrane produce very few disulfide-bonded proteins, although interestingly they do express proteins capable of catalyzing disulfide bond formation). There is a significant difference between the redox states of the cytosol (which is highly reducing) and the periplasm (which is oxidizing), and it was originally thought that this difference was sufficient to allow disulfide bond formation to occur at a reasonable rate.

However, proteins have been identified in the periplasm of *E. coli* that catalyze both the formation and removal of disulfide bonds and thus speed up the rate at which proteins that contain such bonds are folded.

The two major proteins in *E. coli* involved in this process are both periplasmic proteins and are called DsbA and DsbC (Fig. 6). DsbA is a powerful oxidant, catalyzing the formation of disulfide bonds between cysteine residues in proteins that have been newly secreted to the periplasm, and itself becoming reduced as a consequence. Although bacteria lacking DsbA are viable, they show enhanced turnover of periplasmic proteins (owing to the presence of proteases in the periplasm that degrade unfolded proteins), lack of motility (owing to the failure of flagella to assemble), and high sensitivity to reducing agents. DsbA apparently binds proteins in a hydrophobic groove, rather analogous to the mechanism seen in some of the

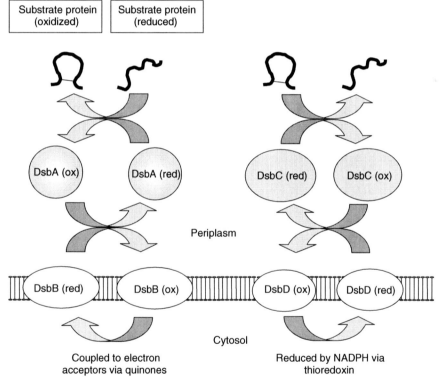

Fig. 6 Mode of action of some of the periplasmic Dsb proteins in *E. coli*. DsbA acts as an oxidase to introduce disulfide bonds into proteins with free thiols. In doing this, it is itself reduced, and must be reoxidized by oxidized DsbB, which in turn becomes reduced. DsbB becomes reoxidized by donating electrons to quinines, which flow to oxygen or other electron acceptors. DsbC in its reduced form acts either as a reductase (the net effect of which is to oxidize DsbC) or an isomerase (where no net redox reaction occurs). In the former capacity, it must be rereduced before regaining its activity; this is done by DsbD, which in turn derives its reducing power ultimately from cytosolic NADPH. Other Dsb proteins also exist in the periplasm of *E. coli* but have been omitted for clarity.

chaperones discussed above, but it is not significantly specific about which cysteines it oxidizes, so both correct and incorrect isomers of the same protein can be formed. The reduced DsbA must be reoxidized for further activity, a process that is mediated by a membrane protein called DsbB, which itself interacts directly with intermediates of the oxidative electron transport pathway to transfer the electrons liberated in disulfide bond formation to quinones and ultimately to either oxygen or, under anaerobic conditions, to other electron acceptors.

The formation of incorrect disulfide-bonded isomers by DsbA points to the need for an isomerase protein that can catalyze the breakage and reformation of disulfide bonds in proteins until the correct isomeric (and active) form is reached. In the *E. coli* periplasm, this role is performed by the protein DsbC. Like protein disulfide isomerase (PDI) (see below), this protein has an innate chaperone activity, in that it can recognize and bind unfolded proteins even when such proteins lack any cysteine residues. It seems likely that DsbC acts to bind incorrectly folded proteins, and then breaks the disulfides if they are present. Disulfides may then be reformed either by subsequent action of DsbA, or by spontaneous reoxidation of the reduced cysteines. In contrast to DsbA, therefore, DsbC must be maintained in a reduced state, and this is mediated by another membrane protein called DsbD. DsbD is in turn maintained in a reduced state by interacting with the cytosolic protein thioredoxin, which itself is kept reduced by metabolically generated NADPH.

The site for disulfide bond formation and isomerization in eukaryotes is the lumen of the endoplasmic reticulum. As is the case with bacteria, disulfide bond formation in newly imported proteins is an enzyme-catalyzed process, the enzyme in this case being protein disulfide isomerase. PDI is by far the most abundant chaperone in the ER. As with bacteria, a membrane protein (Ero1p) exists to restore the active (oxidized) state of PDI after the formation of disulfides has taken place. Reduced Ero1p in turn is reoxidized directly by oxygen, though its activity is also modulated by FAD. In addition to acting as an oxidase, PDI can shuffle the disulfide bonds in a protein after they have formed (hence its name), so it effectively combines the roles of DsbA and DsbC. As with DsbC, it is capable of showing chaperone activity, and so presumably can recognize proteins in which the disulfide bonds are not in the correct position for complete folding. The catalytic site and chaperone site are not the same and indeed are on different domains of the protein. This activity is more important in eukaryotic cells than in prokaryotic ones, as eukaryotic proteins have much larger numbers of disulfide bonds on average, and hence the possibility of forming incorrect disulfide bonds is correspondingly higher.

5.1.2 Hsp33: A Chaperone with a Redox Switch

An intriguing chaperone protein is found in *E. coli*, with homologs in many other species, which is heat shock induced but active only when oxidized. This protein, Hsp33 or HslO, binds very tightly to zinc ions when reduced, coordinating the zinc with four cysteine residues. Under relatively mild oxidizing conditions, the zinc is released and the cysteines form two disulfide bonds. This release causes a substantial change in the conformation of the protein, which significantly increases the amount of exposed hydrophobic area of the protein. In this state, Hsp33 acts as a "holder" rather like the sHSPs, and has been shown to be capable of protecting a number of substrates from aggregation, whether induced by heat shock or by chemical denaturation. However, it has not yet been possible to demonstrate refolding of the bound protein substrates *in vitro*, implying that *in vivo* Hsp33 may interact either with one of the known chaperone proteins, or with a cofactor which is yet to be identified. The likely function of Hsp33 is protection of proteins, which may be particularly susceptible to oxidative damage *in vivo*, and the rapid activation of the chaperone activity of Hsp33 that takes place when the protein is oxidized, makes it an excellent first line of defense against oxidative stress, which can act until the cell has had time to induce a range of other defenses that will tend to act by restoring the normal redox potential of the cytosol.

5.2 Chaperone Networks and Protein Quality Control in the Cell

As research into molecular chaperones has progressed, it has moved to embrace considerations not only of how the members of individual chaperone families function, but also how they interact with other chaperone families within the cell. Much emphasis is now placed on understanding these interactions or networks. One such network has already been discussed briefly above: the interaction of Hsp70 and Hsp90 proteins in the activation of steroid hormone receptors. Chaperone networks have evolved to enable cells to maximize the efficiency with which they produce folded and active proteins, a process that faces two particular hurdles: the inefficiency implicit in the folding process itself (i.e. the extent to which proteins may finish up becoming misfolded or aggregated), and the fact that folding can be reversed by stresses such as heat shock. These networks also have to be integrated with the cell's mechanisms for disposing of proteins that are not functional. In the next two sections, we will look at two of the best understood of these: the system of protein quality control in the *E. coli* cytosol, and the system in the ER of eukaryotic cells. Neither of these, it must be emphasized, is yet completely characterized.

5.2.1 Chaperone Networks and Protein Quality Control in the *E. coli* Cytosol

It has already been pointed out in the text above that DnaK (one of the Hsp70 protein homologs in *E. coli*) and the unrelated protein trigger factor cooperate in chaperoning proteins that are still in the process of being synthesized on the ribosome. This is an essential process, and the trigger factor and DnaK must significantly overlap in the roles that they play in it, as demonstrated by the fact that loss of both genes together is lethal to the cell, whereas loss of either gene individually is not. It is likely that in normal cells, they act sequentially, with the trigger factor binding to proteins as they first emerge from the ribosome, and DnaK interacting with a subset of these later during synthesis.

Other interactions may exist between the DnaK and the GroEL chaperone systems. *In vitro* studies using purified substrate proteins have demonstrated that proteins can be transferred between these two systems effectively in either direction. Detailed *in vivo* evidence for a network between these two proteins is lacking, however, except inasmuch as a few proteins have been identified by coimmunoprecipitation experiments as being substrates of both systems.

More direct evidence exists to link the sHSPs of *E. coli* with the Hsp70 and Hsp60 major chaperone families. *In vitro* experiments show that unfolded proteins can be held in a nonaggregated but nonactive state bound to the sHSPs, and subsequently transferred to the DnaK and GroEL families for folding. *In vivo* experiments also support the existence of such interactions, although these interactions cannot be essential for viability as even under fairly extreme conditions, loss of the sHSPs from *E. coli* does not cause a severe phenotype. It may be that, as was seen in the case of trigger factor and DnaK, the ability for the cell to "hold" transiently unfolded proteins in a refoldable form is so important that more than one protein can act as the "holder," and to date the other proteins that can do this have not been identified.

A third important interaction that exists between chaperones in *E. coli* is that between the Hsp70 protein DnaK and the Hsp100 protein, ClpB. Here, the evidence for an important interaction is much more compelling. It is clear both from *in vitro* and *in vivo* experiments that proteins aggregated by a stress such as heat shock can be disaggregated by the combined actions of DnaK and ClpB, and that the two proteins have very different roles in the process as they cannot substitute for each other. It is likely that ClpB acts upon protein aggregates first, altering their structure to a point where DnaK is able to promote their refolding. ClpB has thus been classified by some workers as an "unfolder," and it is not able to assist the refolding of proteins on its own. Its mode of action is not yet understood, but it is a member of a large family of proteins called the AAA proteins, many of which have a major role in protein quality control. This leads us into the complex area of the overlap between protein refolding and protein degradation.

E. coli (and all other organisms) contains a large number of different proteases whose job it is to remove inactive or damaged proteins from the cell. This process becomes particularly important at high temperatures. A large number of these proteases are, along with ClpB, members of the AAA family of proteins, and some have features reminiscent of molecular chaperones. For example, several of them have an ATP-dependent ability to unfold proteins, and exist as ring-shaped oligomers (usually hexamers). For most of these, unfolding occurs as an essential step prior to degradation of the unfolded protein, either by another site on the same polypeptide that mediated the unfolding, or by another protein that acts in concert with the "unfoldase." The unfoldase activity can, in some cases, give rise to a chaperone-like activity in *in vitro* experiments, although only in the case of ClpB does it appear to have a true role in helping proteins to refold *in vivo*. Degradation by these proteases is reduced by mutations in the DnaK and GroEL chaperone families, implying an interaction, indirect or direct, between the chaperone proteins and the proteases.

Thus, when proteins are unfolded in the *E. coli* cell, they may be captured by "holders" for subsequent refolding, or form aggregates that can be disaggregated by ClpB before refolding by DnaK (Fig. 7). In addition, they may be targeted to cellular proteases. Both these processes will help ensure that only active proteins are present in the cell, but the way the balance between them is achieved (i.e. what determines whether a given protein is degraded or refolded) is not known. Clearly, refolding is a preferential option to degradation if the cell is not to lose the investment in ATP that it has already made in producing the protein; equally clearly, refolding is not

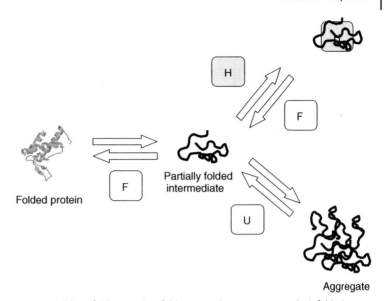

Fig. 7 Holders, folders, and unfolders in a chaperone network. A folded protein may become partially unfolded (for example, due to heat shock). It may refold spontaneously from this state, or may require the assistance of a "folder" chaperone (F), such as the GroEL/ES or DnaK/J systems. The partially unfolded state may interact further with other proteins in the same state to form an aggregate. This can be reversed by the action of an "unfolder" (such as ClpB). Alternatively, a "holder" chaperone (such as an sHSP) may hold the intermediate in a nonactive but nonaggregated state until a folder chaperone can refold it.

always an option, otherwise there would be no need for the cellular proteases to exist.

5.2.2 Protein Quality Control in the Endoplasmic Reticulum

The ER is of particular importance in the quality control of protein folding. It is the site for several important covalent modifications that are key to protein structure, including disulfide bond formation and glycosylation, and proteins pass through the ER en route to the plasma membrane and beyond, where they will be inaccessible to any further modification by chaperones. Thus, a substantial machinery is found within the ER to ensure that proteins are correctly folded and modified, and that leads to the degradation of those that are not. The components of this machinery include proteins that form and isomerize disulfide bonds, a key Hsp70 homolog, several proteins with DnaJ domains, and two proteins called calnexin and calreticulin. Other proteins with chaperone activity also exist in the ER (for example, an Hsp90 homolog called Grp94), although genetic evidence suggests they play a relatively minor role.

The Hsp70 protein homolog (also referred to as BiP, Grp78, and Kar2p) can be shown to associate with incompletely folded proteins using the same rules for binding as other members of the Hsp70 family. BiP has a C-terminal amino acid signal that retains it in the ER by cycles of retrieval from the Golgi body, and so any protein that has bound to it is also kept in the ER until the sites that BiP recognizes are buried in the folded protein (or until the protein is retrotranslocated back to the cytosol for degradation, a process in which BiP also has a major role). Coimmunoprecipitation and cross-linking experiments have shown that BiP may bind early in the folding pathway, as do other Hsp70 homologs such as DnaK, but other studies support a role later during folding. Complexes recovered by such experiments often include other ER resident chaperones as well as BiP, supporting the model of a network of chaperone proteins acting in concert to promote correct folding. BiP can be acted upon by several different membrane proteins with DnaJ domains (all of which face the lumenal side of the membrane) and these may all be involved in BiP's various functions. As with classic Hsp40 proteins, they can act to stimulate the ATPase activity of their Hsp70 partner (BiP) and thus improve the efficiency of the cycling on and off the complex of partially folded protein. BiP is not the only chaperone in the ER that can interact with partially folded proteins: the PDI shares this property, and thus may be involved in retention of incorrectly folded proteins as well as in the formation and isomerization of disulfide bonds.

A more complex system exists to ensure that glycosylated proteins are in their correct conformation before further transport through the secretory pathway. Glycan chains are added at selected asparagine residues to proteins as they pass into the ER through the membrane. These chains have the formula Glc_3-Man_9-$(GlcNAc)_2$, where Glc is glucose, Man is mannose, and GlcNAc is N-acetyl-glucosamine. Shortly after these chains have been added, two of the terminal glucoses are clipped off by the enzymes glucosidase I and II, leaving a single glucose. The presence of this glucose makes the protein bind to one of the two chaperones calnexin and calreticulin, which are respectively transmembrane and lumenal proteins, thus preventing their further transport from the ER. In this state, they can also be acted upon by another PDI called ERp57, which itself binds to both calnexin and calreticulin. The final glucose residue on the bound substrate is removed by glucosidase II, which enables the protein's release from calnexin or calreticulin. If it is correctly folded, it will now exit the ER, but if it is incorrectly folded, a terminal glucose is added back to the protein by another enzyme (UDP-glucose:glycoprotein glucosyl transferase), which only acts on proteins that contain areas that are not folded, and the protein will be rebound by calnexin and calreticulin, thus reinitiating the cycle of protein folding on these chaperones. If after several cycles of this reaction the protein has still failed to fold correctly, it can be targeted for retrotranslocation and degradation in the cytosol.

5.3
Examples of Substrate-specific Chaperones

The chaperones that have been considered above are all examples of what are often referred to as "broad-spectrum" chaperones: that is, they have a range of different substrate proteins with which they interact. Not all chaperones are broad range, however; some interact with only one or a few substrates. There are many examples of these in the literature, two of which are considered here.

5.3.1 Hsp47 and Collagen Assembly
In addition to the various chaperones present in the ER and described in the

previous section, another chaperone protein called Hsp47 is also present. This protein is required for the chaperoning of one class of proteins only: the collagens. Collagens are the most abundant proteins in mammals, and the importance of the action of Hsp47 is shown by the fact that mice with both copies of the *hsp47* gene disrupted do not survive to term and show many abnormalities in collagen processing, which has numerous lethal effects on tissue integrity. How does Hsp47 act? To answer this question, we need to know something about collagen assembly.

Although there are large numbers of different types of collagen, they share certain common features. Collagen is synthesized as procollagen, which contains domains at both the N- and C-termini, which are removed in the formation of the mature collagen molecule. These N- and C-termini form globular domains while the rest of the procollagen molecule forms alpha-helices, three of which coil around each other to form the characteristic collagen triple helix. The alpha-helices consist of repeats of the motif G-X-Y, where Y is often proline (or hydroxyproline, formed in the ER from proline). This molecule is then transported to the Golgi body and ultimately to the cell surface, where the N- and C-terminal domains are removed and collagen triple helices become associated via unknown mechanisms into long fibrils. Hsp47 has been shown to bind to the triple-helical procollagen molecule, and this binding requires the presence of at least one arginine residue in the triple helix. The precise role of the Hsp47 binding is not yet clear, but one promising hypothesis is that it prevents the premature aggregation of procollagen molecules into larger fibrils, that would not be exportable from the ER. Intriguingly, Hsp47 has been shown to be capable of forming trimers, and one speculation is that it may form rings around the triple-helical procollagen molecules that could serve to prevent lateral aggregation with other procollagens.

5.3.2 PapD and the Assembly of Bacterial Pili

The interactions of chaperones and substrates generally lack the degree of specificity that is commonly associated with protein–protein interactions, and it is this lack of specificity that contributes to the broad range of substrates on which chaperones can act. However, the substrate-specific chaperones, such as Hsp47 above, must be able to recognize and act on particular substrates. In one group of proteins, the so-called PapD chaperones, the molecular basis for this recognition is understood, and this gives us useful insights into how such chaperones can act.

PapD chaperones are members of a superfamily of proteins, found in the periplasm of different bacteria, which are involved in the assembly of pili and other surface structures. This gives them considerable applied interest because pili and fimbrae play a major role in the attachment of bacteria to surfaces and in many cases this makes them important virulence factors in disease. More than 30 members of this superfamily are known. PapD is one of the best studied of these and was distinguished as the first chaperone (albeit one whose action is highly specific) for which a crystal structure was obtained.

Certain strains of *E. coli* produce so-called P pili, which are complex structures consisting of an adhesion protein (PapG) at the tip of the pilus, joined via an adaptor protein (PapF) to a string of subunits (of the PapE protein) in an open helical conformation, which in turn are linked by another adaptor protein (PapK) to another string of subunits (PapA) in a helical cylinder. All these proteins are produced in the cytosol of E. coli and exported to the periplasm, and their subsequent assembly into a complete pilus requires the action of the PapD chaperone. In the absence of PapD, the various subunits are degraded by periplasmic proteases. PapD is not, however, part of the assembled pilus, and so it admirably fulfills the definition of a molecular chaperone.

Solving the structure of PapD, both alone and in complexes with proteins (PapK and PapE) whose folding it assists, has demonstrated how this chaperone works. The pilus subunits possess an Ig-fold, which normally consists of seven antiparallel beta sheets. In the pilus subunits, one of these helices is missing, which exposes a deep cleft in the protein. When the protein is in a complex with PapD, an N-terminal beta sheet from PapD fills the space that would normally be occupied by this missing region, thus stabilizing the protein by hiding the hydrophobic residues that would otherwise be exposed. This is referred to as "donor strand complementation," since the completion of the stable Ig-fold is made possible by the "donation" of a β-strand from the chaperone. Indeed, it is likely to be the case that the folding of the individual pilus subunits only takes place in the presence of PapD. Given that this complex is now stabilized by the pilus-chaperone interaction, how does subsequent assembly of the pilus subunit into the growing pilus take place? This assembly happens by a process referred to as "donor strand exchange," where the place occupied by the donated strand of the PapD chaperone is taken by a donated beta strand from the adjacent pilus subunit. This process takes place at an outer membrane pore made up of the protein PapC (sometimes referred to as an "usher"). PapD, complexed to the various different pilus subunits, arrives at the PapC, and donor strand exchange takes place in order to add each new pilus subunit to those already there, which will thus be progressively extruded from the outer membrane of the cell. The order in which the subunits are found is presumably determined by the relative affinities of the different pilus subunits for each other, although this has not been directly proven. Recent results show that the role of PapD in this entire process is far from passive. PapD primes pilus subunits for assembly by holding the groove in an open conformation, which is closed during donor strand exchange, effectively locking the N-terminal extension of each adjacent pilus subunit in place on its neighbor. The interaction between PapD and its substrates is highly specific and, moreover, PapD can be said to be providing steric information required for the correct folding of these substrates.

6
Conclusion

This review has focused on the roles of molecular chaperones in the cell and described some instances of the way in which these roles can be understood in terms of the structures and reaction mechanisms of individual molecular chaperones. It must be emphasized that this is an area of research that has expanded very rapidly over the past decade and a half, and many aspects of molecular chaperone biology have had to be neglected in such a brief discussion. More attention is now being paid to the networks in which molecular chaperones operate and how these overlap with other key areas of the cell's activity, particularly in the whole area of "quality control" – the optimization of protein folding activity under both normal and stressed conditions, and the rapid disposal of proteins which fail to meet the cell's requirements. Attention is also shifting to examining the roles of different chaperones in disease, most notably the protein-folding diseases such as the prion diseases, Alzheimer's, Parkinson's, and others. The possibility that a more complete understanding of the ways in which chaperones are involved in such diseases may lead to new therapies is an exciting one and a powerful incentive to further research in this field.

Acknowledgments

I am grateful to Professor R John Ellis, FRS, for his many useful comments on an earlier draft of this article.

See also Circular Dichroism in Protein Analysis.

Bibliography

Books and Reviews

Buchner, J. (1999) Hsp90 and Co. – a holding for folding, *Trends Biochem. Sci.* **24**, 136–141.

Bukau, B. (1999) *Molecular Chaperones and Folding Catalysts: Regulation, Cellular Functions and Mechanisms*, Taylor & Francis, London.

Graf, P.C.F., Jakob, U. (2002) Redox-regulated molecular chaperones, *Cell. Mol. Life Sci.* **59**, 1624–1631.

Hartl, F.U., Hayer-Hartl, M. (2002) Protein folding – molecular chaperones in the cytosol: from nascent chain to folded protein, *Science* **295**, 1852–1858.

Haslbeck, M. (2002) sHSPs and their role in the chaperone network, *Cell. Mol. Life Sci.* **59**, 1649–1657.

Kadokura, H., Katzen, F., Beckwith, J. (2003) Protein disulfide bond formation in prokaryotes, *Annu. Rev. Biochem.* **72**, 111–135.

Lund, P.A. (2001) *Molecular Chaperones in the Cell*, Oxford University Press, London.

Saibil, H. (2000) Molecular chaperones: containers and surfaces for folding, stabilising or unfolding proteins, *Curr. Opin. Struct. Biol.* **10**, 251–258.

Schrag, J.D., Procopio, D.O., Cygler, M., Thomas, D.Y., Bergeron, J.J.M. (2003) Lectin control of protein folding and sorting in the secretory pathway, *Trends Biochem. Sci.* **28**, 49–57.

Walter, S., Buchner, J. (2002) Molecular chaperones – cellular machines for protein folding, *Angew. Chem., Int. Ed.* **41**, 1098–1113.

Primary Literature

Anfinsen, C.B. (1973) Principles that govern the folding of protein chains, *Science* **181**, 223–230.

Bardwell, J.C.A., McGovern, K., Beckwith, J. (1991) Identification of a protein required for disulfide bond formation *in vivo*, *Cell* **67**, 581–589.

Beckmann, R.P., Mizzen, L.A., Welch, W.J. (1990) Interaction of hsp70 with newly synthesized proteins – implications for protein folding and assembly, *Science* **248**, 850–854.

Braig, K., Otwinowski, Z., Hegde, R., Boisvert, D.C., Joachimiak, A., Horwich, A.L., Sigler, P.B. (1994) The crystal-structure of the bacterial chaperonin GroEL at 2.8- angstrom, *Nature* **371**, 578–586.

Brinker, A., Pfeifer, G., Kerner, M.J., Naylor, D.J., Hartl, F.U., Hayer-Hartl, M. (2001) Dual function of protein confinement in chaperonin-assisted protein folding, *Cell* **107**, 223–233.

Buchberger, A., Schroder, H., Hesterkamp, T., Schonfeld, H.J., Bukau, B. (1996) Substrate shuttling between the dnaK and GroEL systems indicates a chaperone network promoting protein folding, *J. Mol. Biol.* **261**, 328–333.

Buchner, J., Schmidt, M., Fuchs, M., Jaenicke, R., Rudolph, R., Schmid, F.X., Kiefhaber, T. (1991) GroE facilitates refolding of citrate synthase by suppressing aggregation, *Biochemistry* **30**, 1586–1591.

Burston, S.G., Ranson, N.A., Clarke, A.R. (1995) The origins and consequences of asymmetry in the chaperonin reaction cycle, *J. Mol. Biol.* **249**, 138–152.

Chappell, T.G., Welch, W.J., Schlossman, D.M., Palter, K.B., Schlesinger, M.J., Rothman, J.E. (1986) Uncoating ATPase is a member of the 70 kilodalton family of stress proteins, *Cell* **45**, 3–13.

Chen, S., Roseman, A.M., Hunter, A.S., Wood, S.P., Burston, S.G., Ranson, N.A., Clarke, A.R., Saibil, H.R. (1994) Location of a folding protein and shape changes in GroEL-GroES complexes imaged by cryoelectron microscopy, *Nature* **371**, 261–264.

Deshaies, R.J., Koch, B.D., Wernerwashburne, M., Craig, E.A., Schekman, R. (1988) A subfamily of stress proteins facilitates translocation of secretory and mitochondrial precursor polypeptides, *Nature* **332**, 800–805.

Deuerling, E., Schulze-Specking, A., Tomoyasu, T., Mogk, A., Bukau, B. (1999) Trigger factor and DnaK cooperate in folding of newly synthesized proteins, *Nature* **400**, 693–696.

Ehrnsperger, M., Graber, S., Gaestel, M., Buchner, J. (1997) Binding of non-native protein to hsp25 during heat shock creates a reservoir of folding intermediates for reactivation, *EMBO J.* **16**, 221–229.

Ellis, R.J. (1994) Molecular chaperones – opening and closing the Anfinsen cage, *Curr. Biol.* **4**, 633–635.

Ellis, R.J. (1999) Chaperonins, *Curr. Biol.* **9**, R352–R352.

Ellis, R.J. (2001) Macromolecular crowding: obvious but underappreciated, *Trends Biochem. Sci.* **26**, 597–604.

Ellis, R.J. (2001) Molecular chaperones: inside and outside the Anfinsen cage, *Curr. Biol.* **11**, R1038–R1040.

Ellis, R.J., Hartl, F.U. (1996) Protein folding in the cell: competing models of chaperonin function, *FASEB J.* **10**, 20–26.

Ewalt, K.L., Hendrick, J.P., Houry, W.A., Hartl, F.U. (1997) In vivo observation of polypeptide flux through the bacterial chaperonin system, *Cell* **90**, 491–500.

Farr, G.W., Scharl, E.C., Schumacher, R.J., Sondek, S., Horwich, A.L. (1997) Chaperonin-mediated folding in the eukaryotic cytosol proceeds through rounds of release of native and nonnative forms, *Cell* **89**, 927–937.

Flynn, G.C., Chappell, T.G., Rothman, J.E. (1989) Peptide binding and release by proteins implicated as catalysts of protein assembly, *Science* **245**, 385–390.

Frydman, J., Nimmesgern, E., Erdjumentbromage, H., Wall, J.S., Tempst, P., Hartl, F.U. (1992) Function in protein folding of TRiC, a cytosolic ring complex containing TCP-1 and structurally related subunits, *EMBO J.* **11**, 4767–4778.

Goloubinoff, P., Christeller, J.T., Gatenby, A.A., Lorimer, G.H. (1989) Reconstitution of active dimeric ribulose bisphosphate carboxylase from an unfolded state depends on 2 chaperonin proteins and Mg-ATP, *Nature* **342**, 884–889.

Harrison, C.J., Hayer-Hartl, M., DiLiberto, M., Hartl, F.U., Kuriyan, J. (1997) Crystal structure of the nucleotide exchange factor GrpE bound to the ATPase domain of the molecular chaperone DnaK, *Science* **276**, 431–435.

Hebert, D.N., Foellmer, B., Helenius, A. (1995) Glucose trimming and reglucosylation determine glycoprotein association with calnexin in the endoplasmic-reticulum, *Cell* **81**, 425–433.

Hemmingsen, S.M., Woolford, C., Van der vies, S.M., Tilly, K., Dennis, D.T., Georgopoulos, C.P., Hendrix, R.W., Ellis, R.J. (1988) Homologous plant and bacterial proteins chaperone oligomeric protein assembly, *Nature* **333**, 330–334.

Houry, W.A., Frishman, D., Eckerskorn, C., Lottspeich, F., Hartl, F.U. (1999) Identification of *in vivo* substrates of the chaperonin Groel, *Nature* **402**, 147–154.

Jakob, U., Gaestel, M., Engel, K., Buchner, J. (1993) Small heat-shock proteins are molecular chaperones, *J. Biol. Chem.* **268**, 1517–1520.

Jakob, U., Muse, W., Eser, M., Bardwell, J.C.A. (1999) Chaperone activity with a redox switch, *Cell* **96**, 341–352.

Kozutsumi, Y., Segal, M., Normington, K., Gething, M.J., Sambrook, J. (1988) The presence of malfolded proteins in the endoplasmic-reticulum signals the induction of glucose-regulated proteins, *Nature* **332**, 462–464.

Langer, T., Lu, C., Echols, H., Flanagan, J., Hayer, M.K., Hartl, F.U. (1992) Successive action of Dnak, Dnaj and Groel along the pathway of chaperone-mediated protein folding, *Nature* **356**, 683–689.

Lee, G.J., Roseman, A.M., Saibil, H.R., Vierling, E. (1997) A small heat shock protein stably binds heat-denatured model substrates and can maintain a substrate in a folding-competent state, *EMBO J.* **16**, 659–671.

Liou, A.K.F., Willison, K.R. (1997) Elucidation of the subunit orientation in CCT (chaperonin containing TCP1) from the subunit composition of CCT micro-complexes, *EMBO J.* **16**, 4311–4316.

Llorca, O., McCormack, E.A., Hynes, G., Grantham, J., Cordell, J., Carrascosa, J.L., Willison, K.R., Fernandez, J.J., Valpuesta, J.M. (1999) Eukaryotic type II chaperonin CCT interacts with actin through specific subunits, *Nature* **402**, 693–696.

Llorca, O., Smyth, M.G., Carrascosa, J.L., Willison, K.R., Radermacher, M., Martin, J., Horwich, A.L., Hartl, F.U. (1992) Prevention of protein denaturation under heat-stress by the chaperonin Hsp60, *Science* **258**, 995–998.

Martin, J., Langer, T., Boteva, R., Schramel, A., Horwich, A.L., Hartl, F.U. (1991) Chaperonin-mediated protein folding at the surface of GroEL through a molten globule-like intermediate, *Nature* **352**, 36–42.

Martin, J., Mayhew, M., Langer, T., Hartl, F.U. (1993) The reaction cycle of GroEL and GroES in chaperonin-assisted protein-folding, *Nature* **366**, 228–233.

Mogk, A., Tomoyasu, T., Goloubinoff, P., Rudiger, S., Roder, D., Langen, H., Bukau, B. (1999) Identification of thermolabile *Escherichia coli* proteins: Prevention and reversion of aggregation by DnaK and ClpB, *EMBO J.* **18**, 6934–6949.

Munro, S., Pelham, H.R.B. (1986) An Hsp70-like protein in the ER – identity with the 78 kd glucose-regulated protein and immunoglobulin heavy-chain binding-protein, *Cell* **46**, 291–300.

Prodromou, C., Panaretou, B., Chohan, S., Siligardi, G., O'Brien, R., Ladbury, J.E., Roe, S.M., Piper, P.W., Pearl, L.H. (2000) The ATPase cycle of HSP90 drives a molecular 'clamp' via transient dimerization of the N-terminal domains, *EMBO J.* **19**, 4383–4392.

Prodromou, C., Roe, S.M., O'Brien, R., Ladbury, J.E., Piper, P.W., Pearl, L.H. (1997) Identification and structural characterization of the ATP/ADP-binding site in the Hsp90 molecular chaperone, *Cell* **90**, 65–75.

Queitsch, C., Sangster, T.A., Lindquist, S. (2002) Hsp90 as a capacitor of phenotypic variation, *Nature* **417**, 618–624.

Rietsch, A., Belin, D., Martin, N., Beckwith, J. (1996) An *in vivo* pathway for disulfide bond isomerization in *Escherichia coli*, *Proc. Natl. Acad. Sci. U.S.A.* **93**, 13048–13053.

Rudiger, S., Germeroth, L., SchneiderMergener, J., Bukau, B. (1997) Substrate specificity of the Dnak chaperone determined by screening cellulose-bound peptide libraries, *EMBO J.* **16**, 1501–1507.

Rye, H.S., Burston, S.G., Fenton, W.A., Beechem, J.M., Xu, Z.H., Sigler, P.B., Horwich, A.L. (1997) Distinct actions of *cis* and *trans* ATP within the double ring of the chaperonin GroEL, *Nature* **388**, 792–798.

Rye, H.S., Roseman, A.M., Chen, S.X., Furtak, K., Fenton, W.A., Saibil, H.R., Horwich, A.L. (1999) GroEL-groES cycling: ATP and non-native polypeptide direct alternation of folding-active rings, *Cell* **97**, 325–338.

Schroder, H., Langer, T., Hartl, F.U., Bukau, B. (1993) DnaK, DnaJ and GrpE form a cellular chaperone machinery capable of repairing heat-induced protein damage, *EMBO J.* **12**, 4137–4144.

Shtilerman, M., Lorimer, G.H., Englander, S.W. (1999) Chaperonin function: folding by forced unfolding, *Science* **284**, 822–825.

Sternlicht, H., Farr, G.W., Sternlicht, M.L., Driscoll, J.K., Willison, K., Yaffe, M.B. (1993) The T-complex polypeptide-1 complex is a chaperonin for tubulin and actin in vivo, *Proc. Natl. Acad. Sci. U.S.A.* **90**, 9422–9426.

Tasab, M., Batten, M.R., Bulleid, N.J. (2000) Hsp47: a molecular chaperone that interacts with and stabilizes correctly-folded procollagen, *EMBO J.* **19**, 2204–2211.

Tasab, M., Jenkinson, L., Bulleid, N.J. (2002) Sequence-specific recognition of collagen triple helices by the collagen-specific molecular chaperone Hsp47, *J. Biol. Chem.* **277**, 35007–35012.

Teter, S.A., Houry, W.A., Ang, D., Tradler, T., Rockabrand, D., Fischer, G., Blum, P., Georgopoulos, C., Hartl, F.U. (1999) Polypeptide flux through bacterial Hsp70: Dnak cooperates with trigger factor in chaperoning nascent chains, *Cell* **97**, 755–765.

Trent, J.D., Nimmesgern, E., Wall, J.S., Hartl, F.U., Horwich, A.L. (1991) A molecular chaperone from a thermophilic archaebacterium is related to the eukaryotic protein t-complex polypeptide-1, *Nature* **354**, 490–493.

Tu, B.P., Ho-Schleyer, S.C., Travers, K.J., Weissman, J.S. (2000) Biochemical basis of oxidative protein folding in the endoplasmic reticulum, *Science* **290**, 1571–1574.

Weissman, J.S., Hohl, C.M., Kovalenko, O., Kashi, Y., Chen, S.X., Braig, K., Saibil, H.R., Fenton, W.A., Horwich, A.L. (1995) Mechanism of GroEL action – productive release of polypeptide from a sequestered position under GroES, *Cell* **83**, 577–587.

Weissman, J.S., Kashi, Y., Fenton, W.A., Horwich, A.L. (1994) GroEL-mediated protein-folding proceeds by multiple rounds of binding and release of non-native forms, *Cell* **78**, 693–702.

Wiech, H., Buchner, J., Zimmermann, R., Jakob, U. (1992) Hsp90 chaperones protein folding *in vitro*, *Nature* **358**, 169–170.

Xu, Z.H., Horwich, A.L., Sigler, P.B. (1997) The crystal structure of the asymmetric GroEL-GroES-(ADP)(7) chaperonin complex, *Nature* **388**, 741–750.

Yaffe, M.B., Farr, G.W., Miklos, D., Horwich, A.L., Sternlicht, M.L., Sternlicht, H. (1992) TCP-1 complex is a molecular chaperone in tubulin biogenesis, *Nature* **358**, 245–248.

Zhu, X.T., Zhao, X., Burkholder, W.F., Gragerov, A., Ogata, C.M., Gottesman, M.E., Hendrickson, W.A. (1996) Structural analysis of substrate binding by the molecular chaperone DnaK, *Science* **272**, 1606–1614.

10
Motor Proteins

Charles L. Asbury[1,3] *and Steven M. Block*[1,2]
[1] *Department of Biological Sciences, Stanford University, Stanford, California, USA*
[2] *Department of Applied Physics, Stanford University, Stanford, California, USA*
[3] *Present Address: Department of Physiology and Biophysics, University of Washington, Seattle, Washington, SA*

1	Introduction	323
2	**Classic Molecular Motors**	**324**
2.1	Muscle Myosin: Tilting Cross-bridges Drive Contraction	325
2.2	Seeing is Believing: Motility Assays Demonstrate Motor Activity	326
2.3	Kinesin: Intracellular Porter	326
2.4	Ciliary Dynein: The Dark Horse	329
2.5	Processivity Allows Kinesin to Work Alone	330
2.6	Rowers Versus Porters: Duty Ratio Makes a Difference	330
2.7	Molecular Tug-of-war: Applying Force to Individual Motors	332
2.8	Motors Move in Discrete Steps	332
2.9	Different Strokes: Variation Within and Across Motor Families	333
2.10	Fuel Economy and Energy Efficiency	334
2.11	Walk This Way: Processive Mechanoenzymes Move Hand-over-hand	334
2.12	Coordination is Required	335
2.13	The Kinesin Cycle and Working Stroke	336
2.14	Under the Hood, Motors are Still a Mystery	336
3	**Nontraditional Molecular Motors**	**337**
3.1	Nucleic Acid Enzymes	337
3.2	More Than a Motor: Multitasking by RNA Polymerase	337
3.3	RNA Polymerase Structure	338
3.4	What Causes Pauses?	338
3.5	ATP Synthase	339
3.6	The Rotary Motor of Bacterial Flagella	340

Proteins. Edited by Robert A. Meyers.
Copyright © 2007 Wiley-VCH Verlag GmbH & Co. KGaA, Weinheim
ISBN: 978-3-527-31608-3

3.7	Polymers that Push and Pull 342
3.8	Microtubule Ends and Dynamic Instability 343
3.9	Motility Assays with Cytoskeletal Filaments 343
4	Conclusion 344

Bibliography 345
Books and Reviews 345
Primary Literature 345

Keywords

Allostery
Pertaining to or involving a change in conformation caused by the attachment of a ligand or substrate.

Mechanoenzyme
A catalytic enzyme that produces motion and force.

Processivity
A measure of the number of catalytic cycles an enzyme undergoes before detaching from the substrate. In the context of motors, processivity is proportional to the average distance over which the enzyme translocates on its filamentous substrate before detaching.

Substrate
A molecule on which an enzyme acts. In the context of molecular motors, either the fuel source (e.g. NTPs), or the force-generating partner (e.g. actin filament).

Working Stroke
A conformational change that occurs during a single round of catalysis, and which drives motion and force production. The working stroke length, or working distance, is the maximal distance associated with the stroke. The working distance can be different from the distance between consecutive attachment sites on the partner filament.

Cellular motions have fascinated biologists during the 400 years since the invention of the optical microscope first allowed them to be seen. Today, we know that motions underlying the most essential processes of life – such as cell division, energy transduction, muscle contraction, DNA replication, transcription, and translation – are generated by molecular motors. A molecular motor is a protein, or a complex of proteins and nucleic acids, that produces motion and force. For fuel, many molecular motors consume nucleotide triphosphates, breaking an energy-rich phosphate bond to release chemical energy, and then converting this into mechanical work. Other motors tap electrochemical gradients that exist across membranes within bacteria, mitochondria, and chloroplasts. Motor proteins are Nature's nanomachines, and they often function with efficiency that far exceeds the best human-engineered machines.

1
Introduction

Producing motion and force is the primary role of the "classic" molecular motors, myosin, kinesin, and dynein. These *mechanoenzymes* all hydrolyze ATP as a source of energy and drive motion along protein filaments. Myosin generates motion along filamentous actin, and is well known for its role in muscle contraction. The seemingly simple act of flexing ones arm requires $\sim 10^{17}$ myosins working together to slide $\sim 10^{15}$ actin filaments toward one another. Kinesin and dynein move along microtubule filaments. An essential role of kinesin is to haul vesicles across neurons. This can be a 6-day haul since the longest neurons are more than a meter, and vesicle transport proceeds at only $2\,\mu m\,s^{-1}$. Dynein causes the beating of flagella and cilia, such as those lining the lungs, by sliding microtubules past one another.

Besides the classic motors, there are many "nontraditional" motor proteins. In some cases, motion and force production are byproducts rather than a primary function. The main role of DNA and RNA polymerases is to copy and transcribe the genetic code. In order to do so, they move along their nucleic acid templates, sometimes with amazing endurance. A single RNA polymerase molecule can transcribe all 2.5 million bases of the human dystrophin gene in 14 h, at roughly 50 bases per second. Another nontraditional motor is F_1F_0-ATP synthase, which is responsible for replenishing the entire pool of ATP in all cells using energy derived from metabolism. As this enzyme toils, it also spins – it is a rotary motor. A flow of protons causes a shaft within the motor to rotate continuously, and shaft rotation is then coupled to the synthesis of ATP from ADP and phosphate. A third type of nontraditional motor activity is driven by the cytoskeletal polymers, actin, and tubulin. In addition to their roles as structural cables and girders for maintaining cell shape, and as highways for motor proteins to move along, these polymers are themselves dynamic machines that produce force. The leading edges of macrophages and other crawling cells are pushed outward by polymerizing actin filaments. Microtubule depolymerization generates tension that pulls chromosomes apart prior to cell division.

This chapter is a survey of the main classes of molecular motors. It begins with a discussion of the classic mechanoenzymes, myosin, kinesin, and dynein. These motors have been the subject of biophysical research for decades, and our understanding of their function serves as a foundation for the study of other motors. The chapter then turns to nontraditional motors, focusing on a handful of key examples, including nucleic acid enzymes (RNA polymerase), rotary motors

(F_1F_0-ATP synthase, and the bacterial flagellar motor), and protein polymers (actin and tubulin). A central goal of research on motor proteins is to determine how underlying biochemical events, such as ATP hydrolysis, are coupled to mechanical action. Progress toward this goal is chronicled throughout the chapter through description of experiments with classic and nontraditional motors.

2
Classic Molecular Motors

Myosin, kinesin, and dynein are founding members of large families of proteins whose primary function is to generate motion and force. Owing to their structural resemblance, the motors of each family operate in a manner similar to the founding proteins. However, they drive a wide variety of different cellular motions beyond the stereotypical roles of muscle contraction, vesicle transport, and the beating of cilia. There are at least 15 classes of myosin (traditionally denoted with roman numerals I through XV), and only a handful are involved in muscle contraction. Some of the other types are implicated in vesicle budding, cytokinesis, and organelle transport along actin cables. Likewise, kinesin-like proteins and cytoplasmic dyneins are essential for the formation and positioning of the mitotic spindle, chromosome separation prior to cell division, and organelle

---

Fig. 1 Structures of some classic and nontraditional molecular motors. (a) Muscle myosin consists of two heads connected to a common coiled-coil tail. The heads bind actin and carry ATP hydrolysis activity. A rodlike portion of each head (light gray) functions as a lever-arm, tilting $\sim 70°$ relative to the remainder of the head (dark gray) upon attachment to an actin filament. The tail promotes bundling of myosin molecules into thick filaments. (b) Kinesin also has two heads, connected through short polypeptides called *neck linkers* to a common coiled-coil stalk. The heads bind microtubules and carry ATP hydrolysis activity. A conformational change of the neck linkers may drive kinesin motion. The tail binds cargo. (c) Each dynein molecule consists of a donut-shaped head with two rodlike structures, the stem and stalk, emanating from the head. The head contains four ATP-binding sites, only one of which is catalytically active. The tip of the stalk binds microtubules. To drive motion, the head and stalk rotate relative to the stem, which attaches to cargo and can also bundle two or three dynein heads together. (d) RNAP is shaped like a claw, which opens to allow a DNA template to enter, and then wraps completely around the DNA during transcription. While transcribing, RNAP separates a portion of the DNA duplex called the *transcription bubble*, and maintains registration of a short section of hybrid RNA:DNA duplex. Nucleotides enter through a channel, leading to the active site, where they are incorporated into the nascent mRNA chain. (e) F_1F_0-ATP synthase consists of two rotary motors, connected to a common shaft, which act as a motor-generator pair. The F_0 portion taps a proton gradient across the inner mitochondrial membrane to drive spinning of the rotor (c_{12}) relative to the stator (ab_2). The F_1 portion sits directly above F_0, and contains a shaft ($\gamma \varepsilon$) that is rigidly fixed to the rotor of F_0, and also a ring $(\alpha\beta)_3$ that is rigidly fixed to the stator of F_0. Spinning of the shaft relative to the ring drives ATP synthesis. The isolated F_1 portion is known as F_1-ATPase because in the presence of ATP it will spin in reverse, catalyzing ATP hydrolysis. (e) Cross-sectional view of the rotary motor of bacterial flagella. The motor core contains a stack of rings (rotor), embedded in the multilayered cell wall that rotates as a single unit about an axis (dashed line) perpendicular to the surface of the bacteria. Rotation is driven by torque-generating units composed of MotA and MotB proteins. The MotA/B complex is anchored to fixed structures (peptidoglycan) within the cell wall. Protons flow through a channel within MotA/B, where protonation and deprotonation of MotB induces conformational changes in MotA, which attaches and detaches from the base of the rotor and drives its rotation.

movement along microtubules. The discussion here centers on the founding proteins, their functional properties, and some of the experiments that uncovered these properties. Particular attention is given to *in vitro* work with single motor molecules.

2.1
Muscle Myosin: Tilting Cross-bridges Drive Contraction

The motor activity of myosin was discovered more than 50 years ago. Electron microscopy revealed that muscle fibers consist of parallel thick and thin filaments that slide past one another during contraction. Tiny structures termed "cross-bridges," connecting laterally between the filaments, were suspected to drive filament sliding. In some images, the cross-bridges projected from the thick filaments at right angles, but in others, they were tilted, depending on tissue preparation conditions. These observations led to the theory, now well established, that cross-bridges drive filament sliding by cyclically attaching to the thin filaments, tilting, detaching, and untilting.

The thick filaments are now known to be bundles of myosin molecules. Each myosin consists of two identical 200-kDa polypeptides, plus two pairs of light chains (20 kDa). The heavy chains fold into twin globular heads connected to a common coiled-coil tail (Fig. 1a). Two light chains bind each head near the head–tail junction. Myosins bundle together by their long tails, and their heads project from the bundles, forming the cross-bridges that drive

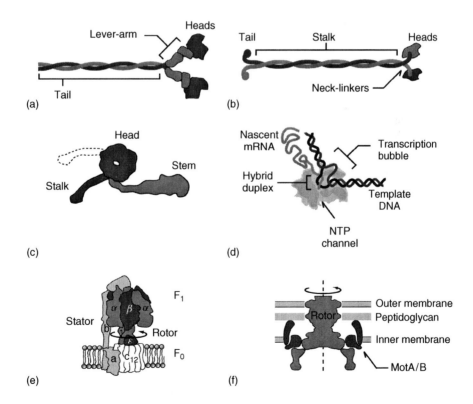

filament sliding. The heads can bind and hydrolyze ATP, and also carry a site that attaches to actin, the main component of the thin filaments, with ATP-dependent affinity. In high-resolution structures, a rodlike portion of the myosin head is found in several different orientations relative to the remainder of the head. Tilting of this "lever-arm," not the entire head, probably drives filament sliding.

The thin filaments of muscle are composed mainly of actin. Actin is a roughly spherical protein that polymerizes into a ropelike structure, with two strands, called *protofilaments*, that twist around one another. The filaments are polar, with a "plus" or "barbed" end, and a "minus" or "pointed" end, which are structurally different. During muscle contraction, the thick filaments slide toward the plus ends of the thin filaments.

2.2
Seeing is Believing: Motility Assays Demonstrate Motor Activity

A wealth of biochemical and structural information supports the tilting cross-bridge model of muscle contraction. However, the most compelling evidence for myosin motility comes from direct observation of motion generated *in vitro*. Myosin and actin are too small to see in an optical microscope. So, *in vitro* motility assays depend on various labeling schemes to render the motion visible. In the earliest assays, micron-sized beads were coated with myosin, and the beads were then observed to move along actin cables in the cytoplasm of the alga *Nitella*, and later along purified actin filaments bound to a glass surface. In an alternate strategy, actin filaments were made visible by fluorescent labeling, and gliding of these labeled filaments on myosin-coated glass surfaces was observed in a fluorescence microscope (Fig. 2a). These important experiments established beyond doubt that actin and myosin alone, without any additional components from muscle cells, were sufficient to generate motion and force. Consistent with the rotating cross-bridge theory, ATP was required for the motility, and the myosins moved toward the plus ends of the actin filaments. Filaments glided *in vitro* at 6000 nm s^{-1} (Table 1), similar to the speed at which thick and thin filaments slide past one another during muscle contraction. As discussed below, these two basic tests – the "bead assay" and the "gliding filament assay" – have been adapted and refined to study a variety of other motors in addition to myosin (Fig. 2b through d).

2.3
Kinesin: Intracellular Porter

Motility assays were instrumental in the discovery of kinesin. Observations of the squid giant axon suggested the existence of motors that consume ATP and haul vesicles at speeds of 1 to 2 µm s^{-1} along the dense array of microtubule filaments within the axon. A putative motor was first isolated by locking the vesicles onto the microtubules using a nonhydrolyzable ATP analog, AMPPNP, followed by purification of the microtubules, and then release of the motor with ATP. A gliding filament assay identical to that developed for myosin confirmed that the purified protein, kinesin, was indeed a motor: glass surfaces coated with kinesin supported the ATP-dependent gliding of microtubules. The gliding velocity, 800 nm s^{-1} (Table 1), closely matched the speed of vesicle transport.

Additional assays, using microtubules marked to reveal their intrinsic polarity, revealed the direction of kinesin-driven

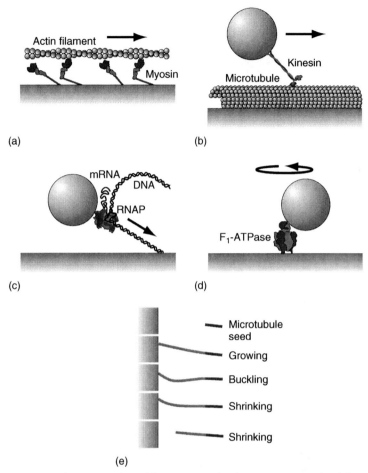

Fig. 2 Motility assays adapted for various molecular motors. (a) In the gliding filament assay, coverslip-bound motors drive filaments to move in a direction parallel to the filament long axis. Here, many myosin heads are shown interacting with a single actin filament. (b) In the bead motility assay, motors are attached to microscopic glass or plastic beads, which are pulled by the motors along coverslip-bound filaments. Kinesin is shown moving along a microtubule. (c) The tethered particle assay was developed to study the motion of nucleic acid enzymes along DNA filaments. Here, a microscopic bead is attached to RNA polymerase, which is transcribing a DNA filament that is attached at one end to the coverslip. Transcription results in a shortening (or lengthening) of the tether (depending on which end of the DNA is surface-bound). (d) In the rotation assay for F_1-ATPase, the motor is attached to a coverslip, and the orientation of the shaft is marked by off-axis attachment of a microscopic bead or filament. Spinning of the shaft causes the bead or filament to rotate. (e) Force generation by polymerization has been demonstrated by growing dynamic microtubule extensions from coverslip-bound seeds. The microtubules continue to elongate even after their growing ends encounter a barricade, which generates enough compressive force to buckle the filaments.

Tab. 1 Properties of selected molecular motors.

Protein Machine	Molecular Weight (kDa)	Force-generating Partner	Energy Source	Maximum Speed (nm s^{-1})[a]	Maximum Force (pN)[a]	Step Size (nm)[a]	Processivity (cycles)
Kinesin, native heterotetramer	340	Microtubule	ATP	1800			
Kinesin, truncated active homodimer	90	Microtubule	ATP	800	6	8	100
Myosin II, native heterohexamer	500	Actin filament	ATP	6000	1.5		
Myosin II, active HMM fragment	110	Actin filament	ATP	8000		6	1
Dynein, inner arm subspecies c	500	Microtubule	ATP	700	1.1	$8n$ ($n = 1, 2, 3 \ldots$)	~10
RNA polymerase, *E. coli* core enzyme	380	dsDNA	NTPs	5	27	0.34	>10 000
F$_1$F$_0$-ATP synthase	540	n/a	Protonmotive				n/a
F$_1$-ATPase, active rotary motor	350	n/a	ATP	150 Hz	40 pN nm	120°	n/a
Bacterial flagellar motor, basal body	9500	n/a	Protonmotive	300 Hz	4600 pN nm		n/a
Microtubule, growing	110 (tubulin dimer)	n/a	Binding	~50	4		n/a
Microtubule, depolymerizing		n/a	GTP	~500			
Actin filament, growing	40 (G-actin monomer)	n/a	Binding	~15			n/a

[a] Different units apply to the rotary motors, F$_1$F$_0$-ATP synthase, F$_1$-ATPase, and the bacterial flagellar motor. For these, the maximum rotation rate, torque, and angular step size are reported in units of Hz, pN nm, and degrees, as noted.

motion. Microtubules are rigid, tube-shaped polymers, composed of tubulin proteins arranged in a lattice, resembling a miniature drinking straw. Like actin filaments, microtubules have two structurally distinct ends, called "plus" and "minus". Polarity-marked filaments driven by kinesin glided with their minus ends leading, implying that kinesin was moving toward the plus ends.

Kinesin and myosin share many structural and functional similarities. Like myosin, each kinesin molecule consists of two identical polypeptides that form twin heads connected to a common coiled-coil stalk (Fig. 1b). The motor activity is carried by the heads. Each head hydrolyzes ATP and attaches to microtubule filaments with nucleotide-dependent affinity. But unlike myosin, the tail of kinesin does not cause bundling, it binds the motor to its cargo. Furthermore, atomic structures of kinesin heads are devoid of any rodlike structure resembling the lever-arm of myosin. Lacking a lever, kinesin's working stroke is likely to be very different from that of myosin.

2.4
Ciliary Dynein: The Dark Horse

Comparatively little functional information is available for the third classic motor, dynein, even though its activity was discovered around the same time as that of myosin. Electron micrographs revealed lateral connections between the parallel microtubules in cilia and flagella, similar to the myosin cross-bridges in muscle. The cross-bridges in cilia and flagella are dynein motors that drive bending motions by sliding microtubules past one another.

Dynein is larger and structurally more complex than kinesin or myosin, but it has many features common to all the classic motors. Depending on the source, dynein consists of one, two, or three large (500 kDa) polypeptides. Each of these forms a donut-shaped head, with two rodlike structures emanating from it, the "stem" and the "stalk" (Fig. 1c). The stem functions similar to the tails of kinesin and myosin, bundling the heads together, and also anchoring them tightly to their cargo. The tip of the stalk binds microtubules in an ATP-dependent manner, like the microtubule binding site within each of kinesin's heads. No atomic resolution structures are available for dynein, but electron microscopy of single dynein particles revealed a conformational change akin to the lever-arm tilting of myosin: under different nucleotide conditions, the stem adopts two different orientations relative to the head and stalk. Thus, dynein may move its stem-bound cargo by cyclically attaching via the stalk to a microtubule, rotating the stalk and head, detaching from the microtubule, and then unrotating. Dynein supports microtubule gliding and bead motion in *in vitro* assays. The direction of dynein-driven motion is toward the minus end of the microtubule, opposite that of kinesin-driven motion.

Dynein's complex structure contains a number of features with unknown functional significance. Dynein motors from different sources have different numbers of heads. Each donut-shaped head consists of six different subdomains arranged in a hexameric ring. Four of these subdomains bind ATP, but only one catalyzes hydrolysis. Nucleotide binding, but not hydrolysis, at one of the other subdomains is essential for motor activity. Uncovering the reasons for this complexity and elucidating dynein's mechanism of action are important frontiers for future research.

2.5 Processivity Allows Kinesin to Work Alone

Soon after its discovery, kinesin was found to possess a tenacity that set it apart from myosin and dynein. Kinesin is highly *processive*, staying attached to the microtubule as it undergoes many catalytic cycles, and translocating over relatively long distances before detaching. This processivity was first demonstrated when gliding filaments or moving beads were found to move long distances (1–2 µm), even when the surface density of motors on the slide or bead was extremely low, ensuring that single kinesin–microtubule interactions were very likely. Several independent lines of evidence now provide very strong evidence of kinesin's processivity (see Fig. 3).

The processivity of kinesin probably evolved as a means to conserve cellular resources. Kinesin's role of transporting vesicles across neurons is a critical task that must be accomplished repeatedly and with high reliability in order for these cells to function. The longest neurons contain millions of vesicles that each take weeks to make the journey from one end to the other. Kinesin's high processivity allows this Herculean task to be completed by just a few motors bound to each vesicle. In principle, the job could also be accomplished by nonprocessive motors, but many more motors would be required. The cell, in turn, would have to devote more energy and resources into producing these additional molecules.

Apart from its biological significance, kinesin's processivity has been a great advantage for experimentalists, allowing the first studies of single motor molecules. Motility assays for myosin relied on hundreds of motors acting together because the tiny tilting motions, or *working strokes*, of the individual heads are too small to see in a conventional optical microscope. However, owing to their high degree of processivity, kinesins generate hundreds of working strokes during each encounter with a microtubule moving distances of ~1 µm. The summation of many strokes renders the motion of individual kinesin motors easily visible.

Results from single molecule motility assays revealed a number of insights about how kinesin moves. The motion of kinesin-driven beads *in vitro* was not random over the microtubule surface, but appeared to follow a path parallel to the protofilaments. Gliding filament assays supplied strong evidence for protofilament tracking, when abnormal microtubules with helical protofilaments were shown to rotate about their long axis as they moved. In bead assays, engineered kinesin proteins with only one head failed to generate highly processive motion, indicating that two heads are, in fact, better than one.

2.6 Rowers Versus Porters: Duty Ratio Makes a Difference

The head domains of myosin and kinesin differ markedly in their duty ratio, the fraction of time during each biochemical cycle that they remain attached to their partner filament. Myosin possesses a low duty ratio that allows groups of molecules to work together efficiently, like the rowers of a large canoe, while kinesin has a high ratio, befitting its role as a lone porter. A myosin head has high affinity for actin just after hydrolysis, when the nucleotide-binding pocket contains either ADP and phosphate, or ADP alone, and low affinity when the pocket is empty or contains ATP. The timing of transitions between these states ensures that the high-affinity

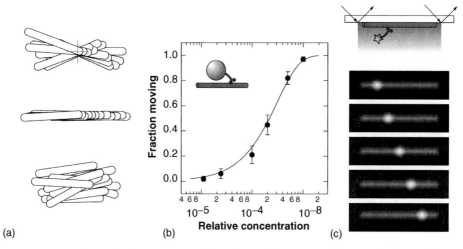

Fig. 3 Strong evidence for the processivity of kinesin. The first evidence for processivity was that kinesin-driven motion persisted *in vitro* even when the surface density of motors was extremely low. (a) Nodal point pivoting in the gliding filament assay also indicated processivity. Microtubules gliding on surfaces decorated sparsely with kinesin rotated erratically about a fixed location, even as they moved through this nodal point. When the trailing end of the microtubule reached the nodal point, it dissociated from the surface and diffused back into solution. A single motor at the nodal point presumably drove the motion (top). Negligible rotation occurred at high motor densities, when multiple motor–filament interactions constrained the filament orientation (middle). Both types of motion were distinct from thermal motion of free filaments in the absence of motor (bottom). [Adapted from Howard, J., Hudspeth, A.J., Vale, R.D. (1989) Movement of microtubules by single kinesin molecules, *Nature* **342**, 154–158.] (b) In the kinesin bead assay, the fraction of moving beads, f, decreased gradually as the relative motor concentration, C, was lowered, as expected if one molecule is sufficient to produce movement. The curve shows a one parameter (λ) fit to Poisson statistics, $f = 1 - \exp(-\lambda C)$. [Adapted from Svoboda, K., Block, S.M. (1994) Force and velocity measured for single kinesin molecules, *Cell* **77**, 773–784.] In the low-density regime, moving beads continued to translocate at normal speeds over distances that were independent of motor concentration (data not shown). (c) A third method used a microscope capable of imaging single fluorophores bound to kinesin (upper panel). The movement of labeled motors along coverslip-bound filaments was directly observed (shown schematically in the lower five panels). [Adapted from Vale, R.D., Funatsu, T., Pierce, D.W., Romberg, L., Harada, Y., Yanagida, T. (1996) Direct observation of single kinesin molecules moving along microtubules, *Nature* **380**, 451–453.] Labeling the motor by fusion to green fluorescent protein avoided chemical modification with reactive dyes, which can damage the motors, and ensured that every motor was labeled with the same number of fluorophores.

states represent <2% of the total cycle time. This low duty ratio is an adaptation that allows the myosin heads to avoid interfering with each other when many are acting on the same actin filament. They detach very quickly after undergoing a working stroke, so the speed of filament sliding is not limited by the hydrolysis rate of the individual heads. In contrast, kinesin heads have a duty ratio >50%, which partially explains the processivity of the motor. Even if the cycles of the two heads were completely uncorrelated, their high duty ratio would ensure that, on average, at least one was always bound to the microtubule.

2.7
Molecular Tug-of-war: Applying Force to Individual Motors

With the development of single molecule assays, it became possible to directly measure the forces generated by individual motors. One method for measuring force production by kinesin was to attach a microtubule filament to a flexible glass fiber, and then hold the fiber near a surface sparsely coated with kinesin. As individual kinesins on the surface bound and moved along the microtubule, they pulled against the glass fiber and caused it to bend. By measuring the amount of bending, the maximum force against which a kinesin motor could move was estimated to be 5 or 6 pN.

Another method for applying force to individual kinesin molecules, which gave a similar estimate of the stall force and also led to a number of other discoveries, was to use an optical trap. An optical trap is made by focusing a laser through the objective lens of a high-magnification microscope, creating a very bright light spot at the specimen. The focused light traps small objects such as micron-sized beads. When the trapped object is moved away from the center of focus, it feels a restoring force pulling it back that is proportional to the distance from the center, as if the trap was a stretched spring pulling on the object. To apply force to kinesin, an optical trap was used to grab beads with single kinesin molecules attached, and to place them near microtubules stuck onto to a glass surface. When the kinesin began moving along the microtubule, it pulled the bead from the trap center, and the trap supplied a restoring force that placed tension on the kinesin. As the bead was pulled gradually away from the trap center, the force increased and the motor speed decreased, halting when the force reached 6 pN.

2.8
Motors Move in Discrete Steps

Kinesin molecules move discontinuously over the microtubule surface, advancing in discrete 8-nm increments and dwelling at well-defined positions between advancements (Fig. 4). Two key innovations allowed the first observation of steps in the motion of kinesin-driven beads. First, tension supplied by an optical trap suppressed the random, thermally driven ("Brownian") motion that would otherwise dominate. Second, the bead position was measured with very high spatial and temporal resolution by monitoring the distribution of scattered light with a photodetector. The 8-nm step size (Fig. 4b) matches the spacing of tubulin dimers in the microtubule lattice. Similar experiments with dynein, which is processive under some conditions, suggest that it also moves stepwise, advancing by multiples of 8 nm.

Optical trapping has been applied to measure the motion of single myosin motors. A different technique than that used for kinesin was required because the interactions between a myosin molecule and an actin filament are fleeting, lasting only a few tens of milliseconds. To resolve these quick attachments, an assay was developed in which an actin filament with beads attached at both ends was suspended between two optical traps and held near a surface sparsely coated with myosin (Fig. 4c). Thermal motion of the beads decreased when a motor became attached to the filament, and the average position of the beads shifted abruptly, due to the tilting of the myosin head, by 6 nm. This tiny

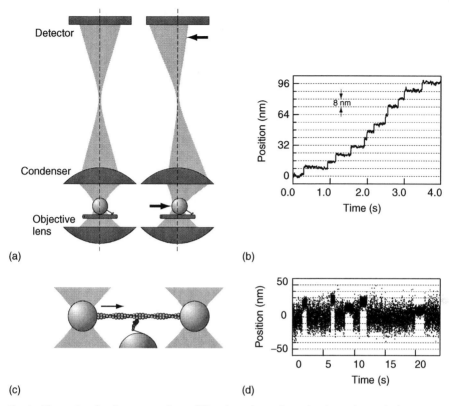

Fig. 4 Measuring the discrete motions of kinesin and myosin molecules using optical traps. Both methods use microscopic beads as handles to apply force, and as markers for the position of the motor or filament. (a) An optical trap applies tension during kinesin-driven movement of a bead along a coverslip-bound microtubule, and this tension reduces thermal motion of the bead so that individual 8-nm steps can be resolved. Bead position is detected with very high spatial resolution by monitoring the distribution of scattered light with a photodetector. (b) Example trace showing 8-nm steps generated by a single kinesin molecule. (c) The three-bead assay developed for measuring working strokes of muscle myosin. An actin filament is pulled taut between two microscopic beads held in optical traps. Binding of a single myosin head to the filament reduces thermal motion of the beads, and also induces a working stroke in the myosin, which causes the beads to deflect by 5 to 15 nm. (d) Example trace showing interactions between a single myosin head and an actin filament. [Data reprinted with permission from Lister, I., Schmitz, S., Walker, M., Trinick, J., Buss, F., Veigel, C., Kendrick-Jones, J. (2004) A monomeric myosin VI with a large working stroke, *EMBO J.* **23**, 1729–1738.]

distance is the maximum sliding distance that a myosin molecule can generate during a single interaction with actin during muscle contraction. More than a million of these interactions are evidently required just to lift a finger.

2.9
Different Strokes: Variation Within and Across Motor Families

In accordance with their diverse roles, motors of the myosin, kinesin, and dynein families often have important differences in the way they move. The cellular role of type V myosin, for example, is more similar to that of kinesin than to muscle myosin. Myosin V acts alone or in small numbers to transport vesicles and organelles along actin cables. It was therefore not surprising to find that this myosin, like kinesin, exhibits processive movement, taking several steps along an actin filament before detaching. The step size of myosin V, 36 nm, is much larger than that of kinesin, and matches the wide spacing of binding sites that occur every half-period in the actin helix. The structure of myosin V explains how it can generate such large movements. Each head contains a lever-arm that is 24 nm long, three times longer than the lever-arm of muscle myosin.

Diversity within motor families invites comparison, which can illuminate important aspects of motor function. Strong evidence supporting the tilting lever-arm model came from comparisons of gliding speeds and stroke lengths generated *in vitro* by myosin-family motors with different lever-arm lengths. The speeds and stroke lengths varied in proportion with the lever-arm length, as predicted by the model. Structural differences between kinesin and Ncd, a related motor that moves toward the opposite end of microtubules, suggested that a "gearbox" region just outside the head domain controls the direction of motion of these motors. This hypothesis was confirmed by testing chimeric motors, made by swapping the gearbox regions of the two proteins, in gliding filament assays.

2.10
Fuel Economy and Energy Efficiency

How many ATP molecules does a motor require to generate a working stroke or step? Is this "fuel economy" always the same? For kinesin, the coupling ratio – the number of ATPs consumed per step – has been measured by comparing the stepping rate in single molecule assays to the rate of ATP hydrolysis. Over a wide range of ATP concentrations and loads, one ATP is hydrolyzed per 8-nm step. This tight, 1 : 1 coupling implies that the energy efficiency of kinesin can be very high. ATP hydrolysis under physiological conditions is worth $\sim$80 pN nm (or $2 \cdot 10^{-23}$ kcal). When kinesin generates 8-nm steps under 5 pN of load, it produces as much as 40 pN nm (10^{-23} kcal), or 50% of the total chemical energy available. So kinesin is more than twice as efficient as the best man-made gasoline engines, which are 24% efficient at full power, and typically achieve only 10 to 15% on the road.

The energy efficiency of cytoplasmic dynein, 10%, is considerably lower than that of kinesin. However, dynein's complex structure may act like an automobile transmission, allowing it to maximize fuel economy. Near dynein's stall force, 1 pN, the motor takes 8-nm steps. At <0.4 pN, however, it seems to advance in larger increments of 24 or 32 nm. Thus, dynein can apparently shift into high gear when carrying a light load. Assuming that the coupling ratio under both conditions is equal, the larger step size will result in proportionally better fuel economy.

2.11
Walk This Way: Processive Mechanoenzymes Move Hand-over-hand

The fact that single kinesin molecules generate hundreds of steps, even under load, together with the earlier findings of motion parallel to the protofilaments and the requirement for two heads, suggested that it might walk – or waddle – from one tubulin dimer to the next. An attractive hypothesis was that the twin heads each take turns, alternately detaching and moving past one another, in a "hand-over-hand" motion resembling that of a person swinging along monkey bars. The same model was also thought to apply to the processive myosin-family motor, myosin V.

Single molecule experiments with myosin V and kinesin have confirmed that both walk hand-over-hand. First, the stride length of myosin V was measured by labeling one of its two heads with a fluorophore, and tracking the label with nanometer resolution. This is like watching a person walking across a field on a moonless night with a flashlight attached to one foot: The person is invisible, but the light moves visibly with every other step. The heads of myosin V took turns making strides that were twice as long as the distance moved by the tail, evidence that strongly supported a hand-over-hand model. Next, optical trapping experiments showed that some kinesin molecules limp along the microtubule, exhibiting a difference in the timing of every other step. Limping implied that kinesin switched between two different configurations as it stepped. The most severe limpers were mutants in which one head hydrolyzed ATP more slowly than the other, arguing for a mechanism in which the two heads swap both mechanical and catalytic activities with each step. Finally, the stride length for one of kinesin's two heads was measured to be 16 nm, double the step size. Taken together, these results make a very strong case for a hand-over-hand mechanism for kinesin and myosin V.

2.12
Coordination is Required

To fully account for the hand-over-hand walking of kinesin and myosin V, coordination between the two heads is essential, and it may be achieved through mechanical tension between the heads. The mechanical cycle of both motors includes a transient state in which both heads are attached to the filament. The motors are probably stretched in this doubly attached state, owing to the relatively large distance between the heads, and the resulting intermolecular strain between the heads could bias their kinetics so that the trailing head nearly always detaches before the leading head. In support of this hypothesis, external loads have been shown to strongly affect the rate of detachment of single myosin V heads, with forward loads accelerating detachment and backward loads slowing detachment.

Coordinated action at distinct sites is a universal requirement for all motors, not just those with high processivity. Automobile engines rely on the carefully timed actions of pistons, valves, and spark plugs. Mechanoenzymes also have critical interacting parts. Consider the mechanochemical cycle of myosin: Within a few milliseconds after attachment of a myosin head to actin, small motions occur in the nucleotide-binding pocket, allowing phosphate release. This triggers lever-arm tilting, which moves the actin filament. Subsequent ADP release triggers a second, smaller motion in certain myosin types (e.g. myosin I and single myosin V heads), but it is not clear if this additional stroke occurs for muscle myosin. ATP binding triggers detachment from actin, and hydrolysis "primes" the motor for the next cycle. For other motors, the specifics and timing are different than for myosin, but this coordinated action is a universal requirement.

2.13
The Kinesin Cycle and Working Stroke

The mechanochemical cycle of kinesin is not as thoroughly understood as that of myosin. Because two heads are involved in the stepping process, the kinesin cycle is necessarily more complex than that of muscle myosin. The specific conformational changes driving motion are poorly defined, and there is uncertainty about which biochemical events are associated with motion. A working hypothesis for kinesin is that all the mechanical action is associated with just one biochemical event. In support of this "one-stroke" hypothesis, reaction schemes with just one force-dependent rate can account for force-velocity and [ATP]-velocity curves measured in single molecule assays. These schemes predict a working stroke after ATP binding, possibly upon ATP hydrolysis. By contrast, a working stroke concomitant with ATP binding is suggested by kinetic measurements, showing that ADP release from one head is stimulated by binding of a nonhydrolyzable ATP analog (e.g. AMPPNP) to the other head.

A conformational change in the structure of kinesin has been discovered that may drive its motion. This putative working stroke is quite different from either the tilting lever-arm of myosin, or the stem reorientation of dynein. In single-headed kinesin constructs, a 15-amino acid peptide known as the "neck linker," which connects each kinesin head to the coiled-coil stalk, undergoes a nucleotide-dependent transition. In the presence of ADP, or when no nucleotide is present, the neck linker is disordered. In this state, it acts like a flexible tether, pivoting about a point on the backside of the head. When the head is attached to a microtubule in the presence of ATP analogs (AMPPNP, ADP-AlF$_4^-$), the neck linker "zips" onto to the surface of the head, and its end points toward the microtubule plus end. In the full two-headed motor, zipping of the neck linker on one head could drive stepping by moving the stalk, and therefore the other head, toward the next attachment site on the microtubule lattice.

2.14
Under the Hood, Motors are Still a Mystery

Even for the best-understood motors, where high-resolution structures are available, and major mechanical steps can be identified with specific biochemical transitions, there are very fundamental questions that remain unanswered. In particular, the atomic-scale motions that transduce small chemical events in the nucleotide pocket and convert them into larger motions elsewhere are poorly understood. Important residues have been identified by comparing sequences of related motors and their precursors (e.g. myosin, kinesin, and G-proteins), and subdomains that move relative to others have been suggested by structural comparison. However, ultimately, these static structures cannot elucidate the timing of movements and the cause-and-effect relationships between the various parts. A myriad of physical and biological techniques will no doubt be essential in this effort, but single molecule techniques that simultaneously record motion and biochemical changes (e.g. through fluorescence) seem particularly valuable in this regard.

3
Nontraditional Molecular Motors

There are many other protein machines that generate force and motion, but which do not fit the classic motor paradigm. A growing number are being studied using *in vitro* assays, like those discussed above, which allow the mechanical output of single motors to be measured.

3.1
Nucleic Acid Enzymes

Some of the most important processes of life are carried out by nucleic acid enzymes, many of which are processive motors that move along DNA. Every human cell stores genetic information in the form of 23 strands of DNA, totaling three billion base pairs, and measuring 1 m in total length. All 23 strands are copied by DNA polymerase enzymes, untangled by topoisomerase enzymes, and packaged into chromosomes by condensins, before cell division. The genetic information contained in the DNA is transcribed into mRNA by RNA polymerase (RNAP), and translated into protein by the ribosome. Each of these nucleic acid enzymes is a protein machine capable of generating force and motion, and each is fascinating in its own right. A discussion of all of them is beyond the scope of this chapter, which will focus on one important example, RNAP.

3.2
More Than a Motor: Multitasking by RNA Polymerase

Even though the size of an RNAP enzyme, by comparison of total molecular weight, is not so different from that of the classic mechanoenzymes, its function is much more complex. Motion is merely a by-product of the biological role of this protein machine, transcribing the genetic code. While moving along a DNA template, RNAP separates a short section of the DNA duplex, the "transcription bubble," and builds a copy of one strand by selecting complementary nucleotides from the surrounding solution and attaching them, one at a time, to the end of the nascent mRNA chain (Fig. 1d). Along the way, it must maintain registration of a short section of "hybrid" RNA:DNA duplex, and also respond to a number of different signals that control the initiation, termination, and elongation rate of transcription. Like other mechanoenzymes, RNAP derives energy from nucleotide hydrolysis, but, in this case, each nucleotide serves a dual role. After hydrolysis, the nucleotide becomes an information-containing subunit incorporated into the growing mRNA. (If automobile engines could make such efficient use of their exhaust, urban air quality would be much improved!)

Motion may not be its primary function, but RNAP is no slouch of a motor. Its motion can be directly observed by attaching a micron-sized bead, and recording bead motion as the enzyme transcribes a DNA template bound at one end to a glass surface (Fig. 2c). Using this "tethered particle" assay in conjunction with optical trapping, the motion of a single RNAP can be tracked with high spatial resolution, and the effect of applied load can be measured. RNAP is slow, moving in these assays at 10 to 15 bp s^{-1}, or just 5 nm s^{-1} (Table 1). This speed is roughly equivalent to the rate of human hair growth, and 160-fold slower than the speed of kinesin. However, RNAP is much more processive than kinesin. In cells, RNAP molecules synthesize mRNA chains of 10^4 (in bacteria) to 10^6 (in mammals) nucleotides. *In vitro*, its processivity is reduced, but the enzyme typically moves across several thousand bases or more before detaching from the template. Movement continues, unhindered at 10 to 15 bp s^{-1}, even when backward loads as high as 27 pN are applied. A high stall force (>5-fold higher than that of kinesin) may be necessary for RNAP to function *in vivo*, perhaps allowing the enzyme to push through "road-blocks" formed by other DNA-binding proteins. RNAP is presumed to move in discrete steps corresponding to the distance between individual bases along the DNA helix, 0.34 nm. This distance is extraordinarily small (20 times smaller than kinesin's 8-nm steps), and steps of this size have not yet been directly observed. However, optical trapping technology is rapidly advancing, and such tiny motions may soon be resolvable.

3.3
RNA Polymerase Structure

RNAP is shaped like a claw. The claw opens to allow a DNA strand to enter, and during transcription it closes, wrapping completely around the DNA. Besides the DNA entry and exit channels, there is also a channel through which the newly synthesized RNA exits, and a pore for nucleotide entry (Fig. 1d). Inside the closed structure, the enzyme makes numerous contacts with the hybrid duplex and the DNA. It is unknown, which parts of RNAP are responsible for generating motion and force production, but many candidate features are apparent in the high-resolution structures. For example, a "bridge helix" located near the site of nucleotide addition may undergo a conformational change that pushes the enzyme to the next site. Determining which portions of RNAP are responsible for its motion is a great challenge for future research.

3.4
What Causes Pauses?

The motion of RNAP along the DNA template is interrupted by pauses, lasting from a few seconds to many minutes. The short-duration pauses (those with lifetimes of seconds) are very frequent, and may result from the enzyme encountering a few GC-rich base pairs of DNA that are tougher-than-average to separate. Occasionally, the enzyme pauses for a much longer duration (20 s to >30 min). These infrequent but long-lived pauses may occur for two reasons, both of which illustrate the sophisticated behavior that RNAP is capable of. First, long pauses can be induced when RNAP encounters a specific sequence in the DNA template it is transcribing. Such sequence-dependent pauses are an important mechanism for regulation of gene expression. By relieving these long pauses, a cell can greatly alter the rate of expression of a pause-containing gene. In some cases, these sequence-dependent pauses occur when the nascent mRNA chain folds into a hairpin structure, which then interacts directly with the RNAP enzyme. In other cases, these pauses occur when the RNAP transcribes a slippery AT-rich sequence, resulting in an unstable RNA:DNA hybrid duplex that allows the enzyme to backtrack. There is a second class of long-duration pauses that are not sequence-dependent, and these probably occur when the enzyme makes a copying error, misincorporating a noncomplementary nucleotide into the mRNA chain. These pauses are also associated with backtracking, and may reflect a "proofreading" activity whereby the enzyme slides backward and then (with the help of accessory factors) cleaves a short section from the end of the mRNA, removing the mistake before resuming elongation. Determining the reasons why RNAP pauses is critical for understanding how gene transcription is controlled. Single molecule experiments will be useful in this effort because the motion of unsynchronized molecules can be followed with very high resolution, in real time.

3.5
ATP Synthase

A rotary machine, F_1F_0-ATP synthase, is at the heart of energy metabolism in plants, animals, and photosynthetic bacteria, where its role is to replenish the cellular store of ATP. The importance of this job is obvious when one considers that a human body contains 100 g of ATP (0.25 moles), each molecule of which gets hydrolyzed 400 times a day to power various cellular tasks. An army of ATP synthase enzymes performs the $\sim 10^{26}$ synthesis reactions required to regenerate the spent ATP. To do so, the enzymes tap into an electrochemical gradient, the protonmotive force, that exists across the inner mitochondrial membrane of animal cells, or across the thylakoid and plasma membranes of plants and eubacteria, respectively.

ATP synthase consists of two separate rotary motors, F_0 and F_1, that work together as a motor–generator pair. Normally, F_0 is the driving motor of the pair. As protons flow through it, down the electrochemical gradient, a portion of F_0 spins like a water wheel. The spinning wheel of F_0 supplies torque that rotates a shaft within the other motor, F_1, causing it to regenerate ATP from ADP and phosphate. The motor–generator pair can also operate in reverse. In this case, ATP hydrolysis by the F_1 motor causes the shaft to rotate backward, which supplies a torque that spins the wheel of F_0 and pumps protons back up the electrochemical gradient.

F_1F_0-ATP synthase is comprised of eight different types of protein subunits (Fig. 1e). The water wheel, or "rotor" portion of F_0 is a ring of 12 identical subunits, c_{12}, that spin in the plane of the membrane relative to a "stator" composed of three other subunits, ab_2. F_1, is a donut-shaped structure made of three pairs of proteins, $(\alpha\beta)_3$, plus two additional proteins, γ and ε, which form the shaft that fits into the center of the donut. Clockwise (CW) rotation of the $\gamma\varepsilon$ shaft (as seen from the F_0 or membrane side) causes ATP synthesis to occur sequentially at three catalytic sites located symmetrically around the $(\alpha\beta)_3$ ring. Because isolated F_1 can function in reverse, catalyzing ATP hydrolysis and counter clockwise (CCW) rotation of the shaft, it is often referred to as F_1-ATPase. In the full, F_1F_0-ATP synthase enzyme, F_1 sits directly above F_0, with the $\gamma\varepsilon$ shaft of F_1 making a rigid connection to the c_{12} ring of F_0, and with the ab_2 stator of F_0 connecting to the $(\alpha\beta)_3$ donut of F_1. Normally, spinning of c_{12} drives rotation of $\gamma\varepsilon$ within $(\alpha\beta)_3$ and hence ATP synthesis.

The ring-shaped structures within F_1F_0-ATP synthase provided the first clues that rotation might be important to its function. As for the classic mechanoenzymes, proof of motion came when an *in vitro* assay was developed, allowing the rotation to be directly observed. F_1-ATPase molecules were attached sparsely to a surface, and the orientations of their $\gamma\varepsilon$ shafts were marked by attaching micron-long, fluorescent-labeled actin filaments. In the presence of ATP, the filaments rotated CCW, indicating shaft rotation. At very low concentrations of ATP, the filaments rotated in discrete steps, dwelling at well-defined orientations in between rapid, 120° reorientations. Rotation rates were one-third of the rate of ATP hydrolysis, implying that each 120° reorientation corresponds to hydrolysis of a single ATP. The filaments used to mark shaft orientation also supplied a drag force acting against the rotation. By calculating the drag on the filaments, F_1 was estimated to deliver 40 pN nm of torque during each 120° reorientation, giving 80 pN nm of mechanical work output per ATP hydrolysis. This is 100% of the available chemical energy, making F_1-ATPase one of the most efficient motors known.

F_1-ATPase is a relative newcomer to the molecular motor scene, but it is quickly becoming one of the best-understood examples of mechanochemical coupling. A series of experiments with single F_1 molecules has revealed that each 120° step occurs in two phases, or substeps, that correspond to particular biochemical transitions in the ATP synthesis reactions occurring at each of the three catalytic sites. Substeps were observed by marking the shaft orientation with 40-nm gold particles (Fig. 2d), rather than the much larger filaments used in earlier work. The small beads resulted in a lower drag force acting on the motor, allowing full speed rotation at 160 revolutions per second (Hz) (Table 1). Capturing the motion with a high-speed video camera showed a substep of 80 to 90° followed by one of 30 to 40°, underlying each of the 120° steps previously observed. Finally, simultaneous observation of shaft orientation and binding and release of a fluorescent nucleotide allowed these biochemical steps to be temporally correlated with the substeps. ATP binding to one of the catalytic sites (site 0) is concurrent with the 80 to 90° substep, and the remaining 30 to 40° substep requires hydrolysis at the site that previously bound ATP (site −1), and release of ADP from the remaining site (−2).

It is unknown whether the events seen during ATP hydrolysis by F_1 are simply the reverse of those occurring during synthesis, but several experiments confirm, at least, that the $\gamma\varepsilon$ shaft rotates in the opposite direction during synthesis. In one experiment, magnetic beads attached to the shaft were used to drive CW rotation in F_1-ATPase, in the presence of ADP and phosphate, and a luciferin–luciferase system that emits a photon upon reacting with ATP was used to verify synthesis. In another experiment, individual, fluorescent-labeled F_1F_0-ATP synthase complexes were embedded in liposomes. A pH difference was created across the membrane by rapid dilution, and rotation was recorded by fluorescence resonance energy transfer (FRET).

3.6
The Rotary Motor of Bacterial Flagella

Bacterial flagella are very different from the flagella and cilia of eukaryotic cells. Flagellated bacteria, such as *Escherichia coli*, swim by rotating a set of four corkscrew-shaped filaments (four, on average) that extend from the cell surface out into the surrounding medium. At the base of each filament is a large protein machine that drives filament rotation. This rotary motor, like F_1F_0-ATP synthase, is powered by the protonmotive force, but it is much larger and structurally more complex than F_1F_0-ATP synthase.

Bacteria swim to find food. They control their swimming behavior by altering the direction of rotation of their flagellar motors. When all four motors rotate CCW (as seen by an observer outside the cell), the cell swims steadily, or "runs," in a relatively straight line parallel to its long axis. When one or more motors rotates CW, the cell "tumbles," erratically moving in place and reorienting itself. The motors switch from CCW to CW at random, so that typical swimming involves runs that last ~1 s, interspersed with tumbles that last a few milliseconds. When the bacteria senses rising nutrient concentrations, it lengthens the runs by increasing the probability of CCW rotation. In this way, the cell moves, on average, toward the food.

The core of the bacterial flagellar motor is a stack of ring-shaped structures, 45 nm in diameter, embedded in the multilayered cell envelope (see Fig 1f). The rings are composed of 20 different types of proteins, but they are all thought to rotate together as a single unit, the "rotor." Rotation of the stack of rings is driven by a circular array of ≤16 "studs" that surround the base of the stack, and which are anchored to the framework of the cell wall. Each stud is composed of two MotA proteins (32 kDa), and one MotB protein (34 kDa). No high-resolution structures are available for the MotA/MotB complex, but both proteins span the cytoplasmic membrane, forming a transmembrane proton channel. MotB has a proton-acceptor site, and MotA contains a site that interacts with the base of the rotor. Protonation and deprotonation of MotB is thought to cause conformational changes in MotA, which probably binds and unbinds from the rotor, driving rotation and torque generation. The minimal torque-generating unit may be composed of two studs (i.e. four MotA subunits plus two MotB subunits). More detailed descriptions of the structure can be found in review articles cited in the bibliography.

Individual flagellar motors can be studied by attaching bacteria to a glass surface by one flagellum. The tail wags the dog in this tethered cell assay – with the flagellum anchored, rotation of the motor causes the whole cell body to spin. To spin the entire cell, the motor must overcome a large viscous drag, so it spins relatively slowly in this assay, at 10 Hz. But it produces an impressive 4600 pN nm of torque. When mutant cells lacking MotB are tethered, they are paralyzed and do not spin. Amazingly, these paralyzed cells can be "resurrected" by expression of MotB from an inducible gene. Resurrected cells begin to rotate within several minutes after induction of MotB expression, and their speed increases in a series of discrete jumps. Each jump in speed represents the incorporation of one additional torque-generating unit into the motor. As many as eight jumps can be seen, implying that the maximum number of torque generators is eight.

Each torque generator is itself a processive, high-duty ratio motor that moves along the surface of the rotor without detaching. The best evidence for processivity is that tethered cells in the resurrection experiment with just one torque-generating unit spin relatively smoothly, and do not freely undergo rotational Brownian motion. Evidently, a single unit is sufficient to prevent the motor from slipping, and each unit remains attached to the rotor during most, or all, of its mechanical cycle. There are twice as many studs (16) as torque-generating units (8), and one hypothesis is that each unit is a co-ordinated pair of studs, possibly moving in a hand-over-hand motion like that of kinesin or myosin V. A single unit is expected to move stepwise over the surface of the rotor, perhaps taking ~26 steps per revolution (the approximate number of subunits composing the base of the rotor), but such steps have not been directly observed. Dividing the maximum torque (4600 pN) by the number of torque-generating units (8), and by their distance from the axis of rotation (20 nm), shows that each unit generates considerable force, up to 29 pN, which is comparable to the stall force of RNAP. Their speed of motion over the rotor surface is also quite high. When the viscous load is minimal, the motor can rotate as fast as 300 Hz (Table 1). This translates into motion of the torque-generators at $38\,000$ nm s^{-1} over the rotor surface, which is >6-fold faster than muscle myosin, and similar to the speed of the fastest myosins (e.g. type XI, responsible for cytoplasmic streaming in algae).

The speed of rotation is proportional to the protonmotive force, as shown by wiring a cell to an external voltage source and watching an inert marker on the motor. To apply voltage, the cell body was drawn halfway into a micropipette, and the membrane permeabilized by chemical treatment. Estimates of the proton flux through the motor suggest that the motor is tightly coupled. Roughly 1200 protons flow through the motor during each complete revolution. By attaching a variety of different-sized latex beads to the filaments and adjusting the viscosity of the surrounding fluid, torque-speed relations have been measured over a wide range of speeds. Forward rotation under assisting torques, and backward rotation under torques above stall (>4600 pN), has been explored by using rotating electric fields or optical traps to apply torque in the tethered cell assay.

3.7
Polymers that Push and Pull

Actin filaments and microtubules are not just static polymers. In addition to their roles as structural cables and girders for maintaining cell shape, and as superhighways for mechanoenzymes to move along, these polymers are also dynamic machines that can produce force. In living cells, the cytoskeletal polymers are in a constant state of flux, and their growth and shrinkage is harnessed to drive many organelle and whole-cell movements. Crawling cells have a dense array of polymerizing actin filaments beneath their leading edge that pushes outward on the plasma membrane and causes protrusion. Similarly, the bacterial pathogen, *Listeria monocytogenes* is pushed by actin polymerization. The bacteria move in graceful arcs through the cytoplasm of a host cell, leaving "comet tails" of polymerized actin in their wake. During mitosis, chromosomes are pushed and pulled by dynamic microtubules whose ends are linked to specialized sites on the chromosomes, the kinetochores. Just before cell division, kinetochore-attached microtubules depolymerize, generating tension that pulls sister chromatids apart.

Both actin and microtubule filaments are composed of protein subunits arranged in a regular lattice. The monomeric form of actin, "G-actin," is a roughly spherical protein, 5 nm in diameter (45 kDa). Like a LEGO block, the surface of an actin monomer has several sites for attachment to other monomers. Each also has a cleft that binds an ATP molecule. Monomers assemble into a ropelike structure, "F-actin," with two strands, called *protofilaments*, that wind around each other with a helical period of 72 nm. The building blocks for microtubules are made of tubulin, a molecule that consists of two nearly identical 50-kDa proteins fused tightly to form a dimer, 8 nm in length. Each dimer has two sites that bind GTP. The dimers assemble into a hollow, tube-shaped structure, 25 nm in diameter, with 13 protofilaments that run parallel to the long axis of the tube. Both types of filaments have fast-growing "plus" ends, and slow-growing "minus" ends.

Nucleotide hydrolysis supplies energy that makes actin and tubulin polymers very dynamic. Actin monomers in solution bind ATP, and have high affinity for one another. After polymerization, the ATP is hydrolyzed and phosphate is released, leaving ADP trapped in the binding clefts of the monomers within a filament. The ADP-containing monomers have reduced affinity, so hydrolysis destabilizes the actin filament, promoting depolymerization. GTP hydrolysis by tubulin has a similar effect. Each tubulin dimer binds two molecules of GTP, one of which is hydrolyzed upon incorporation of the dimer into a microtubule filament. The GDP-containing tubulin dimers have reduced affinity for one another, which destabilizes the lattice and promotes depolymerization. Without hydrolysis, both polymers would simply grow until equilibrium was reached, when the subunit pool was spent. Hydrolysis keeps the filaments out of equilibrium, allowing coexistence of growing and depolymerizing filaments.

3.8
Microtubule Ends and Dynamic Instability

The dynamic behavior of microtubules can be directly observed *in vitro*. In the presence of GTP and pure tubulin, microtubule growth is interrupted by periods of rapid depolymerization. This "dynamic instability" requires GTP hydrolysis. Growth rates are normal in the presence of the nonhydrolyzable GTP analog, GMPCPP, but the growth is uninterrupted. Subunit addition and removal occurs only at the ends of the filaments, which adopt different structures, depending on whether they are in a state of growth or depolymerization. The protofilaments that extend from growing ends are straight, forming sheets that are sometimes hundreds of subunits long. In contrast, the protofilaments at depolymerizing ends become highly curved, peeling away from the lattice. The ends of growing filaments are temporarily stabilized by a "cap" of GTP-containing subunits in which hydrolysis has not yet taken place. The transition between growth and depolymerization, called "catastrophe," is probably triggered when hydrolysis of the cap occurs before more GTP-containing subunits are added.

The curvature of protofilaments at the ends of depolymerizing microtubules, and the fact that the products of depolymerization are often curved oligomers, suggests a structural basis for the coupling of nucleotide hydrolysis and polymerization. Before hydrolysis, the GTP-containing dimers are probably straight, fitting snugly into the growing microtubule lattice. If the GDP-containing subunits are naturally curved, then they would be strained when trapped within the lattice. In this way, energy from hydrolysis could be stored within the lattice as mechanical strain.

3.9
Motility Assays with Cytoskeletal Filaments

Several *in vitro* experiments show that polymerization of pure actin or tubulin, without any additional proteins, can generate pushing force and do mechanical work. Polymerizing actin filaments inside liposomes causes distension of the liposomes, demonstrating that the growing filaments can push outward on the lipid bilayer. Likewise, microtubule filaments grown inside a small chamber can push against the chamber walls with enough force to buckle themselves (Fig. 1e). By analyzing the shapes of buckled filaments, the maximum pushing force of a single microtubule has been estimated at 4 pN.

These important experiments prove that growing filaments can push against an object, but they are incomplete models for the polymer-driven motility that occurs in cells. In cells, a variety of accessory proteins provide spatial and temporal control of filament dynamics, and couple the ends of growing and shrinking filaments to other structures to apply force. *Listeria* promote spatially localized actin polymerization with a nucleation factor, ActA. Likewise, kinetochores contain a host of proteins that modulate microtubule dynamics and maintain attachment to microtubule ends. Understanding the mechanisms of these accessory factors will be key to understanding how cells harness cytoskeletal filaments to produce motion and force.

In vitro motility assays that reconstitute force generation using dynamic filaments coupled to accessory proteins provide more realistic models for filament-based motility in cells. Shrinking microtubules can pull against microscopic beads when the beads are coated with proteins that maintain attachment to the depolymerizing filament ends. This motion is similar to the way chromosomes are pulled apart before cell division, and also to the way the mitotic spindle is positioned inside asymmetrically dividing yeast cells. In a reconstituted assay that closely mimics the motion of *Listeria*, beads coated with ActA protein are pushed around by actin polymerization. The beads follow curved trajectories and leave comet tails of polymerized actin in their wake, just like the bacteria.

4
Conclusion

Motion is fundamental to life. Everyone is familiar with the macroscopic motion of muscle contraction. There are also exquisite motions taking place at the level of cells and molecules. The cells in our immune system crawl around our bodies and engulf invading bacteria. Cilia in our lungs beat to remove inhaled debris. In all these cases, the motion is generated by tiny protein machines, the molecular motors. Molecular motors are ubiquitous, and the list of known motors is growing. Besides the classic motors, myosin, kinesin, and dynein, and the cytoskeletal polymers, filamentous actin and microtubules, there are also protein machines at the heart of energy metabolism, and reading the genetic code. Studies of molecular motors, particularly *in vitro* work with single molecules, have revealed fascinating details about how they convert chemical energy into mechanical work.

While motors are arguably the most machinelike of the biological molecules, they are certainly not the only things inside living cells that remind us of man-made apparatus. The action at a distance that occurs within an allosteric enzyme, for example, is reminiscent of the push rods or levers inside an internal combustion engine. The large, multienzyme complexes that cells use to carry out sequences of reactions remind us of assembly lines. However, molecular motors are an important special case because the motions they produce are large enough to be directly measured. The study of motor proteins offers rare, direct access to address general questions about how a protein's structure dictates it's dynamics and function.

See also Single DNA, RNA and Protein Molecules at Work.

Bibliography

Books and Reviews

Berg, H.C. (2003) The rotary motor of bacterial flagella, *Annu. Rev. Biochem.* **72**, 19–54.

Block, S.M. (1995) Nanometres and piconewtons: the macromolecular mechanics of kinesin, *Trends Cell Biol.* **5**, 169–175.

Cameron, L.A., Giardini, P.A., Soo, F.S., Theriot, J.A. (2000) Secrets of actin-based motility revealed by a bacterial pathogen, *Nat. Rev. Mol. Cell Biol.* **1**, 110–119.

Howard, J. (2001) *Mechanics of Motor Proteins and the Cytoskeleton*, Sinauer Associates, Publishers, Sunderland, MA.

Inoue, S., Salmon, E.D. (1995) Force generation by microtubule assembly/disassembly in mitosis and related movements, *Mol. Biol. Cell* **6**, 1619–1640.

Salmon, E.D. (1995) VE-DIC light microscopy and the discovery of kinesin, *Trends Cell Biol.* **5**, 154–158.

Schliwa, M. (2003) *Molecular Motors*, Wiley-VCH, Weinheim.

Schliwa, M., Woehlke, G. (2003) Molecular motors, *Nature* **422**, 759–765.

Spudich, J.A. (2001) The myosin swinging crossbridge model, *Nat. Rev. Mol. Cell Biol.* **2**, 387–392.

Vale, R.D., Milligan, R.A. (2000) The way things move: looking under the hood of molecular motor proteins, *Science* **288**, 88–95.

Yoshida, M., Muneyuki, E., Hisabori, T. (2001) ATP synthase–a marvellous rotary engine of the cell, *Nat. Rev. Mol. Cell Biol.* **2**, 669–677.

Primary Literature

Asbury, C.L., Fehr, A.N., Block, S.M. (2003) Kinesin moves by an asymmetric hand-over-hand mechanism, *Science* **302**, 2130–2134.

Berliner, E., Young, E.C., Anderson, K., Mahtani, H.K., Gelles, J. (1995) Failure of a single-headed kinesin to track parallel to microtubule protofilaments, *Nature* **373**, 718–721.

Block, S.M., Berg, H.C. (1984) Successive incorporation of force-generating units in the bacterial rotary motor, *Nature* **309**, 470–472.

Burgess, S.A., Walker, M.L., Sakakibara, H., Knight, P.J., Oiwa, K. (2003) Dynein structure and power stroke, *Nature* **421**, 715–718.

Case, R.B., Pierce, D.W., Hom-Booher, N., Hart, C.L., Vale, R.D. (1997) The directional preference of kinesin motors is specified by an element outside of the motor catalytic domain, *Cell* **90**, 959–966.

Diez, M., Zimmermann, B., Borsch, M., Konig, M., Schweinberger, E., Steigmiller, S., Reuter, R., Felekyan, S., Kudryavtsev, V., Seidel, C.A., Graber, P. (2004) Proton-powered subunit rotation in single membrane-bound F0F1-ATP synthase, *Nat. Struct. Mol. Biol.* **11**, 135–141.

Dogterom, M., Yurke, B. (1997) Measurement of the force-velocity relation for growing microtubules, *Science* **278**, 856–860.

Finer, J.T., Simmons, R.M., Spudich, J.A. (1994) Single myosin molecule mechanics: piconewton forces and nanometre steps, *Nature* **368**, 113–119.

Hancock, W.O., Howard, J. (1998) Processivity of the motor protein kinesin requires two heads, *J. Cell Biol.* **140**, 1395–1405.

Henningsen, U., Schliwa, M. (1997) Reversal in the direction of movement of a molecular motor, *Nature* **389**, 93–96.

Howard, J., Hudspeth, A.J., Vale, R.D. (1989) Movement of microtubules by single kinesin molecules, *Nature* **342**, 154–158.

Hua, W., Young, E.C., Fleming, M.L., Gelles, J. (1997) Coupling of kinesin steps to ATP hydrolysis, *Nature* **388**, 390–393.

Ishijima, A., Kojima, H., Funatsu, T., Tokunaga, M., Higuchi, H., Tanaka, H., Yanagida, T. (1998) Simultaneous observation of individual ATPase and mechanical events by a single myosin molecule during interaction with actin, *Cell* **92**, 161–171.

Lang, M.J., Fordyce, P.M., Block, S.M. (2003) Combined optical trapping and single-molecule fluorescence, *J. Biol.* **2**, 6.

Lister, I., Schmitz, S., Walker, M., Trinick, J., Buss, F., Veigel, C., Kendrick-Jones, J. (2004) A monomeric myosin VI with a large working stroke, *EMBO J.* **23**, 1729–1738.

Loisel, T.P., Boujemaa, R., Pantaloni, D., Carlier, M.F. (1999) Reconstitution of actin-based motility of Listeria and Shigella using pure proteins, *Nature* **401**, 613–616.

Lombillo, V.A., Stewart, R.J., McIntosh, J.R. (1995) Minus-end-directed motion of kinesin-coated microspheres driven by microtubule depolymerization, *Nature* **373**, 161–164.

Mallik, R., Carter, B.C., Lex, S.A., King, S.J., Gross, S.P. (2004) Cytoplasmic dynein functions as a gear in response to load, *Nature* **427**, 649–652.

Mehta, A.D., Rock, R.S., Rief, M., Spudich, J.A., Mooseker, M.S., Cheney, R.E. (1999) Myosin-V is a processive actin-based motor, *Nature* **400**, 590–593.

Meyhofer, E., Howard, J. (1995) The force generated by a single kinesin molecule against an elastic load, *Proc. Natl. Acad. Sci. U S A* **92**, 574–578.

Nishizaka, T., Oiwa, K., Noji, H., Kimura, S., Muneyuki, E., Yoshida, M., Kinosita, K. Jr. (2004) Chemomechanical coupling in F1-ATPase revealed by simultaneous observation of nucleotide kinetics and rotation, *Nat. Struct. Mol. Biol.* **11**, 142–148.

Noji, H., Yasuda, R., Yoshida, M., Kinosita, K. Jr. (1997) Direct observation of the rotation of F1-ATPase, *Nature* **386**, 299–302.

Ray, S., Meyhofer, E., Milligan, R.A., Howard, J. (1993) Kinesin follows the microtubule's protofilament axis, *J. Cell Biol.* **121**, 1083–1093.

Rice, S., Lin, A.W., Safer, D., Hart, C.L., Naber, N., Carragher, B.O., Cain, S.M., Pechatnikova, E., Wilson-Kubalek, E.M., Whittaker, M., Pate, E., Cooke, R., Taylor, E.W., Milligan, R.A., Vale, R.D. (1999) A structural change in the kinesin motor protein that drives motility, *Nature* **402**, 778–784.

Ryu, W.S., Berry, R.M., Berg, H.C. (2000) Torque-generating units of the flagellar motor of Escherichia coli have a high duty ratio, *Nature* **403**, 444–447.

Sakakibara, H., Kojima, H., Sakai, Y., Katayama, E., Oiwa, K. (1999) Inner-arm dynein c of Chlamydomonas flagella is a single-headed processive motor, *Nature* **400**, 586–590.

Schafer, D.A., Gelles, J., Sheetz, M.P., Landick, R. (1991) Transcription by single molecules of RNA polymerase observed by light microscopy, *Nature* **352**, 444–448.

Schnitzer, M.J., Block, S.M. (1997) Kinesin hydrolyses one ATP per 8-nm step, *Nature* **388**, 386–390.

Shaevitz, J.W., Abbondanzieri, E.A., Landick, R., Block, S.M. (2003) Backtracking by single RNA polymerase molecules observed at near-base-pair resolution, *Nature* **426**, 684–687.

Svoboda, K., Block, S.M. (1994) Force and velocity measured for single kinesin molecules, *Cell* **77**, 773–784.

Svoboda, K., Schmidt, C.F., Schnapp, B.J., Block, S.M. (1993) Direct observation of kinesin stepping by optical trapping interferometry, *Nature* **365**, 721–727.

Vale, R.D., Reese, T.S., Sheetz, M.P. (1985) Identification of a novel force-generating protein, kinesin, involved in microtubule-based motility, *Cell* **42**, 39–50.

Vale, R.D., Funatsu, T., Pierce, D.W., Romberg, L., Harada, Y., Yanagida, T. (1996) Direct observation of single kinesin molecules moving along microtubules, *Nature* **380**, 451–453.

Veigel, C., Wang, F., Bartoo, M.L., Sellers, J.R., Molloy, J.E. (2002) The gated gait of the processive molecular motor, myosin V, *Nat. Cell Biol.* **4**, 59–65.

Yasuda, R., Noji, H., Kinosita, K. Jr., Yoshida, M. (1998) F1-ATPase is a highly efficient molecular motor that rotates with discrete 120 degree steps, *Cell* **93**, 1117–1124.

Yasuda, R., Noji, H., Yoshida, M., Kinosita, K. Jr., Itoh, H. (2001) Resolution of distinct rotational substeps by submillisecond kinetic analysis of F1-ATPase, *Nature* **410**, 898–904.

Yildiz, A., Tomishige, M., Vale, R.D., Selvin, P.R. (2004) Kinesin walks hand-over-hand, *Science* **303**, 676–678.

Yildiz, A., Forkey, J.N., McKinney, S.A., Ha, T., Goldman, Y.E., Selvin, P.R. (2003) Myosin V walks hand-over-hand: single fluorophore imaging with 1.5-nm localization, *Science* **300**, 2061–2065.

Yin, H., Wang, M.D., Svoboda, K., Landick, R., Block, S.M., Gelles, J. (1995) Transcription against an applied force, *Science* **270**, 1653–1657.

11
DNA–Protein Interactions

Sylvie Rimsky and Malcolm Buckle
Enzymologie et Cinétique Structurale. LBPA. UMR 8113 du CNRS,
Ecole Normale Supérieure, 61 Avenue du Président Wilson, 94235 Cachan cedex
France

1	**Protein–DNA Interactions** 349	
1.1	Structural Composition of Nucleoprotein Complexes 349	
1.1.1	Protein Structural Motifs 349	
1.1.2	DNA Structures Recognized by DNA-Binding Proteins 350	
1.1.3	The Thermodynamics of the Formation of Nucleoprotein Complexes 351	
2	**The Role of Nucleoprotein Complexes in Gene Expression** 353	
2.1	The Core Transcription Apparatus 353	
2.1.1	The Initiation Complex 353	
2.1.2	The Kinetics of Promoter Recognition and Escape 354	
2.2	Bacterial Chromatin Structure 357	
3	**General Methods for Studying Nucleoprotein Complexes** 359	
3.1	Chemical and Enzymatic Footprinting 359	
3.2	Electromobility Shift Assay 361	
3.3	Cross-linking of Nucleoprotein Complexes 363	
3.3.1	Formaldehyde Cross-linking 363	
3.3.2	Laser UV Cross-linking and Photofootprinting 364	
3.4	Surface Plasmon Resonance (SPR) 367	
3.5	Microcalorimetry 371	
3.6	Fluorescence 372	
3.6.1	Fluorescence Intensity Measurements 372	
3.6.2	Fluorescence Anisotropy 372	
3.6.3	Fluorescence Energy Transfer (FRET) 373	
3.7	Electron Microscopy and Atomic Force Microscopy 373	

Proteins. Edited by Robert A. Meyers.
Copyright © 2007 Wiley-VCH Verlag GmbH & Co. KGaA, Weinheim
ISBN: 978-3-527-31608-3

Bibliography 375
Books and Reviews 375
Primary Literature 375

Keywords

Activator
A protein that accelerates transcription, generally at the level of promoter recognition, by binding to specific sites that may be at, near, or distant from the promoter region.

Core Transcription Complex
Machinery responsible for initiating the transcription of one strand of DNA from a promoter sequence into an RNA molecule. This complex contains RNA polymerase.

Genome
The information coded in DNA for the expression of all the proteins specific to the phenotype of a given organism.

Promoter
The minimal sequences needed to initiate correct transcription. In prokaryotes, this consists of consensus -35 and -10 regions, spacer regions (18 bp $\pm$ 2) between the two, UP (upstream) elements, and upstream activator sequences. In eukaryotes, it often contains TATA box elements and initiation sequences but there is more variation with respect to the nature of the putative TATA region (which may, therefore, be CAAT, GC, or octamer), also these promoters are often assisted by enhancer elements.

Transcription Repressor
A molecule that inhibits transcription generally by sterically interfering with the formation of initiation complexes.

Transcription Factors
Proteins required for recruiting RNA polymerase to the transcription start site to form a preinitiation complex.

Proteins interact with DNA to form nucleoprotein complexes that mediate a host of important cellular processes. Nucleoprotein complexes are at the heart of many processes that include DNA recombination and repair, gene transcription, pathology, viral infection, and in defining DNA accessibility through chromosome structure. The interplay between proteins and DNA is dynamic and must be understood in a time-resolved fashion. Whilst a structural analysis of the individual components of such active complexes is essential, it has to be interpreted within the changing environment of a cell as it responds to alterations in the environment or as it transits through a preprogrammed cycle of growth, differentiation, and finally but not always, apoptosis. While the number of processes involving nucleoprotein complexes is large, and the mechanisms by which they carry out these functions are many and complex, there is, however, a restricted known set of the structure–function relationships involved. In this chapter, we will attempt to sketch out some of the more common of these, always relating this to possible dynamic changes that occur during the establishment of the nucleoprotein complex or take place during its function. We will pay particular attention to complexes involved in gene regulation at the level of DNA organization and at the level of the control of transcription where the information contained in the genome is decoded to produce the RNA complement that constitutes the transcriptome and hence, the potential proteome of a cell at any given moment.

We have restricted our analysis to nucleoprotein complexes involved in processes in prokaryotic organisms for reasons of expediency, clarity, and because it falls within our own personal range of interest. However, the molecular basis for the formation of specific nucleoprotein complexes is clearly "system" independent and will be found across the whole range of living organisms; to paraphrase Jacques Monod "the same mechanisms present in *E. coli* will be at work in an elephant." It's clearly logistically more favorable to study the former than the latter!

1
Protein–DNA Interactions

1.1
Structural Composition of Nucleoprotein Complexes

Proteins are not mere collections of amino acids strung together in a series of chains or globular conglomerates but are modular entities consisting of a select set of three-dimensional structures arranged so as to provide a specific range of functions. In the same sense, DNA is most emphatically not a rod consisting of four bases in a sugar-phosphate backbone double helix, but rather possesses a well-defined local and global structure capable of extensive structural flexibility and of engaging in a myriad of interactions with proteins. Although the potential for the types of interactions is enormous, over the years recurring patterns of specific arrangements have been observed.

1.1.1 Protein Structural Motifs

Protein recognition of DNA sequences involves a wide range of contacts including nonpolar, direct, and indirect (water mediated) hydrogen bonding with DNA bases and direct and water mediated contacts with the sugar-phosphate backbone of the DNA. These interactions may occur in the minor groove or in the major groove or indeed in both grooves simultaneously. The protein structural motifs engaged in these interactions include helix-turn helix, basic leucine zipper (coiled coil) and zinc finger structures, using α-helices, β-sheets, and loops to effect the recognition. A relatively recent study showed that of the three types of interactions characterizing complexes involving protein residues and DNA, namely, van der Waals contacts, hydrogen bonds, and water mediated bonds, two thirds fell into the category of van der Waals contacts and the remainder was more or less evenly distributed between the other two modes. This rather surprising observation suggests that the simple definition regarding protein–DNA interactions as being specific or nonspecific needs to be examined in some detail. A step in this direction has been made by taking into account the observation that in general a protein is presented to a nucleic acid so as to promote one or both of two distinct types of interaction:

- indirect readout in which the protein recognizes structural characteristics on the DNA;
- direct readout in which there is generally a direct hydrogen-bond distance interaction between amino acids and a given base on the DNA.

The three types of interactions outlined above, however, are in fact present in both these categories and indeed a given nucleoprotein complex may involve both direct and indirect readout. Thus, the distinction between sequence-specific recognition and structural recognition becomes blurred by the dependence of DNA structure on base sequence. The local structure is going to influence the accessibility of bases to the protein. Van der Waals contacts occur mainly with the phosphate backbone of the DNA; some contacts are made with the sugars and a few directly to bases. Water mediated contacts make up nearly 15% of all contacts observed to date and hydrogen bonds, simple, bidentate, and complex, are found essentially between the phosphate backbone and amino acid residues. A rule of thumb is that while the direct hydrogen bonds reflect specific interactions, water mediated and van der Waals contacts act as stabilizing factors and may be useful in allowing the protein in indirect readout of the DNA. However, to date, there are no data that support a simple general code for protein–DNA recognition.

1.1.2 DNA Structures Recognized by DNA-Binding Proteins

There are many types of DNA structures reflecting the myriad ways in which a rod of DNA can be spatially organized through the disposition of the bases with the sugar-phosphate backbone and the orientation of the major and minor grooves. A typical list of such structures would include bent or curved DNA, four-way cruciform junctions, locally melted "bubbles," forked junctions, and so forth. We are going to concentrate on protein binding to bent or curved DNA sequences to investigate some of the salient features of DNA structure recognition by proteins. The insertion of short runs of adenines (A tracts) phased with the double helix in a DNA molecule generally induces a bend by a mechanism that remains elusive but apparently has to accommodate the more or less universally accepted notion that the center of curvature is toward the minor groove of the A tract and toward the major groove of intervening sequences. In other words, one would expect compression of the minor groove within the A tract and compression of the major groove of adjoining sequences. A key question relates to how this structure lends itself to recognition by certain proteins. Answering this question will require a concerted approach that takes into consideration the thermodynamic, kinetic, and structural aspects of the situation.

1.1.3 The Thermodynamics of the Formation of Nucleoprotein Complexes

Since protein–DNA interactions are competitive and noncovalent, the rates and extents of formation of complexes are affected by macromolecular concentrations and by the concentrations of low molecular mass competitors such as electrolytes. There are essentially five interactions that are important in determining the stability and specificity of nucleoprotein complexes. These are the hydrophobic effect, the polyelectrolyte effect, aromatic ring interactions, ionic interactions, and hydrogen bonding interactions. The hydrophobic effect in which the association of nonpolar surfaces is driven by the release of water molecules, and the polyelectrolyte effects that derive from the reduction of the polyelectrolyte charge density at low salt concentrations essentially help reduce the unfavorable noncovalent interaction between uncomplexed macromolecules and a solvent and hence, effectively promote the formation of macromolecular complexes.

Thermodynamically, the strategies adopted to attain a favorable free energy of binding (ΔG_{bind}) are rather varied. Simply stated, the measure of the change in work energy of a system during a reaction, that is, during the formation of a nucleoprotein complex, involves an enthalpic change ($\Delta H°$) and an entropic change $-T\Delta S°$, given by the Gibbs relationship

$$\Delta G° = \Delta H° - T\Delta S° \qquad (1)$$

The formation of a nucleoprotein complex should be accompanied by a decrease in $\Delta G°$, that is, $\Delta G° < 0$. It is thus obvious that an enthalpically driven reaction ($\Delta H° < 0$) compensates for an unfavorable entropic change ($-T\Delta S° > 0$) and vice versa. An interesting observation shows that whilst $\Delta H°$ and $-T\Delta S°$ values for a number of nucleoprotein complexes involving specific DNA recognition can vary across a broad range (~ 60 kcal mol^{-1}), the resulting $\Delta G°$ values lie within a small range. This observation that $\Delta H°$ and $\Delta S°$ do not vary independently has important repercussions. In Table 1, are collected the various molecular processes that are associated with either $\Delta H°$ or $\Delta S°$. Although this table is incomplete and oversimplified, it illustrates the combinations available in the play-off between $-T\Delta S°$ and $\Delta H°$.

Strain is obviously a major factor to be taken into consideration. Inducing a distortion on the DNA is clearly enthalpically unfavorable. To unstack a single base pair costs between 7 and 11 kcal mol^{-1}, in the

Tab. 1 Generalized enthalpic or entropic favorability of the main groups of interactions in nucleoprotein and protein–protein contacts. Strain means any configuration for a given atom, molecule or group that deviates from the minimal potential energy for that entity in the free protein or DNA.

Process	Effect on $\Delta H°$	Effect on $-T\Delta S°$
Hydrogen-bond contacts	Favorable	–
Ion-pair contacts	Favorable	–
Nonpolar contacts	Favorable	–
Desolvation	Unfavorable	Favorable
Strain	Unfavorable	Favorable

case of open complex formation between RNA polymerase and a promoter sequence involving the unstacking of at least 12 bp this could cost up to 77 kcal mol^{-1}. It is clearly expensive and at first sight unfavorable for proteins to choose a binding process that induces strain in the DNA. An attractive proposition as to why proteins have evolved this mechanism is that the induced distortion in the DNA can then be used for other, related processes. In the case of the formation of open complexes with RNA polymerase, the apparent cost of the distortion may already have been met to some extent by topological distortions induced in the DNA by the prior binding of activators.

It seems, however, that nucleoprotein complexes that distort the DNA benefit from a $\Delta S°$ driven change in the face of an unfavorable $\Delta H°$, whereas protein–DNA complex formation that does not appreciably distort the DNA is driven by a favorable $\Delta H°$ against an unfavorable $\Delta S°$.

How do induced changes in the protein influence the binding characteristics for a given complex? Notwithstanding the many examples that now exist of three-dimensional structures for nucleoprotein complexes, there remains a dearth of concrete data that provide direct insight into this question. Clearly, there is reduced translational and rotational freedom for both the protein and the DNA in the nucleoprotein complex and this will lead to an unfavorable $\Delta S°$. However, in cases in which little distortion of the DNA occurs, then some induced changes in the protein most probably occur and would thus be associated with an unfavorable entropic change. In other words, there may be associated folding changes in the protein during DNA binding, in order to adapt to a given existing structure on the DNA. However, there are no hard and fast rules for predicting the extent of protein rearrangements during binding, and few precise experiments that provide information on this process.

It is worthwhile at this point to examine the phenomena in thermodynamics known as entropy–enthalpy compensation. The fact is that both ΔH and ΔS are temperature dependent. The variation of enthalpy with temperature at constant pressure is given by

$$C_p = \left(\frac{dH}{dT}\right)_p \tag{2}$$

C_p is the molar specific heat capacity for a single compound and is in fact the amount of heat required to raise the temperature of 1 mol by 1 K. Another way of thinking about C_p is to see it as the capacity of a substance to absorb heat energy without increasing the kinetic energy of the substance and thus without increasing the temperature.

Any change in the heat capacity (ΔC_p) between reactants and products in a reaction therefore can be expressed as

$$\Delta C_p = \frac{d(\Delta H)}{dT} \tag{3}$$

Similarly, entropy changes depend on C_p, at equilibrium, $\delta H = T\delta S$, so

$$C_p = \frac{dH}{dT} = \frac{T\,dS}{dT} \tag{4}$$

$$\frac{\Delta C_p}{T} = \frac{d(\Delta S)}{dT} \tag{5}$$

From Eqs. (3) and (5) we can see that integrating from absolute zero to the temperature of interest we can get

$$H = \int C_p\,dT \text{ and } S = \int \frac{C_p}{T\,dT} \tag{6}$$

Thus, if there is a change in heat capacity during a process then both ΔH

and ΔS will be temperature dependent, since $\Delta G° = \Delta H° - T\Delta S°$, then these effects will effectively cancel, thus producing what is called entropy–enthalpy compensation.

It is curious that most protein–DNA interactions are associated with a strong negative ΔC_p change. Since as mentioned above, DNA-binding proteins may or may not distort the DNA and/or undergo some induced fit during binding, any distortion will presumably be associated with a ΔC_p effect, the entropy–enthalpy compensation may thus play a crucial role and the real synergy that exists between the effect of ΔH and ΔS will need to take into consideration the relationships described by Eq. (6). However, to date, insufficient data are available in order to establish the ground rules for a complete thermodynamic understanding of the formation of nucleoprotein complexes but as the following sections illustrate, techniques are developing that should furnish this information.

2
The Role of Nucleoprotein Complexes in Gene Expression

Gene expression requires intimate interactions between proteins and nucleic acids, and therefore represents an ideal context in which to discuss many of the concepts involved in the formation of nucleoprotein complexes. We are going to concentrate on two aspects of this process; first the process whereby RNA polymerases involved in prokaryotic transcription recognize specific regions of DNA to form competent transcription complexes, and secondly how the spatial organization of the DNA affects this process in the cell.

2.1
The Core Transcription Apparatus

2.1.1 The Initiation Complex

Transcription is the process whereby one of the two strands of DNA are copied into an RNA molecule. The decision to transcribe a given region of DNA is at the core of the regulation of gene expression and thus is a fundamental biological process. The enzymes responsible for transcription are the RNA polymerases. In prokaryotes, RNA polymerase is a relatively large multi-subunit enzyme consisting of 4 subunits $((\alpha)_2, \beta', \beta, \omega, \sigma)$ in the core enzyme and 5 subunits $(((\alpha)_2), \beta', \beta, \omega, \sigma)$ in the holoenzyme. The core enzyme (378 784 Da) consists of the two β (150 625 Da) and β' (155 145 Da) subunits, the dimeric form of the α (36 507 Da) subunit involved in interactions with activators and specific regions on the DNA and the ω subunit. The holoenzyme (449 044 Da), the form that actively recognizes promoter sequences responsible for housekeeping genes, contains the σ^{70} subunit (70 260 Da). Recent three-dimensional structures for the core enzyme and holoenzyme in complexes with DNA have shed considerable light on the structural orientation of the RNA polymerase machinery. However, since the process is essentially a dynamic one, the mechanism by which a σ^{70} containing RNA polymerase operates in order to locate its DNA counterpart, locally separate the two DNA strands, and precisely position the catalytic center at the start site for transcription, is not understood at the molecular level. Without a σ-subunit, *Escherichia coli* RNA polymerase is unable to recognize a promoter sequence. In agreement with the direct contact model originally proposed, several genetic studies have indicated that σ^{70} interacts in a bipartite manner via its conserved regions 2.4 and 4.2, with two hexameric consensus sequences (separated by 17 (±1 bp)) located around positions −10 and −35 relative to the transcription start site. More recent biochemical evidence has corroborated the interaction with the −10 region and direct cross-linking data have revealed the presence of intimate contacts made by σ^{70} with the DNA in the −35 region.

2.1.2 The Kinetics of Promoter Recognition and Escape

The kinetics of promoter formation is best characterized in prokaryotic systems and we shall first describe this in some depth. We are using this as an example of the predominantly dynamic aspect of protein–DNA interactions in a very basic and important cellular context. Most processes in the cell are not at equilibrium, and understanding these processes requires an appreciation of the kinetics involved and the evolution of the complexes over time, especially in a cellular context. The basic pathway leading to mRNA production is depicted in the following scheme:

$$R + P \underset{k_{-1}}{\overset{k_1}{\longleftrightarrow}} RP_C \underset{k_{-2}}{\overset{k_2}{\longleftrightarrow}} RP_I \underset{k_{-3}}{\overset{k_3}{\longleftrightarrow}} RP_O$$

$$\overset{NTP}{\longrightarrow} RP_{init} \overset{NTP}{\longrightarrow} RP_E$$

In this scheme R, P, and RP represent the RNA polymerase, the promoter containing DNA, and the various RNA polymerase-promoter complexes respectively. RP_C denotes the initial closed complex. RP_I includes all subsequent species in which the DNA in the promoter region remains closed. RP_O signifies complexes that have undergone promoter opening. RP_{init} denotes those species performing the reiterative synthesis of short abortive products. The ensuing escape from the promoter leads to the formation of a stable elongation complex, RP_E. The formation of the initial closed complex (RP_C) involves the specific recognition of conserved promoter sequences. While the −35 hexamer is certainly involved in the initial recognition process, the upstream portion of the −10 element has been implicated as well, since both regions are specifically recognized by σ^{70} as double-stranded DNA. The rate of formation of the RP_C complex may be accelerated by base nonspecific binding mechanisms that allow for two-dimensional sliding along the DNA, "hopping" or intersegment transfer between juxtaposed regions of DNA. However, the estimated association rate constants remain in line with a process of three-dimensional diffusion. Moreover, the equilibrium between free RNA polymerase (R) and the RP_C complex (K_B) is extremely rapid at many promoters, meaning that the subsequent isomerization step, and not initial complex formation, is rate-limiting.

However, many promoters are not saturated with RNA polymerase *in vivo*, and can thus be activated at K_B through increased recruitment of the RNA polymerase to the promoter. This most often involves the establishment of contacts between DNA-bound activating proteins and the RNA polymerase that serve to increase binding affinity, although additional DNA–RNA polymerase contacts may also serve this purpose. Activators involved in this step often bind to the DNA upstream of the −35 region, at sites centered on the same side of the DNA helix (the activator protein, CRP – cyclic AMP receptor protein – for example). Thus, RNA polymerase may form favorable interactions between its α-CTD and the activator, which increase initial promoter binding. Alternatively, the α-CTD can mediate an increase in recognition via contacts with a UP element. Or, DNA-binding proteins (such as the factor for inversion stimulation, the protein FIS) may serve to induce a DNA bend, which brings upstream sequences closer to the RNA polymerase allowing for additional RNA polymerase-DNA interactions. Intrinsic bends in the upstream DNA region can also serve to facilitate promoter binding, presumably by permitting similar interactions between the RNA polymerase and upstream DNA sequences. Such interactions with the upstream region may be sufficient to cause topological distortion in the DNA, thereby explaining the change in linking number observed during the formation of RP_C. Indeed a change of −0.2 (from −1.6 to −1.4 in linking number, or the number of times that two strands cross over) was measured for this transition, and has been interpreted as resulting from DNA bending or conformational change. The ensuing torsion in the DNA could be exploited by the RNA polymerase later in the transcription cycle. In support of this idea, it has been shown that superhelical density also plays a role in determining K_B at some promoters.

The formation of the RP_I complex is believed to include an induced torque in the DNA and/or bending, which introduces an unwinding twist near the transcription start site. Investigations into the effects of supercoiling on promoter usage have suggested that there is an early topological untwisting of 0.3–0.5 of a helical turn, which takes place during the first isomerization (i.e. before promoter opening). These data have lead to the attractive hypothesis that the formation of RP_I aligns the −35 and −10 hexamers on the same side of the DNA helix. Evidence for DNA bending in this complex is provided by scanning atomic force microscopy, native gel band shift analysis and the DNase I hyperreactivity of bases in the vicinity of −38, −48, and the spacer region.

The transition to an open complex (RP_O) is accompanied by DNA strand separation, which is thought to begin within the −10 region followed by propagation downstream past the transcription start site (+2 or +3). This melting of between 1 and 1.5 turns of DNA thus renders the template strand single stranded and accessible to the initiating NTP (as well as chemical probes such as dimethyl sulfate (DMS), diethyl pyrocarbonate (DEP) and $KMnO_4$). However, the DNA backbone between −10 and +20 remains protected from HO radical attack, and the DNase I footprint is unchanged from that of RP_I, indicating that promoter opening occurs without any additional displacement of the RNA polymerase on the DNA. Also the nontemplate strand remains inaccessible to chemical attack.

No external energy source is necessary for the transition to RP_O, making it likely that the RNA polymerase not only lowers the activation energy for DNA strand separation but forms favorable interactions in the open complex as well. As mentioned previously, it has been postulated that some of the required energy comes from the release of a stressed RNA polymerase-DNA intermediate. This intermediate is presumed to be RP_I. Upon release of this strained species, one would expect conformational changes within the RNA polymerase and/or DNA. In support of this idea, during the RP_I to RP_O transition, there was a change noted in the direction of curvature of the DNA entering and exiting the RNA polymerase and in the linking number (from −1.2 in RP_I to −1.8 in RP_O).

In addition to this role in the facilitation of strand opening, interactions with the RNA polymerase are also believed to stabilize the single-stranded region. Indeed, recent studies have demonstrated that the σ^{70} subunit is competent to recognize specifically single-stranded oligodeoxynucleotides corresponding to the nontemplate strand −10 sequence. The binding affinity measured for this sequence was higher when the region was single stranded than for the analogous double-stranded fragment, thereby implicating a single-strand specificity in this interaction

Once the DNA strands have been locally melted and the stable open complex formed, the template strand becomes accessible to incoming nucleoside triphosphates. The nucleotide at position +1 on the template strand will, by Watson–Crick base pairing, determine the identity of the 5' end of the transcript. The binding of the initiating NTP to the active site serves to stabilize the RNA polymerase–promoter complex, even in the absence of hydrolysis or bond formation. Transcript elongation requires the catalysis of phosphodiester bond formation between the hydroxyl at the current 3' end, and the subsequent NTP, which will be determined by base pairing with the following nucleotide.

However, a significant proportion of the initiating complexes do not immediately go on to synthesize productive transcript. Instead, these complexes repetitively form and then release small transcripts (up to 10 nt) without either leaving the promoter region or releasing RNA polymerase-promoter contacts. This reiterative process, termed abortive initiation, results in the cyclic formation of short nonproductive mRNAs. These small transcripts dissociate from the stably bound polymerase, leaving the active site free once again to bind the initiating NTP. At each early nucleotide position, the decision between transcript release and continued elongation relies on a kinetic competition between the two processes. It follows that the duration of RNA polymerase stalling on a promoter is determined by the strength of RNA polymerase-promoter contacts as well as flanking promoter sequence. In this way, a promoter with sequences optimized for stability at preceding steps can be hindered by these same contacts during promoter clearance. Therefore, each functional promoter must represent a delicate equilibrium between stable complex formation and facile escape.

Productive mRNA synthesis is only achieved after the RNA polymerase ceases this process and moves downstream of the promoter region. This promoter clearance is thought to require the release of the specific RNA polymerase contacts with the promoter region, accomplished by the separation of the core enzyme from σ^{70} during the RP_{init} to RP_E transition. In accordance with this idea, the size of the DNase I footprint produced by RP_E is considerably diminished in comparison to initiating complexes (20–30 bp vs 65–95 bp). The observed decrease in promoter protection coincides with promoter clearance, indicating that core-σ dissociation occurs at this step. The transcription bubble is also enlarged during this transition, from ∼10 to between 17 and 18 nucleotides. The requirement for additional melting of the DNA in the downstream region may also explain the importance of downstream sequence composition in promoter escape.

The resultant elongating complex is incredibly salt tolerant, resisting up to 350 mM NaCl. Thus, the core polymerase in RP_E binds DNA in a fashion that is both extremely stable to dissociation and yet capable of rapid translocation along the DNA. There is an obvious parallel between this RNA polymerase activity and that displayed by the DNA polymerases. However, there is no apparent analog for the "sliding clamp" apparatus for the bacterial RNA polymerases. Numerous studies have probed the conformational changes undergone during the transition by the core subunits themselves. As such it has been proposed that "jaw-like" regions of the β' subunit close around the DNA forming close interactions with the downstream sequences.

2.2
Bacterial Chromatin Structure

In simple terms, the decision to express a given gene is, as we have seen above, taken at the level of recognition of promoter sequences by the transcription machinery. This is clearly a complex process involving a number of proteins, some of which are molecular motors, working in a concerted kinetic fashion. However, in vivo, this dynamic process takes place against the backdrop of the DNA organization within the cell. Although a lot has been achieved in the study of eukaryotic chromosome structure, our general understanding of the organization and expression of genetic information remains incomplete. The progress of research in bacteria has been hampered in part by a low degree of DNA compaction (albeit around 1000-fold) and resultant instability of the bacterial chromatin. Studies conducted in the paradigm organism E. coli revealed that most of the structural proteins of the bacterial nucleoid (Fig. 1) are synthesized in a growth phase-dependent manner.

The relative abundance of several major nucleoid-associated proteins (Fig. 1) in exponential phase is FIS > HU > H-NS > IHF > Lrp, whereas the order of abundance on entry in stationary phase changes to Dps > IHF > HU > H-NS > Lrp > FIS. In organisms like E. coli in which HU is composed of two subunits (α and β), the subunit composition also changes as a function of growth phase. This change in the chromosomal protein composition on transition from exponential growth into stationary phase is accompanied by silencing of genome functions. Thus, these proteins not only define the architecture of the bacterial chromosome but also specifically change the pattern of gene expression in a growth phase-dependent manner. This coupling between the modulation of chromosome structure and the pattern of gene expression is further reinforced by the topological fluctuations of DNA during the growth phase and the strong dependence of bacterial promoter activity on DNA topology.

Although the existence of a link between the metabolic reorganization of the cell and the transitions in the architecture of the bacterial chromosome is widely accepted, how the alterations in cellular physiology are coupled with alterations in the topology of the chromosome remains largely unexplored. We will briefly discuss the role of one of the proteins involved in this process – the H-NS protein – in the light of its DNA-binding properties linked to its function as a transcription repressor.

H-NS is an atypical global regulator involved in the expression of genes, which seem to share the common property of being involved in the cells response to changes in the external environment. Modification of the *hns* gene modifies or completely abolishes the expression of around 200 proteins in E. coli. H-NS is involved in such diverse functions as the response to osmolarity changes, the use of certain carbon sources, antibiotic resistance, ribosomal RNA regulation,

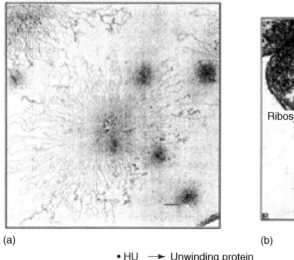
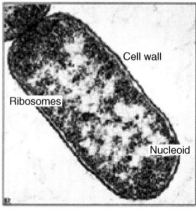

(a) (b)

- HU → Unwinding protein
- IHF → Integration host factor
- FIS → Factor for inversion stimulation
- Lrp → Leucine responsive protein
- H-Ns → Histone-like nucleoid structuring protein
- Dps → DNA binding protein from starved cells

Fig. 1 Structural proteins of the bacterial nucleoid. Electron micrographs of (b) a complete E. coli bacterium showing the DNA condensed (the white material) within the cell) and (a) the same cell exploded after salt treatment showing the DNA now completely unwound, note that the horizontal blue bar in (a) shows the length of the original prior to disruption. The organization of the DNA is mostly carried out by the proteins listed (see color plate p. xv).

pili formation, and the activation of virulence genes in pathogenic bacteria. H-NS affects transcription in a differential manner depending on the growth phase, and consequently is strongly involved in the processes of virulence and silencing in bacteria.

Although H-NS does not exhibit a high DNA sequence specificity, H-NS recognizes and preferentially binds to DNA-containing phased A_n tracts that are intrinsically curved. A number of H-NS responsive promoters have indeed been shown to contain regions of intrinsic DNA curvature located either upstream or downstream of the transcription start point. The available data support the idea that the presence of such DNA sequences *in vivo* renders proximal promoters specifically responsive to transcription modulation by H-NS. A crucial question therefore concerns on the nature of the nucleoprotein complexes involving H-NS. A working model for H-NS binding has been proposed in which differential binding to relatively high and low affinity sites occurs. The formation of an H-NS/DNA complex at a region containing high affinity binding sites constitutes a first step of nucleation; this is followed by a buildup of H-NS binding to lower affinity sites on the DNA leading eventually to total occlusion of the DNA by a process of propagation. The high affinity sites clearly require the presence of a singular structural feature and the helical phasing of the recognized motifs with the periodicity of the DNA. Note, therefore, that specificity of binding is a function of both the intrinsic structure and the deformability of the DNA. An essential feature of this model is that this propagation step is highly dependent on the formation of a "correct" nucleation event. How this binding mode is related to the repression of transcription at proximal or distal promoters remains an object of intensive research; however, a plausible mechanism is that the formation of the nucleation step leads to a facilitated coverage of nonspecific sites on the DNA and that interactions between H-NS bound at "specific" and "nonspecific" sites are responsible for the formation of higher order structures that either physically occlude RNA polymerase from promoter regions (even promoter regions that are not within high affinity H-NS binding-site regions) or provide torsional resistance to the RNA polymerase induced DNA transformations required for the formation of competent transcription complexes. The higher order structure produced by H-NS is most probably not a simple filament that would in general be insufficiently compacted to explain the observed condensation of the DNA, nor would a simple filament explain the phenomenon of silencing in which H-NS is capable of repressing transcription at promoters quite distal to the sites of binding at concentrations incommensurate with the complete coverage of a filamentous DNA. Quite clearly, H-NS is a complex transcriptional regulator exerting its effects through an unusual binding mechanism and forming higher order nucleoprotein complexes. It is to be expected that this mechanism will turn out to be even more intricate when the full nature of interactions between H-NS and other proteins within the bacterial chromatin organization machinery are unraveled.

3
General Methods for Studying Nucleoprotein Complexes

3.1
Chemical and Enzymatic Footprinting

The reactivity of DNA to various chemical and enzymatic reagents has been extensively utilized to probe the local and global structure of DNA in nucleoprotein complexes. The wealth of information that this has yielded has been relatively recently extended by the possibility of applying certain of the techniques in a time-resolved fashion. We will describe the main techniques used in what we call a static (i.e. at equilibrium) and a dynamic (i.e. time resolved) context. We will also try to illustrate attempts to access changes in protein conformation during nucleoprotein complex formation.

Table 2 contains a nonexhaustive list of the reagents that target DNA.

In general, a nucleoprotein complex is treated under specific conditions with the reagent and the products of the reaction examined by denaturing electrophoresis and phosphor imager densitometry. The choice of a specific reagent depends on the nature of the nucleoprotein complex. If knowledge of the position of a protein on a given DNA molecule is required, then DNAse I, exonuclease III, and the more precise hydroxyl radical approach are used. These techniques may also furnish information concerning the quantification of the reaction. The results obtained from attacks by DNAse I are referred to as footprints for the simple reason that the DNA in nucleoprotein complexes will not be accessible to the DNAse I and hence will not be cleaved in those regions of the DNA to which the protein is bound. DNase I, because it cleaves via the minor groove,

Tab. 2 Reagents that target nucleic acids.

Reagent	Nature	Target
Dnase I	Enzymatic	Phosphodiester bond in sugar-phosphate backbone. Attack via binding in the minor groove. Sequence independent, structure dependent
Exonuclease III	Enzymatic	$3'-5'$ exonuclease activity, removes mononucleotides. Sequence dependent, structure independent.
Hydroxy radicals	Chemical	Phosphodiester bond in sugar-phosphate backbone. Sequence independent, structure dependent.
Dimethyl sulfate (DMS)	Chemical	N-7 of guanines in major groove. Also cytosines when unpaired in single-stranded DNA. Sequence dependent, structure dependent.
Diethyl pyrocarbonate (DEP)	Chemical	Carboxyethylates N-7 of imidazole ring of purine residues in distorted DNA (deviation from classical B-form). Sequence dependent, structure dependent.
Potassium permanganate ($KmnO_4$)	Chemical	Oxidizes nonsaturated double bonds to vicinal diols, reacts preferentially with thymines. Sequence dependent, structure dependent.
Osmium tetroxide (OsO_4)	Chemical	Unstacked thymines. Sequence dependent, structure dependent.
1,10-Phenanthroline-copper	Chemical	Cleaves phosphodiester backbone through minor groove. Sequence independent, structure dependent.
Singlet oxygen	Chemical	Generated by irradiation of eosin isothiocyanate in the presence of Tris, singlet oxygen diffuses into the DNA and modifies bases. Sequence independent, structure dependent.
UV laser	Photonic	Photochemical rearrangement of bases, leads primarily to pyrimidine dimer formation and other rearrangements generally involving guanine residues. Sequence dependent, structure dependent.

will produce enhanced cleavage at those regions of the DNA possessing a widened minor groove. Thus, by looking at DNAse I cleavage patterns one can deduce, not only that region occupied by the protein but also those regions in which the DNA has been distorted leading to a widened minor groove.

Sometimes it is useful to probe changes in the local conformation of the DNA. In these cases, reagents that react with unusual configurations are used. Potassium permanganate, for example, which oxidizes unsaturated bonds in pyrimidines, reacts almost exclusively with thymines in which unstacking between a thymine and an adjacent base on the same DNA strand has exposed the 5–6 double bond containing the vicinal diols. Oxidation of thymines produces a labile DNA molecule that may be cleaved at the sugar-phosphate bond adjacent to the modified base. Such modifications may then be visualized, as for DNaseI cleavage, by the use of end-labeled DNA and separation on denaturing polyacrylamide sequencing gels. Consequently, where there is a local change in DNA structure (such as local melting of the DNA strands, characteristic of the formation of an initiating transcription complex with RNA polymerase), the extent of distortion may be determined by the relative reactivity of the thymines to KMnO$_4$. Similarly, the reactivity of cytosines (DMS), adenines (DEP) (Table 2), and thymines (OsO$_4$) may be examined in different nucleoprotein contexts. 1,10-Phenanthroline-copper is a reagent that intercalates in the minor groove of DNA and can, under specific conditions, generate radicals that then cleave the sugar-phosphate backbone. It, therefore, provides information on the relative geometry of the DNA, that is, the minor groove, at specific points on the DNA. One can generate singlet oxygen molecules, which then migrate into the DNA and modify bases; the modified bases render the DNA suitable for cleavage in a similar fashion to the chemical modification (KMnO$_4$ etc.), and thus the site of modification may be simply ascertained.

The use of a reagent is limited by two considerations, the spatial resolution provided by the probe (DNAse I, for example, needs to bind in the minor groove and also requires a specific binding region on the DNA, which therefore limits the spatial resolution of the ensuing footprint), and the degree of perturbation of the nucleoprotein complex by the probe itself. Eosin isothiocyanate, used to generate singlet oxygen, actually very strongly perturbs many nucleoprotein complexes and thus one has to be sure that the structure that is being hit by the reagent is not in fact an artifact due to rearrangements in the nucleoprotein complex induced by the reagent itself.

The judicious use of these types of reagents provides useful information about the local structure of the DNA in these complexes. Indeed, these reagents have provided enormous insights into many, by now classical structures, ranging from cruciforms, and Holliday junctions to transcription initiation complexes.

In cases in which the formation of a nucleoprotein complex occurs over a timescale in excess of tens or hundreds of milliseconds, quite a few of the techniques listed in Table 2 are sufficiently rapid that they may be used to probe changes taking place during this process. This is an extremely important and powerful approach since many of the processes involving nucleoprotein complexes are practically irreversible and knowledge of the kinetics of these processes is essential for a full understanding of the mechanisms involved. Table 3 illustrates how three of the techniques used to probe DNA may be used in a time-resolved fashion.

3.2
Electromobility Shift Assay

The basis of the electromobility shift or more commonly called band shift assay is that protein–DNA complexes remain intact during migration in gel electrophoresis and thus appear as distinctly more slower moving bands than the free DNA fragment. The assay is simple and quick, and the use of radioactive binding-site DNA makes it highly sensitive. It can be used for both highly sequence-specific and nonspecific proteins, such as histones. Band shift assays can be used quantitatively to estimate dissociation constants for protein–DNA complexes. Additionally, band shift experiments can be used to visualize protein–protein interactions between a DNA-binding protein and other non-DNA-binding proteins. The binding of a second protein to a protein–DNA complex to form a triple complex is visualized by a further retardation of mobility that is called a *super shift*. Super shift experiments can

Tab. 3 Some time-resolved methods to study the kinetics of DNA–protein interactions.

Method	Target and spatial resolution	Time resolution	Advantages	Disadvantages
UV laser photofootprinting	van der Waals contacts, T-dimers	~10 ns	ns pulses no need to add any reagents	DNA damage only < 2 Å low quantum yield
DNase I footprinting	Protection and hypersensitive sites	30 ms	Strong signal	Needs Mg^{2+} large footprint
OH X-ray footprinting	Solvent accessibility of bases (amino acids)	3 ms	Single base resolution no need to add any reagents	Sensitive to quenchers such as glycerol and Tris buffer

also be used to assay the binding of a second DNA-binding protein to the DNA or the binding of a second molecule of DNA to the protein.

The band shift assay can be used simply to estimate dissociation constants for nucleoprotein complexes involving just one protein and its binding site. Two methods are generally used; in the first the protein concentration needs to be known accurately, whereas in the second method both the protein and the DNA concentrations are required. In both methods, the amount of free DNA and the amount of DNA bound in the protein–DNA complex have to be estimated either by measuring the intensity of a band in an auto radiograph using techniques such as classical densitometry, or more commonly nowadays, phosphor imaging. An important caveat when using the band shift assay is that association rate constants may be altered by the concentrating influence of the so-called cage effect of the polyacrylamide gel and that dissociation of the complexes may be drastically affected by the change in buffer conditions as the complex enters the gel. The net result may well be important deviations from equilibrium and thus band shift assays are most useful for comparative studies rather than for obtaining absolute equilibrium values.

The first method involves the use of a low DNA concentration (below the dissociation constant) compared to a relatively large range of protein concentration. Since, in a simple situation in which a protein P binds to a DNA-binding site D, the concentration of nucleoprotein complex [PD] is related to the dissociation constant K_D by

$$K_D = \frac{[P][D]}{[PD]}$$ such that when

$[D] \ll K_D$ then $[P]_{free} \cong [P]_{total}$

and so $K_D = ([P]_{total} \cdot [D])/[PD]$ in other words, when the concentration of DNA is equal to the concentration of the protein–DNA complex (50% complex formation) then the dissociation constant is given by the total protein concentration.

The second method for determining dissociation constants requires the preparation of a range of different concentrations of a one-to-one molar mix of protein and DNA. The proportion of DNA in the complex at the different concentrations depends upon the ratio between the total DNA concentration and the dissociation constant. The most convenient way to perform this measurement is to titrate (labeled) DNA at high concentrations (i.e. at least 1000 times the estimated dissociation constant) with increasing amounts of protein. Analysis of a band shift gel allows the determination of the point at which a one-to-one complex is formed. Subsequently, a one-to-one complex, formed under these conditions, is diluted serially and the analysis repeated at different DNA concentrations to estimate the percentage of complex. The concentration of complex is dependent upon the dissociation constant and the concentration of DNA, since in the equilibrium

$$P + D \longleftrightarrow PD$$

$$K_D = \frac{[P]_{free} \cdot [D]_{free}}{[PD]} \quad (7)$$

and

$$[D]_{total} = [PD] + [D]_{free} \quad (8)$$

since the required condition is that

$$[P]_{total} = [D]_{total} \quad (9)$$

then

$$[P]_{free} = [D]_{free} \quad (10)$$

and

$$K_D = \frac{([D]_{total} - [PD])^2}{[PD]} \quad (11)$$

so

$$[PD]^2 - (2[D]_{total} + K_D)[PD] + [D]_{total}^2 = 0 \quad (12)$$

then

$$[PD] = \frac{(2[D]_{total} + K_D) - \sqrt{(2[D]_{total} + K_D)^2 - 4[D]_{total}^2}}{2} \quad (13)$$

It must be emphasized that the apparent K_D values obtained are primarily of use in a comparative sense rather than to allow a rigorous thermodynamic analysis.

3.3
Cross-linking of Nucleoprotein Complexes

3.3.1 Formaldehyde Cross-linking

Formaldehyde is a highly reactive reagent that produces protein–DNA and protein–protein cross-links between macromolecules in close contact. The distance between the cross-linking groups of protein and DNA should be around 2 Å suggesting an interaction at the range of van der Waals radii. Therefore, the formaldehyde cross-linking procedure allows identification of proteins and protein domains in close proximity to the DNA. Formaldehyde cross-linking had been widely used for studies of protein–DNA interactions in chromatin *in vivo* as described in detail. Although the reaction of formaldehyde with nucleotides and DNA is well characterized, the exact mechanism of the cross-linking reaction between proteins and DNA is not well known. It was shown that formaldehyde reacts with the amino groups of cytosines, guanines, and adenines and the imino groups of thymines and probably guanines. In proteins, potential candidates for cross-linking are lysine, arginine, tryptophan, and histidine residues. The first stage includes the reaction with amino or imino groups and results in the formation of methylol derivatives that react with the adjacent second reacting group. In conditions in which short cross-linking times are used and cross-links occur only in the melted DNA regions, the imino groups of thymine (or guanine) bases are highly probable targets of cross-linking reaction. This assumption is based on two known observations:

- Kinetic studies demonstrated slow formaldehyde reaction rates with amino

groups while reaction rates of imino group were fast.
- Reactions with the amino groups of purines can take place with double-stranded DNA as well as with melted DNA regions, while the reaction with imino groups of thymine occurs only in melted DNA regions.

The product of the first reaction step (methylol derivative) is very unstable while the final cross-linked derivatives are stable at 24 °C. Cross-linked complexes of RNA polymerase subunits can be incubated for several hours at room temperature without significant degradation. At 50 °C, slow degradation of protein–DNA cross-links was detected after 20 min of incubation in Tris-HCl containing buffer ($t_{1/2}$ = 90 min). Incubation for 2 min at 95 °C leads to full disruption of the cross-links.

The identification of proteins or subunits cross-linked under different experimental conditions is crucial. Two methods provide simple and direct ways to characterize these contacts. The first method requires polyclonal antibodies to individual proteins or subunits and is a modification of the method used for fractionation of cross-linked chromatin. The method includes precipitation of the cross-linked complexes with subunit-specific antibodies and purification on protein A-Sepharose affinity chromatography. Analysis of bound and nonbound fractions on SDS-PAGE allows the identification of those subunits involved in any particular cross-linked complex. The second method utilizes genetically engineered subunits containing histidine-tags at the carboxyl end or the amino-terminal end. In this case, the cross-linked complexes must be denatured by urea and subjected to fractionation on Ni^{2+}-NTA agarose affinity columns.

3.3.2 Laser UV Cross-linking and Photofootprinting

The intrinsic photoreactivity of DNA and RNA is a useful parameter to measure local DNA structure and identify regions in contact with proteins in given nucleoprotein complexes. In order to satisfy two main criteria of nonperturbation and selectivity, an optimum source of photons must be rapidly delivered to the sample. This is possible by using a UV laser providing high-energy photons (266 nm) in a brief pulse (5 ns) on a small volume of material (10–20 µL). The minimal experimental set up that is required is shown in Fig. 2.

The laser source is a neodymium laser consisting of Nd^{3+} ions at low concentration in yttrium-aluminum-garnet (YAG = $Y_3Al_5O_{12}$) that produces a continuous monochromatic beam at 1064 nm. In Q-switching mode, in conjunction with nonlinear crystals, this beam is quadrupled in frequency to give a homogenous polarized source of photons at 266 nm (approximately 10^{17} photons per pulse proving a dose of around 20–30 mJ per pulse). Since the beam has a diameter of approximately 6 mm, this gives complete irradiation of a solution of 10 to 20 µL at the bottom of an Eppendorf tube. The optimal beam form is a Gaussian rather than a "doughnut" configuration. The final 266 nm beam is selected using an array of dichroic mirrors and directed onto the sample using a Pellin–Brocca Prism (Fig. 2). To all intents and purposes, only nucleotide bases are excited at 266 nm. At the photon fluxes used in the setup described in Fig. 2, each base in a solution of DNA, for example, will see at least one photon carrying approximately 4 eV. Under the conditions used here, the ensuing photochemistry is monophotonic and the DNA is not ionized. This is generally confirmed by the absence of nicks

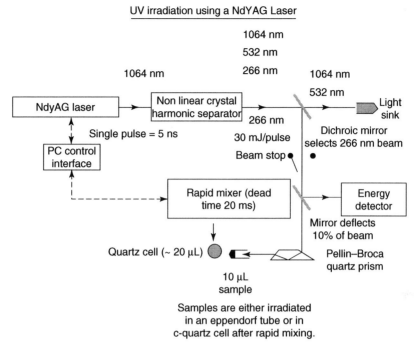

Fig. 2 Typical UV laser set up for the irradiation of nucleoprotein complexes. Complexes may be irradiated in an Eppendorf tube or as they pass through a mixing cell in a stopped flow apparatus.

on DNA strands following irradiation. It should be noted that the ionization energy for thymines is 7.6 eV in water. Furthermore, we can assume that in simple terms, the excited state for the base follows the profile shown in Fig. 3.

Irradiation thus produces excited singlet states of very short (picosecond) duration and a very small proportion of triplet states with longer (µs) lifetimes. In consequence, the photo reactive species are relatively short lived and 98 to 99% are consumed within the 5 ns duration of the laser pulse. What then is the fate of the excited species? There are a number of potential photoreactive products but three in particular are of interest here. The first is the formation of pyrimidine dimers. Adjacent nonsaturated conjugated 5–6 positions in the pyrimidine ring can, following excitation, form covalent cyclic products, notably 4–6 and 5–6 cyclobutanes. The quantum yield for the 4–6 cyclobutanes is around 4%, which essentially means that for a 100 bp DNA, there will be one thymine dimer per molecule following a single pulse at 266 nm. The second event concerns modification of guanine residues. The nature of this reaction is somewhat obscure. The quantum yield may be relatively high and the overall chemistry may require a biphotonic event. The final photochemical event is perhaps the most interesting and involves the formation of covalent links between excited bases and amino acids. The conditions for this cross-linking are that the reactive species are within van der Waals radii distance of each other.

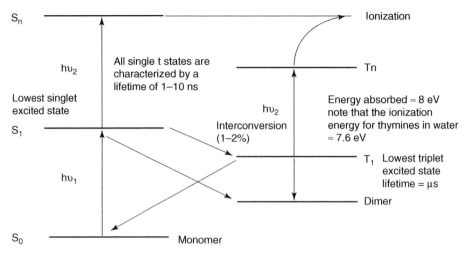

Fig. 3 Electronic states and lifetimes of excited states of atoms in nucleic acids following irradiation. The S1 state is that which is most commonly involved either in cross-linking with amino acids or undergoes internal rearrangement to structures such as pyrimidine dimers, for example.

These limitations arise essentially because of the very short lifetimes of the reactive species (ps). This limitation is of course present for pyrimidine dimer formation and represents one of the major advantages of the technique, namely, that it is extremely rapid, thus reporting configurations that exist before perturbations occur. A simple approach to determine the extent of UV photo-cross-linking to a DNA substrate is to separate the protein–DNA adduct(s) from non–cross-linked DNA by electrophoresis on sodium dodecyl sulfate (SDS)-polyacrylamide gels (SDS-PAGE). Quantification of the separated radio-labeled species is then performed by densitometry, for example, using a PhosphorImager. For most purposes, complexes are formed using 5′-end labeled oligonucleotides (i.e. labeled with $(\gamma\text{-}^{32}\text{P})$-ATP and polynucleotide kinase). After single-pulse UV irradiation, the samples are separated by SDS-PAGE electrophoresis. The gel matrix should be optimized to allow a clear separation between the protein–DNA adduct and the DNA substrate. The electrophoretic migration of the UV-induced adduct will be dependent on its apparent mass, which, in most cases, equates directly to the sum of the covalently linked components (i.e. the apparent mass of the cross-linked DNA strand plus that of the attached polypeptide chain, calculate approximately 325 Da for each nucleotide of the cross-linked strand). If the DNA fragment is too large to be separated clearly, then an alternative methodology must be considered, for example, by first fragmenting the DNA or by degrading it with a nuclease.

When analyzing cross-linking data, it is important to perform measurements over a range of protein concentrations. This is essential when comparing the relative cross-linking efficiencies of different DNA substrates, since it is the saturation isotherm and not the absolute level of DNA cross-linking that contains the information on binding affinity. Generally, titrations of fixed concentrations of the oligonucleotide are performed over a gradient of increasing protein concentration. A typical SDS-PAGE gel (visualized by autoradiography) and resulting binding curve analysis is shown in Fig. 4

The progressive fractional saturation of potential DNA-interacting sites is signaled by the increase in adduct formation, measured in terms of the percentage of the total DNA cross-linked (S_{obs}). Maximal cross-linking (S_{max}) is obtained under saturating concentrations of protein in which all the DNA-binding sites are occupied. Adduct formation may thus be represented simply as

$$S_{obs} = \theta \cdot S_{max} = \theta \cdot \lambda_c \quad (14)$$

where θ ($= S_{obs}/S_{max}$) denotes the fractional saturation of the DNA lattice and λ_c represents the quantum yield of adduct obtained at saturation of the lattice with ligand, a constant factor dictated by the photochemical reactivity of the oligonucleotide sequence occluded and the number and nature of protein–DNA contacts formed. At 266 nm, λ_c is highly dependent on the base composition of the occluded site. Comparison of the data obtained for two independent proteins (gene32 protein and HIV-1 integrase), both of which are capable of binding nonselectively to ssDNA, indicate that the reactivity of each base may be ordered (relative to dT = 1) as T(1) ≫ C(0.04/0.08) > A(0.03/0.04) ≫ G(∼0). Thus, as a general "rule-of-thumb," pyrimidine residues are more reactive than their purine counterparts, with d(T) by far the most efficient at producing protein–DNA cross-links. Of course, in order for an adduct to form, a close contact must exist between an amino acid and the reactive base. While most amino acids appear capable of cross-linking, the photochemistry is poorly understood and the influence of the side chains on the efficiency of the reaction cannot be predicted for a complex nucleoprotein structure. Least-squares analysis of graphical representations of the degree of cross-linking parameter θ as a function of ligand concentration (Fig. 4) then allows the fractional saturation to be derived. This in turn may then be used to determine binding isotherm characteristics from which an apparent equilibrium constant K_{obs} and apparent stoichiometry may be obtained from curves of the type shown in Fig. 4.

A simple technique for establishing those bases having undergone a specific photoreactivity is that of primer extension. Following irradiation, the bases on a given DNA molecule will have been photomodified and the assumption is that for any given DNA molecule, only one photoactive event will have taken place. Figure 4 illustrates the methodology involved. Essentially, a DNA polymerase such as the Klenow fragment is used to carry out primer extension of end-labeled oligomers along an irradiated template. The Klenow fragment terminates extension at position $n - 1$ with respect to a photomodified base. The photomodification may be pyrimidine dimer formation; intrinsic photomodification of a given base (e.g. G) or the presence of a covalent link with a protein. Products may be easily visualized on denaturing sequencing gels and the extent of photomodification determined using phosphor imager densitometry.

3.4
Surface Plasmon Resonance (SPR)

SPR is a technique that paradoxically measures apparent rate constants for the interaction between an immobilized

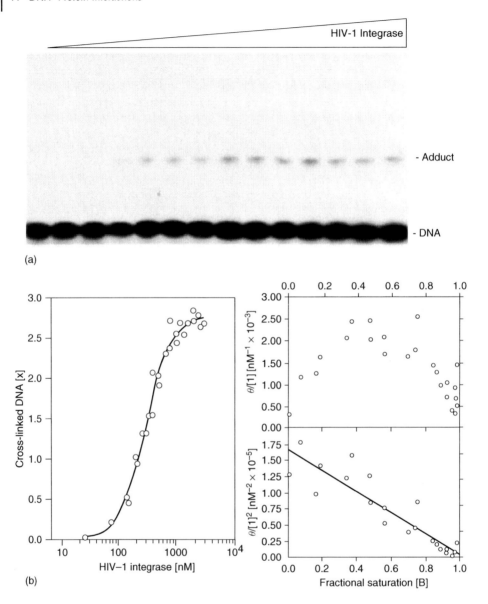

(a)

(b)

ligand and an analyte in solution in order to calculate equilibrium binding constants. The general scope and utility of this approach has been covered in a number of reviews. It is, however, worthwhile here to review the basic principles behind the technique in order to understand its application to studying nucleoprotein complexes. The SPR principles used by the main commercial supplier of this technology (Biacore™) are illustrated in Fig. 5.

Fig. 4 Laser-mediated UV cross-linking of the HIV-1 integrase to DNA. The HIV integrase protein was incubated with a 21-bp duplex DNA molecule (25 nM) for 10 min at 21 °C prior to exposure to a 5-ns pulse of 266 nm UV light. (a) Cross-linked adducts were resolved from un–cross-linked DNA by SDS-polyacrylamide gel electrophoresis. The autoradiograph shows part of a typical titration experiment performed by varying the integrase concentration between 0 and 3 mM. (b) Data points represent the quantification (by phosphor image densitometry) of gels such as the one shown in (a). The sigmoidal curve, shown in semilog form (*panel A*), represents the best fit of the data by nonlinear least-squares procedures in terms of the total HIV-1 integrase concentration and corresponds to a Hill coefficient of 1.90. In *panel B*, the data are transformed, in terms of fractional saturation, to Scatchard form for a first-order (*top*) and second-order (*bottom*) reaction. In this case, the data points are corrected for the bound protein assuming a density of 10 monomers per LTR at saturation. The Hill coefficient is 1.94 after transformation.

Light from a laser source arriving through a prism at a gold surface at the angle of total internal reflection (θ) induces a nonpropagative evanescent wave that penetrates into the flow cell opposite the prism. The intensity of the reflected light is continuously monitored. At a given angle (λ) dependent upon the refractive index of the solution in the flow cell, resonance between the evanescent wave and free electrons in the gold layer results in a reduction in the intensity of reflected light. The change in angle of reduced intensity ($\Delta\lambda$) reflects changes in the refractive index (n) of the solution in the flow cell immediately adjacent to the gold layer. A dextran surface coupled to the gold layer allows immobilization of ligands

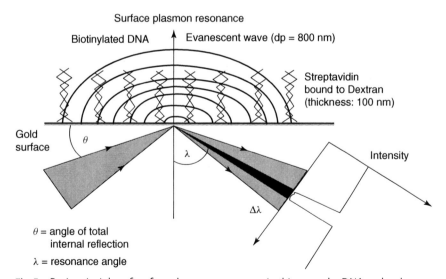

Fig. 5 Basic principles of surface plasmon resonance. In this example, DNA molecules are immobilized through a biotin–streptavidin interaction at the dextran/gold interface. Changes in refractive index as molecules bind to the surface are measured by alterations in the angle λ at which extinction of reflection occurs.

within the evanescent field. In physical terms, the detection system of a typical SPR machine consists of a monochromatic, plane-polarized light source and a photodetector that are connected optically through a glass prism. A thin gold film (50 nm) thick, deposited on one side of the prism, is in contact with the sample solution; this gold film is in turn covered with a long-chain hydroxyalkanethiol, which forms a monolayer (approximately 100-nm thick) at the surface. This layer essentially serves as an attachment point for carboxymethylated dextran chains that create a hydrophilic surface to which ligands can be covalently coupled. Light incident to the back of the metal film is totally internally reflected on to the diode-array detector. A property of this situation is that a nonpropagative evanescent wave penetrates into the solution side of the prism away from the light source. Free electrons in the gold layer enter into resonance with the evanescent wave. In fact, such resonance implies that the amplitude vector characterizing a transversal wave propagating along the gold surface ($\vec{ksp}$) is equal to the component ($\vec{kx}$) of the evanescent wave. Since $\varepsilon = n^2$, then if ω is the frequency of the wave and c the speed of light then

$$|\mathbf{kx}| = \frac{\omega}{c}\sqrt{\frac{\varepsilon_1 - \varepsilon_2}{\varepsilon_1 + \varepsilon_2}} \quad (15)$$

Furthermore, given that for the evanescent wave:

$$|\mathbf{kx}| = \frac{\omega}{c}\sin\theta\sqrt{\varepsilon_g} \quad (16)$$

When resonance occurs, $|\mathbf{ksp}| = |\mathbf{kx}|$ and the intensity of the reflected light decreases at a sharply defined angle of incidence, the SPR angle, given by the simple expression:

$$\sin^2\theta_0 = \frac{\varepsilon_1\varepsilon_2}{\varepsilon_g(\varepsilon_1 + \varepsilon_2)} \quad (17)$$

Thus θ_0, the SPR angle at which a decrease in the intensity of reflected light occurs, measures the refractive index of the solution in contact with the gold surface and is dependent on several instrumental parameters, for example, the wavelength of the light source and the metal of the film. When these parameters are kept constant, the SPR angle shifts are dependent only on changes in refractive index of a thin layer adjacent to the metal surface. Any increase of material at the surface will cause a successive increase of the SPR angle, which is detected as a shift of the position of the light intensity minimum on the diode array. This change can be monitored over time, thus allowing changes in local concentration to be accurately followed. The SPR angle shifts obtained from different proteins in solution have been correlated to surface concentrations determined from radio-labeling techniques and found to be linear over a wide range of surface concentrations. The instrument output, the resonance signal, is indicated in resonance units (RU); in the Biacore configuration, 1000 RU correspond to a 0.1° shift in the SPR angle and for an average protein this corresponds to a surface concentration change of about 1 ng mm^{-2}. The machine measures refractive index changes (Δn) at or near a surface and relates these to changes in local concentration. This relationship is given by the Clausius–Mossotti form

$$\frac{\varepsilon - 1}{\varepsilon + 2} = \frac{N}{3\varepsilon_0}\left(\frac{\alpha + \mu}{kT}\right) \quad (18)$$

of the Debye equation

$$\frac{\varepsilon - 1}{\varepsilon + 2} = \frac{N.\alpha}{3.\varepsilon_0} \quad (19)$$

where ε is the real part of the dielectric constant or permittivity constant related to the refractive index by $\varepsilon = n^2$, N is the number density given by $N_a \rho / M_a$ (N_a = Avogadro's number, ρ = the density and M_a the molecular mass). It is assumed that $\Delta n / \Delta C$ is a constant.

In general, either the DNA or the protein may be immobilized on the surface and the choice will depend on the nature of the interaction to be studied. It is, however, technically more expedient to immobilize DNA. Whilst a variety of techniques exist for the immobilization of DNA on the dextran surface, the most efficient for the majority of protein/DNA interactions is the use of immobilized streptavidin that can then interact with a suitably end-labeled DNA molecule. While it is often possible to obtain useful binding data pertaining to a specific nucleoprotein complex, it is critical to recall that SPR measurements monitor a steady state rather than equilibrium, that is, a reaction involving two components may form an initial binary complex that then "evolves" through intramolecular interactions into a final complex having different dissociation characteristics than that of the intermediate(s) present during the reaction. In many cases now, SPR has furnished useful data concerning specific nucleoprotein complexes and even in complicated cases, such as promoter recognition by RNA polymerase, SPR can furnish useful binding data that correlates well with other footprinting techniques.

An interesting application of this technique toward a more dynamic approach was carried out by using SPR to follow the polymerization of a reverse transcriptase along an immobilized single-stranded DNA template. The resulting change in SPR signal could be closely correlated with the incremental progression of the reverse transcriptase along the immobilized DNA template as it replicated the corresponding RNA or DNA strand. This approach has also been extended to observing RNA polymerases transcribing an immobilized double-stranded DNA template into an RNA molecule and again could be correlated with changes in the population of polymerases on the surface and the appearance of the RNA molecule.

3.5
Microcalorimetry

Microcalorimetry, like SPR, is a noninvasive technique for determining the magnitude of a specific process, such as the formation of a nucleoprotein complex. Microcalorimetry, however, in contrast to SPR, provides access to real thermodynamic measurements of a system at equilibrium. A system at thermodynamic equilibrium is of course defined as one where $\Delta G = 0$. This state is also defined by the equilibrium constant K, where, for a complex AB formed between A and B, $K = [AB]/[A][B]$ (in which case K is known as the *association* or *affinity constant*). This ratio, reflecting the relative probabilities of finding the molecules in a bound or free state, is related to $\Delta G°$ by the simple expression

$$K = e^{\left(\frac{-\Delta G°}{RT}\right)} \quad (20)$$

and since

$$\Delta G° = \Delta H° - T\Delta S°, \quad (21)$$

$$K = e^{\left(\frac{-\Delta H°}{RT}\right)} + e^{\left(\frac{-\Delta S°}{R}\right)} \quad (22)$$

This is a form of the familiar van't Hoff equation that allows calculation of ΔH and ΔS by studying the variation of K as a function of temperature.

Microcalorimetry, however, directly measures heat energy changes during a reaction. There are essentially two approaches in use, differential scanning calorimetry (DSC) and isothermal titration calorimetry (ITC). DSC studies changes that occur as the temperature of a reaction is altered, whereas ITC measures thermal changes during the mixing of reactants. Both approaches provide measurements of ΔC_p and thus by the application of relationships like those given in Eqs. (20), (21), and (22) allow calculation of ΔH and ΔS (and thus ΔG and K).

We mentioned above that nucleoprotein complexes are composed of noncovalent interactions and as such the forces that are involved are therefore relatively weak such that the enthalpy changes involved are of the order of 1–25 kcal mol^{-1}. It is to the credit of calorimetric technology and those who use the approach that they are able to measure the changes in the order of differences of $10^{-6}\,°C$ associated with these weak changes.

3.6
Fluorescence

Fluorescence-based studies are becoming increasingly popular especially amongst the more biophysically oriented research groups. The advantages offered by this technique include the possibility of being able to carry out studies in solution in a true equilibrium fashion and in many cases in a time-resolved fashion. The dynamic range of sensitivity extends from milli to pico molar and the increased availability of diverse reagents extends the versatility of the approach. There are essentially three fluorescence-based strategies that are currently in use with respect to nucleoprotein complexes.

3.6.1 Fluorescence Intensity Measurements

Changes in fluorescence intensity originally relied on changes in the intrinsic fluorescence of protein upon binding to nucleic acids. Alternatively, fluorescence modified bases such as 2-aminopurine and ethenedeoxy-adenosine or extrinsically labeled nucleic acids have also been extensively used. In general, this latter approach has been favored because of the relatively high quantum yield (i.e. the number of reactant molecules that react for every photon absorbed) allowing good sensitivity.

3.6.2 Fluorescence Anisotropy

A phenomenon that is in fact the basis for dichroism is that the absorption of light by a chromophore depends upon the relative orientation of the absorbing chromophore with respect to the polarization of the light. When the incident light is polarized, then absorption will clearly be more likely for those molecules with chromophores whose transition dipole moments are parallel to the direction of polarization. By transition dipole moment we mean that, for a molecule to be able to absorb or emit a photon of frequency v, it must possess, albeit transiently, a dipole oscillating at v. Indeed, it is a property of this phenomenon that the intensity of the transition associated with either absorption or emission of a photon is proportional to the square of the transition dipole moment. In other words, for a transition dipole moment making an angle θ with respect to the direction of light polarization, the molar absorption coefficient ε_0 (also referred to as the extinction coefficient) is given by

$$\frac{\varepsilon_0}{\varepsilon_1} = \cos^2\theta \qquad (23)$$

By using light pulses of short duration and by measuring the fluorescent emission using polarizers, the intensity of emission parallel (I_a) can be compared with the intensity of the emission perpendicular (I_b) to the incident light.

This is very useful since the emission anisotropy may be obtained simply by using the expression

$$A = \frac{I_a - I_b}{I_a + 2I_b} \quad (24)$$

Anisotropy decays exponentially, that is,

$$A_{(t)} = A_{(0)} e^{t/\rho} \quad (25)$$

where ρ is the rotational relaxation time. It is clear that very short ($\sim 10^{-9}$ s) rotational relaxation times may be measured using this technique. Once one has a relaxation time, then it is very straightforward to calculate the hydrated volume V_h for the chromophore-bearing molecule using the simple expression

$$\rho_{(t)} = \frac{\eta V_h}{kT} \quad (26)$$

where η is the viscosity of the solution (J K^{-1} m^{-1} s^{-1}), k is the Boltzmann constant (1.38066 × 10^{-23} J K^{-1}), and T is the temperature in degrees Kelvin.

The rotational diffusion properties of a labeled oligonucleotide alter during complex formation with a protein thus providing information not only about the nature of the complex being formed but also providing the means of carrying out time-resolved analysis of the formation of the nucleoprotein complex. This latter is an increasingly powerful technique.

3.6.3 Fluorescence Energy Transfer (FRET)

FRET is the nonradiative exchange of excitation energy between two chromophores. In fact, an energy donor in an excited state transfers energy to an acceptor by means of intermolecular long-range dipole–dipole coupling. The quantum yield of transfer, E, depends on the distance (R) between the dyes and is given by

$$E = \frac{1}{1 + ([R]/[R]_0)^6} \quad (27)$$

In practice, although this depends to an extent on the nature of the fluorophore, the effective distance at which the phenomena may be useful is of the order of 4 to 6 nm. So one can see that this technique can be used in a nucleoprotein complex to measure distances between select chromophores and indeed has been applied in looking at changes in nucleic acid structure during complex formation or at putative contact points between RNA polymerases and promoter sequences in DNA.

3.7 Electron Microscopy and Atomic Force Microscopy

Electron and near field microscopy possess the capacity to visualize and follow the average behavior of a population of molecules observed individually and in the case of nucleoprotein complexes provide information concerning DNA curvature and flexibility intrinsic or induced by a protein. In classical electron microscopy (EM), molecules are adsorbed onto a carbon film, which may lead to some loss of three-dimensional information but generally will conserve the macromolecules conformational properties. EM generates data in two broad fields: imaging and analysis. Global electron irradiation of a specimen produces images on a fluorescent screen, a photographic film, or a

TV camera. DNA or protein–DNA complexes may be directly adsorbed onto carbon films or mica supports. Visualization is generally carried out by the use of contrasting procedures that do not interfere with the interactions under study. Thus, samples may be either stained using an evaporated metal coating (such as uranyl acetate) or by using an annular dark field in which the electron beam is tilted to provide a tilted illumination of the specimen and the image resulting from those electrons scattered at wide angles by the heavy atoms present in the sample. Contrast is obtained by impeding direct electrons and allowing only scattered electrons through the aperture. In all cases, direct images of DNA molecules or nucleoprotein complexes may be obtained (Fig. 6).

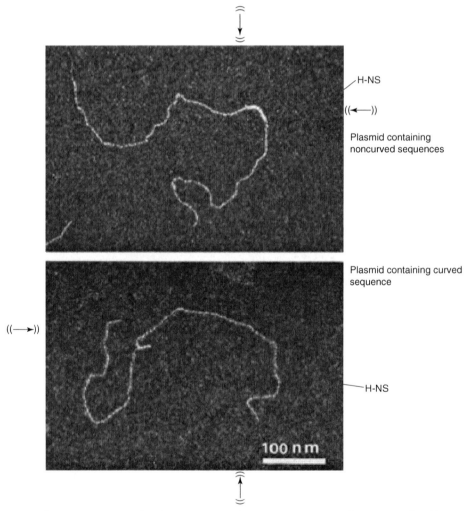

Fig. 6 Electron microscopy of nucleoprotein complexes. H-NS/DNA complexes as visualized by electron microscopy.

A useful parallel technique to EM is cryo-EM, which enables visualization of nucleic acids, for example, without staining. A solution of the macromolecules is rapidly frozen to produce a thin film of vitrified water. By using EM, it is possible, in the correct configuration, to visualize molecules floating in the ensuing amorphous solution.

Scanning force microscopy provides a complementary technique to EM possessing high resolution and contrast. In a simple configuration known as contact mode, a flexible cantilever carrying a sharp tip is scanned across a surface. Deflections of the tip as it pans across the sample on the surface may be translated into a topographic image. A second operating mode known as tapping mode exists in which the cantilever oscillates at a high frequency during the scan. The surface topography is then reconstructed from changes in the oscillating amplitude of the cantilever. Nucleoprotein complexes are generally deposited onto a flat mica surface in a suitable buffer, then rinsed and dried under conditions optimized so as not to disturb the conformation of the complexes under study. In practice, this is a major restriction to the technique.

In general, in images formed by 512 × 512 pixels on a 2 µm scan size, each pixel represents a DNA of 3.9 nm or about 11 bp. In the case of nucleoprotein complexes, changes in the measured contour length of DNA molecules due to the presence of bound protein may be interpreted in terms of specific bending or deformation of the DNA.

See also Footprinting Methods to Examine the Structure and Dynamics of Proteins.

Bibliography

Books and Reviews

deHaseth, P.L., Zupancic, M.L., Record, M.T. Jr. (1998) RNA polymerase-promoter interactions: the comings and goings of RNA polymerase, *J. Bacteriol.* **180**(12), 3019–3025.

Hill, J.J., Royer, C.A. (1997) Fluorescence Approaches to Study of Protein-Nucleic Acid Complexation, in: Braud L., Johnson M.L. (Eds.) *Methods in Enzymology, Fluorescence Spectroscopy*, Vol. 278, Academic Press, pp. 278–416.

Ishihama, A. (1999) Modulation of the nucleoid, the transcription apparatus, and the translation machinery in bacteria for stationary phase survival, *Genes Cells* **4**(3), 135–143.

Kneale, G..G. (1994) DNA-Protein Interactions: Principles and Protocols, *Methods in Molecular Biology*, Vol. 30, Humana Press, Totowa, NJ.

Moss, T. (2001) DNA-Protein Interactions: Principles and Protocols, *Methods in Molecular Biology*, 2nd edition, Vol. 148, Humana Press, Totowa, NJ.

Murakami, K.S., Darst, S.A. (2003) Bacterial RNA polymerases: the wholo story, *Curr. Opin. Struct. Biol.* **13**(1), 31–39.

Ptashne, M., Gann, A. (1997) Transcriptional activation by recruitment, *Nature* **386**(6625), 569–577.

Sauer, R.T. (1991) *Protein-DNA Interactions Methods in Enzymology*, Vol. 208, Academic Press, London, UK.

Travers, A. (1996) Transcription: building an initiation machine, *Curr. Biol.* **6**(4), 401–403.

Travers, A. (1997) DNA-protein interactions: IHF–the master bender, *Curr. Biol.* **7**(4), R252–R254.

Travers, A.A., Buckle, M. (2000) *DNA-Protein Interactions: A Practical Approach*, Oxford University Press, New York.

Primary Literature

Adelman, K., Brody, E.N., Buckle, M. (1998) Stimulation of bacteriophage T4 middle transcription by the T4 proteins MotA and AsiA occurs at two distinct steps in the transcription cycle, *Proc. Natl. Acad. Sci. U.S.A.* **95**(26), 15247–15252.

Azam, T.A., Ishihama, A. (1999) Twelve species of these nucleoid-associated protein from Escherichia coli. Sequence recognition specificity and DNA binding affinity, *J. Biol. Chem.* **274**(46), 33105–33113.

Bloch, V., Yang, Y., Margeat, E., Chavanieu, A., Auge, M.T., Robert, B., Arold, S., Rimsky, S., Kochoyan, M. (2003) The H-NS dimerization domain defines a new fold contributing to DNA recognition, *Nat. Struct. Biol.* **10**(3), 212–218.

Buckle, M., Buc, H. (1989) Fine mapping of DNA single-stranded regions using base-specific chemical probes: study of an open complex formed between RNA polymerase and the lac UV5 promoter, *Biochemistry* **28**(10), 4388–4396.

Buckle, M., Pemberton, I.K., Jacquet, M.A., Buc, H. (1999) The kinetics of sigma subunit directed promoter recognition by E. coli RNA polymerase, *J. Mol. Biol.* **285**(3), 955–964.

Buckle, M., Williams, R.M., Negroni, M., Buc, H. (1996) Real time measurements of elongation by a reverse transcriptase using surface plasmon resonance, *Proc. Natl. Acad. Sci. U.S.A.* **93**(2), 889–894.

Losick, R., Pero, J. (1981) Cascades of sigma factors, *Cell* **25**(3), 582–584.

Luscombe, N.M., Thornton, J.M. (2002) Protein-DNA interactions: amino acid conservation and the effects of mutations on binding specificity, *J. Mol. Biol.* **320**(5), 991–1009.

Mukhopadhyay, J., Kapanidis, A.N., Mekler, V., Kortkhonjia, E., Ebright, Y.W., Ebright, R.H. (2001) Translocation of sigma (70) with RNA polymerase during transcription: fluorescence resonance energy transfer assay for movement relative to DNA, *Cell* **106**(4), 453–463.

Naryshkin, N., Revyakin, A., Kim, Y., Mekler, V., Ebright, R.H. (2000) Structural organization of the RNA polymerase-promoter open complex, *Cell* **101**(6), 601–611.

Pabo, C.O., Nekludova, L. (2000) Geometric analysis and comparison of protein-DNA interfaces: why is there no simple code for recognition? *J. Mol. Biol.* **301**(3), 597–624.

Record, M.T. Jr., Ha, J.H., Fisher, M.A. (1991) Analysis of equilibrium and kinetic measurements to determine thermodynamic origins of stability and specificity and mechanism of formation of site-specific complexes between proteins and helical DNA, *Methods Enzymol.* **208**, 291–343.

Rimsky, S., Zuber, F., Buckle, M., Buc, H. (2001) A molecular mechanism for the repression of transcription by the H-NS protein, *Mol. Microbiol.* **42**(5), 1311–1323.

Travers, A., Muskhelishvili, G. (1998) DNA microloops and microdomains: a general mechanism for transcription activation by torsional transmission, *J. Mol. Biol.* **279**(5), 1027–1043.

Zinkel, S.S., Crothers, D.M. (1990) Comparative gel electrophoresis measurement of the DNA bend angle induced by the catabolite activator protein, *Biopolymers* **29**(1), 29–38.

Part 2
Modeling and Design

Proteins. Edited by Robert A. Meyers.
Copyright © 2007 Wiley-VCH Verlag GmbH & Co. KGaA, Weinheim
ISBN: 978-3-527-31608-3

12
Protein Modeling

Marian R. Zlomislic and D. Peter Tieleman
University of Calgary, Albeta, Canada

1	**Introduction**	**382**
1.1	Primary, Secondary, Tertiary, and Quaternary Structure of Proteins	382
1.2	Relationship between Structure and Function	382
1.3	Experimental Structure Determination Methods	384
1.4	Role of Modeling	385
2	**Structure Prediction Methods**	**385**
2.1	The Protein-folding Problem	385
2.2	*De Novo/Ab Initio* Methods	386
2.3	Fold Recognition Methods	387
2.4	Homology Modeling	388
3	**Structure-based Modeling**	**389**
3.1	Molecular Graphics-based Methods	389
3.2	Poisson–Boltzmann/Electrostatics Calculations	390
3.3	Ligand Docking	391
3.4	Protein–Protein Interactions	392
4	**Simulations of Protein Dynamics**	**392**
4.1	Molecular Dynamics Simulation	392
4.2	Free-energy Calculations	394
4.3	Brownian Dynamics Calculations	394
5	**Example Applications**	**395**
5.1	Acetylcholinesterase	395
5.2	Water Transport Through Aquaporins	396
5.3	ABC Transporters and Multidrug Resistance Proteins	398
6	**Perspectives**	**400**

Proteins. Edited by Robert A. Meyers.
Copyright © 2007 Wiley-VCH Verlag GmbH & Co. KGaA, Weinheim
ISBN: 978-3-527-31608-3

Acknowledgment 401

Bibliography 401
Books and Reviews 401
Primary Literature 402

Keywords

Ab Initio Modeling
The goal of *ab initio* modeling is to predict the structure of a protein from only its amino acid sequence.

Brownian Dynamics Simulation
Brownian dynamics simulations are related to molecular dynamics, but do not include the same amount of detail. Solvent in particular is normally represented as a continuous medium, which significantly decreases the complexity of the models and allows much longer simulations.

CASP, CAFASP, CAPRI
Critical Assessment of Structure Prediction (CASP), Critical Assessment of Fully Automated Structure Prediction (CAFASP), and Critical Assessment of PRedicted Interactions (CAPRI) are large-scale assessments in which hundreds of research groups submit their best structure predictions on the same set of targets, allowing a critical comparison between methods and an assessment of strengths and weaknesses of different methods and the field as a whole.

Docking
The term docking incorporates a group of methods for studying the interactions of small molecules and proteins. The process typically involves extensive searching of possible orientations of the small molecule in a binding site of the protein, and a scoring function that ranks different orientations by their likelihood of occurrence.

Fold Recognition
The goal of fold recognition is to predict which fold, out of ca. 1000 observed protein folds, a given sequence will adopt.

Homology Modeling
Predicting the three-dimensional structure of a protein on the basis of known structure of a related protein.

Molecular Dynamics Simulation
A computer simulation method in which the motions of all atoms in a molecular model, including water and ions are calculated over a period of up to a microsecond.

This computationally expensive method is one of the most detailed ways of studying the dynamics of macromolecules.

Molecular Graphics
Modern computer graphics are used to visualize molecular structures and models. The ability to highlight different features or properties in a three-dimensional model is a powerful way to explore molecules. Stereo hardware and immersive environments further enhance the usefulness of molecular graphics.

Poisson–Boltzmann Equation
The Poisson–Boltzmann equation describes electrostatic interactions between charges in a continuous medium (e.g. water) in the presence of salt. It can be solved numerically by several popular software packages to give insight into electrostatic interactions in proteins, between proteins and their substrates, between proteins and membranes, and in other systems. A common application is the calculation of pK_a-shifts for acidic and basic residues in proteins due to their environment.

Primary, Secondary, Tertiary, and Quaternary Protein Structure
The amino acid sequence of a protein is its primary structure. The α-helices, β-strands, and random coils are typical secondary structures that form between consecutive residues. They are defined by the hydrogen-bonding pattern and dihedral angles observed in the protein backbone. The tertiary structure of a protein describes the relative position of all the protein atoms in three-dimensional space. Elements of secondary structure that were far apart sequentially may now form higher order structures such as parallel or antiparallel β-sheets, or helix bundles. The quaternary structure describes how multiple proteins are arranged together in larger complexes.

QM/MM Simulation
A computer simulation method in which an important part of a system, such as the active site of an enzyme, is described by quantum mechanics, with the rest of the system described by molecular dynamics methods. The part of the system treated by QM includes explicit treatment of electrons.

Structural Genomics
A concerted effort to determine experimentally the high-resolution structure of all proteins in a genome, often with a focus on proteins that are predicted to be structurally different from all structures that are already in the database. This greatly increases the likelihood that a given sequence can be modeled using homology modeling.

> Protein modeling consists of a broad range of computational techniques to understand the properties of proteins and has become an integral part of structural biology and drug design. Modeling can be used to predict the secondary structure or fold of a protein on the basis of its sequence alone, to predict the three-dimensional structure of a protein on the basis of knowledge of the structure of a related protein,

to design new proteins, and to predict properties that depend on the experimentally determined three-dimensional structure of a protein. Examples of such properties include drug binding, protein–protein interactions, and interactions with elements in a protein's environment, including ions, lipids, carbohydrates, and nucleic acids. Conformational changes in proteins can be investigated by molecular dynamics simulations to provide detailed insight into the dynamics of proteins, a crucial aspect of protein function. In recent developments, quantum mechanical calculations are used more and more frequently to study reactions in proteins. With the ever-increasing power of computers, increasingly detailed aspects of protein function can now be investigated by modeling methods, at a scale and level of detail that is often very difficult or impossible to achieve by experimental methods.

In this chapter, the main principles and techniques involved in protein modeling are introduced. A few examples from the literature will highlight how protein modeling can be used in complement with other methods.

1
Introduction

1.1
Primary, Secondary, Tertiary, and Quaternary Structure of Proteins

Proteins are structured at different levels. The primary structure of a protein is its amino acid sequence, encoded by DNA. All proteins are constructed from about 20 common amino acids as well as some amino acids that are formed through chemical modification. The secondary structure of a protein is a sequence of common structural, three-dimensional elements or building blocks. These secondary structure elements include α-helices and β-strands as best-known elements, but there are many other building blocks. The tertiary structure of a protein is its three-dimensional structure in space, which can be thought of as placing the secondary structure elements together in space. The quaternary structure of a protein describes how multiple proteins interact through noncovalent forces to form larger protein complexes. An example of the different levels of structure is shown in Fig. 1, based on the protein barstar. It is shown in a protein complex with barnase to illustrate quaternary structure. The minimal requirement for modeling proteins is knowledge of the primary structure. Since the primary sequence is one of the main results of genome sequencing, this is a very modest requirement. As more experimental structural information becomes available, modeling typically becomes more accurate by using that experimental information to identify plausible models. Any experimental information that puts limitations on possible models is useful.

1.2
Relationship between Structure and Function

A basic assumption in protein science is that the function of a protein follows from its structure. Yet, proteins with very similar three-dimensional structure can have very different functions, so

MKKAVINGEQIRSISDLHQTLKKELALPEYYGENLAALWDCLTGWVEYPLVLEWRQ
FEQSKQLTENGAESVLQVFREAKAEGCDITIILS

(a)

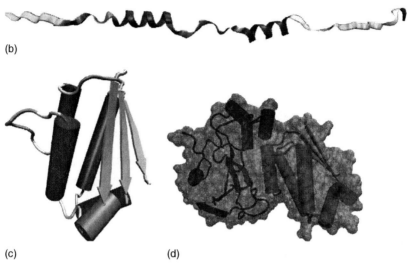

(b)

(c)　　　　　　　　　　(d)

Fig. 1 (a) The primary sequence of the protein barstar in FASTA format. (b) The secondary structure of the first 60 residues of barstar, colored purple for helices, and yellow for β-strands. All are shown in ribbon format. (c) The complete tertiary structure of barstar in cartoon format. The helices are now bundled together, and the β-strands have formed a three-stranded parallel β-sheet. (d) Example of quaternary structure: the barnase/barstar complex. Barstar (red) and barnase (blue) area shown in ribbon and spacefilling formats. Images created with VMD using the PDB entry 1B2U. (See color plate p. xvi.)

precise details of the three-dimensional (tertiary/quaternary) structure are very important. Nonetheless, there is a large amount of information contained in the primary and secondary structure of a protein. Indeed, nature in most cases only seems to need the primary sequence, but computational methods are not yet able to reliably predict the three-dimensional structure of proteins based on primary sequence alone. At the current state of the art, a primary sequence is useful to compare proteins of known function with a similar sequence and sometimes allows assignment of a tentative function. Knowing the secondary structure of a protein significantly narrows down the range of possible three-dimensional structures. It allows a comparison with databases of the secondary sequence of known proteins, which often makes it possible to recognize the three-dimensional fold of a protein on the basis of the secondary structure alone. Interestingly, although in principle the number of possible three-dimensional structures is practically infinite, only about 1000 different "folds" seem to occur in nature. This is particularly useful for whole-genome studies: if we can use the primary sequence and predictions or knowledge of the secondary structure, chances are reasonable that the fold of the protein can be recognized. This is a constructive step toward attaining a three-dimensional structure.

The three-dimensional structure of a protein is the most useful level for molecular modeling. At this level, the interactions between small molecules (drugs, substrate) and an enzyme, permeation properties of ion channels and transporters, activation of receptors by ligand binding, and chemical reactions in enzymes and other processes that require a very detailed knowledge of atomic structure to understand them can be studied. This level is also required to study interactions between protein subunits or for protein–protein or protein–peptide interactions. In practice, the best way to obtain knowledge of the three-dimensional structure is from experimental structure determination, but prediction methods primarily based on homology modeling (see Sect. 2) are increasingly becoming accurate enough to provide useful starting points for further computational studies.

1.3
Experimental Structure Determination Methods

There are several experimental methods to determine the structure of proteins. The primary structure is determined by sequencing, either of the protein itself or of the DNA that encodes the protein. The secondary structure can be measured by spectroscopic methods, which usually determine a percentage of various secondary structure elements. Common techniques to do this are CD (circular dichroism) spectroscopy, IR (infrared) spectroscopy, and NMR (nuclear magnetic resonance) spectroscopy.

As a basis for detailed modeling problems, high-resolution 3-dimensional structures are typically required. The two main methods to experimentally determine 3D structures are X-ray crystallography and nuclear magnetic resonance. Electron microscopy is a third method that has been used to solve the structure of several membrane proteins. Crystallography is a very powerful method that can be used on proteins and protein complexes of any size. The main limitation of the method is that a protein must form regular crystals. Not all proteins do this (especially membrane proteins), and sometimes crystallization forces proteins in structures that are probably not physiologically relevant. NMR experiments can determine the structure of a protein in solution, which is usually a more realistic environment. The main limitations of solution NMR are that it is difficult to apply the method in practice to proteins that are larger than ca. 40 kDa, and that it is usually considerably more labor intensive to determine a structure by NMR compared to crystallography. These limitations have spurred the development of new methods that are catered to work with larger proteins, and to make the process of structure calculation from a measured spectrum more automated. Solid-state protein NMR is a relatively new field that can be used to determine the structure of proteins in ordered systems, such as membranes. Electron microscopy as a method to determine 3-D structures has been mainly used for membrane proteins, which are especially difficult to crystallize for X-ray crystallography. The method is very labor intensive. It does have a significant strength for very large complexes, where the overall arrangement of large subunits can be seen at low resolution. The details of the structure can then be added from crystal structures of individual proteins that make up the complex.

1.4
Role of Modeling

Modeling has many potential uses, at different levels of protein structure. Protein modeling is such a broad field it may be beneficial to distinguish two separate general goals.

One set of goals consists of predicting aspects of the structure of a protein based on less detailed information. If only the primary structure (the sequence) of a protein is known, we can try to predict the secondary structure with reasonable accuracy, often in the 75 to 85% range. On the basis of the degree of hydrophobicity of predicted helices, it is usually also possible to predict transmembrane segments, identifying a protein as a membrane protein, with reasonable accuracy. The secondary structure can be useful to predict the general fold of a protein, which may make it possible to assign a tentative function to the protein. Predicting the tertiary structure directly from the sequence is a very difficult problem with a low success rate. If a high-resolution structure of a related protein is known, then homology modeling can be used to provide a good model of the target protein. In principle, such modeling efforts make it possible to circumvent experimental structure determination. One major goal of protein structure research at the moment is to improve all steps in the process that leads to an accurate three-dimensional model, both by improving *ab initio* prediction methods and by experimentally determining enough protein structure that homology modeling becomes feasible for most proteins in the genome – ideally, one could then obtain structural models for all proteins in an organism directly from its genome sequence. As the number of structures solved experimentally increases, the algorithms to detect homology between remotely related proteins and built homology models will improve, making this goal increasingly feasible.

A second set of goals starts from an experimentally determined high-resolution structure (or a very high-quality homology model) and uses physics-based methods to model aspects of a protein that cannot be easily determined experimentally. This type of approach could include electrostatics calculations to understand how proteins interact with substrates or other proteins, docking calculations to understand differences in binding constants and design new inhibitors for enzymes, or detailed molecular dynamics simulations to investigate the dynamics of proteins. This type of goal usually involves much more biochemical knowledge of the protein and is often centered on very specific questions: why does mutating residue x to alanine change the binding affinity of a particular drug by a factor of 1000? Why is this ion channel selective for potassium over sodium, even though the only difference between potassium and sodium superficially is a small difference in radius?

In the following sections we consider each set of goals in more detail and give specific examples.

2
Structure Prediction Methods

2.1
The Protein-folding Problem

One of the most challenging problems in biophysics is to understand how it is possible that proteins fold rapidly (microseconds to seconds) into a well-defined structure when based on their primary sequence alone, a practically infinite number of structures is possible. Levinthal showed that folding on a realistic timescale could therefore not occur through a systematic search of all possible conformations (Levinthal's paradox). Major research efforts are devoted to understanding protein folding, with a significant focus on the physics of simplified models. Clearly, if it is possible to fold a protein *in vivo* from the sequence only, then with a proper understanding of the laws governing protein folding, we should be able to predict any protein structure computationally. This is still a lofty goal, but in the past years major progress has been made in improving prediction methods at all levels of structure. This progress has been documented strikingly in the proceedings of the semiannual CASP competition (Critical Assessment of Structure Prediction).

CASP is an interesting and exciting venture involving a large scientific community. Its main goal is to obtain an in-depth and objective assessment of our current abilities and inabilities in the area of protein structure prediction (http://predictioncenter.llnl.gov). It involves a neutral organizing committee that collects unpublished, but already solved protein structures from experimental structure determination groups. Specific information, such as the protein's primary sequence, is published on a Web site, and modeling groups can attempt to predict the structures. A panel of assessors judges the predictions against the real structures, and at the CASP scientific meeting the results are discussed, areas of progress and areas of problems are identified, and the progress and directions of the field as a whole are examined. This procedure makes CASP a very fair way of establishing the merits of particular methods and of identifying where future efforts can be most productively focused. CASP6 was held in December 2004 in Italy. In the last few years, CASP has begun to introduce automated prediction software as a separate exercise (CAFASP – Critical Assessment of Fully Automated Structure Prediction). The process has also been adapted to predicting protein–protein interactions (CAPRI – Critical Assessment of PRediction of Interactions), which saw its first edition in 2001. More details about CAPRI can be found at http://capri.ebi.ac.uk. These competitions give a novice in the field of structure prediction an excellent starting point to compare the many protein modeling programs and help a user to decide which program would work best with their requirements. Despite differences in the performance of these programs, many of them draw from the same fundamental approaches to protein structure prediction. The three main approaches are discussed in the following.

2.2
De Novo/Ab Initio Methods

De novo methods are a general designation for methods that predict the structure of a protein on the basis of the primary sequence alone. The ideal method to do this would be based on simple physical laws only. Although this would be a true *de novo* method, it is exceedingly difficult and has only been modestly successful for peptides and small proteins. The difficulty is the extremely large number of possible conformations for a polypeptide chain; predicting the structure of a protein from the sequence only would indeed be equivalent to solving the protein-folding problem. In practice, *de novo* methods incorporate a variety of information derived from a database of existing structures. This can be very direct information, such as taking the conformation of a loop with the same sequence directly from the database, or less direct, such as through heuristical rules derived from statistical analysis of all known protein structures.

2.3 Fold Recognition Methods

Fold recognition methods attempt to identify structures within a new protein (the query) by comparing its sequence to proteins with known structure (the template). This can be called the *inverse folding problem*, since the goal is to find probable folds that might fit the sequence instead of trying to determine how the sequence will fold.

There are two broad approaches to fold recognition methods. The first approach is sequence-based. It relies on finding homologous sequences and then assumes that strong sequence similarity equates to strong structure and function similarity. While this may seem straightforward, current algorithms may not be sensitive enough to recognize distant homologs, and therefore the "correct" template might be missed; as well, mutations between these sequences, including gaps and insertions, sometimes make it difficult to identify the correct sequence alignment. It can be misleading to rely on the sequence alignment alone for fold recognition. It is possible to identify proteins with some sequence identity that do not have any structural or functional similarity to the query, and so, blind sequence alignments may not yield a successful prediction. Classifying protein sequences into families, and then identifying the patterns within the family such as strongly conserved residues (which might be important to function) and the pattern of mutations (For example, is a hydrophobic residue always replaced by another hydrophobic residue? That is an acceptable mutation.) can improve the confidence of your alignment. PSI-BLAST is one of the most popular homology recognition algorithms, although there are a number of others that also perform well.

The second approach to fold recognition is structure-based. "Threading" is a popular term for this approach, which evokes the image of the thread-like primary sequence being "threaded" through the three-dimensional structure of template proteins. Finding the optimal fit of the sequence with each framework is a complex process, particularly if there are gaps and deletions among the sequences of the two proteins. It is very common to use sequence-based methods in conjunction with threading to identify the best matches. Each iteration of the threading process is scored and this score is used to determine the best matches. Following Anfinsen's hypothesis that the native state of the protein is the lowest energy structure, the evaluated energy of the query's fit to the template is a strong predictor of which models are "good" within an experiment. There are a number of threading algorithms available, varying in their choice of protein model (backbone atoms vs just α-carbon atoms, side chains vs interaction sites), their method of alignment, as well as their method of evaluating the energy of the proposed structure, just to name a few differences.

When embarking on a fold recognition study, the state of the art of current fold recognition methods can be judged from results of the previously discussed CASP competition, and also of the Live Bench Project. Unlike CASP, the targets in this exercise are recently deposited structures from the Protein Data Bank. All of the methods assessed are fully automated. An expanding number of fold recognition servers (as well as metapredictors and sequence comparison servers) participate in this weekly assessment exercise, and the most current results can be found on the Web at http://bioinfo.pl/livebench/. It is important to recognize that fold recognition methods are mostly knowledge-based. They are making predictions based only on what is already known. This means that it is impossible to identify any novel folds. Novel fold prediction would fall into the category of an *ab initio* method, where the predictions are based primarily on physical principles.

2.4
Homology Modeling

Homology modeling is the only generally useful method to predict the total three-dimensional structure of a protein. It is based on the observation that the structure of a protein during evolution is much more conserved than the sequence, so that sequences that differ somewhat are still likely to have the same structure. "Somewhat" can be quantified on the basis of the database of known structures: if two proteins with more than ca. 50 residues have a sequence identity of ca. 20 to 30%, they will likely have the same structure. Thus, homology modeling can be used if there is a protein with a known structure (the template) that is sufficiently homologous to the protein whose structure we want to model (the target). In practice, homology modeling follows a series of steps, each of which involve choices by the modeler, or an automated computer server. Following Krieger et al., homology modeling can be thought of as involving seven steps:

1. Template recognition and initial alignment
2. Alignment correction
3. Backbone generation
4. Loop modeling
5. Side-chain modeling
6. Model optimization
7. Model validation

The overall strategy is to begin with aligning the sequence of the target with that of the template, to build the overall chain structure of the target, and then to consider "details." The accuracy of this alignment is the most important determinant of the accuracy of the model. Step 2 is a manual correction of the alignment if there is a compelling reason to do so. For example, in potassium channels there is a nearly universally conserved motif TVGYG that is likely to be aligned even if multiple-sequence alignment algorithms offer alternative alignments. The backbone of the target is typically initially copied from the template, followed by algorithms that allow some flexible adjustments where necessary. The major problems in homology modeling typically involve insertions that are not present in the template. In potassium channels, the loops between transmembrane helices can be very different between channels, even if the transmembrane segments are very similar. Similarly, the loops connecting the transmembrane helices in G-protein coupled receptors convey much of the specificity of different receptors, but this most interesting part generally cannot be modeled by homology modeling. Thus, being able to model loops for the targets that are missing in the template is an important step, and a significant research focus. Once the backbone has been optimized as much as possible, side chains can be modeled and the model tested against experimental information. In many cases, steps have to be repeated.

It is now usually fairly easy to generate homology models, provided a suitable template is available. Commonly used programs include Modeller and WhatIf, as well as automated Web servers such as SwissModel. The key measure of success for homology models is their experimental validation, but less than perfect models can be very useful in interpreting experiments and guiding the design of new experiments.

3
Structure-based Modeling

3.1
Molecular Graphics-based Methods

Over the past 5 to 10 years, the personal computer has begun to replace the large supercomputers and dedicated workstations once needed to use molecular visualization tools. Today, a number of molecular graphics programs are available for use on personal computers, often as freeware. Any program with basic rendering capabilities can be a valuable tool to study a particular protein. A variety of representations, illustrated in Fig. 2 for the protein barnase, can help the modeler understand the link between the protein's structure and its function. By manipulating colors and representations, one can interrogate the structure to see where conserved residues/sequences are in the protein's structure, look at the overall surface of the protein as a guide to potential protein–protein interactions, and look for cavities that might be ligand-docking sites, just to name a few examples. In this section, we aim to point out a few commonly used programs for protein

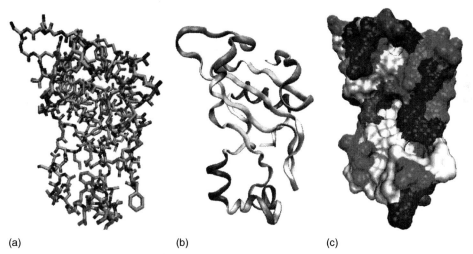

Fig. 2 Barnase rendered in three formats: (a) colored by atom-type (carbon – blue, nitrogen – navy blue, oxygen – red); (b) colored by secondary structure features in ribbon format, where β-sheets are colored yellow, helices are colored purple, and turns and random coil features are colored blue and white; (c) rendered in surface format, colored by residue type, where charged residues are colored red or blue, polar residues are green, and nonpolar residues are white. All figures are rendered with the same view of the protein, looking into the barstar binding pocket. All figures were rendered with VMD. (See color plate p. xvii.)

modeling. Besides those mentioned here, a few more are described by Tate, and even more are listed on the PDB Web site (http://www.rcsb.org/pdb/software-list.html).

Rasmol is one of the most widely used programs for molecular visualization. Besides the basic options in which the display of the protein can be altered from sticks to spacefilling to ribbon format, and so on, there are a variety of tools that can be used through the command line interface. Examples include selection of specific residues of the protein, measuring distances, and highlighting parts of the protein within a certain radius. Rasmol can be used to probe all of the information contained in a ".pdb file," but it has few tools to alter that information. With Swiss-PDB Viewer, protein structures can be manipulated in a number of ways. Residues can be "mutated," and dihedral angles can be modified. This program provides a graphical interface within which several protein structures can be analyzed and manipulated at the same time. Moreover, Swiss-PdbViewer is tightly linked to SwissModel, an automated homology-modeling server. Working with these two programs greatly reduces the amount of work necessary to generate models, as it is possible to thread a protein primary sequence onto a three-dimensional template and get an immediate feedback of how well the threaded protein will be accepted by the reference structure before submitting a request to build missing loops and refine side-chain packing.

VMD is another very powerful visualization program, which can also be used to view trajectories from molecular dynamics simulations. It has a broad range of representations for proteins, including molecular surfaces, electrostatic potential maps, and crystal information. Other useful programs are PyMol and MolMol.

3.2 Poisson–Boltzmann/Electrostatics Calculations

Poisson–Boltzmann theory (PB) has become an important tool for studying biomolecular systems. This type of calculation provides a view of a protein in terms of regions of positive or negative potential. PB calculations are useful for identifying interaction sites on proteins with charged ligands or other proteins. They can also be used for calculating amino acid pK_as in different environments, such as in enzyme active sites where catalytic residues often have large pK_a shifts due to their local environment. PB calculations also provide a basis for Brownian dynamics simulations (see Sect. 4).

PB is based on the Poisson equation, which describes electrostatic interactions in general, but includes implicitly the presence of ions in solution. It is assumed that at equilibrium, the distribution of mobile ions in the system can be approximated by a continuous charge density $\rho_{eq}(\mathbf{r})$, determined by the Boltzmann factor:

$$\rho_{eq}(\mathbf{r}) = \sum_i z_i e n_{0i} \exp\left[-\frac{z_i e \phi(\mathbf{r})}{kT}\right]$$

where i is an ionic species, z_i the charge of ion species i, n_{0i} a reference number density for species i, e is the unit charge, and $\rho(\mathbf{r})$ and $\phi(\mathbf{r})$ are the local equilibrium charge density and the average electrostatic potential, respectively. The electrostatic potential $\phi(\mathbf{r})$ is the solution of Poisson's equation:

$$\varepsilon_0 \nabla \cdot [\varepsilon(\mathbf{r}) \nabla \phi(\mathbf{r})] = -\rho_{eq}(\mathbf{r}) - \rho_{ex}(\mathbf{r})$$

Assuming a 1:1 electrolyte, these two equations can be combined to give:

$$\varepsilon_0 \nabla \cdot [\varepsilon(\mathbf{r}) \nabla \phi(\mathbf{r})] = 2en_0$$
$$\times \sinh\left[\frac{e\phi(\mathbf{r})}{kT}\right] - \rho_{ex}(\mathbf{r})$$

Here the mobile charges represented by ρ_{eq} are in equilibrium, and the fixed charges on the protein or membrane are represented by ρ_{ex}. This is the full nonlinear Poisson–Boltzmann equation, which can be solved numerically. It can be linearized by expanding sinh and retaining only the leading term. When there are no fixed charges and assuming a spatially homogeneous dielectric, this simplifies to:

$$\nabla^2 \phi = \kappa^2 \phi$$

with κ^{-1} is Debye screening length, given by:

$$\kappa^{-1} = \sqrt{\frac{\varepsilon_0 \varepsilon kT}{2e^2 n_0}}$$

The Debye screening length $\kappa^{-1} = 8$ Å for 150 mM salt at 298 K in water with a dielectric constant of 80. The screening of a central charge in bulk solution is about 80% at $r = 3\kappa^{-1}$, or ca. 25 Å. This distance decreases to about 9 Å for 1 M salt solution.

The Poisson–Boltzmann equation, either in its linear or nonlinear form, can be solved for macromolecules by a number of commonly used programs, including DelPhi, UHBD, and APBS.

3.3
Ligand Docking

The binding of small molecules to proteins is a key determinant of many biological processes. The ligand-docking problem is the challenge to predict where, in what orientation, and with what affinity a small molecule binds to a binding site in a protein. In the ultimate state of the art, one would like to predict all three with high accuracy, for a high computational cost if necessary. In addition, we also need methods that allow rapid (computationally inexpensive) screening of a library of millions of compounds to search for drug leads that may interact favorably with potential drug targets. Broadly speaking, ligand docking can be approached from two angles. In the first approach, compounds of known structure are docked into the active site of a protein of interest, and some form of scoring function is a measure of how likely the compound will bind tightly to the protein. In the second approach, starting from the known structure of the protein, a molecule can be designed by fitting functional groups in the active site to optimize the interactions with the protein target. The resulting "computational" molecule can then be synthesized and tested. Several classes of scoring functions exist. One class is based on a combination of interaction energies, with terms that are similar to the potential function used in molecular dynamics simulations. Simpler functions based on, for example, the number of hydrogen bonds or penalties for steric clashes are also useful and can be computationally faster. A second class tries to directly calculate the free energy of binding using a parameterized function or an approximate description of the free energy of binding derived from a potential function. A third class uses statistical information about similar molecules and general features of the binding site to rank ligand–protein complexes. At the moment a combination of methods seems to give the best results, but to our knowledge there is no consensus on which method is the best in any given case.

3.4
Protein–Protein Interactions

Protein–protein interactions are emerging as one of the most important themes in biochemistry. Some interactions are sufficiently stable that a complete complex can be crystallized, but this is relatively rare. More often, proteins interact transiently with other proteins at some point in their lives, often in critical processes such as signaling. The protein database contains mainly single proteins, because to crystallize them they are normally purified to a high degree, thus disrupting all but the strongest noncovalent interactions.

The general goal of protein–protein docking could be described as predicting correctly the structure of a protein–protein complex, given only the structures of the two independent proteins. This assumes that these structures are already known. Whether the proteins are allowed to change their structure somewhat upon binding, which seems realistic, is an additional complexity. There are many approaches in the literature, with two common themes. First, there is a remarkable degree of surface complementarity in most stable complexes. This suggests that it should be possible to draw a molecular surface on both proteins (e.g. see Fig. 1d) and search for all possible relative orientations of the two proteins for the most favorable orientation. This type of search is expensive, and the challenge is to identify the most favorable orientation among many possible orientations. Usually, some form of energy function is used, which gives more favorable values for orientations that optimize interactions such as hydrogen bonding, burying of hydrophobic exposed area, and matching up complementary electrostatic potentials on the surfaces of the proteins that form the complex. A second approach combines experimental data with energy functions to guide the two proteins to a solution that is compatible with experimental data. A recent successful example of this approach is HADDOCK, which combines a molecular mechanics energy function, a simplified form of dynamics to enable flexibility of parts of the interface, and incorporation of "ambiguous interaction restraints." These are "soft" restraints that, by themselves, are insufficient to guide docking but combine to yield more reliable solutions than energy functions alone, provided there is experimental information that can be used as restraints. As mentioned above, progress in the area of protein–protein docking is monitored in the CAPRI challenge (http://capri.ebi.ac.uk).

4
Simulations of Protein Dynamics

Three common simulation methods used to study protein dynamics are explained here.

4.1
Molecular Dynamics Simulation

The most common atomistic simulation technique is molecular dynamics (MD). In MD simulations, the interactions between all atoms in the system are described by empirical potentials. An example of a frequently used potential function is:

$$V(\mathbf{r}^N) = \sum_{bonds} \frac{k_i}{2}(l_i - l_{i,0})^2$$

$$+ \sum_{angles} \frac{k_i}{2}(\theta_i - \theta_{i,0})^2$$

$$+ \sum_{torsions} \frac{V_n}{2}(1 + \cos(n\omega - \gamma))$$

$$+ \sum_{i=1}^{N} \sum_{j=i+1}^{N} \left(4\varepsilon_{ij} \left[\left(\frac{\sigma_{ij}}{r_{ij}} \right)^{12} - \left(\frac{\sigma_{ij}}{r_{ij}} \right)^{6} \right] \right.$$

$$\left. + \frac{q_i q_j}{4\pi \varepsilon_0 r_{ij}} \right)$$

This potential function contains harmonic terms for bonds and angles, a cosine expansion for torsion angles, and Lennard–Jones and Coulomb interactions for nonbonded interactions. The constants k_i are harmonic force constants, l_i, is the current bond length, $l_{i,0}$ the reference bond length, θ_i the current angle, $\theta_{i,0}$ the reference angle, V_n, n and γ are the barrier height, multiplicity, and off-set from the origin for the cosine function used to describe dihedral angles (rotations around a central bond), ε and σ are Lennard–Jones parameters (a different pair for each possible combination of two different atom types), q_i and q_j are (partial) atomic charges, and r_{ij} is the distance between two atoms, and ε_0 (different than ε in the Lennard–Jones potential) is the dielectric constant of the medium.

Using this potential function, the forces (the derivative of the potential with respect to position) on all atoms in the system of interest are calculated and used to solve classical equations of motions to generate a trajectory of all atoms in time. An example of a system studied with MD is shown in Fig. 3: a simulation snapshot of the ABC transporter BtuCD in a realistic environment consisting of lipids, water, and ions.

The primary result of the simulation is a trajectory of all atoms in time, from which specific details of the system can be analyzed. This is an exciting idea, because atoms can be followed as they move in real time on a timescale of up to ca. 100 ns, although longer simulations have also been reported. In principle, any properties that depend on coordinates, velocities, or forces can be calculated, given sufficient simulation time. No assumptions are required about the nature of the solvent, there is no need to choose dielectric boundaries because all atoms are explicitly present, and in principle all interactions (water–ions,

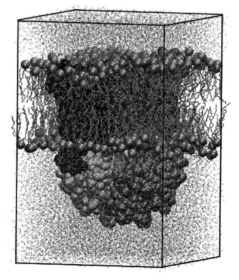

Fig. 3 An orthographic view of the periodic box for the ATP-bound BtuCD simulation, showing water (red and white), lipid with phosphorus atoms enlarged, and the protein. The transporter consists of two transmembrane domains (blue and purple) and two nucleotide binding domains (orange and ochre). The two docked MgATP molecules are partially visible (green and red). Figure courtesy of E. Oloo. See Oloo, E.O., Tieleman, D.P. (2004) Conformational transitions induced by the binding of MgATP to the vitamin B_{12} ABC-transporter BtuCD, *J. Biol. Chem.* **279**, 45013–45019 for more details. Rendered with VMD. (See color plate p. ix.)

water–protein, water–lipid, lipid–protein etc.) are incorporated.

The main limitations of MD are its computational cost, the limited time and length scale that can be treated in a simulation, and technical limitations such as the accuracy of current empirical force fields.

4.2
Free-energy Calculations

Several useful extensions of the basic simulation method make it possible to calculate free-energy differences from simulations. If a reaction coordinate can be identified, processes that are orders of magnitude slower can be studied than would be possible by direct simulation. Such a reaction coordinate could be a concerted conformational change, or a pathway for ion permeation in ion channels. In umbrella sampling, a biasing potential is used to restrict a simulated system to sample phase space within a specified region (called a *window*). By placing windows along the reaction coordinate, one can generate a free-energy profile (also called *potential of mean force*), which will quantitatively describe why one region of space is more favorable than another.

Relative free energies of different side-chain mutations in a protein can be studied using free-energy perturbation by slowly changing one side chain into another within the computer (also known as computational alchemy). This makes it possible to investigate in the computer the effect of mutations on the stability of a protein or on, for example, the binding affinity of a ligand. The theory behind this kind of calculations is well developed and the algorithms have been implemented in many molecular dynamics software packages.

4.3
Brownian Dynamics Calculations

In Brownian dynamics (BD) simulations, the trajectories of individual particles (ions, molecules) are calculated using the Langevin equation:

$$m_i \frac{dv_i}{dt} = -\gamma_i v_i + \mathbf{F}_R + \mathbf{F}_i$$

where m_i, and v_i are the mass and velocity of atom i. Water molecules are not included explicitly, but are present implicitly in the form of a friction coefficient $\gamma = kT/D_i m_i$ (where k is the Boltzmann constant, T is temperature, and D is the diffusion coefficient) and a stochastic force $\mathbf{F}_R$ arising from random collisions of water molecules with ions, obeying the fluctuation-dissipation theorem. $\mathbf{F}_i$ is the force due to other particles in the system as well as external sources, such as an applied electric field. When the friction is large and the motions are overdamped, the inertial term $m_i d\mathbf{v}/dt$ may be neglected, and the simplified form

$$\mathbf{v}_i = \frac{D_i}{kT}\mathbf{F}_i + \mathbf{F}_R$$

may be used. This is the approximation made in Brownian dynamics.

BD simulations require only a few input parameters: in its simplest form, only the diffusion coefficients of the different species of ions and the charge on the ions are needed. However, the model can be refined. To study a protein such as an ion channel, the residues of the protein can be modeled as a set of partial charges, and some form of interaction potential between the mobile ions and the protein must be specified (as seen earlier).

The result of BD simulations is a large set of trajectories for diffusion particles such as ions, proteins, or ligands. By averaging over these trajectories macroscopic properties such as the conductance of ion channels or association rate of protein–protein or protein–ligand encounters can be calculated. In addition, the simulations yield molecular details of the paths through space for the diffusing particles.

Although Brownian dynamics simulations are conceptually simple, replacing solvent by a continuum description and ignoring internal conformational changes in proteins are significant assumptions that require careful consideration. When BD simulations are valid, they are a very powerful method to study biological processes on a timescale that is much longer than can be reached by MD.

5
Example Applications

The literature is now very extensive, and hundreds of papers are published that are based on each of the methods described above. For illustration purposes, we give a few examples of biochemical problems that have been addressed by modeling and simulation methods.

5.1
Acetylcholinesterase

Acetylcholinesterase (AChE) is an extremely fast enzyme that hydrolyzes the neurotransmitter acetylcholine to terminate signaling in cholinergic synapses. It has been studied in great detail by a number of computational methods, including molecular dynamics simulations, continuum electrostatics calculations and Brownian dynamics simulations.

One of the first puzzles posed by the crystal structure of AChE is the question of how the substrate acetylcholine gains access to the active site, since in the crystal structure there is no unobstructed pathway to the active site. This original observation for AChE from the fish *Torpedo californica* has subsequently been reiterated in mouse and human AChE. Simulations have shown how breathing motions in the enzyme facilitate the displacement of substrate from the surface of the enzyme to the buried active site. These motions appear quite complex and spatially extensive, which suggests possible modes of regulation of the activity of the enzyme. Such a mechanism has been observed in other proteins, including hemoglobin and myoglobin, but the fast reaction rates of AChE are hard to reconcile with major structural rearrangements. MD simulations suggest that the primary point of access opens and closes on a timescale that is fast enough not to slow down the entrance of the substrate substantially. Interestingly, the protein appears to have several secondary channels that allow water to enter and leave the active site as the substrate enters.

The reaction rate of AChE is limited by the diffusion of the neurotransmitter acetylcholine. Brownian dynamics are a suitable method to study encounters of acetylcholine with AChE. The acetylcholine diffuses under the influence of intermolecular interactions between acetylcholine and the protein and a random force. The net effect is that the protein "guides" the acetylcholine to its active site through an electrostatic mechanism that enhances the rate of finding acetylcholine beyond the pure diffusional limit. A similar mechanism has since been identified for several fast-reacting proteins.

AChE is the target of the neurotoxin fasciculin, a peptide. Brownian dynamics of the peptide and AChE give insight into

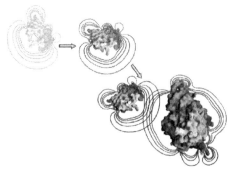

Fig. 4 Schematic illustration of the association of fasciculin (on the left) with acetylcholinesterase. Blue and red contour lines indicate regions of positive and negative electrostatic potential, respectively. Figure courtesy of D. Sept and A. Elcock. See Elcock, A.H., Gabdoulline, R.R., Wade, R.C., McCammon, J.A. (1999) Computer simulation of protein-protein association kinetics: acetylcholinesterase-fasciculin, *J. Mol. Biol.* **291**, 149–162 for more details. (See color plate p. vi.)

the binding kinetics of fasciculin and the structure of the resulting complex. In a series of Brownian dynamics simulations, Elcock et al. investigated the effect of mutations in the protein on association rate constants. The electrostatic interaction between AChE and fasciculin that promotes association is illustrated in Fig. 4. In its simplest form, the Brownian dynamics simulations reproduced the correct order of rate constants for different mutants, although the absolute values were too large by about a factor of 30. In a more accurate treatment of the continuum electrostatics problem, this discrepancy can be reduced greatly. Interestingly, these calculations make it possible to distinguish kinetic and thermodynamic effects, and suggest that for the case of fasciculin-AChE-binding mutations can separately affect the rate of association and the binding constant.

5.2
Water Transport Through Aquaporins

Aquaporins are ubiquitous membrane proteins that aid in the maintenance of crucial osmotic balance in cells, on whose discovery the Nobel Prize in chemistry was awarded in 2003. The road to high-resolution structures of these proteins, and to understanding with atomic detail how they function, is a nice example of the link between experiment, modeling, and simulations.

In early 2000, De Groot and coworkers used 4.5-Å resolution data, insufficient to identify the fold of the protein directly, to predict the fold of aquaporin based on the constraints of helix packing, atomic force microscopy data, and primary sequence. From 1440 possible folds, they were able to identify a maximum of 8 possible folds. Later that year, a 3.8-Å resolution structure of aquaporin from cryoelectron microscopy became available, as well as a 2.2-Å X-ray crystal structure for a homolog of aquaporin, GlpF, the glycerol transport facilitator. The fold of aquaporin is illustrated in Fig. 5(a). It has six transmembrane helices in an "hourglass" arrangement. The protein pore is shaped by two reentrant loops on either side of the membrane that have a conserved coil–NPA–helix motif, in which the Asn-Pro-Ala (NPA) residues of these loops meet in the middle of the channel. The partial helices of this motif are labeled HB and HE. The aromatic rich constriction region (ar/R), labeled in Fig. 5(a), is suspected of contributing to aquaporin's highly specific behavior.

Direct simulation of the AQP structure obtained from cryoelectron microscopy

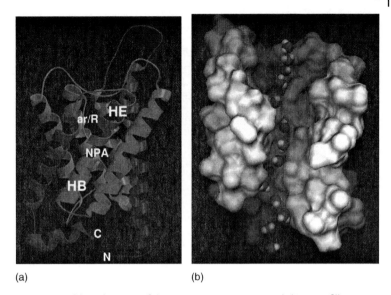

Fig. 5 (a) Ribbon diagram of the aquaporin monomer and (b) spacefilling representation illustrating a water file through the channel pore. (a) Starting at the N-terminus of the monomer, there are two transmembrane helices, followed by the coil–NPA–helix motif. A third transmembrane helix completes the first half of the protein. The second half of the monomer has two transmembrane helices, followed by the coil–NPA–helix motif, followed by another transmembrane helix. (b) The path of the water channel is through the core of the protein, following the path of the coil motifs, which meet at the NPA signature. This is more clearly illustrated in 5(b). The protein is rendered as a molecular surface in white, with the front surface of the protein cut away so that we can clearly see the pore in the middle of the protein. Surfaces colored yellow are those that interact most strongly with passing water molecules. The waterfile displayed is an overlay of a number of snapshots from the 10 ns simulation. The dipole inversion of water at the NPA motif is clearly illustrated here. Figure courtesy of B. de Groot. See De Groot, B.L., Grubmuller, H. (2001) Water permeation across biological membranes: mechanism and dynamics of aquaporin-1 and GlpF, *Science* **294**, 2353–2357. De Groot, B.L., Frigato, T., Helms, V., Grubmuller, H. (2003) The mechanism of proton exclusion in the aquaporin-1 water channel, *J. Mol. Biol.* **333**, 279–293 for more details. (See color plate p. xvi.)

showed a discontinuous water file through the channel and obvious defects in the structure, particularly at the NPA motifs, because of instability observed through the MD simulations. Further homology modeling between the available structures yielded improved stability in the core of the channel. Simulations of the high-resolution GlpF structure, and the improved homology model of AQP1 both observed continuous water files diffusing through the water channel, illustrated in Fig. 5(b), at rates comparable to those observed experimentally.

In late 2001, a 2.2-Å resolution X-ray crystal structure was obtained of human aquaporin-1. With confidence in the three-dimensional structure, the simulation studies that followed were able to focus on measuring physical properties of the channel using classical simulation methods, and also tried to pinpoint the origin of aquaporin's ability to transport water at high diffusion rates, while restricting proton transport. The basis of proton transport is a topic better addressed by quantum mechanical simulations, and so combinations of classical dynamics methods with semiempirical methods have been employed by a number of groups to determine the basis for proton exclusion by these channels.

What is particularly exciting about aquaporin as a case study is that the biological function of this protein, to act as a water channel across the cell membrane, occurs at a rate which is accessible to study by the current state of the art in molecular dynamics simulations. Through simulations, researchers were able to point out deficiencies in the low-resolution crystal structures on the basis of structural instability, and because the observed behavior *in silico* did not agree with that measured experimentally. This observation helped to point out the significance of certain structural motifs in the protein (i.e. the reentrant loops with the NPA motif). The evolution of our understanding of the aquaporin structure through experiments, homology modeling, and simulations has lead to similar methods being applied to other aquaporins with yet unsolved structures.

5.3
ABC Transporters and Multidrug Resistance Proteins

Many human proteins of major medical interest are difficult to obtain in large enough quantities for structural studies. The only exceptions have been proteins that occur in large amounts, such as bovine lens protein rhodopsin or Aqp1 from blood (as seen earlier). An emerging theme in membrane protein structural biology over the past years has been the use of bacterial homologs of human proteins, with, as striking examples, several potassium channel structures and several structures of ABC transporters. ATP-binding cassette (ABC) transporters are modular mechanical machines that couple the hydrolysis of ATP with the transport of molecules across membranes. They consist of two nucleotide binding domains (NBDs), two transmembrane domains, and optional additional domains, organized in a varying number of polypeptide chains. Mutations in the genes encoding many of the 48 ABC transporters of human cells are associated with several diseases, including cystic fibrosis. Increased expression of certain ABC transporters is a major cause of resistance to peptide antibiotics, antifungals, herbicides, anticancer drugs, and other cytotoxic agents. Interestingly, ABC transporters probably form the largest group of homologous proteins and exist in all species. In an exciting recent development, three crystal structures of ABC transporters have been determined: MsbA from *Escherichia coli* at 4.5-Å resolution, MsbA from *Vibrio cholera* at 3.8-Å resolution, and BtuCD from *E. coli* at 3.2-Å resolution. In addition, about 10 structures of nucleotide binding domains (NBDs) have been solved as well as ATP binding domains from other proteins. Despite this impressive progress in structural studies, several of the key questions about the basic mechanism of action of ABC transporters are unresolved. Simulations and homology modeling can be used to make some progress toward understanding the dynamics of ABC transporters and to translating the bacterial structures to human homologs like P-glycoprotein.

Molecular dynamics simulation studies have been used to investigate the "real-time" dynamics of the vitamin B_{12} importer BtuCD from *E. coli*, based on the single snapshot captured in the crystal structure. An example of a simulation model that incorporates BtuCD is shown in Fig. 3. In this model, the crystal structure is incorporated in a realistic environment of phospholipids, water, and ions, and simulated for 20 ns, both in the absence and presence of MgATP in the two binding sites. The results demonstrate that the docking of ATP to the catalytic pockets progressively draws the two cytoplasmic nucleotide-binding cassettes toward each other. Movement of the cassettes into closer opposition in turn induces conformational rearrangement of α-helices in the transmembrane domain. The shape of the translocation pathway consequently changes in a manner that could aid the vectorial movement of vitamin B_{12}. These results suggest that ATP binding may indeed represent the power stroke in the catalytic mechanism. Moreover, occlusion of ATP at one catalytic site is mechanically coupled to opening of the nucleotide-binding pocket at the second site. This may indicate that asymmetric behavior at the two catalytic pockets forms the structural basis by which the transporter is able to alternate ATP hydrolysis from one site to the other. While this remains to be tested, this study is an example of the use of dynamics simulations that build on a single crystal structure to obtain more information about the full process the protein is involved in.

The multidrug resistance P-glycoprotein mediates the extrusion of chemotherapeutic drugs from cancer cells. Characterization of the drug binding and ATPase activities of the protein have made it the paradigm ATP binding cassette (ABC) transporter. Although no high-resolution structure is known, P-glycoprotein has been imaged at low resolution by electron cryomicroscopy and extensively analyzed by disulfide cross-linking. Stenham et al. used an interesting combination of approaches to create a molecular model of P-glycoprotein (shown in Fig. 6) that fits with most experimental data from cross-linking and imaging, helps interpret these experiments, and provides insight into possible mechanisms of drug transport. As described earlier, the only ABC transporter structures whose high-resolution structures are known are MsbA and BtuCD. There is no homology between the transmembrane domains of BtuCD and P-glycoprotein, but the homology between MsbA and P-glycoprotein is 23% sequence identity for the transmembrane domains and 51% sequence identity for the nucleotide binding domains. MsbA is the closest known bacterial relative of P-glycoprotein. Unfortunately, the nucleotide binding domains in MsbA are poorly resolved and the structure has a low resolution. In contrast, the nucleotide binding domains in BtuCD are well resolved. The situation is further complicated by the tertiary and quaternary organization of the domains of MsbA, which is not consistent with either the BtuCD crystal or the extensive cross-linking data already published for P-gp. Nevertheless, Stenham et al. have shown that it is possible to generate atomic scale models of P-gp by combining experimental and theoretical methods. Their homology models are based on an MsbA template but with a tertiary organization that reflects the increasingly accepted consensus NBD dimer interface. This model will be useful in designing and interpreting experimental work on P-gp. In turn, results from experimental work can be used

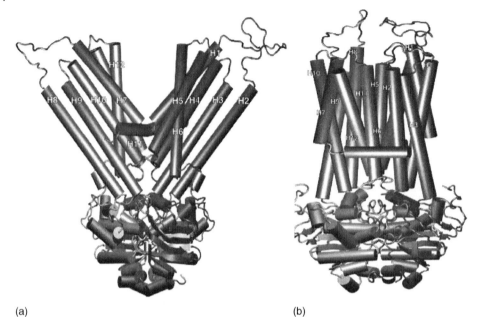

Fig. 6 (a) Construction of a molecular model for P-glycoprotein (P-gp). Each half of P-gp was modeled by homology to the crystal structure of MsbA (PDB code 1JSQ), which had been extended to a full atom representation. The two halves of P-gp were assembled such that the NBDs adopt the ATP-dependent orientation observed in MJ0796 (PDB code 1L2T). The constituent domains are individually colored. (b) Reconciliation of cross-linking data with a P-gp model. P-gp-Model-B was generated by rotation of each NBD with respect to its cognate TMD. The final model contains a parallel TMD:TMD interface (blue and gold subunits) and a consensus NBD:NBD interface (green and purple subunits). Adapted from Stenham, D.R., Campbell, J.D., Sansom, M.S.P., Higgins, C.F., Kerr, I.D., Linton, K.J. (2003) An atomic detail model for the human ATP binding cassette transporter P-glycoprotein derived from disulphide cross-linking and homology modeling, *FASEB J.* **17**, 2287–2289. (See color plate p. xviii.)

directly to improve the model, driving an iterative process that may advance our understanding of medically important ABC transporters.

6
Perspectives

Molecular modeling methods have become standard techniques to address biochemical problems at the level of biomolecular structure. A number of factors have contributed to this success. Computers themselves have become vastly more powerful in the past decades. Initially, simulations were restricted to very simple systems on timescales and levels of detail that were typically not adequate to answer specific questions about proteins. With a 10-fold increase in speed every 5 years a wide range of problems is now accessible directly by computer simulation, such as water transport in aquaporins. The fastest folding proteins can now be studied directly by simulations on the same timescale as the fastest experiments, with unprecedented progress in our understanding of protein folding. Computer graphics hardware has enjoyed a similar increase in speed, enabling complex surface representations of large biomolecular complexes and interactive 3D graphics. A second factor is the improvement in simulation and modeling software, which has led to the development of software that is vastly more user friendly than even a few years ago and runs on commodity hardware instead of exotic (expensive) workstations. A third factor is the improvement in algorithms and parameters that describe interactions between atoms or predict the structure of proteins. With increasing accuracy or more efficient algorithms, molecular modeling is able to answer increasingly more detailed and complicated questions. Finally, the massive advances in experimental studies at the level of biomolecular structure feed back into computer modeling. The expanding database of high-resolution protein structures is a key resource for further development of modeling methods. Advanced single-molecule experiments provide a direct link with simulations that was very rare in the past, and an increasing number of experimental techniques can now be linked to simulation results directly. An example of this is data from nuclear magnetic resonance spectroscopy, which provides both structural and dynamical data on timescales that are comparable to those of simulations.

Because molecular modeling is such a broad area, it is difficult to indicate exactly where most progress can be expected. Technical advances in hardware and software engineering, algorithm and parameter development will continue to increase the number of problems that can be addressed by molecular modeling. These advances will also facilitate incorporation of computationally intense calculations such as QM and QM/MM methods into accessible programs that are user-friendlier to the molecular modeler. The use of Grid technology, distributed computing, and massively parallel commodity-based Beowulf clusters will bring CPU-intensive calculations within reach of an increasing number of researchers, allowing us to study bigger and more complex systems. We expect that important areas of growth will be a stronger link between modeling and experiments, the routine incorporation of modeling and simulation in otherwise mostly experimental studies, and improved methods to study protein–protein interactions and chemical reactions.

Acknowledgment

Tieleman is a Scholar of the Alberta Heritage Foundation for Medical Research and a Sloan Foundation Fellow. Work in his laboratory is supported by the National Sciences and Engineering Research Council and the Canadian Institutes of Health Research.

See also Proteomics; Protein NMR Spectroscopy; Protein Structure Analysis: High-throughput Approaches.

Bibliography

Books and Reviews

Bourne P.E., Weissig, H. (Eds.) (2003) *Structural Bioinformatics*, John Wiley & Sons Inc., New Jersey, MI.

Day, R., Daggett, V. (2003) All-atom Simulations of Protein Folding and Unfolding, in: Eisenberg, D., Kim, P. (Eds.) *Advances in*

Protein Chemistry, Vol. 66, Elsevier Academic Press, New York, pp. 373–403.

Frenkel, D. Smit, B. (2001) *Understanding Molecular Simulation. From Algorithms to Applications*, 2nd edition, Academic Press.

Honig, B., Nicholls, A. (1995) Classical electrostatics in biology and chemistry, *Science* **268**, 1144–1149.

Jorgensen, W.L. (2004) The many roles of computation in drug discovery, *Science* **303**, 1813–1818.

Karplus, M., McCammon, J.A. (2002) Molecular dynamics simulations of biomolecules, *Nat. Struct. Biol.* **9**(9), 646–652.

Leach, A.R. (2001) *Molecular Modeling, Principles and Applications*, 2nd edition, Pearson Education Limited, England.

Petsko, G.A., Ringe, D. (2004) *Primers in Biology: Protein Structure and Function*, New Science Press Ltd.

Schlick, T. (2002) *Molecular Modeling and Simulation – An Interdisciplinary Guide*, Springer, New York.

Primary Literature

Agre, P. (2004). Aquaporin Water Channels (Nobel Lecture). *Angew. Chem. Int. Ed.* **43**, 4278–4290.

Altschul, S.F., Madden, T.L., Schaeffer, A.A., Zhang, J., Zhang, Z., Miller, W., Lipman, D.J. (1997) Gapped BLAST and PSI-BLAST: a new generation of protein database search programs, *Nucleic Acid Res.* **25**, 3389–3402.

Ash, W.L., Zlomislic, M.R., Oloo, E.O., Tieleman, D.P. (2004) Computer simulations of membrane proteins, *Biochim. Biophys. Acta-Biomembranes* **1666**, 158–189.

Baker, N.A., McCammon, J.A. (2003) Electrostatic Interactions, in: Bourner, P.E., Weissig, H. (Eds.) *Structural Bioinformatics*, Wiley & Sons, Inc., New Jersey, MI.

Baker, D., Sali, A. (2001) Protein structure prediction and structural genomics, *Science* **294**, 93–96.

Baker, N.A., Sept, D., Joseph, S., Holst, M.J., McCammon, J.A. (2001) Electrostatics of nanosystems: application to microtubules and the ribosome, *Proc. Natl. Acad. Sci. U.S.A.* **98**, 10037–10041.

Berendsen, H.J.C. (2001) Bioinformatics – Reality simulation – Observe while it happens, *Science* **294**, 2304–2305.

Berman, H.M., Westbrook, J., Feng, Z., Gilliland, G., Bhat, T.N., Weissig, H., Shindyalov, I.N., Bourne, P.E. (2000) The protein data bank, *Nucleic Acids Res.* **28**, 235–242.

Bradley, P., Chivian, D., Meiler, J., Misura, K.M.S., Rohl, C.A., Schief, W.R., Wedemeyer, W.J., Schueler-Furman, O., Murphy, P., Schonbrun, J., Strauss, C.E.M., Baker, D. (2003) Rosetta predictions in CASP5: Successes, failures, and prospects for complete automation, *Proteins: Struct., Funct., Genet.* **53**, 457–468.

Burykin, A., Warshel, A. (2003) What really prevents proton transport through aquaporin? Charge self-energy versus proton wire proposals, *Biophys. J.* **85**, 3696–3706.

Campbell, J.D., Biggin, P.C., Baaden, M., Sansom, M.S.P. (2003) Extending the structure of an ABC transporter to atomic resolution: modeling and simulation studies of MsbA, *Biochemistry* **42**, 3666–3673.

Chakrabarti, N., Tajkhorshid, E., Roux, B., Pomes, R. (2004) Molecular basis of proton blockage in aquaporins, *Structure* **12**, 65–74.

Chance, M.R., Fiser, A., Sali, A., Pieper, U., Eswar, N., Xu, G.P., Fajardo, J.E., Radhakannan, T., Marinkovic, N. (2004) High-throughput computational and experimental techniques in structural genomics, *Genome Res.* **14**, 2145–2154.

Chang, G. (2003) Structure of MsbA from Vibrio cholera: a multidrug resistance ABC transporter homolog in a closed conformation, *J. Mol. Biol.* **330**, 419–430.

Davis, M.E., McCammon, J.A. (1990) Electrostatics in Biomolecular Structure and Dynamics, *Chem. Rev.* **90**, 509–521.

De Groot, B.L., Grubmuller, H. (2001) Water permeation across biological membranes: mechanism and dynamics of aquaporin-1 and GlpF, *Science* **294**, 2353–2357.

De Groot, B.L., Engel, A., Grubmuller, H. (2003) The structure of the aquaporin-1 water channel: a comparison between cryo-electron microscopy and x-ray crystallography, *J. Mol. Biol.* **325**, 485–493.

De Groot, B.L., Frigato, T., Helms, V., Grubmuller, H. (2003) The mechanism of proton exclusion in the aquaporin-1 water channel, *J. Mol. Biol.* **333**, 279–293.

De Groot, B.L., Heymann, J.B., Engel, A., Mitsuoka, K., Fujiyoshi, Y., Grubmuller, H.

(2000) The fold of human aquaporin-1, *J. Mol. Biol.* **300**, 987–994.

Dill, K.A., Chan, H.S. (1997) From Levinthal to pathways to funnels, *Nat. Struct. Biol.* **4**, 10–19.

Dominguez, C., Boelens, R., Bonvin, A.M. (2003) HADDOCK: a protein-protein docking approach based on biochemical or biophysical information, *J. Am. Chem. Soc.* **125**, 1731–1737.

Elcock, A.H. (2004) Molecular simulations of diffusion and association in multimacromolecular systems, *Numerical Comput. Methods, Pt D* **383**, 166–198.

Elcock, A.H., Sept, D., McCammon, J.A. (2001) Computer simulation of protein-protein interactions, *J. Phys. Chem. B* **105**, 1504–1518.

Elcock, A.H., Gabdoulline, R.R., Wade, R.C., McCammon, J.A. (1999) Computer simulation of protein-protein association kinetics: acetylcholinesterase-fasciculin, *J. Mol. Biol.* **291**, 149–162.

Ermak, D.L., McCammon, J.A. (1978) Brownian dynamics with hydrodynamic interactions, *J. Chem. Phys.* **69**, 1352–1360.

Fiser, A.S., Sali, A. (2003) MODELLER: generation and refinement of homology-based protein structure models, *Macromol. Crystallogr., Pt D* **374**, 461–46.

Gabdoulline, R.R., Wade, R.C. (2001) Protein-protein association: Investigation of factors influencing association rates by brownian dynamics simulations, *J. Mol. Biol.* **306**, 1139–1155.

Godzik, A. (2003) Fold Recognition Methods, in: Bourner, P.E., Weissig, H. (Eds.) *Structural Bioinformatics*, John Wiley & Sons, Inc., New Jersey, MI, pp. 525–546.

Gues, N., Peitsch, M.C. (1997) Swiss-model and the Swiss-PDB viewer: an environment for comparative protein modeling, *Electrophoresis* **18**, 2714–2723.

Hartley, R.W. (1989) Barnase and barstar – 2 small proteins to fold and fit together, *Trends Biochem. Sci.* **14**, 450–454.

Holland, I.B., Cole, S.P.C., Kuchler, K., (Eds.) (2002). *ABC Proteins: From Bacteria to Man*. Academic Press.

Honig, B., Nicholls, A. (1995) Classical electrostatics in biology and chemistry, *Science* **268**, 1144–1149.

Humphrey, W., Dalke, A., Schulten, K. (1996) VMD – visual molecular dynamics, *J. Mol. Graph.* **14**, 33–38.

Ilan, B., Tajkhorshid, E., Schulten, K., Voth, G.A. (2004) The mechanism of proton exclusion in aquaporin channels, *Proteins* **55**, 223–228.

Janin, J., Henrick, K., Moult, J., Eyck, L.T., Sternberg, M.J., Vajda, S., Vakser, I., Wodak, S.J. (2003) CAPRI: a critical assessment of predicted interactions, *Proteins* **52**, 2–9.

Jensen, M.O., Tajkhorshid, E., Schulten, K. (2003) Electrostatic tuning of permeation and selectivity in aquaporin water channels, *Biophys. J.* **85**, 2884–2899.

Kollman, P. (1993) Free-energy calculations – applications to chemical and biochemical phenomena, *Chem. Rev.* **93**, 2395–2417.

Krieger, E., Nabuurs, S.B., Vriend, G. (2003) Homology Modeling, in: Bourner, P.E., Weissig, H. (Eds.) *Structural Bioinformatics*, Wiley & Sons, Inc, New Jersey, MI.

Kuhlman, B., Dantas, G., Ireton, G.C., Varani, G., Stoddard, B.L., Baker, D. (2003) Design of a novel globular protein fold with atomic-level accuracy, *Science* **302**, 1364–1368.

Lee, L.P., Tidor, B. (2001) Barstar is electrostatically optimized for tight binding to barnase, *Nat. Struct. Biol.* **8**, 73–76.

Levinthal, C. (1969). How to Fold Graciously. *Mossbauer Spectroscopy in Biological Systems*, University of Illinois Press, Urbana, Illinois, MN.

Locher, K.P., Lee, A.T., Rees, D.C. (2002) The E-coli BtuCD structure: a framework for ABC transporter architecture and mechanism, *Science* **296**, 1091–1098.

MacKinnon, R. (2004) Potassium channels and the atomic basis of selective ion conductance (Nobel lecture), *Angew. Chem.-Int. Ed.* **43**, 4265–4277.

Marti-Renom, M.A., Stuart, A.C., Fiser, A., Sanchez, R., Melo, F., Sali, A. (2000) Comparative protein structure modeling of genes and genomes, *Annu. Rev. Biophys. Biomol. Struct.* **29**, 291–325.

McCammon, J.A., Gelin, B.R., Karplus, M. (1977) Dynamics of folded proteins, *Nature* **267**, 585–590.

McCammon, J.A., Gelin, B.R., Karplus, M., Wolynes, P.G. (1976) Hinge-bending mode in lysozyme, *Nature* **262**, 325–326.

Oloo, E.O., Tieleman, D.P. (2004) Conformational transitions induced by the binding of MgATP to the vitamin B12 ATP-binding cassette (ABC) transporter BtuCD, *J. Biol. Chem.* **279**, 45013–45019.

Rost, B. (2003) Prediction in 1D: Secondary Structure, Membrane Helices, and Accessibility, in: Bourner, P.E., Weissig, H. (Eds.) *Structural Bioinformatics*, John Wiley & Sons, Inc, New Jersey, MI.

Rychelewski, L., Fischer, D., Elofsson, A. (2003) LiveBench-6: large-scale automated evaluation of protein structure prediction servers, *Proteins: Struct., Funct., Genet.* **53**, 542–547.

Sali, A., Blundell, T.L. (1993) Comparative protein modeling by satisfaction of spatial restraints, *J. Mol. Biol.* **234**, 779–815.

Sayle, R., Milner-White, E.J. (1995) RasMol: biomolecular graphics for all, *Trends Biochem. Sci.* **20**, 374.

Schreiber, G., Fersht, A.R. (1995) Energetics of protein-protein interactions – analysis of the barnase-barstar interface by single mutations and double mutant cycles, *J. Mol. Biol.* **248**, 478–486.

Sharp, K.A., Honig, B. (1990) Electrostatic interactions in macromolecules – theory and applications, *Annu. Rev. Biophys. Biophys. Chem.* **19**, 301–332.

Shen, T.Y., Tai, K.H., Henchman, R.H., McCammon, J.A. (2002) Molecular dynamics of acetylcholinesterase, *Acc. Chem. Res.* **35**, 332–340.

Smith, G.R., Sternberg, M.J. (2002) Prediction of protein–protein interactions by docking method, *Curr. Opin. Struct. Biol.* **12**, 28–35.

Smith, G.R., Sternberg, M.J. (2003) Evaluation of the 3D-dock protein docking suite in rounds 1 and 2 of the CAPRI blind trial, *Proteins* **52**, 74–79.

Stenham, D.R., Campbell, J.D., Sansom, M.S.P., Higgins, C.F., Kerr, I.D., Linton, K.J. (2003) An atomic detail model for the human ATP binding cassette transporter P-glycoprotein derived from disulphide cross-linking and homology modeling, *FASEB J.* **17**, 2287–2289.

Tajkhorshid, E., Nollert, P., Jensen, M.O., Miercke, L.J.W., O'Connell, J., Stroud, R.M., Schulten, K. (2002) Control of the selectivity of the aquaporin water channel family by global orientational tuning, *Science* **296**, 525–530.

Tajkhorshid, E., Aksimentiev, A., Balabin, I., Gao, M., Israelwitz, B., Phillips, J.C., Zhu, F., Schulten, K. (2003) Large Scale Simulation of Protein Mechanics and Function, in: Eisenberg, D., Kim, P. (Eds.) *Advances in Protein Chemistry*, Vol. 66, Elsevier Academic Press, New York, pp. 195–247.

Tate, J. (2003) Molecular Visualization, in: Bourner, P.E., Weissig, H. (Eds.) *Structural Bioinformatics*, Wiley & Sons, Inc, New Jersey, MI.

Taylor, R.D., Jewsbury, P.J., Essex, J.W. (2002) A review of protein-small molecule docking methods, *J. Comput. Aided Mol. Des.* **16**, 151–166.

Tieleman, D.P., Marrink, S.J., Berendsen, H.J.C. (1997) A computer perspective of membranes: molecular dynamics studies of lipid bilayer systems, *Biochim. Biophys. Acta-Rev. Biomembranes* **1331**, 235–270.

Van Gunsteren, W.F., Berendsen, H.J.C. (1990) Computer-simulation of molecular-dynamics - methodology, applications, and perspectives in chemistry, *Angew. Chem.-Int. Ed. Engl.* **29**, 992–1023.

Vaughan, C.K., Buckle, A.M., Fersht, A.R. (1999) Structural response to mutation at a protein-protein interface, *J. Mol. Biol.* **286**, 1487–1506.

Vriend, G. (1990) What if – a molecular modeling and drug design program, *J. Mol. Graph.* **8**, 52–55.

Warshel, A., Aqvist, J. (1991) Microscopic Simulations of Chemical Processes in Proteins and the Role of Electrostatic Free Energy, in: Beveridge, D.L., Lavery, R. (Eds.) *Theoretical Biochemistry and Molecular Biophysics*, Vol. 2, Adenine Press, New York, pp. 257.

13
Synthesis of Peptide Mimetics and their Building Blocks

Bruce K. Cassels[1] *and Patricio Sáez*[2]
[1] *Department of Chemistry, Faculty of Sciences, University of Chile, and*
[2] *Faculty of Medical Sciences, University of Santiago, Chile*

1	Introduction	407
2	Azapeptides	408
3	β-, γ-, δ-, ε-, and Related Peptides	411
4	Aminooxy Peptides	412
5	Oligocarbamate and Oligourea Peptidomimetics	414
6	Peptoids	415
7	Variations on the Peptoid Theme	417
8	Proline Substitutes	420
9	Biological Activity and Therapeutic Hints	420
	Bibliography	426
	Books and Reviews	426
	Primary Literature	427

Keywords

Azapeptides
Peptide mimetics in which one or more amino acid residues is (or are) replaced by azaamino acid(s) (i.e. amino acid(s) where the carbon atom bearing the amino group is replaced by nitrogen).

Proteins. Edited by Robert A. Meyers.
Copyright © 2007 Wiley-VCH Verlag GmbH & Co. KGaA, Weinheim
ISBN: 978-3-527-31608-3

Aminooxy Peptides
Analogs of β-, γ- or δ-peptides in which the β-, γ- or δ-carbon is replaced by an oxygen atom.

Oligocarbamate
A compound containing repeating O−CO−NH units separated by two carbon atoms, one of which may bear a side chain.

Oligourea
A compound containing repeating NH−CO−NH units separated by two carbon atoms, one of which may bear a side chain.

β-, γ-, δ-, ε-Peptides
Peptide mimetics based on β-, γ-, δ- or ε-amino acids.

Peptide Mimetics (or Peptidomimetics)
Compounds imitating the structure of bioactive peptides, preserving overall conformational features and retaining some of the chemical groups that interact with biological targets, but incorporating nonamino acid units.

Peptoids
Peptide mimetics in which the amino acid side chains are shifted to the amide nitrogen atoms.

> Very many peptides have the potential of being used as drugs. Nevertheless, they usually suffer from serious limitations stemming from their metabolic lability and unfavorable pharmacokinetics. A promising approach to circumvent these problems is the use of conformationally similar molecules, able to interact with the same biological targets, but less susceptible to metabolic degradation and with better absorption and distribution properties. This may be achieved through the synthesis of peptide mimetics where some or all of the amino acid residues have been replaced by different chemical moieties. This article gives a brief overview of peptide mimetic structures on which medicinal chemists have focused in the last few years, together with the chemical approaches used for their synthesis and of the amino acid replacements that serve as their building blocks. These methods range from classical organic synthetic methods to quite sophisticated catalytic and solid-phase protocols.
>
> In many cases, the target compounds have been tested in biological systems with varying degrees of success. These results should not be construed as an indication of unreliability of the peptidomimetic approach, but rather as a reflection of the limited sample examined to date. Because of the solid foundations on which this therapeutic strategy is based, we think that a growing number of peptide mimetics are likely to enter the clinic in the coming years.

1
Introduction

The large number of endogenous peptides that act as cytokines, hormones, or neurotransmitters, in the regulation of neural function and blood pressure, and in inflammation, continues to increase. Not only are primary gene products important, but in many cases their breakdown also affords smaller peptides with increased potency or with qualitatively different properties. Nevertheless, the development of clinical therapies based on endogenous peptides has been rather intuitive and not actually realized because of pharmacokinetic as well as pharmacodynamic issues, which seem to be difficult to solve properly. In order to overcome these drawbacks, the discovery by screening, the design, synthesis, and pharmacological testing of nonpeptide ligands of peptide receptors – either as agonists or antagonists – has become a very active field, particularly during the last decade.

Peptide mimics, peptide mimetics, or peptidomimetics may be defined as compounds that can substitute for peptides in their interactions with receptors or enzymes. Early examples such as morphine – a mimic of opioid peptides – did not result from any awareness of their mechanism of action. More recently, screening efforts have led to "hits" such as the natural, moderately potent cholecystokinin receptor antagonist asperlicin, described by Chang et al. in 1985, which has been modified to afford a plethora of synthetic analogs based on either the 1,4-benzodiazepine or the quinazolidinone moieties, reviewed by Herranz in 2003, in both of which series peptide analog portions may be discerned.

The design, synthesis, and evaluation of peptide mimics dates back at least to the late 1960s, but the vast majority of papers in the field are less than 20 years old, and the number of publications in the last decade is close to three-quarters of the total. The subject of peptide mimics can be divided into three major areas: modifications of the amino acid side chains, modifications of the peptide backbone, and dipeptide mimics with constrained conformations. The first aspect has been addressed by Williams in the previous edition of this encyclopedia (1996), and will not be considered here. Isosteric replacements of the peptide bond (CONH) by groups such as thioamido (CSNH), ester (COO), ketomethylene ($COCH_2$), methyleneoxy-, methylenethio-, or methyleneamino (CH_2O, CH_2S, CH_2NH), retro-amido (NHCO), or trans C=C double bonds (CH=CH, CF=CH) were reviewed by Goodman and Ro in 1995 in a chapter of Burger's Medicinal Chemistry that is more concerned with conformationally constrained dipeptide mimics. These are a newer development that relies heavily on a firm understanding of protein conformation, which has been reviewed by Hruby and Balse in 2000. Another review volume is that of Abell in 1999. An extensive, five-volume treatise on the subject of peptide and peptidomimetic synthesis was published by Goodman et al. in 2002, and again, as a "workbench edition," in 2004. We will consequently concentrate on advances made in the last few years.

The synthesis of these compounds relies in part on classical peptide chemistry, but almost always incorporating specific modifications to accommodate the different building blocks required. An important general methodology for the construction of peptoids, for example, is the approach of Zuckermann, described below, which continues to be used in most cases.

2
Azapeptides

Azapeptides are peptide mimics in which one or more backbone α-carbon atoms have been replaced by nitrogen. From the structural standpoint, this means introducing one or more urea moieties in the peptide chain. Occasional syntheses of monoazapeptides have been described at least for the last 40 years and were reviewed as far back as 1970. However, this type of peptide mimic was not included in the Goodman and Ro chapter published in 1995, although it has been reviewed in 2002 by Zega and Urleb, so we will discuss them in some detail.

Azapeptide synthesis has generally been achieved by a combination of hydrazine chemistry and peptide chemistry. For example, the reaction of a 1-alkyl-2-Boc-protected hydrazine or an unprotected monoalkylhydrazine with a carboxylic acid ester α-isocyanate affords an N-protected or unprotected monoazadipeptide derivative (Scheme 1). Even in unprotected monoalkylhydrazines, reaction at the substituted nitrogen atom is favored. If deprotected, the resulting semicarbazide can be coupled to the carboxyl group of an N-protected amino acid or peptide using, for example, conventional dicyclohexylcarbodiimide (DCC) coupling, to provide the elongated product (Scheme 2).

A stepwise example of this, requiring purification at each step, but which resulted in potent, selective cysteine protease inhibitors, is that of Wieczerzak et al. in 2002. In this case, peptides were prepared in solution and then an azaglycine residue was appended to the free amino group by coupling *t*-butyl carbazate (Boc-hydrazine) and an amino acid or peptide ester or amide with carbonyldiimidazole, where the key reaction is shown in Scheme 3. Although experimental details are lacking, the Boc-azapeptide could presumably be deprotected and coupled to a Boc-protected amino acid or peptide using conventional methodology (the use of 2-(1-benzotriazol-1-yl)-1,1,3,3-tetramethyluronium tetrafluoroborate – TBTU – or DCC, and 1-hydroxybenzotriazole – HOBt – are mentioned).

The synthesis of a series of Boc-protected alkylhydrazines was described by Dutta and Morley in 1975, and still constitutes a generally useful approach. This is

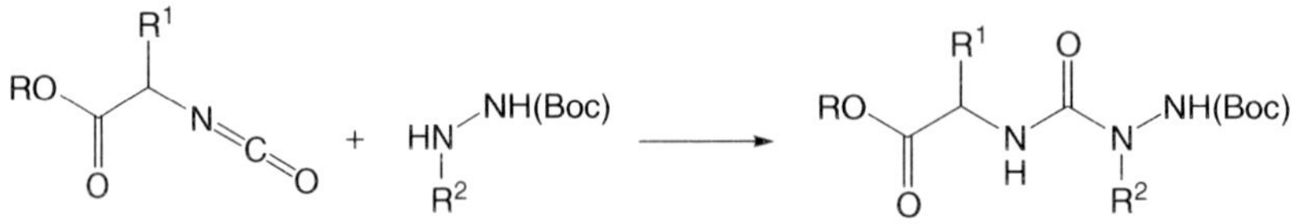

Scheme 1

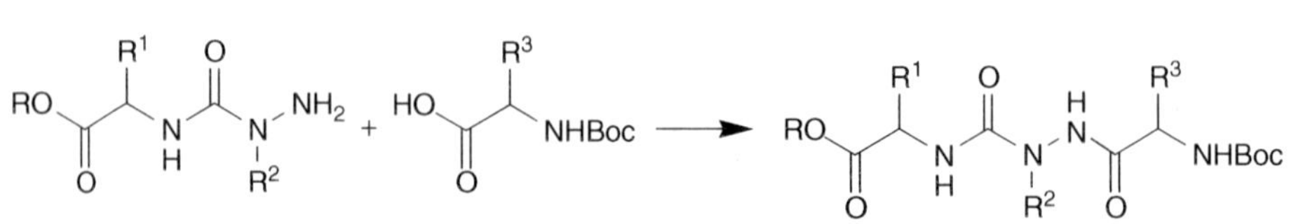

Scheme 2

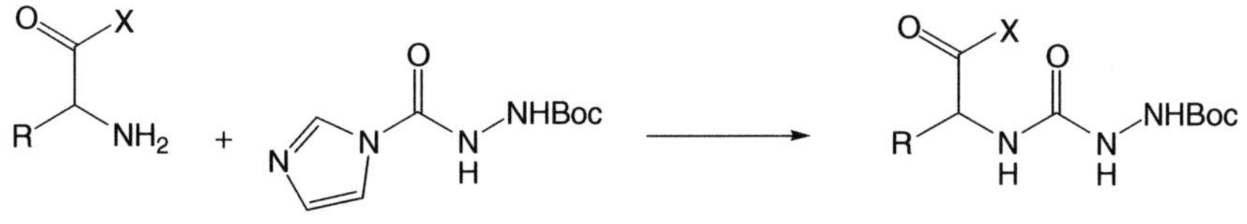

Scheme 3

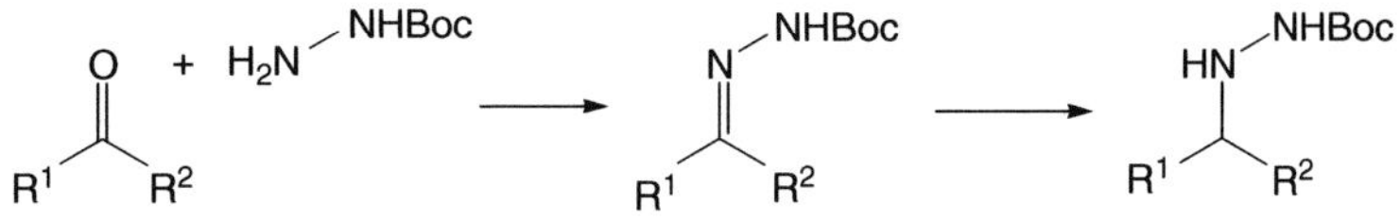

Scheme 4

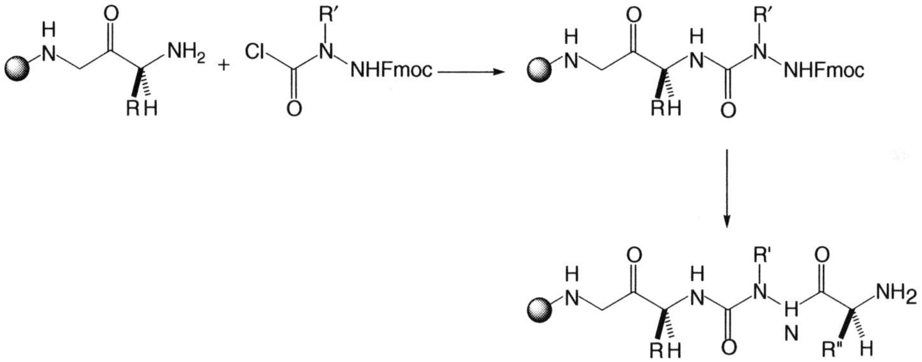

Scheme 5

done by preparing the Boc-hydrazides of suitable aldehydes or ketones and then reducing them by catalytic hydrogenation (Scheme 4).

In 2001, Hart and Beeson in a newer example of compounds introduced important modifications allowing the use of solid-phase Fmoc methodology on Rink amide resin (Scheme 5).

A large series of compounds of this class, based on the sequence of the antigenic hen ovalbumin pentadecapeptide Gln-Ala-Val-His-Ala-Ala-His-Ala-Glu-Ile-Asn-Glu-Ala-Gly-Arg by replacing one amino acid at a time by azaglycine, azaalanine, or azaglutamic acid, bound detectably to the major histocompatibility complex. Some of them, for example, Gln-Ala-Val-His-Ala-Ala-His-Ala-Glu-Ile-Gly[a]-Glu-Ala-Gly-Arg (where the superscript[a] indicates an azaamino acid), were as potent as activators of T cells as the native peptide.

The azaalanine monomer was built starting from methylhydrazine (Scheme 6). Synthesis of the azaglutamic acid started from *tert*-butyl acrylate (Scheme 7).

A recent patent describes an apparently general method for the preparation of azapeptides or azatides that, like Hart and Beeson's procedure in 2001 but unlike earlier methods, is amenable in theory to automated synthesis by modification of existing protocols. An important advantage

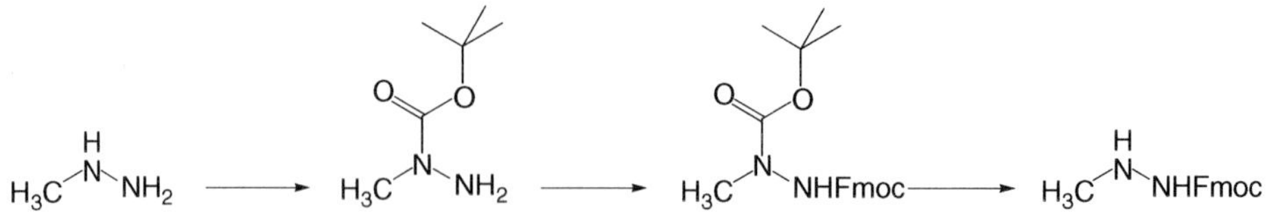

Scheme 6

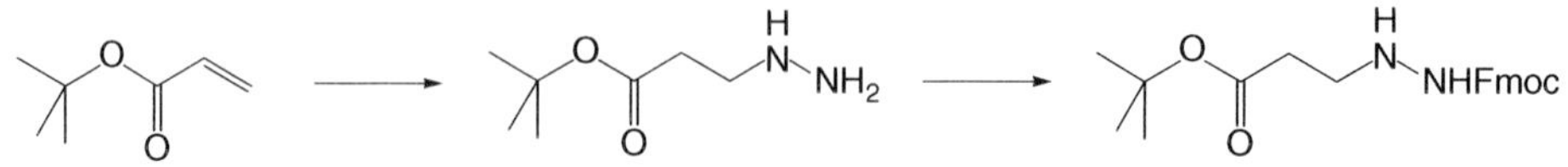

Scheme 7

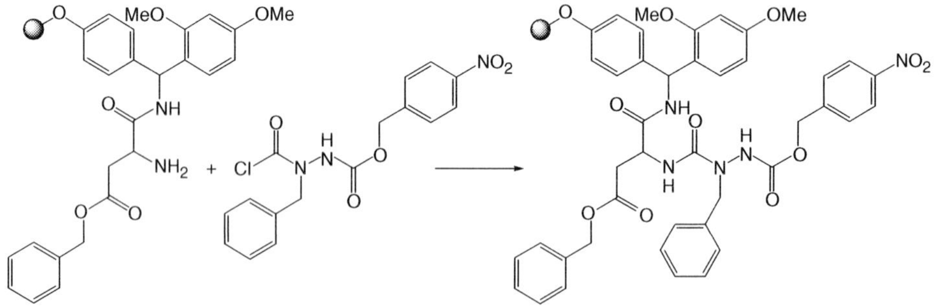

Scheme 8

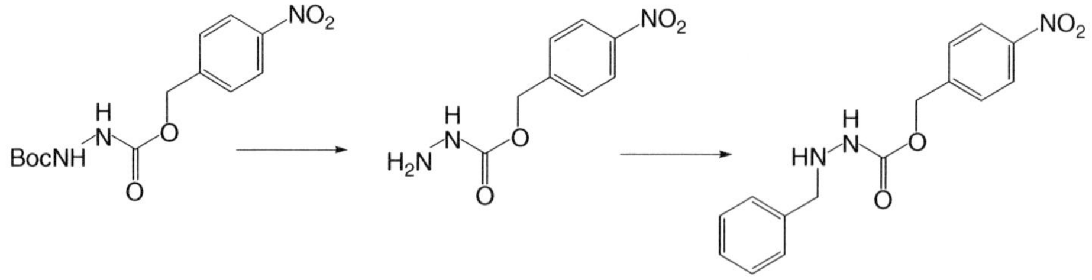

Scheme 9

of this over earlier methods is that it uses the same basic reaction sequence as the Merrifield synthetic strategy. This involves the reaction of an azaamino acid building block, protected with a group such as *p*-nitrobenzyl, with a carbonyl-releasing agent such as phosgene in the presence of base to afford a protected, activated carbonyl azaamino acid. This crude product is subsequently allowed to react with a resin-bound amino acid, peptide, or amino acid or peptide azaanalog (Scheme 8). Subsequent cleavage of the protecting group, for example, with $SnCl_2$, releases the freshly synthesized, resin-bound azapeptide. This methodology requires the synthesis of aza-amino acid building blocks, which can usually be achieved in three steps by reaction

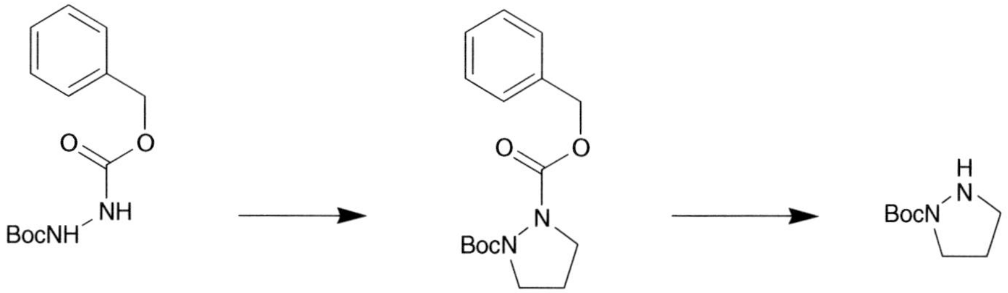

Scheme 10

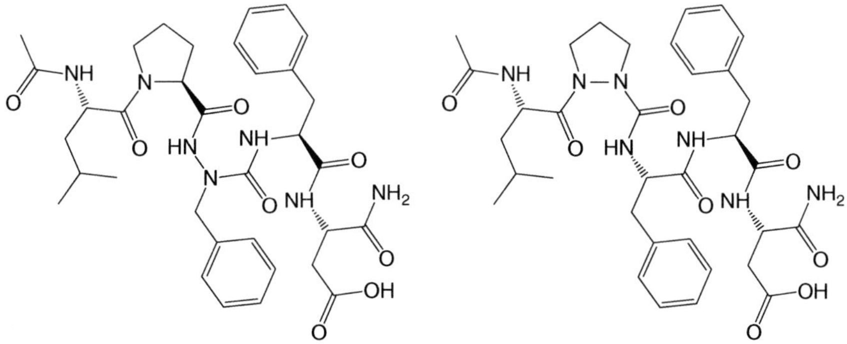

Fig. 1 Two azapeptide analogs of Ac-Leu-Pro-Phe-Phe-AspNH$_2$.

of t-butyl carbazate with p-nitrobenzyl chloroformate, selective removal of the t-butyloxycarbonyl group, and alkylation of the primary amino group (Scheme 9).

A slightly different challenge is posed by the azaproline moiety, which was synthesized according to the sequence shown in Scheme 10. In this particular case, the t-butyloxycarbonyl derivative was used directly in the chlorocarbonylation and coupling steps. This methodology was used to synthesize azapeptides such as Ac-Leu-Pro-Phe[a]-Phe-AspNH$_2$ and Ac-Leu-Pro[a]-Phe-Phe-AspNH$_2$ (Fig. 1).

3
β-, γ-, δ-, ε-, and Related Peptides

β-, γ-, δ- and ε-peptides are oligomers of β-, γ-, δ- or ε-amino acids (Fig. 2). Depending upon their backbone substitution, they may be classified as $β^2$-, $β^{2,2}$-, $β^3$-, and so on (depicted are only $β^2$-, $γ^2$-, $δ^2$- and $ε^2$- peptides). Although sugar amino acids are not covered in this chapter, they may also be viewed as building blocks for conformationally restricted β-, γ-, δ-, and ε-peptides, for example, Fig. 3. These compounds have been reviewed as recently as 2002 by Gruner et al.

Another recent review, concentrating on the structure of β-peptides, but also briefly reviewing their biological properties, is that of Cheng et al. in 2001. Another review by Lelais and Seebach in 2004 with an update of the subject almost to the present day would render any other attempt repetitious at this time.

On several occasions, γ-peptides have been reported to adopt hydrogen bond–stabilized helical conformations, but in

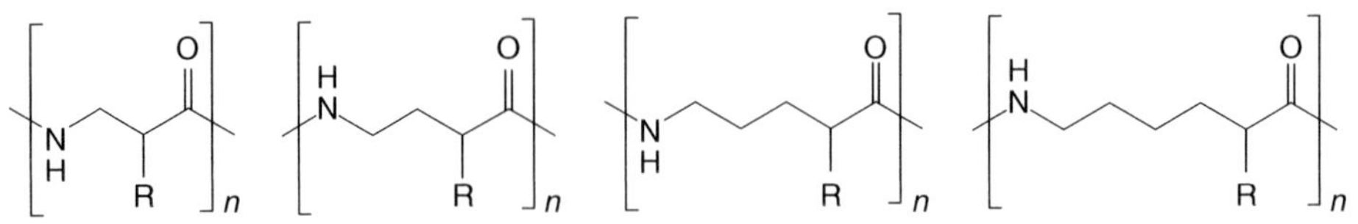

Fig. 2 General structures of β^2-, γ^2-, δ^2- and ε^2- peptides.

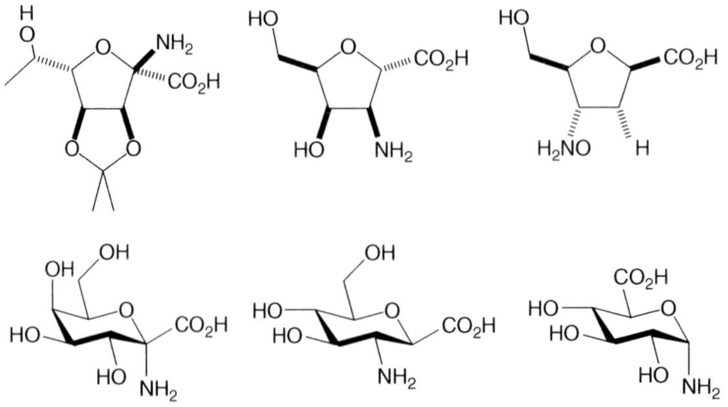

Fig. 3 Examples of sugar amino acid building blocks.

4
Aminooxy Peptides

α-, β- and γ-aminooxypeptides are analogs of β-, γ- or δ-peptides in which the β-, γ- or δ-carbon is replaced by an oxygen atom (Fig. 4). The unusual conformational properties of peptides containing a single α-aminooxy acid residue were first predicted and observed by Yang et al. in

oligomers of γ-amino acid residues, a sheet secondary structure is preferred. Nowick and Brower in 2003 and Zhao et al. in 2004 have constructed bulky N^α-acylornithine-derived δ-peptides that adopt well-defined folded conformations, unlike earlier, presumably more flexible examples. The synthesis of these compounds usually adheres to standard peptide synthetic methodology.

1996. A recent important development in this field was the first solid-phase synthesis of oligomeric α-aminooxy peptides. This was based on the stepwise extension of a resin-bound α-phthalimidoxy acid (Scheme 11). The same group has introduced 2-oxanipecotic acid as a cyclic α-aminooxy acid for conformational studies, for example, Fig. 5. The synthesis of phthaloyl δ-aminooxy acids, the building blocks of chiral δ-aminooxy peptides, has been reviewed and improved upon by Shin et al. in 2000.

Chiral β^3-aminooxy peptides are a recent development. They have been obtained in a straightforward manner by coupling of phthalimidoxy acids with the free amino groups of their partners, under similar conditions to those used previously by the same authors. The chiral β^3-aminooxy amino acid building blocks were synthesized from chiral α-aminooxy

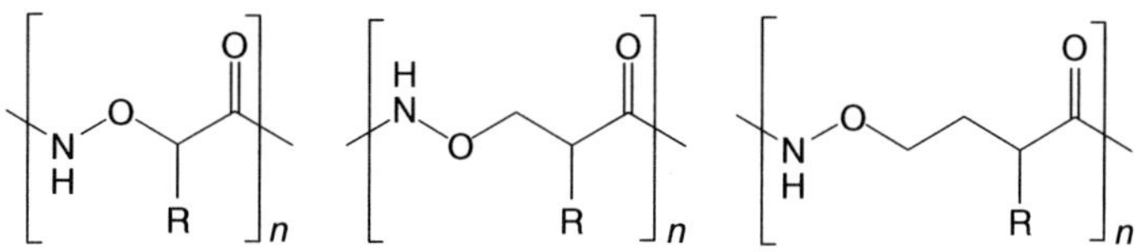

Fig. 4 General structures of α-, β- and γ- aminooxypeptides.

Fig. 5 A 2-oxanipecotic acid dimer moiety.

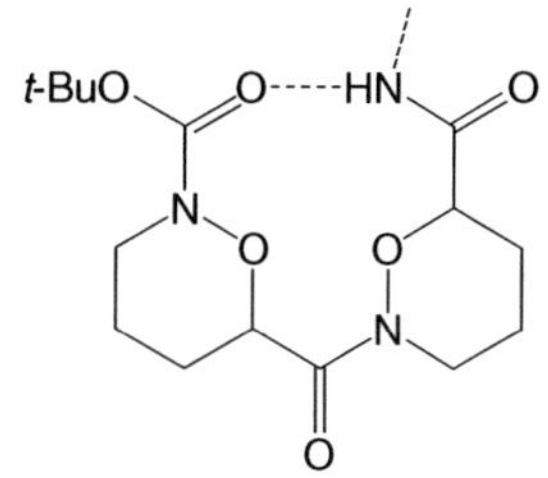

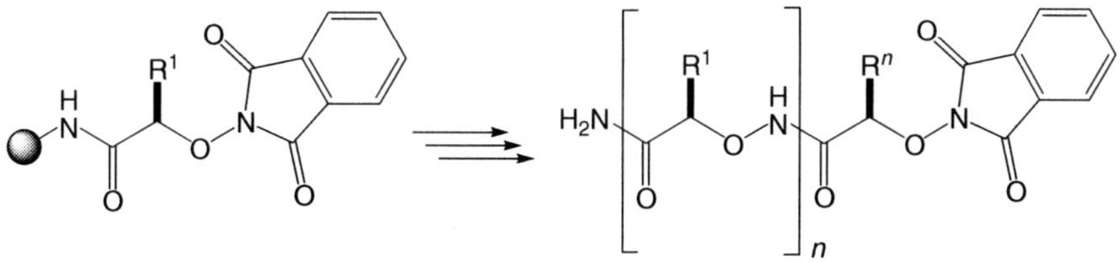

Scheme 11

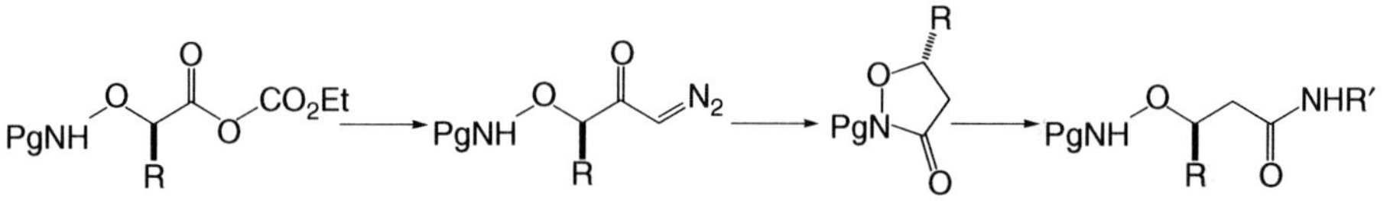

Scheme 12

Fig. 6 A γ^4-aminooxydipeptide.

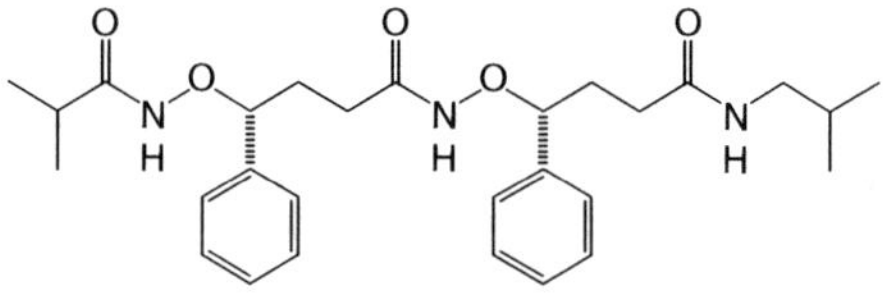

acids by Arndt-Eistert homologation (Pg = protecting group) (Scheme 12) or, preferably, by enantioselective reduction of β-keto esters using either baker's yeast or chiral Ru(II) complexes, followed by reduction of the ester group and amination of the secondary alcohol (Scheme 13).

γ^4-Aminooxy peptides have also been synthesized very recently and have been shown to adopt unique helical conformations, different from those of homochiral oligomers of α- and β-aminooxy acids (Fig. 6). Experimental details of their syntheses have not yet been published.

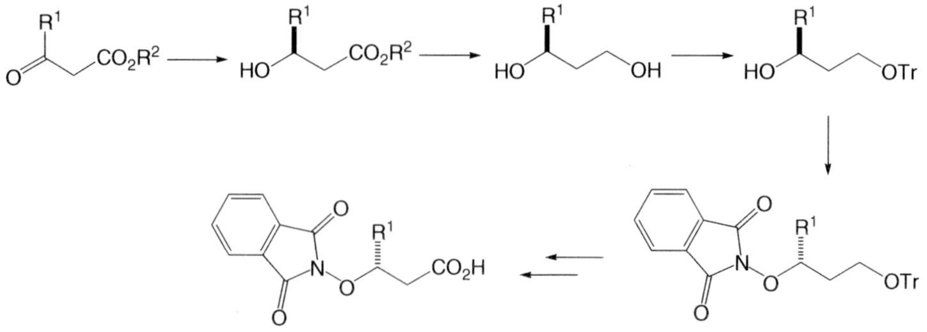

Scheme 13

5
Oligocarbamate and Oligourea Peptidomimetics

Oligocarbamate peptide mimetics were first synthesized as a library by Cho et al. in 1993, using methodology described for the preparation of peptide libraries. Their general structure may be depicted as shown in Fig. 7. The first successful attempts to prepare oligourea peptidomimetics (Fig. 8) using solid-phase technology were published in detail by Burgess et al. in 1997 and Kim et al. in 1996. The carbamate or urea amino acid analogs are generally represented by the usual three- or one-letter amino acid notations with the superscripts c or u, respectively.

The earlier methods were based on phthalimido-protected isocyanates and azido 4-nitrophenyl carbamates as building blocks, which diverge considerably from the standard methodologies using, for example, Fmoc-protected amino acids or amino acid analogs, such as described by Chan and White in 2000. More recent detailed protocols for the solid-phase synthesis of oligourea peptidomimetics and their building blocks using Fmoc protection under conditions mimicking those used in automated peptide synthesis are given in Boeijen et al. (2001). The general procedure is based on the extension of a Rink amide resin-linked starter by successive addition of Fmoc and p-nitrophenyl-protected urea monomers with intermediate capping and deprotection steps (Scheme 14). Most of the monomers were synthesized from Fmoc-protected amino acids by successive conversion to the corresponding alcohols, azides,

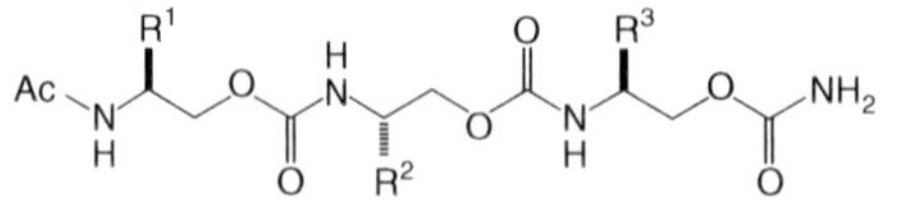

Fig. 7 An oligocarbamate peptide mimetic.

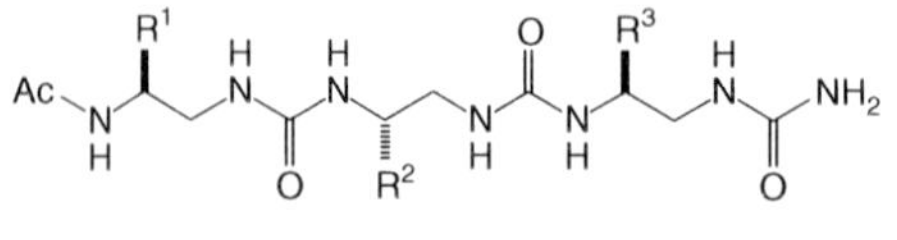

Fig. 8 An oligourea peptide mimetic.

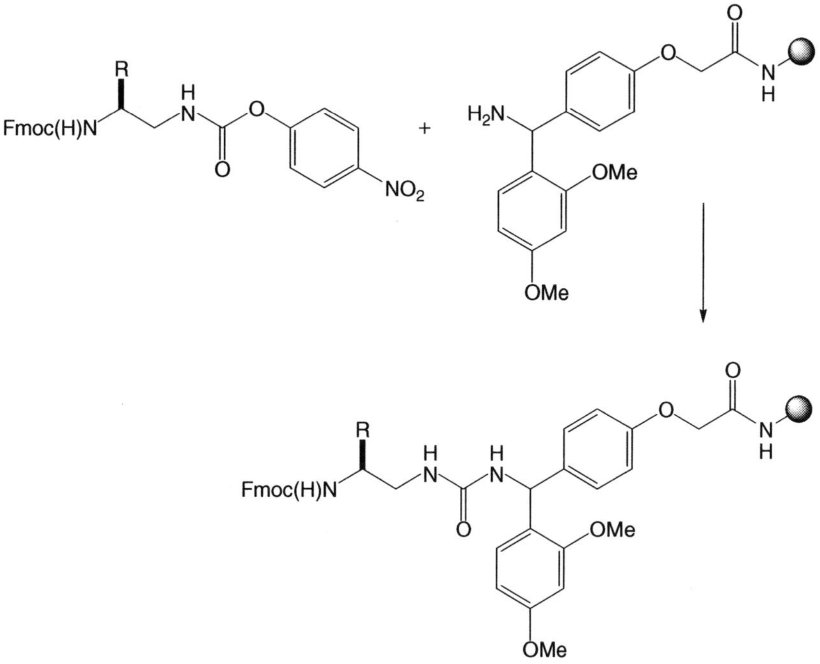

Scheme 14

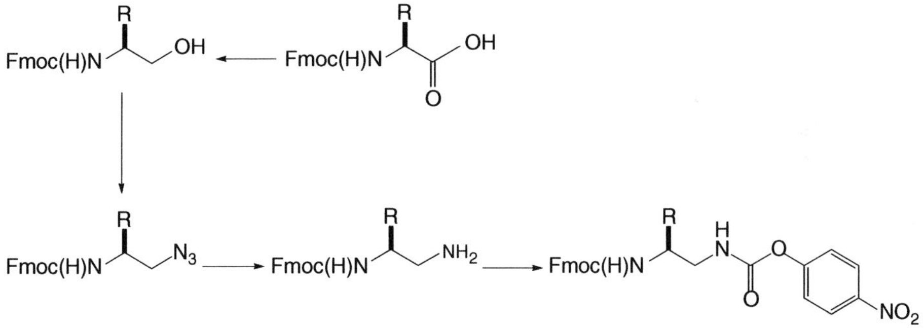

Scheme 15

and amines, which were finally derivatized to the p-nitrophenyl carbamates (Scheme 15). The Fmoc-protected glycine urea monomer, however, was prepared from Boc-ethylenediamine. This methodology was used to synthesize Ac-Gly^u-Gly^u-Phe^u-Leu^u-NH$_2$ and a couple of neurotensin analogs.

6
Peptoids

Peptoids are N-alkylglycine oligomers, that is, peptide mimetics in which side chains are bound to the amide nitrogen atom rather than to the α-carbon atom (Fig. 9). Peptoid synthesis based on commercially

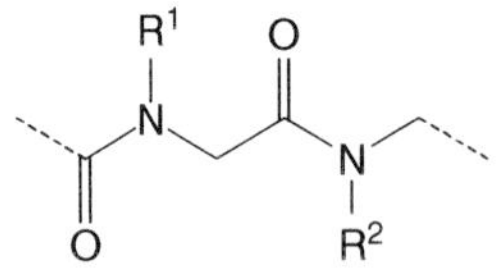

Fig. 9 A dipeptoid moiety.

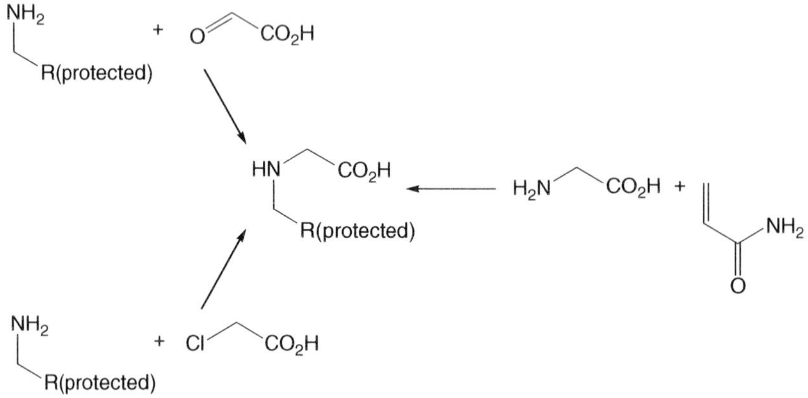

Scheme 16

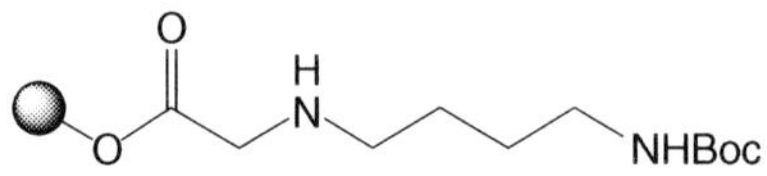

Fig. 10 A resin-bound peptoid residue.

available primary amines is eminently amenable to the preparation of very large libraries using combinatorial approaches. Earlier work in this direction, pioneered by Zuckermann's group has been reviewed by Figliozzi et al. in 1996. A limitation of such processes has been the long reaction time (of up to several hours) required for the coupling of each residue. However, it has been shown in 2002, by Olivos et al. that the solid-phase synthesis of peptoids can be successfully accelerated under microwave irradiation, achieving reaction times of less than one minute. The basic strategy is the approach described by Simon et al. in 1992), which uses standard solid-phase synthesis techniques. The monomers are prepared by different methods from appropriate primary amines (Scheme 16).

Shankaramma et al. in 2003 synthesized tetradecapeptide-peptoid macrocycles containing a single N-substituted glycine residue, allowing a β-hairpin conformation induced by two successive proline residues to be maintained, stabilized here by the array of intramolecular hydrogen bonds between carbonyl and NH groups. The synthesis of the linear precursor was carried out on 2-chlorotrityl chloride-polystyrene resin using Fmoc chemistry, initially binding bromoacetic acid to the resin and then replacing the bromine with a 4-Boc-aminobutylamino group to create the single peptoid residue as a starting point (Fig. 10). This precursor was then released from the resin, cyclized and deprotected to afford the final product.

7
Variations on the Peptoid Theme

Several new variations on the peptoid theme have been synthesized. One of the earliest is the ureapeptoid structure built by Kruijtzer et al. in 1997 and by Wilson and Nowick in 1998 (Fig. 11). Ureapeptoids can be prepared from Boc-protected N-1-substituted ethylenediamines, by conversion to the corresponding isocyanates and condensation of the latter with sarcosine methyl ester (Scheme 17). Subsequent removal of the Boc protective group and condensation with an appropriate p-nitrophenyl carbamate allowed extension of the chain to a ureapeptoid trimer.

A more efficient method starts from Boc-ethylenediamine or oligomers containing this moiety, by chain extension with N-(2-nitrobenzenesulfonyl)-2-imidazolidinone containing the Ns protective group of Fukuyama et al., 1995 followed by N-alkylation and deprotection (Scheme 18).

Retropeptoids, for example, Fig. 12, are peptidomimetics in which the substituted nitrogen atom of the peptide N-substituted glycine is moved up one position to replace the α-carbon atom as described

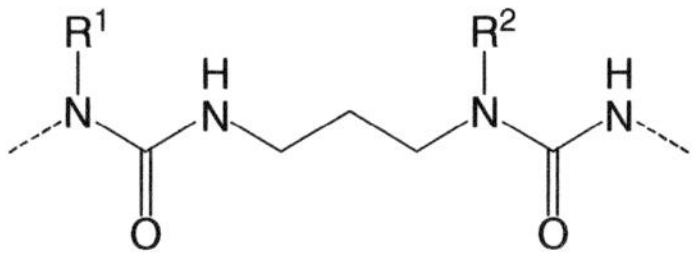

Fig. 11 A ureapeptoid moiety.

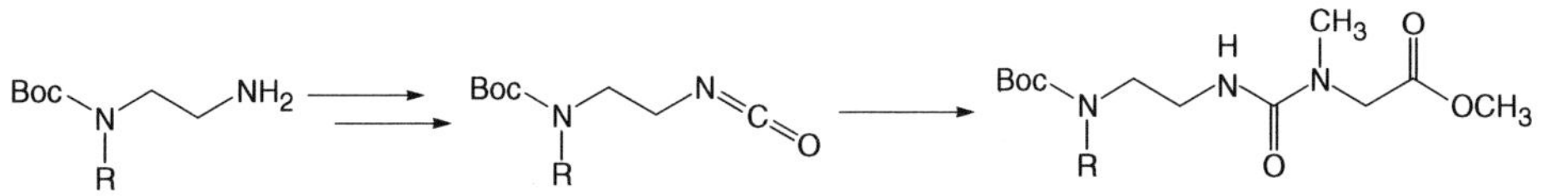

Scheme 17

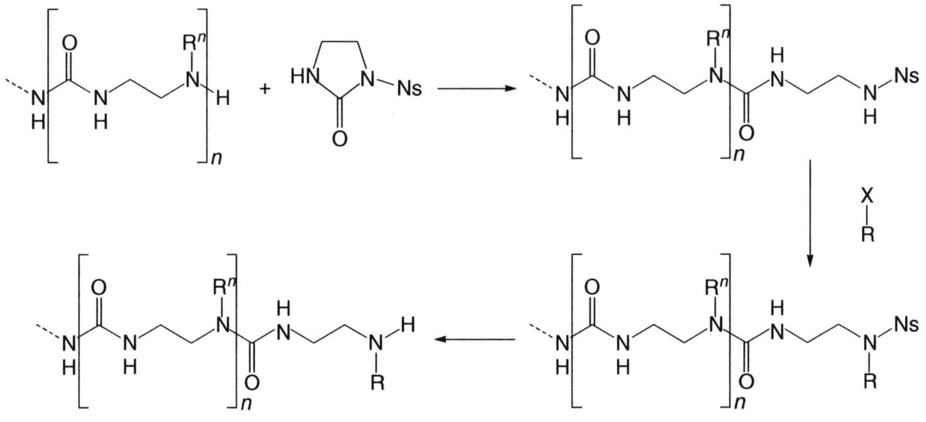

Scheme 18

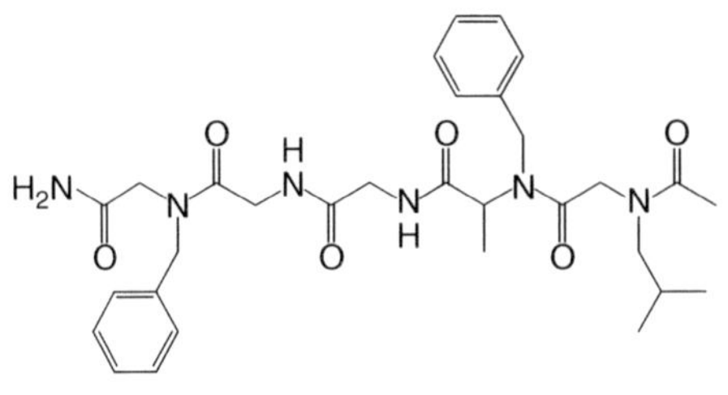

Fig. 12 A retropeptide.

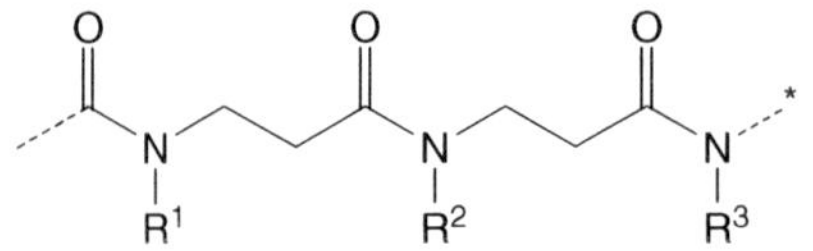

Fig. 13 A β-peptoid moiety.

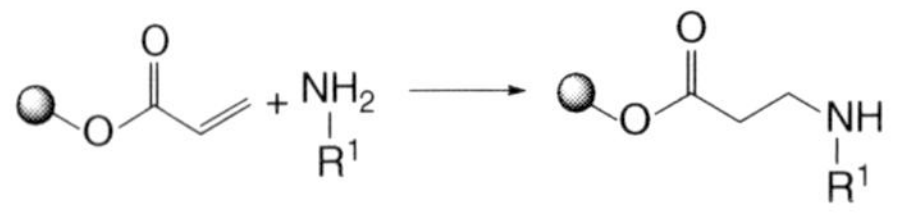

Scheme 19

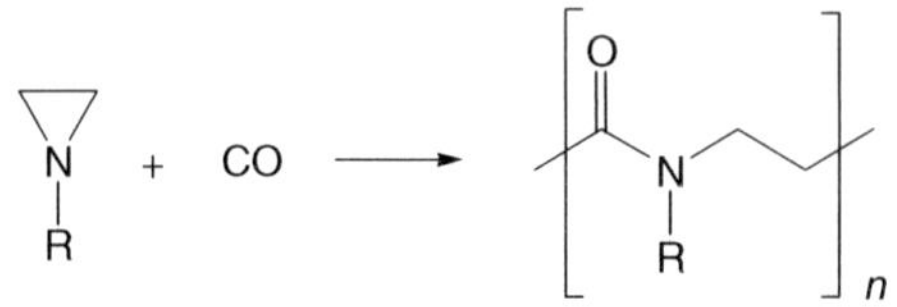

Scheme 20

by Kruijtzer et al. in 1998. β-Peptoids (Fig. 13) are N-substituted oligomers of β-aminopropionic acid. This type of oligomer had been prepared previously by coupling of N-substituted β-amino acids as explained by Seebach et al. in 1996 using standard peptide methodology, and shown to be stable toward pepsin. The Seebach approach suffers from the limitation that it requires a collection of β-amino acid monomers as building blocks. Hamper et al. in 1998 devised a solid-phase strategy, based on the Michael addition of primary amines to an acrylic acid residue bound to Wang resin by treatment with acryloyl chloride (Scheme 19). Successive couplings with acryloyl chloride followed by the addition of primary amines and finally cleavage from the resin led to the synthesis of several di-β-peptoids, as well as trimeric N-benzyl-β-aminopropionic acid, the latter in 67% overall yield.

Poly-β-peptoids are oligomers of a single N-substituted β-aminopropionic acid residue. Approaches to these structures by copolymerization of CO and imines were explored initially by Kacker et al. in 1998 and by Dghaym et al. also in 1998, and have been developed into a successful catalytic method by the group of Jia in 2002 and Darensbourg in 2004. This involves the alternating copolymerization of carbon monoxide and an N-alkylaziridine using a cobalt carbonyl catalyst (Scheme 20).

Hydrazinoazapeptoids are based on the hydrazinopeptide concept (Fig. 14). The synthesis of hydrazinoazapeptoids (as hybrids with a hydrazinopeptide moiety)

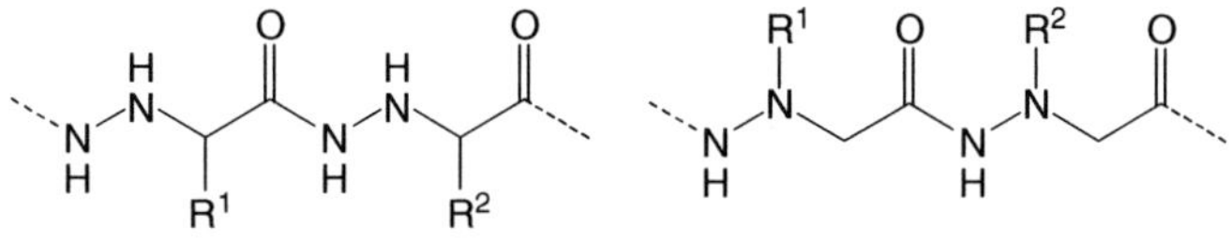

13 Synthesis of Peptide Mimetics and their Building Blocks | 419

Fig. 14 A hydrazinopeptide and a hydrazinopeptoid moieties.

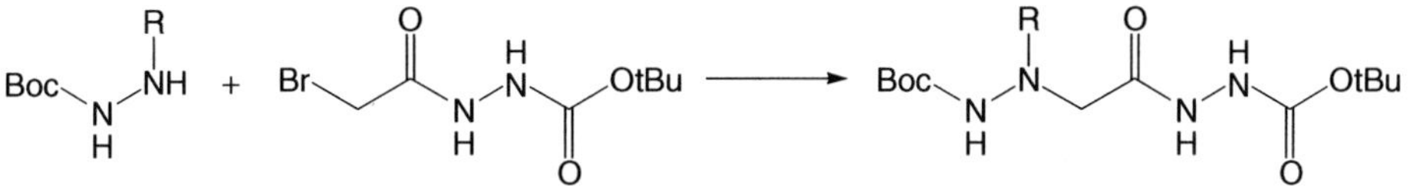

Scheme 21

Scheme 22

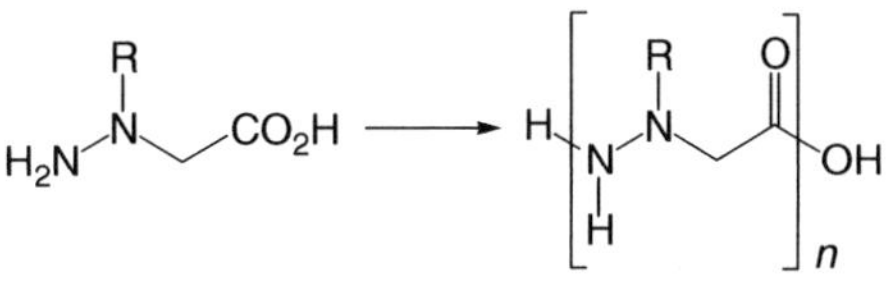

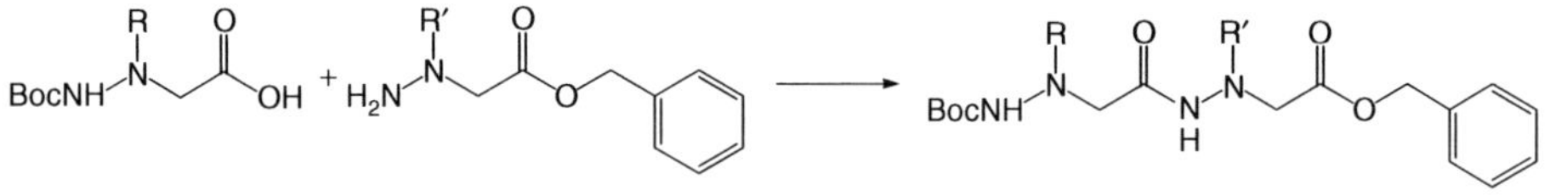

Scheme 23

Fig. 15 An aminooxypeptide and an aminooxypeptoid moieties.

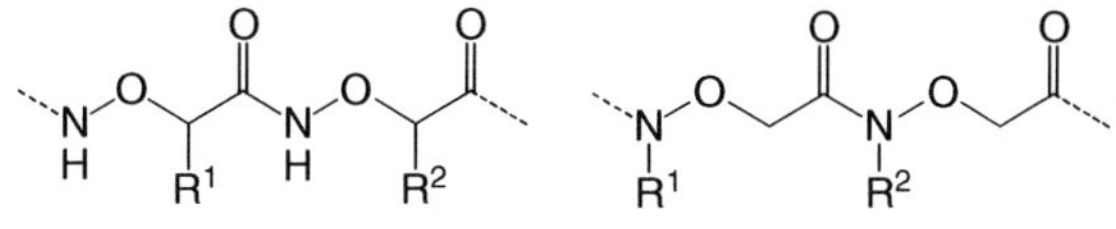

has been carried out in solution, using the submonomer approach, which implies alkylation of one of the nitrogen atoms of a monosubstituted alkyl or arylalkylhydrazine with a bromoacetyl residue on the growing oligomer, requiring the use of protected hydrazine derivatives to insure alkylation at the desired position (Scheme 21).

Aza-β^3-peptides are oligomers of N^α-substituted hydrazinoacetic acid, and may also be regarded as aza analogs of β^3-amino acids (Scheme 22). The monomers can be synthesized as N^β-Boc-protected benzyl or methyl esters, and monomers, dimers, and oligomers all coupled very efficiently under DCC/DMAP (dicyclohexylcarbodiimide/dimethylaminopyridine) activation, followed by deprotection and elongation steps as required (Scheme 23).

Aminooxypeptoids, based on the aminooxypeptide concept, can be envisioned as N-alkyl- or arylalkylhydroxylamine derivatives (Fig. 15). The synthesis of aminooxypeptoids has also been accomplished using the submonomer approach

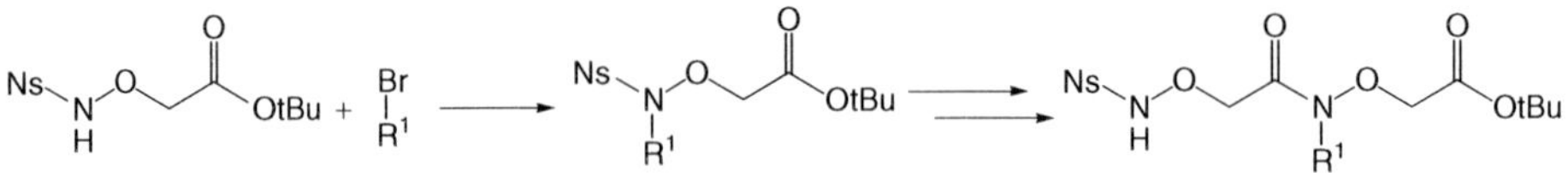

Scheme 24

(Ns represents o-nitrobenzenesulfonyl), but unacceptably low yields were obtained (Scheme 24). Stepwise assembly of the N-alkylated monomers in either the N to C or the C to N direction proved to be more satisfactory, although yields were still low when bulky side chains were present at the terminal positions of the deprotected monomers.

8
Proline Substitutes

The cyclic structure of proline and the similar stabilities of cis- and trans-peptide bonds involving its secondary amine nitrogen, separated by a relatively low potential energy barrier, are features that explain the unique roles of this amino acid. As similar properties are found in other $N^\alpha-C^\alpha$-cyclized amino acids, a number of proline mimics with different ring sizes have been synthesized and incorporated into peptides. These modifications are discussed in the Goodman and Ro review (1995), as well as others in which an additional heteroatom and/or a conformationally stabilizing feature such as a double bond, a bicyclic system, or a *gem*-dimethyl substitution is included in the proline ring. A number of proline substitutes have been biologically characterized (see Sect. 9).

9
Biological Activity and Therapeutic Hints

Although theoretically considered as promising fields to develop innovative therapies for a number of relevant human diseases, the biological characterization of most of the peptide mimetics described here has not been the subject of too much research, as compared with the abundant literature regarding synthetic aspects. Regarding endogenous peptides, these were used rather intuitively. Some classical examples are insulin, human growth hormone, interferons, or erythropoietin. Peptidase inhibition, as illustrated by angiotensin-converting enzyme (ACE) inhibitors, also offers enormous therapeutic opportunities, and the many successful ACE inhibitors currently available may be viewed broadly as mimics of the peptide substrate. Moreover, peptide processing is a crucial step in viral replication, and protease inhibition has become a fundamental aspect of HIV antiretroviral therapy. Increasing knowledge of the roles of peptidases is continually suggesting novel approaches to the treatment of disease. Therefore, the design and synthesis of substrate or transition state analogs to inhibit these enzymes is another fertile field of research.

While these molecules have found some clinical use, the drug potential of many others remains unrealized. One reason for this is that peptides are not bioavailable orally, their transport through biological membranes is usually unsatisfactory, and they are rapidly metabolized. Another is that the inherent conformational flexibility of peptides signifies an entropic penalty for them to be able to adopt appropriate conformations to interact with their targets.

Azapeptides such as Ac-Leu-Pro-Phe[a]-Phe-AspNH$_2$ and Ac-Leu-Pro[a]-Phe-Phe-AspNH$_2$ are effective, proteolytically stable inhibitors of the cytotoxicity of β-amyloid peptide, and are therefore of interest as β-sheet breaking compounds of potential utility in the treatment of diseases such as Alzheimer's, chronic progressive traumatic encephalopathy, and vascular dementia with amyloid angiopathy.

Pure or all-azaaminoacid oligomers or "azatides" are azapeptides in which all the α-carbon atoms have been replaced by nitrogen. Although early attempts to synthesize this kind of analog date back as far as the history of (mono)azapeptides, the first successful azatide synthesis appears to be a report by Gante et al. in 1995 on the preparation of an analog of a peptidic renin inhibitor, which, however, does not seem to have been tested. A year later, Han and Janda described the development of general solution-phase procedures for the coupling of Boc-protected aza-amino acid monomers and the synthesis of these building blocks. This methodology was exemplified by synthesizing the azatide mimic of Leu-enkephalin. Unfortunately, this product did not compete with Leu-enkephalin for an antibody raised against β-endorphin, the only assay carried out as a test of biological activity. This disappointing result, together with the subsequent discovery that newer azapeptide analogs of the original compound, but with one or more natural amino acid residues, also lacked this activity, may explain why this avenue of research does not seem to have been followed more extensively.

On the basis of the sequence of the RNA-binding domain of HIV-1 Tat protein, [48]Gly-Arg-Lys-Lys-Arg-Arg-Gln-Arg-Arg[57], Tamilarasu et al. in 2001 synthesized an oligocarbamate (Phe-Gly[c]-Arg[c]-Lys[c]-Lys[c]-Arg[c]-Arg[c]-Gln[c]-Arg[c]-Arg[c]) and an oligourea (Gly[u]-Arg[u]-Lys[u]-Lys[u]-Arg[u]-Arg[u]-Gln[u]-Arg[u]-Arg[u]-Tyr) analog, which inhibited transcriptional activation by Tat protein in human cells with IC$_{50}$ values of 1 and 0.5 µM, respectively.

The pharmacological activity of peptoids designed as analogs of peptides with recognized bioactivity has often been disappointing. Nevertheless, some peptoid structures lacking any obvious fit to a target exhibit interesting activities. Thus, a combinatorial library of peptoid trimers yielded several new compounds (with a large, lipophilic, nonpeptidomimetic, also antibacterial dehydroabietylamine moiety and unnatural amino acid–like residues) exhibiting activity against both gram-negative and gram-positive bacteria, including some resistant strains to all known antibiotics. The MIC (minimal inhibitory concentration) values were as low as 5 µM in some cases. One of the analogs fully protected *Staphylococcus aureus*-infected mice at 10 and 30 mg kg^{-1} (Fig. 16).

It should be pointed out that dehydroabietylamine itself is no more than an order of magnitude less potent *in vitro* than the most effective peptoid hybrid of this series. Nevertheless, it is inactive in the infection

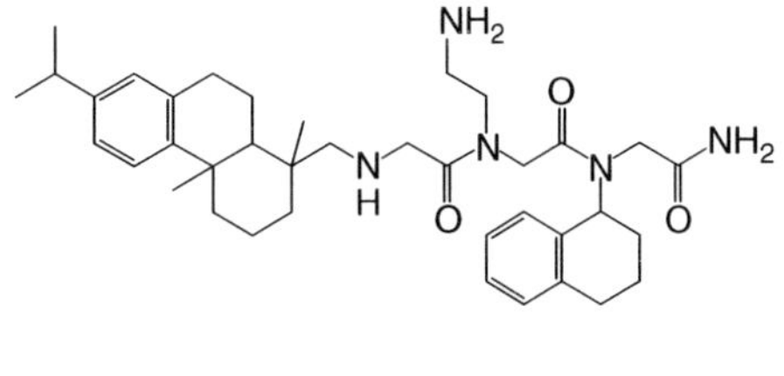

Fig. 16 A dehydroabietyl-aminopeptoid.

model, suggesting that the pharmacokinetics for some peptoids is more favorable. The authors suggest that, because of their broad spectrum of antibacterial activity, their lack of structural specificity, and β-galactosidase and propidium iodide leakage, these substances act on the bacterial cell membrane. Although the more thoroughly studied compound depicted above had some hemolytic activity, this appeared at concentrations greater than the MIC for staphylococci. The screen that led to these compounds showed that dehydroabietylamine, at any of the three positions, although preferably at the N-terminal location, was a frequent (but neither necessary nor sufficient) component of the active analogs. The extension of the antibacterial spectrum to gram-positive organisms seems to be related to the presence of a basic residue (such as the aminoethyl group mentioned earlier) at the intermediate position.

Quite recently, a conjugate of ethylenediaminetetraacetic acid (EDTA) with two 5-aminosalicylic acid (5-ASA) methyl ester moieties (EBAME), was described (Fig. 17). This compound can be viewed as a peptoid in which two glycine conjugates are joined through their α-amino acid nitrogen atoms, combining the anti-inflammatory activities of 5-ASA, with the redox-active metal chelating activity of acid (EDTA), suppressing the gastric irritation mediated by 5-ASA, removing potentially harmful transition metal ions and mimicking superoxide dismutase (SOD) when complexed with Cu^{2+} ions. EBAME was shown to bind to Cu^{2+} in a 1:1 ratio, but the superoxide-destroying activity of this complex was rather low. With Mn^{2+}, however, a very satisfactory SOD-like activity was demonstrated. The authors claim that EBAME has potential as a dual-function anti-inflammatory agent with reduced gastric irritant potential.

A couple of macrocyclic peptide–peptoid hybrids, mentioned in Sect. 6, has also quite recently been shown to have antibiotic activity against gram-positive and – negative bacteria, with low hemolytic activity against erythrocytes, one of the major drawbacks of earlier members of this class (Fig. 18). The replacement of an arginine group N-substituted glycine residue present in a previously synthesized purely peptidic prototype by an N-4-aminobutyl-substituted glycine residue led to a slightly

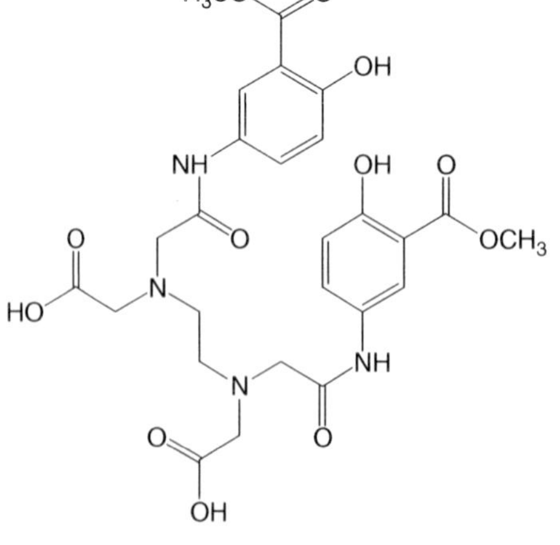

Fig. 17 Ethylenediaminetetraaceptic acid bis-(5-aminosalicylic acid methyl ester) (EBAME).

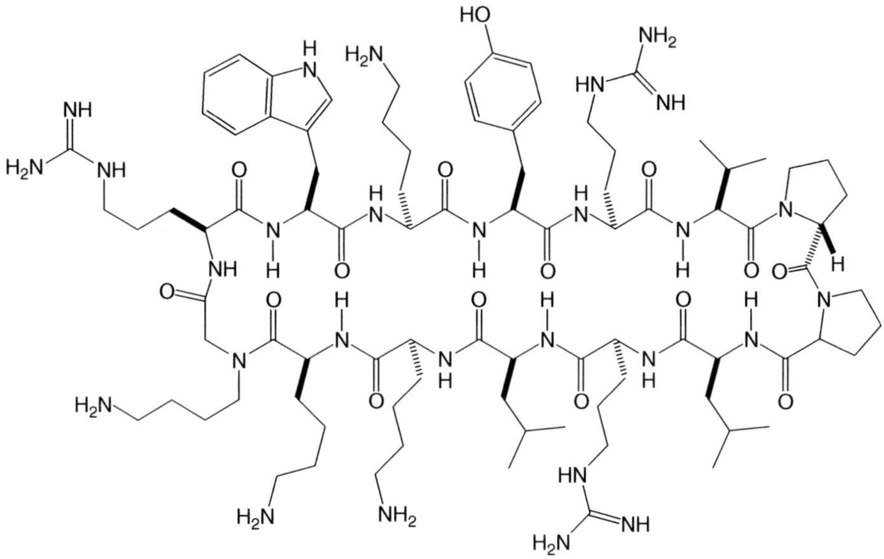

Fig. 18 A macrocyclic peptide–peptoid hybrid.

improved antibiotic potency (MIC between 4 and 64 μg mL^{-1}) and reduced hemolytic activity (0.5% at 100 μg mL^{-1}), while preserving the β-hairpin conformation believed to be important for this biological activity.

When α-chiral, bulky side chains are repeated along a peptoid backbone, stable helical structures arise in spite of the lack of potentially hydrogen-bonding NH groups. The same structural concept has been applied to the design and synthesis of a mimic of lung surfactant protein C, containing 22 N-substituted glycine residues. This product successfully reproduces several desirable characteristics such as the ability to adsorb rapidly to an air–water interface, to reduce and control surface tension and to respread quickly upon surface expansion (Fig. 19).

Several helical, cationic, amphipathic peptoids designed as mimics of the natural magainins have also shown antibacterial activity, for example, Fig. 20.

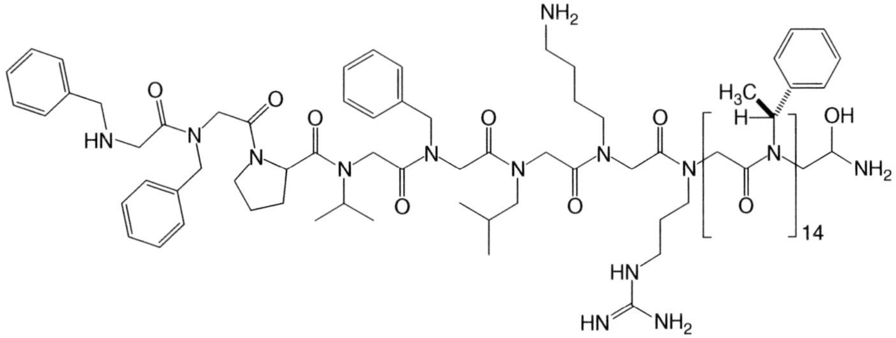

Fig. 19 A peptoid mimic of lung surfactant protein C.

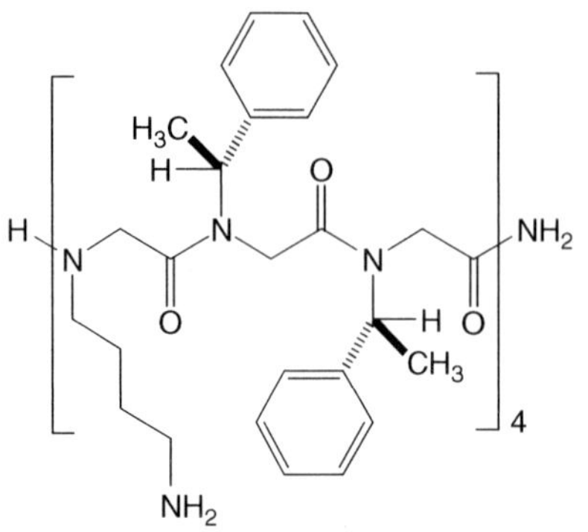

Fig. 20 A peptoid magainin mimic.

These oligomers were prepared from peptoid polyamines, synthesized sequentially on Rink amide resin, which were then guanidinylated using pyrazole 1-carboxamidine (Scheme 25).

The retropeptide Leu-enkephalin amide mimic depicted earlier (Fig. 12) appears not to have been subjected to biological testing. The more complex substance P peptoid and retropeptoid analogs, however, are reported to be *in vitro* agonists in the above-mentioned article, although this observation does not seem to have been followed up by a more thorough pharmacological study. Rather surprisingly, the high resistance of the peptoid analog to degradation by pepsin is mentioned as evidence that the retropeptoid is similarly stable.

Some hydrazinoazapeptoids designed as possible proteasome inhibitors, based on the Ac-Leu-Leu-Norleucinal template, exhibit antiproliferative activity in L1210 cells, for example, Fig. 21, but their potency is weak compared to that of the structurally unrelated, proven proteasome inhibitor bortemozib.

A recent successful example is the synthesis of a potent ($K_i = 1.5$ nM), specific human proline endopeptidase inhibitor

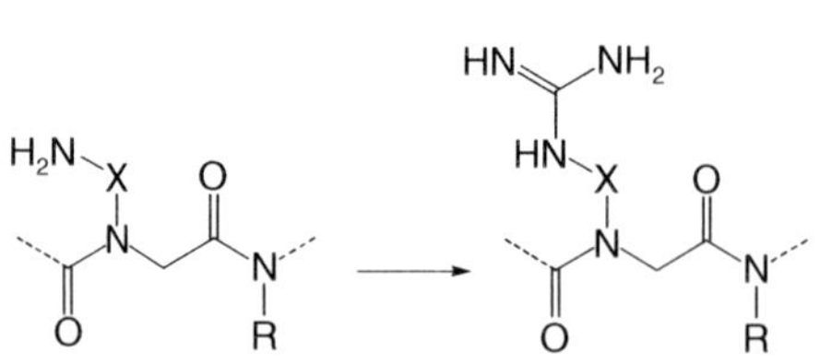

Scheme 25

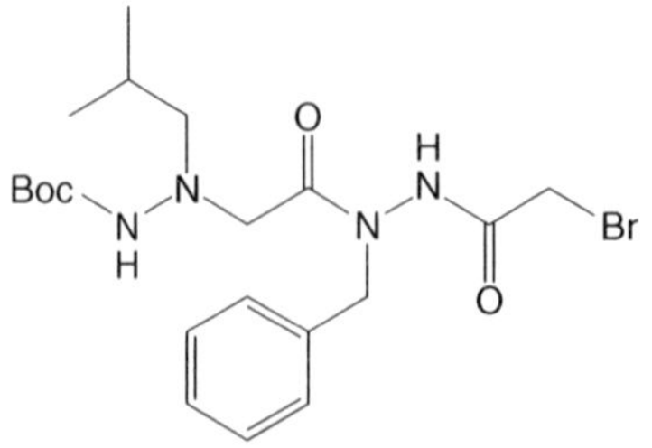

Fig. 21

that was viewed as a probe of the possible role of this enzyme in the production of β-amyloid peptide, a hallmark of Alzheimer's disease (Fig. 22).

A newer development is the replacement of a proline residue by a cyclopent-2-ene-1,2-dicarboxylic acid moiety. Applying this concept, these authors synthesized and evaluated a series of thrombin inhibitors related to inogatran and melagatran (in which the pyrrolidine ring of a proline residue is replaced by a piperidine or an azetidine ring, respectively), some of them with submicromolar IC_{50} values, which, however, are still somewhat unsatisfactory when compared with the 22.4 and 4.7 nM IC_{50} values of the reference compounds. Their most potent example ($IC_{50} = 0.87$ μM) is shown in Fig. 23. For the sake of comparison, the analog lacking the N-ethoxy group has $IC_{50} = 1.51$ μM. Although the (2R)-cyclopent-2-ene-1,2-dicarboxylic acid moiety generally afforded more potent compounds, its replacement by trans-(1S,2S)-cyclopentane-1,2-dicarboxylic acid in the latter example results in slightly enhanced activity ($IC_{50} = 1.32$ μM). This structural modification has been incorporated much more successfully into inhibitors of prolyl oligopeptidase, a serine peptidase that preferentially catalyzes the hydrolysis of peptide bonds at the carboxyl side of proline residues. Thus, the following analog has a 0.3 nM IC_{50}, while the corresponding value for the reference compound with a proline residue in the central position is 0.2 nM (Fig. 24).

Recents efforts in this direction are those of Han et al. in 1999, Avenoza et al. in 2002, and Jenkins et al. in 2004. The former authors used 7-azabicyclo[2.2.1]heptane-1-carboxylic acid as a proline replacement in a boroarginine thrombin inhibitor, and as an inducer of a β-turn in the dipeptides Ser-Pro and Pro-Ser for conformational studies (Fig. 25).

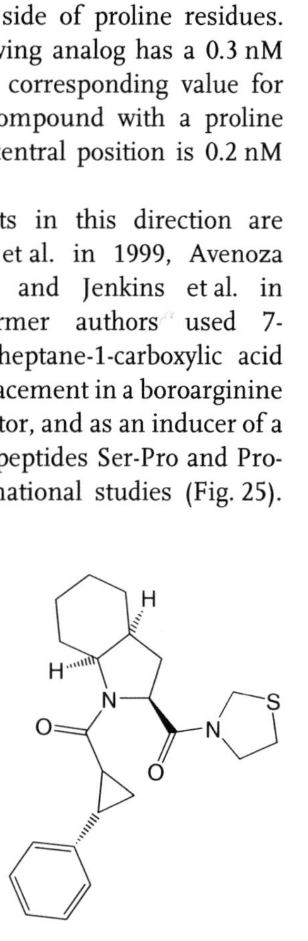

Fig. 22

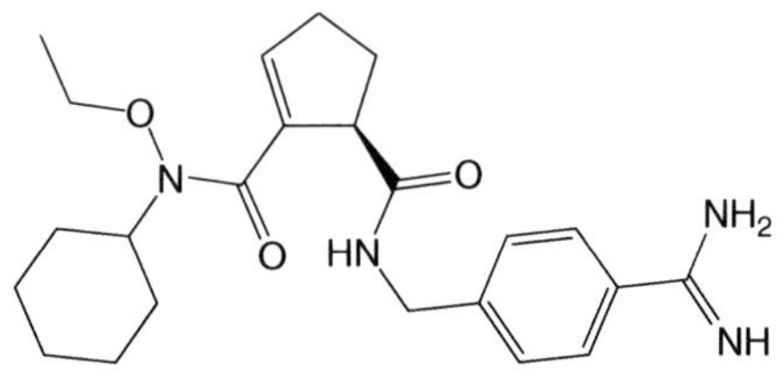

Fig. 23

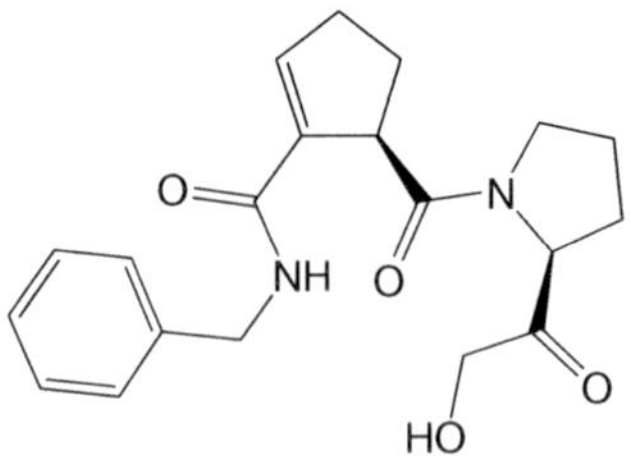

Fig. 24

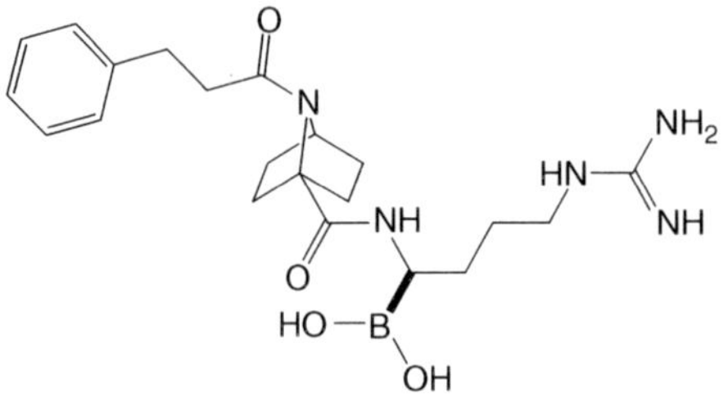

Fig. 25

This substance, synthesized from 7-Boc-1-carboxy-7-azabicyclo[2.2.1]heptane t-butyl ester, inhibited thrombin with $K_i = 2.9$ nM, which compares reasonably well with the parent proline compound ($K_i = 0.10$ nM) suggesting that this replacement mimics the active conformation in Han et al. in 1999.

The more recent paper analyzed the implications of this replacement, with electronegative (HO or F) substituents at C-4, for the conformation of collagen. Jenkins et al. in 2004 concluded that the bridge abolishes the effect of the electronegative substituents on the proline ring on the trans/cis ratio of the peptide bonds, previously observed by the same group with unbridged, electronegatively substituted proline residues.

The unique features of peptide-inspired derivatives pose many synthetic challenges that must still be met. Certain structural aspects, such as the tendency of different peptidomimetic scaffolds to adopt stable conformations in solution, are currently a very active field of research that is also a stimulus for the development of novel synthetic methodologies. In view of the extremely varied therapeutic potential of peptide mimetics, the development of reliable therapies based on these compounds will require further expansion of the broad spectrum of synthetic methods available in order to overcome the current practical limitations of these very versatile moieties, without altering their essential chemical nature. This frontier in the field of peptide mimetics will foreseeably be a major area of research in the coming years.

Bibliography

Books and Reviews

Abell, A. (Ed.) (1999) *Advances in Amino Acid Mimetics and Peptidomimetics*, JAI Press, Greenwich.

Chan, W.C., White, P.D. (Eds.) (2000) *Fmoc Solid Phase Peptide Synthesis: A Practical Approach*, Oxford University Press, Oxford.

Cheng, R.P., Gellman, S.H., DeGrado, W.F. (2001) β-Peptides: From structure to function, *Chem. Rev.* **101**, 3219–3232.

Goodman, M., Ro, S. (1995) in: Wolff, M. (Ed.) *Burger's Medicinal Chemistry and Drug*

Discovery, Vol. 1, 5th edition, John Wiley & Sons, New York, p. 803.

Goodman, M., Moroder, L., Toniolo, C., Felix, A. (Eds.) (2002) *Houben-Weyl, Methods in Organic Chemistry, Vol. E22. Synthesis of Peptides and Peptidomimetics*, Thieme Medical Publishers, Stuttgart and New York.

Hruby, V.J., Balse, P.M. (2000) Conformational and topographical considerations in designing agonist peptidomimetics from peptide leads, *Curr. Med. Chem.* **7**, 945–970.

Simon, R.J., Kania, R.S., Zuckermann, R.N., Huebner, V.D., Jewell, D.A., Banville, S., Ng, S., Wang, L., Rosenberg, S., Marlowe, C.K., Spellmeyer, D.C., Tan, R., Frankel, A.D., Santi, D.V., Cohen, F.E., Bartlett, P.A. (1992) Peptoids: a modular approach to drug discovery, *Proc. Natl. Acad. Sci. U.S.A.* **89**, 9367–9371.

Williams, R.M. (1996) Amino Acid Synthesis, in: Meyers, R.B. (Ed.) *Encyclopedia of Molecular Biology and Molecular Medicine*, Vol. 1, Wiley-VCH Verlag GmbH, Weinheim, p. 52.

Zega, A., Urleb, U. (2002) Azapeptides, *Acta Chim. Slov.* **49**, 649–662.

Primary Literature

Avenoza, A., Busto, J.H., Peregrina, J.M., Rodríguez, F. (2002) Incorporation of Ahc into model dipeptides as an inducer of a β-turn with a distorted amide bond. Conformational analysis, *J. Org. Chem.* **67**, 4241–4249.

Baek, B.-H., Lee, M.-R., Kim, K.-Y., Cho, U.-I., Boo, D.W., Shin, I. (2003) Novel consecutive β- and γ-turn mimetics composed of a-aminooxy acid tripeptides, *Org. Lett.* **5**, 971–974.

Bailey, M.A., Ingram, M.J., Naughton, D.P. (2004) A novel anti-oxidant and anti-cancer strategy: a peptoid anti-inflammatory drug conjugate with SOD mimic activity, *Biochem. Biophys. Res. Commun.* **317**, 1155–1158.

Barelli, H., Petit, A., Hirsch, E., Wilk, S., De Nanteuil, G., Morain, P., Checler, F. (1999) S 17092-1, a highly potent, specific and cell permeant inhibitor of human proline endopeptidase, *Biochem. Biophys. Res. Commun.* **257**, 657–661.

Barré, C., Le Grel, P., Robert, A., Baudy-Floc'h, M. (1994) Design and synthesis of a new series of peptide analogs: the hydrazinopeptides, *J. Chem. Soc. Chem. Commun.* **1994**, 607–608.

Boeijen, A., van Ameijde, J., Liskamp, R.M.J. (2001) Solid-phase synthesis of oligourea peptidomimetics employing the Fmoc protection strategy, *J. Org. Chem.* **66**, 8454–8462.

Bouget, K., Aubin, S., Delcros, J.-G., Arlot-Bonnemains, Y., Baudy-Floc'h, M. (2003) Hydrazino-aza and N-azapeptoids with therapeutic potential as anticancer agents, *Bioorg. Med. Chem.* **11**, 4881–4889.

Bretscher, L.E., Jenkins, C.L., Taylor, K.M., DeRider, M.L., Raines, R.T. (2001) Conformational stability of collagen relies on a stereoelectronic effect, *J. Am. Chem. Soc.* **123**, 777–778.

Burgess, K., Linthicum, D.S., Shin, H. (1995) Solid-phase syntheses of unnatural biopolymers containing repeating urea units, *Angew. Chem., Int. Ed. Engl.* **34**, 907–909.

Burgess, K., Ibarzo, J., Linthicum, D.S., Russell, D.H., Shin, H., Shitangkoon, A., Totani, R., Zhang, A.J. (1997) Solid phase synthesis of oligoureas, *J. Am. Chem. Soc.* **119**, 1556–1564.

Chang, R.S., Lotti, V.J., Monaghan, R.L., Birnbaum, J., Stapley, E.O., Goetz, M.A., Albers-Schonberg, G., Patchett, A.A., Liesch, J.M., Hensens, O.D., Springer, J.P. (1985) A potent nonpeptide cholecystokinin antagonist selective for peripheral tissues isolated from Aspergillus alliaceus, *Science* **230**, 177–179.

Cheguillaume, A., Lehardy, F., Bouget, K., Baudy-Floc'h, M., Le Grel, P.M. (1999) Submonomer solution synthesis of hydrazinoazapeptoids, a new class of pseudopeptides, *J. Org. Chem.* **64**, 2924–2927.

Cheguillaume, A., Salaün, A., Sinbandhit, S., Potel, M., Gall, P., Baudy-Floc'h, M., Le Grel, P.M. (2001) Solution synthesis and characterization of aza-β^3-peptides (N^α-substituted hydrazino acetic acid oligomers), *J. Org. Chem.* **66**, 4923–4929.

Chen, F., Zhu, N.-Y., Yang, D. (2004) γ^4-Aminoxy peptides as new peptidomimetic foldamers, *J. Am. Chem. Soc.* **126**, 15980–15981.

Cho, C.Y., Moran, E.J., Cherry, S.R., Stephans, J.C., Fodor, S.P.A., Adams, C.L., Sundaram, A., Jacobs, J.W., Schultz, P.G. (1993) An unnatural biopolymer, *Science* **261**, 1303–1305.

Darensbourg, D.J., Phelps, A.L., Le Gall, N., Jia, L. (2004) Mechanistic studies on the copolymerization reaction of aziridines and carbon monoxide to produce poly-β-peptoids, *J. Am. Chem. Soc.* **126**, 13808–13815.

Dghaym, R.D., Yaccato, K.J., Arndtsen, B.A. (1998) The novel insertion of imines into a late metal-carbon σ bond: developing a palladium-mediated route to polypeptides, *Organometallics* **17**, 4–6.

Dutta, A., Morley, J. (1975) Polypeptides. Part XIII. Preparation of α-aza-amino-acid (carbazic acid) derivatives and intermediates for the preparation of α-aza-peptides, *J. Chem. Soc., Perkin Trans. 1* 1712–1720.

Figliozzi, G.M., Goldsmith, R., Ng, S.C., Banville, S.C., Zuckermann, R.N. (1996) Synthesis of N-substituted glycine peptoid libraries, *Methods Enzymol.* **267**, 437–447.

Fukuyama, T., Jow, C.-K., Cheung, M. (1995) 2- and 4-nitrobenzenesulfonamides: exceptionally versatile means for preparation of secondary amines and protection of amines, *Tetrahedron Lett.* **36**, 6373–6374.

Gante, J. (1970) Azapeptides, a novel class of peptide analogs, *Angew. Chem., Int. Ed. Engl.* **9**, 813.

Gante, J., Krug, M., Lauterbach, G., Weitzel, R., Hiller, W. (1995) Synthesis and properties of the first all-aza analog of a biologically active peptide, *J. Pept. Sci.* **1**, 201–206.

Gibson, C., Goodman, S.L., Hahn, D., Hoelzemann, G., Kessler, H. (1999) Novel solid-phase synthesis of azapeptides and azapeptoids via Fmoc-strategy and its application in the synthesis of RGD-mimetics, *J. Org. Chem.* **64**, 7388–7394.

Goodson, B., Ehrhardt, A., Ng, S., Nuss, J., Johnson, K., Giedlin, M., Yamamoto, R., Moos, W.H., Krebber, A., Ladner, M., Giacona, M.B., Vitt, C., Winter, J. (1999) Characterization of novel antimicrobial peptoids, *Antimicrob. Agents Chemother.* **43**, 1429–1434.

Gruner, S.A.W., Locardi, E., Lohof, E., Kessler, H. (2002) Carbohydrate-based mimetics in drug design: sugar amino acids and carbohydrate scaffolds, *Chem. Rev.* **102**, 491–514.

Hamper, B.C., Kolodziej, S.A., Scates, A.M., Smith, R.G., Cortez, E. (1998) Solid phase synthesis of β-peptoids: N-substituted β-aminopropionic acid oligomers, *J. Org. Chem.* **63**, 708–718.

Han, H., Janda, K.D. (1996) Azatides: solution and liquid phase syntheses of a new peptidomimetic, *J. Am. Chem. Soc.* **118**, 2539–2544.

Han, H., Yoon, J., Janda, K.D. (1998) Investigations of azapeptides as mimetics of Leu-enkephalin, *Bioorg. Med. Chem. Lett.* **8**, 117–120.

Han, W., Pelletier, J.C., Mersinger, L.J., Kettner, C.A., Hodge, C.N. (1999) 7-Azabicycloheptane carboxylic acid: a proline replacement in a boroarginine thrombin inhibitor, *Org. Lett.* **1**, 1875–1877.

Hart, M., Beeson, C. (2001) Utility of azapeptides as major histocompatibility complex class II protein ligands for T-cell activation, *J. Med. Chem.* **44**, 3700–3709.

Herranz, R. (2003) Cholecystokinin antagonists: pharmacological and therapeutic potential, *Med. Res. Rev.* **23**, 559–605.

Hess, H.-J., Moreland, W.T., Laubach, G.D. (1963) N-[2-Isopropyl-3-(L-aspartyl-L-arginyl)-carbazoyl]-L-tyrosyl-L-valyl-L-histidyl-L-prolyl-L-phenylalanine, an isostere of bovine angiotensin II, *J. Am. Chem. Soc.* **85**, 4041–4042.

Jarho, E.M., Venäläinen, J.I., Huuskonen, J., Christiaans, J.A.M., García-Horsman, J.A., Forsberg, M.A., Järvinen, T., Gynther, J., Männistö, P.T., Wallén, E.A.A. (2004) A cyclopent-2-enecarbonyl group mimics proline at the P2 position of prolyl oligopeptidase inhibitors, *J. Med. Chem.* **47**, 5605–5607.

Jenkins, C.L., Lin, G., Duo, J., Rapolu, D., Guzei, I.A., Raines, R.T., Krow, G.R. (2004) Substituted 2-azabicyclo[2.1.1]hexanes as constrained proline analogs: Implications for collagen stability, *J. Org. Chem.* **69**, 8565–8573.

Jia, L., Sun, H., Shay, J.T., Allgeier, A.M., Hanton, S.D. (2002) Living alternate copolymerization of N-alkylaziridines and carbon monoxide as a route for synthesis of poly-β-peptoids, *J. Am. Chem. Soc.* **124**, 7282–7283.

Kacker, S., Kim, S.J., Sen, A. (1998) Insertion of imines into palladium-acyl bonds: Towards metal-catalyzed alternating copolymerization of imines with carbon monoxide to form polypeptides, *Angew. Chem., Int. Ed. Engl.* **37**, 1251–1254.

Kirshenbaum, K., Barron, A.E., Armand, P., Goldsmith, R., Bradley, E., Cohen, F.E., Dill, K.A., Zuckermann, R.N. (1998) Sequence-specific polypeptoids: a diverse family of heteropolymers with stable secondary structure, *Proc. Natl. Acad. Sci. U.S.A.* **96**, 4303–4308.

Kruijtzer, J.A.W., Lefeber, D.J., Liskamp, R.M.J. (1997) Approaches to the synthesis of ureapeptoid peptidomimetics, *Tetrahedron Lett.* **38**, 5335–5338.

Kruijtzer, J.A.W., Hofmeyer, L.J.F., Heerma, W., Versluis, C., Liskamp, R.M.J. (1998) Solid-phase syntheses of peptoids using Fmoc-protected N-substituted glycines: the synthesis of (retro)peptoids of Leu-enkephalin and substance P, *Chem. Eur. J.* **4**, 1570–1580.

Lee, M.-R., Lee, J., Baek, B.-H., Shin, I. (2003) The first solid-phase synthesis of oligomeric alpha-aminooxy peptides, *Synlett* 325–328.

Lelais, G., Seebach, D. (2004) β^2-Amino acids – Syntheses, occurrence in natural products, and components of β-peptides, *Biopolymers* **76**, 206–243.

López-Areiza, J.J., Rueckle, T., Soto-Jara, C. (2004) Aza-peptides, WO 2004/050689 A2, June 17, 2004.

Murphy, J.E., Uno, T., Hamer, J.D., Dwarki, V., Zuckermann, R.N. (1998) A combinatorial approach to the discovery of efficient cationic peptoid reagents for gene delivery, *Proc. Natl. Acad. Sci. U.S.A.* **95**, 1517–1522.

Ng, S., Goodson, B., Ehrhardt, A., Moos, W.H., Siani, M., Winter, J. (1999) Combinatorial discovery process yields antimicrobial peptoids, *Bioorg. Med. Chem.* **7**, 1781–1785.

Nöteberg, D., Brånalt, J., Kvanrnström, I., Linschoten, M., Musil, D., Nyström, J.-E., Zuccarello, G., Samuelsson, B. (2000) New proline mimetics: synthesis of thrombin inhibitors incorporating cyclopentane- and cyclopentenedicarboxylic acid templates in the P2 position. Binding conformation investigated by X-ray crystallography, *J. Med. Chem.* **43**, 1705–1713.

Nowick, J.S., Brower, J.O. (2003) A new turn structure for the formation of beta-hairpins in peptides, *J. Am. Chem. Soc.* **125**, 876–877.

Olivos, H.J., Alluri, P.G., Reddy, M.M., Salony, D., Kodadek, T. (2002) Microwave-assisted solid-phase synthesis of peptoids, *Org. Lett.* **4**, 4057–4059.

Patch, J.A., Barron, A.E. (2003) Helical peptoid mimics of magainin-2 amide, *J. Am. Chem. Soc.* **125**, 12092–12093.

Seebach, D., Brenner, M., Rueping, M., Schweizer, B., Jaun, B. (2001) Preparation and determination of X-ray-crystal and NMR-solution structures of $\gamma^{2,3,4}$-peptides, *Chem. Commun.* 207–208.

Seebach, D., Overhand, M., Kühnle, F.N.M., Martinoni, B., Oberer, L., Hommel, U., Widmer, H. (1996) β-Peptides: synthesis by Arndt-Eistert homologation with concomitant peptide coupling. Structure determination by NMR and CD spectroscopy and by X-ray crystallography. Helical secondary structure of a β-hexapeptide in solution and its stability towards pepsin, *Helv. Chim. Acta* **79**, 913–941.

Shankaramma, S.C., Moehle, K., James, S., Vrijbloed, J.W., Obrecht, D., Robinson, J.A. (2003) A family of macrocyclic antibiotics with a mixed peptide-peptoid beta-hairpin backbone conformation, *Chem. Commun.* 1842–1843.

Shankaramma, S.C., Athanassiou, Z., Zerbe, O., Moehle, K., Mouton, C., Bernardini, F., Vrijbloed, J., Obrecht, D., Robinson, J. (2002) Macrocyclic hairpin mimetics of the cationic antimicrobial peptide protegrin I: a new family of broad-spectrum antibiotics, *ChemBioChem.* **3**, 1126–1133.

Shin, I., Park, K. (2002) Solution-phase synthesis of aminooxy peptides in the C to N and N to C directions, *Org. Lett.* **4**, 869–872.

Shin, I., Lee, M.R., Lee, J., Jung, M., Lee, W., Yoon, J. (2000) Synthesis of optically active D-aminooxy acids from L-amino acids or L-hydroxy acids as building blocks for the preparation of aminooxy peptides, *J. Org. Chem.* **65**, 7667–7675.

Tamilarasu, N., Huq, I., Rana, T.M. (1999) High affinity and specific binding of HIV-1 TAR RNA by a Tat-derived oligourea, *J. Am. Chem. Soc.* **121**, 1597–1598.

Tamilarasu, N., Huq, I., Rana, T.M. (2001) Targeting RNA with peptidomimetic oligomers in human cells, *Bioorg. Med. Chem. Lett.* **11**, 505–507.

Wang, X., Huq, I., Rana, T.M. (1997) HIV-1 TAR RNA recognition by an unnatural biopolymer, *J. Am. Chem. Soc.* **119**, 6444–6445.

Wender, P.A., Mitchell, D.J., Pattabiraman, K., Pelkey, E.T., Steinman, L., Rothbard, J.B. (2000) The design, synthesis, and evaluation of molecules that enable or enhance cellular uptake: peptoid molecular transporters, *Proc. Natl. Acad. Sci. U.S.A.* **97**, 13003–13008.

Wieczerzak, E., Drabik, P., Lankiewicz, L., Ołdziej, S., Grzonka, Z., Abrahamson, M., Grubb, A., Brömme, D. (2002) Azapeptides structurally based upon inhibitory sites of cystatins as potent and selective inhibitors of cysteine proteases, *J. Med. Chem.* **45**, 4202–4211.

Wilson, M.E., Nowick, J.S. (1998) An efficient synthesis of N,N'-linked oligoureas, *Tetrahedron Lett.* **39**, 6613–6616.

Woll, M.G., Lai, J.R., Guzei, I.A., Taylor, S.J.C., Smith, M.E.B., Gellman, S.H. (2001) Parallel sheet secondary structure in γ-peptides, *J. Am. Chem. Soc.* **123**, 11077–11078.

Wu, C.W., Sanborn, T.J., Zuckermann, R.N., Barron, A.E. (2001) Peptoid oligomers with α-chiral, aromatic side chains: effects of chain length on secondary structure, *J. Am. Chem. Soc.* **123**, 2958–2963.

Wu, C.W., Seurynck, S.L., Lee, K.Y.C., Barron, A.E. (2003) Helical peptoid mimics of lung surfactant protein C, *Chem. Biol.* **10**, 1057–1063.

Yang, D., Zhang, Y.-H., Zhu, N.Y. (2002) $\beta^{2,2}$-Aminoxy acids: A new building block for turns and helices, *J. Am. Chem. Soc.* **124**, 9966–9967.

Yang, D., Ng, F.-F., Li, Z.-J., Wu, Y.-D., Chan, K.W.K., Wang, D.-P. (1996) An unusual turn structure in peptides containing α-aminoxy acids, *J. Am. Chem. Soc.* **118**, 9794–9795.

Yang, D., Li, B., Ng, F.F., Yan, Y.L., Qu, J., Wu, Y.D. (2001) Synthesis and characterization of chiral N-O turns induced by α-aminoxy acids, *J. Org. Chem.* **66**, 7303–7312.

Yang, D., Qu, J., Li, B., Ng, F.-F., Wang, X.-C., Cheung, K.-K., Wang, D.-P., Wu, Y.-D. (1999) Novel turns and helices in peptides of chiral α-aminoxy acids, *J. Am. Chem. Soc.* **121**, 589–590.

Yang, D., Zhang, Y.-H., Li, B., Zhang, D.-W., Chan, J.C.Y., Zhu, N.Y., Luo, S.W., Wu, Y.D. (2004a) Effect of side chains on turns and helices in peptides of β^3-aminoxy acids, *J. Am. Chem. Soc.* **126**, 6956–6966.

Yang, D., Zhang, Y.-H., Li, B., Zhang, D.-W. (2004b) Synthesis of chiral β^3-aminoxy peptides, *J. Org. Chem.* **69**, 7577–7581.

Zhao, X., Jua, M.-X., Jiang, X.-K., Wu, L.-Z., Li, Z.-T., Chen, G.-J. (2004) Zipper-featured δ-peptide foldamers driven by donor-acceptor interaction. Design, synthesis and characterization, *J. Org. Chem.* **69**, 270–279.

14
Design and Application of Synthetic Peptides

Gregory A. Grant
Washington University School of Medicine, St. Louis, MO, USA

1	Overview	432
2	Design and Structure	434
3	Nomenclature	436
4	Synthesis	438
4.1	The Solid-phase Method	439
4.2	Chemistry	439
5	Evaluation	442
6	Applications and Perspectives	445
6.1	Antigenic and Immunogenic Uses and Vaccines	445
6.2	Antisense Peptides	448
6.3	Peptide Libraries	449
6.4	Other Applications	449
	Bibliography	450
	Books and Reviews	450
	Primary Literature	450

Keywords

Peptide Bond
The amide bond formed between the α-carboxyl group and the α-amino group of two adjacent amino acid residues in a peptide.

Proteins. Edited by Robert A. Meyers.
Copyright © 2007 Wiley-VCH Verlag GmbH & Co. KGaA, Weinheim
ISBN: 978-3-527-31608-3

Peptide

The common usage for a shortened form of polypeptide. Peptide refers to any organic compound composed at least partially of amino acids linked by peptide bonds. The distinction between peptides and proteins is arbitrary with the borderline often placed at 50 to 100 amino acids.

Residue

An amino acid residue is an amino acid in a peptide linkage. Its structure is the same as that of the free amino acid except that it has lost a molecule of water consisting of a hydrogen from the α-amino group and a hydroxyl from the α-carboxyl group.

Epitope

The term *epitope* is used to operationally define that portion of a protein that interacts specifically with some component of a biological system. With respect to the immune system, an epitope was initially defined as that portion of a protein that is recognized by an antibody. However, the usage has now been expanded to include portions of proteins that interact in other ways, such as epitopes that interact with T-cell receptors to induce a heightened antibody response toward another epitope.

■ A peptide consists of two or more amino acids linked together by amide bonds. Synthetic peptides are peptides that have been produced with organic, synthetic techniques to either mimic naturally occurring peptides or segments of naturally occurring proteins, or to interact in a specific manner with biological systems. Synthetic peptides are important molecules for biomedical research and medicinal chemistry that are useful for understanding biological systems and that have important applications in the development of vaccines and therapeutic agents and the process of *de novo* drug discovery.

1
Overview

Synthetic peptides are organic compounds, generated by synthetic chemical techniques, which are composed of two or more amino acids linked together by a peptide bond. They are often intended to resemble or duplicate naturally occurring peptides or segments of naturally occurring proteins. Synthetic peptides can be linear (**I**), cyclic (**VII**), or branched (**VI**) (bold roman numerals refer to the structures in Fig. 1) and are usually composed of, but not limited to, the 20 naturally occurring L-amino acids or their D-amino acid counterparts (**IV**). Synthetic peptides may also contain non–amino acid constituents (**III**) as well as chemical modifications of the natural side chains (**VI, IX**) or the polypeptide backbone structure (**V**). Synthetic peptides are important as tools for research in the biological and biomedical sciences, and a significant number

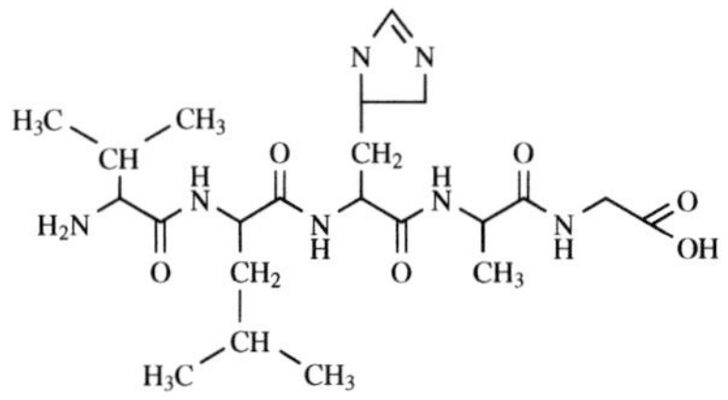

Val — Leu — His — Ala — Gly

I

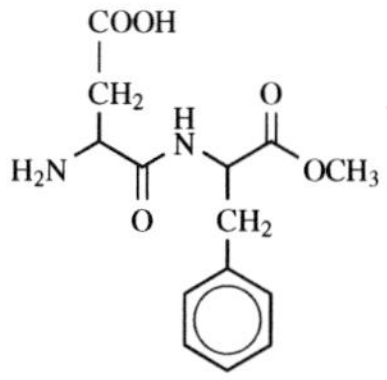

Asp — Phe — OCH_3

II

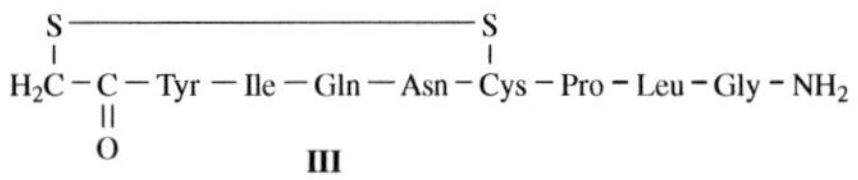

III

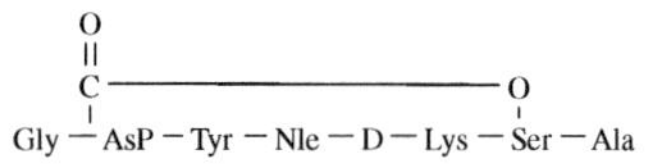

IV

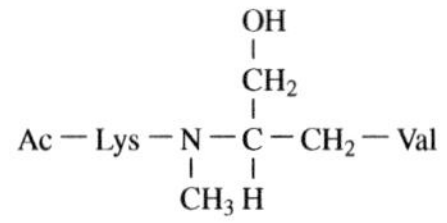

V

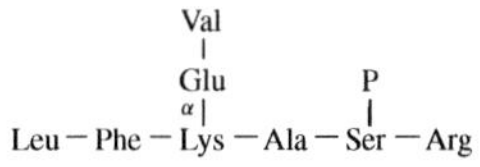

VI

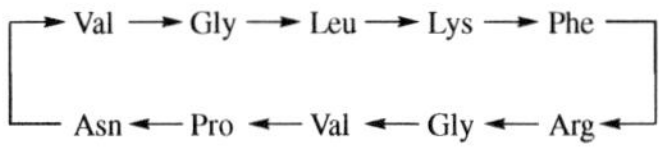

VII

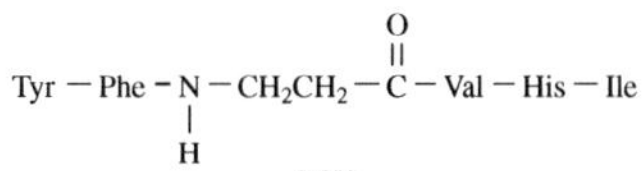

VIII

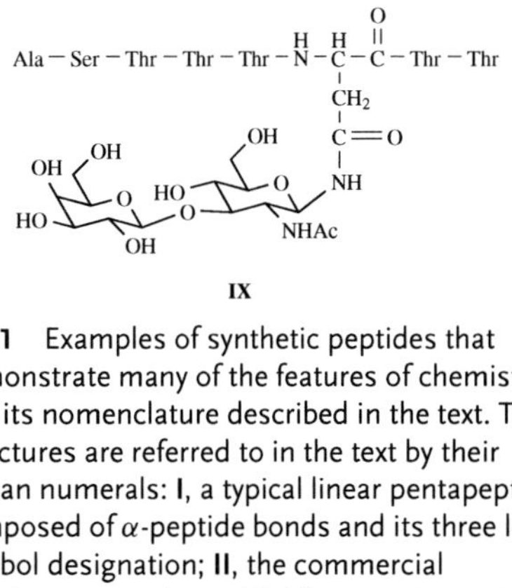

IX

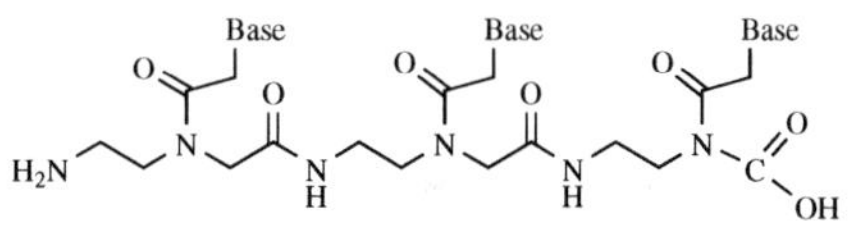

X

Fig. 1 Examples of synthetic peptides that demonstrate many of the features of chemistry and its nomenclature described in the text. The structures are referred to in the text by their roman numerals: **I**, a typical linear pentapeptide composed of α-peptide bonds and its three letter symbol designation; **II**, the commercial sweetener, aspartame; **III**, [β-mercaptopropionic acid[1]] oxytocin containing a terminal non–amino acid constituent and illustrating the disulfide linkage for clarity; **IV**, a heptapeptide containing a nontypical amino acid (Nle, norleucine) a D-amino acid and an intramolecular lactone bond; **V**, a tetrapeptide containing an N-terminal acetyl group, an alkylated α-amino group, and a fully reduced α-carboxyl or alkane constituent as part of the backbone structure; **VI**, a hexapeptide containing a phosphorylated serine residue and a branched chain connected by a peptide bond between the ε-amino group of lysine and the α-carboxyl group of glutamic acid; **VII**, circular depiction of a cyclized decapeptide; **VIII**, a hexapeptide containing a β-peptide bond; **IX**, an octapeptide with a glycosyl group as a sidechain modification; **X**, an example of a peptide nucleic acid (note the unusual nature of the backbone).

of synthetic peptides have pharmaceutical and commercial uses. For instance, synthetic polypeptide hormones have useful clinical applications, and the commercial sweetener aspartame, a dipeptide (N-L-α-aspartyl-L-phenylalanine-1-methyl ester, II), is the basis for an entire industry and is perhaps the best-known synthetic peptide.

Naturally occurring peptides have played crucial roles in life processes since the very beginnings of living organisms. Large peptides, generally referred to as proteins, serve in catalytic, structural, recognitive, and signal transduction processes in both plants and animals. Small peptides have equally important functions as hormones, growth factors, neurotransmitters, and cytokines.

The science of producing and using synthetic peptides is generally recognized as having its beginnings in the late 1800s and early 1900s and is associated mainly with the work of Theodore Curtius and Emil Fischer. Fischer, who is generally regarded as the father of peptide chemistry, was the first to introduce the concepts of peptides and polypeptides and present protocols for their synthesis. Over the years, many investigators have contributed to the advancement of the production and use of synthetic peptides. Perhaps the greatest of these was the introduction of the solid-phase method of synthesis by Bruce Merrifield in 1963. This development, along with improvements in chemistry, instrumentation, and analytical techniques by a large number of other investigators, has made it possible to obtain useful amounts of synthetic peptides on a relatively routine basis.

The genesis of a synthetic peptide can be viewed as being composed of several discrete steps ultimately culminating in its use or application. These steps consist of the design of the peptide to serve a specific purpose, its chemical synthesis, and the evaluation of the integrity of the product, which includes its purification from the initial crude product of the synthesis.

2
Design and Structure

The design of a particular synthetic peptide is often based on its intended use and on considerations of the synthetic chemistry. With modern synthetic techniques it is possible to produce peptides containing in excess of 100 amino acids, although those shorter than 50 are much more common and easier to synthesize. Short peptides tend not to exhibit preferred or stable solution conformations but rather possess a large degree of flexibility. As peptides increase in length, they have a greater tendency to adopt elements of secondary structure, such as helices and β-sheet structures connected by discrete turns, which impart an overall decrease in flexibility. These are structures generally found in naturally occurring proteins and large peptides that are responsible for their specific biological interaction. From a knowledge of the features that contribute to the formation of these structures, peptides can be specifically designed to contain them.

The physical, chemical, and physiological properties of synthetic peptides are determined by the nature of their constituent amino acids. In nature, the range of structure of these amino acids is vast. In addition to the 20 common ribosomally incorporated "protein" amino acids, there are enzymatically formed, posttranslational modifications to these amino acids found in proteins as well as hundreds of naturally occurring, enzymatically synthesized amino acids not found in proteins.

Add to this the imagination of a synthetic chemist, and the possibilities are almost limitless.

In its most basic sense, a synthetic peptide consists of at least two amino acids joined by a peptide bond, which is an amide bond formed between the amino group of one amino acid and the carboxyl group of an adjacent amino acid (Fig. 2). Peptides composed of more than two amino acids possess a number of these amide bonds in series known as the *peptide backbone*. This backbone structure is an important contributor to the overall peptide structure. The resonance properties of the carbon–nitrogen peptide bond give it a substantial double bond character. As a result the amide bond is flat, with the carbonyl carbon, oxygen, nitrogen, and amide hydrogen all lying in the same plane. Thus, no free rotation occurs around the carbon–nitrogen bond. The torsional angle of that bond, called ω, is defined by the atoms $C\alpha-C(O)-N-C\alpha$, and because of the double bond character and the large energy barrier to rotation (25 kcal mol^{-1}), there are two rotational isomers: trans ($\omega = 180°$) and cis ($\omega = 0°$). The trans-isomer possesses the lower energy and is generally found for all peptide bonds not involving proline. The energy of the trans X-Pro bond is somewhat elevated and the barrier to rotation is lowered. Thus, proline-containing peptides will often exhibit cis–trans isomerization.

In addition to ω, the peptide backbone can be completely described with two additional angles, ϕ and ψ. ϕ is the angle described by $C(O)-N-C\alpha-C(O)$ and ψ is described by $N-C\alpha-C(O)-N$. By convention, the angle ϕ is 180° when the two carbonyl carbons are trans to each other and the angle ψ is 180° when the two amide nitrogens are trans to each other.

Peptide and protein structure is organized in several levels. The order of amino acids, or amino acid sequence, is referred to as the primary structure. The secondary structure is the folding up of the primary structure into discrete elements such as helices, sheets, and turns. The tertiary structure describes how the elements of secondary structure found in a single peptide chain interact with each other to produce a final three-dimensional arrangement. The design of synthetic peptides is basically concerned with primary and secondary structure; the concern with secondary structure becoming more important as the size of the peptide increases. With the largest synthetic polypeptides,

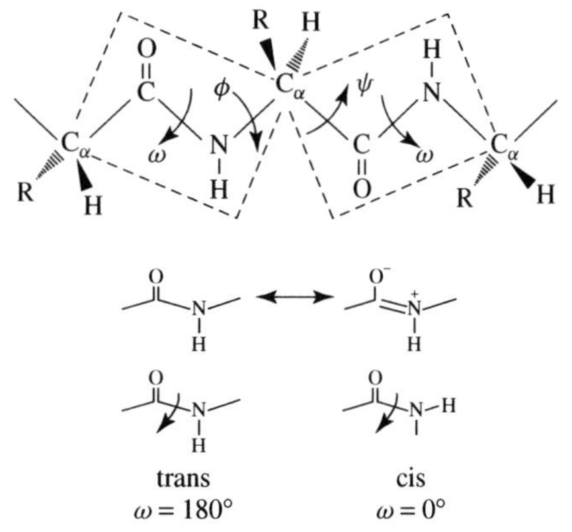

Fig. 2 A stretch of polypeptide demonstrating the planar nature of the peptide bond, the angles that describe the conformation of the polypeptide backbone, and the chirality of L-amino acids. The resonance structures of the peptide bond and cis- and trans-peptide bonds are also shown.

elements of tertiary structure may also be important.

Secondary and tertiary structures are determined by, and can be influenced by a variety of interactions. Basically, secondary and tertiary structures develop from non-covalent interactions such as hydrogen bonds, salt bridges, and hydrophobic interactions resulting from main chain and side chain atoms interacting with each other and with bulk solvent.

The synthetic approach allows the introduction of novel or unusual components such as D-amino acids (**IV**), other unnatural amino acids (**IV, X**), ester or alkyl backbone bonds in place of the normal amide bond (**V**), N- or C-alkyl substituents (**V**), amino acids in which the amino group is not attached to the α-carbon (**VIII, X**), side chain modifications (**VI, IX**), and constraints such as disulfide bridges (**III**) and side chain amide or ester linkages (**IV**). The results of such changes may be higher biological potency, greater stability resulting in longer biological lifetime, or the ability to specifically interact with or covalently label a biological macromolecule or receptor for localization or structure–function relationship studies.

Peptide solubility is another potentially important concern in the design of a synthetic peptide. If the intention is to work with a peptide in aqueous media, it is important that the peptide has sufficient polar character for solubility. This is a point that might especially be overlooked when a peptide is made based on a sequence found internally in a protein. The solubility properties of a particular peptide may be vastly different from the protein when it is removed from the context of the surrounding protein structure. The peptides, composed completely of uncharged amino acids, tend to be less soluble in aqueous environments and their insolubility tends to increase as the size of the peptide increases. Therefore, it is advisable to incorporate charged amino acids into a synthetic peptide when possible. Histidine, lysine, and arginine promote solubility at pHs where their side chains tend to be positively charged and aspartic acid and glutamic acid promote solubility at pHs where their side chains tend to be negatively charged. The contribution of the charge properties of the α-amino and α-carboxyl groups should also be taken into account. In borderline cases, uncharged, polar groups can also promote solubility.

3
Nomenclature

A proper use of nomenclature for describing peptides is important. The International Union of Pure and Applied Chemists (IUPAC), in conjunction with the International Union of Biochemists (IUB), have formulated a set of guidelines for peptide nomenclature (*J. Biol. Chem.* (1985) **260**, 14–42). These guidelines are too extensive to be described here in their entirety, but some of the more common conventions are presented.

The sequence of amino acids within a peptide is usually designated with either the one-letter or three-letter amino acid symbol (Table 1). The amino acid symbols denote the L-configuration unless designated with a D or DL preceding the symbol (**IV**). In the case of linear polypeptides, the convention is to express the peptide left to right from the amino-terminus to the carboxy-terminus (**I**). Thus, a pentapeptide might be represented either as V-L-H-A-G or as Val-Leu-His-Ala-Gly. The α-amino group is assumed to be at

Tab. 1 Common symbols for amino acids in expressing peptide structure.

Residue	Three-letter symbol	One-letter symbol[a]
Alanine	Ala	A
Asparagine[b]	Asn	N
Aspartic acid[b]	Asp	D
Arginine	Arg	R
Cysteine	Cys	C
Glutamic acid[b]	Glu	E
Glutamine[b]	Gln	Q
Glycine	Gly	G
Histidine	His	H
Isoleucine	Ile	I
Leucine	Leu	L
Lysine	Lys	K
Methionine	Met	M
Phenylalanine	Phe	F
Proline	Pro	P
Serine	Ser	S
Threonine	Thr	T
Tryptophan	Trp	W
Tyrosine	Tyr	Y
Valine	Val	V
Hydroxylysine	Hyl	
4-Hydroxyproline	Hyp	
α-Aminobutyric acid	Abu	
Norvaline	Nva	
Norleucine	Nle	
Homoserine	Hse	
Homocysteine	Hcy	
Ornithine	Orn	
Citrulline	Cit	
α-Aminoisobutyric acid	Aib	
γ-Carboxyglutamic acid	Gla	

[a] Symbols for less common amino acids should be defined in the individual publications in which they appear.
[b] Asx or B and Glx or Z are used when the distinction between aspartic acid/asparagine and glutamic acid/glutamine is uncertain. This convention is most commonly applied in sequence analysis.

the left-hand side of the amino acid symbol when using hyphens and at the point of the arrow when using arrows for special cases such as cyclic peptides (**VII**, and see below). Normally, when the termini are free amino or carboxyl groups, no additional designation of this is necessary. However, in order to avoid confusion or ambiguity, the state of the termini is often indicated (**III**). Free termini are designated with a H– at the amino end and an –OH at the carboxyl end so that the above peptide would be written as H-Val-Leu-His-Ala-Gly-OH. Similarly, a C-terminal amide or aldehyde group would be designated by –NH$_2$ or –H respectively.

Occasionally, the names of small peptides will be written out. In this case, the names of all amino acids contributing carboxyl groups to peptide bonds end in -yl, and the amino acid with the terminal carboxyl group retains its original name. The peptapeptide shown above and in **I** would thus be expressed as valylleucylhistidylalanylglycine.

Cyclic peptides are usually expressed on a single line of text, similar to linear peptides, with a line or "extended bond" drawn to connect the joined amino acids that lead to the cyclic nature of the peptide. If the amino acids are joined between side chains, such as in disulfide bonds or side chain lactams, the line starts and ends in the middle of the amino acid name or abbreviation, such as

$$\overline{\text{Gly-Val-Glu-Ala-Leu-Ile-Lys-Ala-Thr.}}$$

If the structure drawn in this manner is ambiguous, additional atoms can be illustrated, as shown in Fig. 1 (**III**, **IV**), to clarify the meaning.

If the connection is between the α-amino and α-carboxyl groups of the termini, the line is drawn from the sides of the names, such as

$$\underline{\text{Gly-Val-Glu-Ala-Leu-Ile-Lys-Ala-Thr.}}$$

In this case, the N→C direction is read left to right. If the peptide is written out in a circular fashion, as if it was written along the circumference of a circle or a loop, then the N→C direction must be designated by arrows (**VII**). Finally, cyclic peptides are sometimes designated as such with the prefix *cyclo*, such as cyclo(Gly-Val-Glu-Ala-Leu-Ile-Lys-Ala-Thr), to indicate the same structure as immediately above.

Synthetic analogs of naturally occurring peptides that have specific names associated with them represent a special case. These include substances such as insulin, oxytocin, enkephalin, vasopressin, and bradykinin to name a few. For example, the polypeptide hormone, oxytocin, has the structure

$$\overline{\text{Cys-Tyr-Ile-Gln-Asn-Cys-Pro-Leu-Gly-NH2,}}$$

where the cysteine residues form a disulfide bridge. If one or more amino acids are replaced in the synthesis of an oxytocin analog, the new amino acids and their positions are placed in brackets before the name. Thus, [Phe^2, Asn^4]-oxytocin refers to an oxytocin analog in which the tyrosine at position 2 and the glutamine at position 4 have been replaced with phenylalanine and asparagine respectively. The prefix *endo* is used to designate the insertion of an extra amino acid into the chain. Thus, *endo*-Ala^{7a}-oxytocin denotes an alanine insertion between Pro^7 and Leu^8. Similarly, deletion of an amino acid is designated by the prefix *des*, for example, *des*-Ile^3-oxytocin denotes that Ile at position 3 has been deleted. Peptide sequences that form fragments of a longer sequence that has a trivial name are designated by the trivial name followed by numbers giving the positions of the first and last amino acids and then a designation describing the number of amino acids in the fragment. For example, the peptide Tyr-Ile-Gln is a segment of oxytocin (see above) and can be designated as oxytocin-(2–4)-tripeptide.

4
Synthesis

Peptides can be synthesized either in solution, which is the classical approach, or by solid-phase methods. With the exception of specialized applications in which solution phase chemistry is used because it is easier, less expensive, or the necessary chemistry is not amenable to the solid-phase method, most peptides made for research purposes are synthesized with the solid-phase method that was introduced by Merrifield in 1963. This technique, for which he was awarded the Nobel Prize in Chemistry in 1984, was primarily responsible for opening the way to the widespread use of synthetic peptides in chemical and biomedical investigations.

Solid-phase methods are readily automated and most solid-phase syntheses performed today are done with the aid of machines. A variety of peptide synthesizers are commercially available for batchwise and continuous flow operations as well as for the synthesis of multiple peptides within the same run.

4.1
The Solid-phase Method

The solid-phase approach to peptide synthesis is presented in Fig. 3. Basically, it consists of anchoring the growing peptide to an insoluble support or resin. This is usually accomplished through the use of a chemical handle that links the support to the first amino acid at the carboxyl terminus of the peptide. Subsequent amino acids are then added, one at a time, in a stepwise manner until the peptide is fully constructed. The solid-phase method allows excess reagents and soluble reaction by-products to be easily removed by filtration and washing, thus doing away with the need to isolate or crystallize the product after each step.

Synthesis of the growing polypeptide chain is accomplished by formation of an amide bond between the free α-amino group of the receiving amino acid (anchored to the support) and the activated α-carboxyl group of the incoming amino acid (in solution). Note that synthesis proceeds from the carboxy end to the amino end. To prevent unwanted side reactions with reactive amino acid side chains and self-coupling of the α-carboxyl group of one incoming amino acid to the α-amino group of another incoming amino acid, these groups must be protected at the appropriate times during synthesis. As depicted in Fig. 3, the group protecting the α-amino group of the amino acid being coupled must ultimately be removed, after coupling is complete, in order to accept the next amino acid during the next cycle. Chemical groups that allow this are referred to as "temporary protecting groups". The reactive amino acid side chains, on the other hand, must remain permanently protected throughout the synthesis. Thus, chemical groups that accomplish this are referred to as "permanent protecting groups". At the end of the synthesis, the completed peptide is cleaved from the resin support and the permanent protecting groups are removed, usually in a single step process, to yield the desired peptide product.

4.2
Chemistry

A number of different chemical approaches for the synthesis of peptides on solid phase supports have been investigated and are in use today. Generally, two levels of protecting group stability are required. The permanent protecting groups must be such that they are stable during the repetitive conditions of removing the temporary protecting groups from the growing peptide chain. However, the permanent protecting groups, as well as the attachment handle holding the peptide to the support, must ultimately be removed under conditions

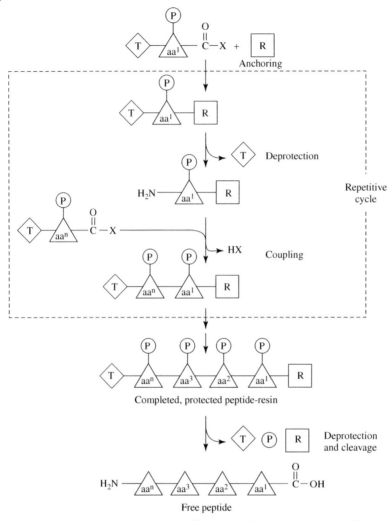

Fig. 3 Stepwise solid-phase synthesis of linear peptides. R, resin or insoluable polymeric support; T, temporary protecting group; P, permanent protecting group; X, denotes activated carboxyl group; $aa^1 \ldots aa^n$, amino acid residues numbered from the C-terminus.

Fig. 4 Common protection schemes for solid-phase peptide synthesis (a) Protection scheme for solid-phase synthesis based on the Boc group utilizing graded acidolysis. The Boc group is removed at each step by trifluoracetic acid (TFA). The permanent side chain protecting groups and the PAM linkage are cleaved simultaneously by hydrofluoric acid (HF) or other strong acid. (b) Mild orthogonal protection scheme for solid-phase synthesis based on the Fmoc group. The Fmoc group is removed at each step by the base catalyzed elimination mechanism, usually using piperidine, as shown. Permanent side chain–protecting groups and the HMP/PAB linkage are cleaved by treatment with TFA. (Figures are reprinted with permission from Freeman, W. H. and Fields, G. B., et al. in Grant, G. A. (Ed.) (2002) *Synthetic Peptides: A User's Guide*, 2nd edition, Oxford University Press, Oxford, New York).

14 Design and Application of Synthetic Peptides | 441

that do not harm the underlying peptide structure. Two general methods have emerged as the most common and are depicted in Fig. 4. They are referred to as either "Boc", which uses the tert-butyloxycarbonyl group, or "Fmoc", which uses the 9-fluorenylmethyloxycarbonyl group for temporary α-amino protection. Today, the Fmoc chemistry is probably used more often than the Boc chemistry largely because the permanent protecting groups and the resin handle can be removed under milder conditions with reagents that do not present as much of a potential hazard as those used for Boc chemistry. However, the Boc chemistry remains a vital and useful chemistry that is still used by many peptide chemists.

A wide variety of permanent protecting groups are available for these two general approaches and a few of them are depicted in Fig. 4 and Table 2. The Boc strategy relies on graded acidolysis to selectively differentiate between temporary and permanent protection. Temporary protecting groups are removed by treatment with trifluoroacetic acid while cleavage from the resin and removal of permanent protecting groups is usually accomplished with anhydrous hydrofluoric acid or other strong acid. The Fmoc strategy is a milder orthogonal approach in that the two classes of protecting groups are removed by different chemical mechanisms such that either class can be removed while preserving the other class. The temporary Fmoc group is removed by a base-catalyzed β-elimination mechanism, usually employing piperidine. Removal of permanent protecting groups as well as cleavage from the support is accomplished by acidolysis, usually with trifluoroacetic acid.

Formation of the peptide bond between the carboxyl group of the incoming amino acid and the amino group on the peptide-resin requires that the carboxyl group be activated chemically. There are currently four major coupling strategies that utilize either active esters, preformed symmetrical anhydrides, acid halides, or generation of the activated acid *in situ* with a variety of different reagents.

Peptides modified at the α-amino group can be easily produced by treatment with an acylating agent, such as acetic anhydride, prior to removal of the permanent protecting groups. Resins are also available for both Boc and Fmoc strategies that will directly yield peptides with either a free acid or an amide function at the carboxy-terminus.

In some instances, larger peptides have been formed from smaller ones by the use of segment condensation and chemoselective ligation techniques. In either case, two or more smaller peptides are synthesized in such a way that they can subsequently be linked together to form a large peptide. However, these procedures are not common and are still relatively difficult to perform.

5
Evaluation

Although the synthesis of any given peptide is often considered routine and can be performed with standard protocols on automated instruments, the chemistry used to produce synthetic peptides is complex and can be influenced by a number of factors that cannot always be foreseen or controlled. As a result, one must never assume that the final product is correct until it is proven to be so.

Homogeneity of the product and correct covalent structure are the two main aims in the production of synthetic peptides. The initial product obtained from the

Tab. 2 Some commonly used side chain protecting groups for solid-phase peptide synthesis.

Amino acid	Compatible with boc protection	Compatible with fmoc protection
Asp/Glu	Benzyl (OBzl)	tert-Butyl (Ot-Bu)
Ser/Thr	Benzyl (Bzl)	tert-Butyl (t-Bu)
Tyr	2-Bromobenzyloxycarbonyl (2-BrZ)	tert-Butyl (t-Bu)
Lys	2-Chlorobenzyloxycarbonyl (2-ClZ)	tert-Butyloxycarbonyl (Boc)
His	Benzyloxymethyl (Bom)	Triphenylmethyl (Trt)
Cys	4-Methylbenzyl (Meb)	Triphenylmethyl (Trt)
Asn/Gln		Triphenylmethyl (Trt)
Arg	4-Toluenesulfonyl (Tos)	2,2,5,7,8-Pentamethyl chroman-6-sulfonyl (Pmc)
Trp	Formyl (CHO) (on N^{in})	

stepwise solid-phase peptide synthesis will contain a variety of by-products in addition to the desired peptide. These products usually consist of deletion sequences generated during synthesis, and chemically modified peptides generated either during synthesis or during the cleavage and deprotection step. A successful synthesis is one in which these products are minimal and the intended peptide is present in sufficient yield to allow its purification from the mixture. Today, virtually all peptide purifications can be accomplished by high-pressure liquid chromatography (HPLC) using either reverse-phase or ion-exchange columns.

A common problem encountered in peptide synthesis is the production of deletion sequences. These come about as a result of either incomplete coupling or incomplete deprotection at any given cycle. Incomplete coupling of the incoming amino acid leaves a free α-amino group on the peptide chain that will persist into the next cycle and be able to couple with the next incoming residue. Incomplete deprotection results in the retention of a protected α-amino group that can subsequently be deprotected in the next cycle and thus be able to receive the incoming amino acid at that cycle. Both cases result in a finished peptide that is missing one amino acid residue. Deletion sequences due to incomplete coupling are probably the most common and the easiest to manage. The elongation of deletion peptides due to incomplete coupling can be halted by the use of a capping step. Capping simply refers to the reaction of the free α-amino group remaining at the end of a coupling step with a reagent such as acetic anhydride that derivatizes the α-amino group in such a way that it can no longer be elongated during subsequent cycles. Although capped peptides represent contaminants that will have to be removed from the final product, they are generally easier to separate from the full-length product than are deletion peptides, which may differ from the full-length product by only one amino acid.

Unwanted chemically modified peptides result either from incomplete deprotection of the side chains or from random adducts formed at some point during synthesis or cleavage and deprotection. As long as they do not represent an overwhelming proportion of the total product, they can usually be separated from the desired product without the need to identify their structure. When unwanted side-products show up as major species, either the chemical protocol or instrument operation needs to be checked and modified. In this regard, it is important to note that the isolation of a single major product from a synthesis does not guarantee that it is the intended product. Facile modifications are not uncommon and there are many instances in which the wrong peptide was isolated and used in experimentation because its structure was not verified.

There are numerous analytical techniques available for the evaluation of synthetic peptides. The techniques that have been commonly used include analytical high performance liquid chromatography (HPLC), capillary zone electrophoresis (CZE), amino acid compositional analysis, UV spectroscopy, sequence analysis, and mass spectrometry.

Today, mass spectrometry is most often used as the primary means of characterizing synthetic peptides. Efficient characterization of synthetic peptides is best obtained by a combination of analytical reverse-phase HPLC and mass spectrometry. Analytical reverse-phase HPLC provides evidence for the homogeneity of the peptide product by presenting a picture of the number of peptide species present based on their chromatographic elution characteristics. Multiple peaks indicate heterogeneity and the area under the peaks is an indication of the relative amounts in each peak. Note, however, that a single main peak suggests relative homogeneity but does not exclude the possibility of comigrating species. Both electrospray ionization mass spectrometry (ESMS) and matrix-assisted laser desorption ionization mass spectrometry (MALDI-MS) can be used for the accurate determination of peptide mass. The position of modifications and deletions, if present, can be identified by sequencing with either chemical methods (Edman chemistry) or tandem mass spectrometry. In this respect, sequencing is most easily accomplished by collision-induced decomposition on ion-trap or tandem instruments such as Quadrupole-Time of Flight (Q-TOF) instruments. Amino acid analysis is the best technique for the quantitation of the product. It specifically measures the amount of amino acids present and is accurate to within ±5 to 10%. Gravimetric techniques will also measure the amount of salt and adsorbed water present in the peptide preparation. This can be substantial, up to 50% of the total, thus causing as much as a twofold error.

6
Applications and Perspectives

Peptides have become an important class of molecules in biomedical research and medicinal chemistry. Already a number of synthetic hormones or their analogs have found use as therapeutic agents. Synthetic peptides show great potential in being effective in regulating such physiological processes as blood pressure, neurotransmission, reproduction, and endocrine functions, to name a few. Synthetic peptide substrates for enzymes such as proteases, kinases, and phosphatases have been used successfully to study enzyme kinetics and mechanism and develop effective inhibitors. Synthetic peptides are seeing great success as antigens for the preparation of polyclonal and monoclonal antisera and offer the potential for more effective vaccines. The process of *de novo* drug discovery is enhanced through the development of peptide libraries. Epitope mapping of proteins with synthetic peptides has identified regions important for biological activity. Synthetic peptides have been instrumental in receptor localization, characterization, and isolation. Finally, even whole enzymes have been produced by synthetic methods. A comprehensive discussion of the many and diverse uses for synthetic polypeptides is obviously beyond the scope of this article. Therefore, the following sections attempt to briefly present some of the more recent and interesting applications in the field.

6.1
Antigenic and Immunogenic Uses and Vaccines

The ability of synthetic peptides to immunologically mimic regions of whole proteins has resulted in their extensive use as antigens and probes of protein epitopes. Antibodies that are raised against a native protein bind to relatively small areas on the surface of that protein and can often recognize and bind to the same antigenic sequence in the absence of the rest of the protein. Thus, highly antigenic sequences or epitopes can be mapped in proteins by synthesizing overlapping peptides, spanning the full sequence of the protein and testing them for binding of the antibody.

Many gene products have been identified and isolated by prediction of antigenic regions in the sequence of a protein that was initially deduced from a gene sequence. Synthetic peptides corresponding to these predicted antigenic regions are used to produce antibodies that in turn are used to specifically interact with the intact protein product for identification and isolation. Although this approach has proven very successful, the main problem in its implementation is the prediction of the protein's antigenic sequences. Although there is no guarantee of success, certain general guidelines for choosing surface epitopes have been described. First, use of a hydropathy profile locates areas that are more hydrophilic. This is based on the assumption that the surface of a protein is exposed to solvent and will thus be hydrophilic in nature. If you only have a short section to work with, choose areas that are charged or polar. In general, the immediate amino and carboxyl ends of a protein may tend to be more solvent accessible than the interior sequence. These regions may be better candidates if you have no other criteria. Secondly, since short stretches of sequence in proteins do not have exposed termini that can become charged, synthesize the peptide with α-amino and α-carboxyl blocking groups such as N-terminal acetyl groups and C-terminal amide groups. Keep in mind that the peptide must be sufficiently polar to be soluble in aqueous solution. Peptides of 10 to 15 residues are usually adequate for the production of antibodies, and including a cysteine residue at either end can be helpful for cross-linking to carrier proteins via heterobifunctional cross-linking reagents (see Fig. 5). The use of a carrier protein is usually necessary because small peptides, in general, are not sufficiently antigenic to produce an immune response on their own. This is because they generally lack T-cell epitopes that can be provided by the carrier. Antibodies are produced by B-cells that recognize or respond to certain epitopes. T-cells also recognize epitopes, but these epitopes are different from those recognized by B-cells. In the presence of T-cell epitopes or antigens, T-cells stimulate B-cells to boost their antibody production. Thus, to produce a good antisera, both B- and T-cell epitopes should be present.

In addition to linking polypeptide antigens to carrier proteins, two purely synthetic approaches have also been shown to be successful in producing antisera (Fig. 5). One approach, using the concept of a template-assembled synthetic protein (TASP) involves synthesizing both B- and T-cell peptide epitopes as branches on a β-sheet peptide backbone template. Another approach, known as the multiple antigen peptide system (MAPS), uses a peptide core of radially branching lysine residues covalently connected to a conventional solid support for peptide synthesis. Peptide antigen sequences are built on the α- and ε-amino groups of the terminal multiple lysine residues. The resultant structure consists of multiple, tightly packed peptide antigen sequences on a single structure. Although no apparent T-cell epitopes are present, MAPS antigens have been shown to produce a strong immunogenic response that results in high titer antisera for the peptide antigen. This is thought to be due to the local high concentration of peptide presented.

One very exciting area that is based on the use of synthetic peptides as antigens is in the development of vaccines.

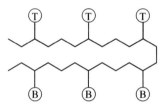

Carrier protein linked via heterobifunctional reagent

(a)

Multiple antigenic peptide system (MAPS)

(b)

Template assembled synthetic protein (TASP)

(c)

Fig. 5 Methods for the use of synthetic peptides as antigens to produce high titer antisera. (a) a synthetic peptide/carrier protein conjugate using the heterobifunctional cross-linking reagent *m*-maleimidobenzoyl-N-hydroxysuccinimide ester (MBS, in brackets). (b) multiple antigenic-peptide system with peptide antigen linked to a polylysine core. (c) incorporation of both T- and B-cell peptide epitopes on a β-sheet peptide template.

Synthetic peptide-based vaccines are actively being pursued as potential antiviral, antibacterial, and antiparasitic agents for such diseases as foot-and-mouth disease, poliovirus, hepatitis B, influenza, HIV, feline leukemia virus, malaria, diphtheria, and cholera to name a few. Prevention of some carcinomas, such as melanoma, is also being explored by this technique on the basis of segments of melanoma cell specific proteins that are presented antigenically to the host's immune system. These vaccines are based on the observations that viruses and microorganisms present major, protein-derived antigenic determinants on their surfaces, that many bacterial diseases stem from protein or peptide toxins secreted by the bacteria, and that cancer cells express antigens that are recognized as foreign to the host. If effective antisera can be raised against these regions, they have the potential to neutralize the virus or toxic agent. Major impediments to developing effective peptide-based vaccines are the inability to definitively recognize the proper determinants to be used as a basis for the synthetic peptide antigen and the problem of serological diversity. With viruses, for example, selecting peptides that will be effective as vaccines is more difficult than just developing antibodies that will react with viruses. Virus particles possess a large number of different epitopes, but not all will produce antibodies that will neutralize virus infectivity. Even when a major neutralizing epitope is identified, diversity within that epitope can be a major problem. For instance, foot-and-mouth disease virus has as many as seven different serotypes throughout the world, and vaccination with one does not necessarily protect against the others. This is exacerbated by the fact that variation can exist within each serotype; many viruses display a much larger number of serotypes, and some display the ability to mutate fairly rapidly. However, if invariant regions can be found among the various subtypes, the number of synthetic peptide-based vaccines needed to protect against the agent can be reduced and vaccine development becomes feasible. The prospect of developing vaccines from synthetic peptides has the distinct advantages that the antigen is chemically well defined, relatively stable, and is free from possible biological contamination that can be a major problem with vaccines from biological sources.

6.2
Antisense Peptides

The area of antisense peptides is very new and still the subject of controversy that stems mainly from the fact that the experimental observations have not yet been adequately explained theoretically.

An antisense peptide is a peptide whose sequence is deduced from the nucleotide sequence that is complementary to the nucleotide sequence which codes for a naturally occurring peptide. In general, although it has not been shown to be the case in all instances, antisense peptides have shown selectivity in binding to their sense counterparts and have displayed unusual or surprising biological activity.

The basis for the unusual interaction between sense and antisense peptides appears to stem from the genetic code. In general, the codons for hydrophobic or hydrophilic amino acids on the coding strand of DNA are complemented by codons for hydrophilic or hydrophobic amino acids, respectively, on the opposite strand. Furthermore, reading of the complementary DNA strand in either direction displays the same pattern of hydrophobicity in the resultant amino acid sequence since the middle base of the codon seems to specify the hydropathic nature of the amino acid.

One of the most stunning successes in the use of antisense peptides is the demonstration that antibodies raised against an antisense peptide of adrenocorticotropin hormone (ACTH) were successful in recognizing and purifying the ACTH receptor. In other words, the antibodies, acting like an image of an image, interacted with the receptor in a manner similar to ACTH. Although many other similar successes have been reported, it is important to note that many failures have also been seen. Thus, the exact nature of the interaction is not well understood but the area of antisense peptides continues to be an exciting and growing field.

6.3
Peptide Libraries

The process of developing new synthetic peptides that are useful for diagnostic or therapeutic purposes usually involves the costly and time-consuming process of synthesizing hundreds of peptide analogs based on the structure of an original active peptide. In spite of recent advances that have improved this process, the task of screening all possible combinations of even short peptides has been impractical. For instance, there are 64 000 000 possible hexapeptides (20^6) for all combinations of the 20 naturally occurring protein amino acids.

Recently, the development of synthetic peptide libraries has greatly aided this task. Basically, the libraries are generated by synthesis of a mixture of a large number of peptides of defined length in which each position, or a defined number of positions, contains a random mixture of amino acids. These heterogeneous mixtures are then screened with some appropriate assay system, and the reactive sequences are isolated and determined. Several different methods for the generation and screening of synthetic peptide libraries have been described.

6.4
Other Applications

One approach to identifying, locating, and isolating receptors to polypeptide ligands such as peptide hormones and opioids is to establish chemical cross-links between the peptide ligand and the receptor subsequent to their binding. An ever-increasing number of heterobifunctional cross-linking reagents, many with photoactivatable reactive groups for *in situ* activation, are being developed for this purpose which can be attached to the biologically active peptides at a variety of locations. Once attached to the peptide, these reagents are capable of producing covalent cross-links between the peptide and the receptor. If the synthetic peptide is labeled, the label can be followed in isolating the receptor and identifying the area of contact in the peptide – the receptor complex.

Peptide nucleic acids are compounds that are synthesized with a peptide-like backbone with DNA- and RNA-derived bases as side chains (**X**). These polyamide oligomers appear to be capable of invading the duplex strand of DNA at specific locations determined by the identity of the bases, causing displacement of one strand, and disrupting transcription elongation of that stretch of DNA. As such, they display potential as antisense and antigene agents that may be superior to some phosphodiester backbone-based antisense agents with regard to stability and bioefficacy.

The use of peptidomimetics is a relatively new but growing area related to the field of synthetic polypeptides. Although the mimetic structure is often not strictly a peptide because it may completely lack peptide bonds, the structure is based on that of a peptide. Since the three-dimensional shape of a peptide determines how it will interact with its biological target, producing specific stable conformations in synthetic peptides is one of the most challenging areas in the design of biologically active peptides. In general, small peptides tend not to exhibit stable or even preferred conformations in solution. The general idea with peptidomimetics is to synthesize a compound that mimics some structural or conformational property that a peptide might assume biologically or that mimics some stable, relatively rigid structure found in proteins, such as β- or γ-turns. The approach in synthesizing peptidomimetics is usually to replace part or all of the normal peptide structure with nonpeptidelike chemical structure so that the resulting conformation is stabilized and mimics the original peptide in space. The final mimic often has very little chemical resemblance to the peptide, but displays stable three-dimensional elements of conformation which is similar to that assumed by the peptide. Some approaches that have been taken in this regard have been to replace peptide bonds with rigid structures that maintain bond distance and side chain orientation or introduction of intramolecular cross-links to stabilize turns or side chain positions.

The area of peptidomimetics has been very actively pursued in rational drug design and pharmaceuticals. Although success has been limited owing to the difficulty in accurately mimicking the exact active biological conformation, some advances have been made and will no doubt continue since peptidomimetics have many potential applications. One successful example is the design of a tricyclic, conformationally constrained lactam that is among the most potent angiotensin converting enzyme (ACE) inhibitors known.

Finally, whole, biologically active proteins have been produced by purely synthetic methods. One striking example of this is the complete chemical synthesis of the human immunodeficiency virus-1 (HIV-1) protease that possesses full biological activity. Even more striking is the demonstration that the HIV-1 protease synthesized from all D-amino acids displays specificity for substrates with opposite chirality to the natural substrate.

The above sections only touch on some of the diverse and potentially very useful applications of synthetic peptides. From what has been accomplished with synthetic polypeptides already, it seems clear that they will continue to play an increasingly larger role in biomedical research and in the treatment of human, animal, and plant disease.

Bibliography

Books and Reviews

Atherton, E., Sheppard, R.C. (1989) *Solid Phase Peptide Synthesis: A Practical Approach*, IRL Press, Oxford.

Bodanszky, M. (1988) *Peptide Chemistry, A Practical Textbook*, Springer-Verlag, Berlin.

Bodanszky, M., Bodanszky, A. (1984) *The Practice of Peptide Synthesis*, Springer-Verlag, Berlin.

Grant, G.A. (Ed.) (2002) *Synthetic Peptides: A User's Guide*, 2nd edition, Oxford University Press, Oxford, New York.

Jones, J. (1991) *The Chemical Synthesis of Peptides*, Clarendon Press, Oxford.

Kates, S.A., Albericio, F. (Eds.) (2000) *Solid-Phase Peptide Synthesis: A Practical Guide*, Marcel Dekker, New York.

Sewald, S., Jakubke, H.-D. (2002) *Peptides: Chemistry and Biology*, Wiley-VCH, Germany.

Smith, B.J. (Ed.) (1997) *Protein Sequencing Protocols*, Humana Press, Totowa, New Jersey, USA.

Stewart, J.M., Young, J.D. (1984) *Solid Phase Peptide Synthesis*, Pierce Chemical Company, Rockford, IL.

Van Regenmortel, M.H.V., Briand, J.P., Muller, S., Plaue, S. (1988) in: Burdon, R.H., Kippenberg, P.H. (Eds.) *Synthetic Polypeptides as Antigens*, In the Laboratory Techniques in Biochemistry and Molecular Biology Series, Elsevier, Amsterdam.

Wieland, T., Bodanszky, M. (1991) *The World of Peptides: A Brief History of Peptide Chemistry*, Springer-Verlag, Berlin.

Primary Literature

Addona, T., Clauser, K. (2002) De novo peptide sequencing via manual interpretation of

MS/MS spectra, *Current Protocols in Protein Science*, John Wiley & Sons, New York, pp. **16**.11.1–**16**.11.19.

Angeletti, R.H., Bonewald, L.F., Fields, G.B. (1997) Six-year study of peptide synthesis, *Methods Enzymol.* **289**, 697–717.

Bergman, M., Zervas, L. (1932) Ueber ein allgemeines Verfahren der Peptid-Synthese, *Ber. Dtsch. Chem. Ges.* **65**, 1192–1201.

Berk, S.C., Rohrer, S.P., Degrado, S.J., Birzin, E.T., Mosley, R.T., Hutchins, S.M., Pasternak, A., Schaeffer, J.M., Underwood, D.J., Chapman, K.T. (1999) A combinatorial approach toward the discovery of non-peptide, subtype-selective somatostatin receptor ligands, *J. Combinatorial Chem.* **1**, 388–391.

Blalock, J.E., Bost, K.L. (1986) Binding of peptides that are specified by complementary RNAs, *Biochem. J.* **234**, 679–683.

Blalock, J.E., Smith, E.M. (1984) Hydropathic anti-complementarity of amino acids based on the genetic code, *Biochem. Biophys. Res. Commun.* **121**, 203–207.

Bodanszky, M., Ondetti, M.A., Levine, S.D., Williams, N.J. (1967) Synthesis of secretin II. The stepwise approach, *J. Am. Chem. Soc.* **89**, 6753–6757.

Bodanszky, M., Williams, N.J. (1967) Synthesis of secretin I. The protected tetradecapeptide corresponding to sequence 14–27, *J. Am. Chem. Soc.* **89**, 685–689.

Bost, K.L., Blalock, J.E. (1986) Molecular characterization of a corticotropin (ACTH) receptor, *Mol. Cell. Endocrinol.* **44**, 1–9.

Carr, S.A., Annan, R.S. (1999) Overview of peptide and protein analysis by mass spectrometry, *Current Protocols in Protein Science*, John Wiley & Sons, New York, pp. **16**.1.1–**16**.1.27.

Crabb, J.W., West, K.A., Dodson, W. Scott, Hulmes, J.D. (1997) Amino acid analysis, *Current Protocols in Protein Science*, John Wiley & Sons, pp. **11**.9.1–**11**.9.42.

Edman, P., Begg, G. (1967) A protein sequenator, *Eur. J. Biochem.* **1**, 80–91.

Fischer, E. (1902) Ueber einige Derivate des Glykocolls Alanins und Leucins, *Ber. Dtsch. Chem. Ges.* **35**, 1095–1106.

Fischer, E. (1903) Synthese von Derivaten der Polypeptide, *Ber. Dtsch. Chem. Ges.* **36**, 2094–2106.

Fischer, E. (1906) Untersuchungen über Aminosäuren, Polypeptide, und Proteine, *Ber. Dtsch. Chem. Ges.* **39**, 530–610.

Fischer, E. (1907) Synthese von Polypeptiden XVII, *Ber. Dtsch. Chem. Ges.* **40**, 1754–1767.

Grant, G.A. (1999) Synthetic peptides for the production of antibodies that recognize intact proteins, *Current Protocols in Protein Science*, John Wiley & Sons, pp. **18**.3.1–**18**.3.14.

Harington, C.R., Mead, T.H. (1935) Synthesis of glutathione, *Biochem. J.* **29**, 1602–1611.

Henzel, W.J., Stults, J.T. (1999) Matrix-assisted laser desorption/ionization time-of-flight mass analysis of peptides. *Current Protocols in Protein Science*, John Wiley & Sons, pp. **16**.2.1–**16**.2.11.

Hirschmann, R., Nutt, R.F., Veber, D.F., Vitali, R.A., Varga, S.L., Jacob, T.A., Holly, F.W., Denkewalter, R.G. (1969) Studies on the total synthesis of an enzyme V. The preparation of enzymatically active material, *J. Am. Chem. Soc.* **91**, 507–508.

Hruby, V.J., Al-Obeidi, F., Kazmierski, W. (1990) Emerging approaches in the molecular design of receptor selective peptide ligands: conformational, topographical and dynamic considerations, *Biochem. J.* **268**, 249–262.

Hruby, V.J., Matsunaga, T.O. (2002) Applications of synthetic peptides, in: Grant, G.A. (Ed.) *Synthetic Peptides: A User's Guide*, Vol. 5, Oxford University Press, Oxford, New York, pp. 292–376.

Milton, R.C., Milton, S.C., Kent, S.B. (1992) Total chemical synthesis of a D-enzyme: the enantiomers of HIV-1 protease show reciprocal chiral substrate specificity, *Science* **256**, 1445–1448.

Moore, R.E., Young, M.K., Lee, T.D. (2000) Protein identification using a quadrupole ion trap mass spectrometer and SEQUEST database matching, *Current Protocols in Protein Science*, John Wiley & Sons, pp. **16**.10.1–**16**.10.9.

Nair, S.A., Kim, M.H., Warren, S.D., Choi, S., Songyang, Z., Cantley, L.C., Hangauer, D.G. (1995) Identification of efficient pentapeptide substrates for the tyrosine kinase pp60c-src, *J. Med. Chem.* **38**, 4276–4283.

Nielsen, P.E. (2001) Peptide nucleic acid: a versatile tool in genetic diagnostics and molecular biology, *Curr. Opin. Biotechnol.* **12**, 16–20.

Nutt, R.F., Brady, S.F., Darke, P.L., Ciccarone, T.M., Colton, C.D., Nutt, E.M., Rodkey, J.A., Bennett, C.D., Waxman, L.H., Sigal, I.S., Anderson, P.S., Veber, D.F. (1988) Chemical synthesis and enzymatic activity of a 99 residue peptide with a sequence proposed for the human immunodeficiency virus protease, *Proc. Natl. Acad. Sci. U.S.A.* **85**, 7129–7133.

Pinilla, C., Appel, J.R., Blanc, P., Houghten, R.A. (1992) Rapid identification of high affinity peptide ligands using positional scanning synthetic peptide combinatorial libraries, *BioTechniques* **13**, 901–905.

Pinilla, C., Appel, J.R., Blondelle, S.E., Dooley, C.T., Eichler, J., Ostresh, J.M., Houghten, R.A. (1994) Versatility of positional scanning synthetic combinatorial libraries for the identification of individual compounds, *Drug Dev. Res.* **33**, 133–145.

Reim, D.F., Speicher, D.W. (1997) N-terminal sequence analysis of proteins and peptides, *Current Protocols in Protein Science*, John Wiley & Sons, pp. **11**.10.1–11.10.38.

Rodda, S.J., Geysen, H.M., Manson, I.J., Schaafs, P.G. (1986) The antibody response to myoglobin-I. Systematic synthesis of myoglobin peptides reveals location and substructure of species dependent continuous antigenic determinants, *Mol. Immunol.* **23**, 603–610.

Schneider, J., Kent, S.B.H. (1988) Enzymatic activity of a synthetic 99 residue protein corresponding to the putative HIV-1 protease, *Cell* **54**, 363–368.

Schwyzer, R., Sieber, P. (1963) Total synthesis of adrenocorticotrophic hormone, *Nature* **199**, 172–174.

Sifferd, R.H., du Vigneaud, V. (1935) A new synthesis of carnosine, with some observations on the splitting of the benzyl group from carbobenzoxy derivatives and from benzylthio ethers, *J. Biol. Chem.* **108**, 753–761.

Tam, J.P., Spetzler, J.C. (1997) Multiple antigen peptide system, *Methods Enzymol.* **289**, 612–637.

Tuchscherer, G., Servis, C. Corradin, G., Blum, U. Rivier, J., Mutter, M. (1992) Total chemical synthesis, characterization, and immunological properties of an MHC class I model using the TASP concept for protein de novo design, *Protein Sci.* **1**, 1377–1386.

du Vigneaud, V., Ressler, C., Swan, J.M., Roberts, C.W., Katsoyannis, P.G., Gordon, S. (1953) The synthesis of an octapeptide amide with the hormonal activity of oxytocin, *J. Am. Chem. Soc.* **75**, 4879–4880.

15
Synthetic Peptides: Chemistry, Biology, and Drug Design

Tomi K. Sawyer
ARIAD Pharmaceuticals, Cambridge, MA, USA

1	**Peptide Chemistry** 456	
1.1	Peptide Structure and Three-dimensional Properties	456
1.2	Peptide Chemical Synthesis and Molecular Diversity	464
2	**Peptide Biology** 466	
2.1	Peptide Biosynthesis and Metabolism	466
2.2	Peptide Biological Functions and Mechanisms of Action	468
3	**Peptide Drug Design** 472	
3.1	Peptide Pathophysiology and Therapeutic Targets	472
3.2	Peptide, Peptidomimetic, and Nonpeptide Drug Discovery	474
	Bibliography 481	
	Books and Reviews 481	
	Primary Literature 482	

Keywords

Amino Acids
The building blocks of peptides (as well as polypeptides and proteins) are amino acids. Chemically, an amino acid consists of a central carbon (termed α-*carbon* or C-α) to which are attached an amino (NH_2) group, a carboxyl (CO_2H) group, a side chain (generically referred to as an *R group*), and a hydrogen (H) atom. Amino acids differ in terms of their R group, and the three-dimensional chemical structure (or configuration) of amino acids also may be differentiated at the central α-carbon, except where R is hydrogen. With the exception of glycine, the configuration of the 20 naturally occurring amino acids is designated as L (unless specified otherwise).

Proteins. Edited by Robert A. Meyers.
Copyright © 2007 Wiley-VCH Verlag GmbH & Co. KGaA, Weinheim
ISBN: 978-3-527-31608-3

First and Second Messengers

Peptide hormones are considered to be first messengers because they initiate a cellular/tissue response by interacting with cell surface receptors (ligand-specific binding molecules). Peptide (exemplifying a class of ligand) binding then leads to an amplified cascade of signal transduction events that increase or decrease production of specific intracellular (second) messengers. Second messengers include cyclic 3′,5′-adenosine monophosphate (cAMP), cyclic 3′,5′-guanosine monophosphate (cGMP), diacylglycerol (DAG), inositol triphosphate (IP3), and calcium ion (Ca^{2+}). Receptor-coupled effector enzymes or transport proteins (e.g. adenylate or guanylate cyclase, phospholipase, calcium channels) regulate intracellular levels of such second messengers.

Peptidomimetics and Nonpeptides

More complete synthetic tailoring of a peptide to yield analogs having essentially no naturally occurring amino acids or dipeptide substructure became popularized during the past two decades on the basis of the premise that such compounds might be better suited for drug development, being low in molecular mass, metabolically stable, and bioavailable by oral administration. Such radically modified peptides are generically referred to as *peptidomimetics* (including peptoids and pseudopeptides), and such compounds vary significantly in terms of the nature of the chemical template and design strategy used. "Nonpeptide" is used here to generically refer to molecules originally derived from screening of natural product sources or chemical files that competitively displace the natural peptide (or synthetic peptide analog) from the native receptor/enzyme target. In contrast to peptidomimetics, which are designed on the basis of chemical information of a peptide lead or its target receptor/enzyme active site, the discovery of nonpeptides from screening strategies may yield compounds that have no obvious chemical similarity to the native peptide (or synthetic peptide analog).

Primary Structure

The primary structure of a peptide is defined as the linear ordered linkage of constituent amino acids from the N terminus to the C terminus. "Primary sequence," although often used, represents incorrect terminology.

Recombinant Peptides

These include naturally occurring peptides as well as analogs containing naturally occurring amino acids. Such peptides may be prepared by biosynthetic methods based on overexpression of a cloned gene to yield high levels of the desired peptide (or precursor) in a host organism or cell line from which the peptide can subsequently be processed and/or purified to homogeneity.

Secondary Structure (Conformation)

The three-dimensional properties of peptides (as well as polypeptides and proteins) are determined by the individual compound's amino acid sequence, the spatial orientation of both its backbone and side chains, and its characteristic specific intramolecular hydrogen-bonding interactions. Peptide conformations include α-helix, β-sheet, and

β- and γ-turns. Secondary structures may be experimentally determined by spectroscopic and/or crystallographic techniques.

Synthetic Peptides
These peptides, prepared by total chemical synthesis or, in some cases, semisynthesis, provide the opportunity for sophisticated structural modification by incorporation of unusual amino acids, dipeptide replacements or surrogates, and cyclization (involving backbone and/or side-chain linkage), among other types of chemical modifications. Synthetic peptide analogs may be chemically modified in a modest or radical manner with respect to backbone and/or side-chain functionalization. Synthetic alterations of a peptide are exemplified by amide substitution with alkylamide, aminomethylene, ketomethylene, or other nonhydrolyzable isosteres. With the exception of the alkylamide modification, such partial transformation of the backbone amide substructure of a peptide results in analogs generally referred to as *pseudopeptides*. Other typical chemical modifications of synthetic peptides are alterations of the backbone hydrocarbon (CH) substructure by alkylation or stereoinversion.

> Peptides are molecules composed of amino acids that are chemically bonded in a specific manner by backbone amide linkage (peptide bond) and, in several instances, by side-chain linkage (e.g. disulfide bond). The discovery and chemical determination of biologically active peptides have become well established over the past few decades. Such work originally needed to overcome two particular challenges: (1) peptides generally exist in very low concentrations, making isolation extremely difficult for early scientists who lacked sophisticated technologies for purification and/or characterization of such molecules as have been more recently developed; and (2) since peptides often consist of a complex molecular framework, early scientists faced major difficulty in accurately determining the structure as well as in proving it by independent chemical synthesis (the latter work, especially, requires highly specific and reversible masking of chemically reactive amine, carboxylate, and other side-chain functionalities of each constituent amino acid).
>
> The rapid development of basic and biomedical research on peptides over the past half-century has been predicated on the premise that peptides are promising molecules for the discovery of new drugs that mimic or block their biological properties. Peptides play key roles in the physiology of all living organisms, as exemplified in humans, where peptides regulate growth, metabolism and development, reproduction, cardiovascular homeostasis, central nervous system (CNS) function, gastrointestinal (GI) function, immune system function, and a plethora of other biological activities.
>
> Over recent years, worldwide attention on peptide chemistry and biological research has been the general theme of international scientific conferences (e.g. biennial symposia of the American, Australian, Chinese, European, and Japanese peptide societies; the Peptide Gordon Conference) and scientific journals dedicated

to such research (e.g. *Journal of Peptide Research, Journal of Peptide Science, Peptide Science (Biopolymers), Peptides, Regulatory Peptides, Neuropeptides,* and *Letters in Peptide Science*). The relative importance of peptide research is evident by the number of Nobel prizes awarded to investigators for their contributions to the science, technology, and/or medicine of these important biomolecules.

Some peptide hormones (e.g. insulin, growth hormone, growth hormone–releasing factor) are being produced on a large scale by recombinant biosynthesis technology to treat specific human afflictions. Human diseases or disorders for which peptide-based drug therapy exists include osteoporosis (calcitonin), diabetes (insulin), prostate cancer and endometriosis (gonadotropin-releasing hormone), acromegaly and ulcers (somatostatin), diuresis and hypertension (vasopressin), hypoglycemia (glucagon), and hypothyroidism (thyrotropin-releasing hormone).

On the basis of the development of the aforementioned naturally occurring or synthetically modified peptide hormones, which are currently marketed as prescription medications, there is hope that such chemically altered peptides (hereinafter referred to as *pseudopeptides, peptidomimetics,* or *peptoids*, according to the type of chemical modification) may also prove useful in therapeutic intervention for still other life-threatening or disabling afflictions, such as immunological dysfunction (e.g. AIDS), cancer, neural dysfunction (e.g. Alzheimer's disease), and cardiovascular dysfunction (e.g. hypertension). Furthermore, the past success of peptide sweeteners (e.g. the dipeptide aspartame) may provide impetus to research and development strategies aimed at the discovery of new applications of peptides.

1
Peptide Chemistry

1.1
Peptide Structure and Three-dimensional Properties

Peptides are composed of two or more amino acids linked together by a peptide bond (Fig. 1). The number of amino acids determine whether the peptide is referred to as a *dipeptide, tripeptide, tetrapeptide,* and so forth. Of the 20 or so genetically coded amino acids, naturally occurring peptides are generally composed of L-stereoisomers (except for Gly). Peptides, polypeptides, or proteins may be structurally differentiated on the basis of the number of constitutive amino acids (see Table 1). For example, the term "peptide" is frequently used to categorize such compounds ranging from 2 to 20 amino acids. Similarly, "polypeptides" often designate compounds ranging from 20 to 50 amino acids, and "proteins" is used for such molecules of more than 50 amino acids (see Table 2). Although a more precise definition may take into consideration other factors (e.g. three-dimensional chemical complexity or biological function), there still exist a substantial amount of personal preference among experts in this field of research. In this article, the generic term "peptide" is used to designate all such chemical agents or messengers.

The precise linear arrangement, amino $\rightarrow$ carboxy (N $\rightarrow$ C) directionality, of amino acids in a peptide is referred to as

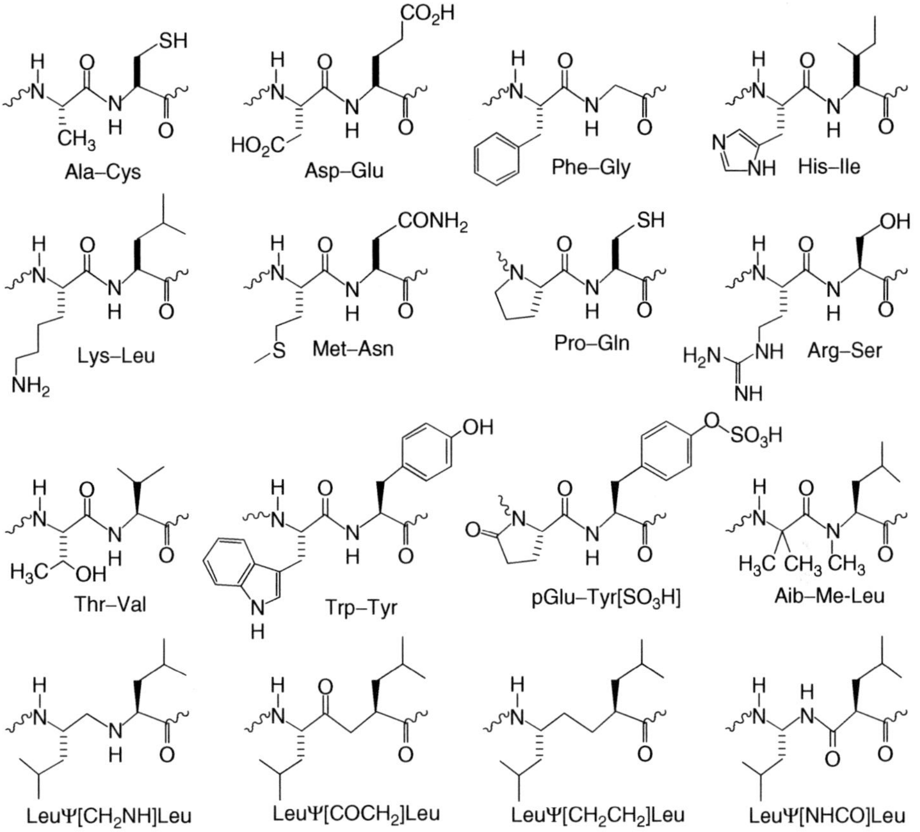

Fig. 1 Chemical structures of dipeptides composed of 20 genetically coded amino acids and some synthetically modified analogs.

its primary structure (Table 3). The three-dimensional folding of the peptide onto itself through covalent (e.g. S−S, disulfide) or noncovalent (e.g. hydrogen and/or ionic) bonding determines the secondary structure of the peptide. Furthermore, the covalent linkage between two or more cysteine residues through disulfide bond formation dictates the transformation of linear peptides to cyclic (or multicyclic) peptides. Such chemically defined restriction of the molecular flexibility of linear peptides does frequently lead toward higher ordered three-dimensional structural complexity and defined spatial arrangement of constitutive amino acid side chains (i.e. tertiary structure). For example, the secondary structure of neuropeptide-Y has been proposed to include a poly-Pro-type helix at the N terminus and an α-helix within its central fragment sequence and a conformationally flexible C terminus. Contributing to such a secondary structure is the tendency of peptides to adopt spatial orientations of both the backbone and side-chain functionalities to bring hydrophobic amino acids (e.g. Trp, Tyr, Phe, Leu, Ile, Val, Met) into proximity to form a hydrophobic surface. Similarly, residues of hydrophilic amino acids (e.g. Lys, Arg,

Tab. 1 Structure determinations of the first 100 peptide hormone, neurotransmitters, growth factors, and cytokines.

Year	Peptide
1951	Oxytocin
1953	Insulin
1954	Adrenocorticotropin, vasopressin
1956	β-Melanotropin, angiotensin
1957	α-Melanotropin, glucagon
1960	Bradykinin
1962	Eledoisin
1964	Gastrin, physalaemin
1965	β-Lipotropin
1966	Secretin, somatotropin
1968	Calcitonin, cholecystokinin, C-peptide
1969	Thyrotropin-releasing hormone, nerve growth factor, prolactin
1971	Chorionic gonadotropin, vasoactive intestinal peptide, gastric inhibitory peptide, lutropin, lutropin-releasing hormone, substance P, bombesin, neurophysin
1972	Follitropin, epidermal growth factor
1973	Motilin, neurotensin, somatostatin, tuftsin
1975	Enkephalin, pancreatic polypeptide, thymopoietin
1976	Delta sleep-inducing peptide, β-endorphin, insulin-like growth factor
1977	Relaxin, granuloliberin
1978	Fibroblast growth factor
1979	Gastrin-releasing peptide, glicentin, α/β-neoendorphin, kyotorphin, thymosin
1980	Interferon-α/β
1981	Corticotropin-releasing factor, dynorphin, oxytomodulin, peptide-HI, calcitonin-gene-related peptide, erythropoietin, sauvagine
1982	Growth hormone-releasing factor, neuropeptide-Y, peptide YY, rimorphin, katacalcin, urotensin, interferon-γ
1983	Neuromedin B, neuromedin K, galanin, melanin-concentrating hormone, atriopeptide, platelet-derived growth factor, transforming growth factor-α/β, interleukin 2
1984	Tumor necrosis factor-α/β, cerebellin, interleukin-3
1985	Valosin, granulocyte/macrophage colony-stimulating factor, interleukin-1α/β, neuropeptide K, neutrophil peptide
1986	Cyclinopeptin, leukopyrokinin, leucokinin, leukomyosuppressin, pancreastatin, galanin-associated peptide, interleukin-6
1987	Follicular gonadotropin-releasing peptide, cardioactive peptide, interleukin-4, interleukin-5, interleukin-8
1988	Corticostatin, interleukin 7, neuropeptide-Y, γ-endothelin

Source: Adapted from Eberle, A.N. (1991) *Chimia* **45**, 145.

Glu, Asp, Thr, Ser, Gln, Asn, His) may be distributed within such three-dimensional structures to form a hydrophilic surface that may provide a high degree of solvation by water. In the case of peptides such as growth hormone–releasing factor (GRF) and glucagon, an α-helical-type secondary structure gives rise to amphiphilicity (one side-chain surface being hydrophobic and the other hydrophilic). Furthermore, peptides may be composed of two amino acid chains, or subunits, forming a so-called

Tab. 2 Physiological actions of some regulatory peptides.

Peptide (abbreviation)	Amino acids[a]	Principal physiological actions[b]
Angiotensin II (AII)	8	↑Vasoconstriction, ↑aldosterone secretion, ↑dipsogenesis (i.e. thirst)
Bradykinin	9	↑Smooth muscle contraction, ↑vasodilation, ↑inflammatory algesia
Calcitonin	32	↓Blood Ca^{2+}
Cholecystokinin (CCK)	8	↑Pancreatic enzyme and electrolyte secretion, ↑gallbladder contraction, ↑satiety, ↑secretion of insulin and glucagons
Adrenocorticotropin (ACTH)	39	↑Adrenal steroidogenesis
Corticotropin-releasing hormone (CRF)	41	↑Corticotropin secretion
Dynorphin	17	↑Sedative analgesia, ↑feeding behavior
β-Endorphin	31	↑Opiate-like activity, ↑central analgesia, ↑respiratory depression, ↑euphoria
Leu-enkephalin	5	↑Spinal analgesia, ↑emotional effects
Epidermal growth factor (EGF)	53	↑Epidermal growth and keratinization,
Follicle-stimulating hormone (FSH)	230	↑Spermatogenesis in males, ↑ovarian follicle growth and estradiol synthesis in females
Gastrin-17	17	↑Pepsin, gastric acid–HCl and pancreatic enzyme secretion, ↑intestine smooth muscle contraction
Glucagon	29	↑Blood glucose, ↑gluconeogenesis, ↑Glycogenolysis
Gonadotropin-releasing hormone (GnRH)	10	↑Luteinizing hormone and follicle-stimulating hormone secretion
Insulin	51	↓Blood glucose, ↑gluconeogenesis, ↑glycogenesis
Luteinizing hormone (LH)	210	↑Testicular androgen synthesis in males, ↑Ovarian estradiol and progesterone synthesis in females
α-Melanotropin (MSH)	13	↑Melanin biosynthesis
Motilin	22	↑Gastrointestinal villous motility

(continued overleaf)

Tab. 2 (Continued)

Peptide (abbreviation)	Amino acids[a]	Principal physiological actions[b]
Nerve growth factor (NGF)	118	↑Sympathetic neurite development
Oxytocin	9	↑Milk ejection, ↑uterine contraction
Parathyroid hormone (PTH)	84	↑Blood calcium
Prolactin	199	↑Milk synthesis, ↑mammary gland development
Secretin	27	↑Pancreatic secretion of H_2O, bicarbonate, electrolyte; ↑stomach secretion of gastrin
Somatomedins	50–80	↑Peripheral nervous system growth and development
Somatostatin-14 (SRIF)	14	↓Somatotropin, thyrotropin, parathyroid hormone, calcitonin, renin, and gastric acid secretion
Somatotropin (growth hormone, GH)	191	↑Hepatic somatomedin synthesis
Growth hormone-releasing factor (GRF)	44	↑Somatotropin secretion
Substance P (NK1)	11	↑Peripheral nervous system pain transmission, ↓central nervous system pain transmission
α_1-Thymosin	28	↑Lymphocyte proliferation and differentiation
Thyrotropin (TSH)	201	↑Thyroid hormone T3 and T4 synthesis and secretion
Thyrotropin-releasing hormone (TRH)	3	↑Thyrotropin and prolactin secretion
Vasoactive intestinal peptide (VIP)	28	↑Vasodilation, ↑bronchodilation, ↑glucagon and insulin secretion, ↑glycogenolysis, ↑lipolysis
Vasopressin	9	↑Renal H_2O adsorption, ↑vasoconstriction

[a] Prevalent form of peptide in humans.
[b] Other biological activities have been described (only principal actions are summarized here).
Source: Adapted from Sawyer, T.K., Smith, C.W. (1990) in: Dulbecco, R. (Ed.), *Encyclopedia of Human Biology*, Vol. 5, Academic Press, Harcourt Brace Jovanovich Publishers, San Diego, CA, pp. 725–735.

Tab. 3 Primary structures of some naturally occurring peptides and synthetic analogs thereof.

Peptide	Primary structure[a]
Aspartame	Asp^1-Phe^2-OMe
Thyrotropin-releasing hormone	<Glu^1-His-Pro^3-NH$_2$
FMRF-amide	Phe^1-Met-Arg-Phe^4-NH$_2$
Enkephalin (Met)	Tyr^1-Gly-Gly-Phe-Met^5
DPDPE	Tyr1-D-Pen-Gly-Phe-D-Pen^5 *Pen, penicillamine
BQ-123	cyclo(Leu-D-Trp-D-Asp-Pro-D-Val)
L-365,209	cyclo(Pro-D-Phe-Ile-D-Dhp-Dhp-D-MePhe) *Dhp, dehydropiperazyl
GHRP	His^1-D-Trp-Ala-Trp-D-Phe-Lys^6-NH$_2$
Ebiratide	Met(O$_2$)^1-Glu-His-Phe-D-Lys-Trp^6-NH-(CH$_2$)$_8$-NH$_2$ *Met(O$_2$), methionine sulfone
Cholecystokinin 8	Asp-Tyr[SO3H]-Met-Gly-Trp-Met-Asp-Phe^8-NH$_2$
Angiotensin II	Asp1-Arg-Val-Tyr-Ile-His-Pro-Phe8
Sandostatin^{TM}	D-Phe^1-Cys-Phe-D-Trp-Lys-Thr-Cys-Thr^8-ol *Thr-ol, threoninol
Oxytocin	Cys^1-Tyr-Ile-Gln-Asn-Cys-Pro-Leu-Gly^9-NH$_2$
Vasopressin	Cys^1-Tyr-Phe-Gln-Asn-Cys-Pro-Arg-Gly^9-NH$_2$
Desmopressin^{TM}	Mpa^1-Tyr-Phe-Gln-Asn-Cys-Pro-D-Arg-Gly^9-NH$_2$ *Mpa, mercaptoproprionic acid
Bradykinin	Arg^1-Pro-Pro-Gly-Phe-Ser-Pro-Phe-Arg^9

(continued overleaf)

Tab. 3 (Continued)

Peptide	Primary structure[a]
HOE-140	Arg[0]-D-Arg[1]-Lys-Arg-Hyp-Thi-Ser-D-Tic-Oic-Arg[9] *Hyp, 4-hydroxyproline; Thi, thienylalanine; Oic, octahydroindole-2-carboxylic acid
Lutropin-releasing hormone	<Glu[1]-His-Trp-Ser-Tyr-Gly-Leu-Arg-Pro-Gly[10]-NH$_2$
Buserelin	<Glu[1]-His-Trp-Ser-Tyr-D-Ser(tBu)-Leu-Arg-Pro[9]-NH-Et
MDL-28050	Suc-Tyr[1]-Glu-Pro-Pro-Glu-Glu-Tyr-Ala-Cha-Gln[10] *Suc, succinyl; Cha, cyclohexylalanine
Substance P	Arg[1]-Pro-Lys-Pro-Gln-Phe-Phe-Gly-Leu-Met[11]-NH$_2$
Cyclosporin A	cyclo(MeLeu-MeLeu-MeThr[4R-4-(E-2-butenyl)-4-methyl]-Abu-Sar-MeLeu-Leu-MeLeu-Ala-D-Ala-MeLeu)
α-Melanotropin	Ac-Ser[1]-Tyr-Ser-Met-Glu-His-Phe-Arg-Trp-Gly-Lys-Pro-Val[13]-NH$_2$
Melanotan-I (NDP-MSH)	Ac-Ser[1]-Tyr-Ser-Nle-Glu-His-D-Phe-Arg-Trp-Gly-Lys-Pro-Val[13]-NH$_2$ *Nle, norleucine
Neurotensin	Glu[1]-Leu-Tyr-Glu-Asn-Lys-Pro-Arg-Arg-Pro-Tyr-Ile-Leu[13]
Somatostatin	Ala[1]-Gly-Cys-Lys-Asn-Phe-Phe-Trp-Lys-Thr-Phe-Thr-Ser-Cys[14]
Endothelin	Cys[1]-Ser-Cys-Ser-Ser-Leu-Met-Asp-Lys-Glu-Cys-Val-Tyr-Phe-Cys-His-Leu-Asp-Ile-Ile-Trp[21]
Glucagon	His[1]-Ser-Gln-Gly-Thr-Phe-Thr-Ser-Asp-Tyr-Ser-Lys-Tyr-Leu-Asp-Ser-Arg-Arg-Ala-Gln-Asp-Phe-Val-Gln-Trp-Leu-Met-Asp-Thr[29]
Galanin	Gly[1]-Trp-Thr-Leu-Asn-Ser-Ala-Gly-Tyr-Leu-Leu-Gly-Pro-His-Ala-Ile-Asp-Asn-His-Arg-Ser-Phe-His-Asp-Lys-Tyr-Gly-Leu-Ala[29]-NH$_2$

Calcitonin	Cys¹-Gly-Asn-Leu-Ser-Thr-Cys-Met-Leu-Gly-Thr-Tyr-Thr-Gln-Asp-Phe-Asn-Lys-Phe-His-Thr-Phe-Pro-Gln-Thr-Ala-Ile-Gly-Val-Gly-Ala-Pro³³-NH₂
Neuropeptide-Y	Tyr¹-Pro-Ser-Lys-Pro-Asp-Asn-Pro-Gly-Glu-Asp-Ala-Pro-Ala-Glu-Asp-Leu-Ala-Arg-Tyr-Tyr-Ser-Ala-Leu-Arg-His-Tyr-Ile-Asn-Leu-Met-Thr-Arg-Gln-Arg-Tyr³⁶-NH₂
Adrenocorticotropin	Ser¹-Tyr-Ser-Met-Glu-His-Phe-Arg-Trp-Gly-Lys-Pro-Val-Gly-Lys-Lys-Arg-Arg-Pro-Val-Lys-Val-Tyr-Pro-Asn-Gly-Ala-Glu-Asp-Glu-Ser-Ala-Glu-Ala-Phe-Pro-Leu-Glu-Phe³⁹
Corticotropin-releasing factor	Ser¹-Glu-Glu-Pro-Pro-Ile-Ser-Leu-Asp-Leu-Thr-Phe-His-Leu-Leu-Arg-Glu-Val-Leu-Glu-Met-Ala-Arg-Ala-Glu-Gln-Leu-Ala-Gln-Gln-Ala-His-Ser-Asn-Arg-Lys-Leu-Leu-Asp-Ile-Ile⁴¹-NH₂
Growth hormone releasing factor	Tyr¹-Ala-Asp-Ala-Ile-Phe-Thr-Asn-Ser-Tyr-Arg-Lys-Val-Leu-Gly-Gln-Leu-Ser-Ala-Arg-Lys-Leu-Leu-Gln-Asp-Ile-Met-Ser-Arg-Gln-Gln-Gly-Glu-Ser-Asn-Gln-Glu-Arg-Gly-Ala-Arg-Ala-Arg-Leu⁴⁴-NH₂
Insulin	Gly¹-Ile-Val-Glu-Gln-Cys-Cys-Thr-Ser-Ile-Cys-Ser-Leu-Tyr-Gln-Leu-Glu-Asn-Tyr-Cys-Asn-(A-chain above; B-chain below) Phe¹-Val-Asn-Gln-His-Leu-Cys-Gly-Ser-His-Leu-Val-Glu-Ala-Leu-Tyr-Leu-Val-Cys-Gly-Glu-Arg-Gly-Phe-Phe-Tyr-Thr-Pro-Lys-Thr³¹
Parathyroid hormone	Ser¹-Val-Ser-Glu-Ile-Gln-Leu-Met-His-Asn-Leu-Gly-Lys-His-Leu-Asn-Ser-Met-Glu-Arg-Val-Glu-Trp-Leu-Arg-Lys-Leu-Gln-Asp-Val-His-Asn-Phe-Val-Ala-Leu-Gly-Ala-Pro-Leu-Ala-Pro-Arg-Asp-Ala-Gly-Ser-Gln-Arg-Pro-Arg-Lys-Lys-Glu-Asp-Asn-Val-Leu-Val-Glu-Ser-His-Glu-Lys-Ser-Leu-Gly-Glu-Ala-Asp-Lys-Ala-Asp-Val-Asp-Val-Leu-Thr-Lys-Ala-Lys-Ser-Gln⁸⁴

[a]Asterisks indicate abbreviations commonly found in the literature.

dimeric structure composed of either homologous or heterologous subunits. The individual subunits may be either covalently (e.g. disulfide) or noncovalently (e.g. ionic or hydrophobic) bonded together. Both interchain and intrachain disulfide bonds may be present within a peptide (e.g. insulin, Table 3). In some cases, such a complex chemical species may be required to establish the biologically active form of a peptide (e.g. nerve growth factor dimer).

The three-dimensional substructure of peptides may be further described in terms of the torsion angles between the backbone amine nitrogen (N-α), backbone carbonyl carbon (C'), backbone hydrocarbon (C-α), and side-chain hydrocarbon functionalization (e.g. C-β, C-γ, C-δ, C-ε of Lys) as depending on the amino acid sequence (Fig. 2). The torsion angle nomenclature is exemplified by the following cases: Ψ, N-α−C-α−C'−N-α; ω, C-α−C'−N-α−C-α; Φ, C'−N-α−C-α−C'; χ_1 N-α−C-α−C-β−X (where X is not hydrogen); χ_2, C-α−C-β−C-γ−X. A Ramachandran plot of Ψ versus Φ for peptides possessing intrinsic secondary structure has further indicated that particular combinations of torsion angles for a helical, reverse-turn, or extended conformation do exist in a predominant fashion. For amide bond torsion angle (ω) the trans geometry is preferred for most dipeptide substructures; when Pro is the C-terminal partner, however, the cis geometry is possible. It is noted that the three-dimensional structural flexibility is directly related to covalent and/or noncovalent bonding interactions within the amino acid sequence of a particular peptide, and synthetic modifications such as N-α alkylation and C-α or C-β alkylation effect significant conformational constraints locally as related to backbone and/or side-chain torsion angles (e.g. conformationally modified Phe analogs, Fig. 2). Finally, it is noted that backbone amide replacements may be defined in terms of a nomenclature system, in which the functional group substitution of the dipeptide substructure is identified as a Ψ[surrogate] (e.g. substitution of the aminomethylene surrogate in Leu-Val would be described as LeuΨ[CH$_2$NH]Val).

1.2
Peptide Chemical Synthesis and Molecular Diversity

The chemical synthesis of peptides and peptide libraries has advanced tremendously over the past century relative to the pioneering efforts of many scientists (including, in some cases, Nobel laureates) such as George Barany, Max Bergmann, Miklos Bodanszky, Louis Carpino, Theodor Curtius, Emil Fischer, Murray Goodman, Mario Geysen, Richard Houghten, Victor Hruby, Stephen Kent, R. Bruce Merrifield, Stanford Moore, Shumpei Sakakibara, William Stein, James Tam, Claudio Toniolo, Vincent du Vigneaud, and Theodor Wieland. In retrospect, the field of peptide chemistry is enriched with a phenomenal record of achievements involving the solution- and solid-phase synthesis of complex peptides, purification and analytical characterization of peptides, and more recently, the development and implementation of specialized technologies that expedite peptide-based drug discovery in diverse areas of research (e.g. immunology; cardiovascular and neurological physiology). The scope of peptide synthesis pervades both naturally occurring peptides and their analogs, as exemplified by a variety of possible side-chain and/or backbone substitutions that have been developed to exploit conformational modeling hypotheses (local or global

15 Synthetic Peptides: Chemistry, Biology, and Drug Design | 465

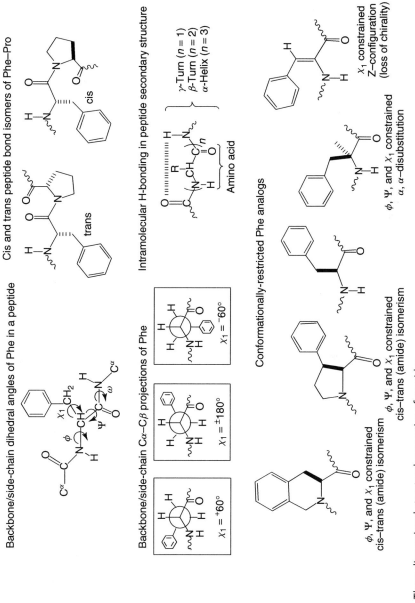

Fig. 2 Three-dimensional structural properties of peptides.

three-dimensional structural constraints), molecular recognition and mechanistic properties at biological targets (receptors, enzymes, antibodies, nucleic acids), and/or metabolism by processing or degrading peptidases. The chemical synthesis of peptides has become a rapidly developed scientific art owing to such factors as automated solid-phase methods, orthogonal protection–cleavage strategies, and a sophisticated database of literature that details possible reagents and potential side reactions to further optimize the task of preparing simple and/or complex peptides. The synthesis of large-sized and/or multicyclic peptides such as interleukin-3, conotoxin, endothelin, transforming growth factor, corticotropin-releasing factor, and calcitonin has been achieved in many research laboratories, and contemporary peptide chemistry methods have been successfully applied to the synthesis of yet more complex macromolecular targets (e.g. HIV protease dimer). Most notably, the chemical synthesis of peptides has expanded to the preparation of combinatorial peptide libraries for which high volume target screening (e.g. receptor binding or enzyme inhibition) is performed to identify compound(s) that possess biological activity. This generally requires the interface of two specialized technologies, namely, synthesis of combinatorial peptide libraries, and their target screening. As opposed to focused analysis of the native peptide with respect to biological activity, such techniques provide tremendous molecular diversity of potential peptide lead compounds to study molecular recognition and mechanistic aspects of peptide interaction with protein (receptors, enzymes), nucleic acid, or carbohydrate targets. Similarly, phage-based peptide libraries have been developed for such applications. The principal advantage of the synthetic peptide or peptidomimetic libraries is the extraordinary molecular diversity of amino acid building blocks that may include unnatural amino acids, pseudodipeptides, and nonpeptide template intermediates.

2
Peptide Biology

2.1
Peptide Biosynthesis and Metabolism

Peptides are biosynthesized on ribosomes, where their specific amino acid sequence is determined (translated) by a specific messenger RNA sequence (codon triplet codes for an amino acid). The nucleotide sequences of the RNA are dictated (transcribed) from specific chromosomal deoxyribonucleotide sequences (DNA genes). Cellular biosynthesis of peptides is believed to proceed very specifically, with error rates of less than 1 in 104 amino acids at a rate of 20 amino acids/ second. The nascent peptides are subsequently released and transported into the cisternae of the rough endoplasmic reticulum and then to the Golgi elements, where they may be posttranslationally modified (e.g. sulfated, glycosylated). Vesicles containing the peptide are pinched off the terminal cisternae of the Golgi apparatus, and then they are targeted to intracellular organelles, the plasma membrane, or are secreted into the extracellular space. In some peptides, the tyrosine residues may be posttranslationally sulfated (Tyr[SO_3H], Fig. 1) as exemplified by the GI peptide hormones cholecystokinin-8 and gastrin. If glutamic acid is present, it may be cyclized into a pyroglutamic acid derivative (<Glu, Fig. 1) as exemplified by thyrotropin-releasing

factor, a hypothalamic peptide. Peptides vary with regard to whether their C terminus is a free carboxylic acid (i.e. $-CO_2H$) or carboxamidated (i.e. $-CONH_2$) and whether the N terminus is a free amine (H_2N-), acetylated (CH_3CONH-), formylated ($HCONH-$), or bearing some other modified amine group (Fig. 1). Some peptides are glycosylated; that is, they are conjugated to one or more carbohydrate groups (e.g. sialic acid). These various examples of posttranslational functionalization are often directly linked to the biological activities of the parent peptides as related to the "bioactive" conformation and/or "message" sequence of the peptide. Such posttranslational "tag"î also serve as biomolecular zip codes to target the peptide to a particular cellular compartment. Finally, it is important to note that the amino acid sequence itself may differ to varying degrees among species. This information is useful in determining the evolutionary relatedness between species. Two or more isoforms of a peptide may also exist within an individual species of animal. These isoforms, although chemically similar, may differ either in primary structure length (e.g. cholecystokinin family of CCK-51, CCK-39, CCK-33, CCK-12, and CCK-8 peptides), or in other chemical aspects (e.g. site-specific amino acid substitutions and/or side-chain modification by glycosylation, sulfation, etc.). Gene duplication followed by single or multiple nucleic acid mutations most likely account for such variations in peptide chemical structure and subsequent evolution of various families of peptides (e.g. the neurohypophyseal hormones such as oxytocin and vasopressin).

Some peptides (e.g. oxytocin, vasopressin) require extensive posttranslational processing because they are not coded directly by DNA. Frequently, N- and/or C-terminally extended amino acid sequences are initially biosynthesized. These propeptides may then be packaged within secretory vesicles along with proteolytic enzymes. Both endopeptidases and exopeptidases (Table 4) can contribute to appropriate processing of the inactive precursor peptide to yield the active peptide. Therefore, peptides may be derived indirectly by way of a propeptide, which itself may be derived from a prepropeptide. For example, proopiomelanocortin is a large inactive peptide (MW 28 500) that is cleaved by specific peptidases into several active peptides, including adrenocorticotropin

Tab. 4 Some examples of endopeptidases and exopeptidases.

Aspartyl class	Cysteinyl class	Metallo class	Serinyl class
Pepsin	Cathepsin B	Peptidyl dipeptidase A	Thrombin
Renin	Cathepsin H	Collagenase	Trypsin
Cathepsin D	Cathepsin L	Endopeptidase 24.11	Chymotrypsin A
Cathepsin E	Cathepsin S	Aminopeptidase M	Elastase
HIV protease	Cathepsin M	Carboxypeptidase A	Kallikrein
	Cathepsin N	Stromolysin	Cathepsin A
	Cathepsin T	Gelatinase A	Cathepsin G
	Calpains	Gelatinase B	Cathepsin R
	Papain		Tissue plasminogen activator (TPA)
	Proline endopeptidase		

(ACTH), β-endorphin, α-melanotropin, and β-melanotropin. Some large plasma proteins also serve as propeptides for peptide production. Renin, an enzyme released from the juxtaglomerular cells of the kidneys, acts on a liver-borne substrate protein (angiotensinogen) to convert it to the decapeptide angiotensin I, which is then processed by another enzyme to yield the active peptide hormone angiotensin II. Other precursor plasma proteins, the kininogens, are similarly converted by specific serine proteases (kallikreins) to kinins, such as bradykinin, which is an important peptide hormone in regulating blood flow in certain vascular beds. Bradykinin and angiotensin II are examples of peptide hormones that are released from liver cells as larger propeptides to be converted into active hormonal peptides within the blood. In contrast, peptidases may also inactivate peptides by splitting the molecules at specific internal peptide bonds. Exopeptidases, both carboxypeptidases and aminopeptidases, cleave off the C-terminal or N-terminal amino acids, respectively. Some peptides may be inactivated by simple deamidation at the C-terminal (if amidated) end of the molecule. Endopeptidase cleavage is generally quite specific, and both exopeptidases and endopeptidases (Table 4) are well characterized in terms of mechanistic properties and substrate specificity. Most importantly, it is well known that the clinical use of some peptides has been limited significantly by their short plasma half-life, which may in part be related to their lability to peptidases. Therefore, knowledge of biodegradation mechanisms has been considered to be an important research objective to facilitate the design of synthetic peptide analogs that may exhibit sustained biological activities (see Sect. 3.2). Some peptides may also undergo inactivation by other chemical transformations such as that exemplified by insulin, in which its cystine-bridged heterodimeric structure is liable to cleavage by means of reduction of the interchain disulfide bonds by the enzyme insulinase. Considerable effort has been devoted to the objective of designing synthetic peptide analogs (including pseudopeptides) and peptidomimetics, which are structurally altered to compromise or eliminate biological cleavage–inactivation by peptidases. Alternatively, there has also been considerable effort devoted to directly inhibiting either processing (cleavage activation) or degrading peptidases to modify the generation or the lifetime of a particular endogenous (or exogenous) peptide, respectively. In particular, the historically important discoveries of peptide and, subsequently, peptidomimetic inhibitors of angiotensin-converting enzyme (ACE) (i.e. teprotide and captopril: see Sect. 3.2, Fig. 4) deserve particular mention because they provided tremendous incentive to rational drug design and natural product lead-finding technologies. Such peptidase targets exemplify processing enzymes, which are required for the generation of endogenous peptides, and inhibition of such peptidases compromises the agonist activity of a particular peptide.

2.2
Peptide Biological Functions and Mechanisms of Action

Peptides serve diverse physiological roles, and their biological functions and mechanisms of action have been the subject of intensive molecular and cell biology, biochemistry and pharmacology research, as exemplified by the pioneering efforts of many scientists (including, in some cases, Nobel laureates) such as Frederick

Banting, Stanley Cohen, Pedro Cuatrecasas, Charles Deber, Lila Gierasch, Alfred Gilman, Roger Guillemin, Mac Hadley, Rita Levi-Montalcini, Frank Porreca, Jean Rivier, Martin Rodbell, Alan Saltiel, Andrew Schally, Solomon Snyder, Donald Steiner, Wylie Vale, and Henry Yamamura. Glutathione is a tripeptide that serves as a cofactor for one or more enzyme systems as related to its oxidation–reduction chemical properties. Ion-binding peptides also constitute a category of transporting molecules (ionophores) as exemplified by valinomycin (Table 3). Ion channel–binding peptides have been identified from natural sources such as venoms, and such peptides may selectively regulate sodium (Na^+), potassium (K^+), or calcium ion (Ca^{2+}) transport. Furthermore, the relationship between calcium ion and peptides extends to the finding that some peptide hormones (α-melanotropin and adrenocorticotropin) apparently require this divalent metal ion to bind and/or effect receptor-mediated signal transduction. Antibiotic peptides include gramicidin (Table 3), actinomycin D, bacitracin, penicillin, and defensins. Peptide toxins include agatoxin, α-bungarotoxin, and ricin. Finally, it is well recognized that a vast number of peptide chemical messengers exist, which manifest their biological activities as regulatory peptide hormones, neurotransmitters, growth factors, and cytokines.

Hormones, which include peptides, steroids, catecholamines, prostaglandins, and related naturally occurring compounds, are chemical messengers released by one cell to act on one or more other cell types to elicit a physiological response. The largest category of hormones consists of peptides, and the biological properties of such peptide hormones is extensive, with well-established examples related to endocrine, cardiovascular, neural, immune, and gastrointestinal system regulation. In terms of functional properties, peptide hormones regulate differentiation, growth, reproduction, blood pressure, glucose homeostasis, and behavior, in addition to performing many other regulatory activities. Peptide hormones normally exist at very low concentrations ($10^{-12}-10^{-9}$ M) in bodily fluids or tissues, in which they persist for short periods of time (1–30 min), since they are rapidly inactivated by peptidases and/or excreted (i.e. cleared) from the body. The transitory actions of peptide hormones are well suited for continuous regulation (i.e. homeostatis) of various physiological systems.

Some of the examples of peptide hormones and their particular regulatory functions summarized earlier (Table 2) are amplified briefly here. Thyroid-stimulating hormone (TSH), which acts on the thyroid gland, is a glycoprotein, as are lutropin and follitropin, which act on the gonads (testes and ovaries). Somatotropin (growth hormone, GH) is necessary for normal growth, and defective or excessive secretion can lead to dwarfism or acromegaly, respectively. Prolactin also possesses growth-promoting activity, most specifically, mammary growth and milk production. All the pituitary peptide hormones are individually controlled by peptide factors (hormones) of hypothalamic origin. These hypophysiotropic peptides either stimulate or inhibit pituitary hormone secretion. The peptide hormones glucagon and insulin, both of pancreatic origin, are responsible for elevating or depressing glucose levels through their actions on the liver and other organs. Glucose homeostasis is therefore very tightly regulated by these two counterregulatory peptides. Similarly, parathyroid hormone (PTH) and calcitonin, derived from the parathyroid and thyroid glands respectively, act in a counterregulatory manner to maintain serum Ca^{2+} levels. Specifically, PTH elevates and calcitonin lowers blood Ca^{2+} levels. The GI peptide hormones regulate the movement, metabolism, digestion, and intestinal absorption of metabolic substrates that are essential for life. Peptide hormones regulate GI smooth muscle motility (and therefore substrate transport) and the pancreatic enzyme secretions that degrade metabolic substrates into absorptive substrates. Gastrin, cholecystokinin (CCK), and vasoactive intestinal peptide (VIP) are examples of a dozen or so recognized peptide hormones that regulate GI, pancreatic, and other processes. A large number of peptide growth factors have been discovered, which play a role in normal as well as neoplastic (cancerous) growth. Factors such as epidermal growth factor (EGF), nerve growth factor (NGF), transforming growth factors, somatomedins, and erythropoietin each affect certain cell types to regulate their normal growth and proliferation. Excess secretion of certain of these peptide growth factors may correlate to neoplastic transformation, tumor growth, and/or metastasis. A most important discovery was that peptides produced by the brain exhibit analgesic activity, as do the opiate drugs (e.g. morphine). This observation led to the discovery of opiate receptors and the demonstration that endogenous morphine-like substances (e.g. Met-enkephalin, β-endorphin, dynorphin) interact with the same receptors to affect analgesia.

Peptide hormones, probably without exception, mediate their initial actions at the level of the cell plasma membrane (phospholipid bilayer). Peptide hormones ("first messengers") interact at the cell surface with protein (or glycoprotein) macromolecular receptors, which bind to the peptide ligand in a very specific manner. This results in activation (signal transduction) of proximally located enzymes or transport proteins, which may be modified in a positive or negative manner to produce within the cell, a "second messenger" (e.g. cAMP, cGMP, IP_3, DAG, Ca^{2+}/K^+; Table 5). Alternatively, the peptide–receptor complex may result in kinase-like activities to phosphorylate other cellular signaling proteins such as those exemplified by growth factor receptor kinases. Many peptide hormones and neurotransmitters effect receptor-mediated stimulation or inhibition of adenylate cyclase via intermediary signaling heterotrimeric proteins known as *G proteins*. Recognition domains of such G protein–coupled peptide receptors (typically, a seven transmembrane–spanning helical protein superfamily) exist for both the peptide ligand and the specific G protein. Three-dimensional models are being developed to provide further insight into the structure–activity relationships of these G protein–coupled receptors as related to protein-coupled receptors as related to probing the molecular basis of peptidergic receptor-mediated biological signaling by site-directed mutagenesis. In cases of oncogene activation, the uncontrolled production or mutation of cellular proteins can mimic, for example, a signaling pathway mediated by a growth factor peptide–receptor complex, which can lead to a pathophysiological (malignant carcinogenic) transformation of that particular cell. Examples of oncogenes that can lead to disregulated signal transduction pathways include peptide growth factors (e.g. *sis*), receptors (e.g. *erb B*), transduction proteins (e.g. *ras*), and transcription factors (e.g. *fos, myc*). Binding

Tab. 5 Peptidergic receptor subtypes and signal transduction pathways.

Peptide	Receptor subtype	Signal transduction second messenger	Peptide//Peptidomimetic/Nonpeptide (Agonists or antagonists)
Angiotensin II	AT_1	↓cAMP, ↑IP_3/DAG	DuP753 (nonpeptide antagonist)
	AT_2	Not determined	PD123177 (nonpeptide antagonist)
Bradykinin	B_1	Not determined	BK_{1-8} (peptide agonist)
			[Leu^8]BK_{1-8} (peptide antagonist)
	B_2	↑IP_3/DAG	[Phe^8(CH_2NH)Arg^9]BK (pseudopeptide agonist)
			d-Arg[Hyp^3, Thi^5, $\underline{D}$-Tic^7, Oic^8]BK (peptide antagonist)
Cholecystokinin	CCK_A	↑IP_3/DAG	A-71623 (peptide agonist)
			Devazepide (nonpeptide antagonist)
	CCK_B	Not determined	Gastrin (peptide agonist)
			CI-988 (peptidomimetic antagonist)
Endothelin	ET_A	↑IP_3/DAG	ET-1 (peptide agonist; nonselective)
			BQ-123 (peptide antagonist)
	ET_B	↑IP_3/DAG	Sarafotoxin S6c (peptide agonist)
Neuropeptide-Y	Y_1	↓cAMP	[Leu^{31}, Pro^{34}]NPY (peptide agonist)
	Y_2	Not determined	NPY_{13-36} (peptide agonist)
Met-enkephalin	δ-Opioid	↓cAMP, ↑K^+ channel	DAMGO (peptide agonist)
			CTOP (peptide antagonist)
β-Endorphin	μ-Opioid	↓cAMP, ↑K^+ channel	DPDPE (peptide agonist)
			ICI-174864 (peptide antagonist)
Dynorphin	κ-Opioid	↓Ca^{2+} channel	U-69593 (nonpeptide agonist)
			Norbinaltorphimine (nonpeptide antagonist)
Substance P	NK_1	↑IP_3/DAG	[Pro^9]SP (peptide agonist)
			CP-96345 (nonpeptide antagonist)
Substance K (NK_A)	NK_2	↑IP_3/DAG	[Lys^5, $MeLeu^9$, Nle^{10}]NKA (peptide agonist)
			L-659877 (peptide antagonist)
Neurokinin-B (NK_B)	NK_3	↑IP_3/DAG	Senktide (peptide agonist)
			[Trp^7, β-Ala^8]NKB_{4-10} (peptide antagonist)
Vasopressin	V_{1A}	↑IP_3/DAG	d(CH_2)$_5$[Tyr(Me)$_2$]AVP (peptide antagonist)
	V_{1B}	↑IP_3/DAG	–
	V_2	↑cAMP	Desamino[$\underline{D}$-Arg^8]AVP (peptide agonist)
			d(CH_2)$_5$[$\underline{D}$-Ile^2, Ile^4]AVP (peptide antagonist)

Source: Adapted from Watson and Abbott, *Trends in Pharmacological Sciences* (TiPS Receptor Nomenclature Supplement – January 1992).

of a peptide to a receptor (or enzyme) results from a multiplicity of noncovalent intermolecular interactions, and these events are fundamentally related to structure–activity studies focused on a plethora of biologically active peptides. Binding is considered to precede transformation of the receptor (or enzyme) from its inactive to active state by molecular mechanisms often not understood. In many cases, it has been determined which substructural features of a peptide are required for binding (i.e. the "address" sequence) versus signaling/activation (i.e. the "message" sequence). In the case of adrenocorticotropin, for example, the central 5–24 amino acid sequence of the native peptide (Table 3) is capable of effecting full steroidogenic agonism, whereas further N-terminal truncation (i.e. $ACTH_{8-24}$ or $ACTH_{11-24}$) gives rise to partial agonism or antagonism. Similarly, corticotropin-releasing factor (CRF) appears to require a significant portion of its C-terminal sequence for binding (i.e. residues 12–41), whereas biological activity requires N-terminal extension (i.e. residues 6–41). The three-dimensional structural features of a peptide that provide the essential functional group ensemble for molecular recognition (binding) at a target receptor or enzyme may be composed of backbone and/or side-chain components, and such a three-dimensional substructure of the peptide has been referred to as the *pharmacophore*. Development of both intuitive peptide pharmacophore models and models derived experimentally (biophysical and computational chemistry) is providing an opportunity to explore the molecular recognition and mechanistic properties of peptide interactions with macromolecular targets. Such studies are of particular significance to peptidomimetic drug design.

3
Peptide Drug Design

A large number of pharmaceutical companies and biotechnology firms have successfully advanced the discovery and development of synthetic or recombinant peptides as well as synthetic peptidomimetic or nonpeptide drugs for numerous medical applications. Relative to synthetic peptide, peptidomimetic, and nonpeptide drug discovery, the integration of drug design using NMR spectroscopy, X-ray crystallography, computational chemistry, and other biophysical and *in silico* methods has been critical to provide insight to the intimate structure–activity relationships and related biological properties necessary to create potent, selective, metabolically stable, safe-acting and, in many cases, orally effective molecules. Of course, knowledge of the pathophysiology of biologically active peptides and their therapeutic target (e.g. receptor, peptidase or signal transduction protein) is essential to identify opportunities for peptide (synthetic or recombinant), peptidomimetic and nonpeptide drug discovery.

3.1
Peptide Pathophysiology and Therapeutic Targets

The pathophysiology related to biologically active peptides may be due to defects in the biosynthesis of the peptide or its target receptor (or enzyme), among a number of other factors. Table 6 lists a variety of endocrine disorders that result from excessive or deficient peptide hormone levels. Amino acid mutations of regulatory peptides are known to exist and to provide the molecular basis for pathophysiological conditions requiring replacement therapy. Causative factors in some cases of diabetes

Tab. 6 Pathophysiological states of known relationship to peptides.

Disease (symptom)	Relationship to peptide
Addison's disease (abnormal carbohydrate metabolism)	Deficiency in adrenocorticotropin
Cushing's disease (increased protein catabolism)	Excess adrenocorticotropin
Secondary hypogonadism	Deficiency in gonadotropin
Secondary hypergonadism	Excess gonadotropin
Hyperglycemia and glucosuria (diabetes mellitus or insulin-dependent diabetes, if resulting from cytotoxic autoantibodies to β-cells)	Deficiency in insulin
Abnormal blood Ca_2 (increased), hyperparathyroidism	Excess parathyroid hormone
Abnormal growth (decreased), Laron-type dwarfism	Deficiency in somatomedin
Abnormal growth (decreased), hypopituitary dwarfism	Deficiency in somatotropin
Abnormal growth (increased) in children, giantism	Excess somatotropin
Abnormal growth (increased) in adults, acromegaly	Excess somatotropin
Abnormal metabolism (increased), Graves' disease antibodies	Excess thyrotropin-mimicking, anti-TSH receptor
Hypovolemia–dehydration (increased), pituitary-type diabetes insipidus	Deficiency in vasopressin

Source: Adapted from Hadley, M.E. (2000) *Endocrinology*, 5th ed., Prentice Hall, Englewood Cliffs, NJ.

mellitus are defects in the primary structure of insulin, either in the active site of the peptide or at sites of cleavage, where the insulin chains are proteolytically cleaved from the prohormone structure. Some, albeit rare, types of diabetes are also due to defects in the amino acid sequence of the insulin receptor that compromise binding of the peptide hormone to the receptor or receptor-mediated signal transduction. In one form of familial (genetic) hypothyroidism, a mutation in one codon of the gene that transcribes for the β-subunit of the hormone thyrotropin (TSH) results in a single amino acid substitution to yield a conformationally abnormal β-subunit that cannot associate with the α-subunit to form a functionally active heterodimeric peptide hormone. Furthermore, since this particular defect in producing endogenous active TSH is familial, and therefore present at birth (congenital), the consequences (e.g. cretinism) are devastating. Within the scope of mimicking or blocking regulatory peptide effects at the level of target receptors, there exists a tremendous research effort in identifying peptidomimetic or nonpeptide derivatives (agonists or antagonists), and such studies have enabled both pharmacological analysis and, in some cases, the discovery of therapeutic agents of synthetic or recombinant origin (see Sect. 3.2). Furthermore, the design of peptide receptor-targeted agonists or antagonists that may be radiolabeled or conjugated to anticancer drugs has made it possible to use such agents in the diagnosis, localization, or chemotherapy of tumors. For example, radioactive indium attached to a somatostatin analog has been developed to localize certain tumors by external scintography.

Enzyme targets that have been determined to be of particular clinical importance as related to the pathophysiology of regulatory peptides include a number of peptidases (Table 4). Representative peptidase targets may be classified as follows: serine peptidases, such as thrombin, tissue plasminogen activator (TPA), and elastase; metallopeptidases, such as ACE, collagenase, and enkephalinase; aspartic peptidases, such as renin, cathepsin D, and HIV protease; and cysteine peptidases, such as cathepsin B, and calpains. As described earlier, peptidases may function primarily as processing or degrading enzymes of regulatory peptides. Within the scope of blood pressure regulation, inhibitors of either ACE or renin may serve to reversibly terminate the stepwise processing of the α_2-macroglobulin angiotensinogen to give rise to the vasoconstrictor angiotensin II (AII). In the case of ACE inhibitors, the clinical efficacy of such drugs as antihypertensives has been well established. In the latter case, the discovery of potent peptidomimetic inhibitors of renin has not readily translated into the development of orally active therapeutic candidates to date. However, such research has catalyzed the identification and design of peptidomimetics that inhibit HIV protease, an aspartyl peptidase having structural and mechanistic resemblance to renin, and such potential drugs may enable a novel chemotherapeutic intervention strategy for the threatening spread and fatal pathogenesis related to AIDS. As a separate case, but still related to processing peptidase targets, peptidomimetic inhibitors of thrombin are being designed in the hope that they might be used to prevent thrombus (clot) formation following surgery. Within the realm of degrading peptidases, the discovery of peptidomimetic inhibitors of neutral endopeptidase (NEP) has been a strategy of interest to advance novel analgesic agents that function via inhibiting the cleavage inactivation of endogenous opioid peptides (e.g. enkephalins). Similarly, peptidomimetic inhibitors of collagenase, a collagen-degrading peptidase, are thought to be of potential utility in the discovery of therapeutic agents effective in pathological conditions such as rheumatoid arthritis.

3.2
Peptide, Peptidomimetic, and Nonpeptide Drug Discovery

Peptide, peptidomimetic, and nonpeptide drug discovery is an intriguing area of research efforts in numerous pharmaceutical and biotechnology companies, and interdisciplinary strategies have emerged over the past two decades that have provided the framework to advance lead compounds and, in some cases, drugs (Fig. 3). Noteworthy to such peptide, peptidomimetic, and nonpeptide drug discovery has been the pioneering work of many scientists from both academia and industry, including Richard DiMarchi, Roger Friedinger, Ralph Hirschmann, Victor Hruby, Isabella Karle, Horst Kessler, Garland Marshall, Henry Mosberg, John Nestor, Daniel Rich, Joseph Rudinger, Tomi Sawyer, Peter Schiller, Robert Schwyzer, and Daniel Veber.

The first recombinant peptide drug commercialized was insulin in 1982. Currently, a number of peptide drugs (or preclinical leads) have been advanced using genetic engineering technologies. Examples are somatotropin, tumor necrosis factor, EGF, erythropoietin, hepatitis B vaccine, interferons, various cytokines, and TPA (Table 7). On the basis of the success of insulin replacement therapy for the treatment of some types of diabetes mellitus, the potential for therapeutic application of other structurally complex peptides has been advanced by sophisticated molecular biology methodologies, which permit biosynthesis of peptides not readily obtained by chemical synthesis.

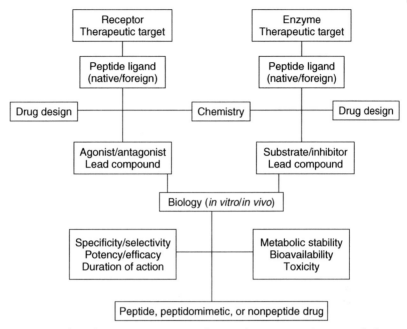

Fig. 3 Interdisciplinary strategies in peptide, peptidomimetic, and nonpeptide drug discovery.

The known chemical diversity of peptide, pseudopeptide, peptidomimetic, and nonpeptide structures that have been advanced as lead compounds or marketed drugs is extensive. Synthetic peptides include native molecules and analogs that incorporate side-chain and/or backbone modifications. In many cases, the resultant synthetic derivatives are more potent, target tissue (receptor) selective, and/or stable to peptidases relative to the native peptide. Such examples include Sandostatin (SRIF analog), Desmopressin (vasopressin analog), Goserelin (GnRH analog), Tetracosactide (corticotropin analog), and Melanotan-I (MSH analog). The discovery of "first-generation" peptide antagonists has, in a majority of cases, been based on chemically modified analogs. Peptide-based competitive antagonists of bradykinin, oxytocin, PTH, enkephalin, vasopressin, glucagon, CCK, gastrin, angiotensin II, GnRH, substance P, and CRF have been reported during the past decade, and such compounds have typically incorporated unusual amino acids (e.g. $\underline{D}$-amino acids, cyclic amino acids, α,β-unsaturated amino acids; Fig. 2) and backbone CONH replacements.

The chemical transformation or replacement of peptide agonists or antagonists by simpler (molecular weight-wise), metabolically stable, and orally bioavailable "second-generation" peptidomimetics or nonpeptides has been accomplished (Table 7). For example, research focused on peptide receptor targets has led to the identification of effective antagonists. To some extent, molecular recognition properties required to mimic the agonist properties of peptides may be more complex and, perhaps, uncompromising relative to those effecting antagonist properties. In fact, the identification of many nonpeptide

Tab. 7 Examples of peptide, peptidomimetic or nonpeptide drugs, clinical candidates or preclinical lead compounds.

Recombinant peptides — ***Proposed/Known therapeutic application***

Atrial natriuretic factor	Potential use in prophylaxis and/or treatment of acute renal failure
Epidermal growth factor	Potential use in skin grafting, eye surgery, and/or treatment of burns or ulcers
Somatotropin	Treatment of growth defects
Insulin	Treatment of type-I diabetes
Glucagon-like insulinotropic peptide	Potential treatment of insulin-insensitive diabetes
Teriparatide	Treatment of osteoporosis
Transforming growth factor	Potential treatment of wound healing and burns
Nerve growth factor	Potential treatment of neural plasticity defects as possibly related to Alzheimer's disease
Interleukin-1α/β	Potential use in cancer therapy and inflammation
Hirudin	Potential use in prevention or treatment of venous blood clots
Tissue plasminogen activator	Potential use as an anticoagulant in heart attacks
Factor VII	Potential use as a blood clotting factor in major forms of hemophiliacs
Erythropoietin	Potential uses in treatment of anemia in kidney dialysis patients, AIDS, and cancer
Platelet-derived growth factor	Potential uses in promoting growth of fibroblasts, keratinocytes, and formation of new blood vessels
Interferon-α	Potential use as an immune stimulant for cancer therapy
Interferon-β	Potential use as an immune stimulant for treatment of viral diseases and multiple sclerosis
Interferon-γ	Potential use as an immune stimulant for treatment of infectious diseases, cancer, and rheumatoid arthritis

Synthetic peptides (relationship to native peptide) — ***Proposed/known therapeutic applications***

Aspartame (dipeptide mimic of glucose)	Artificial sweetener
Cyclosporin A (peptide natural product)	Immunosuppressive drug

Peptide	Application
Sandostatin™ (hexapeptide agonist analog of SRIF)	Symptomatic treatment of acromegaly and carcinoid syndrome
Techtide P829 (radiolabeled SRIF analog)	Detection of tumors
Desmopressin (nonapeptide agonist analog of vasopressin)	Treatment of severe diabetes insipidus
Buserelin (nonapeptide agonist analog of GnRH)	Treatment of prostate cancer
Goserelin (GnRH agonist)	Treatment of cancer and hormonal menstruation disorder
HOE-140 (decapeptide antagonist analog of BK)	Potential treatment of pain, inflammation, rhinitis, and/or asthma
Melanotan-I (tridecapeptide agonist analog of MSH)	Potential use as a stimulant of skin pigmentation
Tetracosactide (peptide agonist analog of ACTH)	Diagnostic agent of adrenal function
BQ-123 (cyclic pentapeptide, natural product antagonist of endothelin)	Potential treatment of hypertension, restenosis, and related disorders
Calcitonin (identical to native peptide)	Treatment of hypercalcemia, Paget's disease, osteoporosis, and pain affiliated with bone cancer
Ebiratide (hexapeptide agonist analog of ACTH)	Potential application for CNS disorders; cognition treatment
Pentigetide (pentapeptide antagonist analog of IgE)	Potential antiallergic agent
MDL28050 (decapeptide antagonist analog of Hirudin)	Potential use as thrombin inhibitor based, anticoagulant agent
Eptifibatide (cyclic heptapeptide)	Potential antithrombotic drug
L365209 (pseudohexapeptide, natural product antagonist of oxytocin)	Potential use as uterine relaxant for prevention of premature labor
Atosiban (oxytocin antagonist)	Potential use for preterm labor
DPDPE (pentapeptide agonist analog of enkephalin)	Potential analgesic lead; δ-selective opioid agonist
GHRP-6 (hexapeptide agonist at gherelin receptor)	Potential GH secretagogue lead
Groliberin (growth hormone-releasing factor analog)	Potential use for treatment of growth hormone deficiency
Enfuvirtide (inhibitor of viral fusion)	Potential antiviral and anti-HIV drug

(continued overleaf)

Tab. 7 (Continued)

Peptidomimetics or nonpeptides (relationship to native peptide)	Proposed/known therapeutic applications
Morphine (peptidoligand natural product agonist at μ-type enkephalin receptor)	Analgesic drug
Captopril, Enalapril (peptidomimetic inhibitors of ACE)	Antihypertensive drugs
DuP-753 (nonpeptide antagonist at AT_1-type AII receptor)	Potential antihypertensive drug
A-72517 (peptidomimetic inhibitor of renin)	Potential antihypertensive drug
Ro-318539, A-75925 (peptidomimetic inhibitors of HIV protease)	Potential anti-HIV/AIDS drug
CP-96345 (nonpeptide antagonist at NK_1 receptor)	Potential use as an antiallergic or analgesic drug
SC-47643 (peptidomimetic antagonist at fibrinogen receptor)	Potential use for prevention of thrombus formation
Integrilin (peptidomimetic antagonist at fibrinogen receptors)	Potential use for myocardial infarction
OPC-21268 (nonpeptide antagonist at V_1 vasopressin receptor)	Potential use as a diuretic agent
Ro-24 9975 (peptidomimetic agonist at TRH receptor)	Potential use as cognitive enhancer agent
CI-988 (peptidomimetic antagonist at CCK_B receptor)	Potential anxiolytic drug
MK-329	Potential use for treatment of pancreatitis (nonpeptide antagonist at CCKA receptor)
L-692429 (nonpeptide agonist at gherelin receptor)	Potential use as a GH secretagogue
SCH-34826 (peptidomimetic inhibitor of metallo-endopeptidase EC 3.4.24.11)	Potential antihypertensive drug; inhibitor of atrial naturetic peptide (ANP) degradation
CT-0543 (peptidomimetic inhibitor of gelatinase)	Potential antimetastatic agent
MD-805 (peptidomimetic inhibitor of thrombin)	Potential antithrombolytic drug
Bivalirudin (thrombin inhibitor)	Potential anticoagulant and antianginal drug
Ampicillin, amoxycillin (nonpeptide inhibitors of bacterial cell wall peptidoglycan synthesis)	Antibacterial penicillin-related drugs

Note: See Table 3 and Fig. 4 for some chemical structures.
Source: Adapted from Sawyer, T.K. (1995) in: Amidon, G., Taylor, M. (Eds.), *Peptide-Based Drug Design: Controlling Transport and Metabolism*, ACS Books, Washington, DC.

antagonists at peptide receptors has been significantly advanced by mass screening assays using chemical collections as well as various natural product sources, thereby illustrating the fact that the peptide ligand is not exclusive toward providing the molecular framework for peptide antagonist drug discovery. Interestingly, the recent development of peptide library technologies (including emerging pseudopeptide and peptidomimetic libraries) has accelerated opportunities to identify novel leads as well as the ability to evaluate analogs in a structure–activity sense to expedite the overall process of drug candidate selection. The application of biophysical chemistry (e.g. NMR spectroscopy and X-ray crystallography) and computational chemistry methods have collectively established powerful drug design technologies for the synthetic transformation of native peptides to chemically modified pseudopeptides or peptidomimetics. In the case of cyclic, conformationally constrained peptides, the synergism of NMR spectroscopy and computational chemistry-based analysis of three-dimensional structural properties and model building points to an extremely promising approach to advancing peptide-based drug design in a rational manner. In the case of X-ray crystallography and computational chemistry, the study of peptide or peptidomimetic interaction with proteins (peptidases, kinases, Src homology proteins, antibodies) exploits high-resolution (1.8–3.0 Å) molecular maps of the complex to provide insight into further design strategies. In this regard, recent work on HIV protease, an aspartyl peptidase, is particularly outstanding in that more than 120 inhibitor–enzyme complexes have been determined by X-ray crystallography. Furthermore, this work provides an excellent example of peptide and peptidomimetic drug design and development strategies that integrate molecular biology, biochemistry, synthetic chemistry, biophysical and computational chemistry, pharmacology, and drug delivery research (Fig. 4).

Among the major challenges of developing peptide drugs has been the mode of administration of the particular compound. Peptide drug delivery exists in many forms, including parenteral (intravenous or intramuscular injection), interstitial (subcutaneous implant), oral, nasal, and percutaneous (transdermal) administration. Perhaps, the most well-known injectable peptide is that of insulin, and with respect to implants the recent success of the gonadotropin-releasing hormone analog Zoladex deserves mention. In the case of oral administration, one of the key reasons for pursuing the chemical transformation of peptides into peptidomimetic derivatives has been to identify prototypic lead compounds that may exhibit oral bioavailability. Such efforts have shown success as exemplified by the development of orally effective ACE inhibitors, renin inhibitors, HIV protease inhibitors, thyrotropin-releasing hormone (TRH) agonists, fibrinogen antagonists, and so forth. Nevertheless, the possibility of alternative modes of administration of peptide (or peptidomimetic) drugs remain an intriguing area of research, and some success has been achieved with respect to nasal (oxytocin, desmopressin, buserelin, and calcitonin), pulmonary, buccal, rectal, and transdermal routes of peptide drug administration. Of particular impact on the oral delivery of peptide drugs is the limitation on passive or active transport of molecules of high molecular weight (typically including tetrapeptides and larger peptides) posed by the intestinal epithelia. Similarly, stability toward gastric acid and degradative enzymes, including

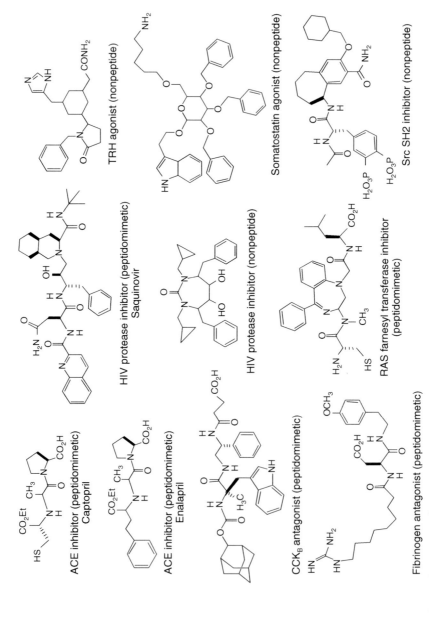

Fig. 4 Chemical structures of some peptidomimetic or nonpeptide drugs, clinical candidates, or preclinical lead compounds.

peptidases associated with the epithelial membrane (luminal side) or within such cells, provides another challenge a peptide-based drug candidate must surpass to eventually access systemic circulation after oral administration.

The apparent medical potential for peptide, peptidomimetic, or nonpeptide drugs is already tremendous, and future opportunities will continue to make this area of pharmaceutical research extremely competitive. There are approximately 100 synthetic and recombinant peptides that are marketed worldwide, and an estimated 200 molecules that are in clinical testing. Nine marketed peptides are responsible for annual sales surpassing the $4 billion mark. Some past and current examples of peptide, peptidomimetic, and nonpeptide preclinical lead compounds, clinical candidates, and marketed drugs are exemplified above in Table 7. Most recently, synthetic and recombinant peptides that have been approved by the FDA include Teriparatide (PTH agonist analog for osteoporosis), Goserelin (GnRH agonist analog for cancer and hormonal menstruation disorder), Integrilin (integrin gpIIb/IIIa antagonist for myocardial infarction) Enfuvirtide (inhibitor of viral fusion to effect antiviral and anti-HIV activity), Eptifibatide (cardiovascular, neuroprotective), Bivalirudin (thrombin inhibitor for anticoagulant and antianginal), Atosiban (oxytocin antagonist for preterm labor), Techtide P829 (radiolabeled somatostatin analog for tumor detection), and Groliberin (GRF analog for treatment of growth hormone deficiency). Past and still existing major peptide and peptidomimetic drugs include insulin (for treatment of diabetes) and dipeptide sweeteners such as aspartame. Without question, the field of peptide, peptidomimetic, and nonpeptide drug discovery has emerged as a superclass of molecules for a plethora of disease as exemplified by the therapeutic scope and molecular diversity as described in this chapter. Future research and drug development holds further promise to combat many existing diseases and to deal with many unmet needs throughout the world.

See also Design and Application of Synthetic Peptides; HPLC of Peptides and Proteins.

Bibliography

Books and Reviews

Amidon, G., Taylor, M. (Eds.) (1995) *Peptide-Based Drug Design: Controlling Transport and Metabolism*, ACS Books, Washington, DC.

Chorev, M., Sawyer, T.K. (Eds.) (2004) Peptide Revolution: Genomics, Proteomics and Therapeutics. *Proceedings of the Eighteenth Peptide Symposium.* American Peptide Society, San Diego, CA.

Gross, E., Meienhofer, J., Udenfriend, S. (Eds.) (1979–1987) *The Peptides: Analysis, Synthesis, Biology*, Vols. 1-IX, Academic Press, New York.

Hadley, M.E. (1992) *Endocrinology*, 3rd edition, Prentice Hall, Englewood Cliffs, NJ.

Hider, R.C., Barlow, D. (Eds.) (1991) *Polypeptide and Protein Drugs: Production, Characterization and Formulation*, Ellis Horwood, Chichester, UK.

Nego-Vilar, A., Conn, P.M. (Eds.) (1988) *Peptide Hormones: Effects and Mechanisms of Action*, Vols. 1–III, CRC Press, Boca Raton, FL.

Ward, D.J. (1991) *Peptide Pharmaceuticals: Approaches to the Design of Novel Drugs*, Open University Press, Buckingham, UK.

Wieland, T., Bodanszky, M. (1991) *The World of Peptides: A Brief History of Peptide Chemistry*, Springer-Verlag, Berlin, Germany.

Williams, W.V., Weiner, D.B. (Eds.) (1993) *Biologically Active Peptides: Design, Synthesis and Utilization*, Technomic, Lancaster, PA.

Primary Literature

Ahn, J.A., Boyle, N.A., MacDonald, M.T., Janda, K.D. (2002) Peptidomimetics and peptide backbone modifications, *Mini Rev. Med. Chem.* **2**, 463–473.

Al-Obeidi, F.A., Hruby, V.J., Sawyer, T.K. (1998) Peptide and peptidomimetic libraries: molecular diversity and drug design, *Mol. Biotechnol.* **9**, 205–223.

Annis, I., Hargittal, B., Barany, G. (1997) Disulfide bone formation in peptides, *Methods Enzymol.* **289**, 198–221.

Burke, T.R. Jr., Yao, Z.J., Liu, D.G., Voigt, J., Gao, Y. (2001) Phosphoryltyrosyl mimetics in the design of peptide-based signal transduction inhibitors, *Biopolymers* **60**, 32–44.

Bursavich, M.G., Rich, D.H. (2002) Designing non-peptide peptidomimetics in the 21st century: inhibitors targeting conformational ensembles, *J. Med. Chem.* **31**, 541–558.

Cheng, R.P., Gellman, S.H., DeGrado, W.F. (2001) Beta-peptides: from structure to function, *Chem. Rev.* **101**, 3219–3232.

Chorev, M., Goodman, M. (1995) Recent developments in retro peptides and proteins—an ongoing topochemical exploration, *Trends Biotechnol.* **13**, 438–445.

Craik, D.J., Simonsen, S., Daly, N.L. (2002) The cyclotides: novel macrocyclic peptides as scaffolds in drug design, *Curr. Opin. Drug Discov. Dev.* **5**, 251–260.

Dawson, P.E., Kent, S.B. (2000) Synthesis of native proteins by chemical ligation, *Annu. Rev. Biochem.* **69**, 923–960.

Eguchi, M., Kahn, M. (2002) Design, synthesis, and application of peptide secondary structure mimetics, *Mini Rev. Med. Chem.* **2**, 447–462.

Fairlie, D.P., Abbenante, G., March, D.R. (1995) Macrocyclic peptidomimetics: forcing peptides into bioactive conformations, *Curr. Med. Chem.* **2**, 654–686.

Fauchere, J.-L. (1986) Elements for the Rational Design of Peptide Drugs, in: Testa, B. (Ed.) *Advances in Drug Research*, Vol. 15, Academic Press, London, UK, pp. 29–69.

Figliozzi, G.M., Goldsmith, R., Ng, S.C., Banville, S.C., Zuckermann, R.N. (1996) Synthesis of N-substituted glycine peptoid libraries, *Methods Enzymol.* **267**, 437–447.

Freidinger, R.M. (1999) Nonpeptidic ligands for peptide and protein receptors, *Curr. Opin. Chem. Biol.* **3**, 395–406.

Goodman, M., Ro, S. (1994) Peptidomimetics for Drug Design, in: Wolff, M.E. (Ed.) *Medicinal Chemistry and Drug Design, Vol. I. Principles of Drug Discovery*, 5th edition, John Wiley & Sons, New York, pp. 803–861.

Holder, J.R., Haskell-Luevano, C. (2004) Melanocortin ligands: 30 years of structure activity relationship (SAR) studies, *Med. Res. Rev.* **24**, 325–356.

Houghten, R.A. (2000) Parallel array and mixture-based synthetic combinatorial chemistry: tools for the next millennium, *Annu. Rev. Pharmacol. Toxicol.* **40**, 273–282.

Hruby, V.J. (2002) Designing peptide receptor agonists and antagonists, *Nat. Rev. Drug Discov.* **1**, 847–858.

Hruby, V.J., Al-Obeidi, F.A., Kazmierski, W. (1990) Emerging approaches in the molecular design of receptor-selective peptide ligands, *Biochem. J.* **268**, 249–262.

Jackson, D.C., Purcell, A.W., Fitzmaurice, C.J., Zeng, W., Hart, D.N. (2002) The central role played by peptides in the immune response and the design of peptide-based vaccines against infectious diseases and cancer, *Curr. Drug Targets* **3**, 175–196.

Kempf, D., Sham, H.L. (1996) HIV protease inhibitors, *Curr. Pharm. Des.* **2**, 225–246.

Lam, K.S., Renil, M. (2002) From combinatorial chemistry to chemical microarray, *Curr. Opin. Chem. Biol.* **6**, 353–358.

Lauer-Fields, J.L., Juska, D., Fields, G.B. (2002) Matrix metalloproteinases and collagen catabolism, *Biopolymers* **66**, 19–32.

Liu, L.P., Deber, C.M. (1998) Guidelines for membrane proteins engineering derived from de novo designed model peptides, *Biopolymers* **47**, 41–62.

Marshall, G.R. (1993) A hierarchical approach to peptidomimetic design, *Tetrahedron* **49**, 3547–3558.

Marshall, G.R. (2001) Peptide interactions with G-protein coupled receptors, *Biopolymers* **60**, 246–277.

Micklatcher, C., Chmielewski, J. (1999) Helical peptide and protein design, *Curr. Opin. Chem. Biol.* **3**, 724–729.

Olson, G.L., Bolin, D.R., Bonner, N.P., Bos, M., Cook, C.M., Fry, D.C., Graves, B.J., Hatada, M., Hill, D.E., Kahn, M., Madison, V.S., Rasiecki, V.K., Sarabu, R., Sepinwall, J., Vincent, G.P., Voss, M.E. (1993) Concepts and progress in the development of peptide mimetics, *J. Med. Chem.* **36**, 3039–3049.

Pavia, M.R., Sawyer, T.K., Moos, W.H., Eds. (1993) The generation of molecular diversity. *Bioorg. Med. Chem. Lett.* **3**, 387–396, Symposium-in-print.

Qian, Y., Sebti, S.M., Hamilton, A.D. (1997) Farnesyltransferase as a target for anticancer drug design, *Biopolymers (Peptide Sci.)* **43**, 25–41.

Quan, M.L., Wexler, R.R. (2001) The design and synthesis of noncovalent factor Xa inhibitors, *Curr. Top. Med. Chem.* **1**, 137–149.

Rich, D.H. (2002) Discovery of nonpeptide, peptidomimetic peptidase inhibitors that target alternate enzyme active site conformations, *Biopolymers* **66**, 115–125.

Ripka, A.S., Rich, D.H. (1998) Peptidomimetic design, *Curr. Opin. Chem. Biol.* **2**, 441–452.

Sadler, K., Tam, J.P. (2002) Peptide dendrimers: applications and synthesis, *J. Biotechnol.* **90**, 195–229.

Sakakibara, S. (1995) Synthesis of large peptides in solution, *Biopolymers* **37**, 17–28.

Sawyer, T.K. (1997) Peptidomimetic and nonpeptide drug discovery: chemical nature and biological targets, in: Reid, R. (Ed.) *Drugs and the Pharmaceutical Sciences*, Vol. 101, Marcel Dekker, New York, pp. 81–114.

Sawyer, T.K. (1999) Peptidomimetic and Nonpeptide Drug Discovery: Impact of Structure-Based Drug Design, in: Veerapandian, P. (Ed.) *Structure-Based Drug Design: Diseases, Targets, Techniques and Developments*, Marcel Dekker, New York, pp. 559–634.

Sawyer, T.K., Bohacek, R.S., Dalgarno, D.C., Eyermann, C.J., Kawahata, N., Metcalf, C.A., Shakespeare, W.C., Sundaramoorthi, R., Wang, Y., Yang, M.G. (2002) Src homology-2 inhibitors: peptidomimetic and nonpeptide, *Mini Rev. Med. Chem.* **2**, 475–488.

Singh, S.B., Lingham, R.B. (2002) Current progress on farnesyl protein transferase inhibitors, *Curr. Opin. Drug Discov. Dev.* **5**, 225–244.

Skiles, J.W., Gonnella, N.C., Jeng, A.Y. (2001) The design, structure and therapeutic application of matrix metalloproteinase inhibitors, *Curr. Med. Chem.* **8**, 425–474.

Songyang, Z., Cantley, L.C. (1995) Recognition and specificity in protein tyrosine kinase-mediated signalling, *Trends Biochem. Sci.* **20**, 471–475.

Sundaram, R., Dakappagari, N.K., Kaumaya, P.T. (2002) Synthetic peptides as cancer vaccines, *Biopolymers* **66**, 200–216.

Yamashita, D.S., Dodds, R.A. (2000) Cathepsin K and the design of inhibitors of cathepsin K, *Curr. Pharm. Des.* **6**, 1–24.

Zhang, Y.Z. (2001) Protein tyrosine phosphatases: prospects for therapeutics, *Curr. Opin. Chem. Biol.* **5**, 416–423.

Part 3
Expression, Synthesis and Degradation

Proteins. Edited by Robert A. Meyers.
Copyright © 2007 Wiley-VCH Verlag GmbH & Co. KGaA, Weinheim
ISBN: 978-3-527-31608-3

16
Ribosome, High Resolution Structure and Function

Christiane Schaffitzel and Nenad Ban
Institute for Molecular Biology and Biophysics, Swiss Federal Institute of Technology, Zürich, Switzerland

1	**A Macroscopic View of Peptide Synthesis**	490
2	**The 30S Ribosome Structure** 493	
2.1	The Path of the mRNA Through the 30S Subunit	493
2.2	Interactions of the 30S Subunit with tRNA	495
2.3	The Decoding Center	495
2.4	Modes of Antibiotic Inhibition	499
2.5	Translation Initiation – Interactions with Initiation Factors	499
3	**The 50S Ribosome Structure** 500	
3.1	The Molecular Mechanism of Peptidyl Transfer	501
3.2	The Exit Tunnel	503
3.3	Antibiotics Targeting the Large Subunit	505
3.4	Regulatory Nascent Chains	506
3.5	The GTPase Factor-Binding Center	507
3.6	Structural Mimicry	507
4	**The 70S Ribosome Structure** 507	
4.1	The 30S–50S Interface	508
4.2	Translocation – The Concerted Movement of tRNAs and mRNA	508
4.3	Movement of tRNAs through the Ribosome	510
4.4	Cotranslational Folding	510
4.5	Protein Secretion and Membrane Insertion	511
5	**Perspectives** 512	

Proteins. Edited by Robert A. Meyers.
Copyright © 2007 Wiley-VCH Verlag GmbH & Co. KGaA, Weinheim
ISBN: 978-3-527-31608-3

Bibliography 512
　Books and reviews   512
　Primary Literature   513

Keywords

Codon and Anticodon
The specific sequence of three nucleotides on the mRNA (codon) and the complementary sequence on the tRNA (anticodon).

Decoding
The mechanism by which the ribosome finds the correct aminoacyl-tRNA based on codon–anticodon interaction.

Elongation Factors
Proteins involved in the elongation of the growing peptide chain.

Exit Tunnel
The tunnel in the large ribosomal subunit for the exit of the nascent peptide chain.

Initiation Factors
Proteins involved in the translation initiation steps, that is, the initial association of the two ribosomal subunits and the binding of the first tRNA.

Molecular Chaperones
Proteins that help other proteins to fold and/or prevent aggregation.

Nascent Peptide
The growing peptide still connected to the ribosome as peptidyl-tRNA.

Peptidyl Transferase Reaction
The elongation of the polypeptide chain by one amino acid.

Release Factors
Proteins required for the release of the nascent peptide when a stop codon is encountered on the mRNA.

Translocation
The movement of the mRNA and the tRNAs through the ribosome during translation.

Translocation Machinery
Accomplishes the transport of the nascent polypeptide through the membrane or its integration in the membrane.

Abbreviations

A, P, E sites:	aminoacyl, peptidyl, exit sites
CryoEM:	cryo-electron microscopy
EF-G:	elongation factor G
EF-Tu:	elongation factor Tu
GTP:	guanosine triphosphate
mRNA:	messenger RNA
rRNA:	ribosomal RNA
SRP:	signal recognition particle
tRNA:	transfer RNA

■ The code for the amino acid sequence of proteins is translated by the ribosome using messenger RNA as the template. Ribosomes are RNA-protein complexes built of two subunits. In prokaryotes, the ribosome consists of a large 50S and a small 30S ribosomal subunit. Both subunits together form the 70S ribosome, the prokaryotic translation machinery. Structures of both ribosomal subunits have been solved at near-atomic resolution. The large ribosomal subunits are from the halophilic archaebacterium *Haloarcula marismortui* and from the eubacterium *Deinococcus radiodurans*. The small subunit has been solved from the thermophilic eubacterium *Thermus thermophilus*. Structures of the ribosomal subunits were also determined in complex with various antibiotics, substrate analogs, and two translation initiation factors. Finally, the entire 70S ribosome from *T. thermophilus* with mRNA and tRNAs is available at 5.5 Å as a molecular model. This wealth of structural data allows for a detailed understanding of the molecular mechanism of peptide synthesis by the ribosome.

The two most important steps in protein synthesis are the recognition of the cognate aminoacyl-tRNA and the peptidyl transfer reaction. These two functions are allotted to the two subunits of the ribosome. The small subunit mediates the interactions between tRNAs and the mRNA and selects for the correct tRNA in the decoding center. The large subunit comprises the peptidyl transferase center and provides the exit tunnel for the growing nascent polypeptide chain. The crystal structures of the separated subunits reveal many mechanistic details of the peptidyl transfer and decoding. The structures show unequivocally that mainly ribosomal RNA is present in the peptidyl transferase center as well as in the decoding center, indicating that it is in fact ribosomal RNA that is responsible for all aspects of peptide synthesis. In other words, the ribosome is a ribozyme.

Here, we analyze the crystal structures of prokaryotic 50S and 30S subunits and the intact 70S ribosome with respect to their function and describe how our understanding of the mechanism of translation has been influenced by these recent structures. Furthermore, we outline the impact of cryo-electron microscopic studies (cryoEM) on our knowledge of ribosome conformational flexibility and ribosome complexes with translation factors.

1
A Macroscopic View of Peptide Synthesis

Aminoacyl-tRNAs are the substrates for protein synthesis. They all have an L-shaped structure with the anticodon loop at the end of the long L-arm. The anticodon sequence is part of the anticodon loop and complementary to the cognate mRNA base triplet (codon). At the 3′ end of the short L-arm, the acceptor stem has a terminal CCA sequence to which individual amino acids are attached by aminoacyl-tRNA synthetases. In the cell, a specific aminoacyl-tRNA synthetase exists for each amino acid, catalyzing the formation of an ester bond between the carboxyl group of the amino acid and the 2′ or 3′ hydroxyl group of the 3′ adenosine of tRNA with the appropriate anticodon.

Both ribosomal subunits have three binding sites for tRNA, called the *A (aminoacyl), P (peptidyl) and E (exit) sites*. During protein synthesis, the tRNA anticodons interact with the mRNA and the small ribosomal subunit, while the respective aminoacyl-acceptor stems extend into the large subunit where the peptidyl transfer reaction occurs (Fig. 1). In the *decoding center* of the small ribosomal subunit,

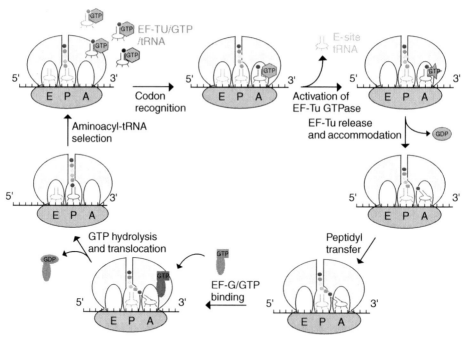

Fig. 1 Overview of the translation cycle. The cognate aminoacyl-tRNA is selected in the A site of the small subunit. The tRNA binds as a ternary complex together with EF-Tu and GTP. Binding of the cognate aminoacyl-tRNA induces GTPase activity of EF-Tu. The release of EF-Tu and GDP is followed by a conformational change of the aminoacyl-tRNA such that it binds in the A site of the peptidyl transferase center. The elongating chain is then transferred from the P-site peptidyl-tRNA to the amino group of the A-site aminoacyl-tRNA. The peptide chain leaves the peptidyl transferase center through the ribosomal tunnel, which spans the large subunit. EF-G catalyzes GTP hydrolysis and thereby drives translocation, that is, transport of the mRNA, the deacylated tRNA from the P to the E site and the peptidyl-tRNA from the A to the P site.

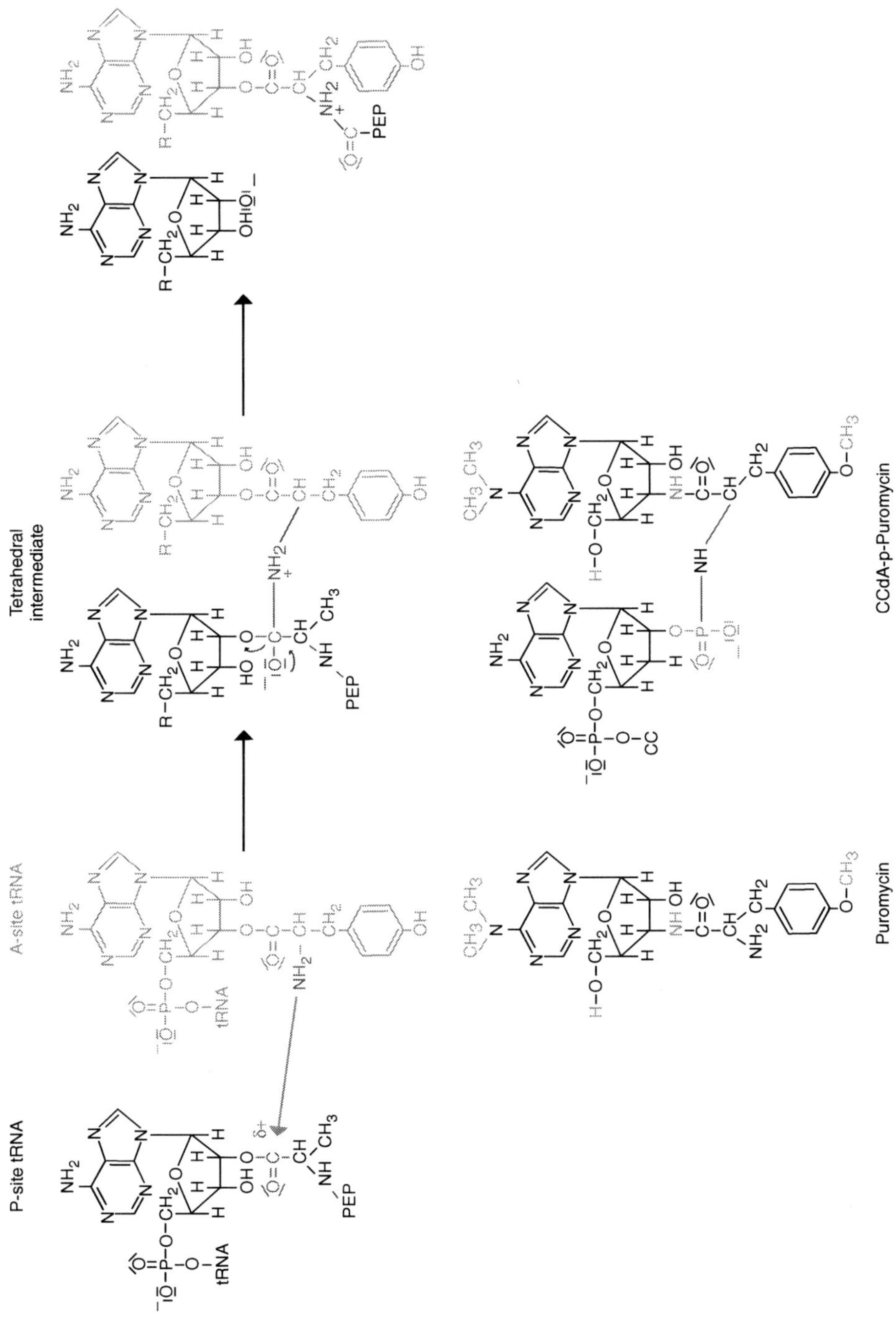

Fig. 2 (See the legend on next page)

selection of the correct aminoacyl-tRNA occurs based on the interaction of the mRNA base triplet with the anticodon triplet of the aminoacyl-tRNA. Aminoacyl-tRNAs bind to the ribosome in complex with elongation factor Tu (EF-Tu) and GTP. These ternary complexes bind tightly to the ribosome only if the anticodon matches the mRNA codon (*initial selection*). Then, GTP is hydrolyzed, resulting in release of phosphate and EF-Tu in complex with GDP (Fig. 1). Concomitantly, a large conformational change of the A site aminoacyl-tRNA occurs, referred to as *accommodation*. This positions the 3′ end of the aminoacyl-tRNA in the *peptidyl transferase center* adjacent to the P site–bound peptidyl-tRNA. Subsequently, the α-amino group of the aminoacyl-tRNA bound in the A site attacks the ester group of the peptidyl- (or N-formyl methionine initiator) tRNA in the P site at its carbonyl atom (Fig. 2). This nucleophilic attack results in a tetrahedral intermediate. A new peptide bond is formed by displacement of the P-site tRNA hydroxyl group from this intermediate (Fig. 2). As a result, the tRNA in the A site carries the nascent peptide chain, which is extended by one amino acid. The synthesized nascent peptide chain diffuses from the peptidyl transferase center into the *tunnel* and exits at the opposite side of the large ribosomal subunit (Fig. 1).

Before the next cycle of peptidyl transfer can occur, the peptidyl-tRNA has to be transported from the ribosomal A site to the P site. Furthermore, the deacylated tRNA in the P site is moved to the E site of the ribosome. This complex concerted movement of tRNAs and mRNA through the ribosome is termed *translocation*. It occurs with high accuracy and involves both subunits. The energy for translocation is provided by GTP hydrolysis catalyzed by elongation factor G (EF-G, Fig. 1). The tRNA bound in the E site interacts still with both ribosomal subunits; it is only released when the next aminoacyl-tRNA is accommodated in the ribosomal A site. A cartoonlike overview of the catalytic centers of both ribosomal subunits is given in Fig. 3.

Fig. 2 Mechanism of peptide bond formation catalyzed in the peptidyl transferase center of the large subunit. (a) The ribosomal peptidyl transferase reaction. The substrates of the peptidyl-transferase reaction are aminoacyl-tRNA and peptidyl-tRNA binding to the ribosomal A site and P site respectively. The α-amino group of the aminoacyl-tRNA in the A site of the ribosome attacks the carbonyl carbon atom of the ester bond that links the nascent polypeptides to P site–bound tRNAs. Thereby, a tetrahedral intermediate is formed at the carbonyl group with a chiral carbon atom (D) and a negatively charged oxygen. The 2′ (or 3′) OH group of the tRNA 3′-adenosine is the leaving group of the tetrahedral intermediate, and a new peptide bond (amide bond) is generated, resulting in a peptidyl-tRNA in the A site extended by one amino acid and a deacylated tRNA in the P site.
(b) Puromycin resembles the 3′ terminus of tyrosyl-tRNA, and it binds to the A site of the ribosome without the need for any elongation factors. Subsequently, the peptide is transferred to the puromycin and the nascent chain is released. (c) The Yarus inhibitor (CCdA-p-Puromycin) is formed by coupling the 3′ OH group of CCdA (in the P site) to the amino group of puromycin (in the A site) via a phosphoramide group, which mimics the tetrahedral carbon intermediate. The inhibitor binds tightly to the ribosome and inhibits peptidyl transferase activity. Important differences between the CCdA-p-Puromycin and the tetrahedral carbon intermediate formed during translation are the nonchiral phosphorus atom, the delocalized negative charge, and the missing 2′ OH group of the adenine in the P site.

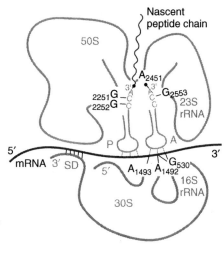

Fig. 3 Schematic overview of the ribosomal active sites. The Shine-Dalgarno (SD) sequence of the mRNA forms a helix with the complementary 3′ terminal tail of 16S rRNA. The anticodons of the A- site and P- site tRNA pair with the mRNA. In the A site decoding center of the small subunit, the codon–anticodon interaction is monitored by 16S rRNA via tertiary interactions. The tRNA acceptor stems point into the peptidyl transferase center of the large subunit. The 3′ CCA ends of the acceptor stems base-pair with 23S rRNA A-site and P-site RNA loops. The adenine of the central loop of 23S rRNA is suggested to play a role in peptidyl transfer catalysis. The nascent peptide chain is extending into the exit tunnel of the large subunit.

2
The 30S Ribosome Structure

The small subunit of the ribosome is responsible for binding mRNA and selection of the correct aminoacyl-tRNA. The structure of the small ribosomal subunit from a thermophilic bacterium *T. thermophilus* was recently determined. On the basis of electron microscopic studies, the 30S subunit is classically divided into head, body, neck, shoulder, and platform. It consists of one 16S rRNA chain and 20 ribosomal proteins (S1–S20) (Fig. 4). The shape of the small subunit is mostly determined by ribosomal RNA. The proteins are distributed over the top, sides, and back of the 30S subunit, while interface with the 50S subunit is mostly formed by rRNA (Fig. 4). Only protein S12 is located near the decoding site, whereas other proteins are found at the periphery of the subunit interface allowing them to contact the 50S subunit. Besides a globular domain many ribosomal proteins have extensions or tails that penetrate far into the rRNA core and closely interact with it, thereby stabilizing the RNA domains. The mRNA binding site and the 3′ end of 16S rRNA are situated on the platform. The tRNA binding sites are located in a cleft formed by the platform and the head (Fig. 4).

2.1
The Path of the mRNA Through the 30S Subunit

The exact path of mRNA through the ribosome was determined by X-ray crystallography of 70S ribosomes at 7 Å resolution. About 30 nucleotides of mRNA are shown to be in contact with the ribosome, predominantly with 16S rRNA. The mRNA wraps around the neck of the 30S subunit, starting with its 5′ end between head and platform and exiting on the opposite site between head and shoulder. The Shine-Dalgarno sequence, that is, the ribosome binding side provides for proper reading of the mRNA by forming a helix with the complementary 3′ terminal tail of 16S rRNA. The Shine-Dalgarno helix is located in the 30S subunit cleft between the head

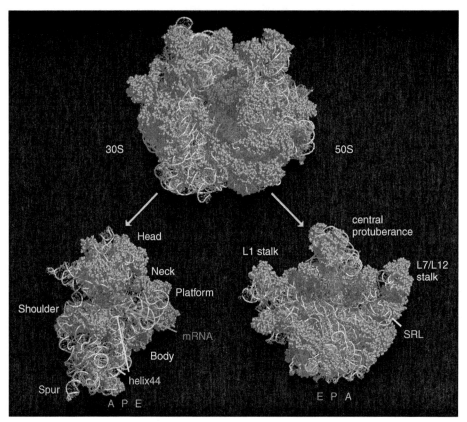

Fig. 4 Complete 70S ribosome is modeled by docking high-resolution 50S and 30S structures onto the low-resolution 70S structure. The 30S structure is from *T. thermophilus* and 50S structure from *H. marismortui* with tRNA positions determined from lower resolution 70S studies. Ribosomal RNA is depicted in gray and proteins in yellow. The 70S is shown from the side with the 30S subunit on the left and the 50S subunit on the right. In the subunit interface cavity, the tRNAs are visible. The A-site tRNA is depicted in blue, the P-site tRNA in magenta and the E-site tRNA in green. The anticodon arms of the tRNAs point to the 30S subunit, while the acceptor stems point into a large cleft in the large subunit. On the left-hand side, the 30S subunit is shown from the interface with the A-, P- and E-tRNA anticodon stem-loops. The anticodon loops of the A and P tRNAs contact the mRNA. Architectural characteristics and important features are labeled. On the right-hand side, the large subunit with tRNAs docked onto the *Haloarcula* 50S structure is shown from the interface as well. The acceptor stems of A- and P-site tRNAs point into the peptidyl transferase center. The sarcin–ricin loop (SRL) is a central part of the GTPase factor binding center. The three characteristic protuberances of the large subunit are labeled. (See color plate p. xx).

and the back of the platform. From there, the mRNA enters a short tunnel, which is situated between head and platform. This region comprises the E site, which does not have full access to the interface. As a consequence, codon–anticodon interaction is not possible to the full extent for E site tRNA. Downstream of the E site, the mRNA reaches the interface by a sharp turn. On the interface, only about eight nucleotides including the A- and P-site codons are presented to the tRNAs. There is a sharp kink in the mRNA between the codons in the A site and in the P site. This kink could play a role in maintaining the reading frame during translocation. A- and P-site codons occupy the major groove of the penultimate helix 44, which is the longest helix of the small subunit. It expands from the head to the bottom of the body and protrudes to interact with the large subunit (Fig. 4). Downstream of the A site responsible for decoding, the mRNA passes through a tunnel between head and shoulder of the subunit. Subsequently, it reaches a ring formed by proteins S3, S4, and S5 that functions probably as a helicase to remove mRNA secondary structure before entering the decoding region. During translation initiation, binding of mRNA to the small subunit requires an opening and closing of at least one of the 30S subunit tunnels at the 3′ or 5′ site of the mRNA.

2.2
Interactions of the 30S Subunit with tRNA

The three tRNAs are aligned on the 30S subunit with the anticodon loops bound in the 16S rRNA groove between the head, body, and platform (Fig. 4). In the *T. thermophilus* 30S ribosome structure, the interaction of mRNA and tRNA in the P site is mimicked through fortuitous interactions between neighboring subunits in the crystal. The 3′ end of 16S rRNA, which is complementary to the Shine-Dalgarno sequence, folds back and imitates mRNA in the P site while a stem-loop structure (the spur, Fig. 4) of a neighboring molecule in the crystal mimics the anticodon loop of the tRNA. As mentioned above, the P-site codon of the mRNA binds to the major groove of a well-conserved region of helix 44. The anticodon stem-loop of the tRNA forms several hydrogen bonds to 16S rRNA. Furthermore, the ribosomal proteins S13 and S9 interact with the anticodon loop of P-site tRNA.

The *E site*, unlike the A and P sites, consists mostly of ribosomal proteins as visualized in the *T. thermophilus* 70S structure. The E site tRNA is tightly bound and interacts with 16S rRNA as well as with a number of ribosomal proteins, mainly S7 and S11. Codon–anticodon interactions are not observed in the E site.

Recently, 30S subunit structures have been solved in complex with mRNA and cognate as well as near-cognate tRNA anticodon stem-loops in the A site, revealing important aspects of the decoding mechanism. The A site is much shallower and wider than the P and E sites, consistent with lower affinity for tRNA. It is mainly composed of 16S rRNA with a small contribution by ribosomal proteins S13 and S12, which contact the mRNA. Importantly, parts of the 16S rRNA, that is, the 3′ major domain and helix 44 (the penultimate helix), are responsible for the functioning of the decoding center.

2.3
The Decoding Center

The error rate of translation is 10^{-3} to 10^{-4}. Thus, the ribosome must be able to discriminate cognate from near and noncognate tRNAs by screening all three codon positions for correct interactions. This high translation fidelity cannot be explained by the difference in free energy between correct and incorrect base pairing of the codon–anticodon helix, since the free energy of some mismatches is very similar to the base pairing of cognate tRNAs. For instance, the gain of free energy of GU base pairing is very similar to the gain of AU pairing. In addition, AU base pairing is energetically less favorable than GC pairing, that is, the gain of free energy is not the same for all cognate codon–anticodon interactions.

Moreover, the difference in free energy does not explain why the recognition of the third base pair is sometimes less stringently recognized than the preceding two. Consequently, a *wobble* can be observed in the pairing of the third base of the codon, giving rise to part of the degeneracy of the genetic code. Therefore, some tRNAs can recognize two or three codons with differences in the third base. It is clear that the *difference in free energy* of correct and incorrect base pairing may well contribute to the accuracy of translation, but by itself it cannot provide a satisfactory explanation.

As a second source of selectivity, the ribosome is capable of recognizing deviations from the Watson-Crick *base-pair geometry*, that is, the suboptimal shape of the sugar-phosphate backbone or the minor and major grooves in mismatched pairs. This sensitivity of the ribosome with regards to base-pair geometry is in analogy to enzyme *substrate recognition*.

The complex between 30S subunit, mRNA and the tRNA anticodon stem-loop switches from the open to the closed conformation if the codon–anticodon interaction is cognate. This conformational change involves flipping out of two adenines, A1492 and A1493, of helix 44. Theses adenines then tightly pack in the minor groove of the first two base pairs of the codon–anticodon helix. Both adenines form hydrogen bonds with the 2′ ribose hydroxyl groups of the codon–anticodon base pairs: a motif termed "A-minor" (Fig. 5). These types of tertiary interactions involving the two adenines are not possible if the codon–anticodon base pairing is noncognate, that is, not a Watson-Crick pair. Similarly, DNA polymerase and RNA polymerase use specific arginine and glutamine residues in their active center, which recognize the shape of Watson-Crick base pairs via the minor groove. A third residue, guanine530, is positioned on a loop that belongs to the 30S subunit shoulder and interacts with the second and the wobble-position base pair (Fig. 5). Consistent with the wobble base pairing in this position, the interaction of G530 with the third base pair is less sensitive to mismatches than the interaction described above involving the two adenines. The structure of 30S subunit, mRNA, and a cognate anticodon stem-loop shows that upon binding of the cognate tRNA, G530 flips from a *syn* to an *anti* conformation (Fig. 6). Near-cognate tRNA (with a single mismatch) does not cause flipping out of the adenines or G530. Consistent with their important role in decoding, A1492, A1493, and G530 are universally conserved and essential for cell viability.

In summary, the ribosome precisely monitors the minor groove geometry of the codon–anticodon interaction at the first two positions in the A site. Therefore, codons that differ in the first or second base but encode the same amino acid must pair with different tRNAs. The third position allows mismatches, which explains the frequent appearance of inosine in the anticodon at this position. Inosine can pair with cytosine, adenine, and uridine, thus maximizing the number of codons recognized by the specific tRNA molecules.

The close contacts of the cognate codon–anticodon pair with A1492, A1493, and G530 induce a characteristic closing of the 30S subunit around the A site. This locking up involves conformational changes of the overall 30S architecture. In particular, these are rotations of the

Fig. 5 Monitoring of the codon–anticodon interactions in the decoding center. (a), (b), and (c) show the minor groove recognition at the first (a), second (b) and third base position (c) respectively. (a) Adenine1493 binds into the minor groove of the first AU base pair. It forms hydrogen bonds to the 2′OH group of the tRNA with its N1 atom and to the 2′ OH group of the mRNA base with its 2′OH group. An additional hydrogen bond is formed between the 2′OH group of A1493 and the O2 group of uracil in the mRNA. (b) In the second position, A1492 and G530 both bind into the minor groove of the codon–anticodon helix and form hydrogen bonds with their N1 atoms. A1492 interacts with the 2′OH group of the mRNA via its N3 and 2′OH group. G530 also forms two hydrogen bonds with its 2′OH and N3 and the 2′OH group of the tRNA respectively. (c) The third base pair is monitored less stringently (wobble position). O6 of G530 forms a hydrogen bond to the 2′OH group of the mRNA. A second, magnesium ion-mediated interaction occurs between the 2′OH group of the mRNA and O2 of C518, as well as the main chain carbonyl of proline48 of protein S12. The GU mismatch is not selected against at this third position.

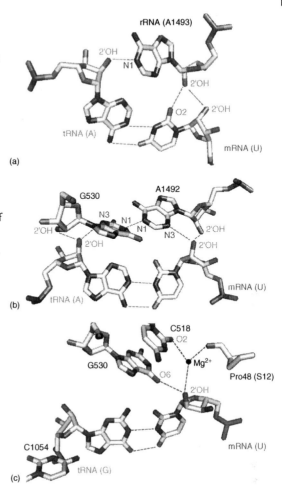

head toward the shoulder and the subunit interface, as well as movement of the shoulder toward the interface and the platform. Particularly well characterized is the "helix 27 accuracy switch" during the transition to the closed conformation. In this case, the packing of helix 27 to helix 44 changes because of a disrupted RNA turn. Thus, the structure of the decoding center is altered. Noncognate and near-cognate tRNA binds to the open conformation and cannot induce closing of the 30S subunit. In the closed conformation, the aminoacyl-tRNA cannot leave the A site. In conclusion, *induced fit* is a further source of accuracy, in addition to the A1492-, A1493-, and G530-based proofreading mechanism (substrate recognition) and free energy gains of codon–anticodon interaction.

During *initial selection*, the specific aminoacyl-tRNA is still bound by EF-Tu, and the acceptor stem of the aminoacyl-tRNA is not in contact with the peptidyl transferase center of the large ribosomal subunit. In cryoEM studies of the ternary complex bound to the 70S ribosome stalled with kirromycin resembling the

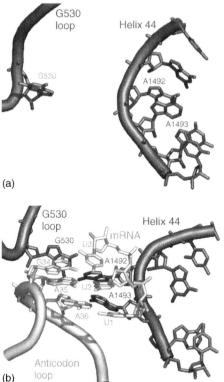

Fig. 6 (a) The empty decoding center of the native 30S subunit. The two 16S rRNA loops in the catalytic center are depicted in gray. The bases A1492 and A1493 are stacked in the helix 44 loop interior. G530 is in the *syn* conformation. (b) The decoding center in the presence of mRNA and the cognate tRNA anticodon loop. The mRNA and the tRNA anticodon loop are shown in white. A1492 and A1493 are now flipped out of helix 44 and pack tightly into the minor grove of the codon–anticodon helix monitoring the first two codon positions. G530 changes its conformation to *anti* and binds in the minor groove of the codon–anticodon helix monitoring the second and the third codon position.

GTPase activated state, tRNA adopts an unusual conformation that appears to facilitate codon–anticodon interaction. Cognate tRNA induces the closed conformation of the 30S subunit and triggers GTP hydrolysis by EF-Tu bound to the large subunit. Subsequently, the 3'CCA end of the aminoacyl-tRNA is released by the EF-Tu·GDP complex and can then rotate into the A site of the peptidyl transferase center. This *accommodation* process needs the full-size tRNA molecule, indicating that tRNA may play an active role in intersubunit signaling. Indeed, in the cryoEM studies of the kirromycin stalled ternary complex, the elbow region of the aminoacyl-tRNA could contact the GTPase associated center of the 50S subunit (while EF-Tu binds the sarcin–ricin loop) during initial selection and thus could induce GTP hydrolysis by a conformational change. However, since the data is at low resolution, it can also support a different model of GTPase activation triggered by conformational changes in both ribosomal subunits transmitted from the 30S subunit to the 50S subunit through intersubunit bridges. In fact, a significant movement of the two subunits relative to each other is observed during accommodation Accommodation can lead to dissociation of the aminoacyl-tRNA from the ribosome. This provides a *second proofreading step* during translation because the interactions between the ribosome and the codon–anticodon minor groove must be sufficiently strong to prevent the dissociation of the A site tRNA in the absence of EF-Tu. In summary, two selection stages exist, which are separated by irreversible GTP hydrolysis.

Clearly, there is a sensitive equilibrium between the open and closed conformation of the 30S subunit during decoding, and stabilization of one or the other conformation seriously affects translation accuracy. The transition into the closed conformation reduces polar interactions present at the S4/S5 interface. Destabilizing mutations in S4 and S5 residing between the shoulder and platform facilitate the transition to the closed conformation. This results in a higher affinity for tRNA and thus in an *error-prone ribosome*. In the closed conformation, S12 can form new salt bridges to helices 27 and 44. Destabilizing mutations of the involved residues result in reduced affinity for tRNA. Thus, the open conformation is favored and the ribosome is in a *hyperaccurate state*.

2.4
Modes of Antibiotic Inhibition

Structures of antibiotics bound to both ribosomal subunits provided detailed information on antibiotic action and the mechanistic basis of antibiotic resistance. In addition, further insights into the translation process were derived from such analyses. About 20 structures of antibiotics bound to the ribosome are available to date. Most of the antibiotics examined bind to functionally important parts of the 30S subunit at or near the decoding center or at locations that are important for conformational changes during translocation. Here, we focus on two 30S subunit binding antibiotics, paromomycin and streptomycin, because the mechanism of their action can be explained with the obtained structures.

The binding of the error-inducing aminoglycoside antibiotic *paromomycin* has been structurally characterized by both, NMR using a fragment of 16S rRNA and by crystallography of the 30S subunit in complex with paromomycin, mRNA, and a near-cognate tRNA anticodon stem-loop. Paromomycin binds to helix 44 inducing a conformational change where crucial bases of the decoding center, A1492 and A1493, are flipped out of the helix and interact directly with the minor groove of the codon–anticodon helix. This is the same conformation otherwise observed upon binding of cognate tRNAs. Thus, paromomycin decreases ribosome's selectivity.

Streptomycin acts during the accommodation step by stabilization of the overall closed conformation of 30S subunit, leading to error-prone translation. Streptomycin binds tightly to helix 27, helix 44, the 530 loop, and protein S12. Resistance to streptomycin is conferred by mutations in S12 that destabilize the closed conformation of the 30S subunit, leading to a hyperaccurate state.

2.5
Translation Initiation – Interactions with Initiation Factors

The initiation factors IF1 and IF3 bind to the small ribosomal subunit. Their structures have been solved in complex with the 30S subunit. IF1 prevents tRNA binding to the A site on the 30S subunit. In the IF1–30S complex, IF1 binds to a site that includes the A site of the small subunit, that is, helix 44, the 530 loop, and protein S12. The ribosomal binding site of IF1 contains many basic residues whereas the solvent-exposed site is rich in negative charges. A loop of IF1 interacts with the minor groove of helix 44. The two adenine bases, A1492 and A1493, are flipped out of helix 44 upon IF1 binding and are buried in IF1. In addition, IF1 induces a conformational change of helix 44 and shifts the relative orientations of the 30S domains. The head, platform, and the shoulder rotate toward the A site, adopting a conformation that probably resembles an intermediate between subunit association and dissociation.

IF2 binds to the acceptor stem of the initiator tRNA, and it interacts with IF1. The IF2 GTPase domain presumably binds to the GTPase factor binding site on the 50S subunit, but direct location of IF2 has not been determined yet.

IF3 binds tightly to the 30S subunit and interferes with its association to the 50S subunit. In addition, it helps in selecting the N-formyl methionine-tRNA by excluding other aminoacyl-tRNAs from the P site. IF3 has two distinct domains, of which the C-terminal domain has been shown to be sufficient for 70S dissociation. The location of the C-terminal domain of IF3 in the X-ray structure of the 30S–IF3 complex is not the same as observed by cryoEM nor is it as derived from biochemical data obtained by hydroxyl radical cleavage. In the cryoEM structure, the N-terminal domain of IF3 binds to the tRNA binding cleft of the 30S subunit, whereas the C-terminal domain is found at the interface side of the platform and the neck of the 30S subunit, which is the E site. Thus, IF3 is spanning the interface region from the platform to the neck of the 30S subunit. At this position, the C-terminal domain IF3 would be adjacent to the P site and would directly interfere with subunit association. IF3 has been shown to bind to three base pairs of the anticodon stem-loop that are unique to initiator tRNA. Therefore, the C-terminal domain is probably responsible for recognizing and selecting the initiator tRNA without interfering with the codon–anticodon interaction in the P site. In contrast to biochemical and cryoEM data, IF3 was visualized at the upper end of the platform on the solvent-exposed side of the 30S subunit in the X-ray structure of the 30S–IF3 complex. In that case, the negative effect of IF3 on subunit association would have to be much more indirect, for instance, by influencing the conformational state of the 30S subunit.

In summary, IF1 binds to the A site on the small subunit, IF2 binds over the A site in complex with the initiator tRNA which occupies the P site, and IF3 binds to the ribosomal E site. Thus, in the initiation complex, all tRNA binding sites are occupied and IF2 and IF3 both select for the correct initiator tRNA.

3
The 50S Ribosome Structure

The large ribosomal subunit is mostly composed of 23S and 5S rRNA and contains approximately 31 to 35 proteins (L1–L35); the exact amount of protein is species dependent. The ribosome core is well conserved in all species and consists mainly of ribosomal RNA. The 50S structure shows three characteristic protuberances: the L1 stalk, the central protuberance, and the L7/L12 stalk (Fig. 4). The proteins are not distributed equally over the ribosome but cluster at the solvent-exposed surface. Their extensions play an important role in the stabilization of the RNA tertiary structure. Similar to the structure of the small subunit, proteins are largely absent from the functionally important sites, namely, the peptidyl transferase center and the interface. 23S rRNA forms a compact, single hemispherical domain, which is the body of the large subunit (Fig. 4). The protuberances consist of 23S rRNA, proteins L1, L7, L10, L11, and L12 and 5S rRNA in case of the central protuberance. The stalks are involved in bridges to the small subunit or interact with tRNAs or GTP-binding translation factors and thus are likely to be dynamic elements of the 50S subunit.

The first 50S X-ray structure obtained from the archaebacterium *H. marismortui* was resolved to 2.4 Å resolution (Fig. 4). The structure highlighted two paramount features of the large subunit, the peptidyl transferase center and the exit tunnel. It provided a detailed overview of the overall architecture and important information about RNA-protein interactions. Furthermore, water and metal ion positions as well as base modifications could be identified, all of which may contribute to the folding and stabilization of the ribosome. Subsequently, the *D. radiodurans* 50S structure was solved at 3.1 Å resolution. While being very similar to the archaeal large subunit, this structure provided additional information on some regions that were disordered in the *Haloarcula* 50S subunit. The disordered regions were proteins L1, L10, L11, L12, that is, components of the two lateral protuberances, the L11 RNA region, and some RNA stem-loops contacting the small ribosomal subunit. These regions are likely to be flexible in the isolated 50S subunit and could adopt a more defined conformation upon association with their binding partner. Stabilizing elements of the large ribosomal subunit are proteins, water, metal ions, and tertiary RNA interactions named A-minor motifs (as observed in the decoding center).

3.1
The Molecular Mechanism of Peptidyl Transfer

The peptidyl transferase center is located at the bottom of a large cleft in the center of the subunit interface (Fig. 4). Although there was mounting evidence that ribosomal proteins are not involved in peptidyl transfer, unambiguous demonstration that the peptidyl transferase center is entirely composed of 23S rRNA was provided with the *Haloarcula* 50S structure in complex with CCdA-p-Puromycin (the Yarus inhibitor). In fact, there is no ribosomal protein present in the proximity of the peptidyl transferase center within a distance of 18 Å. The catalytic site is formed by the highly conserved central loop (the peptidyl transferase loop) of domain V, which is located at the bottom of the cleft.

Once the cognate aminoacyl-tRNA enters the ribosomal A site of the large subunit, the peptidyl transfer occurs spontaneously, that is, without additional energy input. On the basis of the structure of the 50S subunit in complex with CCdA-p-Puromycin and a number of substrate and product analogs, several roles of the ribosome in catalysis were suggested.

First, the ribosome closely brings the two tRNAs together and orients them precisely (*substrate orientation*). The acceptor stems of the aminoacyl-tRNA (mimicked by an aminoacylated minihelix) and the peptidyl-tRNA (mimicked by CCA phenylalanine caproic acid biotin (CCApcb)) point into the cleft, and their 3'CCA ends are proximate to each other (Fig. 7). The 3'CCA sequence of the A-site tRNA interacts with the A loop (helix 92 of 23S rRNA) in the catalytic center, and the 3'CCA sequence of the P-site tRNA base-pairs with the P loop (helix 80 of 23S rRNA) (Fig. 7). The CCA sequence alone can bind to the large subunit, indicating that this is the most important interaction of tRNA with the 50S subunit. By base-pairing with the CCA ends, both RNA loops position the two acylated tRNAs precisely such that the

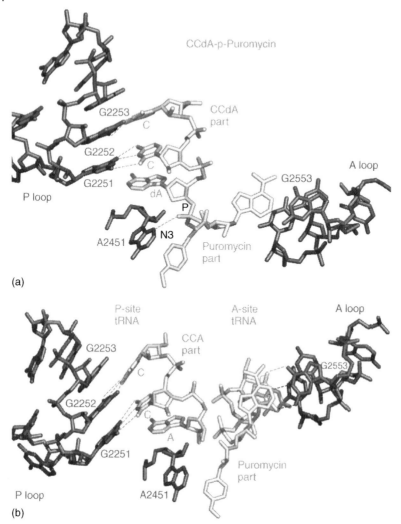

Fig. 7 (a) The structure of the peptidyl transferase center of the 50S ribosomal subunit. The A and P loop of 23S rRNA are shown in gray, the CCdA-p-Puromycin in white. The inhibitor occupies both the tRNA binding sites. The phosphoramide of the inhibitor is close to N3 of adenine2451. A2451 is part of the central loop in the peptidyl transferase center and may play a role in catalysis. (b) The peptidyl transferase center with a P-site and an A-site substrate analogs. In the P site, the cytosines of the CCA pair with two guanines G2251 and G2252 of the P loop. In the A site, the second base of CCA, the cytosine pairs with guanine 2553 of the A loop. The two substrate analogs mimic the situation in the active center immediately after peptidyl transfer.

α-amino group of the A-site aminoacyl-tRNA is close enough to the carbonyl carbon of the P-site peptidyl-tRNA for the nucleophilic attack (Fig. 2).

Second, *transition state stabilization* could lower the activation energy barrier as suggested by the 50S structure in complex with the CCdA-Puromycin inhibitor, which resembles a reaction intermediate. CCdA-p-Puromycin (Fig. 2) binds tightly to the P and A site loops and inhibits peptidyl transfer (Fig. 7a). Nevertheless, this compound is not a good analog because of the nonchiral phosphorus atom, the delocalized negative charge, and the missing 2'OH group on the CCdA part.

The third contribution to catalysis, suggested based on the *Haloarcula* structure with CCdA-p-Puromycin, could be the N3 of A2451 (*E. coli* numbering), which may participate in *acid–base catalysis*. This adenine is closest to the tetrahedral intermediate mimic of CCdA-p-Puromycin (Fig. 7a). Thus, N3 of A2451 could hydrogen-bond and abstract a proton from the amino group of the tetrahedral intermediate facilitating the release of the leaving group.

The pH-dependence of the peptidyl transfer reaction was analyzed by quench flow assays in which purified ribosomal complexes containing radioactive peptidyl-tRNA in the P site were incubated with puromycin at saturating concentrations. At pH 7.5, reaction rates of $50\,s^{-1}$ were obtained, comparable with *in vivo* protein synthesis. At lower pH, the peptidyl transfer was 100-fold slower. Taking into account the pK_a of puromycin, the data indicated the existence of a single titratable group in the catalytic center with a pK_a of 7.5. The pH effect disappeared when an A2451U mutant was analyzed and the reaction rate was decreased by a factor of 130, consistent with the idea that A2451 acts as a general base. Nevertheless, alternative explanations involving protonation of neighboring residues are also possible.

Although the role of A2451 in transition state stabilization and acid–base catalysis is unclear, it certainly plays an important role in peptidyl transfer by substrate orientation and hydrogen bond formation to the α-amino group of aminoacyl-tRNA.

3.2
The Exit Tunnel

The structure of the large ribosomal subunit reveals a tunnel spanning its entire body through which the peptide emerges while still connected to the peptidyl transferase center (Fig. 8). The tunnel is approximately 100 Å in length and has a diameter of about 15 Å. The polypeptide exit tunnel is largely formed by 23S rRNA domains I and II but shows significant contributions from ribosomal proteins L4 and L22 (Fig. 8). Approximately 28-Å apart from the tunnel, exit proteins L4 and L22 fence the most constricted part of the tunnel with a diameter of 10 Å (Fig. 9). The opening of the exit tunnel is at the bottom of the back side of the 50S subunit. The exit is encircled by proteins L19, L22, L23, L24, L29, and L31 (Fig. 10).

The tunnel can accommodate 30 to 40 amino acid residues of a growing polypeptide chain and can therefore protect it from protease digestion or immunoprecipitation with antibodies specific for a nascent polypeptide. In fact, these initial experiments led to the proposal of the existence of a ribosomal tunnel about 35 years ago. Little is known about the conformation of the nascent polypeptide in the tunnel. Since the diameter of the tunnel is 10 Å at its narrowest part, the peptide could adopt an α-helical structure,

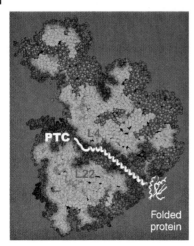

Fig. 8 The polypeptide exit tunnel. The 50S subunit is cut in half such that the tunnel is shown in its entire length. The ribosome atoms are shown in a space-filling representation. All RNA atoms that do not contact solvent are shown in white, and surface atoms are color-labeled with oxygen red, carbon yellow, and nitrogen blue. Proteins are depicted in green, and surface residues are in the same color code as RNA surface atoms. A model of a nascent chain polypeptide passing the tunnel is shown in white ribbon. The polypeptide is still connected to the peptidyl transferase center and can adopt an α-helical conformation in the tunnel. Outside the tunnel, the polypeptide chain can fold cotranslationally adopting its tertiary structure. The narrowest part of the tunnel is constricted by the two proteins L4 and L22. (See color plate p. xix).

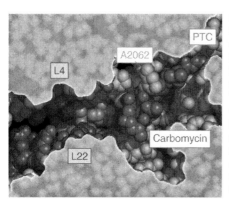

Fig. 9 A section of the ribosomal tunnel with Carbomycin A bound. The first and most constricted half of the tunnel is shown in a longitudinal section. All rRNA atoms are shown in yellow, and protein atoms of L4 and L22 are colored blue in a space-filling representation. The cutting plane is gray. The macrolide antibiotic Carbomycin A (red) is bound to the 50S tunnel blocking the passage of the nascent chain, and its disaccharide isobutyrate extension reaches to the peptidyl transferase center (PTC). A covalent bond is formed between N6 of A2062 of 23S rRNA and an ethylaldehyde substituent of Carbomycin A. (See color plate p. xxi).

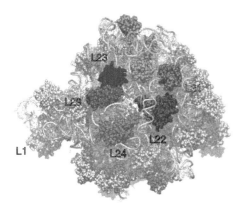

Fig. 10 The tunnel exit of the 50S subunit. Proteins at the rim of the exit tunnel are shown in the different colors. Some of the proteins are suggested to interact with the translocation machinery. L23 and L29 are implied to play a role in protein folding and transport across membranes based on their interaction with chaperones and SRP. The 23S rRNA and other ribosomal proteins are shown in white. (See color plate p. xix).

which would be approximately 6 to 10 Å in diameter (Fig. 8). However, it appears to be unlikely that tertiary protein folding can occur within the tunnel although it widens toward the exit. 23S rRNA can be cross-linked to the nascent chains, although the surface of the tunnel should prevent tight binding of unfolded nascent polypeptides. The tunnel surface is largely composed of hydrophilic, noncharged groups, but no large hydrophobic patches that could form extensive interactions with exposed hydrophobic amino acids are found. Thus, because of selective pressure, the tunnel should allow the passage of any nascent protein with naturally occurring sequence.

3.3 Antibiotics Targeting the Large Subunit

The large ribosomal subunit is targeted by many antibiotics and substrate analogs. These compounds bind to the functionally important regions of the subunit such as the GTPase factor binding center, the peptidyl transferase center and the nascent polypeptide exit tunnel. Because of structural studies, detailed information of their mode of function is currently available for some of the peptidyl transferase inhibitors and the macrolide class of antibiotics.

Chloramphenicol prevents the accurate binding of aminoacyl-tRNA to the ribosomal A site. In contrast to puromycin that acts as an aminoacyl-tRNA analog, chloramphenicol interferes with the exact positioning of the aminoacyl moiety in the A site. Therefore, the formation of the tetrahedral transition state during the peptidyl transfer reaction may be hampered. In addition, the electronegative dichloromethyl group of chloramphenicol is close to the amino group of A site–bound aminoacyl-tRNA, and this may interfere with its nucleophilic attack on the carbonyl carbon atom of the ester bond that links the nascent polypeptide to the P-site anchored tRNA. The other antibiotic that interferes with substrate binding is *Clindamycin*. Binding of this compound overlaps with both A- and P site–bound substrates.

Macrolide antibiotics such as erythromycin block the diffusion of the nascent polypeptide into the tunnel. The exact binding site is found between the peptidyl transferase center and the narrowest part of the tunnel fenced by protein L4 and L22 (Fig. 9). The lactone ring of erythromycin binds with its hydrophobic face directed toward the tunnel wall forming van der Waals contacts with the tunnel surface, whereas its hydrophilic side is exposed. The sugar moieties of the macrolide form hydrogen bonds to the rRNA. Erythromycin binding reduces the diameter of the tunnel from 18 to 19 Å to less than 10 Å, thereby interfering with the path of the elongating peptide. Conversely, if the nascent chain is already in the tunnel, protein synthesis is not inhibited by erythromycin. Mutations in L4 and L22 causing an enlargement or a further constriction of the tunnel confers resistance to erythromycin. These mutations allow the passage of the polypeptide even though erythromycin is bound, or they prevent macrolide binding through conformational changes in the rRNA. Erythromycin binding does not inhibit the peptidyl transferase reaction, and the ribosome can synthesize short tri- or tetrapeptides in the presence of erythromycin. There is a clear correlation between the length of the macrolide substituent at the C5 position of the lactone ring extending toward the active center and the maximal length of the polypeptide that can still be translated. *Carbomycin A*, a macrolide antibiotic with an extended disaccharide isobutyrate substituent on the lactone ring, interferes directly with the peptidyl transfer reaction since its extension reaches to the peptidyl transferase center. In addition to the van der Waals interactions and the hydrogen bonds, some macrolides like carbomycin A carrying an ethylaldehyde substituent on the C8 position can form a reversible covalent bond with N6 of A2062 (Fig. 9).

3.4
Regulatory Nascent Chains

It was recently discovered that certain polypeptide chains interact with the ribosomal exit tunnel. This contradicts the idea that the tunnel is a neutral environment for the exit of polypeptides and indicates that there is interaction between certain nascent peptides and the tunnel, which can regulate the rate of translation. Tight interaction between the nascent chain and the ribosomal tunnel leads to stalling of protein synthesis.

Such a stalling sequence motif was discovered at the C-terminus of the SecM (secretion monitor) protein of *E. coli* by mutational studies. It was determined that a particular sequence motif of 17 amino acids FxxxxWIxxxxGIRAGP is sufficient to cause ribosomal stalling. A genetic screen for stalling suppressor mutants revealed that the SecM peptide interacts tightly with 23S rRNA at the entrance of the exit tunnel and with protein L22 at the narrowest constriction of the tunnel, which is interestingly the same region in which macrolide antibiotics bind (Fig. 9). Accordingly, the stalling motif can be transferred from SecM to other proteins and causes translational arrest as well.

The real function of SecM is regulation of SecA expression through its N-terminal export signal. In the presence of abundant SecA, an ATPase, which is part of the translocation machinery, the stalled SecM peptide is "pulled" out of the ribosomal tunnel and delivered to the translocon. The basis of this regulation is a dicistronic mRNA encoding SecM and SecA, with a strong RNA stem-loop occluding the ribosome binding site of the secA gene. When SecA is missing, translation of the upstream secM gene causes ribosomal stalling, which resolves mRNA stem-loop and permits initiation of secA translation. In the presence of abundant SecA, translational arrest is abolished and the Shine-Dalgarno sequence of the secA gene is inaccessible for the ribosome, preventing SecA synthesis. Remarkably, interactions between the stalling sequence and the ribosome is so strong that overexpression of the SecM stalling sequence without the N-terminal signal sequence, recognized by SecA, is lethal for *E. coli*.

These results demonstrate that the exit tunnel is not a passive environment for the nascent chain. In fact, there are increasing number of effector sequences that either interfere with protein elongation or translation termination. The individual effector sequences share little sequence homology, although several sequences have positively charged residues, which could interact with ribosomal RNA, and the C-terminal residue is frequently a proline.

A distinct mechanism of translational arrest is mediated by binding of the signal recognition particle (SRP). When the translated polypeptide chain has a signal sequence at the N terminus, this sequence is recognized by SRP. The binding of the SRP to the ribosome-nascent chain complex causes a translation arrest, which is resolved only when the peptide binds to the translocon. The stalling effect could either result from a conformational change of the active site of the GTPase factor binding center upon SRP binding to the ribosome, or it could be transmitted in a more subtle way through the peptide itself.

3.5
The GTPase Factor-Binding Center

The elongation factors EF-Tu and EF-G, as well as the other GTP-binding translation factors such as initiation factor IF2 and release factor RF3 interact with the large ribosomal subunit. The location of the GTPase domain binding site, named factor binding center, is known from cryoEM studies. The factor binding center includes the highly conserved *sarcin–ricin loop* (stem-loop 95 of 23S rRNA, Fig. 4), which is absolutely essential for factor binding. Ribosomes can be inactivated by the cleavage of a single covalent bond in the sarcin–ricin stem-loop. The rest of the factor binding region is formed by proteins L14, L23, and L6 on the body of the 50S subunit and proteins L11, L10, and L7/L12 on the flexible stalk. Conformational changes of the ribosome and the elongation factors upon GTP-hydrolysis were visualized by cryoEM. Sequence comparisons of EF-Tu, EF-G, IF2, and RF3 suggest that they all contain two domains with a similar fold that are responsible for binding to the GTPase activating center. This implies that they all bind to the ribosome in a similar fashion.

3.6
Structural Mimicry

Crystallographic studies revealed a *structural mimicry* of the ternary complex comprising tRNA, EF-Tu and GTP and EF-G:GTP. The tRNA shape is almost perfectly mimicked by the three domains of EF-G. This mimicry was further supported by cryoEM data of EF-G bound to the A site on the ribosome.

In fact, several proteins have been identified with a similar shape as the tRNA including release factors eRF1 and RF2, the ribosome recycling factor (RRF), and the tetracycline resistance protein. Clearly, there is also *functional mimicry* between aminoacyl-tRNAs and release factors RF1 and RF2, since both bind to the ribosomal A site, recognize specific mRNA codons, and simultaneously interact with the peptidyl transferase center. In prokaryotic RF1 and RF2, triplets of aminoacids have been identified that are responsible for the RF specificity. In eukaryotic eRF1, which recognizes all three stop codons, a highly conserved sequence motif has been identified as the "anticodon" serving region. It is assumed that the decoding regions of RFs directly interact with the mRNA codon in the ribosome. However, the detailed decoding mechanism of stop codons and the discrimination against sense codons by RFs is still unclear. The catalytic mechanism of peptide release is another unsolved question. Genetic studies suggest that the mechanism of peptide release is distinct from peptide formation but involves 23S rRNA as well.

The final step in translation termination is the disassembly of the ribosomal complex and recycling of the ribosomal subunits. This step is supported by EF-G and RRF, which is a nearly perfect mimic of tRNA.

4
The 70S Ribosome Structure

The *T. thermophilus* 70S structure at 5.5 Å, containing mRNA and three tRNAs, sheds light on the interactions between the 50S and 30S subunit (Fig. 4). In comparison to the isolated subunits, 70S ribosomes show certain conformational differences, most of them stem directly from intersubunit association. Some flexible regions of the 50S structure are ordered in the 70S ribosome, indicating that these elements may adopt a defined geometry upon 30S association.

4.1
The 30S–50S Interface

The 70S structure reveals a network of intermolecular bridges involving RNA–RNA and RNA–protein interactions as well as a single protein–protein interaction. Mobile *intersubunit bridges* are suggested to play a major role in translocation because they allow movement of the ribosomal subunits with respect to each other. Viewed from the interface, ribosomal proteins are mainly located at the periphery of the 30S and 50S subunits (Fig. 4). The interface consists predominantly of RNA: twelve individual intersubunit bridges were identified, most of which are RNA helices. These contacts mostly involve minor groove–minor groove interactions. The RNA bridges are located in the center of the interface, on the 30S platform and helix 44, adjacent to the tRNA binding sites. Contacts involving proteins are peripheral to the functional sites. One of the most important intersubunit contacts seems to be the centrally located flexible bridge element formed by interaction of helix 69 of 23S rRNA with helix 44 of 16S rRNA at the base of the decoding center. Breakage and formation of many of these interface bridges probably accompanies translocation.

4.2
Translocation – The Concerted Movement of tRNAs and mRNA

After peptidyl transfer, the empty P-site tRNA and the peptidyl-tRNA in the A site have to move in concert with the mRNA. Several models have been proposed for translocation. Widely accepted is the *hybrid state model*, which involves independent tRNA movement with respect to the two subunits (Fig. 11). Pursuant to this model, tRNAs can occupy different tRNA binding sites on the small and large subunit, corresponding to a hybrid state. On the small subunit, the mRNA moves in complex with the tRNA anticodon stem-loops, while the acceptor stems of the tRNA are fixed in their respective binding sites on the large subunit (Fig. 11).

Chemical footprinting and cryoEM was used to monitor the movement of tRNAs through the ribosome. Following peptidyl transfer, it was found that the 50S A-site tRNA footprint disappeared and a 50S E-site tRNA footprint emerged instead. At the same time, the tRNA footprints on the small subunit did not change. The hybrid state model suggests a transition into a state in which the former A-site tRNA (A/A) still occupies the A site on the small subunit but its acceptor stem moved into the P site on the large subunit (A/P hybrid state). Concomitantly, the former P-site tRNA (P/P) still forms codon–anticodon contacts in the P site of the small subunit but, at the same time, moves away from the peptidyl transferase center on the large subunit into the E site (P/E hybrid state). This rearrangement is independent of EF-G. In cryoEM, only a small movement of the A-site acceptor stem toward the P site is visible after the peptidyl transfer reaction. In the *Haloarcula* 50S structure, the peptidyl-tRNA remained bound to the A loop in the peptidyl transferase center (the loop that binds A-site aminoacyl-tRNA, Fig. 7b). The P/E hybrid state has been observed by cryoEM, but only under stringent buffer conditions.

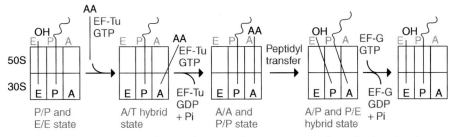

Fig. 11 The hybrid state model for translocation. The tRNAs move independently within the active sites of the two ribosomal subunits. Translocation from the A/T hybrid state into the A/A state depends on GTP hydrolysis by EF-Tu. Translocation into the P/E and A/P hybrid state occurs spontaneously after peptidyl transfer. In contrast, the translocation of the tRNAs and mRNA within the small subunit from the P/E : A/P hybrid states to the E/E : P/P state requires GTP hydrolysis by EF-G.

Upon incubation of the A/P and P/E hybrid state with EF-G and GTP, the tRNAs move to the P/P (occupying the P site on both subunits) and E/E site (Fig. 11), as evidenced by chemical footprinting. This means that upon GTP hydrolysis, EF-G induces tRNA movement with respect to the 30S subunit whereas movement with respect to the 50S subunit occurs spontaneously after peptidyl transfer. Curiously, GTPase activity of EF-G is not essential since translocation can still occur in the presence of nonhydrolyzable GTP analogs, although at a slower rate.

A third type of hybrid state (the *A/T hybrid state*, T for EF-Tu) is observed when the ternary complex of EF-Tu, GTP, and tRNA binds to the ribosome (Fig. 11). In chemical footprinting studies, this complex of tRNA EF-Tu and GTP-protected 16S rRNA in the 30S A site, but there was no tRNA footprint in the 50S A site. Instead, bases in the sarcin–ricin loop in the GTPase factor binding center of the 50S subunit were protected in a similar fashion as upon EF-G binding. After EF-Tu GTP hydrolysis and accommodation, the 50S A site becomes occupied by tRNA.

A prerequisite for a hybrid state is the mobility of the two subunits relative to each other or at least of parts of the subunits. Indeed, it has been observed by cryoEM that the two subunits can rotate 6° relative to each other upon EF-G binding. Following EF-G dependent GTP hydrolysis, translocation occurs, coupled with EF-G release and simultaneous back-rotation of the 30S subunit relative to the 50S subunit. In addition to subunit rotation, regions of the ribosomal subunits, such as intersubunit bridges, could move together with the tRNAs during translocation.

What exactly could be the driving force for the translocation in the 50S subunit? First, the peptidyl transferase reaction could directly provide the free energy for spontaneous translocation from the A/A state to the A/P state. Second, the tRNA binding sites on the large subunit could have different substrate specificities. In that case, the A site would have highest affinity for aminoacyl-tRNA, the P site a higher affinity for peptidyl-tRNA and the E site for deacylated tRNA. The latter possibility would explain the directionality of the process. There is evidence that peptidyl transfer and translocation occur in a sequential manner on the large subunit. Contrariwise, the translocation of tRNAs and mRNA on the small subunit requires energy input. The movement of the tRNA from the A/A state to the A/P hybrid state could decrease the affinity of the tRNA–mRNA helix for the small subunit. This would result in lower activation energy for translocation on the small subunit, provided that the affinity of tRNAs and the bound mRNA for 16S rRNA really presents an energy barrier for the movement between the tRNA binding sites on the small subunit. Moreover, if EF-G binding and GTP hydrolysis depend on a hybrid state of the ribosome, the coupling of peptidyl transfer reaction and translocation would be ensured.

4.3
Movement of tRNAs through the Ribosome

In the 70S structure, the tRNAs are in A/A, P/P, and E/E states rather than in any hybrid state. The large subunit makes contacts with the conserved structures of the tRNA, that is, the tRNA D-stem, elbow, and acceptor arm. The A and P site tRNAs are closest to each other at their acceptor stems, which are within 5-Å vicinity. In the small subunit, the anticodon ends of the tRNAs and the mRNA move only 20 to 28 Å between the A and P site or P and E site respectively. At the same time, the acceptor stem moves over much larger distances in the large subunit. For instance, the E site acceptor stem is rotated away from the P-site tRNA acceptor stem and directed into a separate cleft nearly 50 Å away from the peptidyl transferase cavity (Fig. 4). The L1 stalk and protein S7 seem to lock the deacylated tRNA in the E site as evidenced by the 70S X-ray structure. In the free 50S structure, the L1 stalk has an open conformation, in which it would not prevent release of E-site tRNA. A large rotational movement of tRNA is also necessary during its accommodation from the A/T hybrid state into the A site of the peptidyl transferase center.

4.4
Cotranslational Folding

Upon leaving the tunnel, the peptide chain can fold into its tertiary structure while still being bound to the ribosome. Cotranslational folding has been demonstrated, *in vitro* and *in vivo*, as well as in eukaryotic and prokaryotic cells. For large multidomain proteins, folding in a sequential manner may be beneficial to the folding process to avoid interference between individual protein domains during early folding steps. In the tunnel, the nascent polypeptide can adopt α-helical conformation (Fig. 8), but the diameter of the tunnel does not allow significant tertiary structure formation beyond that. Depending on the amino acid sequence of the nascent polypeptide, a different extent of secondary structure in the tunnel is acquired as indicated by different patterns of cross-linking to 23S rRNA. Moreover, depending on the sequence, different lengths of nascent polypeptides (30–72 amino acids) are protected by the ribosome, suggesting a varying extent of secondary structure formation. The opening of the tunnel is surrounded by ribosomal proteins L19, L22, L23, L24, L29, and L31 and regions of 23S rRNA. These elements may assist in protein folding (Fig. 10).

During translation, the ribosome-nascent peptide chain complexes can interact with various proteins involved in protein targeting and folding, for instance, molecular chaperones like the trigger factor, DnaK/DnaJ/GrpE in *E. coli*, and their homologs in eukaryotes. *Trigger factor*, a cis/trans prolyl isomerase that has protein-folding activity, was shown to be associated with the large ribosomal subunit when binding the nascent chain polypeptide. The N-terminal domain of trigger factor was identified as a ribosome binding domain. It could be cross-linked to proteins L23 and L29, which are located adjacent to each other at the tunnel exit site, exactly where the nascent chain emerges from the tunnel (Fig. 10). By mutational studies, it was shown that ribosome association is essential for trigger factor chaperone activity *in vitro* and *in vivo* and that an exposed, conserved glutamate (E18) of L23 protein is critical for the binding. Thus, ribosomal protein L23 directly links protein synthesis with chaperone-assisted folding.

4.5
Protein Secretion and Membrane Insertion

The ribosome participates directly in the transport of nascent chains into or across membranes to the periplasm or endoplasmic reticulum. Most of the secreted proteins contain an N-terminal hydrophobic signal sequence recognized by the *SRP*. Helicity as well as hydrophobicity of the *signal sequence* are necessary for the productive interaction with SRP and the translocation machinery. SRP binds and targets the ribosome-nascent chain complex to its receptor in the membrane. Subsequently, SRP hands over the ribosomal complex to its receptor, which coordinates the release of SRP from the ribosome-nascent chain complex and the insertion of the nascent chain into the translocation channel, which is built up by the Sec61 complex in eukaryotes and by SecYEG in prokaryotes.

In eukaryotes, SRP is composed of a 7S RNA and six proteins. The 54 kD protein SRP54 has been identified to bind the signal peptide and to interact with the signal receptor. SRP54 can be cross-linked to the large ribosomal subunit in the presence and absence of a nascent chain. Thus, this is clearly the ribosome binding domain of SRP. The two adjacent ribosomal proteins L23 and L35 (L29 in *E. coli*), both located at the rim of the tunnel exit, are the binding partners of SRP54, (Fig. 10). SRP proteins SRP9 and SRP14 are required for the SRP induced translational arrest (as discussed earlier). The nascent chain is released when SRP54 binds GTP and interacts with its receptor. The SRP receptor induces changes in the SRP binding to the ribosome and permits access of the translocation channel to the ribosome-nascent chain complex, such that the signal peptide is handed over to the Sec61 complex and translation continues. In the presence of SRP, direct binding of the Sec61 complex to the ribosomal complexes is blocked.

The structure of the eukaryotic ribosome-nascent chain complex and the eukaryotic *translocation machinery* Sec61 has been studied by cryoEM. The Sec61 complex consists of three transmembrane subunits α, β, and γ, comprising 12 transmembrane helices altogether. Three or four Sec61 complexes assemble to form the functional ringlike translocon with a central pore or indentation. The ringlike Sec61 complex binds to the large ribosomal subunit such that its central pore is aligned with the opening of the protein exit tunnel. Thereby, a path for the nascent chain is formed from the peptidyl transferase center directly into the endoplasmic reticulum. Four connections were identified between the Sec61 complex and the ribosome-nascent chain complexes in the 15-Å cryoEM structure. They involve mostly ribosomal RNA (28S) but also involve the eukaryotic homologs of the ribosomal proteins L19,

L23, L24 and L29 (Fig. 10); the region next to the tunnel exit is relatively conserved in prokaryotes and eukaryotes. The most substantial connection is found for proteins L23, L29, and helix7 of the rRNA. It is interesting to note that L23 and L29 are also implied to play a role in trigger factor and SRP binding, and this may be the reason why deletion of L23 is lethal.

The binding of the translocation machinery to the ribosome leaves a gap between ribosome and the transmembrane channel of at least 15 Å. This finding contradicts earlier postulations that the translocon/ribosome interface should form an ion-tight seal. The cryoEM structure suggests that the channel itself is the permeability barrier between cytoplasm and lumen.

Cotranslational folding also occurs during translocation of proteins into the endoplasmatic reticulum. The minimum distance between the folded domain and the peptidyl transferase center was determined to be 64 amino acids. This agrees well with the protection of 70 amino acids from proteolysis and the minimal peptide length of 70 amino acids needed for glycosylation by the glycosylase in the ER at the exit of the translocon pore. All biochemical and structural data support the idea that the translocon channel extends the ribosomal exit tunnel, even though there appears to be a gap between them. The translocon is probably a dynamic complex, since it allows lateral diffusion of the nascent chain into the membrane bilayer, which probably requires opening of the channel ring.

5
Perspectives

Recent crystal structures improved our understanding of the mechanism of protein synthesis. The peptidyl transferase reaction and the accuracy of aminoacyl-tRNA selection can now be studied with detailed structure-based experiments (Fig. 3). However, many questions remain to be addressed on a molecular level. Especially since information about intersubunit interactions, conformational changes during translation and interaction with translation factors originate from low-resolution structures. Ultimately, all relevant intermediates, complexes, and conformational states of the ribosome should be characterized at high resolution to completely understand translation and associated processes like cotranslational folding and translocation across the membrane.

Functionally important regions of the ribosome including the decoding center, the peptidyl transferase center, and the tunnel are highly conserved. Therefore, it is likely that the fundamental mechanism of translation is the same in all kingdoms of life. Most differences in the architecture of the eukaryotic 80S ribosome compared to the 70S ribosomes are generated by additional proteins and RNA insertions into the periphery of the structure. Nevertheless, eukaryotic translation is much more complex and subjected to a more diversified regulation. Extracting structural information on the eukaryotic ribosome will be undoubtedly difficult, but the task is well worth the extraordinary effort.

Bibliography

Books and reviews

Frank, J. (2001) Cryo-electron microscopy as an investigative tool: the ribosome as an example, *BioEssays* **23**, 725–732.

Kramer, G., Ramachandiran, V., Hardesty, B. (2001) Cotranslational folding–omnia mea mecum porto? *Int. J. Biochem. Cell Biol.* **33**, 541–553.

Moore, P.B., Steitz, T.A. (2002) The involvement of RNA in ribosome function, *Nature* **418**, 229–235.

Nakamura, Y., Ito, K. (2003) Making sense of mimic in translation termination, *Trends Biochem. Sci.* **28**, 99–105.

Nissen, P., Kjeldgaard, M., Nyborg, J. (2000) Macromolecular mimicry, *EMBO J.* **19**, 489–495.

Noller, H.F. (1991) Ribosomal RNA and translation, *Annu. Rev. Biochem.* **60**, 191–227.

Noller, H.F., Yusupov, M.M., Yusupova, G.Z., Baucom, A., Cate, J.H. (2002) Translocation of tRNA during protein synthesis, *FEBS Lett.* **514**, 11–16.

Ramakrishnan, V. (2002) Ribosome structure and the mechanism of translation, *Cell* **108**, 557–572.

Rodnina, M.V., Wintermeyer, W. (2001) Fidelity of aminoacyl-tRNA selection on the ribosome: kinetic and structural mechanisms, *Annu. Rev. Biochem.* **70**, 415–435.

Tenson, T., Ehrenberg, M. (2002) Regulatory nascent peptides in the ribosomal tunnel, *Cell* **108**, 591–594.

Primary Literature

Ban, N., Nissen, P., Hansen, J., Moore, P.B., Steitz, T.A. (2000) The complete atomic structure of the large ribosomal subunit at 2.4 A resolution, *Science* **289**, 905–920.

Bashan, A., Agmon, I., Zarivach, R., Schluenzen, F., Harms, J., Berisio, R., Bartels, H., Franceschi, F., Auerbach, T., Hansen, H.A.S., Kossoy, E., Kessler, M., Yonath, A. (2003) Structural basis of the ribosomal machinery for peptide bond formation, translocation, and nascent chain progression, *Mol. Cell* **11**, 91–102.

Beckmann, R., Spahn, C.M., Eswar, N., Helmers, J., Penczek, P.A., Sali, A., Frank, J., Blobel, G. (2001) Architecture of the protein-conducting channel associated with the translating 80S ribosome, *Cell* **107**, 361–372.

Blobel, G., Sabatini, D.D. (1970) Controlled proteolysis of nascent polypeptides in rat liver cell fractions. I. Location of the polypeptides within ribosomes, *J. Cell Biol.* **45**, 130–145.

Carter, A.P., Clemons, W.M., Brodersen, D.E., Morgan-Warren, R.J., Wimberly, B.T., Ramakrishnan, V. (2000) Functional insights from the structure of the 30S ribosomal subunit and its interactions with antibiotics, *Nature* **407**, 340–348.

Carter, A.P., Clemons, Jr., W.M., Brodersen, D.E., Morgan-Warren, R.J., Hartsch, T., Wimberly, B.T., Ramakrishnan, V.V. (2001) Crystal structure of an initiation factor bound to the 30S ribosomal subunit, *Science* **4**, 4.

Choi, K.M., Brimacombe, R. (1998) The path of the growing peptide chain through the 23S rRNA in the 50S ribosomal subunit; a comparative cross-linking study with three different peptide families, *Nucleic Acids Res.* **26**, 887–895.

Dallas, A., Noller, H.F. (2001) Interaction of translation initiation factor 3 with the 30S ribosomal subunit, *Mol. Cell* **8**, 855–864.

Frank, J., Agrawal, R.K. (2000) A ratchet-like inter-subunit reorganization of the ribosome during translocation, *Nature* **406**, 318–322.

Hansen, J.L., Schmeing, T.M., Moore, P.B., Steitz, T.A. (2002) Structural insights into peptide bond formation, *Proc. Natl. Acad. Sci. U. S. A.* **99**, 11670–11675.

Hansen, J.L., Ippolito, J.A., Ban, N., Nissen, P., Moore, P.B., Steitz, T.A. (2002) The structures of four macrolide antibiotics bound to the large ribosomal subunit, *Mol. Cell* **10**, 117–128.

Harms, J., Schluenzen, F., Zarivach, R., Bashan, A., Gat, S., Agmon, I., Bartels, H., Franceschi, F., Yonath, A. (2001) High resolution structure of the large ribosomal subunit from a mesophilic eubacterium, *Cell* **107**, 679–688.

Katunin, V.I., Muth, G.W., Strobel, S.A., Wintermeyer, W., Rodnina, M.V. (2002) Important contribution to catalysis of peptide bond formation by a single ionizing group within the ribosome, *Mol. Cell* **10**, 339–346.

Kowarik, M., Kung, S., Martoglio, B., Helenius, A. (2002) Protein folding during co-translational translocation in the endoplasmic reticulum, *Mol. Cell* **10**, 769–778.

Kramer, G., Rauch, T., Rist, W., Vorderwulbecke, S., Patzelt, H., Schulze-Specking, A., Ban, N., Deuerling, E., Bukau, B. (2002) L23 protein functions as a chaperone docking site on the ribosome, *Nature* **419**, 171–174.

Malkin, L.I., Rich, A. (1967) Partial resistance of nascent polypeptide chains to proteolytic digestion due to ribosomal shielding, *J. Mol. Biol.* **26**, 329–346.

McCutcheon, J.P., Agrawal, R.K., Philips, S.M., Grassucci, R.A., Gerchman, S.E., Clemons, Jr., W.M., Ramakrishnan, V., Frank, J. (1999) Location of translational initiation factor IF3 on the small ribosomal subunit, *Proc. Natl. Acad. Sci. U.S.A.* **96**, 4301–4306.

Menetret, J.F., Neuhof, A., Morgan, D.G., Plath, K., Radermacher, M., Rapoport, T.A., Akey, C.W. (2000) The structure of ribosome-channel complexes engaged in protein translocation, *Mol. Cell* **6**, 1219–1232.

Nakatogawa, H., Ito, K. (2002) The ribosomal exit tunnel functions as a discriminating gate, *Cell* **108**, 629–636.

Nissen, P., Hansen, J., Ban, N., Moore, P.B., Steitz, T.A. (2000) The structural basis of ribosome activity in peptide bond synthesis, *Science* **289**, 920–930.

Nissen, P., Ippolito, J.A., Ban, N., Moore, P.B., Steitz, T.A. (2001) RNA tertiary interactions in the large ribosomal subunit: the A-minor motif, *Proc. Natl. Acad. Sci. U.S.A.* **98**, 4899–4903.

Ogle, J.M., Murphy, F.V., Tarry, M.J., Ramakrishnan, V. (2002) Selection of tRNA by the ribosome requires a transition from an open to a closed form, *Cell* **111**, 721–732.

Ogle, J.M., Brodersen, D.E., Clemons, Jr., W.M., Tarry, M.J., Carter, A.P., Ramakrishnan, V. (2001) Recognition of cognate transfer RNA by the 30S ribosomal subunit, *Science* **292**, 897–902.

Piepenburg, O., Pape, T., Pleiss, J.A., Wintermeyer, W., Uhlenbeck, O.C., Rodnina, M.V. (2000) Intact aminoacyl-tRNA is required to trigger GTP hydrolysis by elongation factor Tu on the ribosome, *Biochemistry* **39**, 1734–1738.

Pioletti, M., Schluenzen, F., Harms, J., Zarivach, R., Gluehmann, M., Avila, H., Bashan, A., Bartels, H., Auerbach, T., Jacobi, C., Hartsch, T., Yonath, A., Franceschi, F. (2001) Crystal structures of complexes of the small ribosomal subunit with tetracycline, edeine and IF3, *EMBO J.* **20**, 1829–1839.

Pool, M.R., Stumm, J., Fulga, T.A., Sinning, I., Dobberstein, B. (2002) Distinct modes of signal recognition particle interaction with the ribosome, *Science* **297**, 1345–1348.

Schluenzen, F., Zarivach, R., Harms, J., Bashan, A., Tocilj, A., Albrecht, R., Yonath, A., Franceschi, F. (2001) Structural basis for the interaction of antibiotics with the peptidyl transferase centre in eubacteria, *Nature* **413**, 814–821.

Schluenzen, F., Tocilj, A., Zarivach, R., Harms, J., Gluehmann, M., Janell, D., Bashan, A., Bartels, H., Agmon, I., Franceschi, F., Yonath, A. (2000) Structure of functionally activated small ribosomal subunit at 3.3 angstroms resolution, *Cell* **102**, 615–623.

Schmeing, T.M., Seila, A.C., Hansen, J.L., Freeborn, B., Soukup, J.K., Scaringe, S.A., Strobel, S.A., Moore, P.B., Steitz, T.A. (2002) A pre-translocational intermediate in protein synthesis observed in crystals of enzymatically active 50S subunits, *Nat. Struct. Biol.* **4**, 4.

Stark, H., Rodnina, M.V., Wieden, H.J., Zemlin, F., Wintermeyer, W., van Heel, M. (2002) Ribosome interactions of aminoacyl-tRNA and elongation factor Tu in the codon-recognition complex, *Nat. Struct. Biol.* **9**, 849–854.

Valle, M., Sengupta, J., Swami, N.K., Grassucci, R.A., Burkhardt, N., Nierhaus, K.H., Agrawal, R.K., Frank, J. (2002) Cryo-EM reveals an active role for aminoacyl-tRNA in the accommodation process, *EMBO J.* **21**, 3557–3567.

Welch, M., Chastang, J., Yarus, M. (1995) An inhibitor of ribosomal peptidyl transferase using transition-state analogy, *Biochemistry* **34**, 385–390.

Wimberly, B.T., Brodersen, D.E., Clemons, Jr., W.M., Morgan-Warren, R.J., Carter, A.P., Vonrhein, C., Hartsch, T., Ramakrishnan, V. (2000) Structure of the 30S ribosomal subunit, *Nature* **407**, 327–339.

Yusupova, G.Z., Yusupov, M.M., Cate, J.H., Noller, H.F. (2001) The path of messenger RNA through the ribosome, *Cell* **106**, 233–241.

Yusupov, M.M., Yusupova, G.Z., Baucom, A., Lieberman, K., Earnest, T.N., Cate, J.H., Noller, H.F. (2001) Crystal structure of the ribosome at 5.5 Å resolution, *Science* **292**, 883–896.

17
Ubiquitin-Proteasome System for Controlling Cellular Protein Levels

Michael H. Glickman[1] and Aaron Ciechanover[2]
[1]*Department of Biology, Israel Institute of Technology, Haifa, Israel*
[2]*Faculty of Medicine, Israel Institute of Technology, Haifa, Israel*

1	**Mechanisms of Ubiquitination and Protein Degradation**	517
1.1	Ubiquitination 517	
1.2	Selection of Proteins for Degradation 519	
1.3	Degradation 520	
1.4	Regulation of the Ubiquitin System 521	
2	**Biological Processes Regulated by the Ubiquitin System**	522
3	**Modification by Other Ubiquitin-like Proteins**	524
4	**Ubiquitination in Health and Disease**	526
4.1	Cancer 526	
4.2	Neurological Disorders 528	
4.3	Cystic Fibrosis 529	
4.4	Immune Response and Inflammatory Disorders 530	
5	**Summary and Outlook**	531
	Bibliography	**531**
	Books and Reviews 531	
	Primary Literature 531	

Keywords

Deubiquitination
It is the process of removing a ubiquitin tag from a target protein. Deubiquitination is carried out by ubiquitin-specific proteases (also known as deubiquitinating enzymes or

Proteins. Edited by Robert A. Meyers.
Copyright © 2007 Wiley-VCH Verlag GmbH & Co. KGaA, Weinheim
ISBN: 978-3-527-31608-3

DUBs) that hydrolyze the peptide or isopeptide bond following the last residue of ubiquitin.

ER-associated Degradation (ERAD)
It removes proteins from the ER membrane or ER lumen. ERAD involves retrotranslocation of the target protein from the ER into the cytoplasm, ubiquitination, and proteolysis by the proteasome.

Proteasome (a.k.a. 26S proteasome)
It is a large, multisubunit ATP-dependent protease found in all eukaryotic cells responsible for most regulatory proteolysis. The proteasome is composed of a cylindrical *20S core particle* encompassing the proteolytically active subunits, and a *19S regulatory particle* that recognizes polyubiquitinated proteins and prepares them for proteolysis in the 20S core particle.

Proteolysis or Protein Degradation
It is the hydrolysis of peptide bonds in proteins yielding short peptides or amino acids.

Ubiquitin
(Ub) is a highly conserved, compact, and stable 76-residue protein found in all eukaryotes that can be covalently attached at its free carboxyl-terminus to amino groups on other proteins. This posttranslational modification of proteins by ubiquitin is called *ubiquitination* and is carried out by ubiquitinating enzymes.

Ubiquitinating Enzymes
They are a modular assembly of enzymes that are responsible for tagging a designated protein with *ubiquitin*. Ubiquitinating enzymes include an ATP-utilizing E1 that activates the free carboxyl-terminus of ubiquitin, a number of E2s that transfer the activated ubiquitin to the target protein, and numerous E3s that select unique substrates and mediate the action of the proper E2s.

Ubiquitination
It is the posttranslational modification of proteins with ubiquitin or a chain of multiple ubiquitin units. Usually, the outcome of ubiquitination is recognition and proteolysis by the *26S proteasome*. In some cases, ubiquitination does not lead to degradation but serves to alter intrinsic properties of the target protein, or direct its subcellular localization.

Ubiquitin-like Proteins
(Ubl) are small proteins that share homology with ubiquitin either structurally or in sequence. Similar to ubiquitin, most can be attached to the amino group of a lysine side chain on a target protein.

Cells contain many different kinds of proteins, each fulfilling structural, functional, or regulatory roles. Monitoring the state of all these proteins, as well as continuously adjusting their levels to suit demands is paramount to survival. The presence of damaged or mutated proteins, as well as altered levels of normal proteins could cause pathological conditions and even cell death. To exercise such quality control, cells are continuously spending energy both to synthesize new proteins, and to simultaneously degrade them, even though many may still be functional. An important characteristic of regulatory degradation is that it is specific; only the correct proteins are removed in a time-coordinated manner. Such extraordinary specificity is achieved by a modular system that identifies the proteins to be degraded, marks them by covalently attaching ubiquitin to an amino residue, and finally proteolyses the substrate by the 26S proteasome. Recognition of target proteins is carried out by a specific ubiquitin-protein ligase, called an E3. This protein recognizes the substrate and usually directs a ubiquitin-conjugating enzyme, an E2, to attach ubiquitin, a small 76 amino acid protein, onto the substrate. Ubiquitin molecules are often added to one another as well as to the substrate, resulting in chains of ubiquitin extending from the protein targeted for degradation. These polyubiquitin conjugates are then shuttled to the 26S proteasome, a large proteolytic complex, where they are degraded. Interestingly, ubiquitination is a reversible process, with deubiquitinating enzymes able to remove ubiquitin from the target before it can be recognized by the proteasome. Hence, transfer of the polyubiquitinated conjugate to the proteasome must happen swiftly or be shielded from these enzymes. The balance of these processes allows the ubiquitin-proteasome system to control the cellular levels and half lives of thousands of proteins making it a key player in basic biological pathways such as cell division, differentiation, signal transduction, trafficking, and quality control. Not surprisingly, aberrations in the system have been implicated in the pathogenesis of many diseases, certain malignancies, neurodegenerative disorders, inflammation, and immune response. Understanding the underlying mechanisms involved is important for the development of novel, mechanism-based drugs.

1
Mechanisms of Ubiquitination and Protein Degradation

1.1
Ubiquitination

The cellular levels of many proteins are kept in check by regulated degradation via the ubiquitin-proteasome pathway. Substrates are first tagged by covalent attachment of multiple ubiquitin molecules. These tagged proteins are then proteolysed by the 26S proteasome complex simultaneous with release of the ubiquitin tag. This last process is mediated by deubiquitinating enzymes (DUBs), a number of which are attached to the proteasome itself and work together to define the efficiency of the overall process.

Ubiquitin is a 76-residue protein and one of the most (if not the most) evolutionarily conserved proteins; its sequence and structure being almost identical in all eukaryotes. Conjugation of ubiquitin to the protein substrate requires the participation of a slew of ubiquitinating enzymes that are usually broken down into three distinct classes. Initially, the ubiquitin-activating enzyme (UBA; also known as E1), activates ubiquitin in an ATP-dependent reaction to generate a high-energy thiol ester intermediate between the carboxyl-terminus of ubiquitin and the active site cysteine of E1 (E1-S∼ubiquitin). One of several E2 enzymes (ubiquitin-conjugating enzymes; UBCs) transfers the activated ubiquitin from E1, via an additional high-energy thiol ester intermediate, E2-S∼ubiquitin, to the substrate that is specifically bound to a member of the ubiquitin-protein ligase family, E3 (Fig.1). E3s catalyze the last step in the conjugation process: covalent attachment of ubiquitin to the substrate.

The protein to be degraded is recognized by a specific E3. This protein directs the addition of ubiquitin from the E2 onto the substrate. There are several classes of E3 enzymes. Members of the RING finger-containing E3s, the largest family of ubiquitin ligases, mediate the transfer of the activated ubiquitin directly from the E2 to the E3-bound substrate. In the case of another class of E3, the so-called HECT (*Homologous to the E6-AP C-Terminus*) domain E3s, the ubiquitin is transferred from the E2 enzyme to an active site cysteine residue on the E3 to generate yet a third high-energy thiol ester intermediate, ubiquitin-S∼E3. Ubiquitin is then transferred from the E3 to the substrate protein that is bound to the ligase.

At the end of the ubiquitination cascade, the ubiquitin molecule is attached by

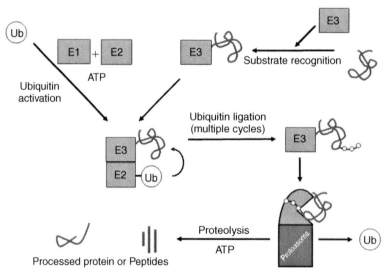

Fig. 1 General scheme of the ubiquitin system. Ubiquitin is activated by the ubiquitin-activating enzyme, E1, followed by its transfer to a ubiquitin-conjugating enzyme, E2. E2 transfers the activated ubiquitin to the protein substrate that is bound to a ubiquitin ligase, E3, which is specific for the substrate. Successive conjugation of ubiquitin moieties to one another generates a polyubiquitin chain that serves as the recognition signal for the 26S proteasome. The substrate is degraded by the 26S proteasome into short peptides, while ubiquitin is released by deubiquitinating enzymes (DUBs) for reuse.

its carboxyl-terminus to an amino group in the substrate protein. Generally, this is the amino group of an internal lysine side chain (an ε-NH_2 group), resulting in a covalent amide bond also known as an *isopeptide bond*. In some cases, ubiquitin is conjugated to the primary amino-terminus forming a linear amide bond; essentially a peptide bond. The ubiquitination reaction continues successively, adding an activated ubiquitin molecule to an internal lysine residue on the previously conjugated ubiquitin in each round, creating a polyubiquitin chain. In some instances, chain extension may require an additional auxiliary ligase called an *E4*. In this scenario, the first ubiquitin moiety is conjugated to the substrate by one E3, while chain elongation is catalyzed by a different ligase, the E4. It is thought that chains containing more than four ubiquitin moieties can bind to the proteasome with particularly high affinity, thus serving to target the protein to which they are conjugated for proteasomeal degradation.

1.2
Selection of Proteins for Degradation

A major task for the ubiquitin system is how to achieve high specificity and selectivity toward its numerous substrates. How to guarantee that the correct substrate will be removed whereas another, often similar protein will remain unscathed? In other words, why are certain proteins stable in the cell, while others are extremely short-lived? Complicating matters, some proteins are inherently short-lived and constitutively degraded while others are stable under most other conditions and degraded only upon a signal such as extracellular stimuli or a particular point in the cell cycle. Within the ubiquitin system, substrate selection is carried out predominantly by the E3 ligases. As such, E3s are the most diverse component of the pathway; to date, over 1000 different E3s or subunits of E3 ligase complexes have been identified in the human genome based on specific, commonly shared structural motifs. Each E3 is specific for a substrate or class of substrates.

Substrates must be recognized by a specific E3 and bind to it as a prerequisite to their ubiquitination. In most cases, however, the substrates are not recognized in a constitutive manner but must undergo posttranslational modification such as specific phosphorylation, dephosphorylation, or oxidation that renders them susceptible for recognition by an E3. In some cases, recognition of the target proteins by the E3 depends on association with an auxiliary protein (such as molecular chaperones) that serves to link the substrate with the appropriate ligase. In fact, many E3 ligases are themselves multisubunit complexes with each subunit participating in a unique network of protein–protein interactions. Certain subunits detach from, or dock to, the core E3 platform tweaking the ligase properties. Such modular behavior allows a limited set of proteins to target a much larger pool of substrates, each at the correct time and cellular context. Well-studied examples are the cullin-based E3 ligases, which regulate many steps of the cell cycle. The cullin subunit and the ring-finger component Rbx1 (plus a few additional subunits) serve as a stable platform while the substrate-recognition domain is a modular component that can be replaced as needed thus directing the cullin-based E3 to ubiquitinate a wholly new subset of substrates. In still other targeting mechanisms, substrates have to dissociate from a complexed form in order to be susceptible for E3 recognition and ubiquitination. This is a common feature of transcription factors that usually have to dissociate from their DNA segment before they can bind to an E3 ligase to be ubiquitinated. Stability of yet other proteins depends on oligomerization. Thus, modifying enzymes (such as kinases), ancillary proteins, or DNA sequences to which substrates bind, also play an important role in the recognition process, in addition obviously to the E3s. In some instances, it is the E3 that must "be switched on" by undergoing posttranslational modification in order to yield an active form that recognizes the target substrate more efficiently.

1.3
Degradation

Degradation of polyubiquitinated substrates is carried out by the 26S proteasome. This multicatalytic protease degrades polyubiquitinated proteins to short peptides and releases ubiquitin for reuse by the ubiquitination machinery. There are a few reports indicating that the 26S proteasome – *in vivo* as well as *in vitro* – can also proteolyse certain (usually unstable or misfolded) proteins that are not covalently modified with the ubiquitin tag. However, the proteasome is the only enzyme that can proteolyse polyubiquitinated proteins. It is composed of two subcomplexes: a 20S core particle (CP) that contains the catalytic sites and proteolytic activity, and a 19S regulatory particle (RP) that is essential for recognition and treatment of ubiquitinated substrates.

The 20S CP is a barrel-shaped structure composed of four stacked rings, two identical outer α-rings and two identical inner β-rings. The eukaryotic α- and β-rings are each composed of seven distinct subunits, giving the 20S complex the general structure of $\alpha_{1-7}\beta_{1-7}\beta_{1-7}\alpha_{1-7}$. Some (though not all) of the β-subunits are proteolytically active, with the remainder serving structural roles in forming the barrel structure. The proteolytically active β-subunits are threonine proteases, members of a unique N-terminal nucleophile hydrolase family. The active site nucleophile is the N-terminal threonine (the protein undergoes activation by posttranslational cleavage *after* it properly assembles into the 20S structure), while the proton acceptor is the N-terminus of the protein itself. This unique active site structure has drawn efforts to design proteasome-specific inhibitors that would not affect other cellular proteases. A number of proteasome inhibitors that target the 20S CP β-subunits are indeed available, with one, Valcade^{tm}, recently receiving FDA approval for treatment of multiple myeloma in humans.

Substrates must enter the lumen of the 20S CP where proteolysis occurs by transverseing through the α-ring at either of the two ends of the 20S barrel. It appears that they can enter from both ends simultaneously enhancing the proteolytic capacity of the proteasome. The α-ring at either of the two ends of the 20S barrel can be capped by a 19S RP. One important function of the 19S RP is to recognize ubiquitinated proteins and other potential substrates of the proteasome. Indeed, a number of polyubiquitin-binding subunits have been identified in the 19S RP, and are thought to recognize, transfer, bind, and anchor the tagged protein to the 19S RP. However, the interplay between these ubiquitin-binding subunits and how each contributes to proteasome function has yet to be deciphered. A second function of the 19S RP is to open an orifice in the α-ring of the 20S CP that will allow entry of the substrate into the internal proteolytic chamber. Even in its opened-state, the passage through the α-ring is too narrow to allow folded proteins through. It is assumed that the

19S regulatory particle unfolds substrates and inserts the unstructured polypeptide chain into the 20S CP. Both the channel opening function and the unfolding of the substrate require metabolic energy, and indeed, the 19S RP contains six different ATPase subunits. Biochemical characterization and mutational analysis have shown these ATPases to be involved in the unfolding and channel gating properties.

Following degradation of the substrate, short peptides derived from the substrate are released from the proteasome. The bulk of these peptides are further degraded in the cytoplasm into free amino acids by cytosolic amino- and carboxypeptidases. A small fraction of product peptides is transported by the TAP (transporter associated with antigen processing) across the ER membrane, further trimmed by specific proteases within the ER, bind to the MHC class I complex, and are carried to the cell surface where they are presented to cytotoxic T cells. In this manner, the proteasome is the major producer of antigenic peptides. Since the proteasome degrades both natural proteins as well as foreign (viral or mutated) proteins, both "self" and "nonself" antigens are produced. When these peptides are derived from a "nonself" protein, the T cell lyses the presenting cell. In rare instances, the proteasome may even splice nonconsecutive segments together, creating a new peptide with a novel sequence not found "as is" in the protein it was derived from. If substantiated as a general phenomenon, peptide splicing will have far reaching implications to the study of antigen presentation.

It should be emphasized that proteasomal degradation does not always continue to completion. In some cases, the proteasome can carve the ubiquitinated substrate in a limited manner, releasing a stable and functional truncated product. The best-studied example is the case of the NF-κB transcriptional regulator: an active subunit (p50 or p52) is generated by the proteasome from a longer inactive precursor (p105 or p100).

In addition to proteolysing the substrate (partially or in full) in the 20S CP, a totally unrelated proteolytic activity is present in the 19S RP and is intimately linked to proteasome function. Proteasomes hydrolyze the peptide or isopeptide bond immediately following the final residue of ubiquitin, thus severing the link between ubiquitin and the substrate or one ubiquitin unit and another in a polyubiquitin chain. This ubiquitin-specific protease activity has been mapped to a subunit of the 19S RP, Rpn11/POH1, as well as to a number of tightly associating components such as UCH37 and ubp6/USP14. The combined action of these DUBs releases ubiquitin, sparing it from being degraded along with the target protein and allowing it to be reused by the ubiquitin system.

1.4
Regulation of the Ubiquitin System

The ubiquitin-proteasome pathway can be regulated at the level of ubiquitination or at the level of proteasome activity. Since conjugation and proteasomal degradation are required for numerous cellular functions, regulation must be delicately executed. In a few cases, general regulation of the entire pathway is observed in response to physiological signals that call for increased or decreased proteolysis. For example, in fungi and plants, a dramatic increase in levels of ubiquitin, ubiquitinating enzymes, and proteasome subunits is observed under many stress conditions such as heat shock or oxidative damage. In animals, upregulation of the entire pathway is observed during massive degradation of skeletal muscle proteins that occurs under normal fasting, and also under pathological conditions such as malignancy-induced cachexia, severe sepsis, metabolic acidosis, or following denervation. In general, these conditions are correlated with accumulation of damaged or misfolded proteins, which should be efficiently removed for cell survival; hence, the need for increased rates of proteolysis. Conversely, it has recently been shown that some conditions require less proteolysis for survival of the cell. During long-term starvation or prolonged and severe heat exposure, proteasome activity is inhibited and general proteolysis rates are slowed down in yeast cells.

In most cases, however, regulation is specific targeting defined substrates that are recognized by specific ligases. By far, the greatest degree of molecular complexity in the ubiquitin system is among the E3 ligases. There are hundreds of ubiquitin ligases encoded in the genomes of multicellular eukaryotes, in addition to hundreds of substrate-recognition subunits that function together with the E3 ligases to target specific substrates. The modular nature of these ligase complexes ensures that slight changes in ratios of the components will direct the pathway to degradation of specific pathways. Thus, many substrate-recognition components of E3 ligases are either differentially expressed or stabilized only upon certain physiological signals.

E3 activity or specificity can also be regulated by posttranslational modifications of defined motifs within the ligase or the substrate. The targeting motif can be a single amino acid residue (e.g. the N-terminal residue), a specific sequence (the destruction box in cyclins), or a domain (such as a hydrophobic patch) that is not normally exposed. In other cases, the motif is generated by a posttranslational modification such as phosphorylation (two neighboring serine residues in the case of IκBα, or a single serine residue in the case of p27), or oxidation (hydroxyproline in the case of the hypoxia-inducible factor-1α — HIF1α). Interestingly, phosphorylation is required also to activate certain ligases, such as c-Cbl. Many proteasome subunits, both in the 20S CP and in the 19S RP can be modified by phosphorylation or glycosylation. These modifications alter proteasome stability or activity, though it is still unclear whether the effect is general toward all substrates, or more interestingly, influence the efficiency of degradation of a subset of substrates.

Eukaryotic genomes also usually encode several dozen different deubiquitinating enzymes. Such a large number suggests that some of them may act only on particular substrates, and raises the possibility that the rate of degradation of some proteins is determined by how fast ubiquitin is removed from them rather than by how fast it is added. Furthermore, substrate delivery to the 26S proteasome probably requires accessory or adaptor molecules, which may be differentially expressed to promote degradation or some ubiquitinated substrates over others.

2
Biological Processes Regulated by the Ubiquitin System

The list of cellular proteins that are targeted by ubiquitin is growing rapidly. By removing these cellular proteins, ubiquitin-mediated proteolysis plays an important role in regulating many basic cellular processes. Among these processes are cell cycle, cell division, differentiation, organ development, cellular response to stress and extracellular effectors, morphogenesis of neuronal networks, modulation of cell surface receptors ion channels and secretory pathways, DNA repair, transcriptional regulation, transcriptional silencing, long-term memory, circadian rhythms, regulation of the immune and inflammatory responses, and biogenesis of organelles.

The paradigm of regulated degradation, where involvement of the ubiquitin-proteasome pathway is well understood, is the cell cycle. Numerous cell cycle regulators such as cyclins, cyclin-dependent kinase inhibitors, proteins involved in sister chromatid separation, tumor suppressors, transcriptional activators and their inhibitors are all short-lived proteins that are rapidly proteolysed upon the correct cues. For example, mitosis must follow duplication of the chromosomes, so certain factors required for DNA synthesis are degraded at the end of S phase and are stabilized only once mitosis has been completed. Progression of mitosis requires that certain inhibitors of mitosis are quickly degraded only once the chromosomes are aligned and are ready to be separated to opposite poles, allowing the cell cycle to advance. The SCF family of ubiquitin-protein ligases is largely responsible for protein ubiquitination in the G1/S phase and the related APC/cyclosome complex performs similar functions in G2/M.

Signal transduction at numerous levels is kept in check by specific ubiquitination and degradation of key components from cell surface receptors, through kinases or phosphatases and down to the ultimate transcription factors. Receptor downregulation (such as growth-hormone receptor) by ubiquitin-dependent degradation is an important aspect of signal transduction. Downstream adaptor proteins such as c-Cbl have been shown to be ubiquitinated, as well. Recently, a number of kinases from the MapK, or other signaling pathways have been shown to be substrates of ubiquitinating enzymes. The synthesis of proteins and transcriptional regulation is also regulated in part by the ubiquitin-proteasome system. Numerous transcription factors are rapidly degraded, usually when dissociated from their DNA sequence. An incomplete list of short-lived transcription factors that are substrates of specific E3 ligases includes members of the AP-1, Id, HIF, P53, NF-κB families. Finally, metabolic pathways are under strict regulation as well. For example, the rate-limiting enzymes in both the sterol and unsaturated fatty acid biosynthetic pathways are located in the membrane of the ER, and their levels are regulated by ubiquitin-dependent degradation. Ubiquitin-dependent degradation of these enzymes is crucial for reducing their activity when levels of sterols or unsaturated fatty acids, respectively, become too high.

Proteasome-dependent proteolysis is a vital component of combating viral infections and aiding the immune response. Upon viral infection, Interferon-γ causes mammalian cells to synthesize new catalytic subunits of the 20S proteasomes with a preference for cleaving peptide bonds between hydrophobic amino acids. These βi subunits are incorporated into newly assembled proteasomes – the so-called *immuno-proteasomes* – with an increased ability to generate protein fragments appropriate for MHC class I presentation. Interferon-γ also induces expression of the 11S regulator, which facilitates the production of antigenic peptides for optimal binding to MHC class I molecules and accelerating the cytotoxic lymphocyte response.

The ubiquitin-proteasome system plays a critical role in heat shock and oxidative damage response by removing mutated, denatured, or misfolded proteins. It appears that in some cases, chaperones (such as Hsp70) cooperate with the ubiquitinating machinery to recognize and ubiquitinate unfolded proteins. By selectively removing these proteins, the system plays a key role in protection against aggregate buildup and in the quality control of the proteome.

All secretory proteins and most transmembrane proteins are inserted into or pass through the membrane of the ER. These proteins must fold, assemble into complexes with other proteins, and undergo various posttranslational modifications. If these proteins fail to fold or assemble properly, or lack appropriate glycosylation, they are prevented from proceeding to the Golgi and instead they are sent back to the cytosol for proteasomal degradation in a process known as *ER-associated degradation* (ERAD). Elaborate machinery exists on the cytoplasmic side of the ER membrane to recognize target proteins, to polyubiquitinate, to extract from the membrane in an ATP-dependent manner and to shuttle to the proteasome for proteolysis. Extraction from the membrane into the cytosol is probably carried out mainly by a hexameric ATPase ring called *Cdc48* or *p97*. This complex is located on the cytoplasmic surface of the membrane, can bind to polyubiquitinated substrates and undergo ATP-dependent conformational changes that ratchet the substrate through the ER channel.

Finally, it is important to note that ubiquitination also plays nondegradative roles in regulating certain biological processes. For example, histone H2A is mono-ubiquitinated in many places, though not proteolysed. Histone ubiquitination is thought to influence chromatin structure and accessibility. Ubiquitination can also serve to target proteins to certain localizations in the cell. In one case, monoamine oxidase A is ubiquitinated and inserted into the outer membrane of the mitochondria though there is no evidence so far that it is targeted for proteasome proteolysis on the way. Modification by polyubiquitin chains that polymerase via lysine 63 on each ubiquitin molecule (rather than the more common polymerization via Lys48) plays a role in a variety of processes, including endocytosis, postreplicative DNA, stress response, mitochondrial DNA inheritance, ribosomal function. Even though such chains probably do bind the proteasome, experimental evidence so far suggests that this type of modification does not involve proteolysis of the target substrate, but rather the modification plays a role in the activation/inactivation of the target protein.

3
Modification by Other Ubiquitin-like Proteins

In parallel to the ubiquitin-conjugating system, there are additional protein conjugation systems. The modifiers are always small globular proteins structurally related to ubiquitin and are known as *ubiquitin-like modifiers* (Ubls). These include SUMO, Nedd8/Rub1, ISG15, Apg8, and Apg12. In all the cases so far, the conjugation process has been carried out in an analogous manner to ubiquitination, utilizing conjugating enzymes that are similar to (yet distinct from) ubiquitinating enzymes. Modification is always on a lysine side chain of the modified proteins to which the carboxyl-terminus of the Ubl is covalently attached. Thus,

SUMO (small ubiquitin modifier), for example, is conjugated by a heterodimeric E1 – Aos1/Uba2, and the E2-conjugating enzyme Ubc9. Though Ubc9 can recognize the SUMOylation motif and transfer SUMO directly to certain substrates, for some proteins, specific E3 enzymes have been described that mediate these processes. A notable distinction between ubiquitination and modification by Ubls, however, is that the latter modification is monomeric, with only one Ubl molecule attached to the target. Recently, evidence has been provided that the ubiquitin-like modifiers SUMO-2 and SUMO-3 may form short chains. Similarly to ubiquitination, modification by Ubls is reversible on account of ubiquitin-like proteases (Ulps) that can shave the modifier off its target. A rapidly increasing number of Ulps have been identified, usually with clear homology to deubiquitinating enzymes.

Both enzymes and substrates of the ubiquitin system as well as other proteins have been found to be modified by ubiquitin-like proteins. The process serves many functions, such as activation of enzymes and transcriptional regulators, subcellular localization, or protecting proteins from ubiquitination. One role of modification by Ubls is to modulate the ubiquitin system. Modification of E3 ubiquitin ligases by ubiquitin-like molecules affects their activity. For example, conjugation of NEDD8/Rub1 to the cullin component of the SCF ligase complex increases its affinity to bind the E2 component of the conjugation machinery, thus increasing overall efficiency of conjugation. In the case of substrates, since modification occurs on lysines – the same residue to which ubiquitin is conjugated – modification can block ubiquitination and consequently enhance cellular stability. For example, in the case of IκBα, the inhibitor of the transcriptional regulator NF-κB, modification by SUMO-1 was shown to protect the substrate from ubiquitination. The activity of yet other transcription factors such as C-Jun, p53, c-Myb to name a few is also altered upon SUMOylation. In a completely different case, SUMOylation of RanGAP1 targets the protein to its final subcellular destination in the nuclear pore complex, NPC. Many SUMOylated proteins are found in promyelocytic leukemia (PML) nuclear bodies suggesting that SUMO may serve as a targeting signal to these or other locals. Complicating matters is that p53 can also be modified by NEDD8, though the exact relationship between ubiquitination, SUMOylation, and NEDDylation of this protein is not yet understood. Modification by the Ubl proteins Apg12 and Apg8 participates in autophagy, while modification by ISG15 functions in interferon-γ signaling.

A completely different group of ubiquitin-like proteins are those that are not involved in protein modification. Thus, certain proteins contain an intrinsic domain that folds into a ubiquitin-like domain. Many are proteins that are part of the ubiquitin system: among them are some E3 ligases, deubiquitinating enzymes, and polyubiquitin-binding proteins. For example, Parkin is an E3 ligase that contains a ubiquitin-like domain that probably facilitates its interaction with other components of the system. hHR23/Rad23, hPLIC/Dsk2, and Ddi1 are ubiquitin-binding proteins that all have an N-terminal Ubl. These proteins have all been implicated in the recognition and turnover of substrates targeted for degradation. They probably shuttle polyubiquitinated substrate to the proteasome to which they bind via their Ubl domain.

4
Ubiquitination in Health and Disease

Inactivation of the key junctions in the ubiquitin–proteasome pathway, such as the enzyme E1 or the proteasome is lethal. Yet, mutations or acquired changes in enzymes or recognition motifs in substrates that do not affect vital pathways or that only partially affect proteolysis may result in a broad array of diseases. Pathological cases include those that result from loss of function – stabilization of a substrate that is normally targeted for degradation by evading the ubiquitin system or due to a mutation in a ubiquitin system enzyme – and those that result from gain of function such as abnormal or accelerated degradation of the protein target. Following are a number of well-studied cases that tie the ubiquitin system to pathogenesis of malignancies, neurodegenerative disorders, genetic diseases, and disorders of the immune and inflammatory responses.

4.1
Cancer

Alterations in ubiquitination and deubiquitination reactions have been directly implicated in the etiology of many malignancies. In general, cancers can result from *stabilization* of oncoproteins, or *destabilization* of tumor suppressor gene products. Some of the natural substrates of the system are oncoproteins that if not properly removed from the cell, can promote malignant transformation. For instance, ubiquitin targets N-Myc, c-Myc, c-Fos, c-Jun, Src, and the adenovirus E1A proteins. Similarly, inadvertent destabilization of tumor suppressors such as p53 and p27 has also been implicated in the pathogenesis of malignancies.

In the case of uterine cervical carcinoma, the level of the tumor suppressor protein p53 is extremely low. Most of these malignancies are caused by high-risk strains of the human papillomavirus (HPV). Detailed studies have shown that the suppressor is targeted for ubiquitin-mediated degradation by the virally encoded oncoprotein E6. Degradation is mediated by the normal and naturally occurring HECT domain E3 enzyme E6-associated protein – E6-AP, where E6 serves as an ancillary protein that allows recognition of p53. E6-AP will not recognize p53 in the absence of E6. E6 associates with both the ubiquitin ligase and the target substrate and brings them to the necessary proximity that is assumed to allow catalysis of conjugation to occur. Removal of the suppressor by the oncoprotein is probably an important mechanism used by the virus to transform cells. While the nature of the native substrates of E6-AP is still elusive, a mutation in the enzyme has been implicated directly as the cause of Angelman Syndrome characterized by mental retardation, seizures, out of context frequent smiling and laughter, and abnormal gait. It is possible that the protein, which accumulates as a result of the mutation, is selectively toxic to the developing neuronal cells.

Similar to the case of p53, low levels of the cyclin-dependent kinase inhibitor p27^{Kip1} have been demonstrated in colorectal, prostate, and breast cancers. p27 acts as a negative growth regulator/tumor suppressor that binds and negatively regulates CDK2/cyclin E and CDK2/cyclin A complexes and thus does not allow cell cycle progression from G1 and entrance into the S phase. As noted, its level is markedly reduced in several cancers, and in many of these cases there was a strong correlation between the low level of p27 and the aggressiveness of the disease – tumor grading, clinical staging, and poor prognosis of the patients. Dissection of the mechanism that underlies the decrease in p27 revealed that the protein is of the WT species and it is probably abnormal activation of the ubiquitin system, and, in particular, dysregulated overexpression of its cognate E3, the F-box protein Skp2, that leads to rapid removal of the suppressor.

β-catenin plays an important role in signal transduction and differentiation of the colorectal epithelium, and aberrations in ubiquitin-mediated regulation of its levels may play an important role in the multistep development of colorectal tumors. In the absence of signaling, casein kinase I (CKI) and glycogen synthase kinase 3β (GSK3β) are active and promote phosphorylation, recruitment of the β-TrCP ubiquitin ligase, ubiquitination, and subsequent degradation of β-catenin. Signaling promotes dephosphorylation, stabilization, and subsequent activation of β-catenin via complex formation with otherwise inactive subunits of transcription factor complexes. β-Catenin interacts with the 300 kDa tumor suppressor APC (adenomatous polyposis coli) and Axin to generate a complex that appears to regulate, in a yet unknown manner, its intracellular level. Aberrations in degradation of β-catenin that lead to its stabilization, accumulation, and subsequent oncogenic activation, can result from two distinct mechanisms: (1) mutations in the phosphorylation motif of the protein by CKI and GSK3β, and (2) mutations in the targeting APC/Axin machinery.

Mutations in components of the ubiquitination machinery can also cause malignancies. Mutations in one germline copy of *VHL* predisposes individuals to a wide range of malignancies, including more then 80% of sporadic cases of renal cell carcinoma, pheochromocytoma, cerebellar hemangioblastomas, and retinal angiomas. A hallmark of $VHL^{-/-}$ tumors is a high degree of vascularization that arises from constitutive expression of the α-subunit of the master switch transcription factor hypoxia-inducible factor-1 (HIF-1), which results in overexpression of hypoxia-inducible gene products including the crucial vascular endothelial growth factor (VEGF). It has been recently shown that pVHL is a subunit in a ubiquitin ligase that is involved in targeting of HIF-1α for ubiquitin- and proteasome-mediated degradation. HIF-1α heterodimerizes with the constitutively expressed HIF-1β to generate the active transcription factors. Under normoxic, and obviously hyperoxic, conditions, HIF-1α is hydroxylated specifically on Pro residue 564 to generate a hydroxyproline derivative. This hydroxylated proline residue is recognized by the pVHL E3 complex that targets the molecule for ubiquitination and subsequent degradation. Under hypoxic conditions, HIF-1α is stable, as the efficiency of the hydroxylation reaction under these conditions is extremely low. Loss of VHL function stabilizes HIF-1α, which can explain the stimulation of vascular growth in tumors in which VHL is mutated or lacking. Since overexpression of VEGF alone or many of the other known target proteins of HIF does not lead to malignant transformation, and since WT VHL can restore normal growth control in these malignant cells, researchers assume that pVHL and/or HIF must have additional, yet unknown substrates.

Defects in deubiquitinating enzymes can also lead to aberrations in growth control and to tumorigenesis. Overexpression of the ubiquitin proteases (UBP) Unp has been found in lung carcinoma in humans and in mouse tumors. Unp may have a role in regulating the degradation of specific, yet unidentified, substrates in these tissues, though it does not seem to be crucial for general proteolysis. An interesting deubiquitinating enzyme is CYLD that is involved in deubiquitinating K63-Ub from NEMO (NF-kB essential modifier), TRAF2 [tumor necrosis factor receptor (TNFR)-associated factor 2] and TRAF6. Ubiquitination of NEMO [a regulator of the IKK (IκB kinase) signaling complex] and the TRAF proteins (which are ubiquitin ligases), and generation of polyubiquitin chains linked via Lys63 of the ubiquitin moiety, does not target these proteins for degradation, but results in their activation. Thus, inhibition of deubiquitination of NEMO, by mutation in CYLD, for example, may lead to uncontrolled activation of the IKK complex with increased activity of NF-κB. Indeed, CYLD was found mutated in familial cylindromatosis, a rare pathology characterized by predisposition to multiple tumors of the skin appendages and the salivary gland.

4.2
Neurological Disorders

Accumulation of ubiquitin conjugates and/or inclusion bodies associated with ubiquitin, proteasome, and certain disease-characteristic proteins, have been reported in a broad array of chronic neurodegenerative diseases, such as the neurofibrillary tangles of Alzheimer's disease (AD), brainstem Lewy bodies (LBs) – the neuropathological hallmark in Parkinson's disease (PD), and nuclear inclusions in CAG repeat expansion (polyglutamine extension) disorders such as occurring in Huntington's disease. However, in all these cases, a direct pathogenetic linkage to aberrations in the ubiquitin system has not been established. One factor that complicates the establishment of such linkage is the realization that many of these diseases, such as Alzheimer's and Parkinson's, are not defined clinical entities, but rather syndromes with different etiologies. Accumulation of ubiquitin conjugates in Lewy inclusion bodies in many of these cases may be secondary, and reflects unsuccessful attempts by the ubiquitin and proteasomal machineries to remove damaged/abnormal proteins. While the initial hypothesis was that inclusion bodies are generated because of the inherent tendency of the abnormal proteins to associate with one another and aggregate, it is now thought that the process maybe more complex and involves active cellular machineries, including inhibition of the ubiquitin system by the aggregated proteins. This aggregation of brain proteins into defined lesions is emerging as a common but poorly understood mechanistic theme in sporadic and hereditary neurodegenerative disorders.

The case of Parkinson's disease highlights the complexity of the involvement of the ubiquitin system in the pathogenesis of neurodegeneration. Mutations in several proteins such as α-synuclein that plays a role in synaptic vesicles dynamics and that leads to its stabilization and accumulation, or in the deubiquitinating enzyme UCH-L1, have been described that link the ubiquitin system to the pathogenesis of the disease. One important player in the pathogenesis of Parkinson's disease is Parkin, which is a RING-finger E3. Mutations in the gene appear to be responsible for the pathogenesis of autosomal recessive juvenile Parkinsonism (AR-JP), one of the most common familial forms of Parkinson's disease. Parkin ubiquitinates and promotes the degradation of several substrates, such as CDCrel-1 (cell division control related protein), a synaptic vesicle-enriched septin GTPase or the Pael receptor, a putative G protein-coupled ER transmembrane polypeptide. It is possible that aberration in the degradation of one of these substrates that leads to its accumulation is neurotoxic and underlies the pathogenesis of AR-JP. However, it should be noted that such a protein has not been identified with any certainty. Complicating the situation is the recent finding that inactivation of the Parkin gene in mice does not lead to any symptoms that are even reminiscent of Parkinson's disease.

Alzheimer's disease is characterized by accumulation/association of ubiquitin with the phosphorylated form of Tau in neurofibrillary tangles and senile plaques, two lesions that are characteristic of the neuronal abnormalities associated with the disease. However, the role of Tau and other putative target proteins in the pathogenesis of the disease is still not clear. A more direct, relationship between the ubiquitin system and pathogenesis of Alzheimer's disease was established with the discovery of a frameshift mutation in the ubiquitin transcript, which leads to extension of the molecule with 20 amino acid residues [Ub(+1)], and, which has been selectively observed in the brains of Alzheimer's disease patients, including those with late onset, nonfamilial disease. Ub(+1) is an efficient acceptor for polyubiquitination, though it cannot be activated by E1 (as it lacks the essential G76 residue) and be transferred to a substrate or to another ubiquitin moiety. The resulting polyubiquitin chains are refractory to disassembly by deubiquitinating enzyme, in particular, isopeptidase T that requires for its activity an exposed G76 residue at the proximal ubiquitin moiety. The accumulated polyubiquitin chains block proteasomal degradation, which results in neuronal apoptosis. Thus, expression of Ub(+1) in the brain, that increases apparently with aging, can potentially result in dominant inhibition of the ubiquitin-proteasome system, leading to accumulation of toxic proteins with neuropathologic consequences. Since (Ub(+1)) was described also in other disorders such as Down's syndrome or supranuclear palsy, it is clear that it is not entirely specific to Alzheimer's disease and a major problem of how the mutation leads to distinct pathologies in different patients remains unsolved.

4.3
Cystic Fibrosis

The cystic fibrosis gene encodes the CF transmembrane regulator (CFTR) that is a chloride channel. Only a small fraction of the protein matures to the cell surface, whereas most of it is degraded from the endoplasmic reticulum (ER) prior to its maturation by the ubiquitin system. One frequent mutation in the channel is ΔF508. The mutation leads to an autosomal recessive inherited multisystem disorder characterized by chronic obstruction of airways and severe maldigestion due to exocrine pancreatic dysfunction. Despite normal ion channel function, $CFTR^{\Delta F508}$ does not reach the cell surface at all, and is retained in the ER from which it is degraded. It is possible that the rapid and efficient degradation results in complete lack of cell surface expression of the F508 protein, and therefore contributes to the pathogenesis of the disease.

4.4 Immune Response and Inflammatory Disorders

A wide array of immune and inflammatory disorders can be caused by untoward activation of the immune system central transcription factor NF-κB. Activation of the factor stimulates transcription of many cytokines, adhesion molecules, inflammatory response, and stress proteins, and immune system receptors. The factor is activated by the ubiquitin system via a two-step proteolytic mechanism: (1) limited processing of the precursor protein p105 to yield the active subunit p50, and (2) signal-induced phosphorylation and subsequent degradation of the inhibitor IκBα that enables translocation of the factor into the nucleus where it initiates specific transcriptional activity. An interesting case in that respect involves mutations in NEMO. Mutations in the protein lead to a series of diseases that affect the skin, among them incontinentia pigmenti (IP), hypohidrotic/anhidrotic ectodermal dysplasia, but also, as expected, immune deficiency. The immunological and infectious features observed in patients result from impaired NF-κB signaling, including cellular response to LPS, and a variety of cytokines.

As described above, the HPV evolved a mechanism for proteolytic removal of p53 that enables continuous replication and propagation of the virus under conditions of DNA damage (insertion of the viral genome into the cellular DNA) that normally would have ended with p53-induced apoptosis. Two other viruses evolved mechanisms that also utilize the ubiquitin system, here to escape immune surveillance. In one case, the Epstein-Barr Virus (EBV) EBV nuclear antigen 1 (EBNA1) persists in healthy virus carriers for life, and is the only viral protein regularly detected in all EBV-associated malignancies, such as Burkitt's lymphoma. Unlike EBNAs 2–4 that are strong immunogens, EBNA1 is not processed and cannot elicit a cytoxic T-cell response. The persistence of EBNA1 contributes, most probably, to some of the pathologies caused by the virus. An interesting structural feature common to all EBNA1 protein is a relatively long and unusual glycine–alanine repeat at the N-terminal domain of the molecule. These Gly–Ala repeats constitute a cis-acting element that inhibits processing of the protein by the 26S proteasome and subsequent presentation of potential antigenic epitopes. Thus, the evolution of the Gly–Ala repeat enabled the virus to evade proteolysis and subsequent presentation to the immune system. An additional interesting observation involves the pathobiology of the human cytomegalovirus (CMV). The virus genome encodes two ER proteins, US2 and US11, that downregulate the expression of MHC class I heavy chain molecules. The MHC molecules are synthesized, transported to the ER where they are glycosylated, but shortly thereafter, in cells expressing US2 or US11, are transported back to the cytosol, deglycosylated, and degraded by the proteasome following ubiquitination. It appears that the viral products bind to the MHC molecules and escort\dislocate them to the translocation machinery where they are transported back into the cytosol.

Thus, it appears that evolution of many viruses has involved intimate recognition with a variety of proteolytic processes. This enabled the evolution of viral mechanisms that enhance the function, via subversion of the normal proteolytic machinery, of the viral replication and propagation machinery.

5
Summary and Outlook

Ubiquitination is involved in the majority of regulated protein degradation in the cell. For proteolytic purposes, usually a polyubiquitin chain is assembled on the protein allowing for tight binding to the 26S proteasome and efficient degradation. However, it should be emphasized that the process is not a unidirectional irreversible process and that ubiquitinated proteins are in a dynamic state balancing further rounds of ubiquitination, deubiquitination, or degradation by the proteasome. Ubiquitin-like proteins have also been found to be regulators of ubiquitination enzymes, or of intracellular targeting and localization. Many questions are still unanswered. Is ubiquitin a strict requirement for proteasome-dependent intracellular degradation, or can some substrates be degraded in an ubiquitin-independent manner? Furthermore, does the ubiquitin-proteasome pathway play additional nonproteolytic roles, such as protein remodeling, transcriptional activation or altering protein–protein interactions or with nucleic acids?

Because of the central role the ubiquitin system plays in such a broad array of basic cellular processes, development of drugs that modulate the activity of the system may be difficult. Inhibition of enzymes common to the entire pathway, such as E1 or the proteasome, may inhibit many processes nonspecifically, although at the right dosage they could affect some targets more than others. One example in which such inhibitors have shown to be beneficial is multiple myeloma, a malignancy that affects the bone marrow. Alternatively, development of small molecules that interfere with the recognition step between specific E3s and their substrates, or site-specific inhibitors of deubiquitinating enzymes may be productive approaches as well.

See also Alzheimer's Disease.

Bibliography

Books and Reviews

Amerik, A.Y., Hochstrasser, M. (2004) Mechanism and function of deubiquitinating enzymes, *Biochim. Biophys. Acta* **1695**, 189–207.

Glickman, M.H., Ciechanover, A. (2002) The Ubiquitin-proteasome proteolytic pathway: destruction for the sake of construction, *Physiol. Rev.* **82**, 373–428.

Goldberg, A.L., Elledge, S.J. Wade, J. (2001) *The Cellular Chamber of Doom*, Scientific American, January, 68–73.

Hicke, L., Dunn, R. (2003) Regulation of membrane protein transport by ubiquitin and ubiquitin-binding proteins, *Annu. Rev. Cell Dev. Biol.* **19**, 141–172.

Hilt, W., Wolf, D.H. (Eds.) (2000) *Proteasomes: The World of Regulatory Proteolysis*, Eurekah.com/LANDES BIOSCIENCE Publishing Company, Georgetown, TX.

Pickart, C.M. (2001) Mechanisms of ubiquitination, *Annu. Rev. Biochem.* **70**, 503–533.

Pickart, C.M., Cohen, R.E. (2004) Proteasomes and their kin: proteases in the machine age, *Nat. Rev. Mol. Cell Biol.* **5**, 177–187.

Schwartz, D.C., Hochstrasser, M. (2003) A superfamily of protein tags: ubiquitin, SUMO and related modifiers, *Trends Biochem. Sci.* **28**, 321–328.

Weissman, A.M. (2001) Themes and variations on ubiquitylation, *Nat. Rev. Mol. Cell Biol.* **2**, 169–179.

Wolf, D.H., Hilt, W. (2004) The proteasome: a proteolytic nanomachine of cell regulation and waste disposal, *Biochim. Biophys. Acta (BBA) – Mol. Cell Res.* **1695**, 19.

Primary Literature

Bloom, J., Pagano, M. (2003) Deregulated degradation of the cdk inhibitor p27 and malignant transformation, *Semin. Cancer Biol.* **13**, 41–47.

Boyd, S.D., Tsai, K.Y., Jacks, T. (2000) An intact HDM2 RING-finger domain is required for nuclear exclusion of p53, *Nat. Cell Biol.* **2**, 563–568.

Buschmann, T., Fuchs, S.Y., Lee, C.G., Pan, Z.Q., Ronai, Z. (2000) SUMO-1 modification of Mdm2 prevents its self-ubiquitination and increases Mdm2 ability to ubiquitinate p53, *Cell* **101**, 753–762.

Chen, Z., Hagler, J., Palombella, V.J., Melandri, F., Scherer, D., Ballard, D., Maniatis, T. (1995) Signal-induced site-specific phosphorylation targets I κ B α to the ubiquitin-proteasome pathway. *Genes Dev.* **9**, 1586–1597.

Ciechanover, A., Hod, Y., Hershko, A. (1978) A heat-stable polypeptide component of an ATP-dependent proteolytic system from reticulocytes, *Biochem. Biophys. Res. Commun.* **81**, 1100–1105.

Ciechanover, A., Heller, H., Katz-Etzion, R., Hershko, A. (1981) Activation of the heat-stable polypeptide of the ATP-dependent proteolytic system, *Proc. Natl. Acad. Sci. U.S.A.* **78**, 761–765.

Ciechanover, A., Elias, S., Heller, H., Hershko, A. (1982) "Covalent affinity" purification of ubiquitin-activating enzyme, *J. Biol. Chem.* **257**, 2537–2542.

Ciechanover, A., Heller, H., Elias, S., Haas, A.L., Hershko, A. (1980) ATP-dependent conjugation of reticulocyte proteins with the polypeptide required for protein degradation, *Proc. Natl. Acad. Sci. U.S.A.* **77**, 1365–1368.

Elsasser, S., Chandler-Militello, D., Muller, B., Hanna, J., Finley, D. (2004) Rad23 and Rpn10 serve as alternative ubiquitin receptors for the proteasome, *J. Biol. Chem.* **279**, 26817–26822.

Etlinger, J.D., Goldberg, A.L. (1977) A soluble ATP-dependent proteolytic system responsible for the degradation of abnormal proteins in reticulocytes, *Proc. Natl. Acad. Sci. U.S.A.* **74**, 54–58.

Finley, D., Ciechanover, A., Varshavsky, A. (1984) Thermolability of ubiquitin-activating enzyme from the mammalian cell cycle mutant ts85, *Cell* **37**, 43–55.

Finley, D., Ozkaynak, E., Varshavsky, A. (1987) The yeast polyubiquitin gene is essential for resistance to high temperatures, starvation, and other stresses, *Cell* **48**, 1035–1046.

Finley, D., Bartel, B., Varshavsky, A. (1989) The tails of ubiquitin precursors are ribosomal proteins whose fusion to ubiquitin facilitates ribosome biogenesis, *Nature* **338**, 394–401.

Fraser, J., Luu, H.A., Neculcea, J., Thomas, D.Y., Storms, R.K. (1991) Ubiquitin gene expression: response to enviormental changes, *Curr. Genet.* **1–2**, 17–23.

Galan, J., Haguenauer-Tsapis, R. (1997) Ubiquitin Lys63 is involved in ubiquitination of a yeast plasma membrane protein, *EMBO J.* **16**, 5847–5854.

Geyer, R.K., Yu, Z.K., Maki, C.G. (2000) The MDM2 RING-finger domain is required to promote p53 nuclear export, *Nat. Cell Biol.* **2**, 569–573.

Giles, R.H., van Es, J.H., Clevers, H. (2003) Caught up in a Wnt storm: Wnt signaling in cancer, *Biochim. Biophys. Acta* **1653**, 1–24.

Glotzer, M., Murray, A.W., Kirschner, M.W. (1991) Cyclin is degraded by the ubiquitin pathway, *Nature* **349**, 132–138.

Goebl, M.G., Yochem, J., Jentsch, S., McGrath, J.P., Varshavsky, A., Byers, B. (1988) The yeast cell cycle gene CDC34 encodes a ubiquitin-conjugating enzyme, *Science* **241**, 1331–1335.

Goldknopf, I.L., Busch, H. (1977) Isopeptide linkage between nonhistone and histone 2A polypeptides of chromosomal conjugate-protein A24, *Proc. Natl. Acad. Sci. U.S.A.* **74**, 864–868.

Goldstein, G., Scheid, M.S., Hammerling, V., Boyse, E.A., Schlesinger, D.H., Niall, H.D. (1975) Isolation of a polypeptide that has lymphocyte-differentiating properties and is probably represented universally in living cells, *Proc. Natl. Acad. Sci. U.S.A.* **72**, 11–15.

Gong, L., Yeh, E.T. (1999) Identification of the activating and conjugating enzymes of the NEDD8 conjugation pathway, *J. Biol. Chem.* **274**, 12036–12042.

Gray, D.A., Inazawa, J., Gupta, K., Wong, A., Ueda, R., Takahashi, T. (1995) Elevated expression of Unph, a proto-oncogene at 3p21.3, in human lung tumors, *Oncogene* **10**, 2179–2183.

Gross-Mesilaty, S., Hargrove, J.L., Chiechanover, A. (1997) Degradation of tyrosine amino-transferase (TAT)via the ubiquitin-proteasome pathway, *FEBS Lett.* **405**(2), 175–180.

Haas, A.L., Rose, I.A. (1982) The mechanism of ubiquitin activating enzyme, *J. Biol. Chem.* **257**, 10329–10337.

Haas, A.L., Rose, I.A., Hershko, A. (1981) Purification of the ubiquitin activating enzyme required for ATP-dependent protein breakdown, *Fed. Proc.* **40**, 1691.

Heessen, S., Masucci, M.G., Dantuma, N.P. (2005) The UBA2 domain functions as an intrinsic stabilization signal that protects Rad23 from proteasomal degradation, *Mol. Cell* **18**, 225–235.

Hershko, A., Tomkins, G.M. (1971) Studies on the degradation of tyrosine aminotransferase in hepatoma cells in culture, *J. Biol. Chem.* **246**, 710–714.

Hershko, A., Ciechanover, A., Rose, I.A. (1979) Resolution of the ATP-dependent proteolytic system from reticulocytes: a component that interacts with ATP, *Proc. Natl. Acad. Sci. U.S.A.* **76**, 3107–3110.

Hershko, A., Ciechanover, A., Rose, I.A. (1981) Identification of the active amino acid residue of the polypeptide of ATP-dependent protein breakdown, *J. Biol. Chem.* **256**, 1525–1528.

Hershko, A., Eytan, E., Ciechanover, A., Haas, A.L. (1982) Immunochemical analysis of the turnover of ubiquitin-protein conjugates in intact cells, *J. Biol. Chem.* **257**, 13964–13970.

Hershko, A., Heller, H., Elias, S., Chiechanover, A. (1983) Components of ubiquitin-protein ligase system, *J. Biol. Chem.* **258**, 8206–8214.

Hershko, A., Ciechanover, A., Heller, H., Haas, A.L., Rose, I.A. (1980) Proposed role of ATP in protein breakdown: conjugation of proteins with multiple chains of the polypeptide of ATP-dependent proteolysis, *Proc. Natl. Acad. Sci. U.S.A.* **77**, 1783–1786.

Hicke, L., Riezman, H. (1996) Ubiquitination of a yeast plasma membrane receptor signals its ligand-stimulated endocytosis, *Cell* **84**, 277–287.

Honda, R., Tanaka, H., Yasuda, H. (1997) Oncoprotein MDM2 is a ubiquitin ligase E3 for tumor suppressor p53, *FEBS Lett.* **420**, 25–27.

Huang, D.T., Walden, H., Duda, D., Schulman, B.A. (2004) Ubiquitin-like protein activation, *Oncogene* **23**, 1958–1971.

Huang, L., Kinnucan, E., Wang, G., Beaudenon, S., Howley, P.M., Huibregtse, J.M., Pavletich, N.P. (1999) Structure of an E6AP-UbcH7 complex: Insights into ubiquitination by the E2-E3 enzyme cascade, *Science* **286**, 1321–1326.

Hunt, L.T., Dayhoff, M.O. (1977) Amino-terminal sequence identity of ubiquitin and the nonhistone component of nuclear protein A24, *Biochem. Biophys. Res. Commun.* **74**, 650–655.

Ivan, M., Kaelin, W.G. (2001) The von Hippel-Lindau tumor suppressor protein, *Curr. Opin. Genet. Dev.* **11**, 27–34.

Jensen, D.E., et al. (1998) BAP1: a novel ubiquitin hydrolase which binds to the BRCA1 RING finger and enhances BRCA1-mediated cell growth suppression, *Oncogene* **16**, 1097–1112.

Kishino, T., Lalande, M., Wagstaff, J. (1997) UBE3A/E6-AP mutations cause Angelman syndrome, *Nat. Genet.* **15**, 70–73.

Kitada, T., et al. (1998) Mutations in the parkin gene cause autosomal recessive juvenile parkinsonism, *Nature* **392**, 605–608.

Klotzbucher, A., Stewart, E., Harrison, D., Hunt, T. (1996) The 'destruction box' of cyclin A allows B-type cyclins to be ubiquitinated, but not efficiently destroyed, *EMBO J.* **15**, 3053–3064.

Kovalenko, A., Chable-Bessia, C., Cantarella, G., Israël, A., Wallach, D., Courtois, G. (2003) The tumour suppressor CYLD negatively regulates NF-kB signalling by deubiquitination, *Nature* **424**, 801–805.

Lorick, K.L., Jensen, J.P., Fang, S., Ong, A.M., Hatakayama, S., Weissman, A.M. (1999) RING finger mediate ubiquitin-conjugating enzyme (E2)-dependent ubiquitination, *Proc. Natl. Acad. Sci. U.S.A.* **96**, 11364–11369.

Lueders, J., Demand, J., Hoehfeld, J. (2000) The ubiquitin-related BAG-1 provides a link between the molecular chaperones Hsc70/Hsp70 and the proteasome, *J. Biol. Chem.* **275**, 4613–4617.

Lyapina, S., et al. (2001) Promotion of NEDD-CUL1 conjugate cleavage by COP9 signalosome, *Science* **292**, 1382–1385.

Ma, C.P., Slaughter, C.A., DeMartino, G.N. (1992) Identification, purification, and characterization of a protein activator (PA28) of the 20S proteasome, *J. Biol. Chem.* **267**, 10515–10523.

Ma, C.P., Vu, J.H., Proske, R.J., Slaughter, C.A., DeMartino, G.N. (1994) Identification, purification, and characterization of a high molecular weight ATP-dependent activator (PA700) of the 26S proteasome, *J. Biol. Chem.* **269**, 3539–3547.

Mahaffey, D., Rechsteiner, M. (1999) Discrimination between ubiquitin-dependent and ubiquitin-independent proteolytic pathways by the 26S proteasome subunit 5a, *FEBS Lett.* **450**, 123–125.

Maniatis, T. (1999) A ubiquitin ligase complex essential for the NF-kB, Wnt/Wingless, and

Hedgehog signaling pathways, *Genes Dev.* **13**, 505–514.

Margottin, F., et al. (1998) A novel human WD protein, human b-TrCp, that interacts with HIV-1 Vpu, connects CD4 to the ER degradation pathway through an F-box motif, *Mol. Cell* **1**, 565–574.

Marunouchi, T., Yasuda, H., Matsumoto, Y., Yamada, M. (1980) Disappearance of a basic chromosomal protein from cells of a mouse temperature-sensitive mutant defective in histone phosphorylation, *Biochem. Biophys. Res. Commun.* **95**, 126–131.

Matsumoto, Y., Yasuda, H., Marunouchi, T., Yamada, M. (1983) Decrease in uH2A (protein A24) of a mouse temperature-sensitive mutant, *FEBS Lett.* **151**, 139–142.

Matsumoto, Y., Yasuda, H., Mita, S., Marunouchi, T., Yamada, M. (1980) Evidence for the involvement of H1 histone phosphorylation in chromosome condensation, *Nature* **284**, 181–183.

Maxwell, P.H., et al. (1999) The tumour suppressor protein VHL targets hypoxia-inducible factors for oxygen-dependent proteolysis, *Nature* **399**, 271–275.

Mitch, W.E., Goldberg, A.L. (1996) Mechanisms of muscle wasting: The role of the ubiquitin-proteasome pathway, *N. Engl. J. Med.* **335**, 1897–1905.

Mizushima, N., et al. (1998) A protein conjugation system essential for autophagy, *Nature* **395**, 395–398.

Moazed, D., Johnson, D. (1996) A deubiquitinating enzyme interacts with SIR4 and regulates silencing in S. cerevisiae, *Cell* **86**, 667–677.

Molinari, M., Milner, J. (1995) p53 in complex with DNA is resistant to ubiquitin-dependent proteolysis in the presence of HPV-16 E6, *Oncogene* **10**, 1849–1854.

Orian, A., et al. (2000) SCF-b-TrCP ubiquitin ligase-mediated processing of NF-kB p105 requires phosphorylation of its C-terminus by IkB kinase, *EMBO J.* **19**, 2580–2591.

Orlowski, M., Cardozo, C., Michaud, C. (1993) Evidence for the presence of five distinct proteolytic components in the pituitary multicatalytic proteinase complex. Properties of two components cleaving bond on the carboxyl side of branched chain and small neutral amino acids, *Biochemistry* **32**, 1563–1572.

Ortega, J., Bernard Heymann, J., Kajava, A.V., Ustrell, V., Rechsteiner, M., Steven, A.C. (2005) The axial channel of the 20 S proteasome opens upon binding of the PA200 activator, *J. Mol. Biol.* **346**, 1221.

Osaka, F., et al. (2000) Covalent modifier NEDD8 is essential for SCF ubiquitin-ligase in fission yeast, *EMBO J.* **19**, 3475–3484.

Panaro, F.J., Na, E., Dharmarajan, K., Pan, Z.Q., Valenzuela, D.M., DeChiara, T.M., Stitt, T.N., Yancopoulos, G.D., Glass, D.J. (2001) Identification of ubiquitin ligases required for skeletal muscle atrophy, *Science* **294**, 1704–1708.

Pickart, C.M., Rose, I.A. (1985) Functional heterogeneity of ubiquitin carrier proteins, *J. Biol. Chem.* **260**, 1573–1581.

Pugh, C.W., Ratcliffe, P.J. (2003) The von Hippel-Lindau tumor suppressor, hypoxia-inducible factor-1 (HIF-1) degradation, and cancer pathogenesis, *Semin. Cancer Biol.* **13**, 83–89.

Rabinovich, E., Kerem, A., Frohlich, K.U., Diamant, N., Bar-Nun, S. (2002) AAA-ATPase p97/Cdc48p, a cytosolic chaperone required for endoplasmic reticulum-associated protein degradation, *Mol. Cell. Biol.* **22**, 626–634.

Saeki, Y., Saitoh, A., Toh-e, A., Yokosawa, H. (2002) Ubiquitin-like proteins and Rpn10 play cooperative roles in ubiquitin-dependent proteolysis, *Biochem. Biophys. Res. Commun.* **293**, 986–992.

Saeki, Y., Tayama, Y., Toh-e, A., Yokosawa, H. (2004) Definitive evidence for Ufd2-catalyzed elongation of the ubiquitin chain through Lys48 linkage, *Biochem. Biophys. Res. Commun.* **320**, 840–845.

Scheffner, M., Huibregtse, J.M., Vierstra, R.D., Howley, P.M. (1993) The HPV-16 E6 and E6-AP complex functions as a ubiquitin-protein ligase in the ubiquitination of p53, *Cell* **75**, 495–505.

Scheffner, M., Werness, B.A., Huibregtse, J.M., Levine, A.L., Howley, P.M. (1990) The E6 oncoprotein encoded by human papillomavirus types 16 and 18 promotes the degradation of p53, *Cell* **63**, 1129–1136.

Shamu, C.E., Story, C.M., Rapoport, T.A., Ploegh, H.L. (1999) The pathway of US11-dependent degradation of MHC class I heavy chains involves a ubiquitin- conjugated intermediate, *J. Cell Biol.* **147**, 45–58.

Shimura, H., et al. (2000) Familial Parkinson disease gene product, Parkin, is a ubiquitin-protein ligase, *Nat. Genet.* **25**, 302–305.

Tsvetkov, L.M., Yeh, K.H., Lee, S.J., Sun, H., Zhang, H. (1999) p27Kip1 ubiquitination and degradation is regulated by the SCFSkp2 complex through phosphorylated Thr187 in p27, *Curr. Biol.* **9**, 661–664.

Turner, C.T., Du, F., Varshavsky, A. (2000) Peptides accelerate their uptake by activating ubiquitin-dependent proteolytic pathway, *Nature* **405**, 579–583.

Vassilev, L.T., Vu, B.T., Graves, B., Carvajal, D., Podlaski, F., Filipovic, Z., Kong, N., Kammlott, U., Lukacs, C., Klein, C., Fotouhi, N., Liu, E.A. (2004) In vivo activation of the p53 pathway by small molecule antagonists of MDM2, *Science* **303**, 844–848.

Ward, C.L., Omura, S., Kopito, R.R. (1995) Degradation of CFTR by the ubiquitin-proteasome pathway, *Cell* **83**, 121–127.

Wigley, W.C., Fabunmi, R.P., Lee, M.G., Marino, C.R., Muallem, S., DeMartino, G.N., Thomas, P.J. (1999) Dynamic association of proteasomal machinery with the centrosome, *J. Cell Biol.* **145**, 481–490.

Wilkinson, K.D., Urban, M.K., Haas, A.L. (1980) Ubiquitin is the ATP-dependent proteolysis factor I of rabbit reticulocytes, *J. Biol. Chem.* **255**, 7529–7532.

Xie, Y., Varshavsky, A. (2001) RPN4 is a ligand, substrate, and transcriptional regulator of the 26S proteasome: a negative feedback circuit, *Proc. Natl. Acad. Sci. U.S.A.* **98**, 3056–3061.

Xirodimas, D., Saville, M., Bourdon, J., Hay, R., Lane, D. (2004) Mdm2-mediated NEDD8 conjugation of p53 inhibits its transcriptional activity, *Cell* **118**, 83–97.

Yaron, A., et al. (1997) Inhibition of NF-kB cellular function via specific targeting of the IkB-ubiquitin ligase, *EMBO J.* **16**, 6486–6494.

Ye, Y., Meyer, H.H., Rapoport, T.A. (2001) The AAA ATPase Cdc48/p97 and its partners transport proteins from the ER into the cytosol, *Nature* **414**, 652–656.

Zhang, Y., Chang, C., Gehling, D.J., Hemmati-Brivanlou, A., Derynck, R. (2001) Regulation of Smad degradation and activity by Smurf2, an E3 ubiquitin ligase, *Proc. Natl. Acad. Sci. U.S.A.* **98**, 974–979.

Zhang, Y., Gao, J., Chung, K.K., Huang, H., Dawson, V.L., Dawson, T.M. (2000) Parkin functions as an E2-dependent ubiquitin-protein ligase and promotes the degradation of the synaptic vesicle-associated protein, CDCrel-1, *Proc. Natl. Acad. Sci. U.S.A.* **97**, 13354–13359.

18
Plant-based Expression of Biopharmaceuticals

Jörg Knäblein
Schering AG, Berlin, Germany

1   Introduction   539

2   Alternative Expression Systems   539

3   History of Plant Expression   541

4   Current Status of Plant-based Expression   542
4.1   SWOT Analysis Reveals a Ripe Market for Plant Expression Systems   542
4.2   Risk Assessment and Contingency Measures   544

5   The Way Forward: Moving Plants to Humanlike Glycosylation   548

6   Three Promising Examples: Tobacco (Rhizosecretion, Transfection) and Moss (Glycosylation)   550
6.1   Harnessing Tobacco Roots to Secrete Proteins   550
6.2   High Protein Yields Utilizing Viral Transfection   551
6.3   Simple Moss Performs Complex Glycosylation   553

7   Other Systems Used for Plant Expression   556

8   Analytical Characterization   557

9   Conclusion and Outlook   558

     Acknowledgments   559

     Bibliography   559
     Books and Reviews   559
     Primary References   559

Proteins. Edited by Robert A. Meyers.
Copyright © 2007 Wiley-VCH Verlag GmbH & Co. KGaA, Weinheim
ISBN: 978-3-527-31608-3

18 Plant-based Expression of Biopharmaceuticals

Keywords

GMP
Good Manufacturing Practice (GMP) was established by WHO in 1968 to guarantee the optimum degree of quality during production and processing of pharmaceuticals (cGMP means under the current regulations of the authorities).

Transgenic
Organisms that have externally introduced foreign DNA/genes stably integrated into their genome to, for example, produce desired substances like human insulin.

Plant-based Expression
Transgenic plants can be genetically modified with a gene of interest to produce a biopharmaceutical of interest.

Glycosylation
It is the addition of polysaccharides to a certain molecule such as a protein. The majority of proteins are synthesized in the rough endoplasmic reticulum (ER) where they undergo glycosylation.

Bioreactor
It is a vessel in which a (bio)chemical process that involves organisms or biochemically active substances (e.g. enzymes) derived from such organisms is carried out.

Biopharmaceuticals are currently the mainstay products of the biotechnology market and represent the fastest growing and, in many ways, the most exciting sector within the pharmaceutical industry. The term "biopharmaceutical" was originated in the 1980s, when a general consensus evolved that it represented a class of therapeutics produced by means of modern biotechnologies. Already a quarter of a century ago, "humulin" (recombinant human insulin, produced in *E. coli* and developed by Genentech in collaboration with Eli Lilly) was approved and received marketing authorization in the United States of America in 1982. Since then the market for biopharmaceuticals has been steadily growing and currently nearly 150 biopharmaceuticals have gained approval for general human use (EU and USA). Over this period it became obvious that production capacities for biopharmaceuticals with "conventional" bioreactors would be a bottleneck and that worldwide fermentation capacities are limited. One exciting solution to these "capacity crunches" is the use of transgenic plants to produce biopharmaceuticals. This article describes different plant expression systems, their advantages and limitations, and concludes by considering some of the innovations and trends likely to influence the future of plant-based biopharmaceuticals.

1
Introduction

Biopharmaceuticals, which are large molecules produced by living cells, are currently the mainstay products of the biotechnology industry. Indeed, biologics such as Genentech's (Vacaville, CA, USA) human growth factor somatropin or Amgen's (Thousand Oaks, CA, USA) recombinant erythropoietin (EPO) have shown that biopharmaceuticals can benefit a huge number of patients and also generate big profits for these companies at the same time. The single most lucrative product is EPO and combined sales of the recombinant EPO products "Procrit" (Ortho biotech) and "Epogen" (Amgen) have reportedly surpassed the $6.5 billion mark. But it has also become obvious over the last couple of years that current fermentation capacities will not be sufficient to manufacture all biopharmaceuticals (in the market already or in development), because the market and demand for biologics is continuously and very rapidly growing; for antibodies alone (with at least 10 monoclonal antibodies approved and being marketed), the revenues are predicted to expand to US$3 billion in 2002 and US$8 billion in 2008. The 10 monoclonal antibodies on the market consume more than 75% of the industry's manufacturing capability. And there are up to 60 more that are expected to reach the market in the next six or seven years. Altogether, there are about 1200 protein-based products in the pipeline with a 20% growth rate and the market for current and late stage (Phase III) is estimated to be US$42 billion in 2005 and even US$100 billion in 2010. But, there are obvious limitations of large-scale manufacturing resources and production capacities – and pharmaceutical companies are competing (see ref Knäblein (2004), review).

To circumvent this capacity crunch, it is necessary to look into other technologies rather than the established ones, like, for example, *Escherichia coli* or CHO (Chinese hamster ovary) cell expression. One solution to avoid these limitations could be the use of transgenic plants to express recombinant proteins at low cost, in GMP (good manufacturing practice) quality greenhouses (with purification and fill finish in conventional facilities). Plants therefore provide an economically sound source of recombinant proteins, such as industrial enzymes, and biopharmaceuticals. Furthermore, using the existing infrastructure for crop cultivation, processing, and storage will reduce the amount of capital investment required for commercial production. For example, it was estimated that the production costs of recombinant proteins in plants could be between 10 and 50 times lower than those for producing the same protein in *E. coli* and Alan Dove describes a factor of thousand for cost of protein (US dollar per gram of raw material) expressed in, for example, CHO cells compared to transgenic plants. So, at the dawn of this new millennium, a solution is imminent to circumvent expression capacity crunches and to supply mankind with the medicines we need. Providing the right amounts of biopharmaceuticals can now be achieved by applying our knowledge of modern life sciences to systems that were on this planet long time before us – plants.

2
Alternative Expression Systems

Currently, CHO cells are the most widely used technology in biomanufacturing because they are capable of expressing eukaryotic proteins (processing, folding, and posttranslational modifications) that cannot be provided by E. coli. A long track record exists for CHO cells, but unfortunately they bring some problems along when it comes to scaling up production. Transport of oxygen (and other gases) and nutrients is critical for the fermentation process, as well as the fact that heat must diffuse evenly to all cultured cells. According to the Michaelis–Menten equation, the growth rate depends on the oxygen/nutrient supply; therefore, good mixing and aeration are a prerequisite for the biomanufacturing process and are usually achieved by different fermentation modes (see Fig. 1). But the laws of physics set strict limits on the size of bioreactors. For example, an agitator achieves good heat flow and aeration, but with increased fermenter size, shear forces also increase and disrupt the cells – and building parallel lines of bioreactors multiplies the costs linearly. A 10 000-L bioreactor costs between US$ 250 000 to 500 000 and takes five years to build (conceptual planning, engineering, construction, validation, etc.). An error in estimating demand for, or inaccurately predicting the approval of, a new drug can be incredibly costly. To compound the problem, regulators in the United States and Europe demand that drugs have to be produced for the market in the same system used to produce them for the final round of clinical trials, in order to guarantee bioequivalence (e.g. toxicity, bioavailability, pharmacokinetics, and pharmacodynamics) of the molecule. So, companies have to choose between launching a product manufactured at a smaller development facility (and struggling to meet market demands) or building larger, dedicated facilities for a drug that might never be approved!

Therefore, alternative technologies are used for the expression of biopharmaceuticals, some of them also at lower costs involved (see Fig. 2). One such alternative is the creation of transgenic animals ("pharming"), but this suffers from the disadvantage that it requires a long time to establish such animals (approximately 2 years). In addition to that, some of the human biopharmaceuticals could be detrimental to the mammal's health, when expressed in the mammary glands. This is why ethical debates sometimes arise from the use of transgenic mammals for production of biopharmaceuticals. Although there are no ethical concerns involved with plants, there are societal ones that will be addressed later. Another expression system (see Fig. 2) utilizes transgenic chicken. The eggs, from which the proteins are harvested, are natural protein-production systems. But production of transgenic birds is still

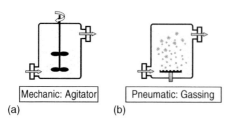
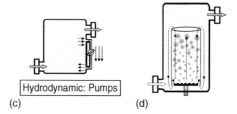

Fig. 1 Different fermentation modes for bioreactors. In order to achieve best aeration and mixing and to avoid high shear forces, different fermentation modes are applied. (a) mechanical, (b) pneumatical, (c) hydrodynamic pumps, (d) airlift reactor. Source: Knäblein J. (2002) *Transport Processes in Bioreactors and Modern Fermentation Technologies*, Lecture at University of Applied Sciences, Emden, Germany.

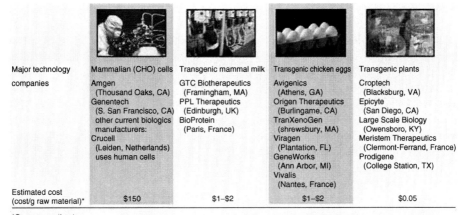

	Mammalian (CHO) cells	Transgenic mammal milk	Transgenic chicken eggs	Transgenic plants
Major technology companies	Amgen (Thousand Oaks, CA) Genentech (S. San Francisco, CA) other current biologics manufacturers: Crucell (Leiden, Netherlands) uses human cells	GTC Biotherapeutics (Framingham, MA) PPL Therapeutics (Edinburgh, UK) BioProtein (Paris, France)	Avigenics (Athens, GA) Origen Therapeutics (Burlingame, CA) TranXenoGen (shrewsbury, MA) Viragen (Plantation, FL) GeneWorks (Ann Arbor, MI) Vivalis (Nantes, France)	Croptech (Blacksburg, VA) Epicyte (San Diego, CA) Large Scale Biology (Owensboro, KY) Meristem Therapeutics (Clermont-Ferrand, France) Prodigene (College Station, TX)
Estimated cost (cost/g raw material)*	$150	$1–$2	$1–$2	$0.05

*Company estimates

Fig. 2 Companies and technologies in biomanufacturing. A comparison of different expression systems shows the big differences in terms of costs, ranging from 150 US$ per gram for CHO cells to 0.05 US$ per gram for transgenic plants. Source: Dove, A. (2002) Uncorking the biomanufacturing bottleneck, Nat. Biotechnol. **20**, 777–779.

several years behind transgenic mammal technology. Intensive animal housing constraints also make them more susceptible to disease (e.g. Asia 1997 or Europe 2003: killing of huge flocks with thousands of chicken suffering from fowl pest). In the light of development time, experience, costs, and ethical issues, plants are therefore the favored technology, since such systems usually have short gene-to-protein times (weeks), some are already well established, and as mentioned before, the involved costs are comparatively low. This low cost of goods sold (COGS) for plant-derived proteins is mainly due to low capital costs: greenhouse costs are only US$ 10 per m^2 versus US$ 1000 per m^2 for mammalian cells.

3
History of Plant Expression

Plants have been a source of medicinal products throughout human evolution. These active pharmaceutical compounds have been primarily small molecules, however. One of the most popular examples is aspirin (acetylsalicylic acid) to relieve pain and reduce fever. A French pharmacist first isolated natural salicin (a chemical relative of the compound used to make aspirin) from white willow bark in 1829. Advances in genetic engineering are now allowing for the production of therapeutic proteins (as opposed to small molecules) in plant tissues. Expression of recombinant proteins in plants has been well documented since the 1970s and has slowly gained credibility in the biotechnology industry and regulatory agencies. The first proof of concept has been the incorporation of insect and pest resistance into grains. For example, "Bt corn" contains genes from *Bacillus thuringensis* and is currently being grown commercially. Genetic engineering techniques are now available for the manipulation of almost all commercially valuable plants. Easy transformation and cultivation make plants suitable for production of virtually any recombinant protein.

Plants have a number of advantages over microbial expression systems, but one of them is of outmost importance: they can produce eukaryotic proteins in their native form, as they are capable of carrying out posttranslational modifications required for the biological activity of many such proteins (see Fischer Schillberg (2004), books). These modifications can be acetylation, phosphorylation, and glycosylation, as well as others. Per se, there is no restriction to the kind of proteins that can be expressed in plants: vaccines (e.g. pertussis or tetanus toxins), serum proteins (e.g. albumin), growth factors (e.g. vascular endothelial growth factor (VEGF), erythropoietin), or enzymes (e.g. urokinase, glucose oxidase, or glucocerebrosidase). However, enzymes sometimes have very complex cofactors, which are essential for their catalytic mode of action, but cannot be supplied by most expression systems. This is why, for the expression of some enzymes, expression systems with special features and characteristics need to be developed. Another very important class of proteins is the antibodies (e.g. scFv, Fab, IgG, or IgA). More than 100 antibodies are currently used in clinical trials as therapeutics, drug delivery vehicles, in diagnostics and imaging, and in drug discovery research for both screening and validation of targets. Again, plants are considered as the system of choice for the production of antibodies ("plantibodies") in bulk amounts at low costs. Since the initial demonstration that transgenic tobacco (*Nicotiana tabacum*) is able to produce functional IgG1 from mouse, full-length antibodies, hybrid antibodies, antibody fragments (Fab), and single-chain variable fragments (scFv) have been expressed in higher plants for a number of purposes. These antibodies can serve in health care and medicinal applications, either directly by using the plant as a food ingredient or as a pharmaceutical or diagnostic reagent after purification from the plant material. In addition, antibodies may improve plant performance, for example, by controlling plant disease or by modifying regulatory and metabolic pathways.

4
Current Status of Plant-based Expression

4.1
SWOT Analysis Reveals a Ripe Market for Plant Expression Systems

When I analyzed the different expression systems regarding their strengths, weaknesses, opportunities, and threats (SWOT), the advantages of plants and their potential to circumvent the worldwide capacity limitations for protein production became quite obvious (see Fig. 3). Comparison of transgenic animals, mammalian cell culture, plant expression systems, yeast, and bacteria shows certain advantages for each of the systems. In the order in which the systems were just mentioned, we can compare them in terms of their development time (speed). Transgenic animals have the longest cycle time (18 months to develop a goat), followed by mammalian cell culture, plants, yeast, and bacteria (one day to transform *E. coli*). If one looks at operating and capital costs, safety, and scalability, the data show that plants are beneficial: therefore, in the comparison (see Fig. 3), they are shown on the right-hand side already. But even for glycosylation, multimeric assembly and folding (where plants are not shown on the right-hand side, meaning other systems are advantageous), some plant expression systems are moving in

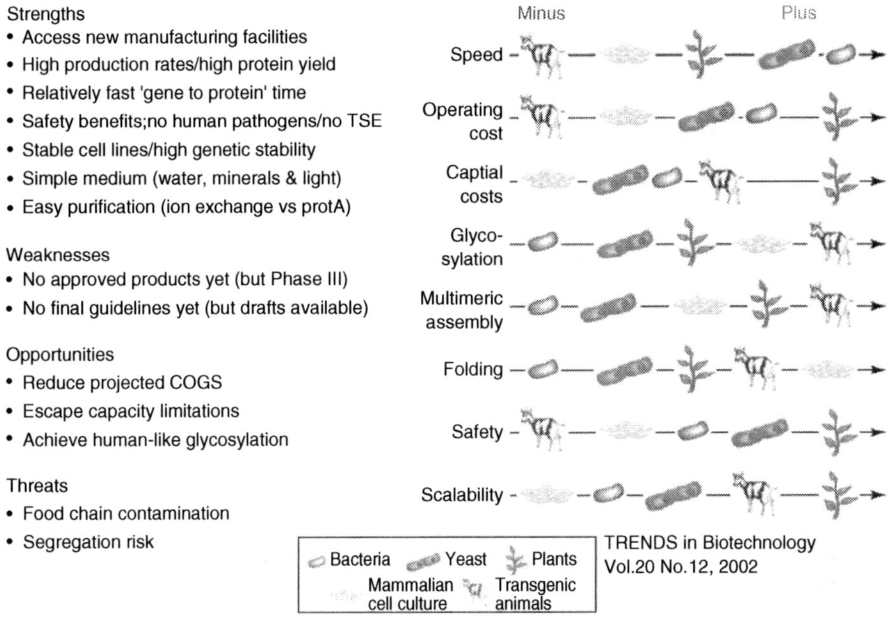
Strengths
• Access new manufacturing facilities
• High production rates/high protein yield
• Relatively fast 'gene to protein' time
• Safety benefits;no human pathogens/no TSE
• Stable cell lines/high genetic stability
• Simple medium (water, minerals & light)
• Easy purification (ion exchange vs protA)
Weaknesses
• No approved products yet (but Phase III)
• No final guidelines yet (but drafts available)
Opportunities
• Reduce projected COGS
• Escape capacity limitations
• Achieve human-like glycosylation
Threats
• Food chain contamination
• Segregation risk
Minus
Plus
Speed
Operating cost
Captial costs
Glyco-sylation
Multimeric assembly
Folding
Safety
Scalability
Bacteria
Yeast
Plants
Mammalian cell culture
Transgenic animals
TRENDS in Biotechnology
Vol.20 No.12, 2002

Fig. 3 SWOT analysis of plant expression systems. Plant expression systems have a lot of advantages (plus) over other systems and are therefore mostly shown on the right-hand side of the picture (Raskin, I., Fridlender, B., et al. (2002) Plants and human health in the twenty-first century, *Trends Biotechnol.* **20**, 522–531). Herein different systems (transgenic animals, mammalian cell culture, plants, yeast, and bacteria) are compared in terms of speed (how quickly they can be developed), operating and capital costs, and so on, and plants are obviously advantageous. Even for glycosylation, assembly, and folding, where plants are not shown on the right-hand side (meaning other systems are advantageous), some plant expression systems are moving in that direction (as will be shown exemplarily in the section on moss). Also the weaknesses and threats can be dealt with, using the appropriate plant expression system. Source: Knäblein J. (2003) *Biotech: A New Era In The New Millennium – From Plant Fermentation To Plant Expression Of Biopharmaceuticals*, PDA International Congress, Prague, Czech Republic.

that direction. An example of this is the moss system from the company greenovation Biotech GmbH (Freiburg, Germany), which will be discussed in detail in the example section. This system performs proper folding and assembly of even such complex proteins like the homodimeric VEGF. Even the sugar pattern could successfully be reengineered from plant to humanlike glycosylation.

In addition to the potential of performing human glycosylation, plants also enjoy the distinct advantage of not harboring any pathogens, which are known to harm animal cells (as opposed to animal cell cultures and products), nor do the products contain any microbial toxins, TSE (Transmissible Spongiform Encephalopathies), prions, or oncogenic sequences. In fact, humans are exposed to a large, constant dose of living plant viruses in the diet without any known effects/illnesses. Plant production of protein therapeutics also has advantages with regard to their scale and speed of production. Plants can be grown in ton quantities (using existing plant/crop technology, like commercial greenhouses), be extracted with industrial-scale equipment, and produce kilogram-size yields from a single plot of cultivation. These economies of scale are expected to reduce the cost of production of pure pharmaceutical-grade therapeutics by more than 2 orders of magnitude versus current bacterial fermentation or cell culture reactor systems (plus raw material COGS are estimated to be as low as 10% of conventional cell culture expenses).

Although a growing list of heterologous proteins were successfully produced in a number of plant expression systems with their manifold advantages, there are also obvious downsides. One weakness is that no product has been approved for the market yet (but will be soon, since some are in Phase III clinical trials already, see Table 1). The other weakness is that no final regulatory guidelines exist. But as mentioned before, regulatory authorities (Food and Drug Administration (FDA), European Medicine Evaluation Agency (EMEA), and Biotechnology Regulatory Service (BRS) and the Biotechnology Industry Organization (BIO) have drafted guidelines on plant-derived biopharmaceuticals (see Table 2) and have asked the community for comments. The FDA has also issued several PTC (Points To Consider) guidelines about plant-based biologics, and review of the July 2002 PTC confirms that the FDA supports this field and highlights the benefits of plant expression systems – including the absence of any pathogens to man from plant extracts. The main concerns of using plant expression systems are societal ones about environmental impacts, segregation risk, and contamination of the food chain. But these threats can be dealt with, using nonedible plants (nonfood, nonfeed), applying advanced containment technologies (GMP greenhouses, bioreactors) and avoiding open-field production.

Owing to the obvious strengths of plant expression systems, there has been explosive growth in the number of start-up companies. Since the 1990s, a number of promising plant expression systems have been developed, and in response to this "blooming field" big pharmaceutical companies have become more interested. Now, the plant expression field is "ripe" for strategic alliances, and, in fact, the last year has seen several major biotech companies begin partnerships with such plant companies. The selection of several such partnerships shown in Table 1 clearly demonstrates that, in general, there has been sufficient experimentation with various crops to provide the overall proof of concept that transgenic plants can produce biopharmaceuticals. However, and this can be seen in the table as well, the commercial production of biopharmaceuticals in transgenic plants is still in the early stages of development and yet the most advanced products are in Phase III clinical development.

4.2
Risk Assessment and Contingency Measures

For a number of reasons, including the knowledge base developed on genetically modifying its genome, industrial processes for extracting fractionated products and the potential for large-scale production, the preferred plant expression system has been corn. However, the use of corn touches on a potential risk: some environmental activist groups and trade associations are concerned about the effect on the environment and possible contamination of the food supply. These issues

18 Plant-based Expression of Biopharmaceuticals | 545

Tab. 1 Plant-derived biopharmaceuticals in clinical trials.

Company	Partner	Protein/indication	Host	Stage
Monsanto	Guy's Hospital London	Anticaries antibody	Corn	Phase III
Large Scale Biology	Own product	scFv (non-Hodgkin)	Tobacco	Phase IIIs
Meristem Therapeutics	Solvay Pharmaceuticals	Gastric lipase	Corn	Phase II
Large Scale Biology	ProdiGene, Plant Bioscience	Anti-ideotype antibody	Tobacco	Phase I
Monsanto	NeoRx	Antitumor antibody	Corn	Phase I
ProdiGene	Own product	TGEV vaccine	Corn	Phase I
Epicyte Pharmaceutical	Dow, Centocor	Anti-HSV antibody	Corn	Phase I
Crop Tech	Immunex	Enbrel (arthritis)	Tobacco	Preclinical
Crop Tech	Amgen	Therapeutic antibodies	Tobacco	Preclinical
AltaGen Bioscience Inc.	U.S. Army 3 + biotechs	Antibodies	Potato	Preclinical
Meristem Therapeutics	CNRS	Human lactoferrin	Corn	Preclinical
MPB Cologne GmbH	Aventis CropScience	Confidential	Potato	Preclinical

Tab. 2 Drafted guidelines on plant-derived biopharmaceuticals.

Agency	Guideline	Status
BRS (Biotechnology Regulatory Services)	"Case study on plant-derived biologics" for Office of Science and Technology Policy/Council on Environmental Quality	Released: Mar 5, 2001
BIO (Biotechnology Industry Organization)	"Reference Document for Confinement and Development of Plant-made Pharmaceuticals in the United States"	Released: May 17, 2002
BIO (Biotechnology Industry Organization)	"BIO Position on Geographic Restrictions for Plant-made Pharmaceuticals and Industrials"	Released: Oct 22, 2002
EMEA (European Medicine Evaluation Agency)	"Concept Paper on the Development of a Committee for Proprietary Medicinal Products (CPMP) Points to Consider on the Use of Transgenic Plants in the Manufacture of Biological Medicinal Products for Human Use"	Released: Mar 01, 2001
FDA (Food and Drug Administration)	"Drugs, Biologics, and Medical Devices Derived from Bioengineered Plants for Use in Humans and Animals"	Issued: Sep 6, 2002
EMEA (European Medicine Evaluation Agency)	"Points To Consider Quality Aspects of Medicinal Products containing active substances produced by stable transgene expression in higher plants"	Issued: Mar 13, 2002

are reflected in the regulatory guidelines and have been the driving force to investigate other plants as well. While many mature and larger companies have been working in this area for many years, there are a number of newcomers that are developing expertise as well. These smaller companies are reacting to the concerns by looking at the use of nonedible plants that can be readily raised in greenhouses. All potential risks have to be assessed and contingency measures need to be established. Understanding the underlying issues is mandatory to make sophisticated decisions about the science and subsequently on the development of appropriate plant expression systems for production of biopharmaceuticals.

Ongoing public fears from the food industry and the public, particularly in Europe ("Franken Food") could have spillover effects on plant-derived pharmaceuticals. Mistakes and misunderstandings have already cost the genetically enhanced grain industry hundreds of millions of dollars. The only way to prevent plant expression systems from suffering the same dilemma is to provide the public with appropriate information on emerging discoveries and newly developed production systems for biopharmaceuticals. Real and theoretical risks involve the spread of engineered genes into wild plants, animals, and bacteria (horizontal transmission). For example, if herbicide resistance was transmitted to weeds, or antibiotic resistance was to be transmitted to bacteria, superpathogens could result. If these genetic alterations were transmitted to their progeny (vertical transmission), an explosion of the pathogens could cause extensive harm. An example of this occurred several years ago, when it was feared that pest-resistant genes had been transmitted from Bt corn to milkweed – leading to the widespread death of Monarch butterflies. Although this was eventually not found to be the case, the public outcry over the incident was a wake-up call to the possible dangers of transgenic food technology. To avoid the same bad perception for biopharmaceuticals expressed in plants, there is the need for thorough risk assessment and contingency planning. One method is the employment of all feasible safety strategies to prevent spreading of engineered DNA (genetic drift), like a basic containment in a greenhouse environment. Although no practical shelter can totally eradicate insect and rodent intrusion, this type of isolation is very effective for self-pollinators and those plants with small pollen dispersal patterns. The use of species-specific, fragile, or poorly transmissible viral vectors is another strategy. Tobacco mosaic virus (TMV), for example, usually only infects a tobacco host.

It requires an injury of the plant to gain entry and cause infection. Destruction of a field of TMV-transformed tobacco requires only plowing under or application of a herbicide. These factors prevent both horizontal and vertical transmission. In addition, there is no known incidence of plant viruses infecting animal or bacterial cells. Another approach is to avoid stable transgenic germlines and therefore most uses of transforming viruses do not involve the incorporation of genes into the plant cell nucleus. By definition, it is almost impossible for these genes to be transmitted vertically through pollen or seed. The engineered protein product is produced only by the infected generation of plants. Another effective way to reduce the risk of genetic drift is the use of plants that do not reproduce without human aid. The modern corn plant cannot reproduce without cultivation and the purposeful planting of its seeds. If a plant may sprout from grain, it still needs to survive the wintering-over process and gain access to the proper planting depth. This extinction process is so rapid, however, that the errant loss of an ear of corn is very unlikely to grow a new plant. Another very well-known example of self-limited reproduction is the modern banana. It propagates almost exclusively through vegetative cloning (i.e. via cuttings).

Pollination is the natural way for most plants to spread their genetic information, make up new plants, and to deliver their offspring in other locations. The use of plants with limited range of pollen dispersal and limited contact with compatible wild hosts therefore is also very effective to prevent genetic drift. Corn, for example, has pollen, which survives for only 10 to 30 min and, hence, has an effective fertilizing radius of less than 500 m. In North America, it has no wild-type relatives with which it could cross-pollinate. In addition to being spatially isolated from nearby cornfields, transgenic corn can be "temporally isolated" by being planted at least 21 days earlier or 21 days later than the surrounding corn, to ensure that the fields are not producing flowers at the same time. Under recent USDA (U.S. Department of Agriculture) regulations, the field must also be planted with equipment dedicated to the genetically modified crop. For soybeans, the situation is different, since they are virtually 100% self-fertilizers and can be planted in very close proximity to other plants without fear of horizontal spread. Another option is the design of transgenic plants that have only sterile pollen or – more or less only applicable for greenhouses – completely prevent cross-pollination by covering the individual plants. One public fear regards spreading antibiotic resistance from one (transgenic donor) plant to other wild-type plants or bacteria in the environment. Although prokaryotic promoters for antibiotic resistance are sometimes used in the fabrication and selection of transgenic constructs, once a transgene has been stably incorporated into the plant genome, it is under the control of plant (eukaryotic) promoter elements. Hence, antibiotic-resistance genes are unable to pass from genetically altered plants into bacteria and remain functional. As stated earlier, another common fear is the creation of a "super bug." The chance of creating a supervirulent virus or bacterium from genetic engineering is unlikely, because the construction of expression cassettes from viral or bacterial genomes involves the removal of the majority of genes responsible for the normal function of these organisms. Even if a resultant organism is somewhat functional, it cannot compete for long in nature with normal, wild-type bacteria of the same species.

As one can see from the aforementioned safety strategies, considerable effort is put into the reduction of any potential risk from the transgenic plant for the environment. In general, the scientific risk can be kept at a minimum, if common sense is applied – in accordance with Thomas Huxley (1825–1895) that "Science is simply common sense at its best." For example, protein toxins (for vaccine production) should never be grown in food plants.

Additionally, the following can be employed as a kind of risk management to prevent the inappropriate or unsafe use of genetically engineered plants:

- An easily recognized phenotypic characteristic can be coexpressed in an engineered product (e.g. tomatoes that contain a therapeutic protein can be selected to grow in a colorless variety of fruit).

- Protein expression can be induced only after harvesting or fruit ripening. For example, CropTech's (Blacksburg, VA, USA) inducible expression system in tobacco, MeGA-PharM, leads to very efficient induction upon leaf injury (harvest) and needs no chemical inducers. This system possesses a fast induction response and protein synthesis rate, and thus leads to high expression levels with no aged product in the field (no environmental damage accumulation).
- Potentially antigenic or immunomodulatory products can be induced to grow in, or not to grow in, a certain plant tissue (e.g. root, leaf/stem, seed, or pollen). In this way, for example, farmers can be protected from harmful airborne pollen or seed dusts.
- Although no absolute system can prevent vandalism or theft of the transgenic plants, a very effective, cheap solution has been used quietly for many years now in the United States. Plots of these modified plants are being grown with absolutely no indication that they are different from a routine crop. In the Midwest, for example, finding a transgenic corn plot among the millions of acres of concurrently growing grain is virtually impossible. The only question here is, if this approach really helps facilitating a fair and an open discussion with the public. Asking the same question for the EU is not relevant: owing to labeling requirements, this approach would not be feasible, as, in general, it is much more difficult to perform open-field studies with transgenic plants.

5
The Way Forward: Moving Plants to Humanlike Glycosylation

As discussed earlier, plant production of therapeutic proteins has many advantages over bacterial systems. One very important feature of plant cells is their capability of carrying out posttranslational modifications. Since they are eukaryotes (i.e. have a nucleus), plants produce proteins through an ER (endoplasmatic reticulum) pathway, adding sugar residues also to the protein – a process called *glycosylation*. These carbohydrates help determine the three-dimensional structures of proteins, which are inherently linked to their function and their efficacy as therapeutics. This glycosylation also affects protein bioavailability and breakdown of the biopharmaceutical; for example, proteins lacking terminal sialic acid residues on their sugar groups are often targeted by the immune system and are rapidly degraded. The glycosylation process begins by targeting the protein to the ER. During translation of mRNA (messenger RNA) into protein, the ribosome is attached to the ER, and the nascent protein fed into the lumen of the ER as translation proceeds. Here, one set of glycosylation enzymes attaches carbohydrates to specific amino acids of the protein. Other glycosylation enzymes either delete or add more sugars to the core structures. This glycosylation process continues into the Golgi apparatus, which sorts the new proteins, and distributes them to their final destinations in the cell (see Fig. 4). Bacteria lack this ability and therefore cannot be used to synthesize proteins that require glycosylation for activity. Although plants have a somewhat different system of protein glycosylation from mammalian cells, the differences usually prove not to be a problem. Some proteins, however, require

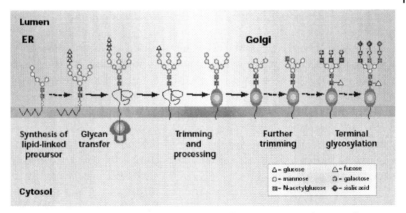

Fig. 4 The glycosylation pathway via ER and Golgi apparatus. In the cytosol, carbohydrates are attached to a lipid precursor, which is then transported into the lumen of the ER to finish core glycosylation. This glycan is now attached to the nascent, folding polypeptide chain (which is synthesized by ribosomes attached to the cytosolic side of the ER from where it translocates into the lumen) and subsequently trimmed and processed before it is folded and moved to the Golgi apparatus. Capping of the oligosaccharide branches with sialic acid and fucose is the final step on the way to a mature glycoprotein. Source: Dove, A. (2001) The bittersweet promise of glycobiology, *Nat. Biotechnol.* **19**, 913–917.

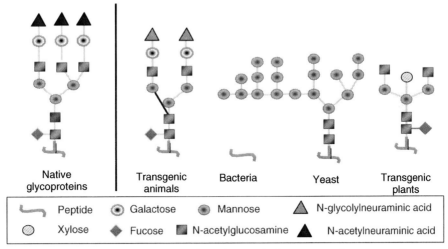

Fig. 5 Engineering plants to humanlike glycosylation. The first step to achieve humanlike glycosylation in plants is to eliminate the plant glycosylation pattern, that is, the attachment of β 1–2 linked xylosyl- and α 1–3 linked fucosyl sugars to the protein. Because these two residues have allergenic potential, the corresponding enzymes Xylosyl- and Fucosyl Transferase are knocked out. In case galactose is relevant for the final product, Galactosyl Transferase is inserted into the host genome. Galactose is available in the organism so that this single gene insertion is sufficient to ensure galactosylation. Source: Knäblein J. (2003) *Biotech: A New Era In The New Millennium – Biopharmaceutic drugs manufactured in novel expression systems*, DECHEMA-Jahrestagung der Biotechnologen, Munich, Germany, 21.

humanlike glycosylation (see Fig. 5) – they must have specific sugar structures attached to the correct sites on the molecule to be maximally effective. Therefore, some efforts are being made in modifying host plants in such a way that they provide the protein with human glycosylation patterns. One example of modifying a plant expression system in this way is the transgenic moss, which will be discussed in the next section.

6
Three Promising Examples: Tobacco (Rhizosecretion, Transfection) and Moss (Glycosylation)

To further elaborate on improving glycosylation and downstream processing, three interesting plant expression systems will be discussed. All systems share the advantage of utilizing nonedible plants (nonfood and nonfeed) and can be kept in either a greenhouse or a fermenter to avoid any segregation risk. Another obvious advantage is secretion of the protein into the medium so that no grinding or extraction is required. This is very important in light of downstream processing: protein purification is often as expensive as the biomanufacturing and should never be underestimated in the total COGS equation.

6.1
Harnessing Tobacco Roots to Secrete Proteins

Phytomedics (Dayton, NJ, USA) uses tobacco plants as an expression system for biopharmaceuticals. Besides the advantage of being well characterized and used in agriculture for some time, tobacco has a stable genetic system, provides high-density tissue (high protein production), needs only simple medium, and can be kept in a greenhouse (see Fig. 6). Optimized antibody expression can be rapidly verified using transient expression assays (short development time) in the plants before creation of transgenic suspension cells or stable plant lines (longer development time). Different vector systems, harboring targeting signals for subcellular compartments, are constructed in parallel and used for transient expression. Applying this screening approach, high expressing cell lines can rapidly be identified. For example, transgenic tobacco plants, transformed with an expression cassette containing the GFP (Green Fluorescent Protein) gene fused to an *aps* (amplification-promoting sequence), had greater levels of corresponding mRNAs and expressed proteins compared to transformants lacking *aps*. Usually, downstream processing (isolation/extraction and purification of the target protein) is limiting for such a system, for example, if the protein has to be isolated from biochemically complex plant tissues (e.g. leaves), this can be a laborious and expensive process and a major obstacle to large-scale protein manufacturing. To overcome this problem, secretion-based systems utilizing transgenic plant cells or plant organs aseptically cultivated *in vitro* would be one solution. However, *in vitro* systems can be expensive, slow growing, unstable, and relatively low yielding. This is why another interesting route was followed. Secretion of molecules is a basic function of plant cells and organs in plants, and is especially developed in plant roots. In order to take up nutrients from the soil, interact with other soil organisms, and defend themselves against numerous pathogens, plant roots have evolved sophisticated mechanisms based on the secretion of different biochemicals (including proteins

Phytomedics (tobacco):

- Root secretion, easy recovery
- Greenhouse contained tanks
- High density tissue
- Salts and water only
- Tobacco is well characterized
- Stable genetic system

Fig. 6 Secretion of the biopharmaceuticals via tobacco roots. The tobacco plants are genetically modified in such a way that the protein is secreted via the roots into the medium ("rhizosecretion"). In this example, the tobacco plant takes up nutrients and water from the medium and releases GFP (Green Fluorescent Protein). Examination of root cultivation medium by its exposure to near ultraviolet-illumination reveals the bright green-blue fluorescence characteristics of GFP in the hydroponic medium (left flask in panel lower left edge). The picture also shows a schematic drawing of the hydroponic tank, as well as tobacco plants at different growth stages, for example, callus, fully grown, and greenhouse plantation. Source: Knäblein J. (2003) *Biotech: A New Era in the New Millennium – Biopharmaceutic Drugs Manufactured in Novel Expression Systems*, DECHEMA-Jahrestagung der Biotechnologen, Munich, Germany, 21. (See color plate. p. xxi)

like toxins) into their neighborhood (rhizosphere). In fact, Borisjuk and coworkers could demonstrate that root secretion can be successfully exploited for the continuous production of recombinant proteins in a process termed *"rhizosecretion."* Here, an endoplasmic reticulum signal peptide is fused to the recombinant protein, which is then continuously secreted from the roots into a simple hydroponic medium (based on the natural secretion from roots of the intact plants). The roots of the tobacco plant are sitting in a hydroponic tank (see Fig. 6), taking up water and nutrients and continuously releasing the biopharmaceutical. By this elegant set up, downstream processing becomes easy and cost-effective, and also offers the advantage of continuous protein production that integrates the biosynthetic potential of a plant over its lifetime and might lead to higher protein yields than single-harvest and extraction methods. Rhizosecretion is demonstrated in Fig. 6, showing a transgenic tobacco plant expressing GFP and releasing it into the medium.

6.2
High Protein Yields Utilizing Viral Transfection

ICON Genetics (Halle, Germany) has developed a protein-production system that relies on rapid multiplication of viral vectors in an infected tobacco plant (see Fig. 7). Viral transfection systems offer

ICON Genetics (tobacco):
- Viral transfection
- Fast development
- High protein yields
- Coexpression of genes

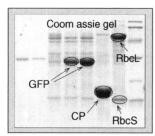

Fig. 7 Viral transfection of tobacco plants. This new generation platform for fast (1 to 2 weeks), high-yield (up to 5 g kg^{-1} fresh leaf weight) production of biopharmaceuticals is based on proviral gene amplification in a nonfood host. Antibodies, antigens, interferons, hormones, and enzymes could successfully be expressed with this system. The picture shows development of initial symptoms on a tobacco following the Agrobacterium-mediated infection with viral vector components that contain a GFP gene (a); this development eventually leads to a systemic spread of the virus, literally converting the plant into a sack full of protein of interest within two weeks (b). The system allows to coexpress two proteins in the same cell, a feature that allows expression of complex proteins such as full-length monoclonal antibodies. Panels (c) and (d) show the same microscope section with the same cells, expressing Green Fluorescent Protein (c) and Red Fluorescent Protein (d) at the same time. The yield and total protein concentration achievable are illustrated by a Coomassie gel with proteins in the system: GFP (protein of interest), CP (coat protein from wild-type virus), RbcS and RbcL (small and large subunit of ribulose-1,5-bisphosphate carboxylase). Source: Knäblein J. (2003) *Biotech: A New Era in the New Millennium – Biopharmaceutic Drugs Manufactured in Novel Expression Systems*, DECHEMA-Jahrestagung der Biotechnologen, Munich, Germany, 21. (See color plate. p. xxii)

a number of advantages, such as very rapid (1 to 2 week) expression time, possibility of generating initial milligram quantities within weeks, high expression levels, and so on. However, the existing viral vectors, such as TMV-based vectors used by, for example, Large Scale Biology Corp. (Vacaville, CA, USA) for production of single-chain antibodies for treatment of non-Hodgkin lymphoma (currently in Phase III clinical trials, see Table 1), had numerous shortcomings, such as inability to express genes larger than 1 kb, inability to coexpress two or more proteins (a prerequisite for production of monoclonal antibodies, because they consist of the light and heavy chains, which are expressed independently and are subsequently assembled), low expression level in systemically infected leaves, and so on. ICON has solved many of these problems by designing a process that starts with an assembly of one or more viral vectors inside a plant after treating the leaves with agrobacteria, which deliver the necessary viral vector components. ICON's proviral vectors provide advantages of fast and high-yield amplification processes in a plant cell, simple and inexpensive assembly of expression cassettes *in planta*, and full control of the process. The robustness of highly standardized protocols allows the use of inherently the same safe protocols for both laboratory-scale as well as industrial production processes. In this system, the plant is modified transiently rather than genetically and reaches the speed and yield of microbial systems while enjoying posttranslational capabilities of plant cells. De- and reconstructing of the virus adds some safety features and also increases efficiency. There is no "physiology conflict," because the "growth phase" is separated from the "production phase," so that no competition occurs for nutrients and other components required for growth and also for expression of the biopharmaceutical at the same time.

This transfection-based platform allows the production of proteins in a plant host at a cost of US$1 to 10 per gram of crude protein. The platform is essentially free from limitations (gene insert size limit, inability to express more than one gene) of current viral vector-based platforms. The expression levels reach 5 g per kilogram of fresh leaf tissue (or some 50% of total cellular protein!) in 5 to 14 days after inoculation. Since the virus process (in addition to superhigh production of its own proteins, including the protein of interest) leads to the shutoff of the other cellular protein synthesis, the amount of protein of interest in the initial extract is extremely high (Fig. 7). It thus results in reduced costs of downstream processing. Milligram quantities can be produced within two weeks, gram quantities in 4 to 6 months, and the production system is inherently scalable. A number of high-value proteins have been successfully expressed, including antibodies, antigens, interferons, hormones, and enzymes (see Klimyuk, Marillonnet, Knäblein, McCaman, Gleba (2005), books).

6.3
Simple Moss Performs Complex Glycosylation

Greenovation Biotech GmbH (Freiburg, Germany) has established an innovative production system for human proteins. The system produces pharmacologically active proteins in a bioreactor, utilizing a moss (*Physcomitrella patens*) cell culture system with unique properties (see Fig. 8). It was stated before that posttranslational modifications for some proteins are crucial to gain complete pharmacological activity. Since moss is the only known plant system that shows a high frequency of homologous recombination, this is a highly attractive tool for production strain design. By establishing stable integration of foreign genes (gene knock-out and new transgene insertion) into the plant genome, it can be programmed to produce proteins with modified glycosylation patterns that are identical to animal cells. The moss is photoautotrophic and therefore only requires simple media for growth, which consist essentially of water and minerals. This reduces costs and also accounts for significantly lower infectious and contamination risks, but in addition to this, the system has some more advantages:

- The transient system allows production of quantities for a feasibility study within weeks – production of a stable expression strain takes 4 to 6 months.
- On the basis of transient expression data, the yield of stable production lines is expected to reach 30 mg L^{-1} per day. This corresponds to the yield of a typical fed-batch culture over 20 days of 600 mg L^{-1}.
- Bacterial fermentation usually requires addition of antibiotics (serving as selection marker and to avoid loss of the

Greenovation (moss system):
- Simple medium (photoautotrophic plant needs only water and minerals)
- Robust expression system (good expression levels from 15 to 25°C)

- Secretion into medium via human leader sequence (broad pH range: 4-8)
- Easy purification from low salt medium via ion exchange

- Easy genetic modifications to cell lines
- Stable cell lines / high genetic stability

- Codon usage like human (no changes required)
- Inexpensive bioreactors from the shelf

- Nonfood plant (no segregation risk)
- Good progress on genetic modification of glycosylation pathways (plant to human)

Fig. 8 Greenovation use a fully contained moss bioreactor. This company has established an innovative production system for human proteins. The system produces pharmacologically active proteins in a bioreactor, utilizing a moss (*Physcomitrella patens*) cell culture system with unique properties. Source: Knäblein J. (2003) *Biotech: A New Era in the New Millennium – Biopharmaceutic drugs Manufactured in Novel Expression Systems*, DECHEMA-Jahrestagung der Biotechnologen, Munich, Germany, 21.

expression vector). For moss cultivation, no antibiotics are needed – this avoids the risk of traces of antibiotics having a significant allergenic potential in the finished product.
- Genetic stability is provided by the fact that the moss is grown in small plant fragments and not as protoplasts or tissue cultures avoiding somaclonal variation.
- As a contained system, the moss bioreactor can be standardized and validated according to GMP standards mandatory in the pharmaceutical industry.
- Excretion into the simple medium is another major feature of the moss bioreactor, which greatly facilitates downstream processing.

As discussed in detail, the first step to get humanlike glycosylation in plants is to eliminate the plant glycosylation, for example, the attachment of β-1-2-linked xylosyl and α-1-3-linked fucosyl sugars to the protein, because these two residues have allergenic potential. Greenovation was able to knockout the relevant glycosylation enzymes xylosyl transferase and fucosyl transferase, which was confirmed by RT-PCR (reverse transcriptase PCR). And indeed, xylosyl and fucosyl residues were completely removed from the glycosylation pattern of the expressed protein as confirmed by MALDI-TOF (matrix assisted laser desorption ionization time of flight) mass spectroscopy analysis (see Fig. 9).

A very challenging protein to express is VEGF because this homodimer consists of two identical monomers linked via a disulfide bond. To produce VEGF in

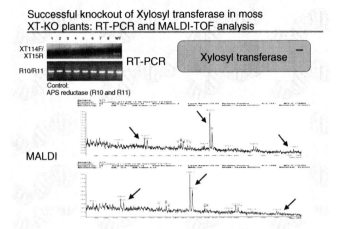

Fig. 9 Knockout of Xylosyl Transferase in moss. To avoid undesired glycosylation, greenovation knocked out the Xylosyl and Fucosyl Transferase, as confirmed by RT-PCR. MALDI-TOF results show that indeed, xylosyl- and fucosyl-residues were completely removed from the glycosylation pattern of the expressed protein (data for knockout of Fucosyl Transferase not shown). Source: Knäblein J. (2003) *Biotech: A New Era in the New Millennium – Biopharmaceutic Drugs Manufactured in Novel Expression Systems*, DECHEMA-Jahrestagung der Biotechnologen, Munich, Germany, 21.

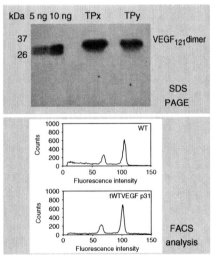
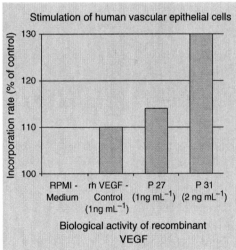

Fig. 10 Greenovation could successfully express the biopharmaceutical VEGF. This growth factor is a very complex protein consisting of two identical monomers linked via a disulfide-bond. To produce VEGF in an active form, the monomers need to be expressed to the right level, correctly folded, assembled, and linked via the disulfide-bond. The analytical assays clearly show that expression in moss yielded completely active VEGF. Source: Knäblein J. (2003) *Biotech: A New Era in the New Millennium – from Plant Fermentation to Plant Expression of Biopharmaceuticals*, PDA International Congress, Prague, Czech Republic.

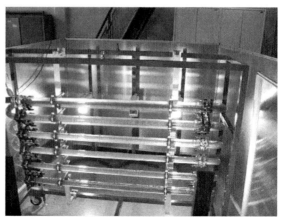

30 L pilot reactor for moss   Two weeks after incubation

Fig. 11 Scaling of photobioreactors up to several 1000 L. The moss bioreactor is based on the cultivation of *Physcomitrella patens* in a fermenter. The moss protonema is grown under photoautotrophic conditions in a medium that consists essentially of water and minerals. Light and carbon dioxide serve as the only energy and carbon sources. Cultivation in suspension allows scaling of the photobioreactors up to several 1000 L. Adaptation of existing technology for large-scale cultivation of algae is done in cooperation with the Technical University of Karlsruhe. Source: greenovation Biotech GmbH (Freiburg, Germany) and Professor C. Posten, Technical University (Karlsruhe, Germany).

an active form, the following need to be provided:

- Monomers need to be expressed to the right level.
- Monomers need to be correctly folded.
- Homodimer needs to be correctly assembled and linked via a disulfide bond.
- Complex protein needs to be secreted in its active form.

And in fact, all this could be achieved with the transgenic moss system as shown in Fig. 10. These results are very promising because they demonstrate that this system is capable of expressing even very complex proteins. In addition to that, the moss system adds no plant-specific sugars to the protein – a major step toward humanlike glycosylation. Furthermore, moss is a robust expression system leading to high yields at 15 to 25 °C and the pH can be adjusted from 4 to 8 depending on the optimum for the protein of interest. Adapting existing technology for large-scale cultivation of algae, fermentation of moss in suspension culture allows scaling of the photobioreactors up to several 1000 L (see Fig. 11). Finally, the medium is inexpensive, since only water and minerals are sufficient.

7
Other Systems Used for Plant Expression

Several different plants have been used for the expression of proteins in plants. All these systems have certain advantages regarding edibility, growth rate, scalability, gene-to-protein time, yield, downstream processing, ease of use, and so on, which I will not discuss in further detail here. A selection of different expression systems is listed:

Alfalfa	Ethiopian mustard	Potatoes
Arabidopsis	Lemna	Rice
Banana	Maize	Soybean
Cauliflower	Moss	Tomatoes
Corn	Oilseeds	Wheat

Some of these systems have been used for research on the basis of their ease of transformation, well-known characterization, and ease to work with. However, they are not necessarily appropriate for commercial production. Which crop is ultimately used for full-scale commercial production will depend on a number of factors including

- time to develop an appropriate system (gene-to-protein);
- section of the plant expressing the product/possible secretion;
- cost and potential waste products from extraction;
- "aged" product/ease of storage;
- long-term stability of the storage tissue;
- quantities of protein needed (scale of production).

Depending on the genetic complexity and ease of manipulation, the development time to produce an appropriate transgenic plant for milligram production of the desired protein can vary from 10 to 12 months in corn as compared to only weeks in moss. Estimates for full GMP production in corn are 30 to 36 months and approximately 12 months for moss. Expression of the protein in various tissues of the plant can result in a great variation in yield. Expression in the seed can often lead to higher yields than in the leafy portion of the plant. This is another explanation for the high interest in using corn, which has a relatively high seed-to-leaf ratio. Extraction from leaf can be costly as it contains a high percentage of water, which could result in unavoidable proteolysis during the process. Proteins stored in seeds can be desiccated and remain intact for long periods of time. The purification and extraction of the protein is likely to be done by adaptations of current processes for the extraction and/or fractionation. For these reasons, it is anticipated that large-scale commercial production of recombinant proteins will involve grain and oilseed crops such as maize, rice, wheat, and soybeans. On the basis of permits for open-air test plots issued by the USDA for pharmaceutical proteins and industrial biochemicals, corn is the crop of choice for production with 73% of the permits issued. The other major crops are soybeans (12%), tobacco (10%), and rice (5%).

In general, the use of smaller plants that can be grown in greenhouses is an effective way of producing the biopharmaceuticals and alleviating concerns from environmental activist groups that the transgenic plant might be harmful to the environment (food chain, segregation risk, genetic drift, etc.).

8
Analytical Characterization

Validated bioanalytical assays are essential and have to be developed to characterize the biopharmaceuticals during the production process (e.g. in-process control) and to release the final product for use as a drug in humans. These assays are applied to determine characteristics such as purity/impurities, identity, quantity, stability, specificity, and potency of the recombinant protein during drug development. Since the very diverse functions of different proteins heavily depend on their structure, one very valuable parameter in protein characterization is the elucidation of their three-dimensional structure. Although over the last couple of years a lot of effort was put into a method for improving the elucidation of protein structures (during my PhD thesis, I was also working in this fascinating field together with my boss Professor Robert Huber, Nobel Prize Laureate in 1988, "for the determination of the three-dimensional structure of a photosynthetic reaction centre"), it is still very time consuming to solve the 3-D structure of larger proteins. This is why despite the high degree of information that can be obtained from the protein structure, this approach cannot be applied on a routine basis. Therefore, tremendous efforts are put into the development of other assays to guarantee that a potent biopharmaceutical drug is indeed ready for use in humans.

9
Conclusion and Outlook

The production of protein therapeutics from transgenic plants is becoming a reality. The numerous benefits offered by plants (low cost of cultivation, high biomass production, relatively fast gene-to-protein time, low capital and operating costs, excellent scalability, eukaryotic post-translational modifications, low risk of human pathogens, lack of endotoxins, as well as high protein yields) virtually guarantee that plant-derived proteins will become more and more common for therapeutic uses. Taking advantage of plant expression systems, the availability of cheap protein-based vaccines in underdeveloped countries of the world is possible in the near future. The cost of very expensive hormone therapies (erythropoietin, human growth hormone, etc.) could fall dramatically within the next decade because of the use of, for example, plant expression systems. Fears about the risks of the plant expression technology are real and well founded, but with a detailed understanding of the technology, it is possible to proactively address these safety issues and create a plant expression industry almost free of mishaps. For this purpose, the entire set up, consisting of the specific plant expression system and the protein being produced, needs to be analyzed and its potential risks assessed on a case-by-case basis. As plant-derived therapeutics begin to demonstrate widespread, tangible benefits to the population and as the plant expression industry develops a longer safety track record, public acceptance of the technology is likely to improve continuously. Plants are by far the most abundant and cost-effective renewable resource uniquely adapted to complex biochemical synthesis. The increasing cost of energy and chemical raw materials, combined with the environmental concerns associated with conventional pharmaceutical manufacturing, will make plants even more compatible in the future. With the words of Max Planck (1858–1947) "How far advanced Man's scientific knowledge may be, when confronted with Nature's immeasurable richness and capacity for constant renewal, he will be like a marveling child and must always be prepared for new surprises," we will definitely discover more fascinating features of plant expression systems. But there is no need to wait: combining the advantages of some technologies that we already have in hand could lead to the ultimate plant expression system. This is what we should focus on, because, then, at the dawn of this new millennium, this would for the first time yield large-enough amounts of biopharmaceuticals to treat everybody on our planet!

Acknowledgments

I would like to thank the companies greenovation Biotech GmbH (Freiburg, Germany), ICON Genetics (Halle, Germany), and Phytomedics (Dayton, NJ, USA) for providing some data and figures to prepare this manuscript.

Bibliography

Books and Reviews

Fischer, R., Schillberg S. (Eds.) (2004) *Molecular Farming: Plant-made Pharmaceuticals and Technical Proteins*, Wiley, ISBN: 3-527-30786-9

Fischer, R., et al. (2004) Plant-based production of biopharmaceuticals, *Curr. Opin. Plant Biol.* 7(2), 152–158.

Horn, M.E., Woodard, S.L., Howard, J.A. (2004) Plant molecular farming: systems and products, *Plant Cell Rep.* 22(10), 711–720.

Klimyuk, V., Marillonnet, S., Knäblein, J., McCaman, M., Gleba, Y. (2005) Production of Recombinant Proteins in Plants, in: *Modern Biopharmaceuticals – Design, Development and Optimization*, Wiley-VCH, in press.

Knäblein, J. (2004) Biopharmaceuticals Expressed in Plants – A New Era in the New Millennium, in: Müller, R., Kayser, O. (Eds.) *Applications in Pharmaceutical Biotechnology*, Wiley-VCH, ISBN 3-527-30554-8.

Ma, J.K., Drake, P.M., Christou, P. (2003) The production of recombinant pharmaceutical proteins in plants, *Nat. Rev. Genet.* 4(10), 794–805.

Stoger, E., et al. (2004) Antibody production in transgenic plants, *Methods Mol. Biol.* 248, 301–318.

Primary References

Arakawa, T., Chong, D.K.X., Langridge, W.H.R. (1998) Efficacy of a food plant-based oral cholera toxin B subunit vaccine, *Nat. Biotechnol.* 16, 292–297.

Arakawa, T., Yu, J., Chong, D.K., Hough, J., Engen, P.C., Langridge, W.H. (1998) A plant-based cholera toxin B subunit-insulin fusion protein protects against the development of autoimmune diabetes, *Nat. Biotechnol.* 16, 934–938.

Arthur D. Little, Inc. (ADL), *AgIndustries Research*, Cambridge, MA, Copyright© 2002.

Artsaenko, O., et al. (1998) Potato tubers as a biofactory for recombinant antibodies, *Mol. Breeding* 4, 313–319.

Beachy, R.N., Fitchen, J.H., Hein, M.B. (1996) Use of Plant Viruses for Delivery of Vaccine Epitopes, in: Collins, G.B., Sheperd, R.J. (Eds.) *Engineering Plants for Commercial Products and Applications*, New York Academy of Sciences, New York, pp. 43–49.

Boothe, J.G., Parmenter, D.L., Saponja, J.A. (1997) Molecular farming in plants: oilseeds as vehicles for the production of pharmaceutical proteins, *Drug Develop. Res.* 42, 172–181.

Borisjuk, N.V., Raskin, I., et al. (1999) Production of recombinant proteins in plant root exudates, *Nat. Biotechnol.* 17, 466–469.

Borisjuk, N.V., Raskin, I., et al. (2000) Tobacco ribosomal DNA spacer element stimulates amplification and expression of heterologous genes, *Nat. Biotechnol.* 18, 1303–1306.

Cabanes-Macheteau, M., et al. (1999) N-glycosylation of a mouse IgG expressed in transgenic tobacco plants, *Glycobiology* 9, 365–372.

Chance, R.E., Frank, B.H. (1993) Research, development, production and safety of biosynthetic human insulin, *Diabetes care* 16(3), 133–142.

Chaudhary, S., Parmenter, D.L., Moloney, M.M. (1998) Transgenic Brassica carinata as a vehicle for the production of recombinant proteins in seeds, *Plant Cell Rep.* 17, 195–200.

Conrad, U., Fiedler, U. (1994) Expression of engineered antibodies in plant cells, *Plant Mol. Biol.* 26, 1023–1030.

Conrad, U., Fiedler, U., Artsaenko, O., Phillips, J. (1998) High-level and stable accumulation of single-chain Fv antibodies in plant storage organs, *J. Plant Physiol.* 152, 708–711.

Cramer, C.L., Boothe, J.G., Oishi, K.K. (1999) Transgenic plants for therapeutic proteins: linking upstream and downstream strategies, *Curr. Top. Microbiol. Immunol.* 240, 95–118.

Cramer, C.L., et al.. (1996) Bioproduction of Human Enzymes in Transgenic Tobacco, in: Collins, G.B., Sheperd, R.J. (Eds.) *Engineering Plants for Commercial Products and Applications*, New York Academy Of Sciences, New York, 62–71.

Dalsgaard, K., et al. (1997) Plant-derived vaccine protects target animals against a viral disease, *Nat. Biotechnol.* **15**, 248–252.

Davies, L., Plieth, J. (2001) The challenge of meeting the escalating demand for proteins, *Scr Mag* **10**, 25–29.

Della-Cioppa, G., Grill, L.K. (1996) Production of Novel Compounds in Higher Plants by Transfection with RNA Viral Vectors, in: Collins, G.B., Sheperd, R.J. (Eds.) *Engineering Plants for Commercial Products and Applications*, New York Academy of Sciences, New York, pp. 57–61.

Dieryck, W., et al. (1997) Human haemoglobin from transgenic tobacco, *Nature* **386**, 29–30.

Doran, P.M. (2000) Foreign protein production in plant tissue cultures, *Curr. Opin. Biotechnol.* **11**, 199–204.

Dove, A. (2001) The bittersweet promise of glycobiology, *Nat. Biotechnol.* **19**, 913–917.

Dove, A. (2002) Unkorking the biomanufacturing bottleneck, *Nat. Biotechnol.* **20**, 777–779.

Drake, P.M., Chargelegue, D., Vine, N.D., Van Dolleweerd, C.J., Obregon, P., Ma, J.K. (2002) Transgenic plants expressing antibodies: a model for phytoremediation, *FASEB J.* **16**(14), 1855–1860.

Drug & Market Development Publications, *Antibody Engineering: Technologies, Applications and Business opportunities*, Westborough, MA, Copyright© 2003.

Evangelista, R.L., Kusnadi, A.R., Howard, J.A., Nikolov, Z.L. (1988) Process and economic evaluation of the extraction and purification of recombinant glucouronidase from transgenic corn, *Biotechnol. Prog.* **14**, 607–614.

Fischer, R., Emans, N. (2000) Molecular farming of pharmaceutical proteins, *Transgenic Res.* **9**, 279–299.

Fischer, R., Hoffmann, K., Schillberg, S., Emans, N. (2000) Antibody production by molecular farming in plants, *J. Biol. Regul. Homeost. Agents* **14**, 83–92.

Ganz, P.R., et al.. (1996) Expression of Human Blood Proteins in Transgenic Plants: The Cytokine GM-CSF as a Model Protein, in: Owen, M.R.L., Pen, J. (Eds.) *Transgenic Plants: A Production System for Industrial and Pharmaceutical Proteins*, John Wiley & Sons, London, UK, 281–297..

Garber, K. (2001) Biotech industry faces new bottleneck, *Nat. Biotechnol.* **19**, 184–185.

Giddings, G., Allison, G., Brooks, D., Carter, C. (2000) Transgenic plants as factories for biopharmaceuticals, *Nat. Biotechnol.* **18**, 1151–1155.

Goddijn, O.J.M., Pen, J. (1995) Plants as bioreactors, *Trends Biotechnol.* **13**, 379–387.

Hamamoto, H., et al. (1993) A new tobacco mosaic virus vector and its use for the systematic production of angiotensin-I-converting enzyme inhibitor in transgenic tobacco and tomato, *Biotechnology* **11**, 930–932.

Hiatt, A., Cafferkey, R., Bowdish, K. (1989) Production of antibodies on transgenic plants, *Nature* **342**, 76–78.

Hood, E.E., Jilka, J.M. (1999) Plant-based production of xenogenic proteins, *Curr. Opin. Biotechnol.* **10**, 382–386.

Johnson, E. (1996) Edible plant vaccines, *Nat. Biotechnol.* **14**, 1532–1533.

Knäblein, J. (2003) Biotech: A new era in the new millennium – fermentation and expression of biopharmaceuticals in plants, *SCREENING – Trends Drug Discov* **4**, 14–16.

Knäblein, J., McCaman, M. (2003) Modern biopharmaceuticals-recombinant protein expression in transgenic plants, *SCREENING – Trends Drug Discov* **6**, 33–35.

Knäblein, J., Huber, R., et al. (1997) [Ta6Br12]2+, a tool for phase determination of large biological assemblies by X-ray crystallography, *J. Mol. Biol.* **270**, 1–7.

Kumagai, M.H., et al. (1993) Rapid, high-level expression of biologically active alpha-trichosanthin in transfected plants by an RNA viral vector, *Proc. Natl. Acad. Sci. USA* **90**, 427–430.

Kusnadi, A., Nikolov, Z.L., Howard, J.A. (1997) Production of recombinant proteins in transgenic plants: practical considerations, *Biotechnol. Bioeng.* **56**, 473–484.

Ma, J.K.C. (2000) Genes, greens, and vaccines, *Nat. Biotechnol.* **18**, 1141–1142.

Ma, J.K.C., Hein, M.B. (1995) Plant antibodies for immunotherapy, *Plant Physiol.* **109**, 341–346.

Ma, J.K.C., Hein, M.B. (1996) Antibody Production and Engineering in Plants, in: Collins, G.B., Sheperd, R.J. (Eds.) *Engineering Plants for Commercial Products and Applications*, New York Academy of Sciences, New York, pp. 72–81.

Ma, J.K.C., Hiatt, A. (1996) Expressing Antibodies in Plants for Immunotherapy, in: Owen, M.R.L., Pen, P. (Eds.) *Transgenic Plants:*

A Production System for Industrial and Pharmaceutical Proteins, John Wiley & Sons, London, UK, pp. 229–243.

Ma, J.K.C., Vine, N.D. (1999) Plant expression systems for the production of vaccines, *Curr. Top. Microbiol. Immunol.* **236**, 275–292.

Ma, J.K.C., et al. (1998) Characterization of a recombinant plant monoclonal secretory antibody and preventive immunotherapy in humans, *Nat. Med.* **4**(5), 601–606.

Ma, S.W., et al. (1997) Transgenic plants expressing autoantigens fed to induce oral immune tolerance, *Nat. Med.* **3**, 793–517.

McCormick, A.A., et al. (1999) Rapid production of specific vaccines for lymphoma by expression of the tumor-derived single-chain Fv epitopes in tobacco plants, *Proc. Natl. Acad. Sci. USA* **96**, 703–708.

McGarvey, P.B., et al. (1995) Expression of the rabies virus glycoprotein in transgenic tomatoes, *Biotechnology* **13**, 1484–1487.

Moloney, M.M. (1995) "Molecular farming" in plants: achievements and prospects, *Biotechnol. Eng.* **9**, 3–9.

Morrow, K.J. (2002) Economics of antibody production, *Genet. Eng. News* **22**, 34–39.

Mushegian, A.R., Shepard, R.J. (1995) Genetic elements of plant viruses as tools for genetic engineering, *Microbiol. Rev.* **59**, 548–578.

Parmenter, D.L., et al. (1995) Production of biologically active hirudin in plant seeds sing oleosin partitioning, *Plant Mol. Biol.* **29**, 1167–1180.

Pen, J. (1996) Comparison of Host Systems for the Production of Recombinant Proteins, in: Owen, M.R.L., Pen, J. (Eds.) *Transgenic Plants: A Production System for Industrial and Pharmaceutical Proteins*, John Wiley & Sons, London, UK, pp. 149–167.

Ponstein, A.S., Verwoerd, T.C., Pen, J. (1996) Production of Enzymes for Industrial Use, in: Collins, G.B., Sheperd, R.J. (Eds.) *Engineering Plants for Commercial Products and Applications*, Vol. 792, New York Academy of Sciences, New York, pp. 91–98.

Raskin, I., Fridlender, B., et al. (2002) Plants and human health in the twenty-first century, *Trends Biotechnol.* **20**, 522–531.

Richter, L.J., Thanavala, Y., Arntzen, C.J., Mason, H.S. (2000) Production of hepatitis B surface antigen in transgenic plants for oral immunization, *Nat. Biotechnol.* **18**, 1167–1171.

Ruggiero, F., et al. (2000) Triple helix assembly and processing of human collagen produced in transgenic tobacco plants, *FEBS Lett.* **469**, 132–136.

Sijmons, P.C., et al. (1990) Production of correctly processed human serum albumin in transgenic plants, *Biotechnology* **8**, 217–221.

Smith, M.D. (1996) Antibody production in plants, *Biotechnol. Adv.* **14**, 267–281.

Smith, M.D., Glick, B.R. (2000) The production of antibodies in plants, *Biotechnol. Adv.* **18**, 85–89.

Stoger, E., et al. (2000) Cereal crops as viable production and storage systems for pharmaceutical scFv antibodies, *Plant Mol. Biol.* **42**, 583–590.

Tacket, C.O., Mason, H.S. (1999) A review of oral vaccination with transgenic vegetables, *Microbes Infect.* **1**, 777–783.

Tacket, C.O., et al. (1998) Immunogenicity in humans of a recombinant bacterial antigen delivered in a transgenic potato, *Nat. Med.* **4**, 607–609.

Technology Catalysts International Corporation, *Biopharmaceutical Farming*, Falls Church, VA, Copyright© 2002.

Thanavala, Y., et al. (1995) Immunogenicity of transgenic plant-derived hepatitis B surface antigen, *Proc. Natl. Acad. Sci. USA* **92**, 3358–3361.

The Context Network, Biopharmaceutical Production in Plants, *Biopharma Prospectus*, West Des Moines, IA, Copyright© 2002.

Tomsett, B., Tregova, A., Garoosi, A., Caddick, M. (2004) Ethanol-inducible gene expression: first step toward a new green revolution? *Trends Plant Sci.* **9**(4), 159–161.

Valdes, R., et al. (2003) Hepatitis B surface antigen immunopurification using a plant-derived specific antibody produced in large scale, *Biochem. Biophy. Res. Commun.* **310**, 742–747.

Vandekerckhove, J., et al. (1989) Enkephalines produced in transgenic plants using modified 2S storage proteins, *Biotechnology* **7**, 929–932.

Whitelam, G.C. (1995) The production of recombinant proteins in plants, *J. Sci. Food Agric.* **68**, 1–9.

Whitelam, G.C., Cockburn, W. (1996) Antibody expression in transgenic plants, *Trends Plant Sci.* **1**, 268–272.

Zhong, G.Y., et al. (1999) Commercial production of aprotinin in transgenic maize seeds, *Mol. Breeding* **5**, 345–356.

19
Cell-free Translation Systems

Takuya Ueda, Akio Inoue and Yoshihiro Shimizu
The University of Tokyo, Kashiwa, Japan

1	**History** 564	
1.1	Beginning 564	
1.2	Determination of the Genetic Code 565	
1.3	Elucidation of Factors Involved in the Translation Process 565	
2	**Applications** 566	
2.1	Protein Production 566	
2.2	Synthesis of Artificial Protein 568	
2.3	Linkage between Genotype and Phenotype 570	
3	**Perspectives** 572	
	Bibliography 572	
	Books and Reviews 572	
	Primary Literature 572	

Keywords

Genetic Code: Relationship between Codons and Amino Acids
CFCF (continuous-flow cell-free) translation system: cell-free translation system developed for protein production by supplementing low molecular weight substrates to cell-free translation.
CECF (continuous-exchange cell-free translation): improved cell-free translation system by the dialysis method.

Suppressor tRNA
tRNA decoding termination codons.

Proteins. Edited by Robert A. Meyers.
Copyright © 2007 Wiley-VCH Verlag GmbH & Co. KGaA, Weinheim
ISBN: 978-3-527-31608-3

Unnatural Amino Acid
Amino acids that are not used by living organisms.

***In vitro* Virus**
The selection method by RNA molecules covalently bound to the protein product.

Ribosome Display
The selection method for mRNA by the function of the polypeptide displayed on ribosome stalling during the translation process.

> A disrupted cell is capable of synthesizing protein from amino acids depending upon the template RNA. This discovery, half a century ago, set the stage for biochemical studies on the gene expression process occurring in cells. The cell-free translation system successfully revealed the function of a number of key molecules participating in the translation system, such as ribosome, mRNA, tRNA, and protein factors and, moreover, elucidated the very complicated mechanism of protein synthesis. Although still frequently utilized in these experimental studies, cell-free translation is beginning to make a mark as an attractive tool for protein production as a potential alternative to an *in vivo* expression system. Moreover, the cell-free translation system is prospective as a method for the synthesis of protein with unnatural amino acids and for the selection of genotypes from peptide libraries.

1
History

1.1
Beginning

Proteins are translated on ribosomes from mRNAs transcribed from DNA. The outline of how translation proceeds had been elucidated by both *in vitro* and *in vivo* approaches for two decades after the discovery of the double-helix structure of DNA. In the 1950s, most studies focusing on the gene expression process were carried out on the basis of a genetic approach using bacteria and phages. The hypothesis of the central dogma, DNA makes the RNA that makes protein, had been evidenced by various observations obtained from *in vivo* experiments. However, the precise mechanism of translation, such as deciphering information on mRNA, had to await a series of verifications by the *in vitro* translation system using disrupted cells. The first demonstration of protein synthesis using cell-extract was described by Zamecnik's group in 1952. The finding that rat liver extract was capable of incorporating amino acids into polypeptides allowed us to elucidate the involvement of individual molecular players in the gene expression process. Thereafter, it was shown in 1955 that the ribonucleoprotein complex, which was designated as ribosome, catalyzes the peptide bond formation. Up to 1960, the construction of bacterial cell-free translation systems had been achieved by several groups. The addition of viral genomic RNA resulted in a large increase of amino acid incorporation into polypeptide, concluding that RNA serves as a template for protein.

1.2
Determination of the Genetic Code

One of the most remarkable contributions of the cell-free translation system in molecular biology was the determination of the genetic code in the 1960s. Matthaei and Nierenberg demonstrated that the addition of exogenous artificial RNA to the cell-free translation system enhanced the incorporation of particular amino acids into polypeptides. The first demonstration was that, in the presence of the polyurdylic acid (poly U), in *Escherichia coli*, the cell-free translation system was capable of synthesizing polymers of only phenylalanine at higher magnesium concentrations than physiological conditions. This finding in 1960 lead to the determination of the correspondence of the triplet codon to amino acid over the next five years. For instance, it was immediately shown that poly C was able to stimulate proline incorporation into polypeptide, indicating that the codon CCC corresponded to proline. By the various synthesized RNAs consisting of randomized two or three kinds of nucleotides, the stimulation of the incorporation of amino acid into the polypeptide was examined, and the relationship between codons and amino acids was determined using these polynucleotides. However, this approach came to a standstill because statistical speculation resulted in ambiguity.

This problem was solved by the elegant experiment called the *triplet-binding experiment*, in which aminoacyl tRNA charged with cognate amino acids binds synthesized trinucleotide on ribosome and determined the correspondence of amino acids to triplet codons. Simultaneously, the chemically synthesized oligonucleotide with a defined sequence was employed for examination of the incorporation of amino acid into polypeptides. Together with these results, the genetic code table had been established by 1966.

Another contribution of the cell-free translation system was the direct verification that tRNA is an adaptor molecule linking amino acids to codons on mRNA. The existence of an adaptor molecule had been proposed by Crick, which had not been evidenced with any experimental results. In 1957, Hoagland and Ogata independently found that the small soluble RNA molecule, which would later be named tRNA, was charged with amino acids. Nierenberg's triplet-binding experiment leads us to the conclusion that this tRNA molecule plays the role of the adaptor molecule that deciphers comma-less triplet codons on mRNA. The determination of the primary structure of tRNA by Holley demonstrated that tRNA folded in cloverleaf structure fulfills the decoding function based upon the interaction between the anticodon and codon.

1.3
Elucidation of Factors Involved in the Translation Process

To evaluate the role of protein factors involved in the translation process, the cell-free translation system has appeared to be the most useful system. The translation process proceeds in three steps: initiation, elongation, and termination, in which specific protein factors govern a respective refined process. These factors were first discovered to enhance protein synthesis using cell-free translation, and the detailed behavior was elucidated by a well-defined system capable of verifying the elementary steps. In the case of the elongation process, Elongation Factor (EF)-Tu was first found to be a factor that stimulates poly U-dependent polyphenylalanine synthesis and thereafter it appeared to convey aminoacylated tRNA to the ribosome as a GTP-binding form. EF-Ts were shown to catalyze the exchange of GTP to GDP on EF-Tu. EF-G appeared to catalyze the hydrolysis of the triphosphate of GTP depending on the ribosome and appeared to be involved in the translocation of tRNA from the A-site to the P-site.

In the case of the initiation process, initiation factor-2 (IF-2) was shown to carry initiator tRNA attached by formymethionine onto a 30S subunit of ribosome. Initiation factor-3 (IF-3) was indicated to dissociate 70S ribosome resulting from the termination reaction. Initiation factor-1 (IF-1) was found to enhance the disassembling activity of IF3 by entering an inter-subunit space. Regarding the termination process, release factor-1(RF-1) and release factor-2 (RF-2) enter the A-site of the ribosome depending on the termination codons to catalyze the hydrolysis of peptidyl-tRNA on the P-site of ribosome. RF-1 recognizes UAG and UAA codons, while RF-2 is responsible for UGA and UAA. While the function of these two RFs was verified at an early stage of the study on translation, RF-3 and the ribosome recycling factor (RRF) had remained less understood. Very recently RF-3 was shown to be involved in the release of RF-1 and RF-2 from the ribosome. RRF was first discovered as a factor enhancing the activity of the cell-free translation system and shown to promote the dissociation of mRNA and tRNA from the ribosome.

Similarly, eukaryotic translational factors have been studied by the cell-free translation system. Although most essential factors coincide with bacterial factors, finding a number of additional proteins enhancing the cell-free translation system, especially in the initiation process, suggests that the eukaryotic system appeared to be much more complicated than the bacterial system.

2
Applications

2.1
Protein Production

Although having long served as a useful system for the elucidation of the translation mechanism, the cell-free translation system had been paid only slight attention as a tool for the preparation of protein. Most protein productions at present are based upon the expression system of cloned genes in living cells, in which we sometimes undergo failure, particularly in cases of toxic or unstable proteins. Therefore, it was desired that an *in vitro* gene expression system using the cell-free translation system would compensate for the *in vivo* expression system. However, the low yield of protein product had hindered the utilization of the cell-free translation system. The conventional cell-free translation system using crude extracts, such as *E. coli* S30, rabbit reticulocyte lysate and wheat germ extracts, produces translation reactions for no longer than 1 h in the batch system and synthesized protein products to such a small degree as to be detected only by labeling with radioactive amino acids. To improve productivity, transcription/translation, coupled with using phage RNA polymerases such as T7 or SP6 RNA polymerases, and energy regenerating systems for ATP recycling such as creatine phosphate/creatine kinase or phosphoenolpyruvate/pyurvate kinase systems, had been developed, though both showed limited results.

However, the short lifetime of the cell-free translation system was overcome by the development of the continuous-flow cell-free (CFCF) system by A Spirin's group in 1988. In CFCF, prolongation of the reaction was achieved by the continuous feeding of amino acids and energy components, and continuous removal of reaction products. The reaction was carried out in a chamber connected to a reservoir to supply the substrates, ATP, GTP, and amino acids, by pumping, and the reactor was separated at the outlet by a membrane facilitating permeation of the synthesized protein and a low molecular weight reaction product (Fig. 1a). By this system, it appeared that the reaction using wheat germ extract proceeds for up to 100 h and was able to synthesize 120 to 240 µg of protein in a 1-mL reaction chamber. The most plausible explanation for this long sustainability is that the shortage of an energy source such as ATP and GTP during the reaction is compensated by the supplement of a feeding solution, considering that in the batch system using cell extracts, the ATPase inevitably dwelling in the extract rapidly consumes ATP. The simpler system without the aid of pumping was developed on the basis of the dialysis method of low molecular weight compounds and designated as the continuous-exchange cell-free (CECF) system (Fig. 1b).

The improvement of productivity and reproducibility of cell-free translation has also been achieved by examining the reaction condition and preparation procedures of crude cell-extract. Preparation of crude extract from polished wheat germ ensures constant production of proteins and prolongation of reaction in the CFCF system. By washing the embryo, endogenous inhibitors of translation, including ribosome inactivating proteins (RIP) such as ricin toxin, are successfully eliminated from extracts, which may result in such drastic

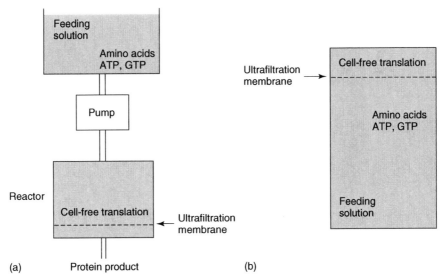

Fig. 1 Schematic illustrations of (a) CFCF and (b) CECF translation systems. A feeding solution containing ATP, GTP, and amino acids is supplied to the reaction chamber in which cell-free translation proceeds, by pumping (CFCF) or diffusion (CECF).

extensions of reaction and high yields of proteins. More recently, a reconstruction of the cell-free translation system using purified components was successfully achieved. Three initiation factors, three elongation factors, three release factors, RRFs, 20 individual aminoacyl tRNA synthetases and methionyl-tRNA transformylase were cloned from the *E. coli* genome for efficient expression in *E. coli* cells and purified. This reconstituted system, designated as the Protein synthesis using recombinant elements (PURE) system, is capable of synthesizing approximately 0.1 mg of proteins in 1-mL volume per hour without any supplemental system such as CFCF and CECF.

2.2
Synthesis of Artificial Protein

One application of cell-free translation is the incorporation of unnatural amino acids into polypeptides. The amino acid–comprising proteins are canonically confined to 20 kinds, barring exceptions such as selenocystein, and therefore *in vivo* systems are unable to synthesize proteins with other amino acids. In 1989, Shultz's group developed a method to incorporate unnatural amino acids at a desired position in polypeptides using the cell-free translation system. In this system, as illustrated in Fig. 2, synthesized suppressor tRNA are first attached by a chemical or enzymatic method to unnatural amino acids and subjected to cell-free translation for DNA templates possessing amber codons at the desired position. The attachment of unnatural amino acids is mostly carried out as follows. An unnatural amino acid is first attached to 3′ OH of adenosine of pCpA dinucleotide by a chemical method. Later, the resultant aminoacylated pCpA is ligated to an amber suppressor tRNA lacking a 3′-terminal CA sequence using

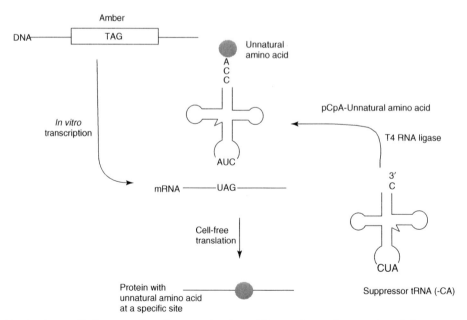

Fig. 2 Strategy for incorporation of unnatural amino acid into polypeptide using the cell-free translation system.

an RNA ligase. If unnatural amino acids are good substrates for aminoacyl tRNA synthetases, they can be attached to tRNA by an appropriate aminoacyl tRNA synthetase. The unnatural amino acid ligated to a suppressor tRNA is further subjected to the cell-free translation system and the codon-forming base pairing with the anticodon of suppressor tRNA is translated into a proteinogenic amino acid.

Although the incorporation of an unnatural amino acid into a polypeptide using cell-free translation is highly advantageous compared to an *in vivo* system, there are several problems to be solved. One problem is the aminoacylation of the suppressor tRNA after one reaction cycle of translation. The suppressor tRNA recharged by an aminoacyl tRNA synthetase during cell-free translation inserts a canonical amino acid at the amber codon, which hinders the efficient incorporation of unnatural amino acid. To solve this problem, tRNAs that are not recognized by an endogenous aminoacyl tRNA synthetase has been developed. For instance, taking advantage of the fact that yeast tRNA^{Tyr} and tRNA^{Gln} do not serve as a substrate for the *E. coli* cognate aminoacyl tRNA synthetase, the suppressor tRNA having the framework of yeast tRNAs is utilized as an adaptor tRNA for unnatural amino acids in an *E. coli* cell-free translation system to prevent recycling of tRNA. *E. coli* tRNA^{Tyr}, which is not recognized by the eukaryotic tyrosyl-tRNA synthetase, can be utilized in wheat germ extract and *vice versa*. Another problem is that the presence of RF-1 corresponding to the amber codon in the extract caused the termination to produce an incomplete peptide, reducing the percentage of peptide containing unnatural amino acid. Using the cell-free translation system in which all the components are in a purified state, this competition is circumvented by eliminating RF-1 from the system, achieving almost full suppression efficiency.

Another problem in the amber suppressor strategy is that only one codon in the open reading frame is available for the introduction of unnatural amino acids. To allow the incorporation of several varieties of unnatural amino acids into a single polypeptide, the four-base codon method was developed. In this approach, tRNA having a four-base anticodon is synthesized by transcription reaction using RNA polymerase, and chemically aminoacylated, followed by cell-free translation. By the four-base codon–anticodon method, one is enabled to incorporate multiple unnatural amino acids at plural codons. Obviously, due to competition with endogenous tRNAs, incorporation of an unnatural amino acid into the polypeptide never reaches full yield, but by employing the four-base codons competing with minor three-base codons, the efficiency of incorporation is easily improved, depending on the amount of endogenous tRNA. This artificial genetic code can be expanded by exploiting the five-base anticodon tRNA.

The expansion of the genetic code in cell-free translation was challenged by the utilization of unnatural nucleobases in codon and anticodon correspondence. The utilization of one additional artificial base pairing besides A-U and G-C pairing in the translation process is able to, in principle, extend the canonical 64 codons to 216 ($6 \times 6 \times 6$) codons in the genetic code, which gives rise to a vast possibility for the incorporation of various unnatural amino acids into polypeptides. First, isoC and isoG (Fig. 3a) were introduced in the codon and anticodon, respectively, and the resultant artificial codon corresponded to the unnatural amino acid. Recently,

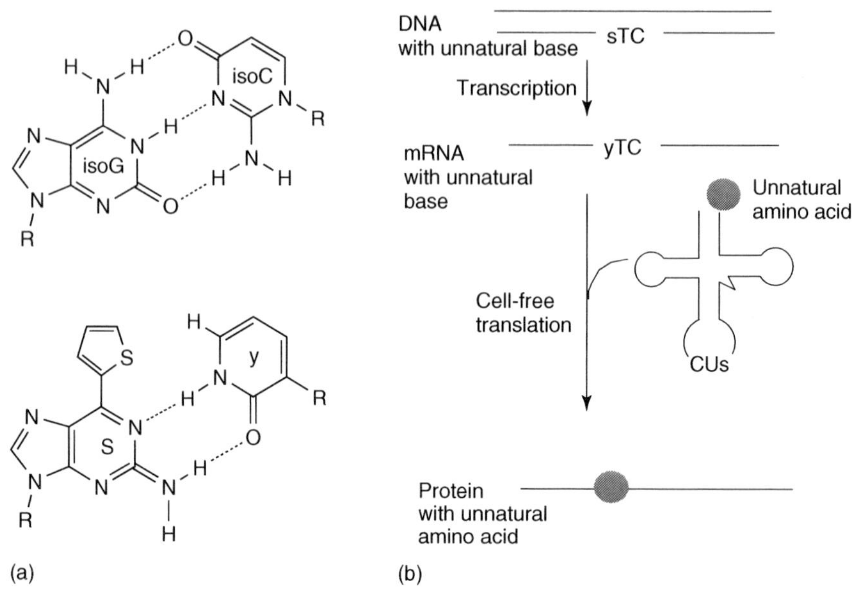

Fig. 3 Structures of unnatural bases utilized in (a) cell-free translation for expansion of genetic code and (b) design of transcription/translation coupled system using unnatural nucleotides for incorporation of unnatural amino acids.

the incorporation of 3-chlorotyrocine was successfully inserted into the desired position of protein utilizing base pairing between 2-amino-6-(2-thienyl) purine (s) and pyridin-2-1 (y) (Fig. 3b) as follows. First, a DNA fragment in which the s-nucleotide was located at the first letter of the desired codon was synthesized for the synthesis of mRNA by transcription reaction in the presence of the y-nucleotide. Simultaneously, yeast tRNA^{Tyr} with the s-nucleotide at the third position of the anticodon was prepared by ligation of RNA fragments and was aminoacylated with 3-chlorotyrosine using tyrosyl-tRNA synthetase. Using the E. coli cell-free translation system, 3-chlorotyrosine was successfully incorporated at the desired position of a target protein. Although this approach looks very expansible, technical difficulties in site-specific introduction of appropriate artificial nucleobases into mRNA and tRNA remain to be solved for the development of a concise and widely distributed system.

2.3
Linkage between Genotype and Phenotype

Cell-free translation provides the potential methodology for linkage between genotype and phenotype. In vitro screening of DNA corresponding to RNAs and proteins with certain functions is of significant importance in the field of molecular biology. In the case of functional RNA, such as ribozymes, it is easy to convert to DNA using the reverse transcriptase after the selection procedure designed for the desired function. In contrast, we have no direct system to convert from polypeptide to mRNA. However, the complex formation of nucleic acid and its protein product enables one to perform pseudoreverse translation.

In order to link polypeptide and mRNA, two approaches using cell-free translation are now available: (1) the "*in vitro* virus" method and (2) the ribosome (or polysome)-display method.

In the "*in vitro* virus" method, creation of a covalent linkage of bonding between the mRNA and the protein is executed by cell-free translation depending on mRNA tagged with puromycin, an antibiotic with a structure resembling the 3′ end of aminoacyl tRNA (Fig. 4a). Emerging nascent polypeptide at the P-site is transferred to mRNA-linked puromycin at the A-site, giving rise to a covalent-bonded molecule of the protein product and responsible mRNA. Consequently, through the selection toward a polypeptide portion, mRNA is concurrently isolated, followed by an RT-PCR reaction of the mRNA fragment.

Meanwhile, the ribosome (or polysome)-display method is based upon a noncovalent complex formation of mRNA, ribosome, and polypeptide (Fig. 4b). The mRNA libraries are translated in the cell-free translation system and the reaction is hampered by antibiotics, for instance, chloramphenicol for the system derived from *E. coli*, or simply by cooling on ice. The stalling ribosomes display nascent polypeptides emerging from translating mRNA, and the selection is directed toward arresting the mRNA-ribosome-polypeptide complex. The isolated complex is dissociated with EDTA by splitting the ribosomal subunit, and the eluted mRNA is subjected to the RT-PCR reaction.

To ensure an efficient ribosome display, the stability of the mRNA–ribosome–polypeptide complex should be sustained

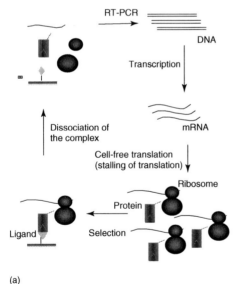
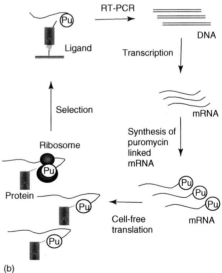

Fig. 4 Schematic drawing of two-selection method of mRNA by protein function. Ribosome display (a) is based on the selection toward attached polypeptide on stalling ribosome. The RNA-ribosome-polypeptide complex is screened by immobilized ligand and RNA derived from a selected complex was amplified by RT-PCR. *In vitro* virus (b) is synthesized puromycin-linked mRNA, which attacks nascent polypeptides attached to tRNA on the ribosome. The covalent complex of mRNA and polypeptide is selected by the ligand and subjected to RT-PCR.

against the presence of RFs in the cell-extract of the cell-free translation system. RFs in cell-free extracts hydrolyze polypeptidyl-tRNA on the P-site, responding to the occupation of the termination codon at the A-site, causing a problematical release of the complex. To prevent this instability, mRNA lacking the termination codon should be designed. The degradation of mRNA during cell-free translation sometimes reduces recovery of the objective mRNA. Thus, nuclease-free cell-free translation systems are eagerly anticipated.

Despite these problems, these two methods allow the mRNA (genotype) to be fished out depending upon the protein function (phenotype) *in vitro*, relatively more conveniently and expeditiously than an *in vivo* system. Moreover, *in vitro* systems are capable of dealing with a huge genetic library comprising 10^{13} molecules, which is an advantage over *in vivo* systems. Particularly, RNA fragments coding for polypeptides with binding properties toward certain ligands, such as antibodies, are enriched in a cloning-independent way by affinity selection from a variety of genetic pools. The multiple cycles of *in vitro* selection and amplification of functional proteins and genes, starting from a randomized pool, can be designated as an *in vitro* evolution system.

3
Perspectives

At present, most protein production processes are carried out utilizing an expression system in living cells. However, we are now able to amplify DNA molecules in a test tube by PCR technology using a DNA polymerase and to synthesize RNA by *in vitro* transcription reaction using an RNA polymerase. Considering that these cell-free systems definitely accelerated an advance in molecular biology, the development of the cell-free translation system will also greatly contribute to studies on the structures and/or functions of proteins. Because of simplicity compared to an *in vivo* expression system, the cell-free translation system is well suited for high-throughput protein production in a postgenome era. Moreover, modification of the translation process is so easily achieved that we will be able to develop an artificial system in addition to synthesis of protein with unnatural amino acids, the *in vitro* virus method and ribosome-display method.

Bibliography

Books and Reviews

Jermutus, L., Ryabova, L.A., Pluckthun, A. (1998) Recent advances in producing and selecting functional proteins by using cell-free translation, *Curr. Opin. Biotechnol.* **9**, 534–548.

Primary Literature

Hirao, I., Ohtsuki, T., Fujiwara, T., Mitsui, T., Yokogawa, T., Okuni, T., Nakayama, H., Takio, K., Yabuki, T., Kigawa, T., Kodama, K., Nishikawa, K., Yokoyama, S. (2002) An unnatural base pair for incorporating amino acid analogs into proteins, *Nat. Biotechnol.* **20**, 177–182.

Noren, C.J., Anthony-Cahill, S.J., Griffith, M.C., Schultz, P.G. (1989) A general method for site-specific incorporation of unnatural amino acids into proteins, *Science* **244**, 182–188.

Piccirilli, J.A., Krauch, T., Moroney, S.E., Benner, S.A. (1990) Enzymatic incorporation of a new base pair into DNA and RNA extends the genetic alphabet, *Nature* **343**, 33–37.

Shimizu, Y., Inoue, A., Ueda, T. (2001) Cell-free translation reconstituted with purified components, *Nat. Biotechnol.* **19**, 751–755.

Sisido, M., Hohsaka, T. (2001) Introduction of specialty functions by the position-specific incorporation of nonnatural amino acids into proteins through four-base codon/anticodon pairs, *Appl. Microbiol. Biotechnol.* **57**, 274–281.

Spirin, A.S., Baranov, V.I., Ryabova, L.A., Ovodov, S.Y., Alakhov, Y.B. (1988) A continuous cell-free translation system capable of producing polypeptides in high yield, *Science* **242**, 1162–1164.